Principles
of
Weed Science

Second Edition

V.S. Rao
International Consultant, Weed Science
Santa Clara, California, USA

CRC Press
Taylor & Francis Group
Boca Raton   London   New York

CRC Press is an imprint of the
Taylor & Francis Group, an informa business

Science Publishers, Inc.
Enfield (NH), USA   Plymouth, UK

Reprinted 2009 by CRC Press

CRC Press
6000 Broken Sound Parkway, NW
Suite 300, Boca Raton, FL 33487
270 Madison Avenue
New York, NY 10016
2 Park Square, Milton Park
Abingdon, Oxon OX14 4RN, UK

SCIENCE PUBLISHERS, INC.
Post Office Box 699
Enfield, New Hampshire 03748
United States of America

Internet site: *http://www.scipub.net*

sales@scipub.net (marketing department)
editor@scipub.net (editorial department)
info@scipub.net (for all other inquiries)

Library of Congress Cataloging-in-Publication Data

Rao, V.S. (Vallurupalli Sivaji), 1940-
   Principles of weed science: (a reference-cum-textbook)/ V.S.
Rao. — 2nd ed.
      p. cm.
   Includes bibliographical references.
   ISBN 1-57808-069-X
   1. Weeds-Control.   2. Herbicides.   I. Title.
SB611.R35 1999
632'.5-dc21
                                                                    99-39247
                                                                          CIP

      Reprinted   2001
      Reprinted   2002

© 2000, Copyright Reserved

ISBN 1-57808-069-X

Dedicated to my loving wife Vallurupalli Nirmala Devi

Preface

Currently, weed science, like the entire field of agricultural science, is going through a new era of technological change. As a consequence, weed scientists are facing newer challenges, particularly in the light of emergence of weeds resistant to herbicides, shifts in weed flora and concerns about herbicide residues in food, soil, groundwater and atmosphere. This necessitated the need to explore newer avenues of weed management technology involving preventive, cultural, ecophysiological, biological and biotechnological methods. The need to look for alternative strategies does not mean that the herbicide technology, the pivot of current weed management strategies, is expected to be abandoned. On the other hand, it necessitates increasing the levels of sophistication in research on herbicide technology which can be utilized in conjunction with other management practices.

In order for the weed scientist to do this job competently, he/she needs to gain adequate knowledge and training in agronomic, soil, genetical, biological and economic aspects of plant growth, behaviour and production, and utilize them to evolve sound weed management practices.

An effort has been made in the second edition of PRINCIPLES OF WEED SCIENCE to present and discuss the various advances made in the field of weed science since the publication of the first edition in 1983. This edition includes an up-to-date information on newer approaches (ecophysiological and biological) in weed management, newer herbicides, bioherbicides, herbicide action mechanisms and transformations in plants, herbicide persistence and behaviour in soil and environment and interaction of herbicides with other agrochemicals. It also contains a chapter on herbicide resistance and genetic engineering. The chapters on the practical aspects of weed science, covered in the first edition, have been retained, but updated and expanded to include the latest developments.

The second edition, like the very popular first edition, has been designed to bring out a comprehensive reference-cum-textbook on the fundamentals and basic principles of weed science. This edition, like the first edition, is expected to serve the students of weed science at undergraduate and graduate (post-graduate) levels as well as teachers, researchers and extension personnel working or interested in weed science in universities, research institutions, governmental and non-governmental organizations, herbicide industry, etc. The subject has been presented in crisp, clear-cut and simple language. As it is impossible to review the entire research done in each of the areas of weed science, I chose to include only the more useful material to bring relevance and objectivity to the subject under discussion. I also chose to exclude some topics because of space constraints.

I accept complete responsibility for choice of information, presentation, interpretation and discussion. The weed management practices suggested for different weeds and in various crop, aquatic and non-crop situations are based on published information and experience. I cannot assume responsibility for success or otherwise of a weed management practice at a given place, crop and situation because environmental conditions, weed spectra, cropping methods, experimental techniques, application methods, etc. vary from one place to another. In view of this, the researcher may need to test a particular weed management practice under local conditions before recommending it to the user. Weeds and herbicides have been referred by the names standardized by Weed Science Society of America. If trade names of herbicides have been mentioned, it was only as a matter of convenience to the reader but not as an endorsement of a particular product. I generally followed British orthography while writing this book, except in case of certain words and common names of weeds and herbicides.

During the course of working on this edition, I have drawn heavily on the computer expertise of my beloved son Srinivas R. Vallurupalli and beloved daughter Madhavi Lata Rajavasireddy and I profusely thank them for their help and assistance. I also appreciate the encouragement and support given by my wife Nirmala Devi who ungrudgingly endured the partial loss of my company when I was busy working on the manuscript.

Santa Clara, California **V.S. RAO**
USA
October 5, 1999

Acknowledgement

I reproduced in my book certain copyrighted material, including certain figures and discussions on them, published in other books and journals. I gratefully acknowledge the permission granted by various publishers mentioned below.

1. Cambridge University Press, New York, NY, USA: To reproduce:
   (a) in Chapter 7 (Fig. 7-5) of this edition Figure 2 `Scheme illustrating chlorophyll biosynthesis; A. Summary of overall pathway' from a chapter on `Carotenoids and Chlorophylls: Herbicidal Inhibition of Pigment Biosynthesis' authored by G. Britton, P. Barry and A.H. Young and published, in 1989, in a book entitled `HERBICIDES AND METABOLISM' edited by A.D. Dodge, and
   (b) in Chapter 7 (Figs. 7-12 and 7-13) of this edition Figure 1 `The metabolism of nitrogen within higher plants' and Figure 2 `The glutamate synthase cycle of ammonia assimilation' from a chapter on `Glutamine Synthetase and its Inhibition' authored by Peter J. Lea and Stuart M. Ridley and published, in 1989, in a book entitled `HERBICIDES AND METABOLISM' edited by A.D. Dodge.

2. John Wiley & Sons, Inc., New York, NY USA: To reproduce in Chapter 2 (Fig. 2-1) of this edition, Figure 7-1 `A schematic presentation of the stair-step technique for the study of allelopathy' published, in 1997, in a book entitled `WEED ECOLOGY: IMPLICATIONS FOR MANAGEMENT', authored by Steven Radosevich, Jodie Holt and Claudio Ghersa.

3. Prentice-Hall, Inc., Upper Saddle River, NJ, USA: To reproduce and adapt, in Chapter 7 (Fig. 7-1) of this edition, Figure 16-8 `Z Scheme, showing reduction potentials and electron flow during photosynthesis' published, in 1996, in a book entitled `PRINCIPLES OF BIOCHEMISTRY', second edition, authored by H. Robert Horton, Lawrence A. Moran, Raymond S. Ochs and K Gray Scrimgeour.

4. Springer-Verlag Heidelberg, Hiedelberg, Germany: To reproduce in Chapter 8 (Figs. 8-15, 8-16 and 8-17) Figure 1 `Metabolic inactivation of nicosulfuron in maize, Figure 3 `Metabolic inactivation of triasulfuron in wheat foliage' and Figure 5 `Metabolic inactivation of proherbicidal sulfonylurea herbicide DPX-H6564' from a chapter on `Recent Advances in Sulfonylurea Herbicides' authored by H.M. Brown and J.C. Cottermann and published, in 1994, in a book entitled 'CHEMISTRY OF PLANT PROTECTION', Vol. 10.

5. Weed Science Society of America, Lawrence, KS, USA: To reproduce the following figures:
   (a). In Chapter 12 (Fig. 12-3) of this edition: Figure 2 `Mechanisms responsible for the synergism of herbicides by other agrochemicals in higher plants' and Figure 3 `Mechanisms responsible for the antagonism of herbicides and other agrochemicals in higher plants' from a chapter on `Interactions of Herbicides with other Agrochemicals in Higher Plants', authored by Kriton K. Hatzios and Donald Penner and published, in 1985, in REVIEWS OF WEED SCIENCE, Vol. 1.

1

Introduction

WEEDS: DEFINITION

Weeds are unwanted and undesirable plants which interfere with the utilization of land and water resources and thus adversely affect human welfare. They can also be referred to as plants out of place. Usually, this means that weeds grow where we either want other plants to grow or where we want no plants at all. In cropland and forests, weeds compete with the beneficial and desired vegetation, reducing the yield and quality of produce. Undesirable vegetation also flourishes in aquatic systems, forestry and non-cropped areas such as industrial sites, roadsides, railway lines, airfields, landscape plantings, water tanks, waterways, etc. Thus, all plants may become weeds in particular situations.

Weeds are an important factor in the management of all land and water resources, but their effective impact is greatest on agriculture. There is no reliable study of worldwide damage due to weeds. However, it is widely known that losses caused by weeds exceed the losses from any category of agricultural pests, such as insects, nematodes, diseases, rodents, etc. Of the total annual loss of agricultural produce from various pests, weeds account for 45%, insects 30%, diseases 20% and other pests 5%.

HARMFUL EFFECTS CAUSED BY WEEDS

Reduction in Crop Yield

Weeds compete with crop plants for nutrients, soil moisture and sunlight. The intensity of weed competition depends upon: a) type of weed species, b) severity of weed infestation, c) duration of weed infestation, d) competing ability of crop plants and e) climatic conditions which affect weed and crop growth.

Reduction in crop yield has a direct correlation with weed competition. Generally, an increase in one kilogram of weed growth corresponds to a reduction in one kilogram of crop growth. Weeds remove plant nutrients more efficiently than crop plants. In a drought situation, they thrive better than crop plants. When left undisturbed, some weeds can grow faster and taller than crop plants and inhibit tillering and branching. They can curtail sunlight and adversely affect photosynthesis and plant productivity.

Depending on the degree of competition, weeds reduce crop yields by 10% to 25%. In other words, if weed competition were completely eliminated from the entire cropland, the world's farm production could be increased by 10% to 25%. At a crop yield loss of 15% in a developing country like India, total weed control in the entire farmland would add US $5 billion to its economy at the current price level. This figure would be far greater if the direct and indirect impact of weeds on aquatic systems, forestry and industrial sites are also considered. This loss, if allowed to continue, amounts to a big drain in the nation's economy.

The economic impact of weeds on the US economy equals or exceeds US $20 billion, with the agricultural sector alone accounting for US $15 billion [3]. Parker and Fryer [11] estimated that the world was losing annually 11.5% of the total food production. That is, if all the weeds in food crops were controlled, the current world's food production would be higher by 11.5% or 450 million tons. This is too staggering a figure to be ignored by the world leaders and weed scientists.

Crop yield loss from weeds is highest in the tropics. For example, in rice, the mainstay of Asia's economy, proper weed control increases gain yields by 20% to 75%. In extreme weed situations, weed management could triple the yield of rice.

Reduction in Land Value

Heavy infestation of perennial weeds can make the land unsuitable or less suitable for cultivation, resulting in loss of its monetary value. Millions of hectares of cultivable area in the rice-growing regions of Asia have been abandoned or are not being regularly cultivated due to severe infestation of *Cyperus rotundus* (nutgrass or purple nutsedge), *Cynodon dactylon* (bermudagrass), *Imperata cylindrica* (thatchgrass, cogongrass, or alang-alang) and other perennial weeds.

Limited Choice of Crops

Crops differ in their ability to compete with weeds. In many instances the presence of particular weeds in fields limits the choice of crops to be grown. Very heavy weed infestations render some economically important crops, particularly pulses, vegetables, cotton, jute and forage crops unsuitable or less suitable for cultivation.

Loss of Quality

The quality of leafy and other vegetable crops suffers in the presence of weeds. In addition, *Allium canadense* (wild onion), *Avena fatua* (wild oats) and wild rice may impair the quality of onion, wheat and rice produce respectively. The leaves of *Mikania micrantha* may contaminate tea leaves during plucking and reduce the value of tea. Contamination of other noxious weed seeds greatly reduces the value of crop seed and grain and sometimes even renders them unsaleable. The presence of weeds and weed debris in crop grain and other farm products reduces their market value and causes spoilage during storage.

Reduced Human Efficiency

Weeds reduce human efficiency through physical discomfort caused by allergy and poisoning. Weeds such as *Parthenium hysterophorus* (parthenium), *Rhus radicans* (poison ivy) and *Ambrosia* spp. (ragweed) that cause itching, hay fever and other debilitating allergies contribute markedly to chronic human illness and suffering. The presence of spiny and thorny weeds such as *Amaranthus spinosus*, *Argemone mexicana*, *Polygonum perfoliatum*, etc. restricts movement of farm workers in carrying out, efficiently, farm practices such as fertilizer application (top-dressing), insect and disease control measures, harvesting, etc.

Increased Costs of Insect and Disease Control

Weeds serve as alternative hosts to several crop insects, nematodes and pathogens. Insects such as aphids, thrips, weevils and stem flies survive on *Brassica kaber* (wild mustard), *Daucus carota* (wild carrot), *Ambrosia* spp. (ragweed), *Amaranthus retroflexus* (redroot pigweed), etc. Weeds such as *Avena fatua* (wild oats) and some perennial grasses harbour pathogens of black stem rust of wheat. By harbouring these insects and pathogens, their attack on crop plants is increased. This results in increased cost of their control.

Problems Caused by Aquatic Weeds

Aquatic weeds reduce markedly the flow of water in irrigation and drainage canals, channels and streams. Obstruction to water flow raises water levels in canals and streams, resulting in: a) flooding, b) seepage into adjoining areas, c) breaks in canal banks and d) inadequate delivery of irrigation water to farms located at a distance from the main water source. In addition, reduced velocity of flow causes increased siltation and reduced carrying capacity, necessitating more frequent mechanical cleaning. Weeds cause enormous loss of water through transpiration. Aquatic weeds also obstruct weirs, gates and other water-regulating structures. Algae and fragments of plant material clog irrigation equipment. Aquatic weeds form breeding grounds for obnoxious insects like mosquitoes. They reduce recreational values by interfering with fishing, swimming, boating, hunting and navigation on streams and canals. Weeds on watershed areas and floodplains utilize large amounts of water and thereby deny its use for the production of food, feed and fibre. Some of the prominent aquatic weeds include *Eichhornia crassipes, Typha angustifolia, Hydrilla verticillata, Potamogeton* spp., *Salvinia molesta, Ipomoea aquatica, Nymphaea* spp., etc.

EVOLUTION OF WEED MANAGEMENT

Weeds are no strangers to man. Weeds have been present ever since he started to cultivate crops, around 10,000 B.C. and they were undoubtedly recognized as a problem from the beginning [6]. When man first started to grow crops for food and fibre, he soon learned that yields were much higher when weeds were removed to allow only crop plants to grow. Thus, the concept of weed management is as old as agriculture itself. Later, insects and diseases became of much greater consequence and importance. The universal occurrence of weeds as constant companions of agricultural environment as opposed to the epidemic nature of other pests, delayed recognition of the importance of weed science in crop production. From the beginning of agriculture to the middle of the 20th century, the plough and hoe have been the only widely employed means of weed control. Thus, the man with the hoe became the classical symbol of field agriculture.

As weed infestations began to seriously limit crop production, methods were devised to combat them. Thus came into existence the various manual, mechanical and biological methods and finally the chemical method of weed management. The discovery of many organic herbicides in the past 50 years enabled substantial progress in controlling weeds the world over. The emergence of herbicides has stimulated research in the economic botany of weeds and this eventually led to broader application of the technology to modern weed management.

Hay [6] identified six stages in the evolution of weed control practices, namely: a) 10,000 B.C.—removing weeds by hand, b) 6000 B.C.—the use of primitive hand tools to till the land and destroy weeds, c) 1000 B.C.—animal-powered implements like harrows, d) 1920 A.D.—mechanically-powered implements like cultivators, blades, harrows, finger-weeders, rotary hoes, rod-weeders, etc., e) 1930 A.D.—biological control and f) 1947 A.D.—chemical control, with the commercial development of organic herbicides such as 2,4-D and MCPA.

Thus, through the advances made in weed control since 10,000 B.C. man has succeeded in preventing weed competition and raising crop yield levels to feed and clothe not only himself and his family, but others as well. Holm [7] observed: 'Crop weed control is the most complex, but also the most important dimension of weed control activity, for it is from our immediate efforts here that we may have more food and fibre. Quite unseen and with the dreary stubborn patience of evil, the weeds in crops suck night and day at the soil and moisture to deprive our plants of the vigour and dry matter needed to feed and clothe the world'.

MODERN WEED SCIENCE

Commenting on the discovery of selective action of copper sulphate as a herbicide in 1890, Bolley [2] observed: 'when the farming public has accepted this method of attacking weeds as a regular farm operation, the gains to the country at large will be greater in monetary consideration than that which has been afforded by any single piece of investigation applied to field work in agriculture'. This prophecy came true some 55 years later when 2,4-D was discovered, which revolutionized agriculture so much that food shortages plaguing North America and Europe at the conclusion of World War II disappeared within a few years. Freed [5] observed that without herbicides the world would have been lacking a good many tons of grain and also that the same varieties of wheat under the same climatic conditions yielded 600-900 kg ha^{-1} more simply due to control of weeds. He further observed that wheat could not be produced in the large acreage in eastern Oregon, USA, which was heavily infested with field bindweed (*Convolvulus arvensis*). Within a year or two of the availability of 2,4-D, this acreage was released for wheat production, something that had not been achieved in almost 15 yr of cultivation using cultural-mechanical practices. Thus, herbicides have made a significant contribution not only to agriculture in the USA and Europe in particular, but also to the world's agriculture in general.

Chandler [4] reported significant increases in the yield of rice through the application of herbicide over the unweeded treatment. The increase due to herbicide treatments was approximately equivalent to two hand weedings. In tea, chemical weed management gave significantly higher yield over manual weed control methods [1]. In Colombia, the use of herbicides improved the yield by 13% in cotton and 24% in rice over what was obtained by manual and mechanical weed control practices and the overall improvement in yields due to herbicides (over manual and mechanical methods) was 19.3% [9]. Matsunaka [10] found that weed control cost in transplanted rice was far higher when hand labour was used instead of herbicides and that tremendous savings could be effected when herbicides supplanted hand labour. These and other numerous reports world over indicate that agriculture is far more profitable when chemical weed management is adopted. Thus, even in developing countries, where the application of modern weed management technology has been slow, the man with the hoe has been replaced by the man with the spray lance.

Herbicides are considered almost synonymous with modern weed science technology as they gave a new direction to the farmer to realize the maximum yield potential of the crop at lower production costs, priorly never possible. At the same time, weed science became an intriguing and a fascinating technology. Weed scientists saw enormous scope for research to unravel the mysteries of their science in the field. Today, 52 years after the commercial development of 2,4-D and MCPA (in 1947), weed scientists have made phenomenal progress in understanding the selective action of over 400 herbicides by studying their absorption and translocation patterns, mechanisms of action in plants, degradative and detoxification mechanisms in plant and soil, interactions with other pesticides and chemicals, etc. All this has helped in making more effective, economical and safe recommendations for control of numerous weeds in different crops, aquatic systems and non-crop areas.

Currently, weed science, like the whole of agricultural science, is going through a new era of technological change. Weed scientists are now facing new challenges, particularly in the light of emergence of weeds resistant to herbicides and concerns and questions about herbicide residues in food, soil, groundwater and atmosphere. These have highlighted the need to evaluate the low-rate concept (LRC), deviate from the path of monoculture systems that rely on herbicides excessively in favour of sustainable agricultural systems incorporating the concepts of integrated weed management and explore the possibility of employing biological, ecological and cultural methods of weed management in addition to the chemical method. There is also a need to study the mid-term and long-term consequences of new weed management practices with respect to the fate of herbicides, soil weed seedbank, weed flora shifts, crop yield and quality, costs of potential control and technical feasibility, and finally the economic returns to the farmer.

Farmers are not the same worldwide. The farmer in a developing country is different from the one in a developed country. Even within a country, the farmers of one region are different from those of other regions. Yet, farmers, irrespective of where they are from, realize that they need to depend on chemical weed management technology as their main defence against weed menace, for the very successful weed control methods the world has ever seen are centred on herbicide use. The farmers need adequate technical support from weed scientists in managing their weeds in the best possible manner. Besides, weed scientists are needed to communicate effectively not only with farmers and the public, but also with politicians, government organizations and environmental groups [13] to project solutions to weed problems.

The field of weed science has now become much broader in scope than it ever was. It has evolved during the past 50 years into a separate and major field of agricultural science. The modern weed scientist is required to gain adequate knowledge in agronomic, soil, genetical, biological, ecological, physiological, biochemical, environmental, toxicological and economic aspects of plant growth, behaviour and production and utilize it to evolve a package of Integrated Weed Management (IWM) strategies comprising chemical, biological, cultural, mechanical and manual methods. IWM is described as the application of these methods of weed management in a mutually supportive manner. The IWM package should be effective enough for long-term maintenance of natural resources and agricultural productivity with minimal adverse environmental impact and adequate economic returns to farmers. In order for the weed scientist to achieve these objectives he may need to interact with specialists of other fields such as entomology, plant pathology, biotechnology and bioengineering and plant breeding. Since weeds dictate most of the crop production practices (e.g. tillage,

6

herbicides, cultivation, planting date and pattern, etc.), weed scientists could become the leaders of research on collaborative and integrated approaches to agricultural systems [12]. Thus, weed scientists are not only required to be experts in the science and technology of weed management, but also in the art of tilting the crop-weed balance in favour of crops.

Recently, weed scientists have been using computer science to develop computerized and bioeconomic models and database for weed management. These help in a better understanding of: a) crop yield losses due to season-long competition with one or mixed species of weeds, b) future weed infestations from either the present weed population or propagule numbers, c) aggregation and spatial variation of weed populations, d) crop-weed interactions on weed reproduction and e) interaction of microclimate on weed reproduction [12]. These models also focus on herbicide efficacy on a particular weed species or a spectrum of weeds, profitability of weed management practices, predicting weed resistance to herbicides, etc. Developing the computer database will be useful for organizing existing information, helping to identify gaps in the information and identifying processes that can be characterized and then linked to bioeconomic models, which in turn, can be used to develop and refine new hypotheses and assess weed management strategies [12].

The preceding discussion indicates that modern weed science is an amalgamated product of various disciplines of agricultural science and a vital branch of sustainable agriculture. Efforts need to be made by governments and educational institutions to accord greater recognition and importance to weed science and to train more people in this field so farmers' weed problems can be solved with increased efficacy and sophistication.

REFERENCES

1. Anonymous. 1979-80. Tocklai Experimental Station, Tea Research Association, Jorhat, Assam, India. *Ann. Sci. Rep. 1979-80*.
2. Bolley, H. L. 1908. Weed control by means of chemical sprays. *Bull. North Dakota Agric. Exp. Stn.* **80**: 541-574 (USA).
3. Bridges, D.C. 1994. Impact of weeds in human endeavors. *Weed Technol.* **8**: 392-395.
4. Chandler, R.F., Jr. 1969. New horizons for an ancient crop. XI Int. Bot. Congr. All-Congress Symposium: World Food Supply.
5. Freed, V.H. 1980. Weed science: the emergence of a vital technology. *Weed Sci.* **28**: 621-625.
6. Hay, J.R. 1974. Gains to the grower from weed science. *Weed Sci.* **22**: 439-442.
7. Holm, L.G. 1967. PAN © 13:90-103.
8. Holm, L.G. 1971. The role of weeds in human affairs. *Weed Sci.* **19**: 485-490.
9. Lange, A., J. Cardenas and R. Cruz. 1973. Crop losses by weeds. *Weeds Today* **4**(1): 11, 14, 22.
10. Matsunaka, S. 1970. Weed control in rice. Technical papers of the FAO Int. Conf. on Weed Control, Davis, California, USA, pp. 7-23.
11. Parker, C. and J.D. Fryer. 1975. Weed control problems causing major reductions in world food supplies. *FAO Plant Protection Bull.* **23**: 83-95.
12. Wyse, D.L. 1994. New technologies and approaches for weed management in sustainable agriculture systems. *Weed Technol.* **8**: 403-407.
13. Zoschke, A. 1994. Toward reduced herbicide rates and adapted weed management. *Weed Technol.* **8**: 376-386.

2

Weed Biology and Ecology

WEED BIOLOGY

Weed biology is related to the study of weeds in relation to their geographic distribution, habitat growth and population dynamics of weed species and communities. Development of an appropriate and effective weed management programme is dependent on the sound knowledge of weed biology. The various aspects related to weed biology are discussed in this chapter. The reader may bear in mind that the discussion on weed biology may sometimes overlap weed ecology.

Weed Classification

Based on lifespan, weeds are classified into three broad groups: *annuals*, *biennials*, and *perennials*. In each group there are both broadleaf weeds (dicotyledonous weeds) and grasses (monocotyledonous weeds).

Annuals complete their life cycle in a year or less. They propagate by seeds. Normally, they are considered easy to control. But they are very persistent because of abundance of seeds which continue to germinate and emerged seedlings that grow fast until conditions become unfavourable. There are two types of annuals: summer annuals and winter annuals.

The summer annuals start germinating in the spring or summer, attain most of their growth during summer, and usually mature and ripen seeds in summer or autumn depending on the length of their life cycle, and then die. The winter annuals germinate in autumn or winter, produce flowers and ripen seeds in the following spring or early summer, and then die. The seeds of winter annuals remain dormant in the soil during summer.

Biennials have a lifespan of two years. They germinate in the spring or summer. The first year's growth may be purely vegetative and this is known as the rosette stage. The taproot is often fleshy and serves as a food-storage organ. During the spring of the second year, a flower stalk arises from the crown; this is known as the bolting stage. After producing seed, the pant dies. The biennials are controlled like annuals by destroying the vegetative growth in the first year.

Perennials live for more than two years and some of them almost indefinitely. They propagate by seeds and underground storage organs like rhizomes, stolons, bulbs, tubers, etc. Some of them strike roots when the nodes come into contact with the soil. Perennials, with their remarkable capacity for both vegetative reproduction and prolific seed production, are very aggressive and competitive weeds. In many cases, no seed is produced in the first year but seeding occurs every year thereafter for the life of the plant. On the basis of vegetative reproduction, perennials are classified into simple perennials, bulbous perennials and creeping perennials; these represent the basic differences in systems of vegetative propagation.

Simple perennials reproduce mostly by seeds. Vegetative reproduction occurs when roots and stems are cut mechanically. Each cut piece may then give out roots and become a plant. Bulbous perennials propagate through the underground parts like bulbs, bulbils and tubers as well as by seeds. Creeping perennials spread by stolons, the lateral extensions of the creeping aboveground horizontal stems, rhizomes (creeping underground stems, with nodes and internodes), roots or seeds. In some creeping perennials, new shoots may develop from both roots and rhizomes.

Thus, this classification of weeds is based only on their lifespan for one year, two years, or more. This behaviour is not always constant, however, as the duration of weeds is sometimes determined by climatic factors. Many weeds that are annuals or biennials in severe climates may act as biennials or perennials respectively in milder climates or in mild winters. Most weeds belong to the group of perennials, followed by annuals. Biennials comprise only a small percentage of weeds. In the USA, 45% of the weeds are perennials, 34% annuals, and only 7% biennials [51], while in India the ratio is 43%, 40%, and 6% respectively. Most of the weeds produce seeds. Among the weeds which do not produce seed, pteridophytes (the ferns) and fern allies predominate.

Weed Propagation

Annuals and Biennials

Annual and biennial weeds depend on seed production as the sole means of propagation and survival while perennial weeds are less dependent on this mechanism. Production of abundant and small seeds is a common adaptation that ensures a high probability of dispersal and reinfestation. A single plant of an annual weed can produce enough seeds in one season to cover an entire area of one acre with this weed species next year. For example, one plant of *Sisymbrium altissimum* (tumble mustard) produces over a half million (511,208) seeds which are enough, if evenly scattered, to sow at least 11 seeds ft^2 in an acre of land or 3200 seeds on every acre of a 160-acre farm [63]. Similarly, *Amaranthus retroflexus* (redroot pigweed), *Portulaca oleracea* (common purslane), and *Solanum nigrum* (black nightshade) produce 196,405, 193,213 and 178,000 seeds plant respectively, while *Brassica nigra* (black mustard) produces 58,363 seeds plant^{-1} [63].

Some weed species have the ability to produce seeds between intervals of normal disturbance associated with a crop situation. *Avena fatua* (wild oats) germinate at the same time the wheat crop is sown and shatter their mature seeds before crop harvest. Many weeds can produce a large number of viable seeds even after having been cut off soon after flowering. A few weed species produce seeds through apomixis, i.e. without fertilization. Weeds such as ferns reproduce by spores rather than seeds.

Weed species differ in their regenerative strategies. Seeds of some species germinate soon after they are shed. Seeds of these species have a short life in the soil and their persistence is dependent on annual seed production and dispersal. In other species, seeds remain in the soil for long periods with intermittent germination of a part of the population. Some of the weed seeds are very long-lived, but these represent a small portion of the total seedbank [92]. In agronomic situations, the majority of the seeds that germinate during the first two years represent a major threat for crop yield loss and control costs.

Perennials

Vegetative reproduction is a common trait of perennial weeds. As discussed above, vegetative propagation includes rhizomes, stolons, tubers, bulbs, corms, roots and stems. Some perennial weeds propagate in one or more ways. Aside from their remarkable capacity for vegetative reproduction, most perennial weeds are also prolific seed producers. Their ability to produce both vegetative propagules and seeds make them very competitive and aggressive in cropping systems. Even if the aerial parts of perennial weeds die, the vegetative propagules continue to live and send up new growth, leading to flowering of shoots and production of a seed crop year after year.

A seedling of a perennial weed species growing from seeds is not a perennial when it first emerges from the soil. For example, *Convolvulus arvensis* (field bindweed) becomes a perennial when it has about 20 true leaves and *Cyperus esculentus* (yellow nutsedge) only 4 to 6 wk after it emerges from seed. Similarly, *Elytrigia repens* (quackgrass) becomes a perennial within 6 to 8 wk of emergence from seeds and *Sorghum halepense* (Johnsongrass) after only 3 to 6 wk of emergence. The population of certain perennial weeds and relative importance of various propagules are presented in Table 2.1.

Table 2.1. Propagation of some important perennial weeds

Weed Species	Common Name	Propagules (in order of importance)
Achillea spp.	yarrow	stolons, terminal rosettes of leaves, seeds.
Agrostis gigantea	redtop	rhizomes (with dormant underground buds),
Allium canadense	wild onion	bulbs and bulbils
Asclepias syriaca	common milkweed	root stocks
Cirsium arvense	Canada thistle	creeping roots
Convolvulus arvensis	field bindweed	creeping roots, seeds
Cynodon dactylon	Bermudagrass	rhizomes, nodes, seeds
Cyperus rotundus	purple nutsedge	tubers, bulbs (nuts), seeds
Elytrigia repens	quackgrass	rhizomes, seeds
Equisetum arvense	field horsetail	rhizomes, spores
Euphorbia esula	leafy spurge	creeping roots, seeds
Imperata cylindrica	thatchgrass, cogongrass	rhizomes, seeds
Oxalis spp.	wood sorrel	bulbils, rhizomes, taproots
Panicum repens	torpedograss	rhizomes, seeds
Pteridium aquilinum	braken fern	rhizomes, spores
Rumex crispus	curly dock	taproots, rosettes of leaves, seeds
Ranunculus repens	creeping buttercup	procumbent stems, seeds
Sonchus arvensis	perennial sowthistle	creeping roots, seeds
Taraxacum officinale	dandelion	taproots, seeds
Veronica filiformis	slender speedwell	creeping stems

Dynamics of Vegetative Propagules

As numerous perennial weeds occur in cropping systems, it is beyond the scope of this book to discuss the dynamics of vegetative propagules for each of the weed species. However, *Cyperus* spp. which occur widely in both tropical and temperate agriculture, have been used as a case study to discuss the subject.

Vegetative propagules serve as storage as well as reproductive organs for perennial weeds. In the case of *Cyperus rotundus* (purple nutsedge or nutgrass) both basal bulbs and tubers store food reserves. They differ primarily by their position in relation to the mother plant [48]. Basal bulbs are directly connected to an aerial shoot. As rhizomes elongate, tubers are produced on them. Tubers consist of rhizomatous tissue with numerous buds. These buds sprout and initiate rhizomatous growth that develops into seedlings and eventually mature plants.

Shoots of purple nutsedge emerge under suitable growing conditions 4 to 7 d after planting tubers. Tuber formation begins from 4 to 6 wk after seedling emergence. More than 95% of purple nutsedge and *Cyperus esculentus* (yellow nutsedge) tubers are formed in the upper 45 cm of soil [9, 87]. In general, more than 80% of tubers occur in the upper 15 cm of soil. Normally, purple and yellow nutsedges are prolific tubers in a very short period of time. Of the two, purple nutsedge is more virulent. Under identical· conditions, purple nutsedge produces larger and a greater number of tubers than yellow nutsedge. Purple nutsedge produced 5700 kg ha^{-1} of rhizomes after 20 wk of planting tubers [43], while yellow nutsedge produced as many as 10 to 15 million tubers ha^{-1} (upper 15 cm of soil) in a growing season [11]. In Israel, established plants of purple nutsedge formed more than 200 new tubers and basal bulbs m^{-2} wk^{-1} [46], while in Georgia, USA they produced 1000 bulbs and 2300 tubers m^{-2} 20 wk after planting, with a dry weight of 3 kg m^{-2} [61].

Production of vegetative propagules, in particular tubers and rhizomes, is influenced by various factors such as excess carbohydrates, growth substances, temperature and photoperiod. Day length has greater influence on tuber formation in yellow nutsedge than in purple nutsedge. Natural photoperiods ranging from 10 to 14 h have no apparent effect on tuberization in purple nutsedge [47], while shorter photoperiods stimulate reproductive growth in yellow nutsedge [51]. Rhizomes differentiate into tubers rather than bulbs in yellow nutsedge plants grown with a photoperiod longer than 12 h [51]. Long photoperiods combined with high nitrogen levels inhibit tuberization in yellow nutsedge while low nitrogen with high temperature promotes tuberization [35].

Vegetative propagules remain dormant in soil for extended periods. Longevity of nutsedge tubers increases as depth in soil increases. Yellow nutsedge tubers have a half-life of 4 to 6 months at 10 to 20 cm in a non-crop environment [83]. Purple nutsedge tubers live longer in the soil than yellow nutsedge tubers. Desiccation can easily kill the tubers or at least reduce their viability. Yellow nutsedge tubers can withstand lower temperatures than purple nutsedge tubers. Generally, purple nutsedge tubers are more tolerant of higher temperatures, while yellow nutsedge tubers are resistant to lower temperatures. Differences to temperature extremes are possibly related to fatty acids, lipids or sugars present in the tubers [84].

Sprouting of tubers is controlled by apical dominance. The apical buds of the nutsedge tuber sprout first, inhibiting development of the distant buds and sprouting of the more basipetal buds. Bud sprouting and shoot emergence are faster from tubers located closest to the soil surface. Sprouting can occur at different times during the growing season, depending on favourable conditions. The first sprouting consumes more than 60% of the dry weight (carbohydrate, oil, starch and protein) of tubers of yellow nutsedge tubers, while the subsequent sproutings use less than 10% of these constituents, each time [82]. Purple nutsedge does not show seasonal dormancy as its tubers can sprout all year round under favourable temperature and moisture conditions. Yellow nutsedge tubers, however, display seasonal dormancy.

Weed Seedbank

The seedbank in the soil is the primary source of new infestations of weeds each year. The species composition and density of weed seed in soil vary greatly and are closely linked to the cropping history of the land. The size of seedbank in agricultural land ranges from near zero to as much as 1 million seed m^{-2} [33]. Seedbanks are generally composed of numerous species belonging to three groups. The first includes a few dominant species accounting for 70 to 90% of the total seed bank [92]. These species represent most of the weed problems in a cropping system. A second group of species, comprising 10 to 20% of the seedbank, generally includes those adapted to the geographic area but not to current production practices. The final group accounts for a small percentage of the total seed and includes recalcitrant seeds from previous seedbanks of the previous crop [93]. This group undergoes constant change due to seed dispersal by humans, other animals, wind and water.

Most of the seeds entering the seedbank come from annual weeds. A characteristic of many weed species is the potential for prolific seed production. Weeds present in agricultural fields usually produce less seed due to competition from the crop, damage from herbicides and other factors. The annual broadleaf weed *Xanthium strumarium* (common cocklebur) produces 7000 seeds plant^{-1} in the absence of crop competition and 1000 seeds plant^{-1} while growing in association with soybean [78]. Seed production of *Abutilon theophrasti* (velvetleaf) was reduced up to 82% by competition with soybean [57]. Shading by crop plants also reduces weed seed population. For example, 76% shade starting when soybean seedlings were 3 wk old reduced velvetleaf seed production as much as 4% [10]. Herbicide doses that do not kill weed plants may help reduce seed production. Sublethal doses of herbicides reduce seed production of several weed species as much as 90% [15].

The position of weed seeds in soil may influence their population dynamics. Seedbank dynamics fluctuate widely depending on the magnitude of new seed introductions and losses. Seedbanks vary from one cropping system to the other. In the Corn Belt of the USA, the average total seedbank densities of annual weeds ranged from 600 to 162,000 viable seeds m^{-2} [34]. The viable seeds that emerged as seedlings ranged from less than 1% for *Barbarea vulgaris* (yellow rocket) to 30% for *Setaria faberi* (giant foxtail). Of the weed species found, redroot pigweed and *Chenopodium album* (common lambsquarters) were the most frequently encountered species. This study also indicated that 50 to 90% of the total seed in the seedbank were non-viable.

Farming practices also influence weed seedbank dynamics and species composition by affecting the quantities of seeds returned to and removed from soil [92]. Cultivation aids in depleting weed populations in arable land. In the absence of input from seed production, the seedbank declines more rapidly with cultivation than without it [12, 18]. Tillage and depth of the tillage alter the seedbank. Bhowmik and Bekech[12] reported that the number of seeds of *Conyza canadensis* (horseweed) declined as tillage depth increased. No germinable seeds were found below 6 cm and at the no-tillage site. Soil tillage by mould-board ploughing turns seed up to the surface while burying the top-surface seeds. Ball [6] reported that weed seedbank declined more rapidly after mould-board ploughing. Row cultivation generally reduced seedbanks of most species.

Weeds are generally more productive in fertile soils, leading to more seeds being added to the seedbank. Redroot pigweed produces more seed under high soil fertility conditions [42]. Farmyard manure can be an important source of seeds for the seedbank. Seeds of many weed species can pass through the digestive tracts of animals and be deposited in a viable state in the dung.

The use of herbicides can drastically reduce weed populations in the soil. Atrazine used in maize monoculture reduced the seedbank by 98% after 6 yr [76]. However, 3 yr after atrazine use had been discontinued in plots treated for 3 yr, the seedbank had rebounded to half of its original density.

Biodiversity of Weeds

In spite of using manual and mechanical methods for thousands of years and adopting modern herbicide technology extensively for over 50 years, weeds have not disappeared from agricultural fields. As man controls certain weed species, new weed species always take their place. Even if control of a particular species is achieved, it is not always complete; certain plants of the same species are left behind, so they can take control of the whole field in the following year(s). In a situation wherein total weed-free crop production is possible for several years, a year of negligence or complacency in using an effective weed control system could lead to the field once again being choked with weeds. One of the most important reasons weeds are so successful, despite numerous and sustained efforts to eliminate them, is their biodiversity [26]. Biodiversity is an inevitable consequence of the struggle an individual weed species undergoes in the presence of neighbours and occupying a physical space in an agrosystem [26]. Biodiversity in weedy populations results from taxonomic diversity, as well as diversity in those traits that affect the survival, mortality, and reproduction of individual weeds [41].

Biodiversity also arises as a result of differential survival mechanisms of individual weed species. Heterogeneous weed populations exploit weaknesses in weed management strategies and adapt well to remain in competition with crop plants. This successful behaviour in a weed population is the aggregate of diverse, individual plant behaviours. Aggregate population behaviour is not always additive, and this non-linear response cannot always be predicted by summation of individual plant responses [27].

Successful weed management requires a thorough knowledge of weed biodiversity. Harper [41] defined five levels of diversity that a plant may meet among its neighbours: 1) genetic variants within a species, 2) somatic polymorphism of plant parts, 3) habitat microsite diversity, 4) age-state diversity within the community and 5) diversity of groupings at a higher level than species. Weed diversity is the adaptive response to these selection pressures. The more important first four levels of biodiversity are briefly discussed below.

Genetic Diversity

Genetic diversity or polymorphism is essentially diversity among and within weed species. Understanding the inherent taxonomic diversity in weeds is a primary need in weed management [26]. Correct weed identification is accomplished in most weed management situations by means of morphological polymorphism, i.e., heterogeneity in the form of a plant. This morphological diversity is utilized to guide weed control tactics as well as to provide insights into weed introduction and spread [26].

New weed genotypes are generated by **mutation, recombination, gene flow** and **segregation distortion** [80]. The genotypic variation often leads to unpredictable phenotypic biodiversity. The phenotype is the product of the interaction of genotype and environment, and the resultant phenotype is frequently unexpected [26]. For example, a single genetic mutation in *Brassica napus* leads to several unexpected pleiotropic (multiple phenotypic expressions) effects in this atrazine-resistant phenotype [25]. A single pair mutation to the **psbA** chloroplast gene, **codon 264**, causes a change in its product, the D-1 protein, a key functional element in PS II electron transport (vide Chapters 7 and 10). This change in the function of the D-1 protein results in a series of other functional changes, pleiotropic effects, in the plant phenotype. These include not only the primary effect of altered photosynthetic function leading to herbicide resistance [55], but also other effects such as changes in membrane lipid structure [69], reorganization of the functional units of chloroplast similar to those in shade-adapted leaves (e.g., changes in thylakoid stacking, loss of starch granules, etc.) [88], changes in stomatal function in leaves [25], changes in the plant's response to temperature [28] and changes in seed dormancy [22, 59]. Thus, a single genetic mutation could result in a dynamic reorganization of the entire organism as the different functional parts become adjusted to each other to reach equilibrium and homeostasis (interdependency among various elements) of the whole [26].

The dynamic reorganization is an example of how the behaviour of weed populations emerges as the aggregate behaviour at lower levels of plant organization [26]. These new phenotypes provide a species with important new sources of functional variants in the population for selection in agroecosystems.

Determining Genetic Diversity

Genetic diversity in a plant population can be determined by an allozyme marker system and two DNA-based marker systems, chloroplast DNA (cpDNA) restriction fragment length polymorphism (RFLP) and random amplified polymorphic DNA (RAPD) analysis.

Allozyme Marker System: **Allozymes** (protein profiles) are useful in documenting genetic variation in many organisms including weed species. This approach is useful when significant allozyme variation occurs. Allozyme marker systems developed for plants have been adapted to new species with minor modifications and provide a relatively inexpensive tool for assessing genetic variation at the molecular level. However, only limited numbers of allozyme loci can be assayed, thus precluding allozymes as a viable marker system.

CpDNA RFLP Analysis: **The Chloroplast DNA Restriction Fragment Length Polymorphism**-based marker technique provides a valuable tool for studying plant genetic diversity, evolution and polygenetics [29]. The chloroplast genome is advantageous over nuclear and mitochondrial genomes because there are multiple genome copies per chloroplast and multiple chloroplasts per cell. The chloroplast genome is relatively small in size and complete cpDNA sequences can be determined in plants. The slow rate of nucleotide substitution in cpDNA molecules is relatively free of large deletion, insertion and inversion events so that cpDNA polymorphisms can be detected within or between closely related plant populations [75]. The cpDNA analysis is an important source of biosystematic information over a wide range of taxonomic levels and a useful system for determining the genetic diversity of weed species.

RAPD Analysis: **The Random Amplified Polymorphic DNA analysis** is a recently developed technique used to determine genetic variation and biosystematic relationships between and within plant populations. It is, in fact, an extension of the methodology known as Polymerase Chain Reaction (PCR) which allows for amplification of specific DNA sequences from a very small amount of total DNA of an organism.

A key feature of RAPD analysis is that primers used in DNA amplification are composed of arbitrary sequences of 9 or 10 bases. This technique, also referred to as **Arbitrary Fragment Length Polymorphism (AFLP)**, is a powerful tool for identification, classification and characterization of closely related genotypes. RAPD markers are being used to identify crop varieties [91], screen genetic resources held in genebanks [65] and somatic hybrids [7]. It is also a useful technique to determine genetic diversity within a plant species [89]. Recently, different biotypes of *Elytrigia repens* were identified by RAPD markers [62].

The RAPD technology is simple to operate and faster for generating data. It does not require filter hybridization and the amount of DNA required is very low [66]. In spite of the several advantages offered by RAPD analysis, it is prone to artifacts caused by contamination of the reaction mixture by foreign DNA.

Although RFLP analysis provides more information than RAPD analysis because bands detected by the RFLP technique are closely related in sequence to the DNA probe used for selection [65], the speed and ease with which RAPD analysis can be done is a significant advantage to measure the degree of nuclear and cytoplasmic diversity in plants.

Role of DNA-based Marker Systems in Weed Biology and Biocontrol

The first step in developing a biological weed control programme involves determining the suitability of a weed as a target of biocontrol. Genetic diversity and genetic relationships between native and introduced populations have a significant bearing on the suitability of a weed for biocontrol. DNA marker data could be used as selection criteria for biocontrol by providing an estimate of a plant's genetic diversity relative to other potential target species [66]. Genetically heterogeneous, sexually reproducing weed populations may require a diverse array of biological events, while less variable, asexually reproducing species may require fewer agents [66]. Thus, weed species that rely on asexual forms of reproduction have a significantly greater chance of being controlled biologically than sexually reproducing species.

The bioagent-weed compatibility is the result of genetic expression of both organisms and environmental conditions. Although determining the molecular basis of bioagent-weed relationship is a desirable goal, studies of this type are time-consuming and costly [66]. The DNA marker systems offer a more efficient and cost-effective method of identifying compatible bioagent-weed relationship.

In studies related to identifying a compatible host-fungal relationship between rice plant and rice blast fungus (*Pyricularia grisea*), RFLP fingerprinting analysis enabled identification of 115 fungal genotypes of six distinct genetic lineages, with each linkage exhibiting virulence on a specific subset of rice cultivars [56]. This led to a better understanding of disease epidemiology and different breeding strategies for a more effective control of the fungal pathogen.

The above example could serve as a guide to improve the efficacy of biological weed control, although it requires a different approach involving identification of the subset of weed biotype that would be most susceptible to a given bioagent. The reader may refer to Chapter 11 for more information on biocontrol of weeds.

Somatic Diversity

Somatic polymorphism occurs when a single genotype develops a variety of phenotypes, each potentially occupying a different niche space [26]. Somatic polymorphism in a single genotype can be expressed by phenotypic variation in plant parts (e.g., different leaf morphologies on the same plant) and whole plant phenotypes (e.g., seed). This somatic differentiation, which is different from genetic segregation, is an adaptation allowing allocation of resources by an individual plant to different ends. This type of somatic diversity can be seen at the seed level (seed polymorphism) in some weed species such as *Avena* spp. [86], *Chenopodium album* [90], *Xanthium* spp. [41], and *Setaria faberi* [40], all of which shed seed from a single plant with different dormancy phenotypes.

Phenotypic plasticity is another type of somatic diversity. Phenotypic plasticity is the capacity for somatic variation in the phenotype as a result of influences of environmental factors on the genotype during development [26]. This type of weed diversity is apparent in the growth and development of *Abutilon theophrasti* (velvetleaf) plants in response to their light environment and the presence of neighbours [1, 2]. This is an important adaptation mechanism of individual weeds, enabling them to adjust their growth over short time periods to the resources available. When in competition with soybean plants, velvetleaf responds to light interference from neighbouring crop plants by shedding lower leaves and increasing internodal length to become taller than soybean plants [1, 2].

Weed competitiveness and crop yield interference can be highly variable due to phenotypic plasticity [26]. An understanding of the diversity of weed responses to neighbouring plants (both crop and weed) and the seasonal variation in environment is essential for assessing weed thresholds more accurately and implementing weed management strategies more effectively.

Habitat Microsite Diversity

Each plant in a field experiences a physical and biotic environment. This spatial (related to occupying a space) diversity of plant habitat varies with each individual plant and is a function of the factors influencing it. Harper [41] conceived that microsite diversity in plant populations is the result of contribution of the following four dimensions: a) diversity dependent on the use of different resources (e.g., differential use of nitrogen by a community of grasses and legumes); b) diversity dependent on lateral heterogeneity of environments (e.g., spatial variability of soil gradients in a field such as pH or moisture availability); c) diversity dependent on vertical heterogeneity of environment (soil depth) and d) diversity dependent on temporal division of the environment (e.g., seasonal variations of growth, flowering, etc.). Spatial variability of weed habitat has become more important now to weed management than in the past, as farmers are becoming more interested in precision farming technology [26].

Age-State Diversity

In an agroecosystem, plants with different life-cycle durations such as annuals (summer and winter), biennials and perennials coexist. The coincidence of growth and reproductive periods of these different life cycles contributes to the overall diversity experienced by an individual weed. The presence of different aged individuals (age-states) within a plant community is one of the important sources of diversity affecting plant biodiversity [41]. The diversity of seeds in a seedbank also constitutes age-state diversity. The longer the viability of seed in soil, the more age-states of that species that might be present in the seedbank and the greater the diversity of seed phenotypes available for infestation by that species in a growing season. This type of weed diversity has implications for weed management [26]. Decisions on weed threshold levels can be based on the age-state diversity potential of a weed population. For example, weed species whose seeds are short-lived (e.g., *Setaria* spp. or foxtails) in the soil seedbank may lead to threshold decisions quite different from those involving long-lived seeds (e.g., velvetleaf) [26].

Relationship of Biodiversity to Weed Management

In a crop-weed situation, crop plants have a lot more homogeneity and uniformity than weeds. In a crop selection programme, the primary criterion is uniformity in individual plants. As a result, the crop population comprises genetically and phenotypically uniform individuals. Such uniformity in crop plants allows no room for the type of plasticity that is needed to adjust to adverse environmental and soil conditions. Furthermore, this lack of diversity in crop population leaves a considerable amount of exploitable resources.

Weeds, on the other hand, are a heterogeneous collection of genotypes and phenotypes that can exploit the many niches left available by crop plants. Selection and adaptation are the special characteristics that weed plants possess and these occur at the level of the individual plant. The crop yield loss caused by weeds, by interfering with crop production, is the aggregate consequence of competition between the less efficient (in terms of utilizing resources and niches left available by crop plants) and homogeneous crop phenotypes and more efficient heterogeneous collection of weed genotypes and phenotypes. In this situation, the weeds will always win. Diversity in weeds ensures their enduring and persistent occupation of a field, and allows them to exploit new and diverse opportunities as they occur in an agroecosystem [26].

Understanding weed diversity in a field is essential for effective management of these weeds. Dekker [26] suggested that the following could aid in the development of sustainable weed management systems and practices.

1. Accurate identification of weed species and weed variants in a production field.
2. Adoption of cropping strategies and practices that increase cropping diversity in order to reduce the niches and resources left available to weeds. These include crop rotation, mixed cropping, herbicide rotation, tillage, etc. These strategies could change relationships between crops and weeds to favour crops.
3. Characterizing and understanding the spatial distribution of weed populations in an individual field, District (county), state, country or continent will provide insights into the factors and forces allowing their spread and enduring occupation. The forces and factors that lead to spots of heavy infestations and areas of no weeds in a field could be manipulated by changes in production practices.

Weed Parasitism

Parasitism is defined as the phenomenon of one living organism living in, on, or with another living organism to complete its life cycle. If a plant survives solely by associating with the living host, it is called an **obligate parasite**. If it survives by living on a living plant or dead plant material (saprophytically) it is termed a **non-obligate parasite**.

Although most parasitic flowering plants occur in about 10 families, the most troublesome parasitic weeds are derived from only four families, viz. Convolvulaceae (*Cuscuta*), Orobanchaceae (*Orobanche*), Scrophulariaceae (*Striga*) and Loranthaceae (*Arceuthobium* spp., *Phoradendron* spp. and *Viscum* spp.) [72]. Parasitic weeds are of greater importance in tropical agriculture than in subtropical and temperate agriculture. Of these, *Striga* (witchweed) and *Orobanche* spp. are root parasites, while *Cuscuta* spp., *Loranthus* spp. and *Arceuthobium* spp. are stem parasites.

Parasitic plants adapt well to find a suitable host plant quickly for their survival. Some parasites, such as *Cuscuta*, have large seeds with large food reserves sufficient to allow the radicle to grow extensively while it is seeking a host plant. Seeds of some other parasitic weeds germinate only when roots of host plants exude certain biochemicals. These include *Orobanche* and *Striga*, which show chemotropic growth of the radicle towards the root of their host plants.

The parasitic species of weeds can be divided into three groups: 1) those that do not have chlorophyll and depend on the host plant for total sustenance (*Orobanche* spp.), 2) those that can synthesize chlorophyll to only a limited degree (*Cuscuta* spp. and *Arceuthobium* spp.) and 3) others that can fix carbon nearly as well as the host plants (*Striga* spp.). *Striga* attaches to the roots of the host plant soon after germination but does not emerge from the soil for several weeks. During this period it is totally dependent upon the host plant. Once they emerge, Striga plants produce chlorophyll and begin to generate their own assimilates, although water and mineral nutrients are drawn from the host plant.

The major organ of parasitic weeds for attachment and penetration of the host tissue is known as the **haustorium**. Although haustoria vary in structure according to species, they all have a similar function, which is attachment and subsequent transport of materials from host plant to parasitic plant. The hyphae of the haustoria contact both xylem and phloem of the host plant for transport of water, minerals and assimilates. These haustoria are also believed to be involved in the transfer of hormones between host and parasite.

Striga is a root parasite on sorghum, millets, maize, sugarcane, etc. There are more than 30 species of *Striga* [47], widely distributed in tropical and subtropical regions of the world (Table 2.2). *Striga asiatica* is the most widely occurring species, particularly in India and Africa. In Africa, *Striga* causes more crop losses than the desert locust. After parasitization, the host crop plants become stunted, yellow and wilted because of loss of nutrients and water. The seeds of *Striga* survive for up to 20 yr in soil and one plant can produce 40,000 to 60,000 seeds, depending on the species. One host crop (maize or sorghum) can support up to 500 *Striga* plants. *Striga* seeds do not germinate in the absence of host-secreted stimulant (e.g., strigol) or an artificial stimulant such as ethylene.

The *Orobanche* parasite, which includes five species, occurs in tobacco, cotton, sunflower, tomato, carrot, soybean, sesame, fingermillet, etc. These species have the narrowest geographic range and broadest host range of the parasite families [94]. The seeds are viable for up to 20 yr in soil. Each *Orobanche* plant can produce up to 500,000 seeds and 1 g of seed contains up to 150,000 seeds. Like *Striga*, germination of *Orobanche* seeds is stimulated by secretions from the host's roots or from roots of non-host plants. The most predominant species, *Orobanche cernua* Loeff., is widely found in India.

Allelopathy

In the plant community, plants exert adverse or depressive effects on their neighbours by releasing toxic substances into the immediate environment. This phenomenon of one plant having a detrimental effect on another through the production and release of toxic chemicals has been termed **allelopathy**. The word **allelopathy** is derived from the Greek **allelo**, meaning "each other" and **patho**, an expression of sufferance or disease. The toxic chemicals may be released from living plant parts or when a plant dies. They are also produced during decay of plant tissue. Allelopathy differs from competition in that the latter is related to the removal or reduction from the environment of some resources (e.g., water, nutrients and light) required by some other plant sharing the habitat. Allelopathy exists in the plant ecosystem and occurs widely in natural plant communities. It is one of the mechanisms by which weeds affect crop growth. Allelopathy is believed to be a significant factor in maintaining the present balance among the various plant species. Exploration of the phenomenon of allelopathy will lead to a better understanding of plant survival and evolutionary strategies, and possibly clues for synthesis and development of newer herbicides.

Chemistry of Allelochemicals

Plants produce numerous chemicals during their growth and development. These compounds are released from the plants as vapour, as leachings from the foliage, as exudates from the roots, or in the course of breakdown or decomposition of dead plant residues.

Table 2.2. Predominant species of *Striga* and their occurrence in different countries (49).

Striga Species	Countries
S. angustifolia (Dem.) Saldanha	India, Indonesia
S. asiatica (L.) Kuntze	India, Indonesia
S. aspera (Willd.) Benth.	Sudan, Sahelian region of Africa
S. bilabiata (Thumb.) Kuntze	Sudan, Sahelian region of Africa
S. baumanii Engler	Congo, Kenya
S. brachycalyx Skn.	Sudan, Sahelian region of Africa
S. crysantha A. Raynal	Congo, Central African Republic
S. densiflora Benth.	India, Indonesia
S. elegans Benth.	Tanzania, Kenya, South Africa
S. forbesii Benth.	Sudan, Eastern and Southern Africa
S. fulgen Hepper	Tanzania
S. gesnerioides (Willd.) Vatke	Mediterranean countries, Yemen, Saudi Arabia, Oman, India, USA
S. hallaie A. Raynal	Congo
S. hermonthica (Del.) Benth.	Zambia, Sudan, Yemen, Saudi Arabia, Senegal
S. junodii Schinz. Skan	South Africa, Mozambique
S. klingii (Engler)	African countries
S. latericea Vatke	Kenya, Tanzania, Ethiopia, Somalia
S. ledormannii Pilger	Cameroon
S. linearifolia (Schumach et Thonn) Hepper	West Africa
S. macrantha Benth.	Western Sudan
S. masuria Benth.	India, China, Indonesia
S. passargei Engl.	Sudan, Sahelian region
S. primuloides Cher.	Nigeria, Ivory Coast
S. publiflora Klotzsch.	Ethiopia, Somalia
S. sulphurea Dalz.	India
S. curviflora Benth.	Australia, Papua New Guinea
S. multiflora Benth.	Australia
S. parviflora Benth.	Australia

Allelochemicals (the secondary metabolites in plant metabolism), chemicals with allelopathic potential, vary from simple molecules such as ammonia to complex conjugated flavonoids such as phlorizin (isolated from apple roots) or the heterocyclic alkaloid caffeine (isolated from coffee). Most of the allelochemicals (Table 2.3) are derived from such diverse chemical groups as organic acids and aldehydes, aromatic acids, terpenoids and steroids, long-chain fatty acids, alcohols, polypeptides, nucleosides, etc. [70, 71]. This diversity reveals the complexity of allelopathic chemistry.

The leaves of black walnut (*Juglans nigra*) produce a chemical called **Juglone** (5-hydroxy-α-naphthaquinone) which when washed off the foliage by dripping water was found to be injurious to plants in the vicinity [23, 81]. Juglone is a powerful toxin when injected into the stems of tomato. Gray and Bonner [36, 37] reported that the leaves of *Encelia farinosa* produced a substance called 3-acetyl-6-methoxybenzal- dehyde, which showed toxicity to many plants competing with *Encelia*. Martin and Rademacher [60] found liberation of scopoletin from roots of oat plants. The leaves of *Camelina alyssum* (flaxweed) produced p-hydroxybenzoic acid and vanillic acid, and these were inhibitory to flax (crop) plants growing alongside the weed [39].

Table 2.3. Allelochemicals found in plants

Chemical	Group	Source (plant/microbe)
Acetic acid	Aliphatic acid	Decomposing, decaying straw
Allyl isothiocyanate	Thiocyanate	Mustard
Arbutin	Phenolic	*Manzanita* shrubs
Bialaphos	Amino acid derivative	Microorganisms
Caffeine	Alkaloid	Coffee
Camphor	Monoterpene	*Salvia* shrubs
Cinnamic acid	Aromatic acid	Guayule plants
Dhurrin	Cyanogenic glucoside	Sorghum
Gallic acid	Tannin	Spurge
Juglone	Quinone	Black Walnut (tree)
Patulin	Simple lactone	*Pencillium* fungus on wheat straw
Phlorizin	Flavonoid	Apple roots
Psoralen	Furanocoumarin	*Psoralea* plants

Production of Allelochemicals

Allelochemicals produced by weeds affect crops and vice versa. They are produced by any plant organ, with roots, seeds and leaves being the most common sources. Flowers and fruits may also produce allelopathic chemicals but these are of less value in terms of allelopathic effect.

Allelochemicals enter the environment through volatilization, foliage leaching or root exudation. They may also result from decomposition of plant residues. The production of allelochemicals is influenced by the intensity, quality, and duration of light, with a greater quantity produced under ultraviolet (UV) light and long days [3]. In a crop-weed situation, the canopy of crop plants is expected to overshadow weeds and to filter the UV light, resulting in production of lower quantities of allelochemicals by weeds. Greater quantities of allelochemicals are produced under conditions of mineral deficiency, drought stress and cool temperatures than at the more optimal growing conditions. Stress condition accentuates the involvement of allelopathy in crop-weed interference and that competition for limited resources may increase the allelopathic potential or sensitivity of the weed, the crop, or both [94]. Although chemicals with allelopathic activity may be present in tissues of many plant species, their mere presence may not mean that allelopathic effects will ensue.

Numerous chemicals, present in plant material, are also released when plant residues are left on the soil surface after harvest or ploughed under. They are released during decomposition of the plant residues. The subsequent plant mortality or growth suppression may not be directly related to these toxic organic substances, but could be as a result of modification in soil microenvironment (e.g., localized alteration in soil pH or other conditions as a result of litter decomposition) [72].

EFFECT OF ALLELOCHEMICALS

Inhibitory effects: There are innumerable reports on the effect of weeds on crop plants. Root exudates of thistles (*Cirsium* spp.) injured oat plants in the field while root exudates of *Euphorbia* and *Scabosia* injured flax [24]. The yield of flax was greatly reduced by the inhibitory substances produced by the leaves of *Camelina alyssum* (flax weed) growing alongside the crop plants [39]. These chemicals were identified as p-hydroxybenzoic acid and vanillic acid [39]. Aqueous extracts of *Convolvulus arvensis* (field bindweed) and *Cirsium arvense* (Canada thistle) inhibited the germination of seeds and growth of seedlings of many crops [44]. In a set of elegant experiments,

Gressel and Holm [38] investigated the effects of aqueous extracts of weed seeds on germination of seeds of different crop species (alfalfa, cabbage, carrot, pepper, radish, tomato, turnip, etc.). The weed species included *Amaranthus retroflexus, Abutilon theophrasti, Chenopodium album, Datura stramonium, Digitaria sanguinalis, Echinochloa crus-galli, Polygonum pennsylvanicum, Portulaca oleracea*, etc. They found that some varieties of carrots, tomatoes and peppers were unaffected while other varieties were highly susceptible. They also found that the inhibitory substances in seeds of *Abutilon* came primarily from the embryo and endosperm. These compounds behaved like amino acids and they were water-soluble, ether-soluble, heat-stable and amphoteric.

Bhowmik and Doll [13] reported soybean yield reductions ranging from 14 to 19% by the extracts of dried residues of several annual species: *Chenopodium album, Amaranthus retroflexus, Abutilon theophrasti* and *Helianthus annuus* (sunflower). Besides, corn yields were also reduced by annual weeds including *Echinochloa crus-galli* and *Setaria viridis* in the field. The inhibitory effects of these weeds were not related to nutrient uptake [14].

Quackgrass (*Elytrigia repens*) is a good example of the allelopathic effect of a weed on crop plants. Osvald [68] observed that the competitive action of quackgrass was partially the result of toxin produced by its roots. Later, Kommedahl *et al.* [53] found that quackgrass was able to inhibit growth of crop seedlings grown in the field on the soil previously used by the weed as also in potted soil containing ground quackgrass rhizomes in the greenhouse. Quackgrass apparently produced a toxin through leaves, roots and seeds. Extracts of 2- and 4-wk-old seedlings of quackgrass showed the same toxicity as leaf extracts from older plants [67]. Experiments conducted by Buchholtz [17] showed that the inhibitory effect of quackgrass grown along with maize was caused partly by a reduction in the availability or absorption of nutrients by maize plants and partly by a systemic effect of quackgrass vegetation on maize. Abundance of moisture did not alleviate the systemic effect. It was possible that chemical compounds exuded by quackgrass rhizomes interfered with the uptake of nutrients by maize.

A similar allelopathic effect is also exhibited by *Imperata cylindrica* (thatchgrass). *Imperata* inhibits the emergence and growth of an annual broadleaf weed *Borreria hispida* in the soil by exuding inhibitory substances through rhizomes. Rhizome extracts, however, do not affect the emergence of *B. hispida* seedlings. Bell and Koeppe [8] reported that the yield reduction of maize by infestations of *Setaria faberi* (giant foxtail) was due to the inhibitory effects of exudates of mature *S. faberi* roots and leachates of dead roots.

Rizvi et al. [74] isolated 1,3,7-trimethylxanthine (1,3,7-T) from *Coffea arabica* and found it inhibiting germination of *Amaranthus spinosus*. Treated seeds showed a marked reduction (about 29.4%) in amylase activity and this could not be counteracted with GA. Similar experiments with seeds of *Phaseolus mungo* showed no effect of 1,3,7-T on germination or amylase activity.

Several workers have found that crop plants show allelopathy on weeds. In a study on the interactions between two weed species, *Chenopodium album* and *Amaranthus retroflexus*, and two crop species, maize and *Lupinus albus*, Dzubenko and Petrenko [31] found that root extractions of crop plants increased catalase and peroxidase activity of weeds and inhibited their growth. Neustruyeva and Dobretsova [64] observed that wheat, oats and peas suppressed growth of *C. album*. Oats exerted an allelopathic effect on *C. album* in addition to their competitive effect. Bell and Koppe [8] suggested that allelopathic effects of crop plants on weed seedlings may be as real as the allelopathic response in the reverse direction and that such phytotoxic compounds might act as natural herbicides.

Stimulatory Effects: Actively growing plants may also produce chemicals that stimulate seed germination and growth of other plant species. Root exudates of some plants contain germination stimulants of seeds of parasitic weeds of the Scrophulareaceae family. For example, dormant seeds of obligate root parasite *Striga lutea* (L.) Kuntze (witchweed) require an exogenous stimulant for germination. Seeds of *Striga* remain dormant in the soil until conditions are favourable for germination and until the seeds come into contact with exogenous stimulants, which are exuded from roots of hosts as well as several other plants [32]. Seeds must be located within a few millimetres of the donor root in order to be exposed to threshold concentrations of the germination stimulant.

Strigol, a highly active stimulant of witchweed seed germination was isolated and identified in root exudates from hydroponically grown cotton, which is not a host of this parasite [21]. Strigol enhanced witchweed germination by 50% at 10^{-11}M concentration and 90-98% at 10^{-10}M. Similarly, maize roots exuded a complex of stimulatory substances which promoted the germination of *Orobanche minor* (broomrape), a parasitic weed in tobacco [85]. Kinetin [6-(2-furfuryl)aminopurine] and certain other 6-substituted aminopurines exuded by the roots of the host (sorghum) stimulate the germination of *Striga*.

Potential of Allelopathy in Weed Management

Allelopathy has the potential to play a prominent role in weed management if it can be properly harnessed. Generally, all plants have allelopathic potential and show some susceptibility to allelochemicals when presented in the right amount, form and concentration at the appropriate time. Adequate work has been done so far to conclude that allelopathy could be utilized for development of a new weed management strategy. However, in order to realize its full potential at the field level much more work is needed in the following areas:

1. Exploration of inhibition of seed germination and seedling emergence and growth of various weed species by potential allelopathic plant (crop) species.
2. Incorporating or enhancing allelopathic activity in crop plants, so that weed suppression could be made more effective. Using plant residues in cropping systems, allelopathic rotational crops or companion plants for allelopathic activity. Crop rotations and multiple cropping should be studied from the point of allelopathic potential. These cropping systems may hold valuable lessons for the development of allelopathy as a useful weed management tool.
3. There are several phytotoxic compounds implicated in allelopathy. Their chemistries may be used to develop newer herbicides. For example, plants produce a phytotoxic substance 1,8-cineole, a terpenoid. It has structural similarity with cinmethylin, a herbicide that controls several annual grasses and suppresses some broadleaf weed species. Although cinmethylin is synthesized, it could have been derived from the known phytotoxicity of the allelochemical 1,8-cineole. Thus, there may be several allelochemicals that could be used to produce and develop new herbicides.

In the years to come, allelopathy may become another weed management tool to be placed in the armoury of farming technology and used in combination with other techniques. It may not, however, be a panacea for all weed problems [94].

Methodologies for Studying Allelopathy

The methodology chosen to determine the existence of naturally occurring toxic substances must be effective and easily reproducible. Allelochemicals may be isolated by using organic solvent extraction or cold-water infusion. The physiological activity of a toxin may be studied by using various bioassay procedures. The test plants or seeds should be exposed or treated with the putative allelochemical. Meaningful interpretation of results can be drawn when only the toxins "released" by allelopathic plants are used to test for physiological activity. Although the extracts of living or dead plant material could be useful in testing for allelopathic effects, they are not the same as toxins naturally released into the environment.

One technique widely used for isolation of inhibitory root exudates is the stair-step system. In this system the donor (allelopathic) and recipient (test) plants are grown separately in sand solution with the pots alternated in a stair-step fashion [72]. The soil solution is circulated from donor plant to recipient plant and back again several times (Fig. 2.1). In a modification of this technique, an exchange column is inserted between donor and recipient plants, so that the substance exuded by roots of the donor can be isolated and bioassayed for phytotoxicity. In these systems, water, nutrients, light and air must be the same at each step of the staircase and never limiting.

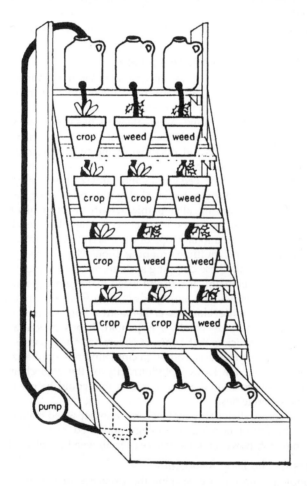

Fig. 2.1 Schematic representation of the stair-step technique for the study of allelopathy [72]

Newer methodologies and instrumentation need to be developed to fully utilize the potential allelopathy offers for effective weed management.

WEED ECOLOGY

Ecology is the interrelationship between organisms and their environment. Weed ecology is thus concerned with growth characteristics and adaptations that enable weeds to survive changes in the environment. The environment includes climatic, edaphic and biotic factors and it determines the distribution, prevalence, competing ability, behaviour and survival of weeds. Man plays an important role in changing the environment by altering the crop husbandry practices and by maintaining weed-free monocrop or multicrop culture. Monocrop cultures (monocultures) utilize moisture, nutrients and light available for plant growth in a way different from multicrop cultures and this influences the ecological requirements for weed growth, behaviour, competition and survival.

Persistence of Weeds

Persistence is a measure of the adaptive potential of a weed that enables it to grow in any environment. In an agricultural situation, the cropping system with its associated habitat management practices determines the persistence of weed species and thus the particular crop-weed association. A persistent weed species will not necessarily pose a hazard if suitable control measures are applied. Persistence of a weed is largely influenced by climatic, edaphic (soil) and biotic factors which affect its occurrence, abundance, range and distribution.

Climatic Factors

The important climatic factors of the environment that affect persistence of weeds are light, temperature, wind and humidity.

Light intensity, quality and duration are important in influencing the growth, reproduction and distribution of weeds. Photoperiod governs flowering and time of seed setting and maturation. Photoperiod also has an important bearing on the evolution of various ecotypes within a weed species. Tolerance to shading is a major adaptation that enables weeds to persist. Competition for light is most under high soil fertility and adequate moisture conditions, conducive for better plant growth and larger foliage area. Weeds with large leaf area indices (LAI) outcompete plants with smaller leaf areas. Successful competitors do not necessarily have more large leaf area but have more foliage positioned at an appropriate angle for greater light interception. Thus, plants with horizontal leaves are more competitive for light than those with upright leaves. Plants with opposite leaves are less competitive than those with alternate leaves. Similarly, weeds that are tall or erect are better competitors for light than short, prostrate plants. Heavily shaded weeds suffer from reduced photosynthesis, resulting in poor growth, a smaller root system, and a reduced ability for water and mineral absorption.

Temperature of atmosphere and soil affects the latitudinal and altitudinal distribution of weeds. Soil temperature affects seed germination and dormancy, which is a major survival mechanism of weeds. The winter survival of underground parts (rhizomes, bulbs, tubers, stolons, etc.) of perennial weeds is very much dependent on their resistance to extremes of freezing temperature in the soil. The effects of temporary temperature extremes on the aboveground parts of weeds in a season are, however, only transitory and of minor significance to the survival and persistence of weed species.

Rainfall and water have a significant effect on weed persistence and distribution. The weed species found in desert conditions differ from those of aquatic environments. Weeds of temperate regions are not always found in the tropics. The distribution pattern of rain is a determining factor in utilization of water supply by the plant, since water shortages at critical stages of growth are often responsible for failure of reproduction and survival. Generally, weeds use about the same amount of water (per unit of plant growth) as do the crop plants with which they compete; but they are better explorers of water from the soil than crop plants. As roots grow more rapidly early in a plant's life, competition for water and nutrients usually begins before competition for light. Competition for water is determined by the relative root volume occupied by competing plants and will be greatest when roots closely intermingle, and crops and weeds try to obtain water from the same volume of soil [78].

Velocity, frequency and direction of wind could also restrict or limit the occurrence and persistence of all plants including weeds. Wind is very much modified by the topographic features of the habitat, such as the altitude, slope and surface. Wind plays a role in stabilizing the oxygen and carbon dioxide balance in the atmosphere. It also modifies the transpirational losses from plants. Wind is also a principal factor in the dissemination of weeds.

Thus, climate has a profound effect on the persistence of weeds, which can adapt to a wide variety of climates. It has an effect on the structure and composition of weeds. Climate can also effect variations in cuticle development, pubescence, vegetative growth, vigour, competitiveness, etc.

Soil Factors

The soil factors which influence weed persistence are soil water aeration, temperature, pH and fertility level, as also cropping system. Weeds are found in soils differing quite widely in physical characters, soil moisture-holding capacities and soil reaction. The fact that they are weeds indicates that they have adaptability to a wide range of soil environment.

It is generally known that some weed species are characteristically 'alkali' plants, also known as **basophiles** (pH range 7.4 to 8.5) which can grow well in alkali soils. Alkaligrass (*Puccinallia* spp.) and *Elytrigia repens* (quackgrass) are the best examples of basophiles. Similarly, weed species such as *Cynodon dactylon* (Bermudagrass), *Digitaria sanguinalis* (large crabgrass), *Rumex acetosella* (red sorrel), *Pteridium* spp., (brackenfern) and *Borreria* spp., (buttonweed) inhabit only acidic soils. These weeds, which have a pH range of 4.5 to 6.5, are termed **acedophiles**. Similarly, there are weeds known as **neutrophiles**, which grow best at a pH range of 6.5 to 7.4. Several species of Compositae and Polygonaceae grow well in saline soils. These weeds, characteristic of soils of different reactions, serve as indicator plants. A shift in soil pH, for example, towards the acid side due to continuous use of ammonium sulphate as a nitrogen source could also cause a shift in the weed spectrum.

Many weeds can adapt and grow well in soils of a fertility level well below that required for optimum crop yields. For example, *Imperata cylindrica* (thatchgrass, cogongrass or alang-alang) grows well in soils of low fertility. But it can adapt well to soils of high fertility. Generally, soils which can support the crop growth are excellent for weed growth as well. Prolific weed growth may indicate the abundance of mineral nutrients in the soil.

Some weed species, e.g. *Commelina benghalensis* (tropical spiderwort), thrive in moist soil conditions while perennial grasses such as *Imperata cylindrica* and *Elytrigia repens*, etc., can persist even in drought conditions. Weeds such as *Typha* spp. live only in waterlogged soils.

Biotic Factors

Plants and animals are among the biotic factors that modify the growth of weeds in a variety of ways that affect weed persistence directly and indirectly. In a cropping situation, the major effects on weeds are those exerted by the crop as it competes for available resources. Once certain weed species are introduced, their persistence in a given crop is determined largely by the degree of competition offered by the crop. The competitive ability of a weed, and hence persistence, depends on its vegetative habit, readiness of seed germination, rate of seedling growth and the extent and nature of root and top growth. Besides, the agricultural practices associated with the growing of a crop may encourage or discourage specific weeds.

Crops that serve as hosts to parasitic weeds and crop-induced stimulants and toxins are examples of other biotic factors.

Survival Mechanism of Weeds

The seed is the primary means of survival mechanism of annual weeds. The vegetative plant propagules such as buds, rhizomes, tubers, bulbs and stolons offer an additional survival mechanism for perennial weeds. The major adaptations for survival of weeds include prolific seed production, survival of vegetative propagules under adverse conditions, seed dissemination and dormancy, and ability of weeds and propagules to resist any detrimental effects of the environment. Of these, dormancy is probably the single most important characteristic of weeds that enable them to survive when crops cannot.

Seed Dissemination

Most weeds are good travellers. They use various forces or agents to transport and scatter themselves from place to place. Of all the agents by which weeds are disseminated, wind, water, animals and man play important roles.

Wind: Many weeds have modifications or adaptations which aid them in being scattered by the wind. Seeds or small fruits with tufts of hair or wing-like appendages are carried by the wind over long distances; the lighter seeds may drift for miles. The various modifications of seed structure that equip them for wind dissemination are termed **saccate, winged comate** (hair-covered), **parachute** and **plumed**. The achenes of dandelions (*Taraxacum officinale*), thistles (*Cirsium* spp.) and species of Compositae floating in the air on a windy day is a common sight. Similarly, the seeds of *Imperata cylindrica* and *Saccharum spontaneum* are carried away by the wind to far away places. The silky floss of *Asclepias syriaca* (common milkweed) is flown away by wind to long distances.

Water: Seeds of many weed species are light or are covered with an oily film, enabling them to float on the water surface. Such seeds are frequently washed into streams and are carried to other fields lower down the valley. Some weeds have air-filled membranous seed envelopes or corky adjuncts of mature fruits which can float on the surface of water. Floodwaters, running streams and irrigation water are important in the spread of weed seeds. Millions of seeds pass a given point of an irrigation canal in a day. Weeds such as *Ambrosia* spp. (ragweed), *Amaranthus* spp. and *Xanthium strumarium* are the best examples of weed dissemination by water. Some weeds can remain viable even after a long time of storage in water.

Animals: Weed seeds are also disseminated by animals. Many seeds pass through the digestive tracts of animals without loss of viability. Birds also consume large quantities of weed seeds and scatter them in droppings. The dispersal of seeds in the form of incompletely digested materials passing through the animals is termed **endozoochory**. The best example of endozoochory is *Prosopis* spp., introduced in certain parts of southern India as a roadside hedge. Seeds also stick to the fur, feathers and muddy feet of birds and animals, and are carried from place to place. This method of dissemination is not, as important however, as the other means.

Man: Seeds of many weeds have specialized structures such as hooks, spines, barbs and awns which tend to cling to man's clothing and footwear or agricultural implements used by man. Fruits of *Tribulus terrestris* (puncturevine) have sharp spines which cling to animals and man and get dispersed far and wide [77]. Similarly, the seeds of *Chrysopogan aciculatus* (silkgrass) have awns and get disseminated by sticking to human clothing. They are carried in packing materials, and in soil and sand or gravel used in construction. Thus, man becomes the most important agent in the dissemination of weeds. He often carries weed seeds of interest from one part of the country to the other and from one country to another. It is due to man's indifference or carelessness that many weeds are 'imported' through various means. The movement of commercial seeds and grain is an important means of weed dispersal by man. A clearcut example is the widespread occurrence in India of *Phalaris minor* (littleseed canarygrass) whose seeds contaminated the wheat grain imported in the 1960s.

Seed Germination

Germination is a critical factor in the establishment of weeds. Germination results only if the environmental conditions required for the mechanism are favourable in a tangible expression of the interactions of factors of climate and soil. Those weed seeds that germinate under the same conditions and at the same time as crop seeds do are the most persistent and successful.

For germination, seeds of both crops and weeds must have adequate soil moisture, favourable temperature and supply of oxygen. Weed seeds, however, possess a variety of special germination mechanisms adapted to changes in temperature, soil moisture, aeration, exposure to light, depth of burial of seeds, etc. When conditions are unfavourable for germination, they can remain dormant or delay germination.

Seeds of many weeds require exposure to light for germination. This is regulated by a bluish-green protein pigment called **phytochrome**, which is responsible for seasonal change in light sensitivity of seeds. Dark germination of photosensitive seeds is promoted by **red (R)** light and this is reversed by an exposure to **far-red (FR)** light. A number of weed species respond to this red-far red reversal mechanism of germination controlled by phytochrome. Germination of the seeds of *Lactuca* spp. (lettuce), *Xanthium* spp. (cocklebur), *Lepidium* spp. (pepper weed), and *Rumex* spp. (sorrels) genera is controlled by phytochrome [50].

Soil disturbance has a beneficial effect on germination due to greater availability and aeration. The quantity of oxygen in the soil is greatly influenced by porosity of the soil and microbial activities. Many weeds germinate under aerobic conditions while some require anaerobic conditions. Soil turnover during ploughing and other land operations exposes the seeds to light and induces germination.

Periodicity of germination is another specialized germination mechanism of weed seeds. Some weeds such as *Amaranthus* spp. have a definite pattern of peaks of germination at regular intervals. Summer annuals favour higher temperatures for germination and one can see flushes of weed seedlings when temperatures reach favourable levels. Winter annuals germinate at lower temperatures and shorter days in the autumn and even in winter. Some weeds can germinate freely throughout the year.

Viability is an integral part of weed infestation. The actual number of seeds that germinate determines the intensity of weed infestation.

Seed Dormancy

Dormancy is a state in which a viable seed fails to germinate even under conditions of moisture, temperature and oxygen favourable for plant growth. It is also a type of resting stage for the seed. It controls the time of the year that a species germinates or delays germination for years, thus guaranteeing viable seed in the soil for several years. When the conditions for germination are not favourable, seeds can become dormant to survive in the soil and persist as serious infestations despite frequent soil disturbances. Many annual weeds produce dormant seeds that germinate under a narrow range of environmental conditions. Seed dormancy is a remarkably efficient survival mechanism of weeds. Seeds of weed species of Boraginaceae, Convolvulaceae, Cucurbitaceae, Leguminosae, and Graminae have a long dormancy period often running into several years [77].

Seed dormancy may be of three types: **inherent, induced** and **enforced**. Inherent or innate dormancy is due to the general background of the plant species and is characterized by: a) rudimentary embryos, b) physiologically immature embryos resulting from an inactive enzyme system, c) mechanically resistant seed coats or scales which prevent embryo expansion, d) impermeable seed coats which prevent entry of water and/or gases and e) excessive presence of inhibitors.

Induced and enforced dormancy depends on the interaction of seed with the environment. Induced dormancy develops when the non-dormant seed becomes dormant after exposure to specific environmental conditions. This type of dormancy usually persists even after the conditions have changed.

In enforced dormancy, limitations of the habitat or environment prevent non-dormant viable seeds from germinating, and germination occurs only when the limitations are removed. Exposure of seeds to sunlight through soil turnover by tillage and optimum temperature and moisture conditions terminate enforced dormancy. Weeds such as *Borreria hispida, Spermacoce ocymoides, Scoparia dulcis,* etc. germinate continuously in the growing season until the environmental conditions become unfavourable (low temperature, drought, etc.).

Induced and enforced dormancy may result when ploughing and other tillage practices bury weed seeds. A shallow tillage later may remove the conditions of enforcement and promote rapid germination of seeds.

MECHANISMS OF SEED DORMANCY

Seed dormancy is controlled by four relatively distinct developmental phases: a) **inductive**, b) **maintenance**, c) **trigger** and d) **germination** [5].

Inductive Phase: Seed dormancy begins with an inductive phenomenon. Certain events during the maturation of seeds may lead to onset of dormancy. These events may be triggered environmentally by light and temperature, or chemically, and are termed **photoinduction, thermoinduction,** and **chemoinduction** respectively.

Dormancy and germination of seeds are regulated by a critical balance of inhibitor-promoter complexes. During seed maturation this balance is shifted in favour of the inhibitor, imposing dormancy. This may happen either through curtailment of synthesis of promoter substances or through build-up of inhibitory components or by a direct antagonism between the two. Abscisic acid (ABA) is antagonistic to gibberellin (GA). Endogenous growth promoters are in high concentration in the early maturation of seeds but decrease with cessation of embryo growth and ripening of fruit. Dormancy may also be induced by the formation of impermeable seed coats, which impose anaerobic conditions on the seed, resulting in build-up of growth retardants.

Maintenance Phase: During the maintenance phase of dormancy the metabolic activity of the seed is very much reduced by blocks at the specific metabolic sites. These metabolic blocks are due to the presence of endogenous inhibitors (e.g. ABA) which are either antagonistic with endogenous promoters (e.g. GA) or interfere with their synthesis. Hormone promoters help in the synthesis and activation of hydrolytic enzymes (α-amylase, protease, lipase, etc.). Thus, dormancy is due to lack of or inactivity of hydrolytic enzymes.

Trigger Phase: In a dormant seed, germination is triggered by a factor or an agent that elicits germination. This **triggering agent** should continue to be present for completion of germination. This agent may be a photochemical one, involving phytochrome-red light mechanism, a thermochemical one involving stratification, the inhibitor-removal mechanism involving scarification and leaching, or the inhibitor-promoter complex favouring the promoter. During this phase the inhibitor is either removed or its effect counteracted by activation or synthesis of the promoter.

Germination Phase: This process involves breaking of seed dormancy and subsequent germination. It is under hormonal control and the naturally occurring hormones (auxins, gibberellins and cytokinins) function as germination agents via the inhibitor-promoter complexes. Of these, gibberellins are the predominant germination agents early in the germination phase during the food-reserve degradation stage. Cytokinins exert their influence later, on the initiation of cell proliferation and expansion.

Vegetative Reproduction

Vegetative or asexual reproduction is a major survival mechanism of persistent weeds, which are generally characterized by deep-root systems and large number of dormant buds, bulbs, tubers, rhizomes, etc. Dormancy and presence of reserve food supplies are common characteristics of these underground plant parts. The relative immunity of the underground organs to destruction by soil disturbance or tillage causes a great problem in eradicating most perennial weeds.

Perennial weeds live for three or more years. The potential for survival is influenced by lifespan, depth of propagule penetration, depth from which regeneration occurs, the age at which the seedlings assume perennial characteristics, the potential for seed production, and the resistance of the plant and its organs to control measures.

Perennial grasses such as *Imperata cylindrica* and *Elytrigia repens* which propagate through rhizomes, have the ability to regenerate even from a small piece of rhizome tissue and have abundant vegetative buds on the rhizomes; these characteristics largely compensate for their vulnerability to soil disturbance. Purple nutsedge (*Cyperus rotundus*) and yellow nutsedge (*Cyperus esculentus*) propagate by seeds and also by means of tubers that have varying degrees of dormancy. The types and mechanisms of tuber dormancy are similar to those that characterize seed dormancy. Increased oxygen availability in the soil promotes sprouting of tubers while high levels of carbon dioxide inhibit sprouting. Some perennial weeds have root systems that extend several feet below the top soil layer and regenerate from the depths. Broadleaf perennials such as *Oxalis* spp. (woodsorrel) propagate by bulbs and bulbils (bulblets) as well as seeds. These bulbils fall to the ground and germinate. Another perennial broadleaf weed,

Asclepias syriaca (common milkweed), propagates by rootstocks and seeds. The dormancy of propagules is dependent on root reserves.

Thus, perennial weeds, through their ability to store food materials in the underground parts and propagule dormancy, can adapt to most cropping systems and survive despite man's efforts to eradicate them. However, perennial weeds are a more serious problem in crops in which soil disturbance is kept to the minimum.

Weed Competition

Weed competition with crops is a part of weed ecology. The word competition comes from the Latin word **competere**, which means to ask or sue for the same things another does. Competition in ecology involves two or more organisms seeking for a particular factor, thing or material when they are in short or limited supply.

Weed competition is complicated because various factors affect the extent to which it occurs. It affects the growth of crop plants. The total effect of the competition as reflected in crop growth and yield results from competition for nutrients, moisture and sunlight. As a general rule, for every unit of weed growth there will be one less unit of crop growth. The plants that germinate first and grow fast tend to exclude others. The first plants that occupy an area have an advantage over the latecomers. If weeds emerge after the crop is well established, they may not pose as serious a problem as those that emerge before the crop plants emerge and establish. Competition between plants is maximum when available resources for crop growth become limiting. Competition between crops and weeds is most severe when the competing plants have similar vegetative habits and demands upon resources.

Many weeds are prolific producers of seeds which remain viable or dormant in the soil depending on the adaptive mechanism used. When the soil is disturbed by tillage, weed seed germination takes place and the weeds emerge before or along with the emergence of crop plants. A vigorous crop or crop variety can enjoy a temporary competitive advantage over weeds, but this advantage is lost in the absence of timely and proper control measures. Under exceptional conditions, an extremely vigorous crop may smother the weeds and remain weed-free for a significant period of time. Thus, dominance of a habitat by the crop or weeds is dependent on the rapidity of germination, seedling establishment and subsequent growth. Differences in photosynthetic area, root development, root growth, etc. determine the competitiveness of plants.

The degree of weed competition is determined by the weed species infesting the area, density of infestation and duration of infestation.

Weed species differ among themselves in competitive ability. For example, the annual grass *Monochoria vaginalis* (monochoria) is a more severe competitor than another annual grass, *Echinochloa crus-galli*, var. *oryzicola* (barnyardgrass) in rice culture [20]. Chang [19] found heaviest losses in rice yield under the infestation of barnyardgrass than of monochoria. The yield losses in rice from barnyardgrass at low and high soil fertility levels were 81% and 88.8%, and from monochoria 64.6% and 58.9%. Similarly, *Avena fatua* (wild oats) was found more competitive in wheat than *Setaria viridis* (green foxtail) [30].

Increase in weed population has a direct effect on crop yield reduction. Alkamper [4] found that oat yield was reduced substantially with an increase in density of weed population. Similar reduction in grain yield with an increase in density of barnyardgrass infestation was also reported in rice by Smith [79] and Chiska [20]. Population density influences the uptake of nutrients.

The duration of weed infestation and the time of weed elimination have a great influence on crop growth and yield. Weeds that are not controlled within two or three weeks of emergence usually affect the yield. In most crops, weed infestation during the first 3 to 8 wk is very critical.

In situations ideal or otherwise for crop and weed growth, weeds always thrive well. Generally, weeds have higher contents of nutrients than crop plants; they grow faster and absorb nutrients earlier, with the result that there may be a lack of nutrients for the crop plants. Further, the weeds are capable of absorbing just as much or even bigger amounts of nutrients than crop plants.

In long term trials conducted by Kolbe [52], it was found that for every 1% increase in yield, weed infestation should be eliminated to the extent of 3% in winter barley, 2% in winter wheat, 0.5% in maize, 1% in potato, 3% in tomato and 8% in apples. This shows that crops differ appreciably in competitive ability with weeds. This is further supported by Kondap *et al.* [54] who reported that soybean had lower competitive ability with purple nutsedge (*Cyperus rotundus*) followed by groundnut (*Arachis hypogea*), blackgram (*Phaseolus mungo*), greengram (*Phaseolus aureus*), horsegram (*Dolichus biflorus*), cowpea (*Vigna sinensis*), sorghum and maize. Yield reductions from nutsedge (nutgrass) infestation were 58% in soybean, 32% in groundnut, 23% in black gram, 23% in green gram, 22% in horse gram, 16% in cowpea, 12% in sorghum and 6% in maize.

In a review on yield losses due to weed competition, Mani et al. [58] reported that yield reduction in wheat ranged from 6 to 35%, in rice 9 to 51%, in maize 29 to 74%, in millets 6 to 82%, in peas 25.3 to 35.5%, in carrot 70 to 78%, in groundnut 30 to 33%, in linseed 31 to 39%, in sugarcane 14 to 72% and in cotton 21 to 61%.

Characteristics of Efficient Weed Competition

The growth and yield of a plant are directly related to its efficiency to assimilate carbon by photosynthesis, which is dependent on many physiological and environmental factors. Black et al. [16] classified plants as efficient and non-efficient on the basis of the following factors:

1) response to light intensity, 2) response to temperature, 3) response to oxygen, 4) presence or absence of photorespiration, 5) pathway of photosynthetic carbon dioxide assimilation and 6) photosynthetic compensation point level.

Black et al. [16] hypothesized that efficient plants are used in agriculture because of their high production and competitiveness. Efficient plants fix carbon even at higher light intensity levels and temperatures, while non-efficient ones cannot do so. In efficient plants, photosynthesis is not inhibited by oxygen. At normal oxygen concentration (21%), photosynthesis is inhibited in non-efficient plants.

Respiration in the presence of light, called **photorespiration**, is wasteful. Photorespiration, found in non-efficient plants, is wasteful because light stimulates oxidation of photosynthetic intermediates to CO_2 and other products. Efficient plants do not exhibit photorespiration.

Carbon fixation by plants, occurs by two different pathways, i.e., the C_3 **cycle** and C_4 **cycle**. In the C_3 cycle, ribulose diphosphate (RuDP) is the CO_2 acceptor, while in the C_4 cycle CO_2 is fixed in a 4-carbon dicarboxylic acid (malic and aspartic) cycle, with phosphoenolpyruvate (PEP) being the CO_2 acceptor. The C_4 cycle is more efficient than the C_3 cycle because PEP has a much higher affinity for CO_2 than RuDP-carboxylase, the enzyme responsible for initial fixation in C_3 plants. Generally, plants do not fix CO_2 by either the C_3 or C_4 cycle. In efficient plants, the C_4 cycle supplements but does not replace the C_3 cycle.

C_4 plants are characterized by their ability to undertake photosynthesis at higher optimal temperature (30-45° C) and light intensity (full sunlight). They are twice as photosynthetically efficient as C_3 plants. Their growth rate is twice faster, producing two to three times more dry matter per unit of water than C_3 plants.

With respect to the level of photosynthetic compensation point, efficient plants have, under normal physiological conditions, a CO_2 compensation point of 5 ppm or less. Non-efficient plants have a compensation point in the range of 30 to 70 ppm CO_2. The compensation point is the concentration of CO_2 below which net carbon assimilation does not occur via photosynthesis. In plants, CO_2 released by respiration is used in photosynthesis with no net oxygen evolution at the compensation point. In plants with a high compensation point less carbon is used in photosynthesis because of the inefficiency of their respiration. Using these criteria, most weeds are categorized as efficient while many crops are termed non-efficient (Table 2.4). Fourteen of the world's 18 worst weeds [45] are C_4 weeds. Of the 76 worst weeds of the world, 42% are C_4. Among the 15 major crops, only 3 (maize, sugarcane and sorghum) are C_4 plants, with the C_3 pathway of photosynthetic fixation prevailing.

Table 2.4. List of efficient and non-efficient crops and weeds [16]

	Efficient	*Non-efficient*
Crops:	Maize, sugarcane, sorghum	Rice, Wheat, Oats, Soybean, Cotton, Tobacco, Sugarbeet, Barley, Lettuce, Spinach, Common bean
Weeds:	*Amaranthus* spp. (pigweed)	*Chenopodium* spp. (lambsquarters)
	Atriplex spp. (saltbrush)	*Abutilon theophrasti* (velvetleaf)
	Cyperus spp. (nutsedge, nutgrass)	
	Echinochloa crus-galli (barnyardgrass)	
	Panicum capillare (witchgrass)	
	Portulaca spp. (purslane)	
	Salsola iberica (Russian thistle)	
	Setaria spp. (foxtails)	
	Sorghum halepense (Johnsongrass)	

The most competitive weeds grow more rapidly, have a shorter vegetative growth period and complete the reproductive stage quickly. They often produce seeds that mature soon after flowering. They may also be tolerant of shade, even if their highest CO_2 assimilation is in full sunlight. They have a high photosynthetic rate, with rapid partitioning of photosynthates into new leaf production. Competitive weeds also develop a large exploitive root system rapidly, capable of tiding over adverse environmental conditions and resource limitations.

REFERENCES

1. Akey, W.C., T.W. Jurik and J. Dekker. 1990. Competition for light between velvetleaf (*Abutilon theophrasti*) and soybean (*Glycine max*). Weed Res. **30**: 403-411.
2. Akey, W.C., T.W. Jurik and J. Dekker. 1991. A replacement series evaluation of competition between velvetleaf (*Abutilon theophrasti*) and soybean (*Glycine max*). Weed Res. **31**: 63-72.
3. Aldrich, R.J. 1984. Weed-Crop Ecology—Principles in Weed Management. Chapter 8: Allelopathy in Weed Management. Breton Pub., N. Scituate, MA.
4. Alkamper, J. 1976. Influence of weed infestation on effect of fertilizer dressings. Pflanzenschtz-Nachrichten Bayer, **29**: 191-235.

5. Amen, R.D. 1968. A model of seed dormancy. Bot. Rev. **34**: 1-31.

6. Ball, D.A. 1992. Weed seedbank response to tillage, herbicides, and crop rotation sequence. Weed Sci. **40**: 654-659.

7. Baird, E., S. Cooper-Bland, R. Waugh, M. DeMaine and W. Powell. 1992. Molecular characterization of inter- and intra-specific somatic hybrids of potato using randomly amplified polymorphic DNA (RAPD) markers. Mol. Gen. Genet. **233**: 469-475.

8. Bell, D.T. and D.E. Koeppe. 1972. Noncompetitive effects of giant foxtail on the growth of corn. Agron. J. **64**: 321-325.

9. Bell, R.S., W.H. Lachman, E.M. Rahn and E.D. Sweet. 1962. Life history studies as related to weed in the Northeast. I. Butgrass. Kingston, RI: Rhode Island Agri. Expt. Station Bull. 36. 33 pp.

10. Bello, I.A., M.D.K. Owen and H.M. Hatterman-Valenti. 1995. Effect of shade on velvetleaf (*Abutilon theophrasti*) growth, seed germination, and dormancy. Weed Technol. **9**: 452-455.

11. Bhowmik, P.C. 1997. Weed biology: Importance to weed management. Weed Sci. **45**: 349-356.

12. Bhowmik, P.C. and M.M. Bekech. 1993. Horseweed (*Conyza canadensis*) seed production, emergence and distribution in no-tillage and conventional-tillage corn (*Zea mays*). Agronomy **1**: 67-71.

13. Bhowmik, P.C. and J.D. Doll. 1992. Corn and soybean response to allelopathic effects of weed and crop residues. Agron. J. **74**: 601-606.

14. Bhowmik, P.C. and J.D. Doll. 1994. Allelopathic effects of annual weed residues on growth and nutrient uptake of corn and soybeans. Agron. J. **76**: 383-388.

15. Biniak, B.M. and R.J. Aldrich. 1986. Reducing velvetleaf (*Abutilon theophrasti*) and giant foxtail (*Setaria faberi*) seed production with simulated-roller herbicide applications. Weed Sci. **34**: 256-259.

16. Black, C.L., T.M. Chen and R.H. Brown. 1969. Biochemical basis for plant competition. Weed Sci. **17**: 338-344.

17. Buchholtz, K. 1968. Use of split-root techniques for study of competition between quackgrass and corn. Abstr. Meeting of Weed Sci. Soc. Am., 90 pp.

18. Buhler, D.D. 1995. Influence of tillage systems on weed population dynamics and management in corn and soybean production in the Central USA. Crop Sci. **35**: 1247-1257.

19. Chang, W.L. 1973. Chemical weed control practice for rice in Taiwan. Pans **19**: 514-522.

20. Chiska, H. 1977. Weed damage to crops: yield loss due to weed competition. *In* J.D. Fryer and S. Matsunaka (eds.). Integrated Control of Weeds. Univ. of Tokyo Press, Tokyo, pp. 1-16.

21. Cook, C.E., L.P. Whichard, M.E. Wall, G.H. Egley, P. Coggon, P.A. Luhan and A.T. Mcphil. 1972. Germination Stimulants. II. The structure of strigol-potent seed germination stimulant for witchweed (*Striga lutea* Lour). J. Amer. Chem. Soc. **94**: 6198-6199.

22. Darmency, H. and J. Pernes. 1989. Agronomic performance of a triazine resistant foxtail millet (*Setaria italica* (L.) Beauv.). Weed Res. **29**: 147-150.

23. Davis, R.F. 1928. The toxic principle of *Juglans nigra* as identified with synthetic Juglone and its toxic effects on tomato and alfalfa plants. Amer. J. Bot. **15**: 620.

24. DeCandolle, M.A.P. 1832. Physiologic Vegetable, Vol. III. Becht Jeune. Lib. Fac. Med. Paris.

25. Dekker, J. 1993. Pleiotropy in triazine resistant *Brassica napus*: leaf and environmental influences on photosynthetic regulation. Zeitschrift Naturforschung **48c**: 283-287.

26. Dekker, J. 1997. Weed diversity and weed management. Weed Sci. **45**: 357-363.

27. Dekker, J., B.I. Dekker, H. Hilhorst and C. Karssen. 1995. Weedy adaptation in *Setaria* spp. IV. Changes in the germination capacity of *S. faberii* embryos with development from anthesis to after abscission. Amer. J. Bot. **83**: 979-991.

28. Dekker, J. and T.D. Sharkey. 1992. Regulation of photosynthesis in triazine resistant and susceptible *Brassica napus*. Plant Physiol. **98**: 1069-1073.

29. Donoghue, M.J. and M.J. Sanderson. 1992. The suitability of molecular and morphological evidence in reconstructing plant phylogeny. *In* P.S. Soltis, D.E. Soltis and J.J. Doyle (eds.). Molecular Systematics of Plants. Routledge, Chapman & Hall, Inc., New York, pp. 340-368.

30. Dryden, R.D. 1974. Losses of nutrients under varying agro-climatical conditions - Weed Control. FAI-FAO Seminar 1974 on 'Optimizing agricultural production under limited availability of fertilizers', New Delhi.

31. Dzubenko, N.N. and N.I. Petrenko. 1971. On biochemical interaction of cultivated plants and weeds. *In* A.M. Grodzinsky (ed.). Physiological-biochemical Basis of Plant Interactions in Phytocenoses, vol. 3. Nukova Dumka, Kiev, pp. 60-66 (in Russian).

32. Egley, G.H. 1986. Stimulation of weed seed germination in soil. Rev. Weed Sci. **2**: 67-89.

33. Fenner, M. 1985. Chapter 4. *In* Seed Ecology. Chapman Hall, New York, NY, pp. 87-104.

34. Forcella, F., R.G. Wilson, K.A. Renner, J. Dekker, R.G. Harvey, D.A. Alm, D.D. Buhler and J.A. Cardina. 1992. Weed seedbanks of the U.S. Cornbelt: magnitude, variation, emergence, and application. Weed Sci. **40**: 636-644.

35. Garg, D.K., L.E. Bendixen and S.R. Anderson. 1967. Rhizome differentiation in yellow nutsedge. Weeds **15**: 124-128.

36. Gray, R. and J. Bonner. 1948a. An inhibitor of plant growth from the leaves of *Encelia farinosa*. Amer. J. Bot. **34**: 52-57.

37. Gray, R. and J. Bonner. 1948b. Structure determination and synthesis of a plant growth inhibitor, 3-acetyl-6-methoxybenzaldehyde found in the leaves of *Encelia farinosa*. J. Amer. Chem. Soc. **70**: 1249-1253.

38. Gressel, J.B. and L.G. Holm. 1964. Chemical inhibition of crop germination by weed seeds and the nature of inhibition of *Abutilon theophrasti*. Weed Res. **4**: 44-53.

39. Grummer, G. and H. Beyer. 1960. The influence exerted by species of *Cammelina* on flax by means of toxic substances. *John L*. Harper, (ed.). The Biology of Weeds. Blackwell, Oxford, pp. 153-157.

40. Haar, M. and J. Dekker. 1994. Somatic polymorphism in *Setaria faberii*: germination-dormancy states at abscission. The 1st International Plant Dormancy Symposium. Corvallis, OR. 214 pp.

41. Harper, J.L. 1977. Population Biology of Plants. Academic Press, San Diego, CA, 892 pp.

42. Hauptli, H. and S.K. Jain. 1978. Biosystematics and agronomical potential of some weedy and cultivated Amaranths. Theor. Appl. Genet. **52**: 177-185.

43. Hauser, E.W. 1962. Development of purple nutsedge under field conditions. Weeds **10**: 315 321.

44. Helgeson, E.A. and R. Konzak. 1950. Phytotoxic effects of aqueous extracts of field bindweed and of Canada thistle. A preliminary report. N. Dak. Agr. Expt. Sta. Bimn. Bul. **12**: 71-76.

45. Holm, L.G., D.L. Plucknett, J.V. Pancho and J.P. Herberger. 1977. The World's Worst Weeds—Distribution and Biology. Univ. Press Hawaii, Honolulu, 609 pp.

46. Horowitz, M. 1965. Data on the biology and chemical control of nutsedge (*Cyperus rotundus*) in Israel. PANS (Pest. Arctic. News Sum.) **11**: 389-416.

47. Horowitz, M. 1972. Growth, tuber formation, and spread of *Cyperus rotundus* from single tubers. Weed Res. **12**: 348-363.

48. Horowitz, M. 1992. Mechanisms of establishment and spreading of *Cyperus rotundus*—the worst weed of warm regions. Proc. First Int. Weed Control Congr. **1**: 94-97.

49. Hosamani, M.M. 1995. Striga—A noxious root parasitic weed. Ayi Graphics, Dharwad, India, 79 pp.

50. Isikawa, S. and T. Fujii. 1961. Photocontrol and temperature dependence of germination of *Rumex* seeds. Plant and Cell Physiol. **2**: 51-62

51. Jensen, L.L. 1971. Morphology and photoperiodic responses of yellow nutsedge. Weed Sci. **19**: 210-219.

52. Kolbe, W. 1977. Long-term studies on relations between weed cover and yield increase with chemical weed control (1967-1976). Pflanzenschutz-Nachrichten Bayer **30**: 121-137.

53. Kommedahl, T., J.B. Kotheimer and J.V. Bernardini. 1959. The effects of quackgrass on germination and seedling development of certain crop plants. Weeds **7**: 1-12.

54. Kondap, S.M., M.R. Reddi, and K. Ramakrishna. 1980. Studies on yield reduction due to *Cyperus rotundus* in different crops. Abst. Conf. Ind. Soc. Weed Sci., p. 25.

34

55. LeBaron, H.M. and J. Gressel (eds.), 1982. Herbicide Resistance in Plants. John Wiley & Sons, Inc. New York, 401 pp.

56. Levy, M., F.J. Correa-Victoria, R.S. Zeigler, S. Xu and J.E. Hamer. 1993. Genetic diversity of the rice blast fungus in a disease nursery in Colombia. Phytopathology **83**: 1427-1433.

57. Lindquist, J.L., B.D. Maxwell, D.D. Buhler and J.L. Gunsolus. 1995. Velvetleaf (*Abutilon theophrasti*) recruitment, survival, seed production, and interference in soybean (*Glycine max*). Weed Sci. **43**: 226-232.

58. Mani, V.S., K.C. Gautam and T.K. Chakraborty. 1968. Losses in crop yield in India due to weed growth. PAN(C) **14**: 142-158.

59. Mapplebeck, L.R., V. Souza-Machado and B. Grodzinski. 1982. Seed germination and seedling growth characteristics of atrazine-susceptible and resistant biotypes of *Brassica campestris*. Can J. Plant Sci. **62**: 733-739.

60. Martin, P. and B. Rademacher. 1960. Studies on the mutual influences of weeds and crop. *In* John L. Harper (ed.). The Biology of Weeds. Blackwell, Oxford. 542 pp.

61. Mercado, B.L. 1979. A monograph on *Cyperus rotundus* L. Biotrop. Bull. **15**: 1-63.

62. Mitra, P.S., P.C. Bhowmik and R. Bernatzky. 1998. DNA profiles of different biotypes of quackgrass (*Elytrigia repens*). Proc. Northeast Weed Sci. Soc. Amer. **52**: 35.

63. Muenscher, W.C. 1936. Weeds. The Macmillan Company, New York, 577 pp.

64. Neustruyeva, S.N. and T.N. Dobretsova. 1972. Influence of some summer crops on white goosefoot. *In* A.M. Grodzinsky (ed.). Physiological-biochemical Basis of Plant Interactions in Phytocenoses. Vol. 3. Naukova Dumka, Kiev, pp. 68-73 (in Russian).

65. Newbury, H.J. and B.V. Ford-Lloyd. 1993. The use of RAPD for assessing variations in plants. Plant Growth Reg. **12**: 43-51

66. Nissen, S.J., R.A. Masters, D.J. Lee and M.L. Rowe. 1995. DNA-based marker systems to determine genetic diversity of weedy species and their application to biocontrol. Weed Sci. **43**: 504-513.

67. Ohman, J.H. and T. Kommedahl. 1960. Relative toxicity of extracts from vegetative organs of quackgrass to alfalfa. Weeds **8**: 666-670.

68. Osvald, H. 1947. Vexternas vapeni kampen on utrymmet. Vaxtodling **2**: 228-303.

69. Pillai, P. and J.B. St. John. 1981. Lipid composition of chloroplast membranes from weed biotypes differentially sensitive to triazine herbicides. Plant Physiol. **68**: 585-587.

70. Putnam, A.R. 1983. Allelopathic chemicals. Chem. Eng. News **61(14)**: 34-35.

71. Putnam, A.R. 1985. Weed allelopathy. *In* S.O. Duke, (ed.). Weed Physiology Vol. 1: Reproduction and Ecophysiology. CHC press, Boca Raton, FL, USA, pp. 131-155.

72. Radosevich, S., J. Holt and C. Ghersa. 1997. Weed Ecology: Implications for Management, 2nd ed. John Wiley & Sons, Inc., New York. 589 pp.

73. Rice, E.L. 1974. Allelopathy. Academic Press, New York, 353 pp.

74. Rizvi, S.J.H., V. Jaiswal, D. Mukherji and S.N. Mathur. 1980. 1,3,7-trimethylxanthine—a new natural weedicide: Its mode of action. Plant Physiol. **65S**: 99.

75. Schaal, B.A., S.L. O'kane and S.H. Rogstad. 1991. DNA variation in plant populations. Trends Ecol. Evol. **6**: 229-233.

76. Schweizer, E.E. and R.L. Zimdahl. 1984. Weed seed decline in irrigated soil after six years of continuous corn (*Zea mays*) and herbicides. Weed Sci. **32**: 76-83.

77. Sen, D.N. and R.P. Bansal. 1978. Seed adaptive mechanisms in arid zone weeds. Abst. All-India Weed Sci. Conf., p. 40.

78. Senseman, S.A. and L.R. Oliver. 1993. Flowering patterns, seed production, and somatic polymorphism of three weed species. Weed Sci. **41**: 412-425.

79. Smith, R.J., Jr. 1968. Weed competition in rice. Weed Sci. **16**: 252-255.

80. Solbrig, O.T. and D.J. Solbrig. 1981. Introduction to population biology and evolution. Addison-Wesley Publ., Reading, MA., 469 pp.

81. Stickney, J.S. and P.R. Hoy. 1881. Toxic action of black walnut. Trans. Wis. State Hort. Soc. **11**: 166-167.

82. Stoller, E.W., D.P. Nema and V.M. Bhan. 1972. Yellow nutsedge tuber germination and seedling development.

83. Stoller, E.W. and L.M. Wax. 1973. Yellow nutsedge shoot emergence and tuber longevity. Weed Sci. **21**: 76-81.
84. Stoller, E.W. and E.J. Weber. 1974. Differential cold tolerance, starch, sugar, protein, and lipid of yellow and purple nutsedge tubers. Plant Physiol. **55**: 859-863.
85. Sunderland, N. 1960. The production of the *Striga* and *Orobanche* germination stimulants by maize roots. J. Exp. Bot. **11**: 236-245.
86. Thurston, J.M. 1957. Morphological and physiological variation in wild oats (*Avena fatua* L. and *A. ludoviciana* Dur.) and hybrids between wild and cultivated oats. J. Agric. Sci. Camb. **49**: 260-274.
87. Tripathi, R.S. 1969. Ecology of *Cyperus rotundus* L. III. Population of tubers at different depths of the soil and their sprouting response to air-drying. Proc. Nat. Acad. Sci., India 39: 140-142.
88. Vaughn, K.C. and S.O. Duke. 1980. Ultraphysiological alterations to chloroplasts in triazine-resistant biotypes. Physiol. Plant. **62**: 510-520.
89. Weining, S. and P. Langridge. 1991. Identification and mapping of polymorphisms in cereals based on the polymerase chain reaction. Theor. Appl. Genet. **82**: 209-216.
90. Welsh, J., R.J. Honeycutt, M. McClelland and B.W.S. Sobral. 1991. Parentage determination in maize hybrids using the arbitrarily primed polymerase chain reaction (AP-PCR). Theor. Appl. Genet. **82**: 473-476.
91. Williams, J.T. and J. L. Harper. 1965. Seed polymorphism and germination. I. The influence of nitrates and low temperatures on the germination of *Chenopodium album*. Weed Res. **5**: 141-150.
92. Wilson, R.G. 1988. Biology of weeds in the soil. *In* M.A. Altieri and M. Liebman, (Eds.). Weed Management in Agroecosystems: Ecological Approaches. CRC Press, Boca Raton, FL, pp 25-39.
93. Wilson, R.G., E.D. Kerr and L.A. Nelson. 1985. Potential for using weed seed content in the soil to predict future weed problems. Weed Sci. **33**: 171-175.
94. Zimdahl, R.L. 1993. Fundamentals of Weed Science. Academic Press, Inc. New York.

3

Traditional, Ecophysiological and Other Approaches in Weed Management

Weed control is a process of reducing or minimizing weed growth to an acceptable level. This concept is narrow in range as it takes into consideration only the effect of a particular method of weed control or elimination, disregarding both its short-term and long-term implications for the cropping system and the environment as a whole. Many of the current weed control measures are narrow in scope and exclude the benefits of other practices. A control measure without adequate efforts to prevent further infestation of weeds could prove to be uneconomical and time consuming.

In weed management, the primary objective is to maintain an environment that is as detrimental to weeds as possible by employing both preventive and post-infested control measures by using various methods either alone or in combination. The crop-weed relationship is manipulated in such a way that the growth of the crop is favoured over that of weeds. The concept of weed management can be applied on a coordinated basis to a single crop or cropping system. The weed spectrum in a crop or a cropping system is dynamic. Control or elimination of a particular weed species or a spectrum of weeds results in an invasion by other weed species. Weeds could become resistant to a particular control measure over a period of time and this necessitates manipulation in the weed management practice. Weed management is essentially a skillful combination of **prevention, control** and **eradication** measures to manage weeds in a crop or environment.

A successful weed management programme takes into account the various crops in rotation and the various crop production practices which bring competitive and ecological pressures on weeds. It takes the entire farm, instead of a small section or a particular area, as a unit to combat weed problems effectively and economically. This type of systems concept of weed management is fast catching the attention of agricultural scientists and environmentalists all over the world.

The various methods of weed management are grouped under three broad groups: traditional, chemical and biological. The chemical and biological approaches in weed management are discussed in later chapters. This chapter deals with the traditional approaches that can be used for effective weed management.

TRADITIONAL APPROACHES

Long before the chemical method became a dominant force in weed management, farmers had been using traditional approaches involving manual, mechanical and cultural measures for centuries. With the availability of herbicides since the late 1940s for almost every weed situation and their wide spread use since then by farmers, particularly of the developed world, most of the traditional weed control practices have been put on the back burner. However, in the light of the environmental and toxicological problems created by certain herbicides, and the rapidity with which herbicide resistance has developed all over the world in the past two decades, it has become necessary to utilize the traditional weed control measures in combination with chemical and, of late, biological methods, to develop a more comprehensive and effective weed management strategy. This type of integration of various weed control measures is an essential component of sustainable agriculture.

The various traditional approaches relevant to weed management are discussed below.

Weed Prevention

Preventive weed control encompasses all measures taken to prevent or arrest the introduction and spread of weeds. No weed control programme can be successful if adequate preventive measures are not taken to reduce weed infestation. Success of a preventive programme depends on the weed species present and the amount of effort made to make it effective. Prolific reproductive capacity, dissemination, dormancy and viability make eradication of a weed species almost impossible. The following preventive control measures may therefore be adopted wherever possible and practicable. The salient points that need to be kept in mind while adopting preventive control measures are as under.

Weed-free Crop Seeds

Seeds of most crops are contaminated with weed seeds. This happens at harvesting time. Some weed species have life cycles similar to that of crops. They set seeds at the same time as crops do. Seeds of some weed species resemble crop seeds in size and shape. Even when seeds are certified with contamination of 1% of weed seeds or less, the consequences are detrimental to crop production. The presence of even a handful of weed seeds may be enough to start a serious infestation. Furthermore, weed seeds planted in the crop rows are the most difficult to control. The following procedures help in producing weed-free crop seeds:

1. Cleaning and testing of all farm-produced crop seeds.
2. Separating crop seeds from an admixture of weed and crop seeds on the basis of physical differences in seed size, weight, shape, surface area, specific gravity, stickiness, pubescence, texture, colour and electrical properties.
3. Using air-screen cleaners and specific gravity separators, which differentiate seeds on the basis of size, shape, surface area and density. Indent-disc and indent-cylinder separators are useful to separate seeds on the basis of seed length. Pneumatic and aspirator separators differentiate seeds on the basis of their set-tling velocity in air. Velvet-roll separators discriminate seeds by the texture of seed coats.
4. Employing clean agricultural equipment, which will help in eliminating the pos-sibility of weed seeds contaminating crop seeds. Weed seeds and vegetative organs adhere to ploughs, harrows, drills, harvesters, weeders and hoes during the pre-planting tillage and post-planting operations and get carried to other seeds. Hence, these implements should be cleaned before using them elsewhere.
5. Using well-decomposed weed-free farmyard manure and compost eliminates the risk of spreading the weed seeds. Animals graze weed areas of pastures and

the weed seeds can pass through their digestive tracks without losing viability. They end up in excretions in the form of dung. Composting the manure destroys most weed seeds. Well-rotted manure is normally free of viable seeds.

6. Adopting measures to prevent carrying of weed seeds and vegetative propagules to the cropping area along with irrigation water. Weed seeds, rhizomes and other propagating materials are light and they readily float or remain suspended and carried long distances by irrigation water. Several million seeds pass a given point at any given time in moving water. Preventive measures require cooperation of farmers and public agencies controlling canal and tank irrigation. Individual farmers can prevent weed infestation by putting appropriate screens across the lateral or feeder channels leading to their fields.

Seed Certification

Seed certification helps in supplying genetically pure seeds and propagating materials of farmers to farmers. It is primarily the responsibility of state and central (federal) governments, which control the certification agencies that set standards for clean seeds.

In India, the National Seeds Corporation of Government of India and the Seed Corporations of various state governments take the responsibility of producing crop seeds and certifying the same produced by seed growers. These agencies standardize inspection and approval procedures for certification. They should, therefore, strictly enforce regulations to produce and supply absolutely weed-free crop seeds. Farmers who grow their own seed or who use the seed of their previous harvest(s) should take adequate precautions to prevent weed seed contamination with crop seeds.

Weed Laws

Weed laws are important in reducing the spread of weed species and in increasing the use of well-adapted high-quality seeds. They help in protecting the farmers from using mislabelled or contaminated seed and also in legally prohibiting seeds of noxious weeds from entering the country.

For the purpose of declaring weed laws, the noxious weeds can be separated into prohibited and restricted weeds. Prohibited noxious weeds are those perennial weeds that not only reproduce by seeds but also by underground roots, stems and other reproductive parts, and they are difficult to control. Restricted noxious weeds are those perennial weeds which are objectionable in cropping areas but which can be controlled.

Success in controlling noxious weeds through the enforcement of weed laws depends on the availability of effective and economical control measures. Each country and state should enact weed laws against the most noxious weeds infesting agricultural and non-agricultural systems and make serious efforts to eradicate them.

Quarantine Laws

Quarantine laws enforce isolation of an area in which a most problematic weed has become established, and prevent movement of the weed into an uninfested area. The normal quarantine laws restrict the entry and interstate and international movement of imported plants capable of spreading plant diseases and insects. But they do not always provide restrictions on import of weed seeds either separately or in the form of admixture with crop seeds.

Post-Infested Weed Management

This includes **control** and **eradication** of weeds already infesting an area. Selection of a proper practice to achieve this is dependent on whether the objective is control or eradication of weeds. Weed eradication involves killing of the existing plants and destroying the viability of all organs of reproduction. This may be desirable when the weed species is extremely noxious and persistent as to make cropping difficult and uneconomical or preclude bringing new area under cultivation. Eradication of perennial weeds cannot be accomplished in one season or year as viable seeds and underground plant propagules may remain dormant for years. On the other hand, when weeds are widespread and difficult or economically not feasible to eliminate, control would be a more practicable measure. A control programme enables the farmer to produce a profitable crop in spite of weeds. Control of weed species is achieved by reducing or killing plant growth and propagules to the point where their presence does not seriously interfere with an area's economic use. Persistent weeds are seldom killed, but when their growth is inhibited, the crop may give a normal yield. As the immense store of seeds in the soil precludes the possibility of eradication, control measures would be more appropriate and economical.

The various methods of post-infested weed control are summarized below.

Manual Methods

HANDWEEDING

Handweeding is done by physical pulling out or removal of weeds by hand or removal by a hand-operated implement, *khurpi*, which resembles the sickle. It is probably the oldest method of controlling weeds and is still a practical method of eliminating weeds from cropped and noncropped lands. It is very effective against annual and biennial weeds, as they do not recover from the pieces of roots left behind in the ground. In the case of perennials, handweeding or pulling usually leaves portions of roots and other propagating parts in the ground and they can regenerate. In such cases, the operation must be repeated at frequent intervals.

Handweeding is also useful in situations where weeds are so scattered that herbicide treatment is uneconomical or herbicide application is not allowed for crop-toxicity reasons or herbicides are not effective on a particular weed species. In places where human labour is cheap and plentiful, this method is widely followed. However, with rapid industrialization and urbanization in developing countries, human labour is rapidly becoming scarce and expensive. Furthermore, under the multiple cropping systems followed in these countries, very little time is available for the slow-paced hand weeding.

DIGGING

Digging is very useful in the case of perennial weeds to remove the underground propagating parts of weeds from the deeper layers of the soil. This practice is more effective than handpulling if the weeds have taproot and deep-rooted vegetative propagules. Digging is accompanied by hand pulling of the plant. Handpulling without proper digging results in the removal of top growth only, allowing regrowth from the leftover plant parts. Generally, digging is a labour-intensive and slow process and hence it is restricted to perennial weed situations where other mechanical and physical methods, and chemical methods are not effective, practicable and economical.

CHEELING

Cheeling is done by hand using a *cheel* **hoe**, similar to a spade, with a long handle. It cuts and scrapes the aboveground weed growth at the ground level and rakes it up. It is useful in the case of annuals and biennials as they do not regrow from the plant parts left behind in the soil. Cheeling is widely used in India in plantation crops, particularly in tea.

SICKLING AND MOWING

Sickling is also done by hand with the help of a sickle, to remove the top growth of weeds, to prevent seed production and to starve the underground parts. It is popular in slope areas where only the tall weed growth is sickled leaving the root system and some growth intact to hold the soil in place and prevent possible soil erosion. In situations wherein excessive weed growth entails using more quantities of postemergence contact herbicides for its control, weeds are sickled down to a particular height and herbicides used. This helps to reduce the quantity of herbicides needed. Sickling is also done to reduce the mature perennial weed growth and facilitate greater effect by postemergence translocated herbicides applied on the subsequent active growth. The machete, another hand-operated implement, is used for slashing the existing weed growth.

For perennial weed control, the best time to sickle or slash is when the underground root reserves are at a low ebb; for control of other weeds it is between full leaf development and the time flowers appear. Mowing is a machine-operated process mostly done along roadsides and on lawns. Its objective is the same as that of sickling. It is ineffective against prostrate or short weeds growing close to the ground. Like sickling, it is, however, a desirable weed control method in areas where bare ground is subject to soil erosion and plant root systems are needed to hold the soil in place.

Mechanical Methods

Mechanical or physical methods of weed control have been employed ever since man began to grow crops. Many agricultural practices are dictated by the need for weed control. Although herbicides have been fast replacing the traditional methods of weed control, mechanical methods are very much needed to make weed control more effective, manageable and economical. Mechanical methods include tillage, hand hoeing, burning, flooding, mulching, etc.

TILLAGE

Tillage removes weeds from the soil, resulting in their death. It may weaken plants through injury or root and stem pruning, reducing their competitiveness or regenerative capacity. Tillage also buries weeds. Besides controlling weeds, tillage affects the physical condition of the soil and turns crop residues under the soil. Tillage is implemented as primary tillage or secondary tillage.

Tillage is done with implements drawn by animals or machines (tractors, tillers, etc.). Before a crop is planted, extensive tillage operations are undertaken to prepare the soil. This preplanting tillage, which includes ploughing, discing, harrowing and levelling, is used to promote the germination of weeds through soil turnover and exposure of seeds to sunlight which can be effectively destroyed later. In the case of perennials, both top and underground growth is injured or destroyed by tillage.

The efficacy of tillage for weed control is largely determined by the type of weeds to be controlled. Annual weeds, biennial weeds without a well-developed taproot and perennial weed seedlings are readily controlled by tillage. Normally, the younger the plants, the more susceptible they are to tillage. As plants become established, they may develop extensive root systems, build up food reserves in the roots, and cover the ground extensively, making tillage operations difficult and less effective. Generally, the greater the disturbance of the weed-soil association, the greater the effectiveness of tillage.

Control of mature perennial weeds by tillage is a difficult proposition. It involves depletion of food reserves through continuous destruction of top growth. Perennial weeds with deep extensive root system and vegetative propagules are more difficult to control than perennial weeds with restricted root system and vegetative propagules. Infestation of stoloniferous and rhizomatous perennial weeds, in fact, spreads more by tillage as the cut pieces of stolons and rhizomes can strike roots and pruning of these propagating parts encourages more intensive weed growth. This requires a continuous programme to destroy new growth after each cultivation. An old infestation of deep-rooted perennials such as *Imperata cylindrica* (thatchgrass, cogongrass or alang-alang)), *Cyperus rotundus* (purple nutsedge or nutgrass), *Elytrigia repens* (quackgrass), etc. which contain food reserves in the underground rhizomes and tubers require more cultivations for thorough eradication than a relatively young infestation. The number and frequency of cultivations required to control perennial weeds vary with a particular species and the soil and climatic conditions. The time intervals between cultivations also vary. The objective, however, should be to kill the top growth precisely at a time when food reserves are no longer depleted from the roots and are beginning to be replenished from the new shoots.

With the introduction of herbicides for control of weeds, the effects of tillage as an agricultural practice independent of weed control are being re-evaluated. Thus came into being the concepts of **conservation tillage, no tillage, minimal tillage, ridge tillage** and **shallow tillage systems**. All these are a part of Integrated Weed Management (IWM). Generally, it is found that tillage other than for weed control is of little benefit in lighter soils. In heavier soils, tillage may offer benefits beyond weed control. The physical condition of the soil normally dictates the number of tillage operations in heavier soils. But in recent years there has been a growing trend towards reducing the amount of tillage. A number of studies conducted around the world in maize, sorghum, rice, etc. indicated that minimal tillage practice of killing the existing weed growth by using translocated and contact postemergence herbicides and planting the crop immediately thereafter, with little or no tillage, compared favourably with conventional methods. These studies also indicated that reduced cultivation is indeed beneficial. The minimal tillage concept is particularly very useful under the present multiple cropping programmes practised across the world, as there is very little time available between harvesting of one crop and planting of the next. Further investigations on minimal tillage, particularly with respect to identification of suitable herbicides, short-term and long- term effects of the practice on physical and microbiological conditions of soil, economic benefits, etc. need to be undertaken before it can become an acceptable and viable practice.

Conservation tillage is defined as any tillage practice that leaves a crop on at least 30% of the soil surface [66]. It includes such practices as **minimal-till, no-till, ridge-till, zone-till,** and **chisel ploughing**. Of these, **minimal-till** and **ridge-till** are currently widely used. These tillage practices reduce soil erosion, lower surface runoff of herbicides and fertilizers, decrease cultivation costs and raise net economic returns of the farm. Generally, crop yields under conservation tillage are comparable to those under conventional tillage. Although increased weed interference caused by reduced tillage may lower crop yields occasionally, the overall benefits outweigh the loss. However, there is potential for changes in the weed spectrum upon implementing a conservation tillage system, as is the case whenever a particular weed management practice is changed. Conservation tillage can affect several aspects of weed population dynamics, including weed seed germination and emergence, intra- and inter-specific competition, resource allocation, reproductive effort, seed production and seed dispersal [10, 64]. Crop residue may suppress and smother weeds such that herbicide use could be reduced. As a consequence, weed population dynamics must be monitored and modelled to help predict changes and develop alternative weed management strategies [65].

In **no-till** practice, the crop is planted in a previously unprepared soil by opening a narrow slot, trench, or band just sufficient in width and depth to obtain proper seed coverage [53]. No other soil preparation is done. However, continuous no-tillage farming can result in a gradual change in the weed spectrum in favour of perennial weeds, which could pose a greater problem than existed before [73]. In such situations, herbicides effective against perennial weeds may be used. A variant of no-till is **minimal-till**, which allows tillage in situations where some tillage is essential. No-till and minimal-till may be used in alternate years to further reduce herbicide application.

In the **ridge-till** system, the crop is planted in parallel, equidistant rows. Ridges are built during interrow cultivation when the crop is 25-40 d old. During harvest, the ridges are not disturbed. The field is left undisturbed until the next season. Before planting the next crop, the stalks are chopped and the field is fertilized. The row crop is planted on the crown of ridges. In ridge-till, the planter follows the old row. The ridge-clearing disc or sweep moves the surface soil, crop residue and many weed seeds out of the row. The ridges are formed during one cultivation and ridging operation [33].

The ridge-till system is considered an ideal crop management system for water and soil erosion control and IWM. In Iowa, USA, a ridge-till watershed lost 1.1 t ha^{-1} of soil compared to a conventionally tilled watershed that lost 10.9 t ha^{-1} [34]. After planting, the ridges must be at least 8 cm higher than the furrows and at harvest they should be 15-20 cm higher [15]. With ridge-till, 84% of the weed seeds are pushed with the 3-6 cm of surface soil from the ridge to the interrow during the planting operation [70]. Weeds in the interrow can be controlled by herbicides or cultivation. Herbicides may be applied PPS (preplant surface), PRE (preemergence) or POST (postemergence). Banding herbicides after planting is a very efficient, cost-saving operation in fields with low weed population [33]. The ridge-till system can be made more effective by applying herbicides 1-3 wk before planting. Row crops like maize, sorghum, millets, soybean, groundnut, etc. are more suitable for the ridge-till system. Integrating the ridge-till system with herbicides could lead to a better and broader spectrum of weed control than either conventional cultivation or herbicides alone.

A shallow tillage followed by herbicide application can effectively eliminate perennial weeds such as *Cyperus* spp., *Elytrigia repens, Cynodon dactylon*, etc. Tillage can bring the underground rhizomes to the soil surface where they get desiccated. Herbicide spraying on the subsequent weed growth will effectively control the weed. Alternatively, an initial herbicide application on the standing weed growth, followed by shallow tillage, and finally another round of herbicide spraying on the new growth can eliminate 90-95% of perennial weed infestation. A POST, translocated herbicide such as glyphosate has proven highly useful in this tillage-herbicide rotation system.

The important primary tillage implements are: the **plough**—mouldboard plough, disc plough, etc. The secondary tillage implements include the **harrow**—disc harrow, spike-toothed harrow, clod crusher, blade harrow, spring-tooth harrow, etc., and the **cultivator**—shovel cultivator, rotary hoe, etc. The type of implements needed for most effective weed control depends on the weed species, type of underground root system, age of the weed, degree of weed infestation, type of soil, crop grown, etc.

HOEING

The hoe has been the most appropriate and widely used weeding tool for centuries. 'The man with the hoe' which symbolized weed control for centuries has been replaced by 'the man with the sprayer'. The hoe, however, is still a very useful implement to obtain results effectively and cheaply. It supplements the cultivator in row crops. Hoeing is particularly more effective on annuals and biennials as weed growth can be completely destroyed. In the case of perennials, it destroys the top growth with little effect on underground plant parts resulting in regrowth. However, it can be more effective on creeping perennials which have a shallow root system.

INTERROW CULTIVATION

Shallow interrow cultivation can be used effectively within the conservation tillage systems [65]. It may be done at a time when weeds are most likely to interfere with the crop. Weeds within the row may be controlled by spraying herbicides in a band over the row. Eadie et al. [17] found that interrow cultivation and banding of herbicides reduced herbicide use by 60% in a ridge-till system and yet maintained crop yields.

ECOFALLOW SYSTEM

The ecofallow system, also known as the chemical fallow system, is a type of reduced tillage practice made possible by herbicides which kill weeds of fallow areas. It conserves soil moisture by minimal disturbance of crop residue and soil and is a dependable practice in low as well as high rainfall areas. In low rainfall areas water is conserved, while in high rainfall areas the soil is conserved. During the interval between two crops, the land is kept fallow by using such POST herbicides as paraquat, glyphosate, paraquat + atrazine, etc. to kill weeds and conserve soil moisture. In the USA, the following crop is planted by direct seeding into the crop residue (or straw) by a special type of planter, the Buffalo no-till planter or the AC no-till planter equipped with flouted coulters [69]. The normal seed drills are not efficient in planting through the residue and uneven ground. The cropresidue acts as an insulator to reduce soil temperature and subsequent evaporation as well as reducing the impact of raindrops and slowing movement of water across the field to allow greater filtration [69]. The ecofallow system helps in reducing herbicide use without affecting the overall efficacy of weed control.

STALE SEEDBED

Stale seedbed is described as a planting system that does not use tillage immediately prior to planting. Tillage is usually performed immediately after crop harvest or any time before the next cropping season begins when soil moisture and environmental conditions allow field operations [60]. This tillage is done to prepare the seedbed well in advance of the next crop or cropping season. Stale seedbed planting cannot be considered a conservation tillage programme in the strictest sense, although the goal of this system is to reduce tillage operations [60]. The stale seedbed system allows timely planting and crop establishment. A key component of a well-prepared stale seedbed is the absence of weeds at planting. In the USA, this concept has been adopted with success on a variety of soils, but it is best suited on poorly drained clay soils [60]. It is useful for such crops as rice, maize, cotton and soybean.

The stale seedbed approach reduces weed seed germination during the cropping season. Sometimes, many weeds in the germination zone germinate and emerge before planting. Foliar herbicides such as glyphosate, paraquat, glufosinate, etc., applied at planting will eliminate these weeds; without additional tillage to bring up new seed fewer weeds will be present later in the season [60]. The stale seedbed system, like other reduced tillage systems, may encourage perennial weed problems in the long run. This can be arrested by using foliage-applied herbicides and herbicide mixtures. Alternatively, deep tillage once every four years may also minimize perennial weed problems.

BURNING

Burning or fire is often an economical and practical means of controlling weeds. It is used to: a) kill off accumulated vegetation, b) destroy dry tops of weeds that have matured, c) kill green-weed growth in situations where cultivation and other common methods are impracticable and d) destroy buried weed seeds and other propagating plant parts. The death of a plant by fire is due to injury to the cells of the leaves and stem. The critical temperature above which cambium cells may be injured is between 45° C and 55° C. In some plants, the cambium layer is protected by an insulating layer of bark even at this temperature range. Succulent weed seedlings, however, succumb to this heat. Dry seeds withstand considerably higher temperatures even at longer exposures.

Fire is used to burn crop residues in cotton, sugarcane, potatoes, maize, soybean, sorghum, castor, etc. When burning is used for selective control of annual broadleaf weeds and grasses in crop rows, it is known as flame cultivation. It then requires carefully controlled flame without injuring crop plants. Cotton plants can resist the flame if it is properly controlled and hence flame cultivation is adopted in cotton. Controlled burning is done with the help of hand-operated vapourizing burners and flame torches.

FLOODING

Flooding is sometimes used for weed control in rice, which is able to grow under flooded conditions. It is done by surrounding the weed infestation with dykes and maintaining water at 15 to 30 cm deep for 3 to 8 wk. The weeds should be submerged in water. Flooding is successful against weed species sensitive to longer periods of submergence in water. Some perennial weeds, e.g. *Convolvulus arvensis*, and other noxious weeds infesting rice can be controlled or even eradicated. Flooding kills plants by reducing oxygen availability for plant growth. The success of flooding depends upon complete submergence of weeds for long periods.

MULCHING

Mulching the soil surface affects the physical, chemical and biological properties and processes of the soil, the transfer of energy and matter between the soil and the atmosphere and crop and weed growth, which in turn influence the need for weed control. Mulch has a smothering effect on weeds by excluding light from the photosynthetic portions of a plant and thus inhibiting the top growth. It is very effective against annual weeds and some perennial weeds such as *Cynodon dactylon*, *Sorghum halepense* (Johnsongrass), etc. Mulching provides an effective barrier to weed emergence.

Mulches are of two types: organic mulches and synthetic mulches. **Organic mulches** consist of plant by-products such as crop residues, paddy husk, groundnut shells, sawdust, grass clippings, wood dust or chips and bark from trees, waste products such as newsprint and crushed rock or small stones. Bark mulches are more effective than straw mulches, as bark releases tannins and phenols which reduce weed growth. **Synthetic mulches** include polyethylene (plastic), polypropylene or polyester sheets or film.

The degree of weed control by mulch is dependent on mulch material, placement, depth and maintenance. Mulches, besides acting as a barrier, enhance herbicide leaching due to greater water infiltration under mulch. To be effective, the mulch should be thick enough to prevent light transmission and eliminate photosynthesis. Mulch with straw could be 10 to 15 cm thick. When plant materials are used as mulch, they encourage soil borne insects like crickets, white ants, etc. Mulching increases soil temperature, which enhances plant growth. In plantation crops like tea and coffee, it encourages proliferation of feeder roots, resulting in efficient uptake of plant nutrients.

When synthetic materials are used, white and black polyethylene plastics are preferred. Synthetic mulches, however, cause soil solarization. Mulching is an expensive proposition and hence its use may be limited to high-value plantation crops such as tea, coffee, etc. In tea, Guatemala grass (*Tripsacum laxum*) which is grown as a soil rehabilitation crop in fallow and non-cropped areas, is extensively used for mulching, particularly in young tea. In some cases citronella grass (*Cymbopogan* spp.) is also used as mulching material.

ECOPHYSIOLOGICAL APPROACHES

Light

In a crop-weed system, light regulates many aspects of weed and crop growth, development and competition. Plants respond variously to the quantity of light (including duration), the spectral quality of light, changing or fluctuating light environments and transient light (sun-flecks).

The growth and photosynthetic rates of plants are affected by the quantity of light. An increase in light leads to greater growth and photosynthetic rate while the same is decreased as light is reduced. Many plants possess plasticity to acclimate to reduced light conditions by redistribution of dry matter, altered leaf anatomy, decreased respiration rates, decreased enzymatic activities and decreased electron transport capacity [30].

Excess light can have a damaging effect on photosynthesis and, therefore, growth. Exposure of photosynthetic apparatus to strong light results in **photoinhibition**. Susceptibility to photoinhibition is highly dependent on genotypic and environmental factors, especially stress [28]. Symptoms of injury from photoinhibition are generally reversible if the inducing light is removed before photodestruction of pigment occurs [36].

The amount of light intercepted by weeds is a major determinant of crop growth and yield. Crop yields can be increased by manipulating canopies so more light is available. Crops and weeds show varying degrees of shade tolerance. In comparison to soybean, the weeds associated with it, such as *Solanum ptycanthum* (eastern black nightshade), *Amaranthus albus* (tumble pigweed) and *Xanthium strumarium* (common cocklebur) were most photosynthetically efficient under low-growth irradiance due to a combination of physiological and morphological adaptations [63]. Many other weeds acclimate under low-growth irradiance by means of plastic responses that reduce the growth-limiting effects of shading and allow restoration of high rates of photosynthesis when the plant is subsequently exposed to high irradiance. Most of the species are capable of adapting to extreme variations in the light environment, i.e., high light condition as well as deep shade.

Most crop canopies are characterized by an exponential decrease in light with increasing depth and heavy shade at lower canopy levels. The presence of weeds in a crop intensifies conditions for light limitation. For example, *Avena fatua* (wild oats) reduces light penetration and growth of wheat by being taller than the crop plants [13]. When wild oat plants were clipped to the height of wheat, light penetration in a mixed canopy was similar to that in monoculture wheat. Similarly, *Abutilon theophrasti* (velvetleaf) intercepted light due to its greater height advantage in soybean [1]. Yield reduction of tomato was greater when grown in competition with *Solanum ptycanthum* (eastern black nightshade) compared to competition with *Solanum nigrum* (black nightshade), due to greater height of the former weed [37]. Photosynthetically active radiation (PAR) at 400-700 nm, reaching the top of a tomato canopy correlated positively with yield and negatively with *S. ptycanthum* density [37]. These and other studies show that weed and canopy architecture, especially plant height, location of branches, and height of maximum leaf area, determine the impact of competition for light and thus have a major influence on crop yield.

Environmental Stress

Environmental stress on plants occurs when the level of an environmental resource adversely affects plant growth through some physiological process or condition. Under normal field conditions, environmental stress prevents any plant from achieving the maximum growth set by its hereditary potential or genotype. It also affects weed susceptibility to insects and diseases as well as the allelopathic potential. Biological factors that influence competition include the species of weed and the density, spatial distribution and duration of growth of both the weed and the crop [7] These factors are modified by physical environment and environmental stress.

Water Stress

Water, or lack of it, is an important environmental factor most often limiting crop growth and yield [5]. Plant water deficit or water stress occurs when transpirational water loss exceeds water absorption through the roots. This lowers the turgor of stomatal guard cells, thereby reducing stomatal conductance and CO_2 uptake. Water stress also reduces photosynthesis by interfering with chlorophyll synthesis, electron transport and phosphorylation, and the synthesis and activity of carboxylating enzymes. Because of its dependence on positive turgor as well as assimilate supply, the process of leaf expansion also is very sensitive to water stress.

Water use efficiency, transpiration rate and response to declining water availability and water stress vary widely among species of crops and weeds, and influence the process of crop-weed competition [49, 55]. The outcome of crop-weed competition for water or any other resource depends upon the relative abilities of weed and crop to obtain the resource and to tolerate a deficit in it [49].

Wiese and Vandiver [71] compared the growth and competitive ability of maize (C_4 plant) and sorghum (C_4) and eight weed species and found that maize produced the most biomass regardless of moisture level. Among the weeds, those species growing in humid regions or in irrigated crops, *Xanthium strumarium* (common cocklebur, C_3), *Echinochloa crus-galli* (barnyardgrass, C_4) and *Digitaria sanguinalis* (crabgrass, C_4) were the most competitive under moist conditions. Weeds typical of dry habitats, *Kochia scoparia* (kochia, C_4) and *Salsola iberica* (Russian thistle, C_4) were more competitive under dry conditions and grew poorly in competition at the high moisture level. Ozturk et al. [44] reported that the productivity of the C_4 species declined at the highest level of water availability while the C_3 species were least productive under dry conditions.

Many weeds appear to be '**water wasters**' in that their stomata are less sensitive to declining leaf water potential than those of the crops with which they compete [49]. If this characteristic is combined with a more extensive root system and/or better physiological tolerance of drought, the weed may quickly exhaust the water supply available to the crop [51]. This characteristic of excessive water use by weeds may be advantageous in the context of crop-weed competition, provided the weeds can survive and reproduce under the resultant soil water deficit [49].

Water stress or moisture availability may also influence the duration of critical period of weed competition for various crops. When competing with natural populations of *Ambrosia artemisiifolia* (common ragweed), the critical period for soybean was 2 wk in a dry year and 4 wk in a wet year [11]. When soil moisture was limiting, soybean competing with natural populations of weeds had a shorter critical period than when soil moisture was adequate [32]. Harrison et al. [27] found that *Setaria faberi* (giant foxtail) reduced soybean growth after only 10 to 15 d when soil moisture was abundant, but significant growth reduction was delayed until 25 d after emergence during a dry year. Differential effects of water stress on the growth of crops and competing weeds probably account for the effects of moisture availability on the duration of critical period [49].

Temperature Stress

Temperature is a major factor governing the seasonal growth of weeds and their geographic distribution. The phenological development of both crops and weeds is closely tied to temperature [4].

Stress imposed by unfavourably high or low temperatures has a significant influence over crop-weed competition. For example, in a comparison with several dicot weeds, maize grew more rapidly at day/night temperatures of 18/12° C, 24/18° C and 30/24° C [26]. Soybean outpaced the weeds in growth rate for the first 10 to 15 d but the weeds subsequently grew more rapidly than soybean. From a study involving three crop and six weed species, Potter and Jones [54] reported significant effects of temperature on relative growth rates (Rw) and relative leaf area expansion rates (Ra), and demonstrated that Rw correlated highly positively with leaf area partitioning. In all nine species, Rw and Ra were greater at 32/21° C than at 21/10° C or 38/27° C. Redroot pigweed (*Amaranthus retroflexus*), a small seeded C_4 dicot, had the highest Rw and Ra regardless of temperature. In other studies, cool temperatures early in the season favoured the growth of maize, upland rice, groundnut and soybean over that of several weeds, but warmer temperatures later in the season accelerated weed growth and weed competition became more severe [40]. Air temperatures of 23° C or greater favoured germination and early growth of *Eleusine indica* (goosegrass) over that of a cultivated millet (*Setaria anceps*), indicating that competitiveness of the millet could be enhanced by earlier planting [29]. These and various other studies demonstrate the role that extremities in temperature can play in crop-weed competition. A sound weed management strategy should take this into consideration.

Soil Solarization

Higher soil temperatures can suppress weed seed germination and kill weed seedlings. Soil temperature can be raised by using a mulch made of plant material or plastic material, as discussed above.

During the process of heating the surface soil, termed **soil solarization**, the soil temperature is raised above the thermal death point for most weed seeds and seedlings. Solarization can be achieved by placing transparent and opaque polyethylene sheets (thickness 0.025-0.050 mm) on moist soil and trapping solar radiation. Its effectiveness for weed control is dependent on warm, moist climate and intense radiation to raise soil temperature, to temperatures high enough to kill weed seeds and seedlings. Besides controlling weeds, solarization inhibits soil-borne diseases and increases crop growth.

During solarization the soil temperature is increased by 8 to 12° C, reaching or exceeding 45° C. Elasticity in soil temperature is, however, dependent on the type of plastic sheet used. Under black plastic sheets, the temperatures do not exceed 45° C, while UV-absorbing plastic sheets raise the temperatures to above 50° C. The effect of solarization is greater at the top 5 to 10 cm layer than at lower layers. This explains the efficacy of solarization on weed seed germination and seedling growth.

Solarization for a period of 5 wk may be adequate for controlling most summer and winter annual weeds, while a period of at least 5 mon is required for such perennial weeds as *Cyperus rotundus*, *Sorghum halepense* and *Cynodon dactylon*. Solarization for 30 d decreased total emergence, excluding *Cyperus rotundus*, by 57 to 83% [8] while the same for 6 wk controlled many annual grasses for a period of up to 3 mon [56]. A month after solarization, *Convolvulus arvensis* (field bindweed), *Sonchus oleraceus* (annual sowthistle), and *Amaranthus blitoides* (prostrate pigweed) covered only 18% of the soil surface in solarized plots as against 85% in non-solarized plots [62]. A one-week period of solarization reduced the percentage of buried seeds of *Sida spinosa* (prickly sida), *Xanthium strumarium* (common cocklebur), *Abutilon theophrasti* (velvetleaf) and *Anoda cristata* (spurred anoda) [18]. Total weed emergence was reduced by 97% one week after removal of plastic sheets and up to 77% for the season [18].

Although soil solarization has been tested and modified since the mid-1970s in more than 38 countries, it has primarily remained an experimental technique under continuing evaluation. However, this hydrothermal process has found successful niches as a commercial crop production practice, especially in greenhouse culture. The principles of solarization may offer a panacea but there are a number of limitations to its current use in large-scale agriculture. These are listed below:

1. Solarization needs large amounts of plastic film for soil mulching. These materials tend to be bulky and difficult to apply. Besides, their disposal is a major problem. Environmental laws prohibit burning plastic waste, and when buried in the soil, plastics are too slow in decomposing.
2. Solarization requires that land be rotated out of production for several weeks and months. This may interfere with the normal cropping patterns.
3. Soil solarization is too expensive a technique for large-scale commercial application. It is a material and labour-intensive practice.

In order to make soil solarization widely acceptable to farmers, improvements in the solarization process may need to be made. New-generation mulching materials, such as photodegradable and biodegradable film may eliminate the disposal problem. Combining solarization with other weed management methods may increase the predictability and effectiveness of the process and allow more widespread use of this method as a viable tool of weed management.

Atmospheric CO_2

Global atmospheric CO_2 enrichment affects crops and weeds, both directly and indirectly, through global warming [49]. Plants with a C_3 photosynthetic pathway are expected to benefit more from CO_2 enrichment than plants with the C_4 photosynthetic pathway [50]. This differential effect on C_3 and C_4 plants has important implications for crop-weed competition because many of the major weeds are C_4 plants while many crops are C_3 plants.

Increasing CO_2 concentration from 350 to 675 ppm increased the competitiveness of a C_3 soybean crop against the C_4 weed *Sorghum halepense* (Johnsongrass), and reduced the competitiveness of Johnsongrass with soybean [52]. Enrichment of CO_2 to 675 ppm improved *Digitaria sanguinalis* (crabgrass) and *Eleusine indica* (goosegrass) [48]. As a result, the weeds, when subjected to drought, produced more dry matter and leaf area at 675 ppm CO_2 than at 350 ppm. When plants were adequately watered, CO_2 enrichment had no effect on growth in these C_4 weed species. Even during drought soybean benefited more from CO_2 enrichment than did the C_4 weeds. Thus the direct effects of CO_2 enrichment seem to favour C_3 plants even when water is limited.

An important indirect effect of global atmospheric CO_2 enrichment may be an increase of 2 to 3° C in the global temperature, with consequent increases in aridity and general instability of weather [67]. These indirect, climatic effects may be more important than direct, physiological effects.

NUTRIENT-BASED APPROACHES

It is well known that plant nutrients clearly promote crop growth. However, in a crop-weed situation, application of nutrients benefits weeds more than crops because of the greater ability of weeds to accumulate minerals. Increased uptake of mineral nutrients in weeds often results in a significant competitive advantage over crop species. By altering the current fertilization strategies, it is possible to simultaneously reduce the harmful effect of weeds on crop growth and yield, and maximize the competitive ability of crops.

Among the major plant nutrients, competition is greatest for nitrogen. When weed densities are low, increased N application helps the crop plants to gain advantage over weeds, leading to minimal competition with weeds and greater crop growth and yield. However, when weed densities are higher, added nutrients favour weed growth than crop growth, leading to little added benefit in crop yield. The increase in weed competition at higher N rates is due to increased nutrient accumulation and use efficiency by weeds.

The ability of weed species to utilize available nutrients better than crop species can also provide an advantage to the former while competing for water and light [9, 35]. Okafor and De Datta [41] found that increasing N in rice benefited *Cyperus rotundus* (purple nutsedge) more than the crop. The subsequent increase in purple nutsedge growth reduced light transmission to the crop, leading to a reduction in rice leaf area index and concomitant decrease in rice grain yield. Liebman and Robichaux [35] reported that a barley-pea mixed cropping was more competitive against *Brassica hirta* (white mustard) when N was not applied, as compared to high N conditions. This corresponded with a reduction in photosynthetic capacity, seed production and shoot biomass in mustard. However, at high N rates, white mustard produced more leaves and was taller than pea.

Strategies to Reduce Weed Competition for Nutrients

Method of Application

One strategy to reduce weed competition is to apply fertilizers in bands in crop row in preference to broadcast application. Otabbong *et al.* [43] compared the effects of weeds on bean yield using three fertilization methods: broadcast application, surface band (5 cm strip) application in seed row, and deep band (7 cm below seed level) application within the seed row. Their results indicated that surface banding in crop row had little beneficial effect on bean yield and weed suppression, and even reduced bean yield in unweeded plots. This was due to increased access by weeds growing in the crop row to concentrated levels of nutrients. In contrast, deep banding of the fertilizer in the crop row significantly increased bean biomass and yield, particularly in unweeded plots, while also suppressing weed biomass by 44%. Thus, beans gained significant competitive advantage when the nutrients were placed below the weed seed level. Similar results were reported for deep placement of fertilizers in rice [38].

An additional advantage of deep banding of fertilizers is the possibility of reducing the quantity of fertilizers needed to maximize crop yields. Otabbong et al. [43] reported that application of fertilizers by deep banding in the bean row not only reduced weed competition with the crop, but also reduced the rate of P_2O_5 (from 100-200 kg ha^{-1} to 50 kg ha^{-1}) required to maximize bean yield. Cochron et al. [12] similarly found no increase in wheat grain yield with N above 45 kg ha^{-1} when deep banding technique was used. At 45 kg ha^{-1}, the wheat grain yield after broadcast application was significantly lower than deep banding.

Deep banding may certainly be beneficial when weeds with shallow root system are predominant in the weed spectrum, but not necessarily when weeds with deeper root system predominate. At the same time, crops with deeper root system may be benefited more from deep band application than those with shallow root system. Thus, extensive studies on placement of fertilizers in relation to root systems of crops and weeds are required before this strategy could be implemented.

Time of Application

The time of fertilizer application may be manipulated to tilt the crop-weed competition for nutrients in favour of crop plants. This could be a valuable strategy in view of varying growth habits, tillering capacities, and root systems of crops and weeds.

In case of wheat, demand for nitrogen (N) is greater before the three-leaf stage when tillering begins [14]. Thereafter, N use efficiency is less in wheat. In contrast, Lolium rigidum (annual ryegrass) produces tillers continuously and is stimulated by late N applications. Therefore, early N application reduces the competitive effect of annual ryegrass on wheat yield. Pandey et al. [46] studied phosphorus (P) uptake in a number of crops and weeds and found that the rate of P uptake varied with age and species. Purple nutsedge (Cyperus rotundus) exhibited rapid P uptake until plants were 24 d old, while the demand for P in other weeds and a number of crops including wheat, pearl millet (Pennisetum glaucum) and chick pea (Cicer arientinum) was higher after 60 d, during the flowering and fruiting stages.

Alternative Nutrient Sources

Crops and weeds vary in their response and sensitivity to various forms of nitrogen. Maize and Amaranthus retroflexus (redroot pigweed) have varying growth responses to nitrate and ammonium [68]. In redroot pigweed, treatment with ammonium (220 mg N kg^{-1}) plus a nitrification inhibitor (EAS at 5 mg ai kg^{-1}) caused leaf chlorosis and crinkling while reducing shoot dry weight, total N accumulation and extractable Mg, malate and oxalate as compared to nitrate (220 mg N kg^{-1}) treatment. Maize was unaffected by either nitrate or ammonium plus EAS. Teyker et al. [68] suggested that enhancing the proportion of N as ammonium may provide more effective weed control in ammonium-sensitive plants, while at the same time minimizing N loss from the crop root zone. Similarly, the infestations of Striga spp. (witchweed) are affected by the form and concentration of N used in fertilizers. Farina *et al*. [21] reported that ammonium dramatically reduced the incidence of witchweed in maize and sorghum. Urea and ammonium are believed to inhibit witchweed germination and root elongation. Thus, shifting the N source from the ammonium-nitrate combination fertilizers to ammonium sulphate and urea could provide more effective control of witchweed in grain crops.

CROP-BASED APPROACHES

Crop-based approaches constitute an expensive but useful traditional method that requires careful thought and planning. In this method all the principles of crop characteristics including stature, growth habits, canopy development, leaf orientation, duration, adaptabilities, plant vigour, etc. are employed to exploit crop abilities and weed deficiencies. The techniques of the crop-based method, also known as the cultural method, are well known to farmers and weed scientists; they are regularly used even in the absence of conscious efforts to manage weeds. These include competitive cultivars, crop rotation, intercropping, companion cropping, plant population, etc.

Competitive Crops and Cultivars

The success of a weed management programme depends on the competitiveness of crops and crop cultivars with weeds. Crops and cultivars with high plant vigour, which grow more rapidly, have an advantage over the slow-growing and late-emerging weeds. They compete better with weeds for plant nutrients, sunlight, soil moisture and carbon dioxide. The major competitive crops include maize, sorghum, soybean, cowpea, etc. They either grow more rapidly and become taller or fill the interrow space with their canopy faster than the weeds. This has a smothering effect on weed growth. Similarly, cultivars which possess rapid growth and development and adaptive abilities perform better in suppressing weed growth and compete well with weeds.

Crop Rotation

Continuous growing of one crop could result in an increase in the population of weeds that characteristically associate with it. This may increase the occurrence of plant diseases and insects, resulting in patchy crop stands which are invaded by weeds.

Some weeds associate with certain crops more than with others. For example, *Echinochloa crus-galli* (barnyardgrass) and *Echinochloa colonum* (junglerice) are common in rice. *Avena fatua* (wild oats) is a common weed in wheat but not in rice. The wild oat infestation can be reduced by growing maize in rotation.

In planning a crop rotation programme, the life cycles, growth behaviour and competitive abilities need to be considered. Crop rotation breaks a weed's life cycle. A good rotation includes crops that reduce weeds that are especially troublesome in succeeding crops. Crops which grow tall may follow or precede one which covers the ground. An ideal crop rotation would not allow undisturbed infestation and development of weeds.

Crop rotation regularly changes the crop in each field, soil preparation practices, subsequent soil tillage and weed control techniques. All these affect weed populations.

In a study involving four cropping systems, Pablico and Moody [45] found that two weeds, viz. *Hedera helix* (itchgrass) and *Amaranthus spinosus* (spiny amaranth) dominated, but their relative magnitude in the cropping systems, on the same soil, was different. In a rice-sorghum rotation, itchgrass dominated, but with continuous sorghum, itchgrass nearly disappeared and spiny amaranth dominated. Thus, different cropping systems affect weed populations by favouring or deterring certain species.

Crop rotation helps in reducing or minimizing the build-up of a weed seedbank in the soil. In long-term studies to determine the effect of different cropping sequences on the population dynamics of winter wild oats (*Avena sterilis* ssp. ludoviciana), Fernandez-Quintanilla *et al*. [25] found that continuous winter cereal cropping (with or without herbicides) increased the winter wild oat seedbank in the soil from 26% to 80% yr^{-1}. With spring barley, the seedbank declined 10% yr^{-1}. When sunflower was used as a summer crop or the field was kept fallow for 12 mon in the rotation to prevent new seed production, the soil seed reserve declined from 57% to 80% annually. There was a great reduction in the size of the soil seedbank of winter wild oats if the cropping programme was other than continuous winter cereals.

The success of rotation systems for weed management is based on the use of crop sequences that employ varying patterns of resource competition, allelopathic interference, soil disturbance and mechanical damage to provide an unstable and frequently inhospitable environment that prevents the proliferation of particular weed species.

Intercropping

Intercropping, also called mixed cropping, is a common cropping system of intensive agriculture, particularly in developing countries. Although intercropping is practised more to maximize land use and minimize risk of crop failure, it has a significant effect in suppressing weed growth. Fewer weeds infest in an intercropping system than in a monoculture system. In pigeonpea-based cropping systems at ICRISAT (International Crop Research Institute for the Semi-Arid Tropics at Hyderabad, India), intercropping with sorghum reduced weed growth by 10%-75% compared to pigeonpea alone) [57]. In a pearlmillet-groundnut intercrop system, an arrangement of one pearl millet row for every three groundnut rows resulted in optimal weed suppression and crop yield [61]. Intercropping cassava and maize is an effective weed control practice [42]. Although intercrops may not always be superior to single crops in weed suppression, the weed biomass accumulation by intercrops is generally lower than that of at least one of the components grown as a single crop.

The effectiveness of intercropping for weed control depends on the crops chosen, their relative proportions and plant geometry. Suppression of weeds during intercropping may have a beneficial effect on a subsequent crop. The benefits derived from intercropping are generally similar to those when crop rotation is followed.

Companion Cropping

Companion crops, also termed cover crops or live mulches, are used as intercrops to suppress weeds. Companion cropping has the potential to become a good weed management technique whose need in a specific situation needs to be determined. Besides providing weed competition, companion crops can build soil organic matter, reduce soil erosion and improve water penetration.

Generally, legumes serve as good companion crops, particularly in tall-growing crops like maize, sorghum, sugarcane and millets, perennial crops such as tea, coffee, etc., and long-term horticultural and tree crops. Akobundu et al. [2] reported that maize grain increased from 1.6 t ha^{-1} in unweeded, no-till plots to 2.7 t ha^{-1} in unweeded, legume live-mulch filled plots. Similarly, clover, a legume, when grown as a companion crop, reduces weed growth and increases maize grain yield. Other legumes such as cowpea, mung bean and crownvetch (*Coronilla* spp.) may also serve as companion crops.

A free floating fern, *Azolla pinnata*, is used in low land and irrigated rice as a sort of companion crop. This fern has a symbiotic relationship with *Azolla anabena*, a nitrogen-fixing blue-green alga. This symbiotic relationship not only contributes up to 100 kg N ha^{-1} but also reduces weed growth by forming an *Azolla* blanket over the water surface in a rice crop. The *Azolla* technique suppresses and controls many annual weeds while leaving out certain annuals which have strong culms (e.g., barnyardgrass) and perennial weeds. The success of this technique largely depends on the ability of the farmers to control water supply and the weed species present. However, continued use of *Azolla* may not only facilitate tolerant weed species spread quickly over the field, but also allow free floating fern to become an aquatic weed in rice, thus complicating the weed problems.

Plant Population

Increasing crop density by using higher seeding rate, narrower row spacing and closer plant-to-plant spacing (within in a row) is an important weed management technique as it enhances crop competitiveness by suppressing or smothering weeds. Higher plant population creates shading which prevents weed seed germination, emergence and establishment. Higher plant population is, however, dependent on growth habit, leaf orientation, duration and other characteristics. This method could be more useful in crops where row spacing is too wide, thus allowing prolific weed growth.

The effects of both row spacing and seed rate were investigated by Moyer et al. [39]. Dry matter of weeds decreased as row spacing decreased from 108 to 36 cm or the broadcast seeding rate of alfalfa increased from 0.33 to 3.0 kg ha^{-1}. The authors found significant herbicide × seeding interaction in one season, with essentially no benefit from herbicide application at the highest plant density and a larger benefit at the lowest density. This trend, however, was absent in the presence of perennial weeds which competed successfully with dense stands of alfalfa.

WEED SEED DORMANCY AND GERMINATION STIMULATION STIMULANTS

As discussed in Chapter 2, weed seed dormancy and germination are regulated by a complex interaction of environmental, edaphic, physiological and genetic factors. Almost all farming practices can affect these factors, primarily by altering the physiological environment of weed seeds in soil, although environmental conditions during embryogenesis and seed ripening may also have an effect [16]. Farming practices influence weed population dynamics and species composition by affecting the quantities of seeds returned to and removed from the soil [72]. Breaking the dormancy of seeds and stimulating their germination constitute a viable weed management tool. Improved tillage practices and chemicals can play a vital role in this respect.

Tillage Practices

Soil tillage often promotes emergence of weed seedlings due to its effect on soil disturbance. This aspect was discussed in Chapter 2.

The size of a weed seedbank can be reduced when reseeding is prevented and the existing seeds are exposed to environmental conditions favourable for their germination. This can be done by a combination of several weed management techniques.

Roberts [58] reported that seed populations can be reduced by stimulating their germination by tillage practices, to approximately 10% of the original amount over a period of several years. Once the population in the plough layer is reduced to a low level, deep tillage may return deeply buried seeds to near the soil surface where they will add to the seed population [59]. Although cultivation will reduce seed population to a certain extent, there will be a large reserve of dormant seeds in the soil for a long time. Inducing germination of these long-lived dormant seeds is a key to solving long-lasting weed problems.

Chemical Stimulants

The chemicals that stimulate germination of both viable and dormant seeds include plant growth regulators (gibberellins, cytokinins and ethylene), plant products (strigol, fusicoccin), respiratory inhibitors (azide, cyanide and hydroxylamine), oxidants (hypochlorite, oxygen), nitrogenous compounds (nitrate, nitrite and thiourea), anesthetics (acetone, ethanol, ethyl ether and chloroform) and miscellaneous compounds [6]. Of these, ethylene, nitrates, strigol and azide have been tested in the field and used for stimulation of weed seed germination in the soil.

Ethylene

Ethylene is an important tool in the control of *Striga* spp. (witchweed). It stimulates witchweed seed germination and reduces the dormant seeds in the soil.

A single injection of ethylene gas at 1.7 kg ha^{-1} at a 20-cm soil depth reduced populations of dormant witchweed by 91% [20]. Ethylene diffused readily through sandy soils. The gas was detected as far as 90 cm below and 120 cm lateral to the application point in the soil. Stimulatory levels of ethylene were retained in the soil for up to 6 h. Under optimum conditions, just 2 h of ethylene contact with witchweed seeds was adequate to trigger germination. The combination of separate applications of ethylene to induce witchweed seed germination and herbicides to kill the emerged witchweed seedlings is an effective strategy to control the parasitic weed very effectively and reduce weed seed populations in the soil substantially. This combined ethylene-herbicide strategy has been used to significantly reduce witchweed infestation in North and South Carolina, USA [19].

Although ethylene stimulates germination of seeds of several weed species under controlled conditions, its effect on weed seeds in the soil so far has been limited. Much work needs to be done on this subject.

Nitrates

Nitrates are well known to enhance germination of many weed seeds, particularly when used in sequence or combination with light, chilling or varying temperatures.

At the field level, ammonium nitrate is known to cause increased emergence of wild oats, *Eleusine indica* (goosegrass), *Echinochloa colonum* (junglerice), *Leptochloa filiformis* (red sprangletop), *Sisymbrium officinale* (hedge mustard) etc. It has no effect on other weeds, however, including *Chenopodium album* (common lambsquarters), *Setaria faberi* (giant foxtail), *Abutilon theophrasti* (velvetleaf), *Datura stramonium* (jimsonweed) and *Amaranthus retroflexus* (redroot pigweed) [22]. Although the nitrate effect was not immediately apparent, long-term effects, such as reduced dormancy and reduced longevity of weeds, might result from the treatment [22].

In spite of the encouraging results obtained at the field level, nitrates appear to be more effective in combination with other germination stimuli, such as light, alternating temperatures and ethylene. As germination stimulants, nitrates offer greater practical utility as nitrate-containing fertilizers can be used widely not only to supply N to crop plants, but also to stimulate weed seed germination and thus reduce weed competition.

Azide

Sodium azide, an inhibitor of aerobic respiration, is known to stimulate weed seed germination. Fay and Gorecki [23] found that sodium azide stimulated wild oats seed germination in the soil and in growth chamber studies. In the field, soil incorporation of sodium azide at 11.2 kg ha^{-1} stimulated a four-fold increase in emergence of seedlings. Because azide is rapidly lost from the soil, sodium azide must be deeply and thoroughly incorporated and soil moisture must be adequate, so the chemical can rapidly act upon the seeds before it is dissipated from the soil. Fay et al. [24] suggested that techniques for controlled release of azide were necessary for the compound to be practical as a germination stimulant in a programme to enhance wild oats control. Besides, the efficacy of sodium azide may be influenced by the time of treatment, the weed species present and the soil environment after treatment [19].

Strigol

Strigol is a chemical stimulant of parasitic weeds, witchweed (*Striga* spp.), orobanche (*Orobanche* spp.), etc. It is a root exudate produced by cotton. It stimulates witchweed seed germination at concentrations as low as 10^{-2} M [31]. At 0.3 to 4.5 kg ha^{-1}, strigol retained activity in moist soil for 21 d, and sufficient quantities of the compound leached down to 30 cm of soil to induce maximum germination of witchweed seeds [31]. Strigol, like ethylene, has the potential to become a highly active soil-applied stimulant to reduce populations of parasitic weed seeds in the soil.

Effective Utilization of Stimulants

The preceding discussion on chemical stimulants indicated that these chemicals have a great potential in breaking dormancy and enhancing weed seed germination. However, their success at field level is dependent on various factors. Egley [19] suggested the following strategies to improve the success of germination stimulants:

a) Application of germination stimulants at peak periods of germination so that fewer non-germinated seeds remain in the soil. Infusion of soil at this time may force germination of those seeds that were poised to germinate or on the verge of germination but unable to complete the germination process.

b) Application of stimulant at the time of the year when the maximum number of seeds are non-dormant and before the seeds pass into secondary dormancy. For summer annuals, a stimulant that is active at low temperatures may be applied to the soil during late winter. To enhance control of winter annuals, a stimulant most effective at high temperatures may be applied to the soil in late summer or early fall, prior to tillage and planting.

These approaches involve determination of time of peak germination of seeds of various weeds and time when they normally break dormancy. This information is necessary to guide timing of application of germination stimulants.

REFERENCES

1. Akey, W.C., T.W. Jurik and J. Dekker. 1990. Competition for light between velvetleaf (*Abutilon theophrasti*) and soybean (*Glycine max*). Weed Res. **30**: 403-411.

2. Akobundu, I.O. 1980. Live mulch: A new approach to weed control and crop production in the tropics. Proc. Br. Crop Protect. Conf. – Weeds, pp. 377-380.

3. Aldrich, R.J. 1994. Weed-Crop Ecology. Breton Publ. North Scituate, MA.,pp. 373-435.

4. Alm, D.M., M.E. McGiffen, Jr. and J.D. Hesketh. 1991. Weed Phenology. *In* T. Hodges (ed.). Predicting Crop Phenology. CRC Press, Boca Raton, FL, USA. pp. 191-218.

5. Begg, J.E. and N.C. Turner. 1976. Crop water deficits. Adv. Agron. J. **68**: 825-827.

6. Bewley, J.D. and M. Black. 1982. Physiology and Biochemistry of Seeds in Relation to Germination. Vol. 2: Viability, Dormancy, and Environmental Control. Springer-Verlag, New York, 375 pp.

7. Bleasdale, J.K.A. 1960. Studies on plant competition. *In* J.L. Harper (ed.). The Biology of Weeds. Blackwell Sci. Publ., Oxford, pp. 133-142.

8. Braun, M., W. Koch and M. Stiefvaler. 1987. Solarization for soil sanitation—possibilities and limitations demonstrated in trial in Southern Germany and the Sudan. Gesunde Pflanzen **39**: 301-309.

9. Carlson, H.L. and J.E. Hill. 1986. Wild oat (*Avena fatua*) competition with spring wheat: Effects of nitrogen fertilization. Weed Sci. **34**: 29-33.

10. Clements, D.R., S.F. Weise and C.J. Swanton. 1994. Integrated weed management and weed species diversity. Phytoprot. **75**: 1-18.

11. Coble, H.D., F.M. Williams and R.L. Ritter. 1981. Common ragweed (*Ambrosia artemisiifolia*) interference in soybeans (*Glycine max*). Weed Sci. **29**: 339-342.

12. Cochron, V.L., L.A. Morrow and R.D. Schirman. 1990. The effect of N placement on grass weeds and winter wheat responses in three tillage systems. Soil Tillage Res. **18**: 347-355.

13. Cudney, D.W., L.S. Jordan and A.E. Hall. 1991. 1991. Effect of wild oat (*Avena fatua*) infestations on light interception and growth rate of wheat (*Triticum aestivum*). Weed Sci. **39**: 175-179.

14. Davidson, S. 1984. Wheat and ryegrass competition for nitrogen. Rural Res. **122**: 4-6.

15. Dickey, E.C., J.C. Siemens, P.J. Jasa, V.L. Hofman and D.P. Shelton. 1992. Tillage system definitions. *In* Conservation Tillage Systems and Management—Crop Residue Management with No-till, Ridge-till, Mulch-till. MidWest Plan Service, Ames, Iowa, USA, pp. 5-7.

16. Dyer, W.E. 1995. Exploiting weed seed dormancy and germination requirements through agronomic practices. Weed Sci. **43**: 498-503.

17. Eadie, A., C.J. Swanton and G.W. Anderson. 1992. Integration of cereal cover crops in ridge-tillage corn production. Weed Technol. **6**: 553-560.

18. Egley, E.H. 1983. Weed seed and seedling reductions by soil solarization with transparent polythene sheets. Weed Sci. **31**: 404-409.

19. Egley, G.H. 1986. Stimulation of weed seed germination in soil. Rev. Weed Sci. **2**: 67-89.

20. Eplee, R.E. 1975. Ethylene: A witchweed seed germination stimulant. Weed Sci. **23**: 433-436.

21. Farina, M.P.W., P.E.L. Thomas and P. Channon. 1985. Nitrogen, phosphorus, and potassium effects on the incidence of *Striga asiatica* (L.) Kuntze in maize. Weed Res. **25**: 443-447.

22. Fawcett, R.S. and F.W. Slife. 1975. Germination stimulation properties of carbamate herbicides. Weed Sci. **23**: 419-424.

23. Fay, P.K. and R.S. Gorecki. 1978. Stimulating germination of dormant wild oat (*Avena fatua*) seed with sodium azide. Weed Sci. **26**: 323-326.

24. Fay, P.K., R.S. Gorecki, and P.M. Fuerst. 1980. Coating sodium azide granules to enhance seed germination. Weed Sci. **28**: 674-677.

25. Fernandez-Quintanilla, C.L. Navarrete, and C. Torner. 1984. The influence of crop rotation on the population dynamics of *Avena sterilis* (L) ssp. *ludoviciana* Dur. in Central Spain. Proc. Third Eur. Weed Res. Soc. Symp. on Weed Problems of the Mediterranean Area, pp. 9-16.

26. Frazee, R.W. and E.W. Stoller. 1974. Differential growth of corn, soybean, and seven dicotyledonous weed seedlings. Weed Sci. **22**: 336-339.

27. Harrison, S.K., C.S. Williams and L.D. Wax. 1985. Interference and control of giant foxtail (*Setaria faberi*) in soybeans (*Glycine max*). Weed Sci. **33**: 203-208.

28. Havaux, M. 1992. Stress tolerance of photosystem I *in vivo*. Antagonistic effects of water, heat, and photoinhibition stresses. Plant Physiol. **100**: 424-432.

29. Hawton, D. 1979. Temperature effects on *Eleusine indica and Setaria anceps* grown in association (I). Weed Res. **19**: 279-284.

30. Holt, J.S. 1995. Plant responses to light: A potential tool for weed management. Weed Sci. **43**: 474-482.

31. Hsiao, A.I., A.D. Worsham and D.E. Moreland. 1983. Leaching and degradation of *dl*-strigol in soil. Weed Sci. **31**: 763-765.

32. Jackson, L.A., G. Kapusta and D.J. Schutte Mason. 1985. Effect of duration and type of natural weed infestations on soybean yield. Agron. J. **77**: 725-729.

33. Klein, R.N., G.A. Wicks and R.G. Wilson. 1996. Ridge-till, an integrated weed management system. Weed Sci. **44**: 417-422.

34. Laflen, J.M., R. Lal and S.A. El-Swaify. 1990. Soil erosion and a sustainable agriculture. *In* C.A. Edwards, R. Lal, P. Madden, R.H. Miller, and G. House (eds.). Sustainable Agricultural Systems. Soil and Water Conser. Serv., Ankeny, Indiana, USA, pp. 569-581.

35. Liebman, M. and R.H. Robichaux. 1990. Competition by barley and pea against mustard: Effects on resource acquisition, photosynthesis, and yield. Ecosystems Environ. **31**: 155-172.

36. Long, S.P. and S. Humphries. 1994. Photoinhibition of photosynthesis in nature. Ann. Rev. Plant Physiol. Plant Molec. Biol. **45**: 633-662.

37. McGiffen, M.E. Jr., J.B. Masiunas and J.D. Hesketh. 1992. Competition for light between tomatoes and nightshades (*Solanum nigrum or S. ptycanthum*). Weed Sci. **40**: 220-226.

38. Moody, K. 1981. Weed-fertilizer interactions in rice. Int. Rice Res. Inst. (IRRI), Int. Rice Paper Ser. 68, pp. 35.

39. Moyer, J.R., K.W. Richards and G.B. Schaalje. 1991. Effect of plant density and herbicide application on alfalfa seed and weed yields. Canadian J. Plant Sci. **71**: 481-489.

40. Noguchi, K. and K. Nakayame. 1978. Studies on weed competition between upland crops and weeds. II. Comparison of early growth of crops and weeds. Jap. J. Crop Sci. **47**: 48-55.

41. Okafor, L.I. and S.K. De Datta. 1976. Competition between upland rice and purple nutsedge for nitrogen, moisture, and light. Weed Sci. **24**: 43-46.

42. Olasantan, F.O., E.O. Lucas and H.C. Ezumah. 1994. Effects of intercropping and fertilizer application on weed control and performance of cassava and maize. Field Crops Res. **39**: 63-69.

43. Ottabbong, E., M.M.L. Izquierdo, S.F.T. Talavera, U.H. Geber and L.J.R. Ohlander. 1991. Response to P fertilizer of *Phaseolus vulgaris* L. growing with or without weeds in a highly P-fixing mollic Andosol. Trop. Agric. **68**: 339-343.

44. Ozturk, M., H. Rehder and H. Zeiglerr. 1981. Biomass production of C3 and C_4 plant species in pure and mixed culture with different water supplies. Oecologia (Berl.) **50**: 73-81.

45. Pablico, P. and K. Moody. 1984. Effect of different cropping patterns and weeding treatments and their residual effects on weed populations and crop yield. Phillipp. Agric. **67**: 70-81.

46. Pandey, H.N., K.C. Misra and K.L. Mukherjee. 1971. Phosphate uptake and its incorporation in some crop plants and their associated weeds. Ann. Bot. **35**: 367-372.

47. Patterson, D.T. 1982. Shading responses of purple and yellow nutsedges (*Cyperus rotundus and C. esculentus*). Weed Sci. **30**: 692-697.

48. Patterson, D.T. 1986. Responses of soybean (*Glycine max*) and three C_4 grass weeds to CO_2 enrichment during drought. Weed Sci. **34**: 203-210.

49. Patterson, D.T. 1995. Effects of environmental stress on weed/crop rotation. Weed Sci. **43**: 483-490.

50. Patterson, D.T. and E.P. Flint. 1980. Potential effects of global atmospheric CO_2 enrichment on the growth and competitiveness of C_3 and C_4 weed and crop plants. Weed Sci. **28**: 71-75.

51. Patterson, D.T. and E.P. Flint. 1983. Comparative water relations, photosynthesis, and growth of soybean (Glycine max) and seven associated weeds. Weed Sci. **31**: 318-323.

52. Patterson, D.T., E.P. Flint and J.L. Beyers, 1984. Effects of CO_2 enrichment on competition between a C_4 weed and a C_3 crop. Weed Sci. **32**: 101-105.

53. Phillips, S.H. and H.M. Young. 1973. No-tillage farming. Reiman Associates, Milwaukee, Wisconsin, USA, 224 pp.

54. Potter, J.R. and J.W. Jones. 1977. Leaf area partitioning as an important factor in growth. Plant Physiol. **59**: 10-14.

55. Radosevich, S.R. and J.S. Holt. 1984. Weed Ecology: Implications for Vegetation Management. John Wiley & Sons, New York, 265 pp.

56 Ragone, D. and J.E. Wilson. 1988. Control of weeds, nematodes, and soil borne pathogens by soil solarization. Alafua Agricultural Bull. **13(1):** 13-20.

57. Rao, M.R. and S.V.R. Shetty. 1976. Some biological aspects of intercropping systems: Crop-weed balance. Indian J. Weed Sci. **8:** 32-34.

58. Roberts, E.H. 1981. The interaction of environmental factors controlling loss of dormancy in seeds. Ann. Appl. Biol. **98:** 552-555.

59. Schweizer, E.E. and R.L. Zimdahl. 1984. Weed seed decline in irrigated soil after six years of continuous corn (*Zea mays*) and herbicides. Weed Sci. **32:** 76-83.

60. Shaw, D.R. 1996. Development of stale seedbed weed control programs for southern row crops. Weed Sci. **44:** 413-416.

61. Shetty, S.V.R. and A.N. Rao. 1981. Weed management studies in sorghum/pigeonpea and pearlmillet/groundnut intercrop systems—some observations. Proc. Intntl. Workshop on Intercropping. 10-13 Jan 1979, ICRISAT, Patancheru, Hyderabad, India, pp. 238-248.

62. Silveira, H.L. and M.L.V. Borges. 1984. Soil solarization and weed control. Proc. Third Eur. Weed Res. Soc. Symp. on Weed Problems of the Mediterranean Area, pp. 345-349.

63. Stoller, E.W. and R.A. Myers. 1989. Response of soybeans (*Glycine max*) and four broadleaf weeds to reduced irradiance. Weed Sci. **37:** 570-574.

64. Swanton, C.J., D.R. Clements and D.A. Derksen. 1993. Weed succession under conservation tillage: A hierarchial framework for research and management. Weed Technol. **7:** 286-297.

65. Swanton, C.J. and S.D. Murphy. 1996. Weed science beyond the weeds: The role of integrated weed management (IWM) in agroecosystem health. Weed Sci. **44:** 437-445.

66. Swanton, C.J. and S.F. Weise. 1991. Integrated weed management: the rationale and approach. Weed Technol. **5:** 657-663.

67. Taylor, K.E. and M.C. MacCracken. 1990. Projected effects of increasing concentrations of carbon dioxide and trace gases on climate. In B.A. Kimball, N.J. Rosenberg and L.H. Allen, Jr. (eds.). Impact of Carbon Dioxide, Trace Gases, and Climate Changes in Global Agriculture. Am. Soc. Agron., Madison, Wisconsin, USA. pp. 1-17.

68. Teyker, R.H., H.D. Hoelzer and R.A. Liebl. 1991. Maize and pigweed response to nitrogen supply and form. Plant Soil. **135:** 287-292.

69. Wicks, G.A. 1976. Ecofallow: A reduced tillage for Great Plains. Weeds Today **7(2):** 20-23.

70. Wicks, G.A. and B.R. Somerhalder. 1971. Effects of seedbed preparation for corn on distribution of weed seed. Weed Sci. **19:** 666-668.

71. Wiese, A.F. and C.W. Vandiver. 1970. Soil moisture effects on competitive ability of weeds. Weed Sci. **18:** 518-519.

72. Wilson, R.G. 1988. Biology of weed seeds in the soil. *In* M.A. Altieri and M. Liebman (eds.). Weed Management in Agroecosystems: Ecological Approaches. CRC Press, Boca Raton, FL, USA.

73. Witt, W.W. and J.W. Herron. 1980. Nitrosamines in Treflan—decisions, decisions. Weeds Today **11(1):** 4-5.

4

Introduction to Chemical Weed Management

CHEMICAL METHOD

Herbicides constitute the principal component of chemical weed management. The word herbicide is derived from the Latin words **herba** (plant) and **caedere** (to kill). In this chemical age, herbicides constitute one of the most vital and essential inputs to maximize crop reduction levels.

Although chemicals (e.g., salts) are reported to have been used in agriculture for several centuries, introduction of the bordeaux mixture in 1896 stimulated interest in chemical weed control. This led to the discovery of copper salts for selective control of broadleaf weeds in cereals. The period between 1896 and 1910 was important for chemical weed control as many chemicals, e.g. sulphuric acid, iron sulphate, copper nitrate, ammonium and potassium salts, sodium nitrate, ammonium sulphate, etc. were discovered for weed control. The important researchers who contributed to the progress of chemical weed control during this period included Martin and Bonnet in France, Bolley in the USA and Schultz in Germany. After this initial interest, research on chemical control of weeds languished for about 30 years, largely due to lack of adequate spraying equipment and the frequent failure of herbicides because of low humidity. The introduction of new and improved agricultural practices like cleaner seeds, vigorous crop varieties, adoption of fallow system, wider usage of gasoline tractors for tillage work, etc. also contributed to the lag in research on chemical control. During this period, particularly in the 1930s, chemicals such as sodium chlorate, carbon bisulphide, sodium arsenite and dinitrophenols were used only occasionally and on a limited scale in the USA and Europe.

The concept of systemic control of weeds through absorption by roots and top growth of the plant and translocation in the plant body and the use of organic chemicals for weed control took root with the introduction of nitrophenols as selective herbicides in 1935. The discovery of 2,4-D in the early 1940s, however, revolutionized the chemical method of weed control. It was in 1941 that Pokorny reported synthesis of 2,4-D which, interestingly, was first tried and found ineffective as a fungicide and insecticide. In 1942, Zimmerman and Hitchcock tested 2,4-D as a growth regulator. The credit for introducing it as a herbicide goes to Marth and Mitchell of the USA, who in 1944 reported selective weed control by 2,4-D in a bluegrass lawn and Hammer and Tukey, also of the USA, who in the same year used 2,4-D successfully in field weed control. Thus, demonstration of the herbicidal properties of 2,4-D in field weed control represented a technological breakthrough that heralded the modern era of weed management.

Research on 2, 4-D established that herbicides could be effective in very small quantities, highly selective and systemic in action. The small quantities of 2,4-D required, as against the huge quantities of inorganic chemicals and oils used before 1944, offered promise as an inexpensive means of weed control. Selectivity was found extremely useful to kill weeds in spite of the presence of crop plants. Systemic action made it possible to kill the underground parts of plants. The immense potential of such herbicides set in motion the development of numerous organic herbicides. Today, over 55 years after the advent of 2,4-D as a selective herbicide, over 400 organic herbicides have been developed and registered in the world for weed control in agricultural and non-agricultural systems. At present, there are few weed problems which herbicides cannot solve.

Herbicide development came at a time when world agriculture was entering an era of increased mechanization and intensive cropping programmes to increase yields and reduce production costs. Since then, herbicides have played a very vital role in augmenting these efforts. Today, sales of herbicides have outstripped those of all other classes of pesticides. Currently, herbicides constitute 55% of the world pesticide market.

CHEMICAL PROPERTIES OF HERBICIDES

Chemical Structure

Most herbicides are organic chemicals, primarily made up of carbon (C) and hydrogen (H) atoms. The carbon atoms of organic molecules bind together to form chains. The chains may be straight, branched or cyclic. Organic compounds composed of only carbon and hydrogen are called **hydrocarbons**. Hydrocarbons are of two types: **saturated** and **unsaturated**. Saturated hydrocarbons (CH_4, C_2H_6, C_3H_8, etc.) are those in which all available bonds are occupied by an atom of C or H. In unsaturated hydrocarbons (acetylene C_2H_2 or benzene C_6H_6), two carbon atoms share more than one bond, resulting in double or triple bond(s). Benzene is a common constituent of many herbicides. Organic chemicals arranged in an unsaturated ring configuration (e.g., benzene) are also called **aromatic hydrocarbons**.

Organic herbicides contain a few elements other than C and H. These elements include oxygen, nitrogen, sulphur, phosphorus and the halogens (bromine, chlorine, fluorine and iodine). Organic compounds that have atoms other than C as part of their ring structure are called **heterocyclic hydrocarbons**. Many herbicides are heterocyclic compounds. Most herbicides contain at least one halogen atom (salt-forming elements) as part of their molecular structure. Alcohols (R-OH), organic acids (R-COOH) and esters (R-O-R) are forms of organic compounds that influence chemical and structural properties. These structures react and influence such chemical and structural properties as water solubility, electrical charge and volatility.

Active Ingredient

The **active ingredient** is that part of a commercially manufactured herbicide that is biologically active. For example, in 2,4-D the active ingredient is the acid form. An active ingredient can be altered slightly by certain chemical processes such as esterification, salt formation, etc. which may improve the biological activity of a herbicide. Herbicides that are derived from alcohols, phenols or organic acids are more soluble in water than those that are not. In contrast, ester forms of herbicides are relatively more soluble in oil and organic solvents and have a tendency to produce vapours. The loss of herbicide as vapour is called **volatility**, which is related to the vapour pressure of the herbicide.

During manufacture of 2,4-D, acid is formed into organic salts and sodium and dimethylamine salt. These organic salts are soluble in water and non-volatile. The ester forms of 2,4-D (ethyl ester, isobutyl ester, isopropyl ester, butoxyethyl ester, isooctyl ester, etc.) are soluble in organic solvents and are more likely to vaporize than the organic salts. The size of ester linkage to the parent 2,4-D acid molecule also influences the degree of volatility of the chemical. For example, the isobutyl ester of 2,4-D is more volatile than the butoxyethyl ester of 2,4-D.

Polarity

Polarity describes the electrical phenomenon of a molecule or an ion. On the basis of polarity, herbicides can be divided into polar and non-polar compounds. The molecules or ions of polar compounds have both electrically positive and negative regions. These polar substances, also known as **hydrophilic** substances, have greater affinity for water and hence are soluble in water and other polar solvents.

Molecules or ions of non-polar compounds do not possess strongly positive and negative areas. These non-polar substances, also known as **lipophilic** substances, have greater affinity for oils and hence are soluble in oil and other non-polar solvents.

Herbicide absorption and translocation by a plant is greatly influenced by these hydrophilic and lipophilic properties. Water molecules attract each other strongly, resulting in a high internal pressure or cohesion and a high surface tension. Because of high surface tension of water molecules, polar solutions (of hydrophilic substances) tend to form large spherical droplets that do not readily wet the waxy cuticle of the leaf surface. This results in poor or improper wetting of the foliage and, as a consequence, the herbicide activity is considerably reduced.

On the other hand, non-polar compounds (lipophilic substances) which have relatively uncharged molecules are held together by van der Waals forces and normally exhibit low water solubility and high oil solubility. Hence, solutions of lipophilic compounds readily wet the waxy cuticle of the leaf surface resulting in better wetting of the foliage. Although lipophilic herbicides penetrate the waxy leaf cuticle better than hydrophilic herbicides, their translocation within the plant is considered to be very slow in the water contiguum. This eventually affects herbicide activity.

HERBICIDE FORMULATION AND APPLICATION

Formulation

Formulation affects the solubility, volatility and specific gravity, and eventually phytotoxicity of a herbicide. A wide variety of herbicide formulations is designed to suit a particular method of application and to achieve increased selectivity and efficacy. Herbicides are usually formulated by combining with a liquid or solid carrier so that they can be applied uniformly. This aspect is discussed in greater detail in Chapter 13.

Application

Time

Time of application is determined by weed species, time of germination of weed and crop plants, and growth stage of weeds. Generally, timing of application falls into three categories: **preplanting**, **preemergence** and **postemergence** in relation to crop and weed.

Preplanting: Preplanting (PPS) treatment refers to application to the soil surface before the crop is planted. Herbicides which have greater toxicity on the emerging crop seedlings are applied before the crop is planted. Similarly, chemicals which need to be incorporated (PPI) into the soil are also applied before planting the crop.

Preemergence: In preemergence (PRE) treatment, herbicides are applied to the soil surface before a crop or weeds or both have emerged. In the case of annual crops, this is normally done after planting the crop but before weeds emerge, and is referred to as preemergence to the crop. In the case of perennial crops, it can be termed as preemergence to weeds.

Postemergence: Herbicide treatment to the weed foliage after emergence of the crop or weed is referred to as postemergence (POST) application. Herbicides may be applied as early POST (within two weeks after crop or weed emergence) or late POST (two weeks after emergence).

Placement

Herbicides are applied either to the soil or foliage. When applied to the soil, they are known as **soil-applied** herbicides; when applied to the foliage of weeds they are termed **foliage-applied** or **foliar** herbicides. Soil-applied herbicides are applied at preplanting or preemergence, while the foliage-applied herbicides are applied at postemergence.

The choice of applying selective herbicides to the soil or foliage depends on the weed-crop situation, type of herbicides, herbicide efficacy, and cost and convenience of application. Soil application may help prevent weed emergence and establishment, thus enabling the crop to grow in a weed-free environment from the beginning of crop emergence. Soil-applied herbicides are placed at or near the root zone. Foliage application ensures better kill of deep-rooted plants.

Herbicides which are either volatile or liable to photodecomposition are incorporated into the soil following soil application. This ensures placement of chemicals below the topsoil layer. Incorporation is done by mixing the upper soil profile and the herbicide to a depth of 1 to 5 cm.

Herbicide application is also done as **blanket** application over the entire soil surface or weed-infested area and as **band** treatment by treating only the narrow strips in the crop rows. Band treatment saves on the chemical. The area not treated by herbicides is normally put under manual methods of weed control. At POST, the band treatment is directed against weeds in the crop rows (Fig. 4.2). Directed spray is done by covering the spray nozzle under a hood or shield or by carefully directing the nozzles to the interrow space. Spot treatment is usually done on small areas of serious weed infestation to kill the weeds and to prevent their spread.

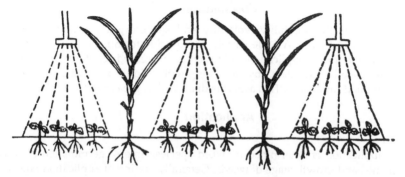

Fig. 4.1. Placement affects herbicide activity. Directed spraying on weeds avoids contact with, and injury to crop plants.

Variations in the placement of a herbicide can affect its activity. A herbicide found toxic to a particular crop under one method of placement may prove safe when applied in a different method. The depth of placement also determines the tolerance or susceptibility of a weed or crop species to a herbicide. If the herbicide is placed below the germinating seeds or roots of weeds it will show greater activity than when placed above them. If the crop seeds are sown below the herbicide layer, the margin of selectivity can be increased in a sensitive crop.

Rate

A herbicide which is selective at a lower rate may become nonselective when applied at higher rates. For every herbicide, there is an optimum rate at which it maintains its selective characteristic; this rate varies from one weed or crop species to another.

HERBICIDE ACTIVITY AND SELECTIVITY

Herbicide **activity** is related to the phytotoxic effects of a chemical on plant growth and development. A herbicide is said to be active, or to possess activity, if it hinders, inhibits or prevents the germination and growth processes of the plant. It is active on sensitive plants and inactive on tolerant or resistant plants. Hence, herbicide activity is determined by the degree of tolerance of the plant to a chemical.

Herbicide **selectivity** refers to the phenomenon wherein a chemical kills the target plant species in a mixed plant population without harming or only slightly affecting the other plants (Fig. 4-2). It is considered to be the greatest single factor that helped chemical weed control to succeed since the advent of 2,4-D. Selectivity enables control of target weed species in spite of the presence of other plants.

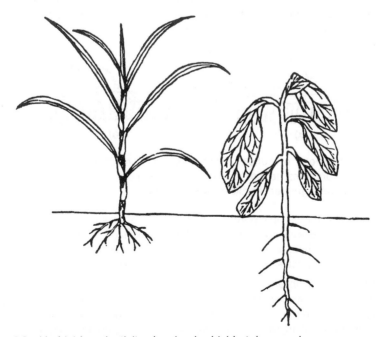

Fig. 4.2. Herbicide selectivity showing herbicide tolerance by a grass species and susceptibility by a broadleaf weed species.

Herbicide activity and selectivity are closely related. Activity refers to the ability of the herbicide to control a weed, while selectivity refers to its ability to control it without affecting the other plants in a mixed stand. Even though a herbicide may exhibit activity, it could affect either selective or non-selective weed control.

Factors Affecting Herbicide Activity and Selectivity

Both activity and selectivity are dependent on a complex chain of events beginning from application to disruption of vital functions of the plant. The chain is mediated mechanically, physically, chemically and metabolically. The various factors that affect herbicide activity and selectivity are briefly discussed below.

Plant Morphology

Plant morphology has a dominant role in determining herbicide activity and selectivity. The amount of spray retained by the foliage influences the amount of chemical available for entry into the plant. The various morphological characteristics of the plant that affect herbicide entry are leaf size and shape, leaf enlargement, pubescence, waxiness of the leaf and thickness of cuticle.

The amount of spray intercepted by the foliage depends on the leaf area and its arrangement with respect to the degree of overlapping and the angle of incidence of the spray. Horizontally arranged leaves intercept more of the spray and retain greater amounts of it than upright leaves. Pubescence in the form of trichomes (leaf hairs) may form an impermeable layer to make the leaf difficult to wet. It prevents contact between the leaf surface and spray droplet. Wax formation on the leaf surface reduces the wettability of the foliage. The wax deposits assume a variety of forms depending on the plant species. The cuticle which varies in thickness from one plant species to another (Fig. 4.3) and with age of the plant, contains waxes, cutin, pectin and cellulose. It is lipoidal in nature which facilitates rapid entry of lipophilic or lipoidal herbicide materials but not hydrophilic or polar herbicide materials.

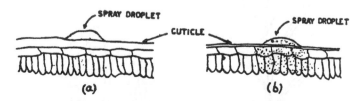

Fig. 4.3. Thickness of cuticle affects herbicide selectivity. Thick cuticle (a) prevents herbicide absorption as against thin cuticle (b) which permits greater herbicide absorption.

The stomata, the microscopic pores on the leaf surface, vary in location number and size in various plant species. In many plants they are located on both upper and lower leaf surfaces and in others only on the lower surface. Plants with large-sized stomata and with a greater number absorb more herbicide and become susceptible.

Stage of Plant Development

Stage of plant development affects the ability of the herbicides to enter the plant. Plants in the germination or young seedling stage (0 to 30 d) are more susceptible to a soil-applied herbicide than in the later stages. Similarly, perennial weeds are more sensitive to foliage-applied herbicides during the active growth period than before or after that stage. The active growth period coincides with active photosynthesis and movement of photosynthates: and this has a direct bearing on herbicide uptake and translocation by the plant. Generally, annual weeds are more susceptible at one- to four-growth development stages and before they reach 20 to 25 cm in height. Application later than these stages may make a herbicide less effective.

Cultivation Practices

The various preplanting and post-planting cultivation practices followed before herbicide application could make a tolerant species susceptible to a particular herbicide after its application. Tillage practices bring the dormant weed seeds to the top soil layer and expose them to sunlight, resulting in germination, thereby increasing their susceptibility to herbicides applied to the soil or foliage. They also bring the roots and stolons of the perennial weeds to the surface so that they can be more easily killed by herbicides applied later. Repeated cultivation also exhausts food reserves of deep-rooted weeds and hence they become more sensitive to herbicides later.

Environmental Factors

Selectivity of an applied herbicide is also influenced by the environmental factors under which the plant grows. The response of a plant to an applied chemical depends on the environmental stresses to which it has been subjected following treatment. A severe stress may become more deleterious to a tolerant species. In field situations, a tolerant crop may be injured under adverse weather conditions. The various environmental factors that affect plant growth and hence herbicide activity and selectivity are temperature, rainfall (water), humidity, light and wind. They largely influence absorption and translocation of herbicides by the plant.

Herbicide Absorption

Once a herbicide is applied, its entry into the plant is facilitated by absorption through the root, shoot and stem. Absorption is influenced by morphological characteristics of the plant and chemical and electrical properties of the plant surface. The rate of absorption and the amount of chemical which enter the plant vary with plant species and stage of plant growth. This differential absorption from one plant species to another determines herbicide activity and selectivity. This aspect will be covered in greater detail in Chapter 6.

Herbicide Translocation

Once a herbicide enters the plant, it must be transported to the site of action for it to disrupt the metabolic activities of the plant. The translocated herbicides move from the site of entry to the site of action via the phloem or xylem. But contact herbicides move very little or not at all from the point of entry. However, under certain conditions, contact herbicides are also subjected to limited movement. Many factors such as plant species, age of the plants, stage of plant development, environmental and soil conditions affecting plant growth, biophysical and biochemical processes within the plant and inherent properties of the chemicals could cause differential movement of herbicides to the sites of action. This determines the activity and selectivity of a herbicide. These, too, will be discussed in greater detail in Chapter 6.

Physiological Differences

After reaching the site(s) of action, a herbicide affects one or many of the metabolic activities related to plant growth and survival. The mechanism by which this is done depends on the herbicide and the physiological tolerance of the plant. Variations in the impediments to herbicidal action form the basis of herbicide selectivity. The physiological tolerance of a plant species, variety or cultivar to a herbicide largely depends on the genetic make-up of the plants and on the taxonomic, morphological and physiological characters related to it. For example, 2,4-D is very effective on dicotyledonous weeds but not on many of the monocotyledonous weeds. Similarly, some cultivars of corn and sorghum are more susceptible to atrazine and propazine respectively than others. This topic will be covered in greater detail in Chapter 7.

Herbicide Metabolism

Any herbicide which enters the plant undergoes metabolism. This depends on the physical and chemical properties of the compound, the chemical composition of the plant and the capacity of the plant to effect metabolic conversion of the applied chemical molecules. It is well known that a tolerant plant inactivates a herbicide in one or many pathways. In certain plant species, detoxification of herbicides takes place at a rate slower than that of accumulation resulting in plant mortality, while in others it is much faster, enabling the plant to be tolerant. This herbicide metabolism and its differences among plant species form the basis for herbicide activity and selectivity. This will be discussed in Chapter 8.

PLANT RESISTANCE TO HERBICIDES

Herbicide resistance is defined as the ability of a plant to withstand the activity of a herbicide at a dosage substantially greater than normally used for its control. The susceptible weed species develops, over the course of time, resistance to the same herbicide(s) to which it used to succumb. Development of herbicide resistance in plants is a worldwide phenomenon, particularly over the last 25 years.

The development of herbicide resistance in weeds is an evolutionary process. In response to repeated treatment with a particular class or family of herbicides, weed populations change in genetic composition such that the frequency of resistance alleles and resistant individuals increase. In this way, weed populations become adapted to the intense selection pressure imposed by herbicides.

Three components contribute to the selection pressure of herbicide resistance: a) efficiency of the herbicide, b) frequency of herbicide usage and c) duration of herbicide effect. The intensity of selection pressure in response to herbicide application is a measure of the relative mortality in target weed populations and/or the relative reduction in seed production of survivors; this will be proportional, in some manner, to herbicide dosage. The duration of selection is a measure of the time period over which phytotoxicity is imposed by a herbicide. Both intensity and duration interact to give seasonal variation in the selection pressure imposed on weed species according to their phenology and growth.

As a result of weed resistance to herbicides, there is a need to change weed management practices, especially the choice of herbicides and herbicide mixtures. This aspect is covered in greater detail in Chapter 10.

5

Classification and Information on Herbicides, Bioherbicides and Herbicide Safeners

HERBICIDE CLASSIFICATION

Over 400 herbicides have been discovered and developed in the past 55 years. There is almost no weed problem that cannot be solved by herbicides. Today, high-yield agriculture heavily depends on herbicides, as they constitute a vital and integral component of weed management practices. Many of the herbicides widely used in the 1960s and 1970s have been phased out and replaced by the newer and more potent herbicides discovered later. Use of some older herbicides has been considerably restricted, reduced and even eliminated in view of environmental and toxicological problems and the availability of more effective and safer chemicals.

Herbicides (organic) are classified on the basis of: a) method of application, b) chemical affinity and structural similarity, and c) mode of action.

Classification on the basis of method of application divides herbicides into two groups: **soil-applied** and **foliage-applied.** All herbicides applied at preplanting (surface or incorporation) and preemergence (to crop, weeds, or both) are included in the soil-applied group; those applied at postemergence on the plant parts are included in the foliage-applied group.

The most common means of discovering new herbicides is by synthesizing the structural variants of a proven herbicide chemical nucleus or family. Thus, herbicides which originate from one chemical nucleus or family have structural similarities and affinities and this provides the basis for chemical classification.

Classification based on the mode of action takes into account differences in the physiological and biochemical actions of the herbicides. On this basis, they are broadly categorized as **systemic** or **translocated** herbicides and **non-systemic** or **contact** herbicides.

Herbicides may also be divided into **selective** and **nonselective** herbicides. Selective herbicides kill or suppress only certain weeds without significantly injuring an associated crop or other desirable plant species. Usually some weeds are also not injured by selective herbicides. In contrast, nonselective herbicides kill or suppress whatever vegetation (including the crop) is treated. They are used with the intention that none of the plants survive.

In recent years, another classification based on the primary site of action has been gaining wider acceptance. It is becoming more important in view of development of herbicide-resistant weeds. Knowledge on the primary site of action is useful in planning a herbicide rotation strategy and reducing the potential for selection of resistant weeds. This classification is included in Chapter 7.

The classification considered in this Chapter includes a combination of mode of application, mode of action and chemical affinity (Fig. 5.1).

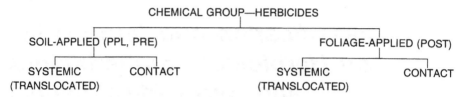

Fig. 5.1. Herbicide classification used in this chapter.

The more commonly used current herbicides belonging to 42 families (groups) are included in this classification (Table 5.1). Herbicides are referred to here by their common names.

Table 5.1. Classification of (organic) herbicides on the basis of chemical affinity, mode of application, and mode of action (herbicides not being widely used or which have been phased out were excluded).

Chemical Family	Soil-Applied Herbicides		Foliage-Applied Herbicides	
	Systemic	Contact	Systemic	Contact
ACETAMIDES	acetochlor, alachlor, dimethenamid metolachlor, napropamide, pronamide, propachlor	—	acetochlor, alachlor, dimethenamid metolachlor, napropamide, propanil	—
ALIPHATICS	—	—	dalapon	acrolein
ARSENICALS	—	—	DSMA, MSMA	cacodylic acid
BENZAMIDES	isoxaben	—	—	—
BENZOICS	dicamba	—	—	—
BENZOTHIA-DIAZOLES	—	—	—	bentazon
BIPYRIDILIUMS	—	—	—	diquat, paraquat
CARBAMATES	—	—	asulam, desmedipham, phenmedipham	—
CINEOLES	cinmethylin	—	—	—
CYCLOHEXA-NEDIONES	—	—	clethodim, cycloxidim, sethoxydim, tralkoxydim	—
DINITROAN-ILINES	benefin, dinitramine, ethalfluralin, fluchloralin, isopropalin, nitralin, oryzalin, pendimethalin, prodiamine, trifluralin	—	—	—
DIPHENYL-ETHERS	fluorodifen, oxyfluorfen	nitrofen	acifluorfen, bifenox, fluoroglycofen fomesafen, 'actofen, oxyfluorfen	—

(Contd.)

Table 5.1 *(Contd.)*

Chemical Family	Soil-Applied Herbicides		Foliage-Applied Herbicides	
	Systemic	*Contact*	*Systemic*	*Contact*
IMIDAZOLID-INONES	buthidazole	—	buthidazole	—
IMIDAZOLIN-ONES	imazapyr, imazaquin, imazethapyr,	—	imazapyr, imazaquin, imazethapyr, imazamethabenz	—
IMINES	—	—	CGA-248757	—
ISOXAZOLID-INONES	clomazone	—	—	—
NITRILES	—	—	—	bromoxynil, ioxynil
OXADIAZOLES	oxadiazon	—	—	—
OXADIAZOLI-DINES	methazole	—	methazole	—
PHENOLS	—	—	—	dinoseb
PHENOXYALK-ANOIC ACIDS				
Phenoxyacetics	2,4-D	—	2,4-D, MCPA	—
Phenoxybutyrics	2,4-DB	—	2,4-DB, MCPB	
Aryloxyphenoxy propionics	—	—	dichlorprop, diclofop, fenoxaprop, fluazifop-P, haloxyfop, mecoprop, quizalofop-P	—
N-PHENYLPHT-HALIMIDES	—	—	flumiclorac	—
PHENYLPYRID-AZINES	—	—	pyridate	—
PHENYL TRIA-ZINONES	sulfentrazone	—	sulfentrazone	—
PHTHALAMATES	naptalam	—	naptalam	—
PYRAZOLIUMS	—	—	difenzoquat	—
PYRIDAZINONES	norflurazon, metflurazon, pyrazon	—	—	—
PYRIDINECARBO-XYLIC ACIDS	—	—	clopyralid, picloram, triclopyr	—
PYRIDINES	dithiopyr, thiazopyr	—	—	—
PYRIDINONES	fluridone	—	fluridone	—
PYRIMIDINYLTH-IO BENZOATES	pyrithiobac	—	pyrithiobac	—
QUNIOLINE CARB-OXYLIC ACIDS	quinclorac	—	quinclorac	—
SULFONYLUREAS	bensulfuron, chlorimuron, chlorsulfuron, halosulfuron, sulfometuron	—	bensulfuron, chlorimuron, chlorsulfuron, halosulfuron, metsulfuron, nicosulfuron, primisulfuron, prosulfuron, sulfometuron, thifensulfuron, triasulfuron, tribenuron	—
TETRAHYDRO-PYRIMIDINONES	experimental compounds	—	experimental compounds	—
THIOCARBAM-ATES	butylate, diallate, EPTC, molinate, pebulate, thiobencarb, triallate, vernolate	—	—	—

(Contd.)

Table 5.1 *(Contd.)*

Chemical Family	Soil-Applied Herbicides		Foliage-Applied Herbicides	
	Systemic	*Contact*	*Systemic*	*Contact*
TRIAZINES	ametryn, atrazine, cyanazine, hexazinone, propazine, prometon, prometryn, simazine	—	atrazine, cyanazine, hexazinone, prometon, prometryn	—
TRIAZINONES	metribuzin	—	—	—
TRIAZOLES	—	—	amitrole, amitrole-T	—
TRIAZOLOPYR-IMIDINE SUL-FONANILIDES	flumetsulam	—	—	—
URACILS	bromacil, terbacil, UCC-C4243	—	UCC-C4243	—
UREAS	diuron, fluometuron, linuron, methabenzthiazuron, metoxuron, isoproturon, monuron, siduron, tebuthiuron	—	diuron, fluometuron, linuron, isoproturon, tebuthiuron	—
UNCLASSIFIED	bensulide, ethofumesate	—	ethofumesate, fosamine, glufosinate, glyphosate	—

INFORMATION ON (ORGANIC) HERBICIDES[1]

In this part of Chapter 5, only the more prominent and widely used organic herbicides are considered. The herbicides are referred to by their common names. While giving chemical names, the structures of herbicides are also illustrated. For each herbicide, its activity on different weed and crop species, rate and time of application, crops for which it is recommended are described. A short description of plant injury symptoms, characteristic of a herbicide group or herbicide, is given. Wherever possible, the persistence of a herbicide in the soil and method of synthesis are given. When trade names are mentioned in this book, it is only in relation to their first commercial release by the companies which originally discovered, developed and manufactured them. A list of some of the common and trade names of herbicides are given in Appendix I.

Herbicides may be divided into two groups: inorganic herbicides and organic herbicides. The inorganic herbicides include copper sulphate, sodium arsenite, sulphuric acid, ferrous sulphate, copper nitrate, ammonium sulphate, ammonium sulphamate, sodium tetraborate, sodium metaborate, amorphous sodium borate, sodium chlorate, sodium nitrate, ammonium thiocyanate, sodium chloride, etc. These were used between 1896 and 1930s. Although they are no more used as herbicides alone, a few of them are mixed with certain organic herbicides. These inorganic herbicides are not discussed here.

Organic herbicides dominate modern agriculture. Although petroleum oils have been used for several years as contact herbicides, the development of organic compounds for weed control really began with the introduction of 3,5-dinitro-O-cresol (a nitrophenol) in 1932. However, the most significant milestone in the history of organic herbicides occurred with the discovery of 2,4-D as a herbicide in 1944.

While referring to the time of application of herbicides, the following terms are abbreviated thus: **preplanting (PPL) ; preplanting surface (PPS); early preplanting surface (E-PPS); preplanting incorporated (PPI); preemergence (PRE); postemergence (POST); early postemergenece (E-POST); late postemergence (L-POST).**

[1]Main Source of Information:
Herbicide Handbook 1994. Weed Science Society of America, Champaign, IL, U.S.A., 352 pp.

Acetamides

Acetamides (acetanilides) constitute a prominent and diverse group of herbicides. Most of them are applied PRE, while others have POST activity. When applied PRE, most susceptible grass and broadleaf weeds fail to emerge. The susceptible monocots that do emerge appear twisted and malformed with leaves tightly rolled in the leaf whorl and unable to roll normally. Leaves do not emerge properly from the coleoptile but may do so underground. Broadleaf weed seedlings may have slightly cupped or crinkled leaves and shortened leaf midribs producing a drawstring effect on the leaf tip, especially under cold conditions. Other injury symptoms exhibited are discussed under individual herbicides.

Acetochlor: 2-chloro-N-(ethoxymethyl)-N-(2-ethyl-6-methylphenyl)acetamide

Acetochlor is applied PPI or PRE at 1.5-3.0 kg ha^{-1} in soybean and maize. A safener (MON 4600 or dichlormid) is required to preclude herbicide injury to maize. Acetochlor controls most annual grasses, *Cyperus esculentus* (yellow nutsedge) and certain broadleaf weeds. It generally provides 8-12 wk weed control, depending on soil type and weather conditions. Acetochlor residues do not persist long enough to injure crops the following season. It is used to make packaged herbicide mixtures with atrazine. Synthesis involves ethoxy methylation of the appropriate acetanilide.

Alachlor: 2-chloro-N-(2,6-diethylphenyl)-N-(methoxymethyl) acetamide

Alachlor was the first chloroacetamide herbicide; developed in 1966, it was commercialized as LASSO in 1969 by Monsanto. It is applied PPI, PRE, or E-POST at 2-4 kg ha^{-1} in maize, sorghum, soybean, groundnut, cotton, etc. It can also be applied E-PPL under no-till conditions and PRE in woody ornamentals. Alachlor controls *Cyperus esculentus*, and many annual grasses such as *Echinochloa crus-galli, Digitaria* spp., *Setaria* spp., *Panicum* spp. and *Eleusine indica*, and certain broadleaf weeds including *Galinsoga* spp., *Chenopodium album, Portulaca* spp. and *Solanum nigrum*. It generally provides 6-10 wk weed control. Alachlor residues do not persist long enough to injure crops the following season. Its field half-life is 21 d. It is used to make herbicide mixtures with atrazine, glyphosate and trifluralin. The synthesis of alachlor involves methoxy methylation of the appropriate acetanilide.

Butachlor: N-(butoxymethyl)-2-chloro-N-(2,6-diethylphenyl)acetamide

Butachlor, also a chloroacetamide herbicide, was first reported in 1970 and commercialized as MACHTE in 1971 by Monsanto. It is applied in irrigated rice within 2-5 d of transplantation of seedlings. It can be applied 10-15 d after sowing in directly seeded irrigated rice. It is also applied PRE in rice under assured rainfall conditions. Butachlor is used in several other crops including groundnut, potato, and soybean and other pulses grown under irrigated or assured rainfall conditions. It is very effective against annual grass weeds such as *Echinochloa crus-galli, Digitaria sanguinalis, Setaria* spp., *Panicum* spp. and certain sedges. Butachlor is the most widely used rice herbicide in the world. The EC (50EC) and granular (5G) formulations are commonly used. The granular formulation is applied in combination with 2, 4-D ethyl ester granular (4G) formulation for a very broad spectrum weed control in irrigated rice. Its field half-life is 13 d. It provides weed control for 4-7 wk and does not injure crops in the following season. Butachlor is synthesized by butoxy methylation of the appropriate acetanilide.

Dimethenamid: 2-chloro-*N*-[(1-methyl-2-methoxy)ethyl]-*N*-(2,4-dimethyl-thien-3-yl)-acetamide

Dimethenamid (SAN 582), a chloroacetamide herbicide, was discovered, synthesized and developed by Sandoz. It was registered as FRONTIER for use in maize in 1993 and soybean in 1994. Susceptible grass and broadleaf weeds fail to emerge. Injury to maize and sorghum can be caused by excessive rates. Affected seedlings show a malformed and twisted appearance. Leaves are tightly rolled in the whorl and may not unroll normally. Injured soybean seedlings show cupped or crinkled leaves. It has no POST activity on established monocot and dicot seedlings. Dimethenamid is applied E-PPS PPI or PRE at 0.8-1.6 kg ha^{-1} in maize and soybean to control many annual grasses such as *Setaria* spp., *Echinochloa crus-galli, Panicum dichotomiflorum* and *Digitaria* spp. as well as *Cyperus esculentus* and certain annual broadleaf weeds including *Amaranthus retroflexus* and *Solanum nigrum*. It can be used as a mixture with atrazine. Dimethenamid can be applied by conventional spraying in water or liquid fertilizer. It can be impregnated on dry bulk fertilizer. Its field half-life is 20 d. It does not injure other crops planted in the following season. A compatibility agent may be required when dimethenamid is mixed with triazine herbicides.

Metolachlor: 2-chloro-*N*-(2-ethyl-6-methylphenyl)-*N*-(2-methoxy-1-methylethyl) acetamide

Metolachlor, was synthesized in 1976 by Ciba-Geigy and made available commercially for use in maize in 1977. It can be applied E-PPS, PPI, PRE, or E-POST at 1.5-4.0 kg ha^{-1} in maize; PRE at 1.0-2.0 kg ha^{-1} in cotton; PPI, POST-plant incorporated or PRE at 1.5-3.0 kg ha^{-1} in groundnut; PRE at 1.5-3.0 kg ha^{-1} in sorghum, safflower and soybean; and PPI, PRE or after hill formation at 1.5-3.0 kg ha^{-1} in potato. Sequential applications allow up to 6.0-7.0 kg ha^{-1}. It is also a useful herbicide in nurseries, landscape planting and turfgrass. Metolachlor controls *Cyperus esculentus*, and many annual grass weeds such as *Setaria* spp., *Echinochloa crus-galli, Digitaria* spp., *Panicum dichotomiflorum, Brachiaria* spp., *Panicum capillare*, etc. It also controls certain broadleaf weed species including *Amaranthus retroflexus, Mollugo verticillata* (carpetweed), etc. It can be applied in liquid or dry bulk fertilizer. Metolachlor can be mixed with atrazine, cyanazine, flumetsulam, metribuzin, etc. When applied alone, it provides weed control for 10-14 wk. Field half-life is 3-5 mon, but residues do not persist long enough to affect crops planted in the following season. The synthesis of metolachlor involves reductive alkylation and subsequent chloracetylation of the alkyl amine.

Napropamide: *N-N*-diethyl-2-(1-naphthalenyloxy)propanamide

Napropamide, a substituted amide, is applied PPI or PRE at 1.0-2.0 kg ha^{-1} in potato, radish, turnip, watermelon, sunflower, safflower, tobacco, groundnut, cabbage, cauliflower, etc. It can also be used PRE (before weed emergence) or E-POST in fruit crops. For best results, it should be incorporated into the soil to a depth of 2-5 cm. Napropamide controls many annual broadleaf and grass weeds including seedling perennial grasses. Seedlings of susceptible weeds fail to emerge after napropamide application. It primarily inhibits root growth, but may also inhibit shoot growth. Napropamide inhibits sprouting buds of *Cyperus rotundus* (purple nutsedge or nutgrass). Its field half-life is 70 d. When incorporated in loamy sand and loamy soils, its half-life is 8-12 wk. The synthesis of napropamide involves reacting 2-(1-naphthoxy)-propionic acid with diethyl amine under dehydrating conditions.

Pronamide: 3,5-dichloro(*N*-1,1-dimethyl-2-propynyl)benzamide

Pronamide is a substituted amide herbicide developed by Rohm and Haas in 1969 (as RH-315 or KERB). It is also known as propyzamide. It can be applied PRE at 1.5-4.0 kg ha^{-1} in cotton, sunflower, groundnut, fruit crops and vegetables. It has PRE activity against many annual broadleaf and grass weeds including *Echinochloa crus-galli, Digitaria sanguinalis, Chenopodium album, Portulaca oleracea,* etc. It also has POST activity on some annual broadleaf and grass weeds. Its field half-life is 60 d. Pronamide does not accumulate from repeated annual applications to the same soil. The synthesis of pronamide involves reaction of 3,5-dichlorobenzyl chloride with 3-amino-3-methylbutyne.

Propachlor: 2-chloro-*N*-(1-methylethyl)-*N*-phenylacetamide

Propachlor, a chloroacetamide herbicide (discovered and developed by Monsanto as RAMROD), is applied PRE at 2.5-6.0 kg ha^{-1} in maize and at 0.5-5.0 kg ha^{-1} in sorghum, millets, soybean, legumes, fruit crops and ornamentals. It controls many annual grass weeds such as *Echinochloa crus-galli, Setaria* spp., *Panicum dichotomiflorum*, and certain broadleaf weeds such as *Amaranthus retroflexus* and *Mollugo verticillata*. Propachlor has short persistence, with half-life of 3 d. The half-life is longer in soils high in organic matter. It does not injure rotational crops in the following season. It can be used to make a package mixture with atrazine. Propachlor is synthesized by reacting *N*-isopropyl aniline with chloroacetyl chloride.

Propanil: 2-chloro-*N*-(1-methylethyl)-*N*-phenylacetamide

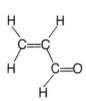

Propanil (an acetamide) was first registered in 1961. It can be applied POST (2- to 4-leaf stage) at 3.0-5.0 kg ha^{-1} in rice and at 1.0-1.25 kg ha^{-1} in wheat. It controls several annual broadleaf and grass weeds including *Setaria* spp., *Echinochloa crus-galli*, *Digitaria* spp., *Eleusine indica* and *Amaranthus retroflexus*. Propanil has a field half-life of 1 d. Its residues do not injure crops planted the following season. It can be used to make mixtures with molinate. Propanil is synthesized by reacting propionic acid and 3,4-dichloroaniline to form *N*-(3,4-dichlorophenyl)propanamide. The residual 3,4-dichloroaniline is reacted with propionic anhydride, followed by vacuum distillation to remove water and residual propionic acid.

Aliphatics

Acrolein: 2-propenal

Acrolein, also known as acrylaldehyde, was commercially developed as a herbicide during the 1930s. It is an aquatic herbicide (1-15 g L^{-100}, injected below the water surface) on submersed and floating weeds and algae in irrigation canals. It kills weeds by contact action. Emerging weeds are not controlled at recommended rates, but floating weeds such as *Pistia stratiotes*, *Eichhornia crassipes* and *Jussiaea* spp., are controlled when applied at 15 g L^{-100} concentration for an extended period. As it is hazardous to use, acrolein is applied only by licenced applicators and trained personnel. It is persistent in water for 2–3 d depending on water temperature and pH. One application can keep a canal clean throughout the season. It is toxic to fish and wildlife and must be kept out of lakes, streams and ponds. Its half-life in water is 175 h at pH 5, 120-180 h at pH 7, and 5-40 h at pH 9. Acrolein is synthesized through vapour phase oxidation of propylene with air or oxygen in the presence of a catalyst.

Dalapon: 2,2-dichloropropionic acid

Dalapon, a POST herbicide developed in 1953, is used for selective control of perennial grasses including *Elytrigia repens* and *Imperata cylindrica*. Repeat applications ensure their better control. It is applied in tea, coffee, sugarcane, orchard crops, etc. It is available as sodium salt.

ARSENICALS

Organic arsenical herbicides have been widely used for over 30 yr even though their herbicidal properties were discovered in 1951. As arsenicals are non-degradable in soils, they provide good control of perennial grasses. They have low to medium mobility on sandy soil and are largely immobile on other soils. The affected plants show foliar chlorosis and necrosis. Arsenicals are essentially foliage-applied herbicides, which include cacodylic acid, DSMA and MSMA. The latter two herbicides are considered for discussion here.

DSMA: Disodium methanearsonate

$$CH_3 — \overset{\overset{\displaystyle O}{\|}}{\underset{\underset{\displaystyle O^-}{|}}{AS}} — O^- \quad 2Na^+$$

DSMA is applied POST at 3.5 kg ha-1 in turf, 2.5 kg ha^{-1} in cotton and 2.5-3.0 kg ha^{-1} in citrus and non-crop areas. It controls *Digitaria* spp., *Paspalum dilatatum*, *Sorghum halepense*, *Cyperus rotundus*, *Setaria* spp., *Ambrosia artemisifolia*, *Xanthium strumarium*, *Amaranthus retroflexus*, etc. A surfactant will enhance its foliage activity. Its field half-life is 180 d. It is strongly adsorbed by soil. DSMA is used to make package mixtures with 2,4-D sodium salt.

MSMA: Monosodium methanearsonate

$$CH_3 — \overset{\overset{\displaystyle O}{\|}}{\underset{\underset{\displaystyle OH}{|}}{AS}} — O^- \quad Na^+$$

MSMA is foliage-applied at 2.5 kg ha^{-1} in cotton, turf and non-crop areas. It controls the weed species listed under DSMA, but more effectively. A surfactant enhances its POST activity. It can be used to formulate mixtures with diuron, fluometuron, dicamba, mecoprop, 2,4-D dimethylamine, prometryn and cacodylic acid. MSMA has moderate to long residual activity in soil, with field half-life of 180 d. It is primarily found in the top 15 cm.

Benzamides

Isoxaben: *N*-[3-(1-ethyl-1-methylpropyl)-5-isoxazolyl]-2,6-dimethoxybenzamide

Isoxaben was discovered in 1979 and first reported in 1982. The susceptible weeds fail to emerge following PRE and POST application. Broadleaf weeds show stunting, reduced root growth, root hair distortion and root clubbing symptoms similar to those caused by dinitroaniline herbicides. Foliar application to susceptible broadleaf weeds causes growth inhibition, swelling and cracking of the stem and petiole within 2 to 3 d. It is applied PPI at 1.5-2.2 kg ha^{-1} before transplanting tobacco. It controls annual grasses such as *Echinochloa crus-galli*, *Digitaria* spp., *Setaria* spp., *Panicum dichotomiflorum*, etc., and broadleaf weeds including *Chenopodium album*, *Amaranthus retroflexus*, *Mollugo verticillata*, etc. Isoxaben has moderate to long residual activity, with a field half-life of 100 d. When applied at recommended rates, it degrades, however, to non-phytotoxic levels during the following season under warm, moist soil conditions. Isoxaben can be mixed with oryzalin and trifluralin.

Benzoics

The herbicidal properties of benzoics were discovered in 1958 with the invention of dicamba. They show strong plant growth-regulating activity. They are relatively strong acids and thus form salts in vitro as well as in plants and soil. Benzoics show symptoms similar to auxinic herbicides on susceptible species. These include twisting and curling of stems and petioles (epinasty), stem swelling (at nodes) and elongation, and leaf cupping. These symptoms are followed by chlorosis, growth inhibition, wilting and necrosis. Dicamba is a major herbicide among benzoics.

Dicamba: 3,6-dichloro-2-methoxybenzoic acid.

Dicamba can be applied PPS, PRE at 0.5 kg ha^{-1} in maize and POST at 0.25 kg ha^{-1} in maize, sorghum, wheat, turf and non-crop areas. It is available as a water-soluble alkali and applied either as a spray or granules. It controls many annual broadleaf weeds. It can be used to make mixtures with 2,4-D, mecoprop, glyphosate, MCPA, atrazine, etc. Dicamba has a field half-life of less than 14 d. It may be leached out of the zone of activity in soils of humid regions in 3-12 wk.

Benzothiadiazoles

Bentazon: 3-(1-methylethyl)-(1*H*)-2,1,3-benzothiadiazin-4(3*H*)-one,2,2-dioxide

Bentazon, first reported in 1968, is available as a sodium salt formulation (BASAGRAN). It is a POST contact herbicide, used for control of certain 2,4-D-tolerant weeds in cereal crops such as maize, wheat, barley, sorghum, rice and soybean, and also to control *Cyperus* spp. Bentazon is effective against weed species of such families as Amaranthaceae, Caryophyllaceae, Compositae, Convolvulaceae, Cruciferae, Chenopodiaceae, Cyperaceae, Ambrosiaceae, Polygonaceae and Solanaceae. It is selective for rice. Its efficacy is improved by adding non-ionic surfactant, oil adjuvant and ammonium sulphate. The sensitive plants show chlorosis within 3-5 d after application, followed by foliar desiccation and necrosis. Foliar bronzing may occur in tolerant crops. Its field half-life is 20 d. Its residues decline to undetectable levels within 6 wk. Bentazon has little or no residual activity. During synthesis of bentazon, anthrinilic acid is reacted with *N*-isopropylamino sulfonyl chloride, followed by ring enclosure with phosgene.

Bipyridiliums

Bipyridiliums, synthesized from pyridine, are related to quaternary ammonium salts, which show phytotoxic properties. Bipyridilium quaternary salts were first evaluated by ICI, UK in 1955 and found to be highly effective, killing plants quickly on contact, since they are rapidly absorbed by the foliage. Even rainfall within 30 min of application has no adverse effect on their activity. Bipyridilium herbicides are available in the form of salts. Diquat and paraquat are the most widely used bipyridilium herbicides.

Diquat: 6,7-dihydrodipyrido[1,2-α:2′,1′-*c*]pyrazinediiium ion

Diquat, first used as a growth regulator in 1955, is available as diquat dibromide salt. It causes rapid wilting and desiccation of plants within several hours of application in full sunlight. Complete foliar necrosis occurs in 1-3 d. It is applied POST, spray-to-wet, at 0.2-0.4 kg ha^{-1} to control several aquatic weeds. It can be applied to ponds, lakes, and drainage ditches for control of algae, submersed aquatic weeds such as *Utricularia* spp., *Ceratophyllum demersum* (coontail) and *Elodea* spp., and floating weeds such as *Centella asiatica*, *Hydrocotyle* spp., *Salvinia* spp., and *Eichhornia crassipes*. Diquat can also be used for tassel control in sugarcane and as a desiccant in cotton, castor, soybean, sorghum, etc. Diquat is strongly bound to clay and hence unavailable to microbes. Its field half-life is 1000 d, but it is not taken up by plants. During synthesis of diquat, 2,2'-bipyridyl is reacted with di-*n*-propyl amine.

Paraquat: 1,1'-dimethyl-4,4'-bipyridinium ion

Paraquat, first used as a growth regulator in 1959, is one of the most widely used herbicides in the world. It is available as dichloride and di(methyl sulphate) salt formulations. Of the two, dichloride formulation is more widely used. It causes rapid wilting and desiccation of plants within a few hours of application, with complete necrosis occurring in 1-3 d. It is a POST, non-selective contact broad-spectrum herbicide in crop and non-crop areas. It is applied POST at 0.25-1.0 kg ha^{-1} to control the existing vegetation at planting in no-till agriculture. It is also widely used in plantation crops such as tea, coffee, rubber, coconut, and in fruit and orchard crops. It can also be used as a post-directed spray in maize, sorghum, sugarcane, pineapple, etc. and as a general spray in non-crop areas. It can be applied as pre-harvest desiccation of potato vines and cotton. It is rapidly and tightly adsorbed by clay particles and becomes completely inactive in soil. Like diquat, paraquat is also biologically unavailable to microbes. Its half-life is 1000 d. Paraquat is synthesized by direct quaternization of 4,4'-bipyridyl with chloromethane under pressure with or without solvent. The iodide salt may be exchanged with the chloride salt or methylsulphate ion by use of ion exchange. Silver methylsulphate gives a quantitative exchange for the halide form.

Carbamates

Carbamates are the esters of carbanilic acid. They were first discovered in 1945 with the discovery of propham. Carbamates are a diverse group of herbicides, which include alkyl- and aryl-substituted derivatives. The methyl carbamates dichlormate, karbutilate and terbutol, covered in the earlier edition, are deleted now as they are not widely used. Among the phenyl carbamate herbicides only asulam, desmedipham, and phenmedipham are discussed here.

Asulam: methyl[(4-aminophenyl) sulfonyl]carbamate.

The herbicidal properties of asulam, commercially available as a sodium salt, were first discovered and reported by May and Baker in 1965. The weeds susceptible to asulam show chlorosis in young leaves and plant stunting, followed by necrosis. The growing points are usually killed within 1-2 wk after treatment, while mature leaves senesce more slowly. Asulam is applied POST at 1.0-3.0 kg ha^{-1} in sugarcane, plantation and orchard crops, and non-crop areas to control many grass weeds including *Sorghum halepense*, *Digitaria sanguinalis*, *Eleusine indica*, and perennial ferns including *Pteridium aquilinum* (bracken fern). Maximum efficacy is obtained when it is applied to vigorous and actively growing immature weeds. The herbicide is translocated to rhizomes to ensure excellent rapid control.

The bracken fern should be in full frond prior to asulam application. The field half-life of asulam in soil is 7 d.

Desmedipham: ethyl{3-[[(phenylamino)carbonyl]oxy]phenyl}carbamate

Desmedipham was first reported in 1969 as a POST herbicide (0.8-1.5 kg ha^{-1}) for sugarbeets to control annual broadleaf weeds such as *Sonchus arvensis, Solanum nigrum, Chenopodium album, Ambrosia artemisiifolia* and *Amaranthus retroflexus.* The affected leaves appear water-soaked within a few days of application, with chlorosis and necrosis following soon thereafter. The field half-life of desmedipham is less than one month in silty loam, sandy loam and silty clay loam soils. Desmedipham is mixed with phenmedipham and ethofumesate.

Phenmedipham: 3-[(methoxycarbonyl)amino]phenyl (3-methylphenyl)carbamate

Phenmedipham was first tested in 1966 and reported in 1967 in Europe. The injury symptoms are similar to those shown by desmedipham. It is applied POST at 0.4-0.7 kg ha^{-1} in sugarbeets and vegetable crops for control of *Setaria* spp., and annual broadleaf weeds including *Solanum nigrum, Portulaca oleracea,* etc. It is used to make package mixtures with desmedipham and ethofumesate. The field half-life of phenmedipahm is 44 d, but it varies with soil temperature and moisture. Soil incorporation slows phenmedipham dissipation.

Cineoles

Cinmethylin: exo-1-methyl-4-(1-methylethyl)-2-[(2-methylphenyl)methoxy]-7-oxabicyclo [2.2.1]heptane

Cinmethylin, a benzyl derivative of the monoterpene 1,4-cineole, was reported in 1985 and later developed by Shell. Although a specific mechanism of action of cinmethylin has not been elucidated, its activity appears to result from the inhibition of mitosis in meristematic regions of susceptible plants. It is a PRE herbicide, effective in transplanted rice at 25-100 g ha^{-1}. It can also be applied PPI or PRE at 0.5-1.5 kg ha^{-1} in soybean, cotton, groundnut, and certain vegetable and tree crops. It primarily controls annual grasses, but is also effective against certain annual broadleaf weeds and sedges. Field half-life of cinmethylin is 25-78 d, increasing with organic matter content. Under anaerobic conditions prevalent in rice, its degradation is reduced because of slower microbial degradation.

Cyclohexanediones (Cyclohexenones)

The cyclohexanedione herbicides were first discovered in the 1980s. They are very effective for the control of many annual and perennial grasses. They are most effective when applied POST. Plant growth ceases soon after application, with young and actively growing leaves being affected first. Leaf chlorosis become brown and mushy at and just above the point of attachment to the node. Older leaves often turn purple, orange, or red before becoming necrotic. When applied PRE, the primary root growth is inhibited and leaves fail to emerge from the coleoptile.

Clethodim: (E,E)-(±)-2-[1-[[(3-chloro-2-propenyl)oxy]imino]propyl]-5-[2-(ethylthio) propyl]-3-hydroxy-2-cyclohexen-1-one

Clethodim was first reported in 1987 and registered in 1991 as SELECT. It is applied POST at 0.1-0.3 kg ha^{-1} in cotton and soybean for the control of many annual and perennial grasses. It has no activity against broadleaf weeds and sedges. It requires an oil adjuvant for maximum activity. Clethodim is non-persistent in soil, with a half-life of 3 d. Its activity is antagonized by bentazon and acifluorfen.

Cycloxidim: 2-[1-(ethoxyimino)butyl]-3-hydroxy-5-(2H-tetrahydrothiopyran-3-yl)-2-cyclohexen-1-one

Cycloxidim (formerly BAS 517H) was discovered by BASF and first reported in 1984. It is commercially available as FOCUS and LASER. Cycloxidim is applied POST at 50-400 g ha^{-1} for control of annual and perennial grasses, with no effect on broadleaf weed species. The emerging grass seedlings are controlled immediately after application. It has a very short residual life, but half-life varies with environmental conditions. Cycloxidim requires an oil adjuvant or surfactant for maximum efficacy.

Sethoxydim: 2-[1-(ethoxyimino)butyl]-5-[2-(ethylthio)propyl]-3-hydroxy-2-cyclohexen-1-one

Sethoxydim was discovered by Nippon Soda (Japan) and developed by BASF. It was first tested in 1978 as BAS 9052 OH. It is commercially available as POAST. Sethoxydim controls annual and perennial grasses in several broadleaf crops. It can be applied POST at 0.10-0.50 kg ha^{-1} in alfalfa, cotton, sunflower, etc. It can also be applied POST at 100-120 g ha^{-1} for a preplant burndown effect in no-till soybean. At higher rates of 350-550 g ha^{-1}, it can be applied POST in ornamental trees, tree crops and non-crop areas for control of shrubs and hard-to-control weeds. Sethoxydim is rapidly degraded in soil, with an average field half-life of 5 d.

Tralkoxydim: 2-[1-ethoxyimino)propyl-3-hydroxy-5-mesitycyclohexen-1-one
Tralkoxydim was developed by ICI, UK. Like other cyclohexanedione herbicides, it is very effective in controlling several annual and perennial grasses when used as a POST herbicide. It is applied at 0.1-0.4 kg ha^{-1} in cotton, soybean and wheat, tea, coffee and other plantation crops, fruit and tree crops, etc. Tralkoxydim is rapidly degraded in soil, with a field half-life of 3-4 d.

Dinitroanilines

Dinitroaniline herbicides were introduced in the 1960s. Trifluralin, the first and most prominent member of this group, was introduced in 1960. The dinitroanilines are orange-yellow coloured compounds with low water solubility. They are volatile and hence need to be incorporated into the soil after application. They are widely used for selective PPI and PRE control of a wide spectrum of grasses and broadleaf weeds. The susceptible weeds fail to emerge, due to inhibition of coleoptile growth or hypocotyl unhooking. Seed germination is not inhibited. Root growth inhibition is a prominent symptom on emerged seedlings and established plants. Roots appear stubby, with tips becoming thickened. The base of grass shoots swell, appearing bulbous, and the hypocotyl may swell in broadleaf weeds. Shoots may be deformed and become brittle. Dinitroaniline herbicides are strongly adsorbed by soil.

Benefin: N-butyl-N-ethyl-2,6-dinitro-4-(trifluoromethyl)benzenamine

Benefin was first described as a herbicide in 1965. It is applied PPI at 1.25-1.60 kg ha^{-1} in groundnut, tobacco, lettuce and turf to control annual grasses such as *Echinochloa crus-galli*, *Brachiaria* spp., *Digitaria* spp., *Panicum dichotomiflorum*, *Setaria* spp. and *Cenchrus* spp., and certain annual broadleaf weeds including *Chenopodium album* and *Amaranthus* spp. It can be applied by conventional spraying or by impregnating on dry bulk fertilizer. Benefin is strongly adsorbed by soil. Its field half-life is 40 d. It controls weeds for a full season. The crops in rotation are not affected by benefin residues. Benefin can be mixed with oryzalin. Its synthesis involves nitration of 4-(trifluoromethyl)chlorobenzene with fuming nitric acid and sulphuric acids, yielding the 2,6-dinitro derivative, which when treated with N-butyl-N-ethylamine produces the final product.

Ethalfluralin: N-ethyl-N-(2-methyl-2-propenyl)-2,6-dinitro-4-(trifluoromethyl) benzenamine

Ethalfluralin was first described as a herbicide in 1974. It is applied PPI at 0.6-1.2 kg ha^{-1} in soybean, groundnut, sunflower and vegetables. It can be applied in standing wheat stubble and incorporated next season for the following crop in conservation tillage system. Ethalfluralin primarily controls annual grasses such as *Setaria* spp., *Echinochloa crus-galli*, *Panicum dichotomiflorum*, *Digitaria* spp., etc. and also certain broadleaf weeds such as *Amaranthus retroflexus*, *Kochia scoparia* and *Solanum nigrum*. Its field half-life is 60 d. Ethalfluralin could injure sensitive crops planted even one year after its application. Such carryover problems can be minimized by deep mouldboard ploughing before planting a sensitive crop.

Fluchloralin: *N*-(2-chloroethyl)2,6-dinitro-*N*-propyl-4-(trifluoromethyl)aniline

Fluchloralin is a PRE herbicide used in cereal crops particularly wheat and rice, groundnut, pulse crops and many vegetable crops for selective control of many annual broadleaf weeds and grasses. The susceptible weeds are affected during germination or seedling emergence. For better results, soil incorporation is needed.

Pendimethalin: *N*-(1-ethylpropyl)-3,4-dimethyl-2,6-dinitrobenzenamine

Pendimethalin was discovered in 1971 by American Cyanamid, first reported in 1974, and registered for cotton, maize, tobacco, sorghum, wheat, peanut, sunflower, rice, sugarcane, etc. It is applied PRE or E-POST at 0.50-2.0 kg ha^{-1} in the aforesaid crops as also potato, onion, vegetables, fruit crops, etc. It controls primarily annual grass and broadleaf weeds mentioned under ethalfluralin. Besides, it can also control *Sorghum halepense*, *Brachiaria* spp., *Abutilon theophrasti*, etc. It is mixed with imazapyr, imazaquin and oxyfluorfen. Pendemethalin has a field half-life of 44 d, but it varies with soil temperature and moisture. Soil incorporation slows pendemethalin dissipation.

Prodiamine: 2,4-dinitro-*N*^{3},*N*^{3}-dipropyl-6-(trifluoromethyl)-1,3-benzenediamine

Prodiamine was developed in 1975. It inhibits root and shoot growth like other dinitroaniline herbicides. It is applied PRE at 0.4-1.7 kg ha^{-1} in established turf and ornamentals. In addition to the weed species listed under isopropalin, prodiamine can also effectively control *Sorghum halepense*, *Eleusine indica* and *Euphorbia humistrata* (prostrate spurge). The field half-life of prodiamine is 120 d. During the synthesis of prodiamine, 2,4-dichlorobenzotrifluoride is subjected to nitration, dipropylamination and ammoniation.

Trifluralin: 2,6-dinitro-*N,N*-dipropyl-4-(trifluoromethyl)benzenamine

Trifluralin is the most prominent member of dinitroanilines. It was first described as a herbicide in 1960. It was widely used as a soil-applied herbicide in 1964. It was a selective herbicide used in maize, pulse crops, cotton, soybean, safflower, castor, sunflower, vegetable crops, horticultural crops, etc. It is applied PPI at 0.56-1.12 kg ha^{-1}. Soil incorporation within 24 h of application prevents vaporization losses. Trifluralin controls several annual grasses and certain small seeded broadleaf weeds. In addition to application by conventional spraying system, trifluralin can be impregnated on dry fertilizer or applied in liquid fertilizer through chemigation (application through irrigation water) systems in maize, soybean, wheat and sorghum at 0.84-2.24 kg ha^{-1}. It can be applied POST via chemigation at 0.84-1.40 kg ha^{-1} in onion, potato, sugarbeets, etc. Trifluralin is mixed with imazaquin, isoxaben, tebuthiuron, benefin, clomazone, metribuzin, triallate and alachlor. Its field half-life is 45 d on moist soils, with < 10% of the applied trifluralin remaining 1 yr after application. Residues of trifluralin applied at 1.1 kg ha^{-1} may persist long enough to injure certain rotational crops in the following season. Its mobility in the soil is low to negligible due to strong adsorption by soil.

Diphenylethers

Since the introduction of nitrofen in 1964 as a first diphenyl ether herbicide (DPE), 10 more have been developed for use in various crops including wheat, maize, rice, soybean, cotton and groundnut. These herbicides include fluorodifen, bifenox, oxyfluorfen, acifluorfen, acifluorfen methyl, fluoroglycofen, fomesafen, lactofen, nitrofluorfen and chlomethoxyfen. Many of these are applied at PRE, while others can be applied both PRE and POST. They are regarded as contact herbicides, even when applied to the soil, because they are not well translocated in plants. The leaves of susceptible plants become chlorotic and then desiccated and necrotic within 1 or 2 d. The youngest expanded leaves of tolerant crops such as soybean and groundnut also may show chlorosis and necrosis, especially at higher rates. Sublethal rates may produce foliar 'bronzing', usually on expanded leaves. Droplet drift may leave bleached spots or flecks on leaves.

Acifluorfen: 5-[2-chloro-4-(trifluoromethyl)phenoxy]-2-nitrobenzoic acid

Acifluorfen is available as sodium salt. It is marketed by American Cyanamid (as SCEPTER) and BASF (BLAZER). Acifluorfen can be applied POST at 0.15-0.40 kg ha^{-1} in groundnut and soybean, and at 0.15 kg ha^{-1} in rice. It controls many annual broadleaf weeds including *Ipomoea* spp. (morningglory), *Solanum nigrum, Xanthium strumarium* (cocklebur), *Ambrosia artemisiifolia, Datura stramonium* (jimsonweed), *Sesbania exaltata* (hemp sesbania), *Chenopodium album, Amaranthus* spp., *Polygonum* spp., *Brassica kaber,* (wild mustard), etc. and several grasses including *Setaria* spp., *Sorghum halepense,* etc. Addition of surfactants and adjuvants enhances weed control efficacy and crop injury. The half-life of acifluorfen is 14-60 d. Its residues in soil do not affect seedling emergence. Factors which promote microbial activity enhance breakdown. Acifluorfen residues do not persist in the environment. It can be used to make herbicide mixtures with bentazon.

Bifenox: methyl 5-(2,4-dichlorophenoxy)-2-nitrobenzoate
Bifenox was first developed in 1973, and later marketed as MODOWN. It is used POST at 0.5-1.0 kg ha^{-1} in grain crops, and PRE or E-POST at 1.5-2.0 kg ha^{-1} in transplanted rice. It is also applied PRE in soybean and sunflower. Bifenox controls many annual broadleaf weeds such as *Kochia scoparia, Amaranthus* spp., *Datura stramonium, Polygonum*

spp., *Brassica kaber, Datura carota, Veronica* spp., etc. and grass weeds such as *Echinochloa crus-galli,* and annual sedges. The field half-life of bifenox is 7-14 d. It can control weeds for 6-8 wk, and residues do not injure susceptible crops planted 1 yr after application.

Fluoroglycofen: carboxymethyl 5-[2-chloro-4-(trifluoromethyl)phenoxy]-2-nitrobenzoate

Fluoroglycofen, also known as benzofluorfen, was developed by Rohm and Haas. It is applied as ethyl ester under the trade name COMPETE. It can be applied POST at 15-40 g ha^{-1} in cereal crops for control of several broadleaf weeds. The half-life of fluoroglycofen is about 1 wk for PRE application and 2-3 wk for POST application. Fluoroglycofen is synthesized by reacting sodium 5-[2-chloro-4-(trifluoromethyl) phenoxy]-2-nitrobenzoate with ethyl monochloroacetate.

Fomesafen: 5-[2-chloro-4-(trifluoromethyl)phenoxy]-N-(methylsulfonyl)-2-nitrobenzamide

The herbicidal activity of fomesafen was first tested in 1978 by ICI, and developed in 1983. It is available as sodium salt under the trade name RE-FLEX. Fomesafen is applied POST at 0.25-0.40 kg ha^{-1} in soybean to control many annual broadleaf weeds including *Ipomoea* spp., *Amaranthus* spp., *Datura stramonium, Brassica kaber,* *Solanum nigrum, Ambrosia artemisiifolia,* etc. Its field half-life is 100 d and residues persist long enough to injure certain susceptible crops such as sugarbeet, sunflower and sorghum for 1 yr after application. It can be used to make a herbicide mixture with fluazifop-P butyl ester. It is synthesized by reacting the appropriate acid chloride with methane sulfonamide.

Lactofen: (±)-2-ethoxy-1-methyl-2-oxoethyl 5-[2-chloro-4-(trifluoromethyl)phenoxy]-2-nitrobenzoate

Lactofen is commercially available as COBRA. It can be applied POST at 70-200 kg ha^{-1} in soybean and as a directed spray in cotton. It controls many annual broadleaf weeds such as *Datura stramonium, Brassica kaber, Solanum* spp. *Ambrosia* spp., *Amaranthus* spp. and *Xanthium* spp. Oil adjuvants, surfactants and fertilizer adjuvants are often added to enhance control. The half-life of lactofen is 3 d. It usually dissipates in and around 7 d in moist soils. Residues do not injure rotational crops. As lactofen is compatible with many herbicides, it is used for making several herbicide mixtures.

Oxyfluorfen: 2-chloro-1-(3-ethoxy-4-nitrophenoxy)-4-(trifluoromethyl)benzene

The herbicidal activity of oxyfluorfen was first reported in 1975 and it was registered in 1976 as GOAL. It is applied at 0.25-2.0 kg ha^{-1} in soybean, groundnut, cassava, leguminous crops, and plantation crops such as tea, rubber and oil palm. It is also a useful herbicide in transplanted rice, vegetables and vineries. As a POST-directed spray, it is effective in maize, cotton, papaya, soybean, fruit trees and plantation crops. Oxyfluorfen controls many annual broadleaf weeds including *Amarathus* spp., *Ageratum conyzoides*, *Borreria hispida*, *Commelina benghalensis*, *Portulaca oleracea*, *Chenopodium album*, etc. and annual grasses such as *Digitaria sanguinalis*, *Echinochloa crus-galli*, *Eleusine indica*, *Paspalum conjugatum*, *Axonopus compressus*, etc. It is compatible with pendemethalin and oryzalin. Oxyfluorfen is moderately persistent in soil with a half-life of 35 d. It is immobile in most soils, but slightly mobile in extremely sandy soils. During the synthesis of oxyfluorfen, 3,4-dichlorobenzotrifluoride is reacted with resorcinol. The resultant intermediate is nitrated and then ethoxylated to produce oxyfluorfen.

Imidazolidinones

Buthidazole: 3-[5-(1,1-dimethylethyl)-1,3,4-thiadazol-2-yl]-4-hydroxy-1-methyl-2-imidazolidinone

Buthidazole is applied PRE and POST for weed control in maize, wheat, alfalfa and other crops. It is effective against numerous broadleaf weeds and grasses including *Elytrigia repens*, *Teraxacum officinale*, *Echinochloa crus-galli*, *Setaria viridis*, *Abutilon theophrasti*, *Chenopodium album*, *Amaranthus* spp., etc.

Imidazolinones

Imidazolinones are a new group of herbicides discovered by American Cyanamid in the late 1970s and 1980s. Growth inhibition by imidazolinone herbicides begins within a few hours after application and injury symptoms usually appear 1-2 wk later. Meristematic areas become chlorotic, followed by a slow foliar chlorosis and necrosis.

Imazapyr: 2-[4,5-dihydro-4-methyl-4-(1-methylethyl)-5-oxo-1H-imidazol-2-yl]-3-pyridinecarboxylic acid

Imazapyr was discovered in the late 1970s, its herbicidal activity found in 1981, and commercial sales begun in 1984 as ARSENAL. Imazapyr acid is formulated as isopropylamine salt. It is applied PRE and POST at 50-150 g ha^{-1} in sugarcane and plantation crops and non-crop areas to control several annual and perennial grasses and broadleaf weeds. A non-ionic surfactant improves POST activity. Field half-life ranges from 25 to 142 d depending on soil type and environmental conditions. Weed control efficacy persists from 3 to 24 mon depending on application rate. It can be mixed with imazaquin, trifluralin, acifluorfen, pendimethalin, etc. During synthesis of imazapyr, 2,3-pyridinecarboxylic acid undergoes dehydration (to form the anhydride), a condensation reaction, and hydrolysis. It is then cyclized (to form the imidazolinone ring) and the end product is precipitated.

Imazaquin: 2-[4,5-dihydro-4-methyl-4-(1-methylethyl)-5-oxo-1H-imidazol-2-yl]-3-quinolinecarboxylic acid

Imazaquin was discovered in 1980, field tested in 1981 and registered in 1984 (as IMAGE and SCECTER). It was the first imidazolinone herbicide widely used. Imazaquin is commercially available as an ammonium salt. It is applied PRE or POST at 50-70 g ha^{-1} in soybean, maize, groundnut, pulses, etc. It can be split-applied to soil PPL or PRE (35 g ha^{-1}) and POST (35 g ha^{-1}). It can control a large range of annual broadleaf weeds including *Amaranthus* spp., *Xanthium strumarium*, *Datura stramonium*, *Kochia scoparia*, *Chenopodium album*, *Ipomoea* spp., *Solanum nigrum*, *Ambrosia* spp., *Euphorbia* spp., *Anoda cristata* and *Abutilon theophrasti* as also some annual grasses such as *Digitaria* spp., *Setaria* spp., *Echinochloa crus-galli*, *Panicum dichotomiflorum*, *Sorghum halepense*, *Sorghum bicolor* (shattercane), and certain nutsedge species. A non-ionic surfactant, oil adjuvant, or fertilizer adjuvant is required for maximum POST activity. The field half-life is 60-90 d and so planting of rotational crops may need to be delayed. It is compatible with trifluralin, imazethapyr and pendemethalin to make herbicide mixtures. During the synthesis of imazaquin, 2,3-quinoline-dicarboxylic acid undergoes dehydration (to form the anhydride), a condensation reaction, and hydrolysis. It is then cyclized to form the imidazolinone ring, and precipitated to obtain the end product.

Imazethapyr: 2-[4,5-dihydro-4-methyl-4-(1-methylethyl)-5-oxo-1H-imidazol-2-yl]-5-ethyl-3-pyridinecarboxylic acid

Imazethapyr was discovered in 1981, field tested in 1987-88, and registered (HAMMER, PURSUIT) in 1989. It can be used in soybean, groundnut, peas, edible beans and imidazolinone-tolerant maize. It can be applied POST at 50-70 g ha^{-1} in soybean, E-PPL and PRE in soybean and imidazolinone-tolerant maize. It can also be used in groundnut and edible legumes. It controls a wide spectrum of annual broadleaf and grass weeds, listed under imazaquin. Soil persistence is similar to that described for imazaquin. It is compatible with trifluralin, pendemethalin, etc. to formulate commercial mixtures. During imazethapyr synthesis, 5-ethyl-2,3-pyridinedicarboxylic acid undergoes dehydration (to form the anhydride), a condensation reaction, and hydrolysis. It is then cyclized to form the imidazolinone ring and precipitated to obtain the end product.

Imazamethabenz: (±)-2-[4,5-dihydro-4-methyl-4-(1-methylethyl)-5-oxo-1*H*-imidazol-2-yl]-4(and 5)- methylbenzoic acid.

m-Isomer

The herbicidal activity of imazamethabenz was first reported in 1982 and registered as AS-SERT by American Cyanamid in 1988. It can be applied POST at 30-50 kg ha^{-1} in wheat and barley and 20-40 g ha^{-1} in sunflower. It controls *Avena fatua*, *Brassica kaber*, *Descurainia sophia* (flixweed), *Sisymbrium* spp., *Polygonum convolvulus* (wild buckwheat), *Thlaspi arvense* (field pennycress), *Asperugo procumbens* (catchweed), *Poa* spp., *Galium* spp., etc. Imazamethabenz can be applied in water or in non-phosphorous liquid fertilizer. Its POST activity can be enhanced by mixing a surfactant or oil adjuvant. Its field half-life is 25-36 d. Planting of susceptible rotational crops may need to be delayed.

Imines

CGA-248757: methyl{[2-chloro-4-fluoro-5-[(tetrahydro-3-oxo-1*H*,3*H*-[1,3,4]thiadiazolo[3,4-a]pyridazin-1-ylidene)amino]phenyl]thio}acetate

CGA-248757, (also referred to as KIH-9201), was jointly developed by Kumiai and Ciba-Geigy. It can be applied POST at 3-15 g ha^{-1} in maize and soybean for control of *Abutilon theophrasti* and certain other broadleaf weeds. Addition of a surfactant or oil adjuvant enhances its efficacy. It has no soil activity. It has a very short persistence in soil, with a half-life of 1-2 d. Synthesis of CGA-248757 involves a multistep process beginning with 2-fluoroaniline.

Isoxazolidinones

Clomazone: 2-[(2-chlorophenyl)methyl]-4,4-dimethyl-3-isoxazolidinone

Clomazone was developed in the early 1980s and commercialized by FMC as COMMAND in 1985. Seedlings of susceptible plants emerge from clomazone-applied soil but are bleached white and become necrotic after several days. When applied POST, susceptible species exhibit foliar bleaching in later growth stages. The clomazone vapours from nearby treated areas may also cause foliar bleaching. It is applied PPI or PRE at 0.56-1.70 kg ha^{-1} in soybean. It is also used PRE or PPI at 0.56-1.12 kg ha^{-1} in vegetable crops. Clomazone controls several annual broadleaf and grass weeds including *Abutilon theophrasti*, *Ambrosia artemisiifolia*, *Chenopodium album*, *Echinochloa crus-galli*, *Digitaria* spp., etc. It can be impregnated on dry bulk fertilizer, besides applying by a conventional sprayer. Its field half-life is 24 d but varies with soil type. It is less persistent in sandy loam than in silt loam or clay loam soils.

Nitriles

Bromoxynil: 3,5-dibromo-4-hydroxybenzonitrile

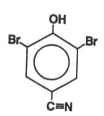

Bromoxynil, a benzonitrile herbicide, was first synthesized in Germany in 1896 and its herbicidal activity reported in 1963. It is applied POST at 0.21-0.56 kg ha^{-1} for weed control in wheat, barley, oats, rye, triticale, maize, sorghum, etc. It is similar to ioxynil but less active. It is effective against certain broadleaf weeds such as *Chorispora tenella, Chenopodium album, Solanum spp., Polygonum convolvulus, Brassica kaber, Kochia scoparia, Xanthium strumarium, Abutilon theophrasti, Capsella bursa-pastoris*, etc. It is also used for weed control on industrial sites, roadsides, railroad shoulders and other non-crop areas. It is applied on weeds at an early growth stage, i.e., 3- to 4-leaf stage. It can be used to make herbicidal mixtures with ioxynil, MCPA, atrazine, etc.

Dichlobenil: 2,6-dichlorobenzonitrile

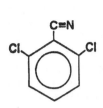

Dichlobenil was discovered in the mid-1950s and its herbicidal properties were reported in 1960. It is a powerful inhibitor of germination. It is applied to the soil, with or without incorporation, at 4-8 kg ha^{-1} in fruit and orchard crops, vineries, forest nurseries, etc. Dichlobenil is very effective against *Cyperus rotundus* and other perennial weeds which propagate through underground plant propagules like rhizomes, tubers and stolons. It also kills young seedlings of both monocot and dicot species. Dichlobenil controls many annual and perennial broadleaf and grass weeds such as *Poa annua, Digitaria* spp., *Portulaca oleracea, Panicum spp., Daucus carota* (wild carrot), *Taraxacum officinale, Cirsium arvense, Euphorbia esula, Elytrigia repens*, etc. It is mixed with pronamide, diuron, etc. Its field half-life is 60 d. Dichlobenil involves a single-step ammoxidation of 2,6-dichlorotoluene to 2,6-dichlorobenzonitrile.

Ioxynil: 4-hydroxy-3,5-diidobenzonitrile

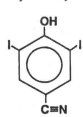

Ioxynil was first synthesized in Germany in 1896 and its herbicidal activity reported in 1963. It is applied POST in wheat, barley and rice. It is available as sodium salt and an oil-soluble amine salt formulation. It is not active through the soil. Treated plants develop blistered necrotic tissues within 24 h of application. This is followed by destruction of leaf tissue. It is normally tank-mixed with bromoxynil (0.2 + 0.2 kg ha^{-1}). Ioxynil controls weeds of Labiatae, Caryophyllaceae, Fumariaceae, Violaceae and Scrophulariaceae families better than bromoxynil. Addition of a surfactant may reduce crop selectivity.

Oxadiazoles

Oxadiazon: 3-[2,4-dichloro-5-(1-methylethoxy)phenyl]-5-(1,1-dimethylethyl)-1,3,4-oxadiazol-2-(3*H*)-one

Oxadiazon, first reported in 1969 and introduced by Rhone-Poulenc, is primarily a PRE herbicide (0.50-4.0 kg ha^{-1}), although it shows some POST activity (0.5-2.0 kg ha^{-1}). It is used in rice, cotton, sugarcane, soybean, groundnut, onion, garlic, potato, sunflower, transplanted vegetables, tea, rubber and banana. It is particularly useful to control weeds that grow from bulbs, rhizomes and other deep-rooted plant propagules, including *Cynodon dactylon, Lolium perenne*, etc. It is effective against many annual and broadleaf and grass weeds such as *Oxalis* spp., *Amaranthus* spp., *Richardia scabra*, *Urtica* spp., *Poa annua*, *Digitaria* spp., *Mollugo verticillata*, *Cenchrus incertus*, *Eleusine indica*, *Setaria viridis*, etc. At POST, it is active during the early growth stage of weeds. Oxadiazon has moderate to long persistence, with a field half-life of 60 d.

Oxadiazolidines

Methazole: 2-(3,4-dichlorophenyl)-4-methyl-1,2,4-oxadiazolidine-3,5-dione

Methazole was synthesized by Sandoz (PROBE) in 1965 and its herbicidal activity reported in 1970. When applied to the foliage, it causes veinal chlorosis followed by general chlorosis and desiccation. It is a PPL, PRE and POST herbicide (0.5-1.5 kg ha^{-1}), effective against a wide range of weed species, both of annual and perennial types. It is used in cotton, sorghum, maize, wheat, oats, soybean, onion, grapes and several fruit crops. The field half-life of methazole is 14 d. It generally degrades within 30-60 d. The residues do not accumulate in soil and do not injure rotational crops in the following season. It is compatible with most herbicides, including MSMA.

Phenols

Organic chemicals were first used as herbicides in 1932 when 4,6-dintro-O-cresol (2-methyl-4,6-dinitrophenol) was introduced. It was widely used in Europe under the name DNOC and in the USA under the name SINOX. Dinitrophenols are widely used as uncouplers of oxidative phosphorylation. Phenols are generally active by contact action.

Dinoseb: 2-*sec*-butyl-4,6-dinitrophenol

Dinoseb was first described as a herbicide in 1945. It is the most toxic of substituted dinitrophenols. Its sodium salt is applied POST for selective weed control in potato, beans, peas, orchard crops, plantation crops and vineyards. It is also used as a desiccant in potato and legume crops for seed. The alkanolamine salt (PREMERGE) is applied PRE to kill germinating weeds in the topsoil layer. It can also be used POST. The parent phenol is mixed with a surfactant and used as a general contact herbicide.

Phenoxyalkanoic Acids

Phenoxyalkanoic acids and their derivatives are a major group of organic herbicides because of their selectivity and outstanding ability to be translocated within the plant.

A. PHENOXYACETICS

These include 2,4-D, MCPA and 2,4,5-T. These chlorophenoxy herbicides are of historical importance because of the discovery of 2,4-D in the 1940s and the impetus given for research on a host of organic herbicides. They have hormonal activity at low rates and bring out growth responses in regions distant from the point of application. At higher rates, they exhibit herbicidal properties. They show a fine degree of selectivity between the susceptible broadleaf weeds and the tolerant grasses, thus facilitating their use in many cereal and grass crops.

The plants susceptible to phenoxyacetics show symptoms typical of other auxin-type herbicides, and include epinastic bending and twisting of stems and petioles, stem swelling and elongation, and leaf curling and cupping. This is followed by chlorosis at the growing points, growth inhibition, wilting and necrosis. The affected plants die slowly, within 2-5 wk. At low concentrations, young leaves may appear puckered and the tips of new leaves may develop into narrow extensions of the midrib.

2,4-D: (2,4-dichlorophenoxy)acetic acid

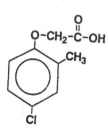

The chemical synthesis of 2,4-D was described in 1941 by Pokorny and its potency as a plant hormone reported in 1942 by Zimmerman and Hitchcock. The herbicidal utility of 2,4-D was reported shortly later by American (E.J. Krause; P.C. Marth and J.W. Mitchell) and British researchers (R.E. Slade, W.G. Templeman and W.A. Secton; T.S. Nutman, H.G. Thornton and J.H. Quaster). It was commercially developed in 1947.

2,4-D is a systemic POST (0.26-2.24 kg ha^{-1}) herbicide used for control of many annual weeds in cereal crops, sugarcane, plantation crops and non-crop areas. Most dicotyledonous crops are sensitive to 2,4-D. It is also effective when applied at PRE (0.26-2.24 kg ha^{-1}) to soil for being absorbed by emerging weed seedlings. It is available in salt and ester formulations. Ester formulations (ethyl ester, butoxyethyl ester, isopropyl ester, isooctyl ester, etc.) are more volatile than salt formulations, which include those of sodium and amine salts. 2,4-D is potentially mobile but rapid degradation in soil and removal from soil by plant uptake minimizes leaching. The persistence of 2,4-D phytotoxicity is 1-4 wk in warm, moist soil. Field half-life is 10 d. It is compatible with most herbicides, making it the most widely used herbicide for making herbicidal mixtures. 2,4-D acid is prepared from 2,4-dichlorophenol and monochloroacetic acid. 2,4-D salts are made by adding the appropriate amine or inorganic hydroxide to the acid. Esters are made by reacting 2,4-D acid with the appropriate alcohols.

MCPA: (4-chloro-2-methylphenoxy)acetic acid

MCPA is an analogue of 2,4-D, used for selective POST (0.25-1.5 kg ha^{-1}) herbicide against broadleaf weeds. It is more selective than 2,4-D at equal rates. It is widely used in cereals, legumes and aquatic areas. It was discovered in 1945. It is available as amines, esters and sodium or potassium salts. The amines and salts are water soluble. MCPA is widely used for making herbicide mixtures with tribenuron, 2,4-D, dicamba, bromoxynil, dichlorprop, etc. It is synthesized by chlorination of O-cresol to produce p-chloro-O-cresol followed by coupling with monochloroacetic acid in an alkane medium.

B. PHENOXYBUTYRICS

2,4-DB: 4(2,4-dichlorophenoxy)butanoic acid
2,4-DB is a POST (0.20-0.45 kg ha^{-1}) herbicide used for selective control of several

O-CH$_2$-CH$_2$-CH$_2$-C-OH (with C=O) on a 2,4-dichlorophenoxy ring structure

broadleaf weeds in forage legume crops. It can also be applied PRE and PPL (0.14-0.25 kg ha^{-1}). It is formulated as dimethylamine salt, isooctyl ester and butoxyethyl ester. The visual injury symptoms are similar to those exhibited by 2,4-D and MCPA. A non-ionic surfactant may be needed for maximum efficacy. In addition to application by conventional spraying, 2,4-DB can be applied through sprinkler irrigation systems. Its field half-life is 5 d for the acid, 7 d for the butoxymethyl ester and 10 d for the dimethylamine salt. It can be mixed with naptalam. 2,4-DB is synthesized by reacting 2,4-dichlorophenol with butyrolacetone in the presence of concentrated NaOH.

C. ARYLOXYPHENOXY PROPIONICS

In the 1970s, a number of aryloxyphenoxy propionic (APP) acids and derivatives were synthesized by Hoechst AG. This soon led to the appearance of several heterocyclic oxypropionic acids from different companies. These compounds were found to be highly active selective herbicides giving control of grass weeds mainly in broadleaf crops. The herbicides are generally more toxic towards graminaceous plants, but tolerated by dicotyledonous and some monocotyledonous plants. They are most effective when applied POST.

The POST treatment of susceptible plants by APPs gives similar symptoms of injury. Newly formed leaves and young and actively growing tissues become visibly chlorotic usually within 2-4 d. Progressive necrosis of the meristematic tissue in the nodes and buds develops within 1-3 wk. The oldest leaves show senescent pigment changes. Growth of roots is also strongly inhibited, with damage appearing in the meristematic region of the root tip. Scattered chlorotic mottling may develop on some susceptible species and may resemble the burning effect caused by contact herbicides. With PRE application, the first leaf may emerge, turn purple and die.

Dichlorprop: 2-(2,4-dichlorophenoxy)propanoic acid

O-CH(CH$_3$)-C-OH (with C=O) on a 2,4-dichlorophenoxy ring structure

Dichlorprop is commercially formulated as butoxyethyl and isooctyl ester, and dimethylamine salt. It is used mainly for control of general vegetation in non-agricultural areas as a POST, general spray-to-wet as well as directed application (5-10 kg ha^{-1}). Dichlorprop controls many hardwood brush weeds. It was introduced as a commercial herbicide in 1961. Its average field half-life is 10 d. It is used with mecoprop, MCPA, 2,4-D and other herbicides to make package mixtures. It is synthesized by coupling 2-chloro-propionic acid with 2,4-dinitrophenol in alkaline solution. A slight molar excess of the phenol and a pH above 9 are needed for good coupling. The acid is obtained by adjusting pH to about 4.9-5.3 and extracting to remove unreacted DCP with toluene. Dichlorprop is then released by adding HCl.

Diclofop: 2-[4-(2,4-dichlorophenoxy)phenoxyl]propanoic acid

structure: Cl and Cl substituted phenyl-O-phenyl-O-CH(CH$_3$)-C-OH (with C=O)

The herbicide activity of diclofop was reported in 1975. Both diclofop acid and its methyl ester formulation, diclofop methyl ester, are herbicidally active. During metabolism, diclofop methyl is hydrolyzed (demethylated) to diclofop in plants, with the former causing leaf membrane damage, and the latter causing inhibition of meristematic activity. Diclofop is applied POST (0.56-1.12 kg ha^{-1}) in wheat and barley to control certain annual grasses such as *Avena fatua*, *Setaria* spp., *Echinochloa crus-galli*, *Lolium* spp., etc. At higher doses, it gives a fairly good control of *Cynodon dactylon*. It may also be applied at PRE (0.75-1.25 kg ha^{-1}). Most broadleaf agronomic and horticultural crops are tolerant to diclofop. The average field half-life of diclofop parent acid is 30 d at pH 7.0.

Fenoxaprop: 2-{4-[(6-chloro-benzoxazolyl)oxy]phenoxy}propanoic acid

The herbicidal properties were first reported in 1982. It is commercially available as an ethyl ester. It is applied at POST (30-100 g ha^{-1}) in soybean and turf. Fenoxaprop by itself may be phytotoxic to wheat and other crops, but it shows selectivity when applied with 2,4-D, MCPA, thifensulfuron and tribenuron which antagonize its activity. It controls many annual and some perennial grass weeds with no injury to broadleaf weeds. The field half-life of fenoxaprop is 5-14 d depending on soil type.

Fluazifop-P: (*R*)-2-[4-[[5-(trifluoromethyl)-2-pyridinyl]oxy]phenoxy]propanoic acid

Fluazifop-P, commercially available as butyl ester formulation (FUSILADE), was discovered by Ishihara Sangyo Kaishi, Ltd., Japan, and developed jointly by ICI (now Zeneca) UK. It was first tested for herbicidal activity in 1981. Fluazifop-P is applied POST (50-200 g ha^{-1}) in cotton, soybean, wheat, tea, coffee, tree crops, etc. for control of several annual and perennial grasses including *Echinochloa crus-galli*, *Digitaria* spp., *Bromus tectorum*, *Panicum* spp., *Setaria* spp., *Cynodon dactylon*, *Imperata cylindrica*, *Sorghum bicolor*, *Elytrigia repens*, *Sorghum halepense*, etc. It has no activity against broadleaf weed species. An oil adjuvant or non-ionic surfactant enhances efficacy of fluazifop-P. Its average field half-life is 15 d. Although several methods are used to synthesize fluazifop-P, one method includes reaction of 4-(5-trifluoromethyl-2-pyridyloxy)-phenol with butyl 2-chloropropionate with a base.

Quizalofop-P: (R)-2-{4-[(6-chloro-2-quinoxalinyl)oxy]phenoxy}propanoic acid

Quizalofop-P was first registered in the late 1980s in soybean and the early 1990s in cotton. It is commercially available as ethyl ester formulation (ASSURE). It provides POST control of annual and perennial grass weeds in soybean (35-80 g ha^{-1}) and non-crop areas (75-100 g ha^{-1}). The grass weeds include *Sorghum halepense, Cynodon dactylon, Elytrigia repens, Muhlenbergia frondosa* (a fern), etc. A non-ionic surfactant or oil adjuvant is required for maximum efficacy. Quizalofop may suppress or control grass weeds germinating after a POST application, but the degree of suppression or control is related to herbicide rate, soil type and environmental conditions. Its field half-life is 60 d.

N-Phenylphthalimides

Flumiclorac: [chloro-4-fluoro-5-(1,3,4,5,6,7-hexahydro-1,3-diaxo-2H-isoindol-2-yl) phenoxy]acetic acid

Flumiclorac was discovered in 1981 by Sumitomo, Japan. It is commercially available as a phenyl ester formulation under its trade names RESOURCE and SUMIVERDE. The susceptible plants exhibit injury symptoms within 1 d under bright sunlight, followed by wilting and bleaching. The leaves then become brown, desiccated and necrotic. It is applied POST at 30-60 g ha^{-1} in soybean and maize to control *Abutilon theophrasti, Chenopodium album, Ambrosia artemisiifolia, Amaranthus* spp., *Euphorbia maculata*, etc. Flumiclorac is extremely non-persistent with half-lives ranging from 1-6 d, with no rotational crop restrictions. During its synthesis, pentyl 2-chloro-4-fluoro-5-aminophenoxyacetate is treated with 3,4,5,6-tetrahydrophthalic anhydride to obtain the phenyl ester of flumiclorac.

Phenylpyridazines

Pyridate: O-(6-chloro-3-phenyl-4-pyridazinyl) S-octyl carbonothioate

Pyridate (TOUGH) was discovered in 1974. It can be applied POST 1.0-1.5 kg ha^{-1} in groundnut, maize, rice, onion, alfalfa, grassland, rapeseed, orchards, vineyards and some vegetables. The susceptible plants show withering at leaf edges and the effect is complete within 5-10 d. It is non-persistent in soil, with a half-life of 7-21 d. During the synthesis of pyridate, 3-phenyl-pyridazone-6 is chlorinated and then saponified to produce 3-phenyl-4-hydroxy-6-chloropyridazine which is further reacted to produce the final product.

Phenyl Triazinones (Aryl Triazinones)

Sulfentrazone: N-{2,4-dichloro-5-[4-(difluoromethyl)-4,5-dihydro-3-methyl-5-oxo-1H-1,2,4-triazol-1-yl]phenyl}methane sulfonamide

Sulfentrazone (F6285, FMC97285) is a new phenyl triazinone herbicide introduced in 1991, field tested from 1991 to 1997 and registered in 1998 for soybean by the FMC corporation. It can be applied PPI and PRE at 60-90 g ha^{-1} for control of several broadleaf weeds including *Ipomoea hederacea* (morningglory), *Cassia occidentalis* coffee senna), *Sida spinosa* (prickly sida), *Cucumis melo* (smell melon), *Acalypha ostryifolia*, etc Other crops tolerant to sulfentrazone include tobacco, sugarcane and some species of turfgrass. It is also used POST (140 g ha^{-1}) to control a broad spectrum of weeds. Several sensitive weeds may be controlled by foliar application of sulfentrazone at as low as 18 g ha^{-1}. Its POST activity can be enhanced by a surfactant to control several weeds including *Cyperus esculentus, Abutilon theophrasti*, etc. at 34-56 g ha^{-1}.

Phthalamates

Naptalam: 2-[1-naphthalenylamino)carbonyl]benzoic acid

Naptalam was discovered by Uniroyal in 1947 and first registered as a herbicide in 1956. It can be applied PRE or POST at 2.0-4.0 kg ha^{-1} to control many annual broadleaf weeds such as *Chenopodium album, Amaranthus spp., Ambrosia artemisiifolia, Xanthium strumarium, Solanum nigrum*, etc. and certain annual grasses including *Setaria faberi*. Susceptible plants show strong epinasty. Naptalam also produces a strong antigeotropic response, abolishing the normal curvature of roots towards the ground and of shoot towards light. It is used in cucumber, melon, watermelon, groundnut, soybean, etc. Its field half-life is 14 d and the residues do not injure crops planted in the following season.

Pyrazoliums

Difenzoquat: 1,2-dimethyl-3,5-diphenyl-1H-pyrazolium

The herbicidal properties of difenzoquat methyl sulphate were first reported in 1973. It is a POST herbicide for wild oats control at 0.7-1.0 kg ha^{-1}. A non-ionic surfactant enhances its efficacy. Plant growth ceases soon after application and injury symptoms appear after 3-7 d. Meristematic regions become chlorotic, leading to necrosis. The field half-life of difenzoquat is < 4 wk.

Pyridazinones

Pyridazinones or fluorinated pyridazinone herbicides were discovered in the 1960s. They are mostly PRE herbicides, which include norflurazon and pyrazon. The susceptible plants show interveinal whitening of leaf and stem tissue. Growth may continue for several days, but seedlings lack green photosynthetic tissue, and eventually turn necrotic and die.

Norflurazon: 4-chloro-5-(methylamino)-2-[3-(trifluoromethyl)phenyl]-3-(2H)-pyridazinone

Norflurazon (SAN 9789H) was first introduced in 1968 by Sandoz. It is a PRE herbicide (2.5-5.0 kg ha^{-1}) used in tree crops, vineries, ornamentals and orchard and plantation crops, besides non-crop areas. It controls many grasses such as *Setaria* spp., *Eleusine indica*, *Panicum dichotomiflorum*, *Digitaria sanguinalis*, *Sorghum halepense*, etc., broadleaf weeds such as *Amaranthus* spp., *Ipomoea*, spp., *Portulaca* spp., *Cirsium arvense*, and many sedges including *Cyperus* spp. Norflurazon has moderate to long persistence in soil, with a half-life of 45-180 d. During the synthesis of norflurazon, trifluormethylaniline is acidified, then nitrified, followed by reaction with sulphites, and then by acidification, to form hydrazine hydrochloride, which is then reacted with mucochloric acid. The brown oil is then reacted with monomethylamine, forming a fine suspension. Base is then added and the solvents are distilled. The norflurazon suspension is filtered and washed.

Pyrazon: 5-amino-4-chloro-2-phenyl-3(2H)-pyridazinone

Pyrazon, also known as chloridazon, was the first pyridazinone herbicide to be developed in 1963. It is commercially available as PYRAMIN. It can be applied PRE or early POST at 3-9 kg ha^{-1} in sugarbeets and onion. Like norflurazon, it controls many annual grass and broadleaf weeds. It provides 4-8 wk weed control, depending on soil moisture and temperature. It is temperature-sensitive, resulting in crop damage under very unusual temperature extremes. In irrigated regions, it is used with sprinkler and drip irrigation. Better results are obtained with PPI and with furrow irrigation. Pyrazon has short to moderate soil persistence, with a field half-life of 21 d. Pyrazon is synthesized by condensing mucochloric acid with phenyl hydrazine, producing 1-phenyl-4,5-dichloropyridazine-6. It is then reacted with ammonia to produce pyrazon.

Pyridinecarboxylic Acids

Clopyralid: 3,6-dichloro-2-pyridinecarboxylic acid

Clopyralid was discovered in 1961. It was first marketed in Europe in 1978, while in the US, it was introduced in 1987-89 under the trade names CURTAIL, STINGER and CONFRONT. It is applied POST at 0.1-0.3 kg ha^{-1} in sugarbeets and maize, and at 0.14-0.56 kg ha^{-1} in pastures and rangelands. It controls many annual and perennial broadleaf weeds including *Cirsium arvense*, *Polygonum* spp., *Xanthium strumarium*, *Datura stramonium*, *Ambrosia artemisiifolia*, *Iva xanthifolia*, etc. Clopyralid is moderately residual, with an average field half-life of 40 d. Residues may injure certain crops such as potato, peas, pulses, etc. planted 1 yr after application.

Picloram: 4-amino-3,5,6-trichloro-2-pyridinecarboxylic acid

Picloram was discovered by Dow Chemical (as TORDON) in 1960 and first reported in 1963. It is available as isooctyl ester and triisopropanolamine salt. The symptoms of picloram injury are typical of other auxin-type herbicides. It is applied to the weed foliage at 0.15-1.0 kg ha^{-1} in forest plantations, non-crop areas, pastures and rangeland. It can be applied on the cut surface (stump, tree injection, or girdle) as treatment for woody species. It controls many annual and perennial broadleaf weeds, vines and woody plants at higher rates. The field half-life of picloram is 90 d. Dissipation is more rapid under warm, humid conditions. It is compatible with 2,4-D and triclopyr to make pre-mix formulations.

Triclopyr: [(3,5,6-trichloro-2-pyridinyl)oxy]acetic acid

Triclopyr, available as a triethylamine salt and butoxyethyl ester, was introduced by Dow Chemical in 1975 as GARLON. It is applied POST at 1.0-10.0 kg ha^{-1} in non-crop areas such as industrial sites, forestry, roadsides, rangeland and permanent pastures. At 0.25-0.40 kg ha^{-1}, it can be used in rice. It can be applied to freshly cut tree stumps, and mixed with oil for bark treatment on young trees. Triclopyr can control many annual hard-to-control broadleaf weeds as well as many tree and brush species. A non-ionic surfactant may increase its efficacy. It is moderately persistent in soil, with a half-life of 30 d, ranging from 10 to 146 d depending on soil type, moisture and temperature. It is compatible with several herbicides such as picloram, clopyralid, 2,4-D, etc. to make herbicidal mixtures.

Pyridines

Pyridine herbicides are mostly soil-applied herbicides, with little or no POST activity. These include dithiopyr and thiazopyr, which cause root growth inhibition and swelling in meristematic regions such as root tips. Susceptible plants may show thickened or swollen hypocotyls or internodes. Seed germination is not inhibited.

Dithiopyr: *S,S*-dimethyl 2-(difluoromethyl)-4-(2-methylpropyl)-6-(trifluoromethyl)-3,5-pyridinedicarbothioate

Dithiopyr was first introduced (as DIMENSION and STAKEOUT) by Monsanto, but purchased by Rohm and Haas in 1994. It is commercially available in liquid and granular formulations. Dithiopyr is applied PRE at 250-500 g ha^{-1} or POST at 120-500 g ha^{-1} in direct-seeded and transplanted rice, established turf, ornamentals, and orchard and other perennial crops. It controls several annual grass and broadleaf weeds such as *Echinochloa crus-galli,* *Digitaria* spp., *Eleusine indica*, *Oxalis* spp., *Euphorbia* spp. It has short to moderate soil persistence, with a half-life of 17 d. Metabolites get dissipated almost completely within 1 yr.

Thiazopyr: methyl 2-(difluoromethyl)-5-(4,5-dihydro-2-thiazolyl)-4-(2-methylpropyl)-6-trifluoromethyl)-3-pyridinecarboxylate

Thiazopyr, like dithiopyr, was first introduced by Monsanto and later purchased by Rohm and Haas in 1994. It is commercially available as VISOR and SPINDLE in EC and granular formulations. Thiazopyr is soil-applied at 0.5-2.0 kg ha^{-1} in several crops including maize, groundnut, soybean, vineries, sugarcane, sunflower, potato, vegetables, alfalfa, tree crops, and in non-crop areas to control several annual and perennial grasses and broadleaf weeds. Its field half-life is 64 d, with a range of 8 to 150 d, depending on temperature, moisture, soil pH and organic matter content. Wheat and sorghum are susceptible to soil residues of thiazopyr.

Pyridinones

Fluridone: 1-methyl-3-phenyl-5-[3-(trifluoromethyl)phenyl]-4(1H)-pyridinone

Fluridone was first described as an aquatic herbicide in 1976. It is applied at 0.06-0.09 mg ai L^{-1} (ppm ai) in ponds, 0.075-0.15 mg ai L^{-1} in lakes and reservoirs, or at 2.24 kg ai ha^{-1} of treated surface in drainage canals, irrigation canals and rivers. Fluridone can be applied on the water surface or subsurface, or as a bottom application just above the hydrosoil. It controls most submerged and emerged aquatic weed species including *Utricularia* spp., *Ceratophyllum demersum*, *Elodea* spp., *Myriophyllum* spp., *Najas* spp., *Potamogeton* spp, *Hydrilla verticillata*, *Brachiaria mutica*, etc. The affected weeds give white, pink and chlorotic appearance in 7-10 d time, with complete kill requiring more than 30 d. The half-life of fluridone is around 20 d in anaerobic pond water and 90 d in the hydrosoil. Fluridone adheres to sediments, but gradually desorbs into the water and then is subject to photodegradation.

Pyrimidinylthio-benzoates (Benzoates)

Pyrimidinylthio-benzoates, also referred to as benzoates, are a new class of herbicides being developed since the early 1990s. Pyrithiobac is the first herbicide in this group.

Pyrithiobac: 2-chloro-6-[(4,6-dimethoxy-2-pyrimidinyl)thio]benzoic acid, sodium salt.

Pyrithiobac (formerly DPX-PE 350 or KIH-8921) was developed by DuPont. It is commercially available as sodium salt. Pyrithiobac-sodium possesses a unique chemistry and is not a member of any existing herbicide family. Pyrithiobac is applied POST at 70 g ha^{-1} in cotton (1- to 2-leaf stage) for effective control of several broadleaf weeds such as *Xanthium strumarium, Ipomoea* spp., *Abutilon theophrasti, Sesbania exaltata* (hemp sesbania), *Salvia reflexa* (lanceleaf sage), *Sida spinosa* (prickly sida), *Amaranthus* spp., etc. When applied at this rate, wheat, soybean and grain sorghum may be rotated with cotton with no crop toxicity to these crops. Cotton is very tolerant of pyrithiobac even at rates as high as 112 g ha^{-1}. It may also be applied PPI or PRE in cotton to control *Cassia occidentalis*, a troublesome broadleaf weed. Higher rates cause injury to cotton. However, caution may need to be exercised, as insecticides (malathion, fenvalerate, methomyl, chlorpyriphos, etc.) applied along with pyrithiobac, as mixtures, are likely to make cotton sensitive to the herbicide. Antagonism occurs when pyrithiobac is applied with graminicides such as quizalofop-P, fluazifop-P, clethodim, or sethoxydim.

Quinolinecarxoxylic acids

Quinclorac: 3,7-dichloro-8-quinolinecarboxylic acid

Quinclorac was synthesized and discovered by BASF. It is an auxin-type herbicide, used in direct-seeded and transplanted rice to control important annual grass weeds. It is particularly very effective against *Echinochloa crus-galli, Setaria* spp., etc. It is applied PRE, late-PRE, or early-POST to weeds in transplanted rice and direct-seeded rice at 0-25-0.50 kg ha^{-1}. It also controls certain annual and perennial broadleaf weeds including *Convolvulus arvensis*. Its phytotoxicity is characterized by inhibition of shoot growth and chlorosis, and subsequent necrosis of the entire shoot. Quinclorac is persistent in soil, with residues injuring certain subsequent crops planted 1 yr after application.

Sulfonylureas

Sulfonylurea herbicides were developed in the 1980s. They show high herbicidal potency at very low rates, making them environmentally safe herbicides. The sulfonylurea group comprises the most widely used herbicides in present-day agriculture. Affected plants show growth inhibition soon after application. Chlorosis appears 3-5 d after application. This is followed by necrosis of the growing point. Plant death occurs within 7-21 d after application. Certain species occasionally remain green, but are stunted and not competitive with the crop.

Bensulfuron: methyl 2-[[[[[(4,6-dimethoxy-2-pyrimidinyl)amino]carbonyl]amino]
sulfonyl]methyl]benzoate

The herbicidal activity of bensulfuron was reported in 1989 and marketed by DuPont
shortly later as LONDAX. It can be applied at PRE or POST in irrigated rice at 35-70 g
ha^{-1} or POST in dryland rice at 30-40 g ha^{-1}. It controls many annual grasses (*Echinochloa
crus-galli* and *E. Colona*), sedges (*Cyperus rotundus* and other Cyperaceae family weeds)
and certain broadleaf weeds such as *Commelina* spp., *Heteranthera limosa* (ducksalad),
Eclipta alba, Sphenoclea zeylandica, Pontederia cordata, Ammania spp., *Plantago* spp., etc.
Bensulfuron has a field half-life of 5-10 d in irrigated rice. It moves no deeper than the
top 5-7 cm of soil.

Chlorimuron: ethyl 2-[[[[(4-chloro-6-methoxy-2-pyrimidinyl)amino]carbonyl]
amino]sulfonyl]benzoate

Chlorimuron was first synthesized and
field tested in 1982 and made commercially available in 1986 as CLASSIC. It is
applied POST at 8-13 g ha^{-1}, and PRE 30-
90 g ha^{-1} in soybean, groundnut and non-
crop areas. It controls many annual
broadleaf weeds including *Xanthium
strumarium, Datura stramonium, Ipomoea*
spp., *Ambrosia* spp., *Amaranthus* spp.,
Polygonum spp, etc. A non-ionic surfactant
or oil adjuvant is required for maximum foliar activity at POST. Chlorimuron persists
longer in high pH soils, with an average field half-life of 40 d. It is used to make
herbicidal mixtures with thifensulfuron, metribuzin, linuron, etc.

Chlorsulfuron: 2-chloro-N-[[(4-methoxy-6-methyl-1,3,5-triazine-2-yl)amino]carbonyl]
benzenesulfonamide

Chlorsulfuron can be applied PRE or POST
at 10-20 g ha^{-1} in wheat. It is used at higher
rates in non-crop areas to control many
broadleaf weeds including *Kochia scoparia,
Cirsium arvense, Brassica* spp., *Amaranthus*
spp., *Chenopodium album*, etc. It is compatible with most broadleaf herbicides and
fungicides, and certain organophosphate
insecticides. Growth of treated plants is
inhibited within a few hours after application, but injury symptoms appear 1-2 wk
later. Meristematic areas gradually become chlorotic and necrotic, followed by a

general foliar chlorosis and necrosis. The field half-life of chlorsulfuron is 40 d, ranging from 4-6 wk. Chlorsulfuron may injure susceptible crops up to 3 or 4 yr after application in high pH soils. It is mixed with metsulfuron to broaden its herbicidal activity.

Halosulfuron: methyl 5-[[(4,6-dimethoxy-2-pyrimidinyl)amino]carbonylaminosulfonyl]-3-chloro-1-methyl-1-*H*-pyrazole-4-carboxylate

Halosulfuron (MON 1200), introduced in 1991, has been made commercially available by Monsanto. It can be applied early PPS, PPI, or PRE to maize at 70-80 g ha^{-1} in combination with a safener, Mon 13900. It can also be used as a POST herbicide at 36 g ha^{-1} in maize, sorghum, sugarcane and turf. It also controls *Abutilon theophrasti, Xanthium strumarium* and several other broadleaf weeds, and *Cyperus* spp. It has short to moderate persistence in soil, with a field half-life of 6-35 d, depending on type of soil, organic matter content and rainfall.

Metsulfuron: methyl 2-[[[[(4-methoxy-6-methyl-1,3,5-triazin-2-yl)amino]carbonyl] amino]sulfonyl]benzoate

The herbicidal properties of metsulfuron were first reported in 1983. It was first marketed as ALLY and ESCORT. It is applied POST at 40-60 g ha^{-1} in wheat for control of broadleaf weeds such as *Kochia scoparia, Brassica* spp., *Cirsium arvense, Chenopodium album*, etc. It can be used at 40-120 g ha^{-1} in several pasture grass species. It is moderately persistent in soil with a half-life of 30 d, with a range of 1-6 wk. It is mixed with chlorsulfuron.

Nicosulfuron: 2-[[[[(4,6-dimethoxy-2-pyrimidinyl)amino]carbonyl]amino]sulfonyl]-*N*-*N*-dimethyl-3-pyridinecarboxamide

Nicosulfuron was first marketed in 1991 as ACCENT. It is applied POST at 35-70 g ha^{-1} in maize for control of most annual and some perennial grasses, including *Setaria* spp., *Sorghum bicolor* (shattercane), *Eriochloa villosa* (wooly cupgrass), *Panicum miliaceum* (wild proso-millet), *Sorghum halepense* and *Elytrigia repens* as well as certain broadleaf weeds. A surfactant or oil adjuvant is required for maximum efficacy. Its field half-life is 21 d at pH 6.5.

Primisulfuron: methyl 2-[[[[[4,6-bis(difluoromethoxy)2-pyrimidinyl]amino]carbonyl] amino]sulfonyl]benzoate

Primisulfuron was synthesized and discovered by Ciba-Geigy, Basel, Switzerland. It was first field tested in 1983 and became commercially available for use on maize in

1990. Its herbicidal activity was first reported in 1987. Primisulfuron can be applied POST at 20-40 g ha⁻¹ in maize to control certain annual and perennial grass weeds such as *Panicum dichotomiflorum*, *Sorghum bicolor*, *Sorghum halepense and Elytrigia repens*, and some annual broadleaf weeds such as *Amaranthus* spp., *Abutilon theophrasti*, *Polygonum* spp., *Ambrosia artemisiifolia*, *Xanthium strumarium*, *Solanum nigrum*, etc. It is moderately persistent in soils, with an average field life of 30 d. Residues can persist long enough to injure susceptible crops the following year. Its synthesis involves several steps beginning with thiourea and diethylmalonate and culminating with coupling of an aminosulfonyl-benzoic acid with a pyrimidinyl isocyanate.

Prosulfuron: 1-(4-methoxy-6-methyl-triazin-2-yl)-3-[2-(3,3,3-trifluoropropyl)-phenylsulfonyl]-urea

Prosulfuron (CGA-152005) was first field tested in 1988 and registered in the U.S. in 1995. It can be applied POST at 15-40 g ha⁻¹ in maize, sorghum, wheat, barley, sugarcane and certain other graminaceous crops. It controls certain annual broadleaf weeds such as *Xanthium strumarium*, *Kochia scoparia*, *Chenopodium album*, *Amaranthus* spp., *Ambrosia artemisiifolia*, *Abutilon theophrasti*, etc. A surfactant or oil adjuvant is required for maximum POST efficacy. It is used to make herbicide mixtures with primisulfuron, bromoxynil, terbuthylazine, etc. It is moderately persistent in soil. Its field half-life is 10 d. It is synthesized, multistep, beginning with aniline-2-sulphonic acid and ending in condensation with the appropriate triazine.

Sulfometuron: methyl 2-[[[[(4,6-dimethyl-2-pyrimidinyl)amino]carbonyl]amino]sulfonyl]benzoate

Sulfometuron (DPX 5648) can be applied PRE (before weed emergence) or POST at 50-200 g ha⁻¹ in forest areas, roadsides, non-crop areas, etc. to control annual and perennial broadleaf and grass weeds including *Sonchus arvensis*, *Malva* spp., *Rumex crispus*, *Ambrosia tenuifolia*, *Echinochloa crus-galli*, *Setaria* spp., *Aegilops cylindrica*, etc. A non-ionic surfactant improves sulfometuron efficacy but may reduce selectivity. Its field half-life is 20-28 d at pH 6 to 7.

Thifensulfuron: methyl 3-[[[[(4-methoxy-6-methyl-1,3,5-triazine-2-yl)amino]carbonyl]amino]sulfonyl]-2-thiophenecarboxylate

The herbicidal activity of thifensulfuron was first reported in 1984 and the herbicide commercially introduced in 1987 by DuPont as HARMONY. It can be applied POST at

10-18 g ha^{-1} in wheat and soybean to control many annual broadleaf weeds such as *Brassica kaber*, *Amaranthus* spp., *Kochia scoparia*, etc. A non-ionic surfactant is used to maximize its efficacy. Its field half-life is 12 d. It is used to make herbicidal mixtures with chlorimuron, tribenuron, fenoxaprop, etc.

Triasulfuron: 2-(2-chloroethoxy)-*N*-[[(4-methoxy-6-methyl-1,3,5-triazin-2-yl)amino]carbonyl]benzene-sulfonamide

Triasulfuron was synthesized by Ciba-Geigy in 1981 and made commercially available for wheat in 1992 as AMBER. It can be applied POST or PRE at 15-30 g ha^{-1} in wheat for control of many annual broadleaf weeds as *Kochia scoparia*, *Amaranthus* spp., *Ambrosia artemisiifolia*, etc. Triasulfuron is persistent, as its residues can persist long enough to injure certain dicot crops 1-3 yr after application. It has a multistep synthesis starting with chlorophenol and culminating with the coupling of an amino triazine with a benzene-sulfonyl isocyanate.

Tribenuron: methyl 2-[[[[(4-methoxy-6-methyl-1,3,5-triazin-2-yl)methylamino]carbonyl]amino]sulfonyl]benzoate

Tribenuron was field tested in 1983 and made commercially available by DuPont as CHEYENNE in Spain in 1986 and as EXPRESS in the U.S. in 1989. It can be applied POST at 9-18 g ha^{-1} in wheat and barley to control many annual broadleaf weeds including *Brassica* spp., *Chenopodium album*, *Lamium amplexicaule*, *Lactuca serriola* and *Kochia scoparia*. It is also used for making commercial mixtures with thifensulfuron, fenoxaprop and MCPA. It has a short residual effect, with an average field half-life of 10 d at pH 6. Rotational crops can be planted 60 d after application.

Tetrahydropyrimidinones

Tetrahydropyrimidinones (1,3-diaryl-2-pyrimidinones), also called cyclic ureas, is a new class of photo-bleaching herbicides. Herbicides of this family are yet to be commercialized. Chemically, these herbicides combine several structural elements of well-known chlorotic herbicides including fluometuron, fluridone and norflurazon. Fluridone and fluometuron are the parent structures of tetrahydropyrimidinones (THPs). THPs are expected to be used in rice. They are active on a broad spectrum of monocot and dicot weeds and sedges including *Echinochloa crus-galli*, *Cyperus difformis*, *Scirpus juncoides*, *Monochoria vaginalis*, *Cyperus serotinus*, *Eleocharis acicularis*, *Saggitaria pygmaea*, etc. while not affecting rice.

Thiocarbamates

When one oxygen in the carbonic acid molecule is substituted by sulphur, thiocarbamic acid is formed. Similarly, with two sulphur substitutions dithiocarbamic acids are formed. Thiocarbamate herbicides are toxic to germinating weeds following their soil incorporation before sowing. The susceptible grass and broadleaf weeds fail to emerge. Injury to sensitive crops like maize, when applied without a safener, manifests as malformed and twisted seedlings. Leaves are tightly rolled in the whorl and may not unroll normally. Injured dicot grasses may have cupped or crinkled leaves with a thick, leathery texture. As thiocarbamate herbicides are volatile, they need soil incorporation soon after application.

Butylate: S-ethyl bis(2-methylpropyl)carbamothioate

Butylate was first reported in 1962. It is applied PPI at 3.0-6.0 kg ha^{-1} in maize for control of many annual grasses such as *Setaria* spp., *Echinochloa crus-galli*, *Panicum dichotomiflorum*, *Sorghum halepense*, *Sorghum bicolor*, etc. as well as *Cyperus esculentus and Cyperus rotundus*, and a few broadleaf weeds. It must be incorporated into the soil soon after application to prevent volatility losses. It is lost substantially when applied to wet soils without incorporation, but little loss occurs after application to dry soils. Liquid formulations can be mixed with certain liquid fertilizers, impregnated on dry bulk fertilizers, or applied through irrigation water (chemigation). Its field half-life is 13 d. Butylate provides 4-7 wk of weed control and does not injure the succeeding crops. Butylate can be mixed with atrazine and safener dichlormid. It is synthesized by reacting ethyl chlorothiolformate with diisobutyl amine and a base.

Diallate: S-(2,3-dichloro-2-propenyl)bis(1-methylethyl)carbamothioate

Diallate was developed in the late 1950s and commercialized as AVADEX in 1961. It consists of *cis* and *trans* isomers. It is applied PPI or postplant incorporated (2.0-5.0 kg ha^{-1}) in wheat, maize, soybean, sugarbeets, oilseed crops, ornamentals, root crops and fruit trees. Diallate is very effective against wild oats (*Avena fatua*). It has a field half-life of 30 d. Diallate is synthesized by reacting appropriate polychlorinated allyl chloride with the appropriate thiol carbamate salt.

EPTC: S-ethyl dipropyl carbamothioate

EPTC, the first thiocarbamate herbicide developed in 1957, is very effective against many annual grasses and broadleaf weeds. As it is volatile, it needs to be incorporated into the soil to a depth of 5-10 cm. It is readily lost when the soil surface is wet at the time of application. It is applied PPI at 2.0-5.0 kg ha^{-1} in many crops including potato, safflower, vegetables, maize, cotton and orchard crops. It is effective against perennial grasses such as *Elytrigia repens*, *Sorghum halepense*, etc., sedges such as *Cyperus* spp., and many annual grasses. EPTC has good activity against broadleaf weeds such as *Ipomoea purpurea, Solanum villosum, Solanum nigrum, Chenopodium album, Portulaca oleracea, Amaranthus* spp., etc. It can be applied through irrigation systems (sprinkler, furrow and drip) and impregnated on dry fertilizers as well as by conventional sprayer. EPTC can be mixed with the microbial inhibitor dietholate and safener dichlormid. It has a field half-life of 6 d and the residues do not injure susceptible crops planted in the following season. It provides weed control for 4-6 wk. EPTC is synthesized by reacting ethyl chlorothiolformate with *di-n*-propyl amine and base.

Molinate: *S*-ethyl hexahydro-1*H*-azepine-1-carbothioate

Molinate is applied at 3.0-5-g ha^{-1} in irrigated rice after transplantation but before weeds emerge. In dryland rice, it can be applied PPI at 2.0-3.0 kg ha^{-1}. It can also be applied by injecting into irrigation water in direct-seeded irrigated rice. Molinate controls several annual grasses including *Echinochloa crus-galli, Leptochloa* spp., *Brachiaria platyphylla,* etc. It can be mixed with propanil. Its field half-life is 3 wk and residues do not persist long enough to injure susceptible crops planted 1 yr after application. Molinate, like EPTC, is readily lost from wet soil surface when not incorporated immediately after application.

Pebulate: *S*-propyl butylethylcarbamothioate

Pebulate was first reported in 1959. It is a selective PPI and PRE herbicide like EPTC used for control of *Cyperus* spp., *Digitaria sanguinalis, Setaria* spp., *Echinochloa crus-galli, Avena fatua,* and many broadleaf weeds. It is used (3.0-6.0 kg ha^{-1}) in tobacco, sugar beets, and tomato. Pebulate needs overhead or furrow irrigation soon after application to prevent volatility losses. Its field half-life is 2 wk. It is synthesized by reacting *n*-propyl chlorothioformate with *n*-butyl ethyl amine and a base.

Thiobencarb: *S*-[(4-chlorophenyl)methyl]diethylcarbamothioate

Thiobencarb, also known as benthiocarb, was developed in 1969. It is applied PPI or PRE in rice nurseries at 1.0-1.5 kg ha^{-1} and PRE in irrigated rice at 1.5-3.0 kg ha^{-1}. It is very selective for rice. It controls many annual grasses including *Echinochloa crus-galli, Echinochloa colonum, Leptochloa* spp., *Brachiaria platyphylla, Aeschynomene virginica, Commelina* spp. and sedges of the Cyperaceae family. The field half-life of thiobencarb is 30-90 d. It can be used to make mixtures with 2,4-D ethyl ester for broad-spectrum weed control in rice.

Triallate: *S*-(2,3,3-trichloro-2-propenyl) bis(1-methylethyl)carbamothioate

Triallate was commercialized as AVADEX BW in 1962. It is like diallate, except that it has two chlorines on the terminal allyl carbon. It is applied PPI (2.5-5.0 kg ha^{-1}) in wheat, maize, soybean, barley, forage legumes, fruit crops and ornamentals. It is particularly very effective against wild oats. Like diallate, it is active on a wide spectrum of weeds.

Triallate has a field half-life of 82 d. It gives 6-8 wk of weed control. It is mixed with trifluralin and synthesized by reacting polychlorinated allyl chloride with appropriate thiol carbamate salt.

Triazines

The herbicidal properties of the symmetrical triazines (s-triazines) were discovered in 1952 by Geigy Ltd., Basel, Switzerland. Subsequently, numerous triazine derivatives have been synthesized and screened for their herbicidal properties. Triazines became one of the most prominent groups of herbicides to broaden the chemical concept of weed control. They are widely used for selective and non-selective weed control. Their greatest success has been in maize where they are exclusively used for selective weed control. Most of them are soil-applied, while some do exhibit activity through the foliage. Atrazine, simazine, ametryn and prometryn are the triazine compounds that gained major recognition in agricultural and non-agricultural systems.

Various substitutions on the triazine nucleus yield compounds of widely different chemical and biological properties. Accordingly, they exhibit a wide spectrum of selective weed control and biological activities. If the 2-position of the triazine ring is substituted by chlorine, it is termed as **chlorotriazine** (see atrazine structure). If it is substituted by the methoxy group, it is called **methoxytriazine,** and if it is replaced by methylthio (methylmercapto) group (see ametryn structure), it is known as **methylthiotriazine (methylmercaptotriazine).**

The triazines are usually colourless, non-explosive, non-corrosive and stable compounds, with low mammalian toxicity. The methoxytriazines are prepared by reacting the appropriate chlorotriazine with methanol in the presence of an equivalent base. They are more water soluble and readily absorbed, and have different selectivities than chlorotriazines. The methylthio or methylmercapto triazines are prepared by reacting the appropriate chlorotriazines with methylmercaptan in the presence of sodium hydroxide.

The typical triazine-injury symptoms begin with interveinal chlorosis of leaves and yellowing of their margins. These are followed by further chlorosis and necrosis. Older leaves are more damaged than the new growth. Browning of leaf tips can occur, while root growth is not affected. Triazines do not affect seed germination.

Ametryn: N-ethyl-N'-(1-methylethyl)-6-(methylthio)-1,3,5-triazine-2,4-diamine

Ametryn is the methylthio analogue of atrazine. It was first released for field evaluation in 1959 and made commercially available in 1964. It is a selective PRE (2.0-8.0 kg ha^{-1}) and POST (0.5-2.0 kg ha^{-1}) herbicide, depending upon the crop, for control of broadleaf and grass weeds in sugarcane, banana, pineapple and citrus. Ametryn has considerable contact activity and hence its POST application must be directed on weeds. It is more soluble than atrazine. The field half-life of ametryn is 60 d. In tropical conditions, it may provide better weed control and its residual life in the soil is expanded to several months. Ametryn is synthesized by successive N-alkylation and thioalkylation of cyanuric chloride.

Atrazine: 6-chloro-N-ethyl-N'-(1-methylethyl)-1,3,5-triazine-2,4-diamine

Atrazine, a prominent triazine herbicide, is currently one of the most widely used herbicides in world agriculture. It was first released for experimental evaluation in 1956 and was made commercially available in 1958. It is a selective PRE herbicide (1.0-2.0 kg ha^{-1}) for control of many broadleaf weeds and grasses in maize, sorghum, sugarcane, pineapple, turf and orchards. It is also used as a non-selective herbicide in non-crop areas and in minimal or no-tillage programmes in maize and sorghum. Atrazine controls broadleaf weeds including *Amaranthus* spp., *Ipomoea* spp., *Datura stramonium, Polygonum convolvulus, Ambrosia* spp., *Polygonum* spp., *Abutilon theophrasti, Xanthium strumarium, Chenopodium album*, etc. and certain annual grasses such as *Echinochloa crus-galli, Setaria* spp., *Panicum dichotomiflorum*, etc. When mixed with a non-toxic crop oil or surfactant, it is effective at POST (1.0-2.0 kg ha^{-1}) as well, particularly in maize and sorghum. Many vegetable crops are sensitive to atrazine. It is used to formulate package mixtures with several herbicides including metolachlor, bentazon, simazine, cyanazine, alachlor, propachlor, bromoxynil, dicamba, butylate etc. and a safener dichlormid. Atrazine has a field half-life of 60 d. Its persistence is increased by higher soil pH as well as by cool, dry soil conditions. Most rotational crops can be planted one year after application at selective rates, except under arid and semi-arid climatic conditions. Atrazine is synthesized by successive N-alkylation of cyanuric chloride.

Cyanazine: 2-ethyl-N'-(1-methylethyl)-6-(methylthio)-1,3,5-triazine-2,4-diamine

Cyanazine was discovered in 1967 and developed as a herbicide in 1976. It is applied E-PPL, PPI, PRE, E-POST at 1.0-4.0 kg ha^{-1} in maize, cotton, beans, etc. Cyanazine controls many broadleaf weeds including *Ipomoea* spp., *Portulaca oleracea, Ambrosia artemisiifolia, Brassica kaber, Kochia scoparia*, etc. and several annual grass weeds such as *Setaria* spp., *Digitaria* spp., *Echinochloa crus-galli, Panicum capillare*, etc. A surfactant or oil adjuvant may improve its POST efficacy. It is used to formulate mixtures with atrazine and metolachlor. Cyanazine has a short residual activity, with a field half-life of 14 d. It is synthesized by reacting cyanuric chloride and aminoisobutyronitrile.

Hexazinone: 3-cyclohexyl-6-(dimethylamino)-1-methyl-1,3,5-triazine-2,4(1H,3H)-dione

Hexazinone, a PRE (0.62-2.5 kg ha^{-1}) and POST (0.62-1.25 kg ha^{-1}) herbicide introduced for noncrop land weed control in 1975, is used for selective weed control in sugarcane, pineapple and alfalfa. In non-crop areas, including industrial sites, railroad shoulders, storage areas, reforestation sites, etc., it is applied at 2.5-13.5 kg ha^{-1}). It controls many annual and perennial broadleaf and grass weeds, as well as many brush species. Hexazinone is effective against many woody plants and aquatic weeds. A surfactant increases efficacy for foliar applications but may cause undesirable levels of crop injury in some situations. Hexazinone has a moderately long residual activity, with a field half-life of 90 d.

Prometryn: *N,N'*-bis(1-methylethyl)-6-(methylthio)-1,3,5-triazine-2,4-diamine

Prometryn, the methylthio analogue of propazine, became commercially available in 1964. It is applied PPI, PRE, and directed-POST in cotton at 0.5-3.0 kg ha^{-1}. It is also used in soybean and non-crop areas. As prometryn has considerable contact activity, it should be applied as directed spray at POST. It controls several weeds including *Physalis* spp., *Chenopodium album*, *Ipomoea* spp., *Amaranthus* spp., *Sida spinosa*, *Setaria* spp., *Eleusine indica*, etc. It is mixed with MSMA. Its field half-life is 60 d. It is synthesized by successive *N*-alkylations of cyanuric acid followed by reaction with mercaptan to attach the thiomethyl group.

Simazine: 6-chloro-*N,N'*-diethyl-1,3,5-triazine-2,4-diamine

Simazine is one of the first triazine herbicides discovered by Geigy. It was made commercially available in 1958. It is widely used as a selective PRE herbicide (1.0-4.0 kg ha^{-1}) to control many annual broadleaf weeds and some grasses in maize, sugarcane, plantation crops like tea, coffee and rubber and many orchard crops. It is also used as a non-selective herbicide for vegetation control on roadsides, banks of irrigation canals and drainage channels, industrial sites and other non-crop areas. It has little or no POST activity. The field half-lives for simazine are 149 d for a sandy loam soil (OM 1.3%, pH 7.3), 60 d for sandy soil (OM 0.9%, pH 7.8), 55 d for a loam soil with low OM (2.1%) and low pH (5.5), and 186 d for a loam soil containing high OM (6.2%) and high pH (7.9). It is used to make packaged mixtures with metolachlor, prometon, atrazine, etc.

Triazinones

Metribuzin: 4-amino-6-(1,1-dimethylethyl)-3-(methylthio)-1,2,4-triazin-5(4*H*)-one

The triazinone herbicide family was first discovered in 1964. Metribuzin was introduced in 1969 and commercially made available by Bayer in 1973. It is applied E-PPS, PPI, PRE, or POST-directed (0.25-0.80 kg ha^{-1}) in sugarcane, maize, soybean, potato, peas, beans, etc. Metribuzin is very active against annual broadleaf weeds such as *Chenopodium album*, *Amaranthus* spp., *Abutilon theophrasti*, *Datura stramonium*, *Brassica* spp. and *Ambrosia artemisiifolia* along with certain annual grasses. It is used to make package mixtures with chlorimuron, trifluralin and metolachlor. Its half-life is 30-60 d. Generally, the residues of metribuzin do not injure susceptible crops planted one year after application. But certain root crops such as onion, potato, and sugarbeets may show susceptibility.

Triazoles

Amitrole, Amitrole-T: 1*H*-1,2,4-triazol-3-amine
Amitrole was first reported in 1953 and commercially introduced in 1954 by Union

Carbide. The primary symptom of amitrole toxicity is bleaching (albinism) in leaves and shoots, and it is most evident in meristems and developing leaves. Bleached leaves eventually wilt and become necrotic. Amitrole is a POST translocated herbicide, applied spray-to-wet at 2.0-10.0 kg ha^{-1}, in non-crop areas and in locations not used for growing food crops. It controls many annual and perennial grass and broadleaf weeds. Its field half-life is 14 d. Amitrole is synthesized by condensation of formic acid with aminoguanidine.

Amitrole-T is a mixture of amitrole and ammonium thiocyanate. It is more active than amitrole.

Triazolopyrimidine Sulfonanilides

Flumetsulam: N-(2,6-difluorophenyl)-5-methyl[1,2,4]triazolo[1,5-a]pyrimidine-2-sulfonamide

Flumetsulam was discovered in 1984. It is commercially available as BROAD-STRIKE. The weed species most sensitive to flumetsulam are killed before emergence, but some weeds may die after emergence under dry conditions. The emerged sensitive plants exhibit stunting, interveinal chlorosis, veinal discoloration (purpling) and necrosis within 1-3 wk. Flumetsulam is marketed as a package mixture with metolachlor for use in maize and soybean, and with clopyralid for use in maize. It can be applied PRE, PPI, or PPS at 50-75 g ha^{-1}. Flumetsulam controls many annual broadleaf weeds such as *Brassica kaber, Amaranthus* spp., *Kochia scoparia, Chenopodium album, Polygonum* spp., *Solanum nigrum, Abutilon theophrasti,* etc, but has little activity against grasses. Flumetsulam is package-mixed with trifluralin or metolachlor to provide control of several annual grasses. Liquid flumetsulam can be impregnated on dry fertilizer. Its field half-life is 1-3 mon. Residues of flumetsulam do not injure soybean, maize, alfalfa, groundnut, potato, rice, sorghum and tobacco planted one year after application.

Uracils

Substituted uracils were first identified as herbicides in 1962. Bromacil and terbacil are the two uracil herbicides presently used in agriculture and non-agricultural areas. The susceptible plants show foliar chlorosis and necrosis, and inhibition of root and shoot growth.

Bromacil: 5-bromo-6-methyl-3-(1-methylpropyl)-2,4-(1H,3H)pyrimidinedione

Bromacil (lithium salt) can be applied PRE at 5.0-7.0 kg ha^{-1} in citrus, 1.5-3.0 kg ha^{-1} in pineapple, and 1.5-5.0 kg ha^{-1} in non-crop areas. It is good for brush control at higher rates. It controls many annual and perennial grasses, sedges, and broadleaf weeds including *Echinochloa crus-galli, Digitaria* spp., *Setaria* spp., *Cynodon dactylon, Cyperus* spp., *Sorghum halepense, Tribulus terrestris, Cenchrus* spp. and

Chenopodium album. At low rates, it is used for selective weed control in orchard crops. Bromacil is used to make herbicide mixtures with diuron, sodium metaborate and sodium chlorate. Its field half-life is 60 d. At higher rates, the phytotoxic residues persist for more than one year.

Terbacil: 5-chloro-3-(1,1-dimethylethyl)-6-methyl-2,4-(1*H*,3*H*)-pyrimidinedione

Terbacil is a PRE (1.0-2.0 kg ha^{-1}) and POST (1.0-1.5 kg ha^{-1}) herbicide for control of many annual and some perennial weeds in sugarcane, established alfalfa, and orchard crops such as apples, peaches and citrus. The weeds include *Stellaria media, Lamium amplexicaule, Chenopodium album, Descurainia* spp., *Lactuca serriola, Lolium* spp. and *Echinochloa crus-galli*, with a partial control of *Cyperus* spp. It has medium to long persistence in soil, with a field half-life of 120 d. It needs to be applied just before or during the active growth period of weeds. If dense growth is present, top growth should be removed. Control of perennial grasses may be improved by cultivation prior to treatment.

UCC-C4243: 2-chloro-5[3,6-dihydro-3-methyl-2,6-dioxo-4-(trifluoromethyl)-1(2*H*)-pyrimidinyl]benzoic acid

UCC-C4243 is an experimental substituted uracil herbicide that controls a wide spectrum of broadleaf and some annual grass weeds when applied to either foliage or soil at 30-140 g ha^{-1}. It is applied in wheat, which tolerates up to 170 g ha^{-1} when applied to soil. Pea, lentil and barley grown in rotation with wheat are too sensitive to this new herbicide to merit its use. Its herbicide potential is under study. However, it has a potential use as a crop desiccant in potato and cotton. When applied to the foliage, UCC-C4243 causes rapid chlorosis and desiccation.

Ureas

The discovery and development of substituted ureas as herbicides began shortly after World War II. The work led to the discovery of monuron in 1951. Subsequently, several other phenyl substituted ureas including diuorn, linuron, fenuron, chloroxuron, fluometuron, chlorbromuron, siduron, tebuthiuron, etc. were developed. Besides, the heterocyclic substituted ureas like noruron and methabenzthiazuron were also discovered and developed. These substituted ureas are used as selective herbicides in agriculture and non-selective herbicides in non-crop areas. These are primarily used at PRE, although some of them show POST activity. At higher rates, they are useful as soil sterilants, particularly in non-crop areas.

The seedlings of susceptible species, following soil application, emerge but become chlorotic within a few days. This is followed by complete necrosis. Foliar application causes interveinal chlorosis of the leaves and yellowing at the leaf margins. The older leaves are more damaged than the new growth.

Of the numerous substituted ureas developed, only the more prominent ones are discussed below.

Diuron: *N*'-(3,4-dichlorophenyl)-*N,N*-dimethylurea

Diuron was first reported in 1951. It can be applied PRE at 0.8-2.5 kg ha^{-1} for controlling emerging grass and broadleaf weed seedlings in tea, cotton, coffee, grapes, pineapple, apples, pears, citrus and other plantation and orchard crops. It can also be applied as POST-directed spray in maize, sorghum, etc. at 0.2-0.4 kg ha^{-1}. It can be used to formulate POST mixtures with MSMA, paraquat, bromacil, dichlobenil, tebuthiuron, sodium metaborate, sodium chlorate, 2,4-D and imazapyr. Diuron has a field half-life of 90 d. The phytotoxic residues dissipate within a season when applied at lower selective rates. At higher selective rates, residues may persist for more than one year. It is synthesized by reacting 3,4-dichlorophenylisocyanate and dimethylamine.

Fluometuron: *N,N*-dimethyl-*N'*-[3-(trifluoromethyl)phenyl]urea

A CH$_3$ substitution on the meta position of the phenyl ring of urea structure gives fluometuron. As a result, its selectivity is increased. It was first reported in 1964. Fluometuron may be applied PPI, PRE (1.0-1.5 kg ha^{-1}), and E-POST (1.0-2.0 kg ha^{-1}) in cotton, sugarcane, sorghum, citrus and ornamental crops. It controls many broadleaf and grass weeds including *Echinochloa crus-galli*, *Digitaria* spp., *Panicum dichotomiflorum*, *Setaria* spp., *Eleusine indica*, *Brachiaria platyphylla*, *Xanthium strumarium*, *Portulaca* spp., *Ipomea* spp., *Chenopodium album*, *Polygonum* spp., *Cassia obtusifolia* and *Euphorbia esula*. It can be mixed with MSMA. Its field half-life is 85 d. Fluometuron residues often dissipate to non-detectable levels within 4 mon of application. It is synthesized by alkylamination of the appropriate aromatic isocyanate.

Linuron: *N'*-(3,4-dichlorophenyl)-*N*-methoxy-*N*-methylurea

Linuron differs from diuron by having a methoxy group in place of one of the methyl substituents. It was first reported in 1962. It can be applied PRE or E-POST-directed (0.30-2.5 kg ha^{-1}) in soybean, cotton, maize, sorghum, wheat, potato and vegetable crops. It controls annual broadleaf weeds including *Brassica* spp., *Amaranthus* spp., *Portulaca oleracea*, *Polygonum convolvulus*, etc. and annual grasses such as *Echinochloa crus-galli* and *Setaria* spp. Foliar application is effective on a wider spectrum of weeds. A non-ionic surfactant increases its weed control efficacy. It can be mixed with chlorimuron. Linuron has a field half-life of 60 d. Its residues do not injure cover crops planted the following season. It is synthesized by reacting 3,4-dichlorophenyl-isocyanate with methoxymethylamine (dimethyl hydroxylamine).

Tebuthiuron: *N*-[5-(1,1-dimethylethyl)-1,3-thiadiazol-2-yl]-*N,N*-dimethylurea

Tebuthiuron was first reported in 1974. It can be applied PRE or POST (0.8-4.0 kg ha^{-1}) in pastures, rangeland and non-crop areas. It controls certain broadleaf weeds and woody brush species at low rates and several broadleaf, grass and brush species at higher rates. Its field half-life is 12-15 mon, and it is considerably greater in low rainfall areas and in muck soils with high organic matter content. It can be mixed with trifluralin and diuron. Tebuthiuron is synthesized by combining pivalic acid and 4-methyl-3-

thiosemicarbazide (MTSC) and then adding a mixture of sulphuric acid and polyphosphoric acid. Towards the end of the reaction, ammonium hydroxide is added to neutralize pH. The lower aqueous layer is removed and isopropylbenzene urea is added to form an intermediate amine. Then 1,3-dimethyl urea is added with HCl. All reactions are conducted under nitrogen. Finally, solvents are removed under vacuum and the resultant crystals are washed in water and dried.

Unclassified Herbicides

Bensulide: *O,O*-bis(1-methylethyl) *S*-[2[(phenylsulfonyl)amino]ethyl]phosphorodithioate

Bensulide was first reported in 1962 and introduced by Stauffer. It is a PPI and PRE herbicide (5.6-6.7 kg ha^{-1}) used to control annual grass and broadleaf weeds such as *Digitaria sanguinalis, Echinochloa crus-galli, Eleusine indica, Amaranthus* spp., *Chenopodium album, Caspella* spp., *Lamium* spp., etc. It is used in vegetable crops such as cucumber, squash, pumpkin, melon, onion, cabbage, carrot, cauliflower and chillies. Bensulide requires rain or irrigation immediately after application for long-term activity in soil. It inhibits root and shoot growth. Its field half-life is 120 d.

Ethofumesate: (±)-2-ethoxy-2,3-dihydro-3,3-dimethyl-5-benzofuranyl methanesulfonate

The herbicidal properties of ethofumesate were first reported in 1969. It is commercially available as PROGRASS. It can be applied before or after weed emergence at 0.8-2.0 kg ha^{-1} in sugarbeet. It can also be applied PPI or postplant-incorporated at 2.0-4.0 kg ha^{-1}. Ethofumesate controls several weeds including *Solanum nigrum, Stellaria media, Chenopodium album, Kochia scoparia, Amaranthus retroflexus, Cirsium arvense, Polygonum* spp., *Echinochloa crus-galli, Digitaria sanguinalis, Setaria* spp., *Bromus tectorum,* etc. Its half-life ranges from 5 wk under warm, moist conditions to 14 wk under dry, cold conditions. Ethofumesate can be used to make herbicide mixtures with phenmedipham and desmedipham. Its synthesis involves reaction of the enamine from morpholine and 2-methyl propionaldehyde with *p*-benzoquinone to give 2,3-dihydro-3,3-dimethyl-2-morpholinobenzofuran-5-ol, which is mesylated, and the product converted to ethofumesate.

Fosamine: ethyl hydrogen(aminocarbonyl)phosphonate

The herbicidal properties of fosamine were first described in 1974. It is available as ammonium salt under the name KRENITE. In most woody plants, fosamine injury is not apparent before normal leaf senescence in the fall. However, leaf and bud development is severely inhibited the following spring. Leaves that emerge appear abnormally small and spindly. Other plants such as pines and bindweed may show a response soon after application. On moderately susceptible to resistant species, suppression of terminal growth may occur. Fosamine is used to control brush weeds in non-crop areas. It is applied to the foliage at 10-50 kg ha^{-1}. It can be applied as a spray-to-wet treatment at 1.5-3.5 kg (in 100 L water). An oil adjuvant or surfactant may enhance its POST activity. Its field half-life is 8 d. During the synthesis of fosamine, triethyl phosphate is reacted with methyl chloroformate and ammonia.

Glufosinate: 2-amino-4-(hydroxymethylphosphinyl)butanoic acid

$$OH-\overset{\overset{O}{\|}}{C}-CH-CH_2-CH_2-\overset{\overset{O}{\|}}{P}-OH$$
$$\underset{NH_2}{|} \qquad\qquad \underset{CH_3}{|}$$

Glufosinate, a phosphorylated amino acid available as ammonium salt, was first reported as a herbicide in 1981. The injured plants show chlorosis and wilting within 3-5 d after application, followed by necrosis in 1-2 wk. The rate of development of toxicity symptoms is enhanced by bright sunlight, high humidity and moist soil. Seedlings are not injured before emergence. Glufosinate is applied POST at 0.35-1.5 kg ha^{-1} in plantation (tea, coffee, etc.) and tree crops and in non-crop areas. As a non-selective herbicide, it controls a broad spectrum of annual and perennial grass weeds such as *Panicum maximum*, *Paspalum conjugatum*, *Axonopus compressus*, *Setaria* spp., *Imperata cylindrica*, *Digitaria sanguinalis*, etc., annual and perennial broadleaf weeds and ferns. Glufosinate has a field half-life of 7 d.

Glyphosate: N-(phosphonomethyl)glycine

$$OH-\overset{\overset{O}{\|}}{C}-CH_2-NH-CH_2-\overset{\overset{O}{\|}}{P}-OH$$
$$\underset{OH}{|}$$

The herbicidal activity of glyphosate was first discovered in 1971 and introduced as an isopropylammonium salt by Monsanto. It is commercially available as ROUNDUP and under several other trade names. The susceptible plants show growth inhibition within 4-7 d of application, followed by general foliar chlorosis and necrosis. Chlorosis appears first, more pronouncedly on immature leaves and growing points. Foliage may turn reddish-purple in certain species. Regrowth of treated perennial and woody species often appears deformed with whitish markings or striations.

Glyphosate is a broad spectrum selective POST herbicide used (at 0.5-4.0 kg ha^{-1}) for effective control of rhizomatous and deep-rooted perennial weeds including *Elytrigia repens*, *Imperata cylindrica*, *Cynodon dactylon*, *Paspalum conjugatum*, *Setaria palmifolia*, *Saccharum spontaneum*, *Polygonum chinense*, *Pteridium aquilinum* (fern), *Eleusine indica*, *Sorghum halepense*, *Mikania micrantha*, *Digitaria sanguinalis*, *Echinochloa crus-galli*, *Panicum repens*, *Paspalum dilatatum* and many other species. Greater activity of glyphosate is observed when applied on the foliage during the active growth period. Perennial weeds should have greater leaf area for greater effect. It is extensively used in plantation crops such as tea, coffee, rubber, oil palm, etc., orchard crops, vineyards, pineapple, sugarcane and in non-crop areas such as banks of irrigation canals, drainage ditches, roadsides and industrial and recreational sites. It is also used in minimal tillage cropping systems. Glyphosate has moderate persistence in soil, with a half-life of 47 d. All crops can be planted immediately after application due to strong adsorption by soil. Tank mixing with residual herbicides such as substituted ureas and triazines or with POST herbicides such as paraquat, dalapon, MSMA and phenoxy (or other auxin-type) herbicides may reduce glyphosate efficacy.

112

Tridiphane: 2-(3,5-dichlorophenyl-2)-2-(2,2,2-trichloroethyl)oxirane

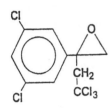

Tridiphane was discontinued as a herbicide in the late 1980s, but is used as a synergist to enhance the POST efficacy of cyanazine and atrazine, particularly against annual grasses such as *Setaria* spp., *Panicum* spp., *Digitaria* spp., etc. It can be applied POST at 0.5-0.8 kg ha^{-1} in combination with cyanazine and atrazine in maize. Its field half-life is 28 d under aerobic and 3 d under anaerobic conditions. Tridiphane does not accumulate with repeated annual applications.

INFORMATION ON BIOHERBICIDES[2]

Currently, three bioherbicide products have been registered for use in certain crops for control of some problem weeds. These are available in commercial formulations. The three bioherbicides BIOMAL®, COLLEGO® and DeVine® are discussed as under.

BIOMAL®

BIOMAL is a POST mycoherbicide useful for the specific control of *Malva pusilla* (round-leaved mallow) in field crops. The BIOMAL® formulation is a wettable powder consisting solely of living spores of *Colletotrichum gloeosporioides* (Penz.) Penz & Sacc. f. sp. *malvae*. It is applied to the actively growing round-leaved mallow plants after the two-leaf stage and preferably before the weeds are 15 cm tall. Although BIOMAL® is effective when applied at any stage of weed growth, control occurs at a slower rate in older and the more mature plants. The most effective stage of application is at an early seedling stage. BIOMAL® is applied by mixing in water, using any conventional field sprayer at a spray volume of 100-150 L ha^{-1}. The suspension should be agitated constantly during spraying to ensure that the spores stay in uniform suspension. The suspension should be applied within 3 h of preparation. High humidity for approximately for 18-24 h after application ensures successful spore germination and infection. Ranfall during or immediately after application facilitates spore germination. If available, sprinkler irrigaion may be used to create more humid conditions after application.

BIOMAL®-infected round-leaved mallow plants show typical anthracnose disease symptoms. Lesions will form on the leaves, petioles and stems of infected plants within 2-4 h after application. As the disease progresses, the stems are girdled by lesions, resulting in plant mortality. The commercial BIOMAL® formulation does not appear to have adverse toxicological effects on humans and the environment. It is manufactured by Philom Bios, Inc, Saskatoon, Saskatchewan, Canada.

COLLEGO®

COLLEGO® is a selective POST mycoherbicide used for the control of *Aeschynomene virginica* (L.) B.S.P. (northern jointvetch) in rice and soybean. The living spores of the fungus *Collectotrichum gloeosporioides* (Penz.) & Sacc. f. sp. *aeschynomene* constitute the actice ingredient of COLLEGO®. It is a two-component product, with component A consisting of a water-soluble spore-rehydrating agent and component B containing a wettable powder formulation of living fungal spores of *C. gloesporioides* f. sp. *aeschynomene*. COLLEGO® should be applied to the emerged northen jointvetch plants that are 20-60 cm tall and have not reached the bloom stage. Rice fields should be flooded before

[2]Source of Information:
Biological Control of Weeds Handbook (Alan K. Watson, ed.). 1993. Weed Science Society of America, Champaign, Illinois, U.S.A., 202 pp.

application. Soybean fields should be applied just before application. Free moisture or relative humidity above 80% and air temperatures of approximately 26°C for at least 12 h are required for development of the highest degree of infection.

C. gloeosporioides f. sp *aeschynomene* causes an anthracnose disease by forming lesions on the above-ground parts of northern joinvetch. Lesions occur principally on the stems and once the stems are girdled, plant parts above the girdle collapse and die. Death of plants may not occur for 4-5 wk after the application of COLLEGO®. This product is not known to be hazardous to humans or to the environment. It is manufactured by Ecogen, Inc., Langhorne, Pennsylvania, U.S.A.

DeVine®

DeVine® is a mycoherbicide used for control of *Morrenia odorata*, strangler or milkweed vine, in citrus plantations. It is a submerged fermentation product containing chlamydospores of the fungus *Phytophthora palmivora* (Butl.) Butl. MWV pathotype. The liquid formulation is available in one pint (U.S. liquid measure, 473.2 ml) containers which must be kept refrigerated (2-8°C) until use. DeVine® may be applied in any type of citrus plantation from May to September after the weed has germinated or is actively growing. It may be applied by using a herbicide boom sprayer, at a spray volume of 124 L ha^{-1}, to achieve uniform coverage of the soil under the tree canopy. The fungus will initiate root infection in milkweed vine plants and kills the vine in 2-10 wk following application, depending on the size and maturity of the vine. The soil surface must be wet at the time of application.

DeVine®-infected milkweed vine shows typical *Phytophthora* rot symptoms. The dying plants are girdled at the soil level and up to an inch above it. The infection progresses until it encompasses all plant roots. Infected plant roots slough the cortex, leaving only the stele when the plant is pulled from the soil. The root rot induces leaf wilt, the leaves wither and eventually fall from the plant. The fungus is not a mammalian pathogen and it presents no imminent hazard to humans upon exposure. DeVine® is manufactured by Chemical and Agricultural Products Division, Abbott Laboratories, Chicago, Illinois, U.S.A.

INFORMATION ON HERBICIDE SAFENERS

Herbicide safeners, earlier known as antidotes, protectants or antagonists, protect susceptible crop plants from herbicide injury. They do not have herbicidal properties. They widen the margin of crop selectivity to non-selective herbicides and permit using higher rates of herbicides than normally possible for more effective weed control. Safeners have been in use since the late 1960s and early 1970s with the discovery and development of 1-8-napthalic anhydride (NA), CDAA and dichlormid. Subsequently several safeners have been developed and their properties and uses are discussed here.

Benoxacor: 4-(dichloroacetyl)-3,4-dihydro-3-methyl-2*H*-1,4-benzoxazine

Benoxacor (CGA-154,281), a morpholine safener, was discovered in 1982 by Ciba-Geigy, and premix formulations with metolachlor were tested in 1986. It protects maize from metolachlor injury that may occur during abnormally wet conditions at the time of germination and emergence. It is packaged and applied in a 30:1 ratio of metolachlor and benoxacor. Thus, it can be soil-applied as an E-PPS, PPI, or PRE along with metolachlor. Benoxacor is synthesized by a multistep synthesis, beginning with 2-nitrophenol and a morpholine closing, followed by acetylation.

Dichlormid: 2,2-dichloro-*N-N*-di-2-propenylacetamide

Dichlormid, earlier referred to as R-25788, was discovered in 1970 as an antidote for preventing EPTC injury to maize. It is now widely used to formulate commercial package mixtures with thiocarbamate (butylate and EPTC) and chloroacetamide (acetochlor, alachlor, etc.) herbicides for use in maize. It is also used with atrazine. It is applied at PPI or PRE at 0.14-0.45 kg ha^{-1} with the herbicide. Dichlormid provides no weed control. It protects maize from injury by acetochlor and butylate. It may also be applied as a soil or seed treatment to protect maize and sorghum from several herbicides.

R-29148: 3-(dichloroacetyl)-2,2,5-trimethyloxazolidine

R-29148, like dichlormid, is also used to make premix formulations with EPTC and butylate for use in maize. It is applied at 0.15-0.30 kg ha^{-1} when used with these herbicides. It was discovered in 1971.

Dietholate: *O,O*-diethyl *O*-phenyl phosphorothioate

Dietholate (R-33865) is a non-phytotoxic microbial inhibitor sold in a commercial package mix with EPTC and butylate for use in maize. By inhibiting soil microbes that degrade EPTC and butylate, dietholate extends the soil residual life of these herbicides, and thereby increases the duration of weed control efficacy. Hence, dietholate is referred to as a herbicide extender. It is applied PPI at 0.75-1.0 kg ha^{-1}. It does not persist in soil.

Flurazole: phenylmethyl-2-chloro-4-(trifluoromethyl)-5-(thiazolecarboxylate)

Flurazole (MON-4606) is used as a seed treatment at 1.25-2.50 kg seed on grain sorghum to protect against injury by alachlor and acetochlor. Like other safeners, flurazole has no weed control activity and does not protect weeds from control by herbicides.

Fluxofenim: 1-(4-chlorophenyl)-2,2,2-trifluoro-1-ethanone *O*-(1,3-dioxolan-ylmethyl) oxime.

Fluxofenim (CGA-133,205), an oxime safener, was first tested in 1982 by Ciba-Geigy. It is also used as a seed treatment, like oxabetrinil, to protect sorghum from injury by chloroacetamide (metolachlor) herbicides applied PRE or PPI. It is applied at 0.4 g (ai) per kg of seed and used like oxabetrinil. Fluxofenim is synthesized by a multi-step condensation of trifluorochloroacetophenone with a halogenated dioxolane.

Oxabetrinil: α-[(1,3-dioxolan-2-ylmethoxy)imino]benzeneacetonitrile

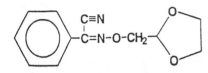

Oxabetrinil (CGA-43,089), an oxime group chemical, was first developed in 1982 by Ciba-Geigy. It is used as a safener to protect sorghum from injury by chloroacetamide herbicides (e.g. acetochlor) applied PPI or PRE. It is applied as a seed treatment at 1.25 g (ai) kg^{-1} of seed, by placing seed in oxabetrinil/water slurry with a binding agent. Sorghum seed must be dried quickly after treatment to avoid imbibition. Good-germinating hybrids/varieties or seed lots show no loss in germination percentage following oxabetrinil treatment. Weaker-germinating hybrids or varieties or seed lots may germinate at slightly lower percentages. Oxabetrinil is synthesized by coupling benzylcyanide with the appropriate halogenated dioxolane.

Other Safeners: There are many safeners which are in various stages of development, and these are listed below. The exception is NA which was one of the first safeners to be developed.

CGA-185,072; HOE-70,542 (Fenchlorazol-ethyl); Fenchlorim (CGA-123,407); Cyometrinil (CGA-43,089); NA (1-8-naphthalic anhydride); MG-191; OTC; MON 13900.

6

Absorption and Translocation of Herbicides

If a herbicide is to become effective on the physiological and biochemical processes of the plant, it must be absorbed by the plant and translocated (except for contact herbicides) in adequate quantity to the site(s) of action. Differential absorption and translocation, which form the basis for herbicide selectivity, determine the tolerance and susceptibility of a plant species to a particular herbicide.

ABSORPTION

Absorption is the process of penetration into the plant tissue. Herbicides are applied either to the soil or plant foliage. Hence, herbicide absorption depends upon the method of application and the plant part with which the chemical comes into contact.

Soil-Applied Herbicides

Herbicides applied to the soil either at preplanting or preemergence are usually taken up by the root or shoot of the emerging seedling. In young plants, the primary absorbing region of a root lies between 5 and 10 cm behind the root tip. Water, salts and water-soluble herbicides are taken up by root hairs and cortex cells of this primary absorbing region (Fig. 6.1). The xylem in the absorbing region is sufficiently differentiated to be functional and the endodermis, a single layer of closely packed cells, is not sufficiently lignified to prevent penetration by solutes. From root hairs and cortex, the molecules migrate via the symplast into the stele where they leak from the symplast to apoplast and ascend into the foliage via the transpiration stream.

Herbicides enter the roots via the same pathways and by similar mechanisms as inorganic ions. They are absorbed by both passive and active mechanisms. The passive entrance is primarily along the absorbed water and the herbicides move with the water throughout the plant in the apoplast system including the xylem. Active absorption requires energy and the herbicides enter into the protoplasm and move via the symplast system. Herbicides may enter the plant and move primarily by one or both of these mechanisms, depending on the chemical and physical properties of the molecule. A nonpolar herbicide can dissolve the suberized casparian strip, which separates cortex and stele at the endodermis, and diffuse through it to reach the stele and move symplastically. However, the polar herbicides cannot do this to enter the symplast; instead, they move inward with water along the apoplast system.

Besides roots, the soil-applied herbicides are also absorbed by the developing shoots and coleoptiles. Shoot entry is considered to be the primary site of entry for some

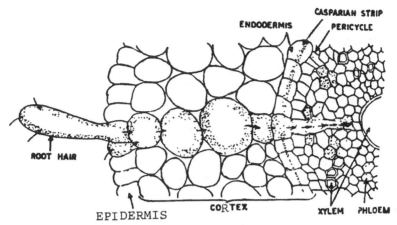

Fig. 6.1. Penetration of herbicide through root hairs and cortex cells in the primary absorbing region of the root (longitudinal section).

herbicides. In the case of EPTC, exposure of roots of barnyardgrass (*Echinochloa crusgalli*) seedlings has little or no effect, while exposure of the shoots results in severe injury. Uptake of EPTC, diallate and propham through the coleoptile is more toxic to oat seedlings than uptake through the roots [1]. Parker [46] suggested that volatility might be a factor in shoot uptake of herbicides such as the thiocarbamates. This volatility factor has also been found to be responsible for the lethal effect of trifluralin on *Setaria italica* and *Panicum miliaceum* when shoots but not roots were exposed to the herbicide [61]. Negi and Funderburk [41] found that vapours of trifluralin injured the shoots of maize drastically but not roots.

Triazine and urea herbicides kill the plants via the shoot. They allow the plants to emerge and effect shoot growth by inhibiting the Hill reaction of photosynthesis. The plants die shortly after emergence.

Seed also plays an important role in the uptake of herbicides. The aqueous solutions of soil-applied herbicides move into the seeds by diffusion. Herbicide absorption by seeds is initially a physical process, independent of water uptake. This process is dependent on herbicide concentration and its solubility in water as also on the size, shape, seed coat hardness and permeability, and oil and protein content of the seed. Phillips et al. [47] investigated the absorption of chlorpropham atrazine, linuron, chloramben and 2,4-D by seeds of 11 soybean strains and found that the total quantity and concentration found in the seeds differed for both herbicides and soybean strains. The total quantity of herbicide absorbed was affected by total oil content or percentage of oil in the seeds. Rubin and Eshel [53] found that seeds of susceptible snapbean (*Phaseolus vulgaris*) absorbed larger amounts of terbutryn and fluometuron from soil than the seeds of resistant cotton. The herbicides accumulated mainly in the seed coat, which is shed following emergence, precluding the possibility of their translocation to the seedling leaves. In snapbean seeds, the herbicides accumulated in the cotyledons, which served as a source of acropetal flow to the developing foliage of the young seedling.

Thus, root, shoot and seed are the major sites of uptake of soil-applied herbicides. The roots may predominate in the uptake of some herbicides, while shoots or seeds predominate in the case of the others. More frequently, the combination of any two or all three sites may effect herbicide absorption.

In the case of soil-applied herbicides, placement of the herbicide in the soil is an important factor in governing the efficacy, selectivity and reproducibility of the weed control obtained. When shallow-rooted weeds are to be removed from the deep-rooted crop, surface or shallow application of a herbicide will ensure greater selectivity. Most of the herbicide applied is usually concentrated in the upper 2 to 8 cm of soil. If the herbicide is not concentrated in the zone where weeds grow, the treatment may prove ineffective or less effective.

Foliage-Applied Herbicides

When a herbicide is applied at postemergence, the spray is usually directed at the plant, most of the chemical reaching the foliage, which then absorbs the herbicide molecules. Before absorption of the foliage is discussed, an understanding of the leaf cuticle is essential.

Cuticle Structure

The herbicide has to pass through a wax layer called the cuticle, which protects the inner cells. The cuticle is a thin, continuous, non-cellular lipoidal membrane covering the plant surface. It covers the guard, mesophyll and epidermal cells adjoining the stomatal chambers and intercellular air spaces in leaves. The cuticle is made up of four different substances: **cutin, cutin wax, pectin** and **cellulose**.

Cutin, the chief structural component of the cuticle, is composed of polymerized long-chain fatty acids and alcohols, and protects the phtosynthetic mechanism of the plant cell. It is not soluble in most organic solvents and is resistant to decay. **Cutin wax**, also referred to as **cutin matrix**, is composed of short-chain esters and alcohols. The optically active cutin waxes are separated by layers of isotropic materials that extend from the cuticular surface to the region of embedded pectic substances. These layers of isotropic material act as pathways of penetration through the impermeable wax layer [42].

The third cuticular substance is **pectin**, which is amorphous and highly hydrophilic. Pectin is composed largely of long chain polygalacturonic acid molecules having side carboxyl groups. As it is hydrophilic it has an important effect on the water-holding capacity of the cell. The fourth substance is **cellulose**, which is a carbohydrate. It is fibrillar in structure, hydrophilic, elastic and has great tensile strength, which is responsible for the high resistance of cell walls to stretching. Cellulose is composed of long-chain molecules, which are relatively stable.

Foliar Penetration

Foliage is the main source of entrance of postemergence herbicides. The effect of a herbicide depends on its ability to penetrate the cuticle, which is a part of the apoplast, and the plasmalemma, which covers all surfaces of the symplast, including plasmodesmata, to enter the symplast.

Foliage-applied herbicides may exist in **polar** form or **non-polar** form. The polar ions are electrically positive as well as negative. Examples of polar compounds are water, carbohydrates, amino acids and herbicides such as 2,4-D salts, PCP salts, TCA, thiocyanates, etc. They are soluble in polar solvents and insoluble in non-polar solvents. In contrast, the non-polar ions are neither strongly positive nor negative, and hence are called non-electrolytes. Non-polar compounds include oils, waxes, 2,4-D acid, 2,4-D esters and most of the organic substances. They dissolve in non-polar solvents such as oil, but not in water.

Foliage-applied herbicides enter the living cells either by a lipoidal route or an aqueous route or both. Fat-soluble and non-polar compounds readily penetrate the cuticle by diffusion. Herbicides such as dinitroaniline and phenoxy compounds penetrate the cuticle by the lipoidal route in the non-polar undissociated form. Polar herbicides take the aqueous pathway, and their penetration is greatly assisted by high humidity and a saturated atmosphere around the leaves. A high water content in the leaf, as a result of high humidity, enables the water system of the leaf cells to become almost contiguous with the outer surface of the cuticle and provides an aqueous pathway from the cuticle to the symplast. Water, polar solutes and polar herbicides readily penetrate the protein and cellulose portions.

Penetration through Trichomes and Stomata

Trichomes or leaf hairs are common features on many plant surfaces and they are involved in foliar absorption of herbicides. Generally, leaf hairs are extensions of epidermal cells. They provide a microclimate which can alter the drying time of aqueous sprays and thus, the pattern of absorption. Trichomes, which contain living protoplasm, have a significant bearing on herbicide absorption.

The leaf contains stomata in the cuticularized epidermis. They allow gaseous exchange. Entry through the stomata takes place when they are open at the time of treatment. Aqueous solutions do not penetrate stomata and hence addition of a wetting agent is required to lower surface tension and improve spreading of a herbicide solution on the leaf surface. Wetting agents destroy the air films between the solution and cuticle that prevent contact of the solution with the cuticle surface. They also improve the spread of herbicide solution along the intercellular spaces, enhance the wetting of the walls of the mesophyll and the bundle sheath cells, and increase penetration between the wax particles of the cuticle. Stomatal absorption of foliage-applied herbicides is greater through the lower than the upper leaf surface. The greater permeability of the lower leaf surface is due to greater number of stomata, trichomes, morphology and chemistry of the cuticle, and amount of waxes present, all of which may have an effect on herbicide penetration. When the stomata are open and conditions favourable, stomatal entry may be comparable with uptake through the cuticle.

Penetration through the Stem

Stems of plants also play a useful role in absorbing herbicides, and in some cases this route of penetration is as effective as foliar penetration. Compared to the foliage, the stem presents a limited target area. However, the stem has an extensive and highly developed transport system. The degree of penetration through the stem depends on the growth and stage of development of the plant. Penetration through a young and immature stem is more rapid than in a mature stem. The mechanism of penetration through the stem is similar to that in the foliage.

Penetration through the bark of woody plants is very difficult. The **periderm** is composed of the **phellogen**, the **phellem** (cork), and the **phelloderm**. The periderm contains tightly packed suberized cells (with no intercellular spaces) which are composed of tannins, fatty acids, lignin, cellulose and terpenes. Hence, it provides a barrier to the penetration of polar herbicides. However, the presence of lenticels, which provide radial channels through the bark layer, could aid herbicide penetration. Herbicide entry through the bark of woody plants is normally done by mechanical cuts or by injections.

TRANSLOCATION

Once a herbicide is absorbed into the plant system, it moves either apoplastically or symplastically (Fig. 6-2). The term apoplast refers to the system of non-living interconnecting cell walls, intercellular spaces, and the water-filled and air-filled xylem elements. Thus, apoplastic movement takes place predominantly through the xylem and intercellular spaces. The word symplast refers to the system of interconnected protoplasm that is connected from cell to cell by means of plasmodesmata, excluding the vacuoles. The sieve tubes are the highly specialized components of the symplast. Generally, the symplast includes the phloem transportation mechanism and the apoplast includes the xylem stream. The apoplast protects the symplast from unfavourable environmental conditions that cause desiccation and abrasion. Any chemical that penetrates the symplast first penetrates the apoplast. The cuticle, which is a part of the apoplast, and the plasma membrane, which covers all surfaces of the symplast including plasmodesmata, are the two main barriers that a chemical must penetrate to enter the symplast.

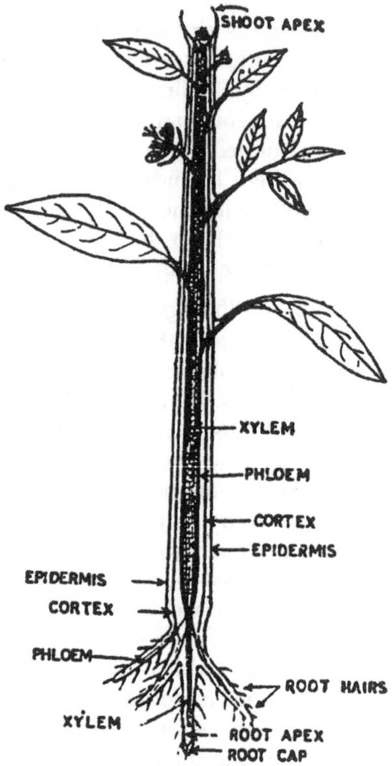

Soil-applied Herbicides

Soil-applied herbicides, taken up by the root or shoot, penetrate the xylem and move upward apoplastically in the transpiration stream. Even highly toxic herbicides can be absorbed from the soil and readily translocated to all parts of the plant without injuring the xylem, which is principally a non-living tissue. Under low soil moisture and high transpiration conditions, water deficits develop within the plant. This situation facilitates a rapid absorption and translocation of herbicides. Triazine, uracil and urea herbicides, which have very low ability to enter the symplast in leaves, move acropetally via the chloroplast to the leaf mesophyll very rapidly with water. They are translocated to the leaves, stems and buds as well.

Fig. 6.2. Longitudinal sectional view of apoplastic (xylem) and symplastic (phloem) movement of herbicides.

Foliage-applied Herbicides

The effect of a translocated herbicide after being absorbed by the foliage depends on the pattern of translocation. The two main transport pathways are the phloem and xylem. The herbicides move along with the assimilates from the leaves in a stream through the phloem utilizing the source-to-sink principle, the main sources being the expanded and growing leaves, and the sinks being the shoot tips, buds, roots, fruits and other plant parts undergoing development and expansion. A strong sink is essential for an active translocation of herbicide from leaves. The direction of movement is from the regions of synthesis of food or hydrolysis of food reserves to regions of food utilization. Herbicides that enter the symplast migrate to the phloem and translocate into the lumina of the sieve tubes in the assimilation stream [15]. Crafts and Yamaguchi [14] reported that compounds such as 2,4-D are subject to export and accumulation in the living cells during migration across the mesophyll along the sieve-tube conduits. Other herbicides, such as dalapon, 2,3,6-TBA, picloram, amitrole, etc. are not accumulated but move freely across the mesophyll along the phloem and into the living growing root tips.

Herbicides such as picloram and dalapon are translocated through the xylem when absorbed by roots, and through the xylem when absorbed by leaves. A herbicide which is translocated primarily through the phloem can also 'leak out' of the symplast and enter the xylem circulation.

Herbicide translocation is, however, a dynamic process that is not restricted to one system but can occur in both the xylem and phloem over an extended period as long as the herbicide remains available. Regardless of their site of entry into the plant, all herbicides can enter the symplast and are potentially available for phloem transport. Some herbicides are predominantly phloem-mobile, while others are almost exclusively xylem-mobile. However, a herbicide can be translocated from the site of entry into the plant in the xylem, then transfer and be translocated in the xylem, with a subsequent retranslocation in the phloem. Maleic hydrazide and amitrole transfer from the phloem to the xylem before being circulated in plants. Glyphosate is translocated apoplastically following phloem translocation out of a leaf. The root-applied picloram first moves to the shoots in the xylem and is then translocated in the phloem before accumulating in young leaves. Imazaquin, following root application and translocation to the shoot in the xylem, is translocated in the phloem of *Xanthium strumarium* L. (common cocklebur) [55]. Dicamba is continuously remobilized, first accumulating in young leaves and then being translocated from these leaves when they have expanded. Thus, most herbicides are **ambimobile** and can be transported in either system, depending on conditions within the plant. Some of these ambimobile herbicides may be translocated primarily in the phloem; its metabolite(s) may be xylem-mobile and translocated apoplastically following metabolism.

Herbicide translocation, particularly in the phloem, depends on the production (or remobilization) and loading of assimilates into the phloem at the source, translocation from the source to sink and unloading at the sink [18]. Any herbicidal action that interferes with these processes can disrupt phloem transport of assimilates and consequently, translocation of the herbicide itself.

A herbicide, applied in a mixture, can reduce the translocation of another herbicide(s) compared to when it was applied alone. On the contrary, translocation of one herbicide may stimulate the mobility of another herbicide. This aspect is discussed in greater detail in Chapter 12.

FACTORS AFFECTING HERBICIDE ABSORPTION AND TRANSLOCATION

A number of factors affect and modify the absorption and translocation of a herbicide, eventually affecting the phytotoxicity of the chemical applied. These factors are discussed in this section.

Plant Factors

Branching Habit

The branching habit of a plant is very critical for foliar-applied herbicides. Most broadleaf species have an open or horizontal branching habit with expanded leaves and exposed growing regions. This facilitates retention of spray droplets and easy surface coverage. Grass species, on the other hand, have minutely ridged surfaces on the leaves. They are often vertically arranged and growing regions are enclosed by sheaths (or bases of older leaves) which serve as a protective cover. This may be the reason why non-polar herbicides are more effective against grasses as they tend to have greater surface coverage.

Plant Surface

The surfaces of leaves and stems are commonly cutinized and suberized by waxy and fatty substances. Hence, aqueous sprays, which are polar in nature, are repelled by most plant surfaces; their droplets are spherical and tend to run off. This is particularly true of aqueous sprays on grasses. This problem can be overcome by the addition of wetting agents to the sprays, resulting in lowered surface tension and better spread. Non-polar herbicides, which have low viscosity and surface tension, have stronger affinities for fatty substances and they wet the plants rapidly and completely; the droplets tend to spread and creep. Under favourable conditions, they even run through open stomata and fill the intercellular spaces of the leaves. After covering the plant, the non-polar herbicide penetrates the cuticle and cell walls and comes into intimate contact with the protoplasm.

Plant Maturity

Herbicides move faster in young plants than in old. Uptake by the foliage and translocation in the phloem are greater and faster in actively growing meristematic tissue. As plants mature, the leaf surfaces thicken and barriers for herbicide penetration become greater. The thickness and nature of the cuticle change with the plant age and maturity. Rapid shoot and root growth also favour rapid herbicide absorption.

In some plants the meristematic tissue, such as leaf primordia, often lacks functional xylem during the early stages of cell differentiation. Therefore, a xylem-translocated herbicide absorbed by roots may move rapidly to the leaves but be prevented from entering leaf primordia.

Plant Species and Varieties

Some plant species, some cultivars within the species, and even some strains within a variety, show differences in absorption and translocation of herbicides. Cotton is moderately tolerant to prometryn while soybean is sensitive. Even though roots of both species absorb equal amounts of prometryn, most of the herbicide in cotton is concentrated in the lysigenous glands and in the root primordia while it is uniformly distributed in soybean plant [56]. Soybean plants had considerably higher concentrations of prometryn in the shoots than the roots whereas the reverse was true in cotton [56]. Similarly, SAN 6706 [4-chloro-5-(dimethylamino)-2(α, α, α-trifluoro-m-tolyl)-3(2H)-pyridazinone] which moves apoplastically in the transpirational stream, accumulated in significant concentrations in lysigenous glands and trichomes of the tolerant cotton plant, while most of it accumulated in the leaves in the case of sensitive maize and soybean plants [60]. Rubin and Eshel [53] found that terbutryn and fluometuron, applied to the roots, accumulated in the root system of both cotton and snapbean (*Phaseolus vulgaris*), but the sensitive snapbean plants absorbed and translocated more herbicide to the shoots than the tolerant cotton plants, followed by rapid distribution in the leaf mesophyll tissue. Such differences in uptake and translocation of herbicides between tolerant and sensitive plants are partially responsible for selectivity.

Environmental Factors

The environment prior to, during and after the herbicide application, has considerable effect on the growth and development of the plant. This eventually affects the absorption and translocation of herbicides. It is difficult to isolate the effect of one environmental factor from another.

Temperature and Humidity

Plant species and varieties vary greatly in their requirement of optimum temperature and humidity. Temperature and humidity have a significant effect on transpiration, respiration and evaporation. High temperature causes injury to plants by altering the balance between water absorbed by roots and that lost through transpiration. When air temperature rises, the rate of transpiration increases, resulting in possible adverse effects on the metabolic and physiological activities of the plant; this may have a critical effect on herbicide absorption and translocation. Respiration is low at lower temperatures and increases with an increase in temperature. High temperature causes rapid evaporation of water, causing water stress in the plant.

Humidity has considerable influence on the development of cuticle, transpiration and water stress. Relative humidity is more important than absolute humidity. Aqueous solutions enter the hydrated cuticle more easily. At high relative humidity, water content of the leaf is high, and this facilitates free movement of the herbicide in the apoplast and the symplast. High turgor pressure in the protoplasm at high humidity leads to more active protoplasmic streaming and more rapid translocation in the phloem sieve tubes. This explains the rapid absorption and translocation of foliar-applied herbicides under high humidity conditions.

An increase in temperature enhances absorption and translocation. Pallas [44] found that at temperatures of 20°C, 25°C and 35°C, less 2,4-D and benzoic acid was absorbed and translocated at lower humidities (34 to 48%) than at higher humidities. The increased absorption and translocation at higher humidity correlated with the degree of stomatal opening. Such an increase in absorption and translocation at higher humidity is due to increased phloem transport.

McWhorter and Wills [38] reported that a constant level of 40% to 100% relative humidity, an increase in air temperature from 22°C to 32°C resulted in a two- or three-fold increase in absorption and a four- to eight-fold increase in translocation of melfluidide in soybean. Similarly, at a constant temperature of either 22°C or 32°C, an increase in relative humidity from 40% to 100%, resulted in less than two-fold increase in absorption and translocation of melfluidide.

Rainfall

Rain occurring soon after application of foliage-applied herbicides reduces the effectiveness of the chemical. This is due to washing of the spray deposits off the foliage. Polar herbicides such as dalapon, 2,4-D salts, maleic hydrazide, etc., which are highly soluble in water, penetrate leaf surfaces very slowly, and can be washed off by rain occurring within 4 to 6 h of application. The absorption of other herbicides, such as paraquat and 2,4-D ester, which penetrate very rapidly, is not seriously affected in most species by rain falling even within 1 h of application. Generally, herbicides formulated in oil or oil emulsions are less affected by rain than the aqueous solutions.

Wind

Wind has a direct effect on evapotranspiration. It aggravates the susceptibility of a plant to a herbicide. Wind can damage the cuticle. At high temperature and low humidity, wind can accentuate the action of a contact herbicide like paraquat. Wind also causes rapid drying of spray solution on the foliage, resulting in reduced absorption.

Light

Light assists herbicide penetration by stimulating stomatal opening. Light also. activates photosynthesis, leading to greater movement of organic solutes and consequently herbicides from the leaf to other parts of the plant. The effect of light varies with duration and intensity. Photoperiod (daylength) is critical for the growth and flowering of many plant species. Variation in intensity of light may cause etiolation, stunting, reduction in flowering and other abnormalities. It also affects photosynthesis, transpiration and stomatal opening.

Soil Factors

Soil factors directly affect the movement of soil-applied herbicides absorbed by roots and shoots. They have an indirect effect on the foliage-applied herbicides through the vigour and growth of the plant.

Soil-Water Stress and Temperature

Transport of a herbicide along its pathway is a function of soil-water stress and soil temperature because of their effects on transport coefficients and herbicide solubility. Soil-water stress and temperature affect the permeability of roots to water and herbicide solution. Under high water stress and high temperatures, the rate of penetration of herbicide molecules will be considerably reduced. Soil-water potential affects transpiration, photosynthesis and root permeability, which in turn influence herbicide uptake, translocation and eventually phytotoxicity. Schrieber et al. [54] observed that a reduction in soil-water potential from –0.35 to –2.50 bars reduced the uptake of bromacil by roots of wheat seedlings, and at this reduced water potential, more of the herbicide absorbed remained in the roots rather than being translocated to shoots. Their study suggested that the rate of uptake was controlled by the effect of soil-water potential on the apoplastic movement of water and solutes in the root cells.

Water stress may also lead to an increased thickness of the cuticle, which results in reduced entry and translocation of foliage-applied herbicides.

Soil pH

Soil pH has profound effects on the uptake of herbicides by roots. Changes in soil pH affect the cation exchange capacity of soils. The ionic nature of certain herbicides is a function of pH. Many acid herbicides have pKa values lower than the pH of the soil to which they are applied, and they exist primarily in ionic form in the soil environment. Generally, herbicides exist in ionic form between pH 4.3 and 7.5, with an increase in ionic concentration being proportional to an increase in pH. The negatively charged functional groups in the soil predominate at pH above 5.3 and repel anionic form herbicides such as 2,4-D and dicamba.

Organic Form and Clay Type

Soil organic matter and type of clay predominant in the soil, which influence movement, availability and absorption of herbicides in the soil, also affect their uptake and transport. This aspect is covered in greater detail in Chapter 9.

Chemical Factors

Herbicide Concentration

Penetration of translocated herbicides has a relationship with their concentration. At supraoptimal concentrations, herbicides may cause physiological injury more rapidly, thus precluding the normal rate of absorption and translocation.

pH of Herbicide Solution

Hydrogen ion concentration plays a significant role in the penetration of foliage-applied herbicides. The effect of pH is dependent on the chemical nature of the herbicide. It has a modifying effect on the membrane potential and the metabolic activity of the cells involved in the uptake and translocation processes.

Chemical Structure

Molecular structure may affect absorption of a herbicide by the plant tissue. Herbicides vary in their lipid solubility. A modification of molecular structure may increase lipid solubility, resulting in greater penetration of herbicides.

Surfactants

Surfactants are non-herbicidal compounds, which enhance the absorption of aqueous solutions of polar herbicides. They reduce the surface tension between the plant surface and spray particles. This aspect is covered below.

ENHANCEMENT OF HERBICIDE ABSORPTION AND TRANSLOCATION

Surfactants

As discussed earlier, polar herbicides tend to be soluble in polar solvents and insoluble in non-polar solvents. When solutions of polar herbicides are sprayed on the waxy cuticle found on plant foliage, the spray droplets become spherical and are repelled, resulting in poor or inadequate wetting of the plant surface. Any substance that brings the polar herbicide into more intimate contact with the leaf surface will improve absorption, and this can be achieved if the chemical has affinity for both water and waxy cuticle. This binding agent is called a **surfactant**, whose primary function is to act as a surface-active-agent between two bodies, i.e. plant surface and spray particle. The surfactant molecule has both the **hydrophilic** (attracted to aqueous or polar sub-stance) and **lipophilic** (attracted to non-aqueous, non-polar or oil substances) por-tions. These surfactants reduce the surface tension between molecules of a liquid.

Surfactants are active at interfaces. The surfactant molecules with their dual nature tend to accumulate at interfaces or surfaces between two dissimilar groups. As a result, the balance of forces that hold a droplet of water on a flat surface can be drastically altered and the droplet then spreads out as a thin layer. The effect of a surfactant can be determined by measuring the angle of contact between a droplet and the surface. In an oil-water mixture, a drop of oil is surrounded by a film of closely packed

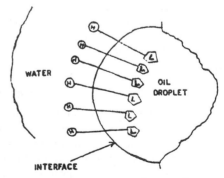

Fig. 6.3. Orientation of a surfactant molecule to an oil droplet at an interface

surfactant molecules, with the lipophilic portion of the surfactant molecule orienting towards the oil and the hydrophilic portion attracting the water. The orientation of lipophilic and hydrophilic portions of a surfactant molecule in the interfaces is shown in Fig. 6.3.

Surfactants provide varying degrees of lipophilic and hydrophilic tendencies. The balance of these two tendencies in a single molecule controls the character of a surfactant. This balance, known as **hydrophilic/lipophilic balance (HLB)** is expressed as an arbitrary number and is used to categorize surfactants. Surfactants with predominantly lipophilic character, i.e., low HLB, tend to promote water-in-oil emulsions, while those with stronger hydrophilic character, i.e. high HLB, promote oil-in-water emulsions.

There are several surfactants which can be used in the herbicide spray to enhance foliage retention, penetration, absorption and, consequently, translocation of postemergence herbicides. They are either non-ionic or ionic in nature, depending on their ionization and dissociation in water. Ionic surfactants may be anionic having negative charge, or cationic with positive charge. In hard water or acidic or alkaline systems, ionic surfactants can be grossly influenced by adverse ratios of ionized to un-ionized molecules as the materials in solution hinder the ionization process of the surfactant. Non-ionic agents have little or no ionization capacity in water. They are non-electrolytes and chemically inactive in the presence of salts. They promote tolerance to electrolytes in the herbicide solution and allow adjustment of the HLB factor. Anionic and non-ionic agents are widely used in agriculture. Generally, anionic agents perform better in cold water and best in soft water, while non-ionic agents work well in warm and hard water. These surfactants are also referred to as adjuvants.

Wetting and Spreading Agents

Wetting agents reduce the interfacial tension and bring one liquid into intimate contact with another. The efficacy of contact herbicides is dependent on complete and uniform wetting of a plant. Waxy leaf surfaces tend to repel water and the addition of a wetting agent to the herbicidal solution increases the adherence of spray particles to the leaf surface. Most wetting agents are also spreading agents, which improve spreading properties of water on the leaf surface. The spreading agents, which are non-ionic in nature, reduce the angle of contact (**contact angle**) between the liquid drop and the leaf surface. They are usually added at a concentration range of 15 to 50 ml per 100 L spray. When excessive volumes of spray are applied, wetting agents may sometimes increase spray run-off because of the very low surface tension and thinness of the wetting film created by spreading. Thus, herbicide selectivity may sometimes be lost by the addition of a wetting and spreading agent. This problem can be avoided by increasing the thickness of the film through use of thickening or viscosity-inducing agents.

Emulsifiers

Emulsifiers occupy the interface between oil and water surfaces by coupling them together, thus maintaining the stability of an emulsion. As a result, tiny dispersed droplets in a stable emulsion are prevented from coalescing. Emulsification is basically a surface action. Many wetting agents also serve as a stabilizer. Most herbicides supplied in liquid formulation are emulsions of oil-in-water type which when mixed with water have about the same viscosity as water.

Thickening and Sticking Agents

If the herbicide spray film is very thin and tends to run off the leaf surface easily, the addition of a thickening agent will be of great help. It improves the viscosity of the spray film and thus increases spray adherence to leaves and reduces the bounce or run-off from spraying. Many thickening agents are also sticking agents (stickers) which, however, do not have to be viscosity-inducers as long as they improve spray adherence to leaves. Some sticking agents are water-soluble polymers which on drying the leaf surface form a water-resistant film. Other sticking agents have both initial contact and run-off-prevention functions. A proper balance of spreader and sticker will markedly improve the activity of herbicides.

Penetrating Agents

Penetration agents improve plant absorption of a herbicide by solubilizing the waxy cuticle or lipoidal portion of the cell wall or membrane of the plant tissue so that penetration is more readily achieved.

Humectants

These substances also improve absorption of herbicides by raising the humidity between the spray film and leaf surface. They operate on the principle that higher humidity aids in rapid and greater absorption.

Dispersing Agents

Dispersing agents reduce cohesion between like particles. They promote separation of like particles and cause deflocculation. Good wetting agents also act as dispersing agents, although these two substances are sometimes incompatible and tend to interfere with each other's action if used together.

Fertilizer Additives

Addition of small quantities of fertilizer materials enhances absorption of aqueous herbicides. The most commonly used fertilizers are ammonium sulphate and dihydrogen phosphate. Fertilizers apparently solubilize the cuticle and increase the permeability across the leaf cells, thus causing enhanced penetration of the chemical. Addition of ammonium sulphate to the spray solution enhances glyphosate activity against *Elytrigia repens* [5, 63] and *Imperata cylindrica* [50], and picloram activity on the seedlings of guava (*Psidium guajava*) and dwarf beans (*Phaseolus vulgaris*) [68]. Similarly, the activity of 2,4-D sodium salt can be enhanced by tank-mixing ammonium sulphate or urea at 0.5% to spray solution [51].

ABSORPTION AND TRANSLOCATION OF INDIVIDUAL HERBICIDES

Acetamides

Alachlor, Propachlor

Both alachlor and propachlor are applied to the soil. They are readily absorbed by the roots and shoots and distributed acropetally throughout the plant parts. Germinating seeds also absorb these herbicides in significant quantities. Chandler et al. [11] found that root uptake of alachlor was greater for wheat, the susceptible species, than for soybean, the resistant species. Older leaves accumulated more alachlor than the younger and actively growing tissue in both species. Foliar application of alachlor to the primary leaves of soybean and wheat resulted in some apoplastic translocation within the leaf of wheat but only a slight translocation in soybean. Absorption of alachlor by the leaf and coleoptile tissue was greater in light than in darkness. These authors concluded that translocation of alachlor was primarily apoplastic, with greater uptake and translocation in wheat (susceptible) than in soybean (resistant).

The main site of uptake of alachlor in *Cyperus esculentus* is the portion slightly above the tuber. When applied to the foliage of *C. esculentus*, alachlor was not translocated in sufficient quantities to the growing point to control the weed [2]. In germinating and emerging seedlings of snap beans (*Phaseolus vulgaris*), root uptake and translocation of alachlor to the shoots were significantly greater at higher (27°C night-32°C day) than lower (16°C night-21°C day) temperature regimes [52].

Propachlor is absorbed primarily by the shoots of germinating seedlings and secondarily by roots. It is translocated acropetally and accumulates throughout the plant, with higher concentrations in vegetative parts than in reproductive parts.

Acetochlor, Butachlor, Dimethenamid, Metolachlor

These herbicides, essentially soil-applied herbicides, are absorbed primarily by the emerging plant shoots, including grass coleoptile, broadleaf hypocotyl or epicotyl. They are also absorbed by the roots of emerging plants. When absorbed by roots, these herbicides are translocated acropetally throughout the shoots, with greater accumulation in vegetative plant parts and lesser accumulation in reproductive parts. As they are phytotoxic only to emerging seedlings, their translocation in established plants is irrelevant to their mechanisms of action.

Propanil

Propanil, a postemergence herbicide, is readily absorbed by leaves. Its translocation from the leaf is quite limited, but when it moves it goes to growing points and then to the other leaves.

Pronamide

Pronamide, a preemergence herbicide, is readily absorbed by plants through the root system, translocated upwards apoplastically and distributed throughout the entire plant. Initial root absorption is a physical process, followed by its dependence on the metabolic processes. Very little pronamide moves through the foliage, but when it does it moves predominantly acropetally in the apoplastic system. The primary reason for restricted foliar activity is lack of cuticular penetration. Placement of pronamide is more phytotoxic to *Elytrigia repens* (quackgrass) when placed in the seed zone; it does not move appreciably within rhizomes after absorption [10].

Aliphatics

Dalapon

Dalapon, primarily a foliage-applied herbicide, is readily absorbed by leaves and translocated symplastically via the phloem along with photosynthates throughout the plant, with greater accumulation in young and meristematic tissues. It also moves acropetally in the apoplastic pathway. It is readily washed off the foliage by rain occurring within 4 to 6 h of application. Dalapon is also readily absorbed from the soil by roots and translocated via the transpiration stream apoplastically and also symplastically. It moves from one system to the other relatively freely. It enters both leaves and roots at a rapid rate initially, followed by a slow and steady rate controlled by the metabolic processes. It accumulates in young tissues. Absorption and translocation of dalapon is dependent on air temperature, relative humidity, light intensity and age of the plant. Surfactants enhance absorption and movement of dalapon.

Arsenicals

DSMA, MSMA

Both these arsenical herbicides, applied to the foliage, move apoplastically as well as symplastically. They apparently follow the source-to-sink pattern, with the amount of herbicide moving being independent of the sink activity. DSMA moves in both acropetal and basipetal directions, but the latter movement is much slower. Duble et al. [19] found that movement of DSMA within a single leaf of *Cyperus rotundus* (purple nutsedge, nutgrass) was not influenced by relative leaf age. It was not redistributed in the shoot or in the tuber system but accumulated in terminal tubers after repeated applications. MSMA, when applied to the foliage, moved primarily in the xylem acropetally, but small amounts moved basipetally. It was translocated rapidly to both the root system and shoot tip. Temperature apparently has a significant effect on translocation of MSMA and DSMA.

Benzamides

Isoxaben

Isoxaben, also a soil-applied herbicide, is readily absorbed by the roots through passive diffusion. Leaf absorption is limited, as only 3% of foliar-applied isoxaben entered the leaves of the highly sensitive *Amaranthus retroflexus* (redroot pigweed) even 3 d after application. When absorbed by leaves, isoxaben moves symplastically in the phloem.

Benzoics

Dicamba

Dicamba, applied to the foliage, is absorbed by leaves and translocated basipetally and acropetally. It accumulates in meristems or young developing parts, indicating symplastic translocation. It is mobile in both the xylem and phloem. Baradari et al. [4] reported that 32% of the foliar-applied dicamba was absorbed; most of it tended to accumulate in young growing leaves. When applied to the soil, dicamba enters roots and moves upwards in the apoplast, and the rate of this transport sometimes exceeds the rate of symplastic transport. Thus, dicamba is readily absorbed by leaves and roots and translocated via both the symplastic and apoplastic systems.

Benzothiadiazole

Bentazon

Bentazon, a foliage-applied herbicide, is absorbed by leaves and translocated primarily acropetally. Wills [66] reported that translocation was affected by leaf maturity, with the greatest movement seen when applied to the mature leaf near the base of the shoot in the case of *Xanthium pensylvanicum* (common cocklebur) and at the youngest fully expanded leaf near the apex of the shoot in the case of soybean. Translocation was greater at high temperature (35°C) and high humidity (96%) than at low temperature (25°C) and low relative humidity (35%). Stomatal penetration plays a significant role in the entry of bentazon into the leaf.

Bipyridiliums

Diquat, Paraquat

Bipyridiliums are contact herbicides applied to the foliage. Their absorption by leaves is so rapid that none is washed off by rain even within 1 h of application. Absorption is physical, involving an ionic attraction between cations of the herbicide and negative charges on the leaf surface.

Diquat and paraquat do not move out of treated leaves in darkness but when the plants are transferred to light they are distributed, via the xylem, throughout the aerial portion. These herbicides are absorbed by leaves into the apoplast, but are not translocated during the dark period. However, when the plants are exposed to light, the herbicides get activated and kill the tissue. The chemicals already accumulated during the dark period are carried with water apoplastically and distributed. Putnam and Ries [49] found greater movement of paraquat in the leaves of *Elytrigia repens* under 6 h of light compared to 6 h of darkness following application. However, a 6 h dark period after application prior to the light period further enhanced movement from the treated leaf.

Relative humidity has a significant effect on the absorption of bipyridilium herbicides. An increase in relative humidity from 50-55% to 93-94% enhances paraquat and diquat uptake 2- to 5-fold. This is attributed to hydration of the cuticle at high relative humidity.

Carbamates

Asulam, Chlorpropham, Propham

Asulam, a postemergence herbicide, is readily absorbed by the foliage and translocated in the phloem symplastically to the growing regions of the plant. Addition of a wetting agent increases the rate of absorption. From the leaves, asulam also moves into the root system, resulting in the death of several dormant buds on the rhizomes. Asulam is also taken up by roots when applied to the soil. Veerasekaran et al. [64] reported that translocation of asulam from the treated fronds of *Pteridium aquilinum* was primarily according to a source-to-sink pattern, with intense accumulation in the metabolically active sinks, viz. Rhizome apices, frond buds, root margin tips and young frond tissue.

Chlorpropham and propham, primarily applied to the soil, have low water solubility, and are rapidly absorbed from the soil. Chlorpropham enters the seed during germination by the physical accumulation process and the penetration is more rapid and independent of water uptake. In some plant species, the seed coat may become a barrier for herbicide penetration. Chlorpropham, taken up by the roots, is distributed apoplastically to all plant parts. Propham, absorbed by the roots and germinating seeds, is translocated throughout the plant. In soybean, it is absorbed by the roots and translocated acropetally [57]. Propham differs in absorption rates and distribution patterns in different plant species [9].

Desmedipham and Phenmedipham

These postemergence carbamate herbicides are absorbed by leaves, and rain falling immediately after application may reduce their effectiveness. Absorption by leaf is quite rapid initially. They move in the direction of transpiration current. The rate of absorption of desmedipham and phenmedipham increases with increasing light intensity and temperature.

Cineoles

Cinmethylin

Cinmethylin, applied primarily to the soil, is absorbed by the roots and shoots of germinating or emerging weeds. It is translocated upward in the xylem.

Cyclohexanediones (Cyclohexenones)

Clethodim, Cycloxidim, Sethoxydim, Tralkoxydim

Clethodim and cycloxidim, the foliage-applied herbicides, are rapidly absorbed into the foliage and translocated symplastically. They eventually accumulate in meristematic regions.

Sethoxydim and tralkoxydim are rapidly absorbed by the roots and leaves, particularly when applied with an oil or other adjuvant. They move across the plasmalemma by passive diffusion. The relatively acidic environment outside the cell allows a significant proportion of the herbicide to remain in the protonated (undissociated) form, which readily diffuses across the plasmalemma into the cell. Once inside the cell, these herbicides translocate both in the phloem and xylem, although primarily in the phloem. They accumulate in meristematic areas of shoots and roots, but the rate of translocation out of the treated leaves is low and the extent of translocation is limited.

Dinitroanilines

Benefin

Benefin, a soil-incorporated herbicide, is absorbed primarily by the emerging shoots (grass coleoptile or broadleaf hypocotyl or epicotyl) and secondarily by seedling roots. Benefin vapours may be adsorbed by the foliage and cotyledons. Plants beyond the seedling stage readily absorb benefin into the roots but acropetal translocation to shoots is limited. Benefin is highly lipophilic and is rapidly partitioned into membranes and other lipid contents of root tissue, thereby limiting translocation. It accumulates more in the roots than in the shoots and this accumulation is linear with time. In resistant species, benefin accumulates in moderate amounts in roots.

Dinitramine, Fluchloralin, Isopropalin, Profluralin

All these herbicides are incorporated into the soil at preplanting or preemergence. Dinitramine and profluralin are absorbed by the roots. But profluralin gets accumulated in the roots while dinitramine moves and gets accumulated in the upper stem tissues and leaves [30]. Hawxby and Basler [30] reported that profluralin was found accumulated in the roots to a greater extent at 16° C than at 38° C; variation in the temperature had no effect on its movement from the roots. Conversely, dinitramine accumulated in the tips of plants to a great extent at 38° C than at 16° C. However, Jacques and Harvey [31] found profluralin and fluchloralin, which were more volatile than dinitramine, accumulated in the shoots when roots were exposed to their vapours. For isopropalin, there was no significant absorption or translocation.

Trifluralin

Trifluralin, incorporated into the soil at preplanting and preemergence, is absorbed by the roots and emerging shoots of germinating seedlings. Shoot exposure inhibits shoot growth more than root exposure, indicating greater absorption by emerging shoots. Movement of trifluralin within the plant is very limited. Strang and Rogers [59] reported that trifluralin accumulated in the roots of cotton and soybean seedlings, due primarily to adsorption by the cuticle, epidermis and, to a lesser extent, cell walls, rather than active accumulation within the living cells. The epidermis was considered to be the major site of entry. Generally, upward translocation is related to the amount of herbicide that can enter the xylem. Trifluralin may actually circulate in the apoplastic and symplastic transpirational systems and accumulate in the areas of high lipid content, i.e. cotyledons [34]. Trifluralin is normally applied to the plant foliage, and wetting agents show little enhancement of postemergence activity.

DiphenylEthers

Acifluorfen, Bifenox, Fluoroglycofen, Fomesafen, Lactofen

Acifluorfen, a foliage-applied herbicide, is readily absorbed by leaves of most plant species. Very little of the foliage-absorbed acifluorfen is translocated basipetally. Root absorption and acropetal translocation to leaves is limited. Bifenox, a soil-applied herbicide, is taken up rapidly by the roots as well as leaves. Foliar application is not easily removed by rainfall. It is translocated apoplastically following either foliar or root application. Fluoroglycofen is rapidly absorbed by the foliage but translocated very little following either root or foliar application. Fomesafen and lactofen are also rapidly absorbed into leaves within 1 h of application. They primarily move in the xylem.

Oxyfluorfen

Oxyfluorfen, primarily a soil-applied herbicide, is readily absorbed by the roots but low relative humidity may reduce absorption. Very little oxyfluorfen is absorbed by the shoots. There is very little basipetal and acropetal translocation following root or shoot absorption. Fadayomi and Warren [20] found little movement of oxyfluorfen from the point of application on leaves.

Imidazolidinones

Buthidazole

Buthidazole, when applied to the soil is absorbed by the roots and emerging coleoptiles and transported to leaf tips apoplastically. Haderlie [25] found that buthidazole was absorbed by the germinating soybean and maize, with the rate of absorption increasing when roots were capable of absorption. Foliar absorption of buthidazole in soybean greatly increased when a non-ionic surfactant was added to the spray solution. There was little or no movement of buthidazole from the treated leaves. Translocation was typical of apoplastic movement. Hatzios and Penner [28] reported that following foliar application, buthidazole moved toward the tip of the treated leaves in maize, a tolerant species, and moved both acropetally and basipetally in Amaranthus retroflexus (redroot pigweed), a sensitive species. They found that buthidazole was rapidly taken up by the roots and translocated to leaves. Hatzios and Penner [29] concluded that buthidazole was absorbed similarly by the leaves and roots of both the tolerant alfalfa and the susceptible Elytrigia repens (quackgrass) and that translocation was evidently apoplastic in both species.

Imidazolinones

Imazapyr, Imazaquin, Imazethapyr, Imazamethabenz

Imazaquin, applied to soil and foliage, is rapidly absorbed by both roots and foliage, with most rapid uptake occurring via the roots. Differential absorption occurs among species, accounting for their tolerance and susceptibility. It is translocated in both xylem and phloem but reduced translocation may be responsible for decreased susceptibility of some weed species. Decreased translocation may be due to its metabolism to non-mobile metabolites.

Imazethapyr, like imazaquin, is rapidly absorbed by the foliage but very slowly by roots. It is translocated in both the xylem and phloem. Ballard et al. [3] found that imazethapyr has both xylem and phloem mobility by translocating both basipetally and acropetally. But a higher percentage of the absorbed imazethapyr accumulated in the lower foliage and roots of the susceptible *Ambrosia trifida* (giant ragweed), than in the tolerant *Ambrosia artemisiifolia* (common ragweed). Imazapyr and imazamethabenz, essentially postemergence herbicides, are rapidly absorbed by shoots and roots. Imazamethabenz methyl ester is highly xylem mobile. Imazamethabenz acid is produced in appreciable amounts only in susceptible weeds and accumulates in meristematic regions.

Imines

CGA 248757

This imine herbicide, applied at postemergence, is absorbed by the foliage and translocated very little basipetally.

Isoxazolidinones

Clomazone

Clomazone, when applied to the soil, is readily absorbed by the roots and emerging shoots (coleoptiles in grass weeds and hypocotyls in broadleaf weeds) and is translocated in the xylem to the foliage. When applied to the foliage, clomazone is poorly absorbed and poorly translocated in the phloem.

Nitriles

Bromoxynil, Ioxynil

Bromoxynil and ioxynil, both applied at postemergence, are readily absorbed by the foliage. They move apoplastically but very slowly. Ioxynil moves down to the base of the petiole of the treated leaf within 5 h, and with increase in time the herbicide moves up the stem and gets accumulated in young leaves at the apex, bypassing the mature leaves [16]. This suggested symplastic movement in sieve tubes. When ioxynil was applied to the roots, there was only slight apoplastic movement to the foliage. Bromoxynil, like ioxynil, has little or no basipetal movement to other plant parts. Bromoxynil and ioxynil have localized contact activity. Retention of spray by the leaves is a major factor in determining selectivity of plant species to ioxynil. Addition of a surfactant increases retention and eliminates selectivity. Ioxynil esters penetrate the leaves more readily than ioxynil salts. Thus, the efficacy of esters is less affected by adverse environmental conditions.

Oxadiazoles

Oxadiazon

Oxadiazon, primarily a preemergence herbicide, is readily absorbed by the shoots of emerging seedlings, but less so by the roots. It is translocated apoplastically and distributed in plant parts. When applied to foliage, it is readily absorbed by leaves. It accumulates in older plant parts as also in roots, rhizomes and buds, with little movement to growing points.

Phenols

Dinoseb

Phenols are contact herbicides, whether applied at preemergence or postemergence. Their rapid contact action destroys the plant organs responsible for apoplastic and symplastic translocation. The preemergence treatment of dinoseb destroys the roots even before the foliage and other parts of the plant are developed. Its movement in the xylem is limited. Environmental factors such as sunlight, high temperature and high humidity hasten contact injury and death of the plant. Dinoseb is non-polar and hence, can penetrate the cuticle.

Phenoxyalkanoic Acids

Phenoxyacetics – 2,4-D, MCPA; Phenoxybutyrics – 2,4-DB, MCPB

Phenoxyacetic and Phenoxybutyric herbicides are primarily postemergence herbicides, but they are active when applied at preemergence as well. Hence, they are absorbed by leaves, stems and roots.

When applied to the foliage, they penetrate the cuticle by diffusion. Since cuticle surface is lipoidal in nature, the sodium and potassium salts of phenoxy herbicides which are polar in reaction and which have low lipid solubility penetrate poorly. But when the pH is lowered, the herbicide molecules get less dissociated and become penetrable. The low pH also lowers the polarity of the cuticle, which then becomes more penetrable. Similarly, addition of surfactants, oil additives, and inorganic ions such as ammonium, nitrate and phosphate, promotes the absorption of sodium and potassium salts of phenoxy herbicides. On the other hand, the esters of phenoxy herbicides penetrate the cuticle very rapidly without assistance from these non-herbicidal compounds.

Once the herbicide has penetrated the cuticle, it moves inside the plant both apoplastically and symplastically. During symplastic transport, the chief pathway of translocation of phenoxy herbicides, the chemical moves with the stream of assimilates from source (photosynthesizing leaves) to sink (sites of active growth in roots, shoots, flowers and fruits). The best effect of the herbicide is observed when it is applied during the periods of vigorous growth when the chemical is drawn with assimilates into the sink. When root growth is inhibited, little movement to roots takes place. Generally, phenoxy herbicides accumulate in the apical meristems.

Increase in temperature has a beneficial effect on the absorption of phenoxy herbicides due to disorganization of the lipid materials on the cuticle and consequent increased membrane permeability. A similar beneficial effect is also obtained at high humidity, which helps in opening of stomata.

Absorption of phenoxyacids into green and succulent stems is similar to absorption into leaves. For aqueous solutions, woody stems are, however, difficult to penetrate. This can be circumvented by using oil as a carrier or by making a cut across the bark and applying the aqueous solutions of the herbicide. Translocation is in the xylem, mainly acropetally, with little downward movement in the phloem.

Aqueous solutions of phenoxyacids are readily absorbed by the roots when applied to the soil. Root absorption requires metabolic energy. Inside the root, the herbicide moves through the external apoplast and into the symplast. Within the symplast, the herbicide migrates from the epiderm cells through the cortex and endodermis into the stele, where it leaks into the symplast. When applied to roots, phenoxy herbicides move poorly than when applied to the foliage, apparently because of greater retention and accumulation by living cells in their passage through the symplast.

Aryloxyphenoxy propionics (APPs) — Dichlorprop, Diclofop, Fenoxaprop, Fluazifop-P, Haloxyfop, Quizalofop-P

Aryloxyphenoxy propionics (APPs), in ester form, are rapidly absorbed by the roots and leaves when applied at preemergence and postemergence. They diffuse readily across the plasmalemma. The acid, by poorly absorbed by the foliage, accumulates in cells by ion trapping, typical of weak acid herbicides (62). This shows that the protonated acid diffuses through the membrane while the dissociated anion is excluded. Once inside the higher pH environment of the cytoplasm, the protonated form of APP herbicides dissociates to the anion and is consequently trapped inside the cell due to its inability to diffuse back through the non-polar membrane.

The ester forms of APP herbicides are lipophilic and hence strongly adsorbed by the epicuticular wax, showing desorption into the apoplast and making removal by leaf wash (by rain) difficult. Various adjuvants, which temporarily dissolve epicuticular waxes, may enhance adsorption to epicuticular waxes, increasing rain-fastness, while slowing movement to the apoplast.

Humid and warm conditions generally favour uptake of APP herbicides by the foliage of certain grass species. Fluazifop absorption by *Cynodon dactylon* (Bermudagrass) is greater at 35° C than at 27° C, while *Sorghum halepense* (Johnsongrass) is unaffected.

Although absorption of APP herbicides by the roots and foliage is good, translocation is considerably reduced and varied. Boldt and Putnam [6] studied translocation of diclofop in five species and found that less than 2% of ^{14}C–diclofop in treated leaves was translocated over a five-day period. Most of the herbicide moved acropetally. Gillespie and Miller [23] reported that less than 2% of diclofop-methyl moved from treated sunflower leaves over a 10-day period, with equal amounts of the herbicide moving to roots, to the shoot above the treated leaf and to the shoot below the treated leaf. Nalewaja et al. [40] found an increase in translocation of diclofop in oats during the three days after application, from about 10% of the absorbed quantity to about 30% and 20% of the absorbed quantity after using petroleum oil and methylated sunflower oil respectively as adjuvants.

Haloxyfop-methyl, like other APP herbicides, though rapidly absorbed, is poorly translocated. Harrison and Wax [27] reported that corn leaf absorbed 43% of the herbicide 5 h after treatment, but only 3% of this quantity was translocated from the treated leaf area. Within the treated leaf, movement was primarily acropetal. However, Whim et al. [65] found almost 50% of the foliar-applied haloxyfop to be translocated within 96 h after treatment.

Translocation of fluazifop-butyl appears to be better than other APP herbicides. The pyridinyloxyphenoxy compounds (of APP group) to which fluazifop-butyl belongs, translocate better than the phenoxyalkanoic analogues which include diclofop. Kells et al. [33] found that up to 25% of the absorbed fluazifop-butyl was translocated, with little difference in translocation in resistant soybean and susceptible quack grass. Most of the herbicide accumulated in the meristem, much of it being in fluazifop acid form and none as fluazifop-butyl. Nalewaja et al. [40] observed that 60% of fluazifop absorbed by oats was translocated both acropetally and basipetally.

The relatively poor translocation of most APP herbicides may be the result of a rapid cellular uptake of the cells and most proximal to the site of application, thus reducing the level of herbicide available for translocation. Generally, lipophilic compounds such as APP esters, though penetrate the foliage readily, are not phloem-mobile and they cross the plasmalemma easily. Once in the symplast, the herbicide is metabolized through conjugation or other alteration mechanism (hydrolysis or demethylation), depending on the plant species and the type of herbicide, resulting in lowering mobility. The de-esterified form, which is less lipophilic, may be the more mobile form of most APP herbicides.

N-Phenylphthalimides

Flumiclorac

Flumiclorac, an N-phenylphthalamate herbicide applied to the foliage, is readily absorbed by the leaves, with little or no basipetal translocation in the phloem into the roots. It has rain-fastness of one hour after application.

Phenylpyridazines

Pyridate

Pyridate, applied to the plant, is rapidly absorbed by foliage, most of it (90%) within 6 h of application. However, its translocation in the phloem is poor. Efficacy of pyridate is unaffected by rain occurring one hour after application.

Phenyl Triazinones (Aryl Triazinones)

Sulfentrazone

Sulfentrazone, a soil-applied herbicide, is readily absorbed by the roots and translocated to the foliage [17]. It moves upwards in the xylem. Dayan et al. [17] reported that there was a difference in uptake and translocation between the suscepti-ble *Cassia occidentalis* (coffee senna) and the relatively tolerant *Senna obtusifolia* (sicklepod). Translocation of root-absorbed sulfentrazone was similar in both species.

Pyrazoliums

Difenzoquat

Difenzoquat is rapidly absorbed by the foliage following postemergence application. It moves primarily in the xylem and consequently translocation is very limited following foliar absorption.

Pyridazines

Metribuzin

Metribuzin, primarily a preemergence herbicide, is readily taken up by the roots through diffusion. It is translocated upwards apoplastically. Its accumulation is highest in roots, stems and leaves, and lowest in fruits and seeds. When applied at postemergence to the young leaves, it moves upward distally. Downward movement apparently does not occur.

Pyridazinones

Norflurazon, Pyrazon

Norflurazon and pyrazon, the preemergence herbicides, are readily absorbed by roots and distributed throughout the plant via the xylem. They are also absorbed by leaves when applied at postemergence, but are not well translocated symplastically. Foliar absorption is increased by a surfactant or oil adjuvant.

Pyridinecarboxylic Acids

Clopyralid, Picloram, Triclopyr

Clopyralid, a postemergence herbicide, is readily absorbed by the foliage and roots. Hall and VandenBorn [26] reported that in sunflower and rapeseed, 97% of the foliar-applied clopyralid was absorbed within 24 h of application. The parent acid of clopyralid is more rapidly absorbed than either the ester or salt forms. Under conditions of low humidity and water stress, absorption of monoethanolamine and K salts is greatly reduced, while the ester and acid forms are unaffected. Herbicide uptake across plant membrane occurs by diffusion of parent acid. Clopyralid is readily translocated primarily via the symplasm, with over 50% of the applied herbicide moving out of the treated leaves of *Cirsium arvense* (Canada thistle) within 24 h of application. It accumulates primarily at the growing points. Bovey et al. [7] found that the salt forms of clopyralid translocated less than the parent acid, but twice as much as the esters.

Picloram, applied at postemergence, penetrates the foliage readily, with 97% of the herbicide being absorbed within 24 h of application [26]. It moves across the plasmalemma through an active protein-mediated process and passive diffusion. After its entry, picloram is translocated primarily by the symplastic pathway, eventually accumulating at the growing points. When taken up by roots, picloram moves to the top apoplastically and accumulates in shoot tips and young leaves.

Triclopyr, a foliage-applied herbicide, readily penetrates the foliage. The butoxyethyl ester is more rapidly absorbed than the acid form. Triclopyr also is readily absorbed by roots. It is translocated primarily via the symplastic pathway, accumulating at the growing points. Lewer and Owen [37] reported that between 40% and 67% of triclopyr penetrating the foliage of barley, wheat and *Chenopodium album* (common chickweed) moved out of the treated leaf within 3 d of application. About 3.6% of the foliar-applied triclopyr accumulates in the roots of *Solanum carolinense* (horsenettle) [24].

Pyridines

Dithiopyr, Thiazopyr

Dithiopyr and thiazopyr, the preemergence herbicides, are primarily absorbed by the roots and to a certain extent by the foliage of susceptible plants. Their translocation is limited but they move symplastically and accumulate primarily in meristematic tissues.

Pyridinones

Fluridone

Fluridone, an aquatic systemic herbicide, is absorbed from water by plant shoots and from the hydrosoil by roots. In cotton, a tolerant species, fluridone is taken up by the roots but with little or no translocation into the shoots. In susceptible species, root-absorbed fluridone is translocated readily into the shoot.

Pyrimidinylthio-benzoates (Benzoates)

Pyrithiobac

Pyrithiobac is absorbed following soil and foliar application. It is translocated within the phloem of treated leaves.

Quinolinecarboxylic Acids

Quinclorac

Quinclorac is absorbed by both leaves and roots. Lamoureux and Rusness [36] found that 20% of the absorbed herbicide remained in the treated leaves for up to 3 wk following treatment and the remainder was translocated basipetally. Once absorbed by the leaves or roots, quinclorac translocates rapidly, but its levels build up more slowly following soil application compared to foliar application. This is probably due to its lower rate of absorption by the roots. The young leaves and apex are strong sinks for quinclorac, regardless of the route of uptake. The root is not an effective sink for quinclorac absorbed through the leaves.

Sulfonylureas

Chlorsulfuron, Primisulfuron, Sulfometuron

Sulfonylurea herbicides are readily absorbed by both the foliage and roots. They are translocated extensively in the xylem, and less so in the phloem, after foliar absorption. When applied to the foliage, sulfonylureas translocate more to shoot-growing points and less to the roots, while root-applied herbicides translocate efficiently to all parts of the shoots, including meristematic regions. Sulfonylureas accumulate predominantly in the meristematic regions.

The uptake of primisulfuron is more rapid than chlorsulfuron, while translocation of nicosulfuron is much faster than primisulfuron [43]. Bruce et al. [8] reported that foliar absorption of nicosulfuron by *Elytrigia repens* (quackgrass) was greater in one-leaf than five-leaf plants, while translocation was similar regardless of growth stage. Its absorption, translocation and accumulation increased with increasing temperatures.

Sulfometuron is readily absorbed by leaves following postemergence application and into the roots from the treated soil. It is translocated in both the xylem and phloem, although not extensively. It moves in the xylem by mass flow along with phloem solutes, accumulating in the meristematic areas of the plant.

Thiocarbamates

EPTC, Butylate, Diallate, Triallate, Pebulate, Thiobencarb

EPTC, a soil-incorporated herbicide, is readily absorbed by the roots and coleoptiles and translocated both upwards and downwards. It is also absorbed by germinating seeds under wet soil conditions. When absorbed by roots, the major site of uptake, EPTC moves upwards to the foliage and is uniformly distributed, with greater accumulation in the youngest tissue. When absorbed by the coleoptiles, it is translocated downwards to the roots. Prendeville et al. [48] placed EPTC in different shoot zones below the soil surface after emergence of barley, oats, wheat and sorghum, and found that wheat, oats and barley were severely injured when treated only at the coleoptilar node. Sorghum was severely injured regardless of the shoot zone exposed. Uptake of EPTC from the soil by the susceptible sorghum was double that of the tolerant wheat.

Diallate and triallate, also soil-incorporated herbicides, are absorbed primarily through the emerging coleoptiles. Application of diallate and triallate 10 to 15 mm above the coleoptilar node caused maximum retardation of growth [45]. Nalewaja [39] reported that when diallate was applied to the coleoptile tips of wheat, barley and wild oat seedlings, it moved through the coleoptiles into the roots. When the shoots were treated, there was acropetal movement to the coleoptile tips and basipetal movement to root tips. Translocation took place primarily in the apoplast.

Pebulate, also applied to the soil, is quickly absorbed by the root system and translocated upwards to the foliage. The amount of uptake from soil increases with the time of germination and is proportionately more with increasing concentration.

The absorption and translocation of butylate and thiobencarb are similar to EPTC and pebulate.

Triazines

Atrazine, Ametryn, Propazine, Prometon, Prometryn, Simazine, Terbutryn

Triazines are primarily soil-applied herbicides. They are readily absorbed by the roots and translocated to leaves. They move rapidly to the shoot via the apoplast. They concentrate first in the veins, then in the interveinal areas and finally in the margins of leaves. The rate of absorption and translocation from roots to shoots is proportional to the amount of water and/or rate of translocation. The inhibition of transpiration may result in reduced absorption and translocation. Grass species, which are less susceptible to triazines, progressively accumulate the herbicide from the tip to the base of the leaf blade following root application. The more susceptible broadleaf species accumulate the herbicide at the margins of the leaves. Generally, resistant species accumulate a far less quantity of the herbicide than susceptible species. Sikka and Davis [56] reported that tolerance of cotton to prometryn was due in part to the poor translocation and accumulation in the lysigenous glands. When applied to the foliage, triazines penetrate and move very poorly, causing contact injury. The waxy leaf surfaces act as a barrier to uptake. Foy [21] studied the foliar absorption and translocation of triazines by several crop species, and found that foliar penetration and acute toxicity appeared to be correlated with the water solubility of the compound, and there was no appreciable movement of any of the herbicides downwards or out of the treated leaf in any of the species. All the compounds moved apoplastically. Addition of a surfactant tended to enhance their activity.

Terbutryn, also a preemergence herbicide, is rapidly absorbed by the roots and translocated acropetally in the xylem. It accumulates in the apical meristems. Terbutryn penetrates the foliage rapidly, minimizing removal by rain, and accumulates in the growing regions. It is also absorbed by germinating seeds. Rubin and Eshel [53] reported that seeds of snap beans (*Phaseolus vulgaris*), the susceptible species, absorbed a larger amount of terbutryn from the soil than did seeds of cotton, the tolerant species. The herbicide accumulated mainly in seed coats (of cotton), which are shed following emergence, with no translocation to the seedling leaves. In snap bean seeds, the herbicide accumulated in the cotyledons, which served as a source of acropetal movement to the developing foliage of seedlings.

Triazoles

Amitrole, Amitrole-T

Both herbicides are applied exclusively to the foliage. Amitrole is readily absorbed by leaves and translocated, within the plant, via the symplastic and apoplastic routes. Following foliar application, amitrole moves either up or down in the stem in the living cells of phloem. Surfactants, paraffin oil and high relative humidity increase foliar absorption and movement of amitrole. Amitrole also moves to the underground rhizomes and tubers of perennial weeds such as *Cyperus rotundus, Sorghum halepense*, etc. Coupland and Peabody [13] found that amitrole was translocated to areas of high meristematic activity such as shoot and rhizome apices and rhizome nodes in *Equisetum arvense* (field horsetail). Addition of ammonium thiocyanate (NH$_4$SCN) to amitrole reduces binding of the herbicide at the site of application. This formulation, amitrole-T, is absorbed and translocated faster than amitrole.

Triazolopyrimidine Sulfonanilides

Flumetsulam

Flumetsulam, a preemergence herbicide, is primarily absorbed by the roots, with some absorption by the emerging shoots. It readily moves apoplastically from roots to shoots and from shoots to roots.

Uracils

Bromacil, Terbacil

Uracils, soil-applied herbicides, are more readily absorbed by the roots, and less so by the foliage and stems. In sensitive species, they rapidly translocate from the roots to the leaves, where they are uniformly distributed in the mesophyll. In tolerant species, they are retained in the roots and their rate of accumulation is slower; most of the herbicide concentrates mainly in and around the veins. Translocation occurs acropetally in the xylem. In perennial grasses, uracils also translocate to the rhizomes, thus inhibiting regeneration. Gardiner et al. [22] found that 83% of the absorbed bromacil remained in the roots, while only 17% moved into the stems and leaves. Jordan and Clerx [32] reported that pineapple and sweet orange (*Citrus sinensis*) roots absorbed twice as much bromacil, and accumulated three times as much in the leaves, as did *Cleopatra mandarin* (*Citrus reticulata*). Uracils can penetrate the foliage when applied in the isoparaffinic oil or when a surfactant is added to the spray solution.

Ureas

Diuron, Fluometuron, Isoproturon, Linuron, Methabenzthiazuron

Diuron, which is primarily applied to the soil, is absorbed by the roots but its movement is restricted to the apoplast. Very little diuron is absorbed by the emerging shoots. It does not accumulate in the phloem. The tolerance of cotton is attributed to accumulation and binding of marked concentrations of the herbicide in lysigenous glands of cotton leaves and in trichomes [58]. Fluometuron is also actively taken up by the roots and translocated apoplastically rather rapidly with varying degrees of distribution patterns. It is primarily confined to the interveinal areas, with very little remaining in the veins. Differential translocation is partly responsible for tolerance and susceptibility of a plant species.

Linuron and isoproturon are most readily absorbed by the root system and rapidly translocated to the shoots following application to the roots. Root absorption is by passive uptake. Foliar absorption of linuron is significantly greatly than that of diuron, monuron and other ureic herbicides. Linuron inhibits the growth of *Setaria faberi* (giant foxtail) when the herbicide is placed in the shoot zone while no inhibition is caused when the chemical is placed in the root zone [35]. When methabenzthiazuron is applied to the foliage, it is not translocated away from the area of application in significant amounts.

Unclassified Herbicides

Bensulide

Bensulide, a soil-applied herbicide, is adsorbed on root surfaces and hence only a small amount is absorbed by the roots. Very little upward translocation occurs as it inhibits root growth.

Ethofumesate

Ethofumesate, when applied at preemergence, is readily absorbed by the emerging shoots and roots and readily translocated to the foliage. When applied postemergence, it is poorly absorbed by maturing leaves with a well-developed cuticle.

Fosamine

Fosamine, a postemergence herbicide, is absorbed slowly by the foliage, with less than 50% of the applied herbicide absorbed in 7 to 8 d after application. Its penetration is good when applied to young stems. Retention and penetration of fosamine is reduced when it is applied to hairy leaf surfaces. Surfactants increase absorption of fosamine. The slow rate of absorption increases potential for wash-off by rainfall.

Glufosinate

Glufosinate, a postemergence herbicide, is absorbed by the roots under field conditions because of rapid microbial breakdown. It is readily absorbed by the foliage but its translocation in the xylem and phloem is limited.

Glyphosate

Glyphosate is a postemergence herbicide absorbed by the leaves and stems. Gross morphology of the target plant may play an important role in glyphosate activity by influencing the amount of herbicide intercepted and retained. The plasmolemma, a fluid bilayer of phospholipids, is an important barrier in the absorption of glyphosate, which is highly hydrophilic. Glyphosate is absorbed passively by diffusion.

Glyphosate is readily and rapidly translocated throughout the actively growing aerial and underground portions of the plant. The underground plant parts like rhizomes, tubers, etc. of perennial species are affected, resulting in failure of regrowth from these propagation sites and subsequent destruction of the plant tissue. It moves in the phloem through the symplast both acropetally and basipetally, following a source-to-sink pattern. Although glyphosate transports through the apoplast, symplastic transport is apparently the main route. Symplastic translocation, from mature leaves, is to both shoots and roots. In perennials, carbohydrate reserves are depleted during early growth and development and replenished later in the life cycle. Claus and Behrens [12] obtained less control of *Elytrigia repens* (quackgrass) when glyphosate was applied to young plants of 11/2- + to 2-leaf stage than when applied later. This was possibly due to a reduced transport of the chemical to the underground vegetative tissue. When glyphosate is applied at, or following, full bloom it would assure larger quantities of glyphosate being translocated to the underground system since the major flow of assimilates is from the mature foliage (source) to the underground storage system (sink).

Zandstra and Nishimoto [69] reported that in younger plants the total activity of glyphosate was higher in roots than in leaves or tubers of *Cyperus rotundus*, as the roots contained many active meristems. As the plants matured, root development decreased and downward movement was more to new rhizomes and tubers, and less to roots. Their results indicated that as plants developed, tubers accumulated a large portion of glyphosate, and the leaves accumulated less glyphosate than tubers at all growth stages when tubers were present. Wills [67] found that the movement of glyphosate in cotton plants was significantly greater following application to the mature lower stem than to the mature lower leaves, immature upper stem, or immature upper leaves of cotton. Claus and Behrens [12] reported that accumulation of glyphosate following foliar application to *Elytrigia repens* was greatest in nodes near the rhizome tip and least in nodes near the mother shoot. Movement of glyphosate is controlled by those factors which influence the assimilate translocation. Therefore, perennial weed control could be enhanced if glyphosate is applied when assimilate translocation into the roots and rhizomes is greatest.

REFERENCES

1. Appleby, A.P., W.R. Furtick and S.C. Fang. 1965. Soil placement studies with EPTC and other carbamate herbicides on *Avena sativa*. Weed Res. **5**: 115-122.
2. Armstrong, T.F., W.F. Meggit and D. Penner. 1973. Absorption, translocation and metabolism of alachlor by yellow nutsedge. Weed Sci. **21**: 357-360.
3. Ballard, T.O., M.E. Foley and T.T. Bauman. 1995. Absorption, translocation and metabolism of imazethapyr in common ragweed (*Ambrosia artemisiifolia*) and giant ragweed (*Ambrosia trifida*). Weed Sci. **43**: 572-577.
4. Baradari, M.R., L.C. Haderlie and R.G. Wilson. 1980. Chlorflurenol effects on absorption and translocation of dicamba in Canada thistle (*Cirsium arvense*). Weed Sci. **28**: 199-200.
5. Blair, A.M. 1975. The addition of ammonium salts or a phosphate ester to herbicides to control *Agropyron repens* (L.) Beauv. Weed Res. **15**: 255-258.
6. Boldt, P.F. and A.R. Putnam. 1980. Selectivity mechanisms for foliar applications of diclofop methyl. I. Retention, absorption, translocation and volatility. Weed Sci. **28**: 474-477.
7. Bovey, R.W., H. Hein, Jr. and F. Nelson Keeney. 1989. Phytotoxicity, absorption and translocation of five clopyralid formulations in honey mesquite (*Prosopis glandulosa*). Weed Sci. **37**: 19-22.
8. Bruce, J.A., J.B. Carey, D. Penner and J.J. Kells. 1996. Effect of growth stage and environment on foliar absorption, translocation, metabolism and activity of nicosulfuron in quackgrass (*Elytrigia repens*). Weed Sci. **44**: 447-454.
9. Burt, M.E. and F.T. Corbin. 1978. Uptake, translocation and metabolism of propham by wheat (*Triticum acestivum*), sugarbeet (*Beta vulgaris*) and alfalfa (*Medicago sativa*). Weed Sci. 36:296-303.
10. Carlson, W.C., E.M. Lignowski and H.J. Hopen. 1975. Uptake, translocation and metabolism of pronamide. Weed Sci. **23**: 148-154.
11. Chandler, J.M., E. Basler and P.W. Santelman. 1974. Uptake and translocation of alachlor in sorghum and wheat. Weed Sci. **22**: 253-258.
12. Claus, J.S. and R. Behrens. 1976. Glyphosate translocation and quack grass rhizome and kill. Weed Sci. **24**: 149-152.
13. Coupland, D. and D.V. Peabody. 1981. Absorption, translocation and exudation of glyphosate, fosamine and amitrole in field horsetail (*Equisetum arvense*). Weed Sci. **29**: 556-560.
14. Crafts, A.S. and S. Yamaguchi. 1964. The autoradiography of plant materials. Agr. Extension Serv. Manual 35, Univ. of California, Berkeley, California, USA, 143 pp.
15. Crafts, A.S. and C.E. Crisp. 1971. Phloem Transport in Plants. W.H. Freeman and Co., San Francisco, USA, 481 pp.
16. Davies, P.J., D.S.H. Drennan J.D. Fryer and K. Holly. 1968. The basis of differential phytotoxicity of 4-hydroxy-3,5-di-iodobenzonitrile. II. Uptake and translocation. Weed Res. **8**: 233-240.
17. Dayan, F.E., J.D. Wette and H.G. Hancock. 1996. Physiological basis for differential sensitivity to sulfentrazone by sicklepod (*Senna obtusifolia*) and coffee senna (*Cassia occidentalis*). Weed Sci. **44**: 12-17.

18. Devine, M.D. 1989. Phloem translocation of herbicides. Rev. Weed Sci. 4: 191-123.

19. Duble, R.L., E.C. Scott and G.C. McBee. 1968. Translocation of two organic arsenicals in purple nutsedge. Weed Sci. 16: 421-424.

20. Fadayomi, O. and G.F. warren. 1977. Uptake and translocation of nitrofen and oxyfluorfen. Weed Sci. 25: 111-114.

21. Foy, C.L. 1964. Volatility and tracer studies with alkylamino-s-triazines. Weeds 12: 103-108.

22. Gardiner, J.A., R.C. Rhodes, J.B. Adams, Jr. and E.J. Soboczenski. 1969. Synthesis and studies with 2-C^{14}-labeled bromacil and terbacil. J. Agric. Food Chem. 17: 980-986.

23. Gillespie, G.R. and S.D. Miller. 1983. Absorption, translocation and metabolism of diclofop by sunflower (*Helianthus annus*). Weed Sci. 31: 658-663.

24. Gorrell, R.M. R.M., S.W. Bingham and C.L. Foy. 1988. Translocation and fate of dicamba, picloram and triclopyr in horsenettle, *Solanum carolinense*. Weed Sci. 36: 447-452.

25. Haderlie, L.C. 1980. Absorption and translocation of buthidazole. Weed Sci. 28: 353-357.

26. Hall, J.C. and W.H. VandenBorn. 1988. The absence of a role of absorption, translocation, or metabolism in the selectivity of picloram and clopyralid in two plant species. Weed Sci. 36: 9-14.

27. Harrison, S.K. and L.M. Wax. 1986. Adjuvant effects on absorption, translocation and metabolism of haloxyfop-methyl in corn (*Zea mays*). Weed Sci. 34: 185-195.

28. Hatzios, K.K. and D. Penner. 1980a. Site of application of ^{14}C-buthidazole in corn (*Zea mays*) and redroot pigweed (*Amaranthus retroflexus*). Weed Sci. 28: 285-291.

29. Hatzios, K.K. and D. Penner. 1980b. Absorption, translocation and metabolism of ^{14}C-buthidazole in alfalfa (*Medicago sativa*) and quackgrass (*Agropyron repens*) Weed Sci. 28: 635-639.

30. Hawxby, K. and E. Basler. 1976. Effects of temperature on absorption and translocation of profluralin and dintramine. Weed Sci. 24: 545-548.

31. Jaques, C.J. and R.G. Harvey. 1979. Vapor absorption and translocation of dinitroaniline herbicides in oats (*Avena sativa*) and peas (*Pisum sativum*). Weeds Sci. 27: 371-374.

32. Jordan, L.S. and W.A. Clerx. 1981. Accumulation and metabolism of bromacil in pineapple sweet orange (*Citrus sinensis*) and Cleopatra mandarin (*Citrus reticulata*). Weed Sci. 29: 1-5.

33. Kells, J.J., W.F. Meggitt and D. Penner. 1984. Absorption, translocation and activity of fluazifop-butyl as influenced by plant growth stage and environment. Weed Sci. 32: 143-149.

34. Ketchersid, M.L. T.E. Bosell and M.G. Merkle. 1969. Uptake and translocation of substituted aniline herbicides in peanut seedlings. Agron. J. 61: 185-187.

35. Knake, E.L. and L.M. Wax. 1968. The importance of the shoot of giant foxtail for uptake of preemergence herbicides. Weed Sci. 16: 393-395.

36. Lamoureaux, G.L. and D.G. Rusness. 1995. Quinclorac absorption, translocation metabolism and toxicity in leafy spurge (*Euphorbia esula*). Pestic. Biochem. Physiol. 53: 210-226.

37. Lewer, P. and W.J. Owen. 1990. Selective action of herbicide, triclopyr. Pestic. Biochem. Physiol. 36: 187-200.

38. McWhorter, C.G. and G.D. Wills. 1978. Factors affecting the translocation of 14C-melfluidide in soybeans (*Glycine max*), common cocklebur (*Xanthium pensylvanicum*) and johnsongrass (*Sorghum halepense*). Weed Sci. 26: 382-388.

39. Nalewaja, J.D. 1968. Uptake and translocation of diallate in wheat, barley, flax and wild oat. Weed Sci. 16: 309-312.

40. Nalewaja, J.D., G.A. Skrzypczak and G.R. Gillespie. 1986. Absorption and translocation of herbicides with lipid compounds. Weed Sci. 34: 564-568.

41. Negi, N.S. and H.H. Funderburk. 1968. Effect of solutions and vapors of trifluralin on growth of roots and shoots. Abstr. Weed Sci. Soc. Amer. pp. 37-38.

42. Norris, R.F. and M.J. Bukovac. 1968. Structure of the pear leaf cuticle with special reference to cuticular penetration. Amer. J. Bot. 55: 975-983.

43. Novosel, K.M. and K.A. Renner. 1995. Nicosulfuron and primisulfuron root uptake, translocation and inhibition of acetolactate synthase in sugarbeet (*Beta vulgaris*). Weed Sci. 43: 342-346.

44. Pallas, J.E., Jr. 1960. Effects of temperature and humidity on foliar absorption and translocation of 2,4-dichlorophenoxy acetic acid and benzoic acid. Plant Physiol. 35: 575-580.

45. Parker, C. 1963. Factors affecting the selectivity of 2,3-dichloroallyl diisopropylthiocarbamate (Di-allate) against *Avena* spp. in wheat and barley. Weed Res. **3**: 259-276.

46. Parker, C. 1966. The importance of shoot entry in the action of herbicides applied to the soil. Weeds **14**: 117-121.

47. Phillips, R.E., D.B. Egli and L. Thompson, Jr. 1972. Absorption of herbicides by sorghum seeds and their influence on emergence and seedling growth. Weed Sci. **20**: 506-510.

48. Prendeville, G.N., L.R. Oliver and M.M. Schreiber. 1968. Species differences in site of shoot uptake and tolerance to EPTC. Weed Sci. **16**: 538-540.

49. Putnam, A.R. and S.K. Ries. 1968. Factors influencing the phytotoxicity and movement of paraquat in quackgrass. Weed Sci. **16**: 80-83.

50. Rao, V.S. and F. Rahman. 1978. Weed control in tea with glyphosate. Two and A Bud. **25**: 71-73.

51. Rao, V.S., F. Rahman, H.S. Singh, A.K. Dutta and M.C. Saikia. 1977. Advances in weed research in tea of North East India. Proc. Weed Sci. Conf., Indian Soc. Weed Sci. pp. 61-73.

52. Rice, R.P. and A.R. Putnam. 1980. Temperature influences on uptake, translocation and metabolism of alachlor in snap beans (*Phaseolus vulgaris*). Weed Sci. **28**: 131-134.

53. Rubin, R. and Y. Eshel. 1977. Absorption and distribution of terbutryn and fluometuron in cotton (*Gossypium hirsutum*) and snapbeans (*Phaseolus vulgaris*). Weed Sci. **25**: 499-505.

54. Schreiber, J.D., V.V. Volk and L. Boersma. 1975. Soil water potential and bromacil uptake by wheat. Weed Sci. **23**: 127-130.

55. Shaner, D.L. and P.A. Robinson. 1985. Absorption, translocation and metabolism of AC 252,214 in soybean (*Glycine max*), common cocklebur (*Xanthium strumarium*) and velvetleaf (*Abutilon theophrasti*). Weed Sci. **33**: 469-471.

56. Sikka, H.C. and D.E. Davis. 1968. Absorption, translocation and metabolism of prometryne in cotton and soybean. Weed Sci. **16**: 474-477.

57. Still, G.G. and E.R. Mansager. 1971. Metabolism of isopropyl-3-chlorocarbanilate by soybean plants. J. Agric. Food Chem. **19**: 879-884.

58. Strang, R.H. and R.L. Rogers. 1971a. A microradiographic study of ^{14}C-diuron absorption by cotton. Weed Sci. 355-362.

59. Strang, R.H. and R.L. Rogers. 1971b. A microradiographic study of ^{14}C-trifluralin absorption. Weed Sci. **19**: 363-369.

60 Strang, R.H. and R.L. Rogers. 1975. Translocation of ^{14}C-SAN 6706 in cotton, soybean and corn. Weed Sci. **23**: 26-31.

61 Swann, C.W. and R. Behrens. 1969. Phytotoxicity and loss of trifluralin vapors from soil. WSSA Abstr. No. 222.

62. Tritter, S.A., F.B. Hall and B.G. Todd. 1987. Diclofop-methyl and diclofop uptake in oat (*Avena sativa* L.) protoplasts. Can J. Plant Sci. **67**: 215-223.

63. Turner, D.J. and M.P. Loader. 1980. Effect of ammonium sulphate and other additives upon the phytotoxicity of glyphosate to *Agropyron repens* (L.) Beauv. Weed Res. **20**: 139-146.

64. Veerasekaran, P., R.C. Kirkwood and W.W. Fletcher. 1977. Studies on the mode of action of asulam in bracken (*Pteridium aquilinum*). I. Absorption and translocation of ^{14}C-asulam. Weed Res. **17**: 33-39.

65. Whim, J.L., W.F. Meggitt and D. Penner. 1986. Effect of acifluorfen and bentazon on absorption and translocation of haloxyfop and DPX-Y6202 in quackgrass (*Agropyron repens*). Weed Sci. **34**: 333-337.

66. Wills, G.D. 1976. Translocation of bentazon in soybeans and common cocklebur. Weed Sci. **24**: 536-540.

67. Wills, G.D. 1978. Factors affecting toxicity and translocation of glyphosate in cotton (*Gossypium hirsutum*). Weed Sci. **26**: 509-515.

68. Wilson, B.J. and R.K. Nishimoto. 1975. Ammonium sulphate enhancement of picloram activity and absorption. Weed Sci. **23**: 289-296.

69. Zandstra, B.H. and R.K. Nishimoto. 1977. Movement and activity of glyphosate in purple nutsedge. Weed Sci. **25**: 268-274.

7

Mechanisms of Action of Herbicides

Herbicides bring about various physiological and biochemical effects on the growth and development of emerging seedlings as well as established plants, either after coming into contact with the plant surface or after reaching the site(s) of action within the plant tissue. The net result is the death of the plant. These physiological and biochemical effects are followed by various types of visual injury symptoms on susceptible plants. These include chlorosis, defoliation, stunting, necrosis, stand reduction, epinasty, morphological aberrations, growth stimulation, cupping of leaves, marginal leaf burn, desiccation, delayed emergence, germination failure, etc. These injury symptoms may appear on any part of the plant, including roots, flowers, fruits, foliage, etc.

The rate of appearance of these effects varies with the characteristic actions of the herbicide and depends on the degree of tolerance or susceptibility of the plant species. Environmental factors and soil conditions affecting plant growth, as well as herbicide formulation and application method significantly influence the effect of herbicides. After reaching the site of action, the herbicide affects the living processes of the plant and these effects are immediately reflected in the growth morphology and physiology of the plant. Herbicides differ in their site of action. They have more than one site of action and of these, the most sensitive ones, the primary sites, are affected first. As the herbicide concentration builds up in the tissue, additional sites, probably the secondary and tertiary sites, may become involved.

The different physiological and biochemical processes that occur within the living plant are briefly discussed here to give background information on the principles of mechanisms of action of herbicides. For more information on these aspects, the reader may refer to books on plant physiology and biochemistry.

PHYSIOLOGICAL AND BIOCHEMICAL PROCESSES IN PLANTS

The various physiological and biochemical processes affected by herbicides are grouped under the following categories. These are:
1. Photosynthesis
2. Mitochondrial activities
3. Protein and nucleic acid biosynthesis
4. Pigment biosynthesis
5. Fatty acid (lipid) biosynthesis
6. Branched-chain amino acid biosynthesis
7. Aromatic compound biosynthesis

8. Ethylene biosynthesis
9. Glutamine biosynthesis
10. Hydrolytic enzyme activities

Most of the herbicides affect at least one or more of these processes.

Photosynthesis

Plant metabolism is fueled by two forms of chemical energy: a) the phosphoryl-group transfer potential of ATP and b) the reducing power of NADH and NADPH. ATP and the reduced forms of nicotinamide cofactors can be produced by the oxidation of carbohydrates via the glycolysis and citric acid cycle, and oxidative phosphorylation, which yields even more ATP. The ultimate source of carbohydrate is produced by plants through a process known as **photosynthesis.**

Photosynthesis is a process that converts atmospheric carbon dioxide and water to carbohydrates, using sunlight as the source of energy. Solar energy, captured in chemical form as ATP and NADPH, is an immediate source of energy for green plants and other photosynthetic autotrophs and the ultimate source of energy for nearly all heterotrophic organisms.

Photosynthesis, usually represented by the following reaction, takes place in two partial reactions.

$$6CO_2 + 6H_2O \xrightarrow{\text{Light}} C_6H_{12}O_6 + 6O_2$$

In the first process, shown in the following reaction (a), light energy is captured by light-absorbed pigments and converted into the chemical energy of ATP and certain reducing agents, particularly NADPH. In this process, protons derived from water are used in the chemiosmotic synthesis of ATP from ADP and Pi, while a hydride ion from water reduces $NADP^+$ to NADPH. This reaction is accompanied by the light-dependent release of oxygen gas produced by splitting water molecules.

$$\text{(a) } H_2O + ADP + Pi + NADP^+ \xrightarrow{\text{Light}} O_2 + ATP + NADPH + H^+$$

In the second process, represented in reaction (b), the energy-rich products of the first reaction, NADPH and ATP, are used as energy source to reduce CO_2 to carbohydrate (glucose). At the same time, NADPH is reoxidized to $NADP^+$ and ATP is broken down to ADP and Pi.

$$\text{(b) } CO_2 + NADPH + H^+ + ATP \longrightarrow \text{Glucose} + NADP^+ + ADP + Pi$$

The first process involving the conversion of light energy into chemical energy is referred to as the **light reaction** or the **light phase** of photosynthesis. The second process, in which carbohydrate (glucose) and other reduced products are formed from CO_2, without depending on light (it requires only a supply of ATP and NADPH), is called the **dark reaction** or **dark phase**. The latter is more appropriately called the light-independent process.

Chloroplast and Chlorophyll

Chloroplasts are the specialized organelles in which photosynthesis occurs in plants. The chloroplast contains a highly folded continuous membrane network called the **thylakoid membrane**, the site of light-dependent reactions that produce NADPH and ATP. The thylakoid membrane is suspended in the aqueous matrix, or **stroma**, of the chloroplast. The stroma contains soluble enzymes that catalyze the reduction of CO_2 to carbohydrate. The aqueous space within the thylakoid membrane is called the **lumen,** into which protons move (across the thylakoid membrane) to create proton-motive force that drives ATP synthesis. The thylakoid membrane is folded into flattened vesicles arranged in stacks, called **grana,** or as single vesicles that traverse the stroma and connect grana. The part of the thylakoid membrane located within grana and not in contact with the stroma is called **granal lamellae.** The part exposed to the stroma is called **stromal lamellae.**

The thylakoid membrane contains pigments that capture light energy for photosynthesis. Chlorophylls, which are green, are the most abundant pigments involved in the photosynthetic process. Chlorophylls include **chlorophyll a (chl a)** and **chlorophyll b (chl b). Chl a** and **chl b** both absorb light in the violet-to-blue region (absorption maximum 400-500 nm) and the orange-to-red region (abosorption maximum 650-700 nm) respectively. In addition to chlorophyll, several accessory pigments such as carotenoids (yellow to brown) and phycobilins, including phycoerythrin (red) and phycocyanin (blue) found in photosynthetic membranes, also serve as receptors of light energy. These accessory pigments which have absorption maxima at wavelengths other than those of the chlorphylls, serve as supplementary light receptors for the portions of visible spectrum not completely covered by chlorophyll, thus broadening the range of light energy absorbed by the photosynthetic pigments. When the light energy is absorbed by these accessory pigments, it must, however, be transferred as excitation energy to chlorophyll molecules before it can be used for photosynthesis.

Photosystems I and II

The functional units of photosynthesis in plants are called **photosystem** I and **photosystem II,** numbered in the order of their discovery. The reaction centre of each photosystem is a complex of proteins, electron-transporting cofactors and two chlorophyll molecules called the **special pair.** The absorption maximum of the special pair of photosystem I (**PS I**) is 700 nm and so these pigments are called **P700.** The special pair of photosystem II (**PS II**) absorb light maximally at 680 nm and these are called **P680.** Besides, pigment molecules also contain **light harvesting complexes (LHCs),** with one fraction bound to PS I, another to PS II and the third, being mobile, can harvest light for either PS I or PS II.

PS I is located in the stromal lamellae and is therefore exposed to the chloroplast stroma, while PS II is located predominantly in the granal lamellae, away from the stroma. The two photosystems have different electron-transferring components.

When a photon of light strikes a pigment molecule capable of absorbing light at a given wavelength, energy is absorbed by some of the electrons, which are thus boosted to high-energy levels. This molecule is then in an energy-rich **excited state.** This excitation of a molecule by light is very rapid. The excited molecule may either return to its low-energy state, the **ground state,** with simultaneous emission of the energy originally absorbed during excitation in the form of light (as fluorescence) or heat or both, or react readily with some other molecule. In such a photochemical reaction, the excited molecule may lose an electron to the other reacting molecule.

Photosynthetic Electron Transport

Photosynthetic electron transport requires **PS I** and **PS II** and three other components embedded or associated with the thylakoid membrane: the **oxygen-evolving complex,** the **cytochrome b6/f complex** and **chloroplast ATP synthase.** The PS II complex is primarily, but not exclusively, localized in the granal lamellae, away from the stroma, while the PS I and ATP synthase are present in the stromal lamellae. The oxygen-evolving complex, which is composed of several peripheral membrane proteins and four manganese ions, is associated with PS II on the luminal side of the thylakoid membrane. The cytochrome b6/f complex is equally located in both granal and stromal lamellae. ATP synthase, which also spans the membrane, islocated exclusively in the stromal lamellae. ATP synthase converts the potential energy of the proton gradient developed during electron transport into high-energy phosphate bond energy in the form of ATP. Photosynthetic electron transport requires all the aforesaid processes and/or components except for ATP synthase.

The path of electron flow during photosynthesis is depicted in a zigzag figure called the **Z-scheme** (Fig. 7.1). The Z-scheme plots the reduction potentials of the photosynthetic electron-transport components. It shows that the absorption of light energy converts P680 and P700 (the pigments that are poor reducing agents) to excited molecules that are good reducing agents.

Oxygen-Evolving Complex

The electrons needed for electron transport are obtained from the oxidation of water catalyzed by the oxygen-evolving complex as illustrated in the following equation:

$$2H_2O \xrightarrow{\text{Light}} O_2 + 4H^+ + 4\bar{e}$$

During this reaction, a) two molecules of water are oxidized, b) four protons (4H+) are released into the lumen, c) one molecule of O_2 is released and diffused to the atmosphere and d) four electrons are produced and transferred, one at a time, to a strong oxidizing agent in PS II, the oxidized form of P680.

Electron Flow

The energy to drive electron flow comes from the absorption of a photon of light by a LHC pigment molecule associated with PS II or PS I, i.e., an antenna pigment or a reaction-centre pigment. The absorbed energy is transferred among pigment molecules to the reaction centre. With each transfer of energy, the promoted electron in the donor pigment molecule returns to the ground state and the electron in the recipient pigment molecule moves to a higher energy level. Because the energy level of an excited reaction-centre chlorophyll is slightly lower than the energy required to excite an antenna pigment, energy transfer proceeds further.

PS II Electron Transport

The PS II complex includes the oxygen-evolving (water-splitting) complex, a reaction-centre complex and the LHC antenna pigment (or proteins).

Electrons are transferred from the oxygen-evolving complex to P680 via the carrier Z, a tyrosine residue of a protein subunit of the PS II reaction centre. The reaction transfer, fueled by light energy, converts P680 to an excited state, designated **P680*** (Fig. 7.1) [72a]. The excited P680* has a more negative reduction potential than P680. Once excited, P680* donates an electron to the next carrier, **pheophytin a (Ph a)** which is identical to **chlorophyll a** except that the magnesium ion of **chlorophyll a** is replaced by two protons. Electron transfer then proceeds from the reduced pheophytin **a** to PQ_A, **a plastoquinone** (an electron acceptor), tightly bound to a PS II polypeptide. Plastoquinone can be reduced by two sequential one-electron transfers. The bound PQ_A transfers two electrons, one at a time, to a second plastoquinone molecule, PQ_B, which is reversibly bound to a PS II polypeptide. The fully reduced PQ_BH_2 is then released into the pool of plastoquinone in the thylakoid membrane. PQ_A and PQ_B are

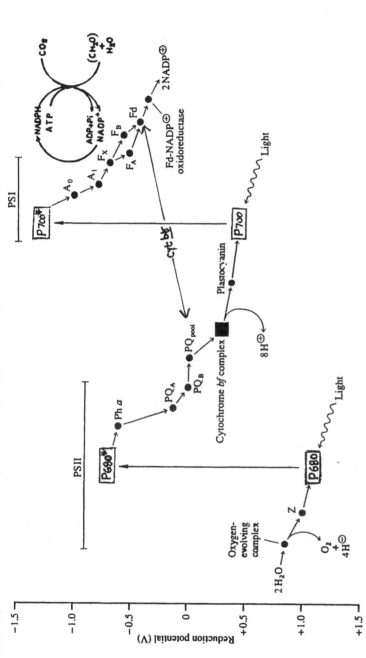

Fig. 7.1. The photosynthetic electron transport scheme (**Z-scheme**) involving the oxygen-evolving complex, reduction potentials and electron flow during photosynthesis, reduction of CO_2 and NADP and phosphorylation. Light energy absorbed by the special pair pigments, P680 and P700, drives electron flow uphill. The reduction potentials of the carriers vary with experimental conditions. Hence, the values shown are approximate. PS II and PS I are connected by a central electron transport chain between pheophytin a, the primary electron acceptor of electrons of P680 of PS II and P700 of PS I. (Reproduced from [72a], but modified, with permission of the publisher).

Abbreviations: P680*: excited form of P680; **Phe *a* :** pheophytin a (**chl *a*** without magnesium); the electron acceptor of PS II; **PQA:** plastoquinone tightly bound to PS II (D_1 protein); **PQB:** reversibly bound (to D_1 protein) plastoquinone undergoing reduction by PS II; **PQpool:** plastoquinone pool made up of PQ and PQH_2 (i.e., plastohydroquinone); **P700:** P I reaction centre; **P700*:** excited form of P700; **Ao:** chlorophyll *a*, the primary electron acceptor and **chl *a*** monomer; **A₁:** phylloquinone (vitamin K-1); **F$_X$, F$_B$, and F$_A$:** iron-sulphur clusters; **Fd:** ferredoxin; **cyt *bf*:** cytochrome *b6/f* complex.

plastoquinone molecules bound to the special niches of the D_2 and D_1 proteins (polypeptides) respectively (Fig. 7.2) [56]. Because of its microenvironment, PQ_A acts as a one-electron carrier and can only be reduced as far as the semi-quinone state. PQ_B acts as a two-electron gate: it accepts two electrons from PQ_A, then accepts two protons from the stroma side of the membrane and then leaves its binding site as **plastohydroquinone (PQH2)**. Another plastoquinone molecule then binds the D_1 protein, replacing the molecule that has left and when bound, is called PQ_B; it then becomes ready to assume the electron transfer role.

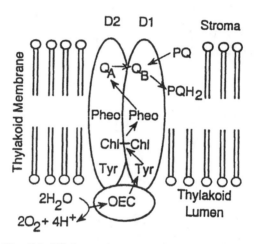

Fig. 7.2. PS II reaction-centre complex scheme.
Abbreviations: OEC: oxygen-evolving complex; **Tyr:** tyrosine amino acid residue 160; **Chl:** chlorophyll *a* dimer, also referred to as P680; **Pheo:** pheophytin; Q_A and Q_B (PQ_A and PQ_B): bound plastoquinone molecules; **PQH$_2$:** plastohydroquinone; D_1 and D_2: proteins [56].

Electron Transport between PS II and PS I

Plastohydroquinone (PQH_2) donates its electrons to plastocyanin (PC) via the cytochrome *bf* (*b6/f*) complex, which contains cytochromes and iron-sulphur proteins (Fig. 7.1) [72a]. During this reaction two protons liberated from PQH_2 move laterally along the thylakoid membrane. Plastocyanin, a primary electron donor, accepts electrons from cytochrome **bf** and shuttles them along the lumen side of the thylakoid membrane to P700, the electron acceptor in PS I. The oxidation of two water molecules to O_2, which generates four electrons, results in two complete rounds of the Q cycle for a net oxidation of two PQH_2 to two PQ and the transfer of eight protons ($8H^+$) into the lumen.

PS I Electron Transport

The PS I complex consists of the components of photosynthetic electron transport that catalyze the photoreduction of ferredoxin, with plastocyanin as the electron donor. PS I is composed of a reaction-centre complex and the light-harvesting chlorophyll antenna proteins which transfer absorbed light energy to P700, the PS I reaction centre. P700, a chl **a** dimer, undergoes a light-induced charge separation resulting in the transfer of electrons and in the process getting P700 excited. The excited P700* is the strongest reducing agent in the chain of electron carriers.

The electron from **P700*** is readily donated to **Ao**, a chl *a* monomer and a primary electron acceptor (P430). The electron is then transferred to phylloquinone (vitamin K_1 or A_1), which donates it to a series of iron-sulphur clusters, first to F_x and then to either F_A or F_B [Fig. 7.1]. F_x is a part of the PS I reaction centre complex, while F_A and F_B are located in a peripheral membrane polypeptide on the stromal side of the thylakoid membrane. F_A and F_B appear to be the intermediate from which the bipyridilium herbicides, paraquat and diquat, accept electrons.

The electron delivered to F_A or F_B is transferred to **ferredoxin (Fd)**, the iron-sulphur protein coenzyme dissolved in the chloroplast stroma. The reduced ferredoxin, one of the most powerful biological reducing agents, reduces $NADP^+$ to NADPH. The reduction of $NADP^+$ is catalyzed by **ferredoxin-$NADP^+$ oxidoreductase**, an enzyme loosely bound to the stromal side of the thylakoid membrane. Two molecules of NADPH are generated for each oxygen molecule formed from water. The formation of NADPH completes the electron transport sequence.

Photophosphorylation

During the photosynthetic electron transport, protons from water are released into the lumen, stromal protons are transported to the lumen during the oxidation of PQH_2 and the uptake of a proton during the reduction of $NADP^+$ lowers the proton concentration in the stroma. In the presence of a transmembrane proton concentration gradient, the membrane-spanning chloroplast ATP synthase catalyzes the formation of ATP from ADP + Pi. The ATP thus produced is essential in the process of photosynthetic CO_2 assimilation and in solute movement in plant cells. Since the process of ATP formation depends on light, it is called **photophosphorylation**.

Photophosphorylation is carried out via cyclic electron transport and non-cyclic electron transport pathways. In **cyclic transport**, the ferredoxin donates its electron not to $NADP^+$ but back to the PQ pool via a specialized cytochrome of the cytochrome *bf* complex, as illustrated in the reaction below (also vide Fig. 7.1). The electron then reduces P700. The cyclic electron transport operates as a cycle to form ATP without the simultaneous formation of NADPH. ATP formation via the cyclic electron transport is called **cyclic photophosphorylation**.

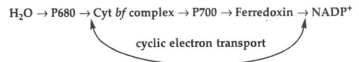

$$H_2O \rightarrow P680 \rightarrow Cyt\ \textit{bf}\ complex \rightarrow P700 \rightarrow Ferredoxin \rightarrow NADP^+$$

cyclic electron transport

In the **non-cyclic electron transport** pathway, NADPH is formed directly from $NADP^+$. For every four electrons transferred to reduce two molecules of $NADP^+$ to NADPH, a proton-motive force develops sufficient to synthesize two molecules of ATP. In **non-cyclic photophosphorylation**, ATP is formed with the simultaneous formation of NADPH.

The rates of cyclic and non-cyclic electron transport are influenced by the ratio of stromal NADPH to $NADP^+$. When the ratio is high, such as in high light intensity, the rate of non-cyclic electron flow is limited, due to the low availability of the $NADP^+$ needed as a substrate for ferredoxin-$NADP^+$ oxidoreductase. In this situation, cyclic electron transport is favoured.

Inhibition of Photosynthesis by Herbicides

Herbicides that inhibit or modify photosynthesis can be classified into: a) **electron transport inhibitors**, b) **uncouplers**, c) **energy transfer inhibitors**, d) **inhibitory**

uncouplers and e) **electron acceptors**. The carbon dioxide fixation and assimilation in photosynthesis is not a primary site of inhibition of herbicides. Of these, only two biochemical mechanisms have been clearly demonstrated to be of primary importance in herbicidal action in photosynthesis: 1) **inhibition of PS II electron transport** and 2) **diversion of electron transfer through PS I**.

PS II electron transport inhibitors belong to a variety of chemical groups. These include phenols, nitriles, pyridazinones, etc. While all these groups inhibit electron transport, they have another mode of action. Nitrophenols and phenols possess uncoupling activity, whereas pyridazinones inhibit lipid and carotenoid biosynthesis.

PS II herbicides block the flow of electrons through PS II and thus also indirectly block the transfer of excitation energy from chlorophyll molecules to the PS II reaction centre. Excited chlorophyll molecules (**singlet chlorophyll**, i.e. ^{1}chl) spontaneously form **triplet chlorophyll** (^{3}chl) through a non-radiative energy transformation of chlorophyll, known as intersystem crossing (**isc**), as shown in Fig. 7.3 [56]. The ^{3}chl (triplet chlorophyll) reacts with molecular oxygen to form singlet oxygen ($^1O_2$).

$$\text{chl} + \text{PSII (hv)} \xrightarrow{\text{isc}} 1\text{chl} \longrightarrow \, ^3\text{chl}$$
$$^3\text{chl} + O_2 \longrightarrow \, ^1O_2 + \text{chl}$$

Fig. 7.3. Reaction involved in the generation of ^{3}**chl** (triplet chlorophyll) and $^1O_2$ (singlet oxygen) by photosystem II transport inhibitors [56]. (**isc:** intersystem crossing).

Both triplet chlorophyll and singlet oxygen can abstract hydrogen from unsaturated lipids, producing a lipid radical and initiating a chain of lipid peroxidation. Lipids and proteins are then oxidized, resulting in loss of chlorophyll and carotenoids and in leaky membranes, which allow cells and cell organelles to dry and disintegrate rapidly.

Mitochondrial Activities

Mitochondria are the cellular organelles that carry out the processes of cellular respiration responsible for the controlled oxidation of the respiratory substrates (organic acids) and the conservation of the energy thus released in forms usable for the energy-requiring functions of the cell. These processes involve: a) synthesis of ATP (adenosine triphosphate) and b) transport of electrons and protons from the substrate to oxygen. During the electron transport of oxygen, ADP (adenosine diphosphate) is phosphorylated to ATP in a process known as **oxidative phosphorylation**. ATP thus synthesized is a common currency for energy expenditure in various other metabolic activities throughout the cell. The Krebs citric acid cycle and the fatty acid oxidation cycle are essentially the preparatory processes for the aerobic regeneration of ATP from ADP and phosphate (Pi) by energy coupling in the respiratory chain. This is the final step in cellular respiration.

Herbicides affect the mitochondrial activities by uncoupling the reactions responsible for ATP synthesis or interfering with electron transport and energy transfer. Uncouplers act on membranes of the mitochondria in which phosphorylation takes place. Electrons leak through the membranes so that the charges that they normally separate are lost. As a result, energy is not accumulated for ATP synthesis. Besides, uncouplers also stimulate respiration of mitochondria, promote hydrolysis of ATP, or inhibit exchange reactions catalyzed by mitochondria in the absence of inorganic phosphate, ADP, ATP and H_2O.

Protein and Nucleic Acid Biosynthesis

Protein synthesis is an active process and a major biological function, which, along with RNA synthesis, is essential for cell elongation. The synthesized proteins are highly organized molecules composed of a specific sequence of amino acids joined by peptide bonds. This process of protein synthesis is an elaborate one requiring the presence and participation of various reactions and components.

Ribosomes are particulate ribonucleoproteins wherein proteins are synthesized. The synthesis of ribosomes requires the synthesis of ribosomal RNA as well as protein particles. These ribosomes are involved in three different stages in the synthesis of soluble proteins: (a) transfer of information from messenger RNA (**mRNA**) to the amino acid sequence, (b) formation of peptide bonds and (c) release and coiling of polypeptide chain. Messenger RNA is the rapidly labelled nucleic acid of nuclear origin. Its function is to carry the genetic information from DNA to the ribosomes for the incorporation of amino acids into polypeptides. It has a nucleotide composition similar to that of DNA. Messenger RNA and ribosomes have been reported to pre-exist in the seed in an active form.

When clusters of ribosomes are held together by mRNA, polyribosomes (polysomes) are formed. A change in the capacity of a biological system for protein synthesis is linked to a change in the polyribosome content. The rate of polyribosome formation in the early stages of seed germination controls the capacity for protein synthesis. Net polyribosome formation is concomitant with amino acid availability. Deprivation of any or all of the essential amino acids results in reduced polyribosome formation.

Transfer RNA (tRNA), also called **soluble RNA (sRNA)**, is required for linking the amino acids, the building blocks for protein synthesis, to form aminoacyl-tRNA with the mediation of activating enzymes, aminoacyl-tRNA synthetases, each of which is highly specific for each of the 20 amino acids for its corresponding tRNA. This enzyme catalyzes a reaction between the carboxyl group of the amino acid and the phosphoryl group of ATP.

Aminoacyl-tRNA is formed as shown below:

(a) Enzyme + ATP + AA_1 $\longrightarrow$ Enzyme—AA_1—AMP + PPi

(b) Enzyme—AA_1—AMP + $tRNA^1$ $\longrightarrow$ AA_1—$tRNA^1$ + Enzyme + ATP

In the first reaction, amino acid is activated by the enzyme to form the enzyme-aminoacyl-adenylate complex which, in the second one, reacts with the tRNA specific for the amino acid in question, resulting in the formation of aminoacyl-tRNA.

The aminoacyl-tRNA so formed is then attached to **polyribosome (mRNA-ribosome)** which is carrying the growing peptidyl-tRNA chain as shown under:

(c) Polyribosome—$tRNA_x$—$AA_{(n)}$ + AA_1— $tRNA^1$ $\longrightarrow$
Polyribosomes + $tRNA^1$—$AA_{(n+1)}$ + $tRNA_x$

At the end of this reaction, the length of the original peptidyl-tRNA chain has grown by one additional amino acid. Repetition of these steps results in the complete growth of the peptide chain from the amino acid terminal to the carboxyl end of the peptide.

Protein synthesis, described briefly above, takes place broadly in three stages: 1) **initiation of the polypeptide chain,** involving formation of the mRNA-ribosome initiation complex and the binding of *N*-**formyl methionyl tRNA (fMet-tRNA),** 2) **chain elongation,** involving transfer of the growing peptidyl chain from the tRNA to which the incoming AA-tRNA is attached and 3) **chain termination,** involving release of the newly synthesized (nascent) polypeptide chains. The scheme of protein synthesis involving various components is given in Fig. 7.4.

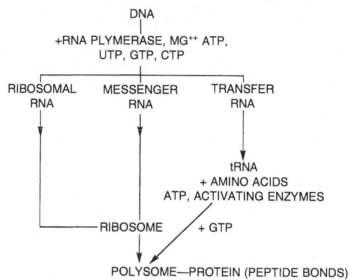

Fig. 7.4. Scheme for protein synthesis

Pigment Biosynthesis

Many herbicides cause either white or yellow chlorosis of leaves as a consequence of the total or partial absence of the normal chloroplast pigments, viz. **chlorophylls** and **carotenoids.** Chlorosis may result from the inhibition of pigment biosynthesis or from destruction of the existing pigment. As a general rule, pigment biosynthesis inhibitors cause chlorosis only in newly developed leaves and they are most effective as herbicides when applied preemergence. On the other hand, inhibition of photosynthesis and photosynthetic electron transport by other herbicides frequently lead to destruction of the existing chloroplast pigments.

The chlorphylls and carotenoids are located specifically in the **pigment-protein complexes (PPC)** within the thylakoid membranes. Each functional PPC has its own distinctive pigment composition. Thus the reaction-centre core complexes of photosystems I and II (PS I and PS II) have chlorophyll **a** and are enriched in β-carotene, whereas the light-harvesting chlorophyll proteins (**LHCP**) associated with PS I and PS II contain both chlorophyll **a** and **b** together with the **xanthophyHs, lutein, violoxanthin** and **neoxanthin.**

Chlorophyll Biosynthesis

The precursor for chlorophyll (chl) is δ-**aminolaevulinic acid (ALA).** In plants, ALA is formed by utilizing the intact carbon skeleton of glutamic acid (glutamate) through the catalytical role of **ALA synthetase** (Fig. 7.5) [22].

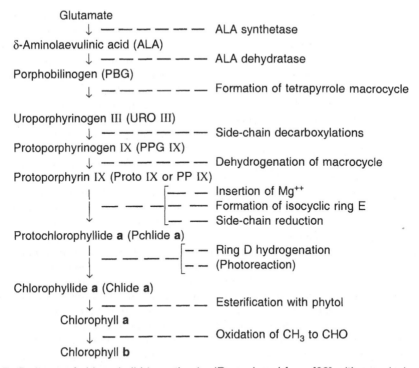

Glutamate
↓ — — — — — — ALA synthetase
δ-Aminolaevulinic acid (ALA)
↓ — — — — — — ALA dehydratase
Porphobilinogen (PBG)
↓ — — — — — — Formation of tetrapyrrole macrocycle

Uroporphyrinogen III (URO III)
↓ — — — — — — Side-chain decarboxylations
Protoporphyrinogen IX (PPG IX)
↓ — — — — — — Dehydrogenation of macrocycle
Protoporphyrin IX (Proto IX or PP IX)
| ⎡— — Insertion of Mg⁺⁺
| — — — ⎢— — Formation of isocyclic ring E
↓ ⎣— — Side-chain reduction
Protochlorophyllide **a** (Pchlide **a**)
| ⎡– – Ring D hydrogenation
| — — — — ⎢
↓ ⎣– – (Photoreaction)
Chlorophyllide **a** (Chlide **a**)
↓ — — — — — — Esterification with phytol
Chlorophyll **a**
↓ — — — — — — Oxidation of CH₃ to CHO
Chlorophyll **b**

Fig. 7.5. Pathway of chlorophyll biosynthesis. (Reproduced from [22] with permission of the publisher).

Condensation of groups of two ALA molecules, catalyzed by ALA dehydratase, results in the formation of **porphobilinogen** (**PBG**), the first pyrrolic intermediate in the chlorophyll biosynthesis pathway. Next, four PBG units are linked together in a head-to-tail sequence and this linear molecule is then enzymatically closed to form **uroporphyrinogen** (**URO III**), the first cyclic tetrapyrrole. URO III is converted to an intermediate coporphyrinogen III (not shown in Fig. 7.5) by decarboxylation of the four acetic acid substituents, leaving methyl groups, and two of the four propionic acid molecules are oxidatively decarboxylated to vinyl groups, forming **protoporphyrinogen IX (PPG IX)**. Removal of six electrons from the porphyrinogen macromolecule leads to the formation of **protoporphyrin IX (Proto IX or PP IX)**. Proto IX is the last common precursor to hemes, bilins and chloropylls.

The chlorophyll branch of the pathway begins with insertion of magnesium into the Proto IX molecules, followed by formation of isocycle ring E and reduction of the vinyl group, on the side chain, to an ethyl group, forming **protochlorophyllide a (Pchlide a)**. Pchlide **a** contains a fifth, isocyclic, ring that is present in all chlorophylls. Pchlide **a** undergoes hydrogenation of ring D in the presence of light to yield **chlorophyllide a (Chlide a)**. **Chlorophyll a** formation involves esterification of the propionic acid on ring D with geranylgeraniol and subsequent reduction of the geranyl geranyl group to phytyl. **Chlorophyll b** is derived directly from chlorophyll **a** by oxidation of the methyl (CH₃) group on ring B to a formyl (CHO) group.

Carotenoid Biosynthesis

Carotenoid biosynthesis is an integral part of the construction of the **pigment-protein complexes** (PPC) in the thylakoids and its regulation is closely linked to that of the formation of other components such as chlorophyll, proteins and lipids. In plants, carotenoids provide essential photoreactive functions, blocking the formation of reactive oxygen species. These pigments also accumulate in chloroplasts, providing the yellow, orange and red colours of many flowers, fruits and storage roots.

The biosynthesis of carotenoids, namely **carotenes** and **xanthophylls**, takes place on a membrane-bound multienzyme complex containing phytoene synthetase, desaturase, cyclase and hydroxylase enzymes. Although the carotenes associated with the reaction centres of PS I and PS II play a peripheral role in electron transport, they play an important and essential role in photosynthesis by protecting chlorophyll against photo-oxidative destruction by singlet oxygen ($^1O_2$; vide Fig. 7.3 and relevant discussion). This highly reactive oxygen species is formed in excess through excited chlorophyll when the photosynthetic machinery (electron transport and subsequent CO_2 assimilation) operates too slowly to accommodate strong light. Protection against singlet O_2 is achieved when nine or more double bonds are present in the carotenoid molecule, as is the case with **lycopene**, β-**carotene** and **xanthophylls**. These molecules physically interact with the triplet state chlorophyll (^{3}chl) and dissipate its energy as heat. If a herbicide prevents carotene and xanthophyll formation, chlorophyll, although formed, will not accumulate either. Consequently, leaves emerging after bleaching-type herbicide treatment will be depleted of all coloured plastidic pigments.

Carotenoid biosynthesis begins with head-to-tail condensation of two molecules of geranylgeranyl pyrophosphate to prephytoene pyrophosphate which, after dephosphorylation, is converted to **15-*cis* phytoene** (Fig. 7.6) [134]. The subsequent pathway involves four desaturation steps (removal of 2H each time) and two cyclization steps to form β-carotene. Desaturation involves successive formation of conjugated double bonds and shifting light absorbance from the ultraviolet (phytoene, peak at 286 nm) to the blue (β-carotene, peak at 448 nm). Phytoene has three conjugated double bonds while β-carotene and the xanthophylls have eleven.

Phytoene is dehydrogenated by **phytoene desaturase (PD)**, a membrane-bound enzyme of the thylakoid. The phytoene desaturase reaction produces ζ**-carotene**, with **phytofluene** as an intermediate.

TARGET SITES FOR HERBICIDAL ACTIVITY

a) **Phytoene synthesis:** Any effect on the early steps of carotenoid biosynthesis, before phytoene synthesis, would have a general effect on the formation of all isoprenoids, both inside and outside the chloroplast. Phytoene synthetase is a likely control point in the biosynthetic pathway, but its specific inhibition is unlikely because the reaction mechanism is similar to that of formation of the sterol precursor, squalene.

b) **Phytoene desaturase:** Desaturation is essential and specific to carotenoid biosynthesis and relatively easy to disrupt. Its inhibition results in the accumulation of phytoene and other intermediates, which have short chromophores and cannot protect against photo-oxidation, so the plants will be rapidly killed by light and oxygen. Blocking the hydrogen abstraction (dehydrogenation) steps between phytoene and lycopene by inhibiting phytoene desaturase is a more effective pathway for the activity of photobleaching herbicides.

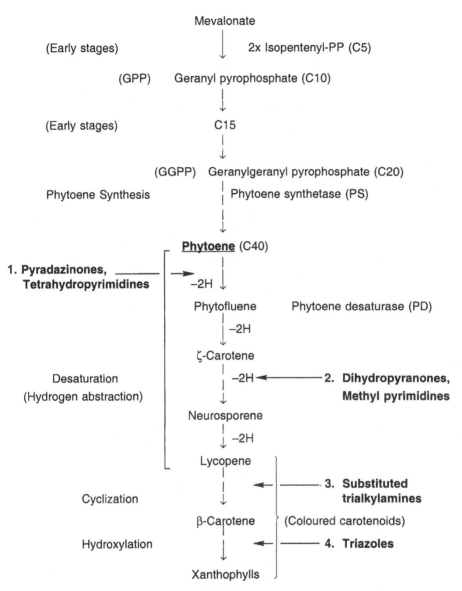

Fig.7.6. Scheme of carotenoid biosynthesis pathway and sites of action of the membrane-bound carotenogenic enzymes (adapted from [134] and modified). The numbers 1, 2, 3 and 4 indicate the sites of action of certain herbicides and inhibitors.

c) **Carotenoid cyclization:** Inhibition of lycopene cyclization to form β-carotene causes lycopene accumulation. The long chromophore of lycopene cannot protect plants *in vivo*, presumably because it cannot be fitted into the thylakoid PPC at the correct sites and in the same orientation as β-carotene and lutein [22]. Herbicides that block carotenoid cyclization may be slower to act than the desaturation of inhibitors.

d) **Carotene hydroxylation:** If hydroxylation is inhibited β-carotenoids would accumulate, but it is unlikely that carotenes would occupy the same specific sites as the xanthophylls and maintain the PPC structures. However, inhibition of carotene hydroxylation may not be specific as it would also affect other oxidative processes such as the formation of squalene epoxide and sterols [22].

Fatty Acid (Lipid) Bisosynthesis

Plastids contain **acetyl-CoA synthetase, acetyl-CoA carboxylase** and **fatty acid synthetase,** key components of fatty acid synthesis in plants [85]. Inhibitors that affect any of the steps involving these enzymes can block glycerolipid and phospholipid synthesis, which results in inhibition of membrane formation [85]. The most affected are young developing leaves and meristematic tissues, which depend on the efficient fatty acid supply.

Acetyl-CoA carboxylase (ACCase) is the major target of inhibitors affecting fatty acid biosynthesis. The enzyme produces malonyl CoA which is a key intermediate in both fatty acid and flavonoid (gibberellins, abscisic acid, carotenoids and other isoprenoids) biosynthesis (Fig. 7.7) [59]. Acetyl CoA, the substrate for ACCase, is derived directly from acetate via the action of acetyl-CoA synthetase or from pyruvate via pyruvate dehydrogenase. Inhibition of ACCase would deprive the plant of a key intermediate (malonyl CoA), essential to both lipid and flavonoid biosynthesis and would lead to phytotoxic effects [85].

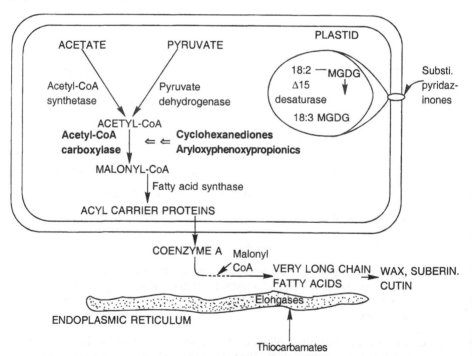

Fig. 7.7. Simplified scheme of fatty acid synthesis in plants and primary targets for cyclohexanediones, aryloxyphenoxypropionics, pyridazinones and thiocarbamates (MGDG: monogalactosyldiacylglycerol (Adapted from [59] and modified).

Four classes of herbicides have a specific effect on fatty acid synthesis. These include pyridazinones, cyclohexanediones, aryloxyphenoxy propionics and thiocarbamates. Of these, cyclohexanediones and aryloxyphenoxypropionics inhibit acetyl-CoA carboxylase. Herbicides of both groups bind to the same region of the target enzyme but they occupy different binding sites. These compounds inhibit lipid and/or flavonoid biosynthesis in susceptible species, suggesting the key role of ACCase in these processes.

Branched-Chain Amino Acid Biosynthesis

Plants die as a direct result of starvation for the branched-chain amino acids in the growing tissues, but not in the mature tissues. Addition of all three branched-chain amino acids, valine, leucine and isoleucine, causes reversal of the toxic effects. In the meristematic tissue, inhibition of **acetolactate synthase**, a prominent enzyme in the biosynthesis of branched-chain amino acids, causes disruption of protein synthesis and cell division as well as increase in translocation of photosynthates to the growing points.

Acetolactate synthase (**ALS**), also referred to as acetohydroxy acid synthase (AHAS), is the first common enzyme in the combined pathway responsible for the biosynthesis of valine, leucine and isoleucine (Fig. 7.8) [66, 145]. ALS is a non-oxidative

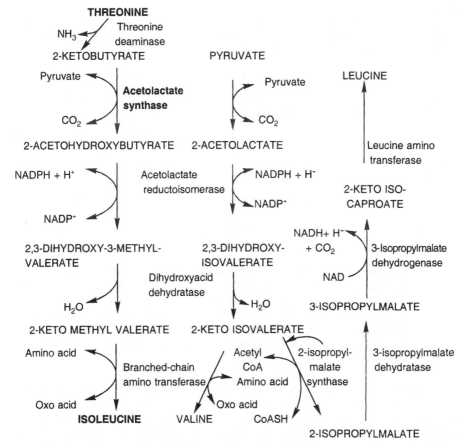

Fig. 7.8. Pathway for the biosynthesis of branched-chain amino acids (adapted from [66, 145]).

thiamine pyrophosphate containing decarboxylase. The initial decarboxylation of pyruvate yields a hydroxyethylamine pyrophosphate intermediate. Four enzymes involved in the three biosynthetic pathways are: **acetolactate synthase, acetolactate reductoisomerase, dihydroxyacid dehydratase** and **branched-chain amino (or amino acid) transferase.** Besides, isoleucine requires one additional enzyme, threonine deaminase and leucine requires three additional enzymes 2-isopropylmalate synthase, 3-isopropylmalate dehydratase and 3-isopropylmalate dehydrogenase [145]. The enzymes involved in the branched-chain amino acid biosynthesis reside in the chloroplast. Regulation of the pathway is caused by inhibition of threonine dehydratase by isoleucine, inhibition of 2-isopropylmalate synthase by leucine and inhibition of ALS by all three amino acids. This suggests that the biosynthetic pathway can be blocked through feedback inhibition from the end-products of the pathway.

Aromatic Compound Biosynthesis

In plants, aromatic compounds such as amino acids (phenyl alanine, tyrosine, tryptophan, etc.), p-aminobenzoic acid, ubiquinone, folic acid, vitamins E and K, lignin, flavonoids, etc. account for up to 35% of the dry weight of higher plants. Herbicides, which inhibit biosynthesis of these aromatic compounds, will effectively block the growth and development of plant tissues.

Biosynthesis of the aromatic compounds involved in plant metabolism proceeds by way of the shikimate pathway, which involves seven steps between erythrose-4-phosphate via shikimate to chorismate (Fig. 7.9).

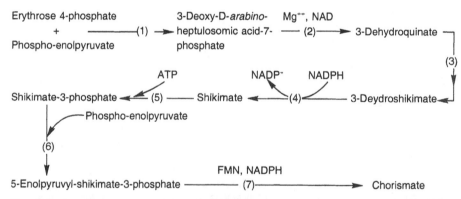

Fig. 7.9. Reactions of the shikimate pathway. [The numbers refer to the following enzymes in the pathway: (1) 3-deoxy-D-*arabino*-heptulosomic acid-7-phosphate (DAHP) synthase; (2) 3-dehydroquinate (DHQ) synthase; (3) 3-dehydroquinase (alternate name: 3-dehydroquinate dehydratase); (4) shikimate dehydrogenase; (5) shikimate kinase; (6) 5-enolpyruvyl-shikimate 3-phosphate (EPSP) synthase (alternate name: 3-phosphoshikimate-1-carboxyvinyltransferase); (7) chorismate synthase].

The shikimate pathway is of key importance in linking primary and secondary metabolism and is initiated by the condensation of of phosphoenolpyruvate (PEP) with erythrose-4-phosphate. Inhibition of EPSP (5-enolpyruvyl-shikimate-3-phosphate) synthase, the enzyme involved in the synthesis of 5-enolpyruvyl-shikimate 3-phosphate (Step 6, Fig. 7.9), will have a resultant inihibitory effect on the formation of chorismate.

Chorismate (chorismic acid), the end-product of this pathway, is the common precursor of all aromatic amino acids and many other important aromatic compounds and a host of secondary metabolites as indicated in Fig. 7.10. In addition, phenylalanine (Phe) feeds into secondary phenolic compound pathways via the important regulatory enzyme phenylalanine ammonia-lyase (PAL) to produce a diverse array of phenolic compounds such as lignin precursors, flavonoids and tannins. Chorismate also gives rise directly to a number of phenolic compounds.

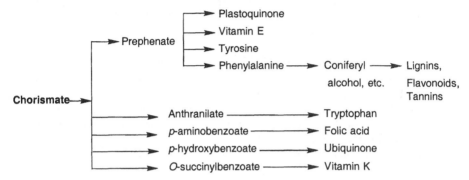

Fig. 7.10. Some of the aromatic compounds derived from chorismic acid in plants.

Ethylene Biosynthesis

Ethylene, the simplest olefin, exists in the gaseous state under normal physiological conditions. It regulates many aspects of plant growth, development and senescence. It is biologically active in trace elements. Ethylene is produced from essentially all parts of higher plants, including leaves, stems, roots, flowers, fruits, tubers and seedlings. Ethylene production is induced during certain stages of plant growth such as germination, ripening of fruits, abscission of leaves and senescence of flowers. Ethylene production can also be induced by a variety of external factors such as mechanical wounding, environmental stresses and certain chemicals, including auxin and other growth regulators. Increased ethylene production can in turn bring out many important physiological consequences.

Methionine is an effective major precursor of ethylene. The pathway of ethylene biosynthesis is illustrated in Fig. 7.11.

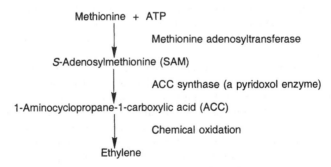

Fig. 7.11. Pathway for biosynthesis of ethylene

The conversion of methionine to ethylene requires ATP and oxygen and is inhibited by an uncoupler of oxidative phosphorylation like DNP. S-adenosylmethionine (SAM) is the first intermediate in ethylene biosynthesis. The conversion of methionine to SAM is catalyzed by methionine adenosyltransferase. In the presence of oxygen, SAM is converted through the catalytic enzymatic action of ACC synthase (a pyridoxal enzyme) to 1-aminocyclopropane-1-carboxylic acid (ACC), the second intermediate, which is later oxidized to form ethylene.

Inhibition of ACC synthase precludes formation of ACC (from SAM), leading to inhibition of ethylene biosynthesis. The conversion of SAM to ACC synthase is a rate-controlling process in ethylene biosynthesis.

Glutamine Biosynthesis

Plants obtain their nitrogen from either nitrate taken up from the soil or by nitrogen fixation carried out by Rhizobium bacteria in legume root nodules. Nitrate reduction to ammonia takes place in the root or shoot. Glutamine synthetase (GS) is the first enzyme involved in the assimilation of ammonia, i.e., the conversion of inorganic nitrogen to an organic form (Fig. 7.13) [92].

GS catalyzes the conversion of glutamate (glutamic acid) to glutamine as shown under:

$$\text{L-glutamate} + \text{ATP} + \text{NH}_3 \xrightarrow{\text{Glutamine synthetase}} \text{L-glutamine} + \text{ATP} + \text{ADP} + \text{Pi} + \text{H}_2\text{O}$$

Following the formation of glutamine (the transaminated form of glutamic acid), the amide nitrogen is transferred to the 2-amino position of glutamate by the enzyme glutamate synthase which exists in two forms: L-glutamate:NAD$^+$ oxidoreductase (transaminating) and L-glutamate:ferredoxin oxidoreductase (transaminating) as shown below:

$$\text{L-glutamine} + \text{2-oxoglutarate} + (\text{NADH}_2 \text{ or ferredoxin}_{red}) \xrightarrow{\text{Glutamate synthase}} \text{2L-glutamate} \\ + (\text{NAD}^\cdot \text{ or ferredoxin}_{ox})$$

Thus, glutamine synthetase (GS) and glutamate synthase act in conjunction to form the glutamate synthase cycle (Fig. 7.12) [92].

Qualitatively, the most important role of GS is in the leaves of C_3 plants, which carry out the process of photorespiration in light. It is essential that the ammonia be rapidly reassimilated by GS so the plant does not die of nitrogen starvation. In plants, the ammonia diffuses through the cytoplasm, in which some assimilation takes place and into the chloroplast, the major site of GS activity. It is the rapid turnover of ammonia in leaves in light that has made GS activity an important target for herbicide action.

The glutamine formed by chloroplast GS is used as an amino donor for a variety of biosynthetic pathways, including amino acid and nucleotide biosynthesis (Fig. 7.13) [92]. Abell [1] observed that only partial inhibition of GS is required for phytotoxicity and the levels of inhibition required can be easily achieved chemically. Inhibition of GS produces not only an increase in ammonia levels, but also a decrease in the rate of CO_2 fixation. In addition to the central role played by GS in nitrogen metabolism, reduction of GS activity has consequences for electron transport and CO_2 fixation under photorespiratory conditions. Inhibition of GS leads to multiple deleterious effects, thus making it an exquisitely toxic site [1].

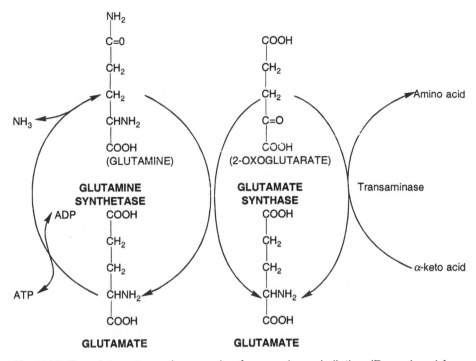

Fig. 7.12. The glutamate synthase cycle of ammonia assimilation (Reproduced from [92] with permission of the publisher and modified).

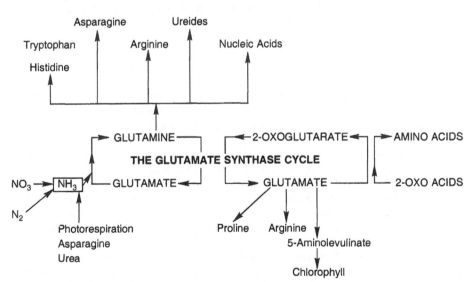

Fig. 7.13. Simplified scheme of nitrogen metabolism in higher plants and the glutamate synthase cycle. (Reproduced from [92] with permission of the publisher).

Hydrolytic Enzyme Activities

During seed germination, enzymes are activated, synthesized and stored, insoluble foods degraded, soluble foods translocated, nutrients mobilized and synthetic reactions related to growth take place [4]. One of the major metabolic processes to take place during this time is the production of hydrolytic enzymes such as amylases, proteases, lipases, phosphatases, esterases, etc. Each of these enzymes degrades the large molecules into subunits: amylases degrade stored carbohydrates into simple sugars, proteases degrade stored proteins into amino acids, lipases act upon fats to form long-chain fatty acids and glycerol compounds, etc.

Of the two types of amylases, α-amylase and β-amylase, the former is predominantly present in the germinating seed. These hydrolytic enzymes, particularly α-amylase, protease, ribonuclease, acid phosphatase and many other enzymes are under the control of gibberellins (GA). The predominant gibberellin is gibberellic acid (GA_3). Aleurone layers of cereal grains are the seat of production of these enzymes. During inhibition of the seed, the embryo supplies the hormone GA_3 to the aleurone, which then induces synthesis of polyribosomes, endoplasm reticulum (ER) and proteins, followed later by hydrolytic enzymes. Thus, production of hydrolytic enzymes requires the synthesis and presence of proteins, polyribosomes and nucleic acids. An effect of the herbicide on any one or more of these events would affect the other events as well, consequently affecting eventual germination of the seed. A significant number of preemergence herbicides, particularly acetamides (acetanilides), thiocarbamates and others, inhibit these metabolic processes vital for seed germination.

HERBICIDE CLASSIFICATION BASED ON PRIMARY SITE OF ACTION

Each herbicide or herbicide family (class or group) has a primary site of action different from the other. Some may have more than one site of action, but the most inhibitory of these will be affected first. The other sites may be considered secondary.

Increased and continuous use of herbicides for weed control has led to the selection of herbicide-resistant weeds for over two decades. As discussed in Chapter 10, resistance is the heritable ability of a weed biotype to survive a herbicide treatment to which the original population was susceptible in the past. There are at least three mechanisms of herbicidal resistance in plants: a) alternation in the site of action, b) changes in the rate of detoxification of the herbicide and c) modifications in the uptake and translocation of the herbicide. Knowledge of the sites of herbicidal action is a key in planning herbicide rotations and mixtures and thereby reducing the potential for selection of resistant weeds. Recognizing this, weed scientists in Australia, Canada, Europe and the USA classified herbicide groups (1, 2, 3, 4, etc. or A, B, C, D, etc.) based on mode of action. The Weed Science Society of America has identified 28 groups [130]. The sites of action of some of these groups (15, 16, 17, 25 and 26) are unknown.

A modified classification, presented in Table 7.1, includes the group number, site of action, chemical family and common names of herbicides. In this classification, the protein and nucleic acid biosyntheses which were excluded in the WSSA classification, but which could possibly serve as primary sites of action of some herbicide families (acetamides, thiocarbamates, pyridine carboxylic acids, phenoxys, benzoics and pyrazoliums), have been included. These sites are designated as I, II and III, representing protein biosynthesis (inhibition), protein and/or nucleic acid biosynthesis (stimulation) and nucleic acid biosynthesis (inhibition), respectively. This classification may be further modified or expanded as and when more chemical families and herbicides with different sites of action are identified. The main purpose of this presentation is to help the reader understand the principles underlying the classification and thereby create awareness of the problem of herbicide-resistant weeds, discussed in Chapter 10. This classification could serve as an aid in the management of herbicide resistance.

Table 7.1. Herbicide classification based on the primary site of biochemical action.

WSSA Group No.	Site of Action	Herbicide Families/Herbicides
	PHOTOSYNTHESIS (Inhibition)	
22	PHOTOSYSTEM I (electron diversion)	Bypiridiliums (diquat, paraquat)
	PHOTOSYSTEM II	
5	— same site; binding behaviour I	Triazines (ametryn, atrazine, cyanazine, prometon, prometryn, propazine, simazine) Triazinones (metribuzin, hexazinone) Uracils (bromacil, terbacil) Pyridazinones (pyrazon) Phenylcarbamates (desmedipham, phenmedipham)
7	— same site; binding behaviour II	Ureas (diuron, fluometuron, linuron, metobromuron, monuron, siduron, tebuthiuron) Acetamides (propanil)
6	— same site; binding behaviour III	Nitriles (bromoxynil, ioxynil) Benzothiadaziole (bentazon) Imidazolidinones (buthidazole) Phenyl pyridazines (pyridate)
	PIGMENT BIOSYNTHESIS (Inhibition)	
12	CAROTENOID BIOSYNTHESIS —phytoene desaturase	Pyridazinones (norflurazon) Tetrahydropyrimidinones
	—desaturation, cyclization, hydroxylation	Pyridinones (fluridone) Nitriles (bromoxynil, ioxynil)
14	PROTOPORPHYRIN IX (PP IX) Protoporphyrinogen oxidase (Protox)	Diphenyl Ethers (bifenox, acifluorfen, fomesafen, lactofen, oxyfuorfen Phenyl triazinones (Aryl triazinones) (sulfentrazone) N-Phenylphthalamides (flumiclorac) Oxadiazole (oxadiazon) Imines (CGA-248757)
11	BLEACHING: Carotenoid Biosynthesis inhibition (unknown target)	Triazole (amitrole, amitrole-T)
13	BLEACHING: Diterpenes (Inhibition)	Isoxazolidinones (clomazone)
	FATTY ACID (LIPID) BIOSYNTHESIS	
1	ACETYL CoA CARBOXYLASE (ACCase) (Inhibition)	Aryloxyphenoxypropionics (diclofop, fenoxaprop, fluazifop-P, haloxyfop, quizalofop-P)

(Contd.)

Table 7.1. *(Contd.)*

WSSA Group No.	Site of Action	Herbicide Families/Herbicides
		Cyclohexanediones (clethodim, cycloxidim, sethoxydim, tralkoxydim)
	ACETYL ELONGASE (ACEase) (Very long Chain Fatty Acids) (Inhibition)	Thiocarbamates (butylate, EPTC, molinate, pebulate, thiobacarb, triallate, vernolate)
		Acetamides (acetochlor, alachlor, butachlor, metolachlor, pronamide, propachlor, dimethenamid)

BRANCHED-CHAIN AMINO ACID BIOSYNTHESIS

WSSA Group No.	Site of Action	Herbicide Families/Herbicides
2	ACETOLACTATE SYNTHASE (ALS) (also called Acetohydroxy-acid Synthase: AHAS) (Inhibition)	Sulfonylureas (bensulfuron, chlorimuron, chlorsulfuron, ethametsulfuron, metsulfuron, nicosulfuron, primisulfuron, rimisulfuron, sulfometuron, triasulfuron, tribenuron, thifensulfuron, triflusulfuron,halosulfuron methyl (MON 1200), MON 37500
		Imidazolinones (imazapyr, imazamethabenz, imazaquin, imazethapyr, AC 263,922, AC 299,263
		Triazolopyrimidine sulfoanilides (flumetsulam)
		Pyrimidinylthio benzoates (pyrithiobac)

AROMATIC COMPOUND BIOSYNTHESIS

WSSA Group No.	Site of Action	Herbicide Families/Herbicides
9	EPSP SYNTHASE (Inhibition)	Glyphosate
10	GLUTAMINE SYNTHETASE (Inhibition)	Glufosinate
4	AUXIN-type action	Phenoxys (2,4-D, MCPA, dichlorprop, MCPA, MCPB)
	(Ethylene Biosynthesis — ACC Synthase	Pyridinecarboxylic acids (clopyralid, fluroxypyr, picloram, triclopyr)
		Benzoic acids (dicamba)
		Quinolinecarboxylic acids (quinclorac)

OTHER BIOCHEMICAL PROCESSES

WSSA Group No.	Site of Action	Herbicide Families/Herbicides
20	CELLULOSE SYNTHASE (Inhibition)	Nitriles (dichlobenil)
		Quinolinecarboxylic acids (quinclorac)
21	CELL WALL synthesis (site B) (Inhibition)	Benzamides (isoxaben)
19	AUXIN-type action (Inhibition)	Phthalamates (naptalam)
23	MITOSIS (Cell division inhibition)	Carbamates (chlorpropham, propham)
		Cineoles (cinmethylin)
		Bensulide
		Acetamides/Amides (napropamide)
3	MICROTUBULE ASSEMBLY (Inhibition)	Dinitroanilines (benefin, ethalfluralin, oryzalin, pendimethalin, trifluralin)
		Pyridines (dithiopyr, thiazopyr)
18	7,8-DIHYDROPTEROATE (DHP) SYNTHASE (Inhibition)	Carbamates (asulam)

(Contd.)

Table 7.1. *(Contd.)*

WSSA Group No.	Site of Action	Herbicide Families/Herbicides
22	UNCOUPLERS (Membrane disruptors) (Inhibition of oxidative phosphorylation)	Phenols (dinoseb) Arsenicals (DSMA, MSMA)
I	PROTEIN SYNTHESIS (Inhibition)	Acetamides (acetachlor, alachlor, butachlor, metolachlor, pronamide, propachlor, dimethenamid, napropamide) Thiocarbamates (butylate, EPTC, molinate, pebulate, thiobencarb, triallate, vernolate)
II	PROTEIN BISOSYNTHESIS and/or RNA BIOSYNTHESIS (RNA Polymerase) (Stimulation)	Pyridinecarboxylic acids (clopyralid, fluroxypyr, picloram, triclopyr) Phenoxys (2,4-D, MCPA, dichlorprop, MCPA, MCPB) Benzoic acids (dicamba)
III	RNA BIOSYNTHESIS (Inhibition)	Pyrazoliums (difenzoquat)

*I, II and III are not WSSA groups. These have been designated by the author.

MECHANISMS OF ACTION OF INDIVIDUAL HERBICIDES

The morphological, physiological and biochemical effects of some of the herbicides are discussed in the remainder of this chapter. The advances made in research on mechanisms of action of herbicides since the publication of the first edition of this book in 1983 have been reviewed for some of the older herbicides. This information is presented under the caption *update*.

Acetamides (Amides)

Alachlor, Propachlor, Prynachlor

These acetanilides, which are applied at preemrgence inhibit seed germination by interfering with the metabolic activities related to it. The seedlings of annual grass species against which they are effective do not emerge following preemergence application.

Alachlor and propachlor inhibited GA_3-induced α-amylase production by seeds of barley, the susceptible species, during germination [40, 76, 111]. Rao [124] and Rao and Duke [126] reported that alachlor, propachlor and prynachlor inhibited α-amylase as well as protease synthesis by seeds of barley. The inhibition of protease was more severe. Dhillon and Anderson [43] found that in the presence of propachlor, storage protein was not digested in squash cotyledons even after 9 d of seedling emergence. Rao [124] and Rao and Duke [126] suggested that these herbicides may be acting as repressors of gene action, preventing the normal expression of the hormonal effect of GA through the synthesis of DNA-dependent RNA. This was confirmed when higher levels of GA_3 overcame alachlor inhibition by removing the repressor effect [126]. Although the inhibition of α-amylase and protease by alachlor was overcome at higher GA_3 concentrations, it did not result in the removal of inhibition on seedling growth. These results suggested that the effects of these acetamides on α-amylase and protease were secondary and these herbicides were possibly acting on biosynthetic reactions required for the synthesis of these hydrolytic enzymes.

Duke [50] and Duke et al. [51, 52] reported that propachlor did not affect ATP formation but inhibited protein and RNA biosynthesis. They concluded that the primary effect of propachlor was on protein biosynthesis and that its effect on nucleic acid biosynthesis was secondary. Moreland et al. [111] also reported that propachlor inhibited protein and RNA synthesis besides the GA-induced production of α-amylase.

Rao [124] and Rao and Duke [125] thoroughly investigated the effects of alachlor, propachlor and prynachlor and found that seedling growth was inhibited by them, but such mitochondrial activities as oxidative phosphorylation leading to ATP synthesis and respiration, *in vivo* or *in vitro*, were not affected. However, these herbicides severely inhibited polyribosome formation (mRNA-ribosome complexing) (Fig. 7.14) by inhibiting ribosome synthesis and nascent peptide synthesis. These effects were observed only in barley, the susceptible species, but not in maize, the tolerant species. They concluded that protein synthesis was the primary site of action and the effects seen on hydrolytic enzymes were only secondary.

Since alachlor, propachlor and prynachlor inhibited the formation of mRNA-ribosome complex (polyribosome), Rao [124] and Rao and Duke [126] concluded that the chain initiation process was the primary target of these herbicides during protein synthesis. This polyribosome complex is required for accepting the aminoacyl-tRNA and increasing the length of the peptide chain. Dhillon [42] speculated that propachlor inhibits amino acid activation and hence prevents formation of aminoacyl-tRNA. The α-halogen of the α-chloroacetamides undergoes nucleophilic displacement with the amino group of methionyl-tRNA [76] and this would prevent peptide chain linkage from occurring [96]. Thus, these acetanilide herbicides could prevent: a) formation of mRNA-ribosome complex, b) activation of amino acid to form aminoacyl-tRNA or c) linkage of the growing peptide chain [polysome-tRNA-AA$_{(n)}$] and the incoming aminoacyl-tRNA, and thereby inhibit chain initiation and eventually protein synthesis.

Deal et al. [37] reported that alachlor and propachlor inhibited protein synthesis in vivo but not the polypeptide elongation process. They suggested that inhibition of protein synthesis could occur a step before translation of the message.

Chandler et al. [26] found that alachlor inhibited or stimulated nitrate ion uptake and nitrate reductase activity, depending on the rate used, order of exposure or the total time of exposure. Alachlor, however, did not inhibit the Hill reaction in isolated chloroplasts. Klepper [86] reported that alachlor was most effective in causing accumulation of nitrites in the germinating mustard seedlings while the effect of propachlor was only moderate. Alachlor and propachlor are also known to inhibit cell division and cell enlargement [36, 43]. Truelove et al. [151] found that alachlor inhibited choline incorporation into phosphotidylcholine during cotton seed germination.

Metolachlor

Metolachlor which is structurally similar to alachlor, inhibits seed germination and early seedling growth. It affects root growth by inhibiting protein synthesis. It also inhibits both cell division and cell enlargement. Deal et al. [37] found that metolachlor inhibited protein synthesis without an effect on the rates of polypeptide elongation or on the synthesis of specific polypeptide.

Truelove et al. [151] reported that metolachlor inhibited incorporation of choline into phosphotidylcholine during germination of cotton seeds. It had no effect on any other phospholipid. Pillai and Davis [120] found that metolachlor inhibited photosynthesis in *Chlorella* and respiration in *Chlorella* and *Phaseolus vulgaris*. However, they did not consider that these effects on photosynthesis and respiration were sufficient to explain the phytotoxicity of metolachlor.

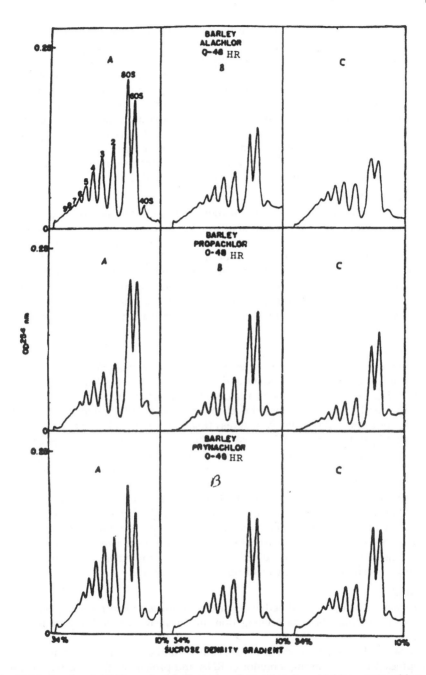

Fig. 7.14. *In vivo* inhibition of polyribosome formation by alachlor, propachlor and prynachlor in 2-day old barley roots. (Ribosome subunits and monosome peaks are numbered 40, 60, and 80s respectively and polyribosome peaks are numbered according to the size of polymer, i.e., 2 for dimer, 3 for trimer, 4 for tetramer, etc.) [124].

On the basis of evidence presented here, it can be concluded that protein synthesis is the primary mechanism of metolachlor action in susceptible species.

Update: Acetanilide herbicides generally inhibit several metabolic processes including protein synthesis and lipid metabolism. Alachlor and metolachlor, however, have little or no effect on lipid biosynthesis (159, 163). On the other hand, **butachlor** and **CDAA** have a strong inhibitory effect on the incorporation of ^{14}C-acetate into lipids in susceptible species. Chloroacetamides (alachlor, **acetochlor**, metolachlor, propachlor, etc.) have been hypothesized to interfere with fatty acid metabolism by alkylating key enzymes involved in fatty acid biosynthesis or by alkylating CoA and thereby interfering with CoA metabolism [55]. CoA plays an important role in lipid metabolism and other metabolic processes that are inhibited by acetamides.

Napropamide

Napropamide caused retardation of root growth [21]. It also reduced bud break, terminal growth and root development in cranberry cuttings [41]. Napropamide is known to inhibit growth by blocking the progression of dividing cells through the cell cycle into mitosis. It reduces cell division and DNA synthesis. This inhibitory effect may be due to an inhibition in the synthesis or activity of cell cycle specific proteins. Mercado [106] found that napropamide inhibited germination of *Cyperus rotundus* (purple nutsedge) tubers and shoot development. It also inhibited α-amylase activity and this inhibition correlated with inhibition of tuber germination, thereby preventing shoot growth.

Pronamide

Pronamide is applied at preemergence as well as postemergence. Foliar application of pronamide decreased the amount and altered the form of epicuticular wax on the leaves [127]. Decrease in wax was associated with wettability, which favoured spore deposition and encouraged fungal infection. Smith et al. [144] found that the nodes of the apex of pronamide-treated quackgrass (*Elytrigia repens*) rhizomes became swollen and necrotic and exhibited an increase in DNA, RNA and protein. These changes in the terminal nodes were accompanied by a dramatic increase in the levels of cellulose and several disruptive changes in the meristematic, nodal and internodal tissue of quackgrass.

Update: Pronamide inhibits mitosis by binding to tubulin and preventing its assembly into microtubules. The underdeveloped microtubules produced in the presence of pronamide make the cells unable to form spindle fibres, resulting in their inability to separate chromosomes to the poles of the cell. Thus mitosis is arrested at prometaphase. Pronamide-treated plants are devoid of cortical microtubules, required to prevent isodiametric cell expansion, thus had club-shaped root tips.

Propanil

Propanil, a foliage-applied herbicide, causes chlorosis and necrosis in sensitive weed species. It also inhibits radicle elongation and destroys permeability of root membranes.

Propanil inhibits the growth of tomato radicles and auxin-induced growth of *Avena* coleoptiles [71]. It is a strong inhibitor of RNA and protein synthesis as well as inhibitor of α-amylase synthesis [111]. It also inhibited oxygen uptake and phosphate esterification in soybean mitochondria [71]. Good [58] and Moreland and Hill [108] found that propanil also inhibited the Hill reaction of photosynthesis and the site of inhibition was the reduction of cytochrome 553 by PS II [116]. Propanil may also destroy chloroplast membranes and increase permeability of red beet membranes [71].

Propanil has a strong inhibitory effect on PS II electron transport. Moreland and Huber [109] reported that propanil interfered with both ATP generation and electron transport in mung bean mitochondria. At higher concentrations, propanil inhibited the state 3 oxidation of malate, exogenous NADH and succinate. At lower concentrations, it acted as an uncoupler.

Aliphatics

Acrolein

Acrolein, a general contact aquatic herbicide, kills the plant through its sulphahydryl reactivity which destroys vital enzyme systems in plant cells. The dead plant tissues disintegrate and flow downstream in the aquatic system.

Dalapon

Dalapon causes leaf chlorosis and necrosis and eventual growth inhibition by systemic action following foliar or root absorption. It interferes with the meristematic activity of the root tip and arrests mitotic activity at prophase. Fawcett and Slife [53] reported that dalapon reduced seed production of Setaria faberi (giant foxtail) and Datura stramonium (jimsonweed), lowered the dormancy of the seeds of Chenopodium album (lambsquarters) and inhibited the vigour of the seedlings of lambsquarters.

Dalapon does not appear to have a precise mechanism or one site of action but it affects many metabolic processes of the plant. It affects glucose metabolism by interfering with glycolysis. Jain et al. [75] reported that dalapon interfered with glucose utilization and that partial inhibition by dalapon may occur at initiation of the glycolytic pathway and within Krebs cycle.

Dalapon also affects nitrogen metabolism. Anderson et al. [6] found that dalapon caused degradation of protein to ammonium compounds and an increase in amides which were further broken down to ammonia in the susceptible Setaria lutescens (yellow foxtail) as well as the tolerant sugarbeets. After some time, the amides and amino acids returned to the normal level in the tolerant species but not in the susceptible species. This abnormal protein metabolism accompanied by unusual accumulations of metabolites may be the cause of dalapon's toxicity as well as its selectivity.

Dalapon markedly reduces the amount of surface wax produced on the leaves of various plants, indicating that it might be disturbing lipid metabolism to interfere with the normal deposition of cuticular components. It, however, does not seem to interfere with respiration or production of metabolic energy. It also does not affect oxidative phosphorylation by mitochondria or the activity of hydrolytic enzymes such as protease, α-amylase and dipeptidase.

Arsenicals

DSMA, MSMA

The first arsenical toxicity symptom is followed by cessation of growth and gradual browning. The plant eventually dies. Arsenicals cause root plasmolysis and leaf wilting followed by discoloration and necrosis of leaf tips and margins. When absorbed, the underground rhizomes also show browning of the storage tissues. Shoot meristematic tissues are probably the major sites of action of cacodylic acid and MSMA [131].

The general effect of arsenicals is lowering the rate of respiration and oxygen uptake. They are uncouplers and hence inhibit transformation of energy to ATP during oxidative phosphorylation. During uncoupling, despite continuation of the oxidative process, the energy produced is dissipated as heat and becomes unavailable for cellular processes. Arsenicals have a denaturing or precipitating effect on the protoplasm. They bind organic sulphahydryl groups, inhibiting enzymes such as pyruvate oxidases and phosphatases so that tissue respiration is inhibited.

Benzamides

Isoxaben

Isoxaben interferes with cell wall biosynthesis in susceptible weeds. It inhibits synthesis of cellulose, a cell wall polysaccharide [31].

Benzoics

Chloramben, Dicamba

Chloramben inhibits elongation and growth of roots and shoots of susceptible species. It promotes cell division in the presence of kinetin, as well as cell enlargement, indicating that it has auxinic properties [81]. It causes growth proliferation, resulting in stem swelling.

Dicamba inhibited oxygen uptake of leaves of *Cyperus rotundus* [101] and of isolated mitochondria [54]. Moreland et al. [111] found that dicamba inhibited GA_3-induced α-amylase synthesis in barley seeds but did not affect RNA and protein synthesis. Gruenhagen and Moreland [62] observed no inhibition of ATP level by dicamba. Watson et al. [160] reported that dicamba reduced ATPase activity. On the other hand, Arnold and Nalewaja [7] suggested that foliar-applied dicamba increases RNA and protein levels in susceptible plants by directly affecting the removal of histone from the DNA template. Quimby [122] found that dicamba influenced the DNA-precipitating properties of histones, suggesting that it may affect normal functioning of the genetic mechanism. Tsay and Ashton [152] reported that dicamba inhibited the development of dipeptidase activity of squash cotyledons.

Benzothiadiazoles

Bentazon

Bentazon, a benzothiadiazole herbicide, inhibits photosynthesis by inhibiting the PQB-binding niche on the D 1 protein of the PS II complex in chloroplast thylakoid membranes, thus blocking electron transport from PQ_A to PQ_B. This stops CO_2 fixation and production of ATP and $NADPH^{++}$, but plant death occurs in most cases by other processes. Inability to reoxidize PQ_A promotes formation of triplet state chlorophyll (^{3}chl) which interacts with ground-state oxygen to form singlet oxygen ($^1O_2$) (Fig. 7.3). Both triplet chlorophyll and singlet oxygen can abstract a hydrogen from unsaturated lipids, producing a lipid radical and initiating a chain reaction of lipid peroxidation. Lipids and proteins are attacked and oxidized, resulting in loss of chlorophyll and carotenoids, as well as leaky membranes; consequently, cell organelles dry and disintegrate rapidly.

Bipyridiliums

Diquat, Paraquat

These herbicides, due to their contact action, cause wilting and rapid desiccation of foliage. In the presence of light, phytotoxic symptoms are enhanced. The presence of both chlorophyll and light are necessary for maximal diquat injury. Paraquat disrupts membrane integrity to possibly cause wilting, necrosis and eventual death of foliage [13]. It causes segregation of the cytoplasm into isolated areas on the inner membrane surface and in the final phase was a rupture of the plasma membrane and collapse of the cell [18]. In the presence of surfactants, diquat altered the electrostatic stability of the membrane, resulting in deterioration of membrane integrity [148].

Bipyridiliums generally affect the plant through their ability to form toxic free radicals by the process of reduction and subsequent reoxidation to yield H_2O_2 and OH. Chloroplasts, oxygen and light are required for the formation of these free radicals. Bipyridiliums competitively inhibit the reduction of NADP to NADPH by removing the electrons from the electron transport system of PS II [167]. Diquat competitively inhibits photophosphorylation catalyzed by phenazine methosulphate (PMS), by acting at the same site as PMS in the electron-transport path of chloroplasts [167]. Thus diquat can shut electrons from ferredoxin to form diquat-free radicals and prevent normal reduction of NADP to NADPH. However, shutting electron transport is not likely to be the sole explanation of the herbicidal activity of diquat and paraquat. The instantaneous reoxidation of the reduced bipyridilium ion to form hydrogen peroxide (H_2O_2) appeared to be vitally important in the activity of diquat and paraquat. H_2O_2 may account for the observed effects of increased permeability of the membranes followed by drastic disruption of the cell organization. This H_2O_2 acts as a toxicant causing rapid injury to plants.

Generally, diquat and paraquat exert their herbicidal effects more rapidly in light than in darkness. However, death of plants could also occur slowly in darkness. Phytotoxicity in darkness is associated with respiration, implying that the herbicides interfere with mitochondrial electron transport.

Update: Bipyridilium herbicides enhance stress directly by generating oxygen radicals [19]. Reoxidation of paraquat leads to the generation of superoxide (O_2^-). When paraquat is applied to the plant as a colourless solution of the divalent cation (PQT^{++}), it is converted to the intensely blue monovalent cation radical ($PQT^{+\bullet}$) when reduced by PS II (Fig. 7.15). The Fa/Fb iron-sulphur centre of PS I is the electron donor to PQT^{++}. The redox potential of PQT^{++} (−446 mV) permits the acceptance of electrons from Fa/Fb (−560 mV).

$$PQT^{++} + PSI\,(\bar{e}) \longrightarrow PQT+$$
$$PQT^{+}. + O_2 \longrightarrow PQT^{++} + O_2^{-\bullet}$$

$$\overset{SOD}{2H^+ + O_2^{-\bullet} + O_2^{-\bullet} \longrightarrow H_2O_2 + O_2}$$
$$H_2O_2 + O_2^{-\bullet} \longrightarrow O_2 + OH + OH^-$$
$$PQT^{++} + H_2O_2 \longrightarrow PQT^{++} + OH^\bullet + OH^-$$
$$H^+ Fe^{++} + H_2O_2 \longrightarrow OH^\bullet + Fe^{+++} + H_2O$$

Fig. 7.15. Reactions involved in the generation of superoxide ($O_2^{-\bullet}$) and hydrogen peroxide (H_2O_2) by PS I electron acceptors during the effect of paraquat on PSI electron transport [56]. (SOD: Superoxide Dismutase).

The reduced paraquat, $PQT^{+\bullet}$, reduces O_2 to superoxide ($O_2^{-\bullet}$). In the process, PQT^{++} is regenerated. The catalytic qualities of PQT^{++} associated with PS I are phytotoxic. The paraquat-generated superoxide ($O_2^{-\bullet}$) is then converted through the catalyzing effect of superoxide dismutase (SOD) to $H_2O_2 + O_2^\bullet$. $PQT^{+\bullet}$ condenses with H_2O_2 to spontaneously produce PQT^{++}, $OH^•$ and OH^-. The $OH^•$ can also be produced by an Fe^{++} catalyst with H^+ and H_2O_2 as reactants [56].

Carbamates

Asulam

In bracken fern (*Pteridium aquilinum*), asulam produces a toxic effect in the frond buds situated in the rhizome branches. This eventually results in either complete death of frond buds, failure of buds to regenerate, or abnormal development of fronds from the affected buds. Veerasekharan et al. (155) found that the foliar application of asulam on bracken fern reduced RNA levels in frond buds and young fronds within 3 d and protein levels after 14 d. It stimulated O_2 uptake but inhibited ^{32}P uptake. Their results suggested that interference with RNA and protein synthesis at the metabolically active sinks (rhizome buds) could be one of the major mechanisms of action of asulam.

Update: Asulam appears to inhibit cell division and expansion of plant meristems by interfering with microtubule assembly or function. It also inhibits 7,8-dihydropteroate synthetase, an enzyme involved in folic acid synthesis which is required for purine nucleotide biosynthesis.

Desmedipham, Phenmedipham

Desmedipham and phenmedipham are postemergence herbicides used mostly in sugarbeets (*Beta vulgaris*). They show marginal or tip leaf burn and chlorosis. At lethal dose, they cause rapid collapse and death of the plant. High temperature and high light levels increase phytotoxicity. Bethlenfalvay and Norris [15] found that desmedipham caused photosynthetic inhibition in sugarbeet plants; this was more evident when the temperature was raised from 10° C to 35° C. High light intensity following foliar application caused greater inhibition of photosynthesis than low light intensity.

Desmedipham and phenmedipham contain the peptide bond structure common to many photosynthetic inhibitors [137]. Phenmedipham inhibits CO_2 assimilation. Both herbicides reduce photosynthesis by inhibiting the Hill reaction.

Cineoles

Cinmethylin

Although a specific mechanism of action for the compound has not been elucidated, cinmethylin, a cineole herbicide, appears to inhibit mitosis in meristematic regions (shoots and roots) of susceptible plants.

Cyclohexanediones (Cyclohexenones)

Clethodim, Cycloxidim, Sethoxydim, Tralkoxydim

Cyclohexanedione herbicides, like aryloxyphenoxypropionics, are efficient graminicides, causing post-emergence control of annual and perennial grasses in broadleaf crops. The treated plants exhibit chlorosis in developing leaves and cessation of growth. Within a few days, necrosis of shoot apex and meristematic regions of leaves and roots occurs.

Sethoxydim, clethodim, cycloxidim and tralkoxydim interfere with lipid metabolism. Burgstahler and Lichthenthaler [23] reported that sethoxydim reduced glycolipid and phospholipid content of maize seedlings, indicating its activity in the early stage in lipid metabolism. This was further corroborated by the work of Ishihara et al. [73]

who found that lipid biosynthesis in maize root tips was more sensitive to sethoxydim than was RNA, DNA, or protein synthesis. Burton et al. [24] reported that sethoxydim inhibited acetyl-coA carboxylase (ACCase), the enzyme which catalyzes *de novo* fatty acid synthesis in chloroplasts of sensitive maize, but not in chloroplasts of tolerant pea.

Dinitroanilines

Benefin, Dinitramine, Fluchloralin, Isopropalin, Pendimethalin, Trifluralin

The most striking symptoms of dinitroaniline herbicide treatment are similar to those obtained by treating seedlings with the classic mitotic disrupter colchicine [153]. Within treated roots, the cells do not proceed further than prometaphase of mitosis. As the cells are arrested at prometaphase, the nuclear membrane reforms around the chromosomes and the chromosome number is doubled. Due to spreading of the chromosomes throughout the cytoplasm, the nucleus assumes a highly lobed morphology [154]. Electron micrographs of the affected cells revealed that no or few spindle and kinetochore microtubules, the cellular structures responsible for chromosome movement during mitosis, occur in cells treated with dinitroanilines [67, 154]. Cortical microtubules, which determine the shape of root cells in the zone of elongation by directing the deposition of cell wall components, are also affected by these herbicides. The end results are isodiametric (square-shaped) rather than elongate (rectangular) cells, contributing to the swollen appearance of root tips.

The bicochemical work progressed so far indicates that dinitroaniline herbicides inhibit polymerization of tubulin, the major protein constituent of microtubules, into microtubules [107, 147]. The herbicides react with free tubulin heterodimers (composed of α and β subunits) in the cytoplasm and, when attached to the growing (or plus) end of the microtubule, prevent further polymerization. Microtubules are dynamic structures, extending at one end through polymerization of the tubulin, and either gradually losing subunits from the minus ends or cataclysmically losing whole microtubules or segments thereof. When polymerization is blocked by these herbicides, the depolymerization process shortens the microtubules continuously until they eventually become undetectable. The net result is arrestation of cell division, formation of polynucleate cells and inhibition of root and plant growth as described earlier.

Diphenylethers

Acifluorfen, Bifenox, Fomesafen, Fluorodifen, Fluoroglycofen, Lactofen, Oxyfluorfen

The most characteristic property of diphenyl ether (DPE) herbicides is that their fast phytotoxic action is strictly light-dependent.

$$
\text{DPE Herbicids} \xrightarrow{\hspace{1cm}} \underset{\substack{\text{protoporphyrin IX}\\ \text{(PP IX)}}}{\text{Accumulation of}} \xrightarrow{\text{Light}} \underset{\substack{\text{destruction of mem-}\\ \text{brane fatty acids}}}{\text{Peroxidative}} \xrightarrow{\hspace{1cm}} \text{Cell lysis}
$$

The first sign of herbicidal injury is water-soaked appearance of the treated tissues, suggesting cell membrane lysis, mainly caused by peroxidative destruction of lipids and other cell constituents including chlorophyll. From their extensive ultrastructural work, Derrick et al. [38] described the following sequence of events after administration of acifluorfen to the leaves of *Gallium aparine* (catchweed bedstraw) in light: a) by

3 h incubation, swelling/distortion of chloroplasts, b) by 3 to 5 h, evagination (unsheathing) and invagination (sheathing) of chloroplast envelopes, some cytoplasmic disturbance and tonoplast perturbation, c) by 15 h, tonoplast disturbance, lysis and/or cytoplasmic vesicle development and d) between 20 and 30 h, a progressive degeneration of membranes and organelles. The sequence of events described above was accompanied by a rapid decline in photosynthetic development. Since chloroplast envelope membranes showed such rapid and dramatic changes which preceded all other ultrastructural changes, Derrick et al. [38] suggested that these membranes may contain the primary target of DPE herbicides. Matringe and Scala [102] reported that the herbicidal activity of DPE herbicides probably resulted from their ability to interfere with the metabolism of tetrapyrroles, involved in protoporphyrinogen formation. Soybean treated with acifluorfen accumulated isoflavins and pterocarpans [32]; this accumulation was preceded by a transient increase in activity of relevant enzymes such as chalcone synthase and phenylalanine ammonia lyase [32, 70]. Other studies showed that the leaf content of phytoalexins and stress metabolites is increased by acifluorfen in various crops [90].

Lehnen et al. [95] found that DPE herbicides were potent inhibitors of the enzyme protoporphyrinogen oxidase (**Protox**) which oxidizes protoporphyrinogen (**PPG IX**) to protoporphyrin IX (**PP IX**), a tetrapyrrole intermediate. In healthy tissues, the enzymatically produced PP IX is normally channelled to the proper metal chelatase (to produce protoheme or Mg-Protox IX), thus preventing the accumulation of photodestructive PP IX. Jacobs et al. [74] suggested that PPG IX destruction may be a mechanism for providing protection from the toxic effects of PP IX accumulation in healthy tissues. Blockage of Protox deregulates the porphyrin pathway by reducing the feedback inhibition by heme [38, 39].

Inhibition of protox also leads to uncontrolled accumulation of PP IX, presumably because the unprocessed PP IX overflows its normal localization in the thylakoid membrane and is oxidized by molecular oxygen to PP IX. The accumulated PP IX can give rise, in light, to toxic species of oxygen and is able to degrade cellular molecules by peroxidative attack, as described in Fig. 7.16 [135]. Light absorption by excited PP IX produces triplet state PP IX which interacts with ground state oxygen to form singlet oxygen. Both triplet and singlet oxygen can abstract a hydrogen from unsaturated lipids (fatty acids), producing lipid radical and initiating a chain reaction of lipid peroxidation (Fig. 7.16). Lipids and proteins are then attacked and oxidized, resulting in loss of chlorophyll and carotenoids and in leaky membranes, eventually leading to rapid disintegration and death of cell and cell organelles.

PP IX accumulation is caused by most DPE herbicides including oxyfluorfen, acifluorfen, bifenox, fomesafen, fluoroglycofen, lactofen, fluorodifen and nitrofen. Matsumoto et al. [103] reported that Protox prepared from the etiolated and greening cucumber (*Cucumis sativa*) cotyledons was very susceptible and equally inhibited by oxyfluorfen. Protox from rice and barrnyardgrass (*Echinochloa crus-galli*) was more susceptible than that from radish (*Raphanus sativus*), cucumber and buckwheat (*Fagopyrum esculentum*). The inhibitory activity of oxyfluorfen and bifenox on protox was higher than that of other DPE herbicides including chlornitrofen and nitrofen. These results suggest that bifenox and oxyfluorfen rapidly inhibited protox in the intact plants and that protox inhibition was indeed the primary site of action of DPE herbicides.

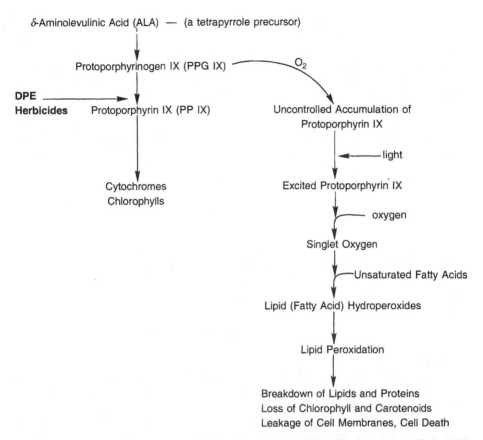

δ-Aminolevulinic Acid (ALA) — (a tetrapyrrole precursor)

Protoporphyrinogen IX (PPG IX) ——— O₂

DPE
Herbicides Protoporphyrin IX (PP IX)   Uncontrolled Accumulation of
Protoporphyrin IX

← light

Cytochromes   Excited Protoporphyrin IX
Chlorophylls

— oxygen

Singlet Oxygen

— Unsaturated Fatty Acids

Lipid (Fatty Acid) Hydroperoxides

Lipid Peroxidation

Breakdown of Lipids and Proteins
Loss of Chlorophyll and Carotenoids
Leakage of Cell Membranes, Cell Death

Fig. 7.16. Primary site of action of diphenyl ether herbicides (Redrawn from Ref. 135)

Imidazolidinones

Buthidazole

Buthidazole disrupts mesophyll chloroplasts and reduces starch in double sheath chloroplasts [65]. It inhibits respiration and photosynthesis.

York et al. [165] found no significant inhibition of PS I-mediated electron transport. York and Arntzen [164] reported that buthidazole inhibited the whole electron transport from water to methyl viologen, PS II-dependent dichlorophenolindophenol reduction, non-cyclic photophosphorylayion and silicomolybdate photoreduction. York et al. [166] observed that the major effect of buthidazole was inhibition of electron transport on the reducing side of PS II at the site of diuron and atrazine inhibition. It also had a secondary site of electron transport inhibition on the oxidizing side of PS II.

Imidazolinones

Imazamethabenz, Imazapyr, Imazaquin, Imazethapyr

The physiological effects of imidazolinones are similar to those of sulfonylureas. Shaner and Reider [139] reported that imazapyr caused no major changes in the rates of lipid biosynthesis (^{14}C-acetate incorporation), protein synthesis (^{14}C-leucine incorporation), leucine synthesis (^{14}C-cysteine incorporation) and RNA synthesis (^{14}C-uridine incorporation). They observed that DNA synthesis is unlikely to be the primary target of imazapyr. Shaner et al. [141] found that imazapyr-applied maize leaves contained increased levels (1.3 to 20-fold) of most amino acids, but markedly reduced levels of valine and leucine. Valine, leucine and isoleucine are linked by a common biosynthetic pathway. Imazapyr inhibits acetolactate synthase (ALS), the first enzyme common to all these three amino acids.

Several workers [112, 140, 146] reported that corn tissue treated with imazapyr contained low levels of extractable ALS, also called acetohydroxyacid synthase (AHAS). Stidham [145] observed that the ALS interaction with imidazolinones is somewhat different from the interaction of the enzyme with sulfonylureas. Though they both inhibit ALS *in vitro*, but only imidazolinones decrease the level of extractable ALS.

Imines

CGA-248757

CGA-248757, like diphenyl ether herbicides, stimulates accumulation of protoporphyrins (PP IX) by inhibiting protoporphyrinogen oxidase (protox) which oxidizes protoporphyrinogen (PPG IX) to PP IX. Accumulation of PP IX leads to peroxidation of membrane lipids in the presence of light and oxygen and irreversible damage to cell membranes and cell function.

Nitriles

Bromoxynil, Ioxynil

These nitrile herbicides are postemergence herbicides. Soon after application they burn the foliage which is followed by necrosis and extensive destruction of the leaf tissue.

Bromoxynil and ioxynil affect a number of essential physiological and biochemical processes in higher plants. Wain [156] reported that ioxynil inhibited the Hill reaction and uncoupled oxidative phosphorylation. Its toxicity was greater in light than in darkness. Paton and Smith [119] found ioxynil inhibiting electron transport, non-cyclic photophosphorylation and CO_2 fixation in leaf chloroplasts. Kerr and Wain [82] observed that ioxynil may depress the uptake of inorganic phosphate (Pi) and oxygen, eventually resulting in a low P/O ratio. Smith et al. [143] suggested that since electron transport is essential for photosynthesis, a blockage of this transport by ioxynil could lead to cessation of ATP synthesis and an eventual death of the plant. They considered the actual site of action of ioxynil to be near plastoquinone (PQ).

Update: Bromoxynil and ioxynil inhibit photosynthesis by binding to the PQ_B-binding niche on the D1 protein of the PS II complex in chloroplast thylakoid membranes, thus blocking electron transport from PQ_A to Q_B. This stops CO_2 fixation and production of ATP and $NADPH^{++}$, all required for plant growth. But plant death may occur by other processes. Inability to reoxidize PQ_A promotes formation of ^{3}chl (triplet chlorophyll) which interacts with ground state oxygen to form $^1O_2$ (singlet oxygen). Both triplet chlorophyll and singlet oxygen extract hydrogen from unsaturated fatty acids, producing a lipid radical and initiating a chain reaction of lipid peroxidation (vide Fig. 7-16). Lipids and proteins are attacked and oxidized, resulting in loss of chlorophyll and carotenoids, as well as leaky membranes; consequently cells and cell organelles dry and disintegrate rapidly.

Oxadiazoles

Oxadiazon

Oxadiazon causes herbicidal injury symptoms similar to those caused by diphenyl ether herbicides, namely rapid loss of pigmentation (bleaching of leaves) and desiccation. These are common symptoms of herbicides that kill by causing rapid peroxidative damage. Oxadiazon can affect protein synthesis, CO_2 fixation, nucleic acid synthesis and lipid synthesis [64]. It specifically inhibits chlorophyll biosynthesis at concentrations that have little effect on lipid peroxidation [48, 49]. Researchers have speculated that this effect is not related to its photobleaching effect. However, a clear link between inhibition of chlorophyll synthesis and lipid oxidation has been established. Duke et al. [46] found that PS II did not appear to be involved in oxadiazon activity. They also found that strong oxidation activity in tissues grown in far-red light (far-red light-grown tissues have no photosynthetic capacity) indicating that photosynthesis was not required for its activity. The oxadiazon-treated tissues accumulated protoporphyrin IX, a photodynamic pigment and reduced protochlorophyllide accumulation of protoporphyrin IX, which acts as a single oxygen-generating pigment [44].

Phenols

Dinoseb

Dinoseb, applied to the foliage, kills the plants rapidly by contact action. Green tissues turn brown and desiccate. The less susceptible species show stunted growth and this may eventually lead to death. The primary mechanism of action of phenols is inhibition of oxidative phosphorylation. They act as an uncoupling agent and inhibitor of oxidative phosphorylation. As uncouplers, they penetrate the membrane of the cell organelles and destroy its permeability. By interfering with oxidative phosphorylation, dinoseb releases respiration from its dependence on a phosphate acceptor. As a result, stimulation of respiration occurs. Failure to form ATP, in spite of continued respiration, rapidly affects the living plant.

Phenoxyalkanoic Acids

Phenoxyacetics (2,4-D, MCPA; Phenoxybutyrics (2,4-DB, MCPB

These chlorophenoxy acids, also known as auxin-type herbicides, cause characteristic phytotoxic symptoms such as epinasty, swelling, twisting and bending of the treated areas and distant plants. The plants stop growing, meristematic cells stop dividing, young leaves stop expanding and the mesophyll tissue ceases development. Roots may lose the ability to absorb water and mineral nutrients and leaves lose the photosynthetic ability. When the tip of the shoot is killed, some lateral buds may begin to grow because of their release from the apical dominance of the shoot tip. Treated plants die slowly, within two or three weeks of treatment. The degree of injury to leaves usually decreases as the leaves mature. Hence, treatment to mature leaves results in less effect. The herbicidal effect is associated with meristematic tissue rather than with expanding tissue.

At herbicidal concentrations, phenoxyacetics and phenoxybutyrics promote various biochemical activities, eventually leading to the destruction of a part or whole of the plant. Chrispeels and Hanson [29] reported that 2,4-D doubled the RNA content in soybean hypocotyls, with over half of the increase appearing in the microsomal fraction and one-quarter in the soluble fraction. Key et al. [83] and Cardenas et al. [25] observed that one of the actions of an auxinic herbicide like 2,4-D was to induce aberrations in the nucleic acid metabolism of sensitive plants. O'Brien et al. [118] found that 2,4-D promoted the synthesis of RNA polymerase in soybean hypocotyls and suggested that this could lead to the read out part of the template not being read by endogenous polymerases, or alternatively there may be a change in DNA sites available for transcription. Thus, 2,4-D could exert its influence at the gene level.

A search into the available literature indicates that 2,4-D affects all types of RNA, i.e. mRNA, rRNA and tRNA, as well as DNA and ribonuclease (RNase), Thus, the primary mechanism of 2,4-D would involve various reactions related to gene action. Chen et al. [27] suggested that in susceptible species auxinic herbicides make more of. the DNA template available for transcription, with a resultant production of more RNA and that the site of action appeared to be at the transcription stage.

Update: The primary action of auxinic herbicides appears to involve cell wall plasticity and nucleic acid metabolism. 2,4-D acidifies cell wall by stimulating the activity of a membrane-bound ATPase-driven proton pump. Reduction in apoplasmic pH induces cell elongation by increasing the activity of certain enzymes responsible for cell-wall loosening. 2,4-D and other auxin-type herbicides stimulate ethylene evolution, which may, in some cases, produce the characteristic epinastic symptoms associated with exposure to these herbicides.

Aryloxyphenoxypropionics (APPs)—Diclofop-methyl, Dichlorprop, Fenoxaprop, Fluazifop-P, Haloxyfop-P

The APP herbicides are used for postemergence control of annual and perennial grasses in certain broadleaf crops. They are readily absorbed and translocated to meristematic regions where they exert their herbicidal activity in grasses. They are applied as esters and rapidly converted to acids via the carboxylesterase activity upon entering the leaf. The acid form of the herbicide, considered to be the active form, is translocated to meristematic regions.

The injury caused by APP herbicides follows the pattern of: a) inhibition of growth of apical, internodal and root meristems, b) chlorosis and c) meristematic necrosis. At the physiological level, these herbicides cause membrane disruption, disruption of auxin action and inhibition of fatty acid biosynthesis. Diclofop methyl causes an increase in leaf cell membrane leakage, which may be direct, due to the result of membrane disruption, and indirect, through a disruption of energy translocation in chloroplasts or mitochondria. There is, however, strong evidence that these herbicides are specific to lipid synthesis and metabolism. Hoppe [72] found that incorporation of ^{14}C-acetate into polar lipids was considerably reduced by pretreatment with doclofop-methyl. Harwood et al. [63] and Wallker et al. [157] reported that when the sensitive wheat and barley plants were sprayed with fluazifop-butyl, subsequent incorporations of radioactivity from ^{1-14}C-acetate into acyl lipids was inhibited, whereas labelling of pigments was unaffected. Other herbicides such as fenoxaprop-ethyl, fenthioprop-ethyl and haloxyfop also reduced lipid synthesis in susceptible species [28, 88]. These results confirmed that APPs inhibit fatty acid synthesis. Cho et al. [28] concluded that their primary site of action is somewhere in the glycolytic pathway.

Much work was done subsequently to identify the exact target site of action of APP herbicides in the fatty acid biosynthetic pathway. Several workers [17, 24, 87, 138, 142] have reported that these herbicides were selective and potent inhibitors of acetyl-CoA carboxylase (ACCase) activity in grasses, indicating that ACCase was indeed the site of action.

ACCase is a multifunctional, biotinylated protein located in the stroma of plastids. It catalyzes the ATP-dependent carboxylation of acetyl-CoA to form malonyl-CoA

(Fig. 7.17). This is the first step in the *de novo* synthesis of fatty acids (vide discussion on Fatty Acid Biosynthesis and Fig. 7.7). The ATP-dependent carboxylation of biotin occurs at the carboxylation site. The biotin prosthetic group then moves to the carboxyltransferase site where that carboxyl group is transferred to acetyl-CoA, forming malonyl-CoA. Of the two partial reactions catalyzed by ACCase, the carboxyltransferase reaction is most sensitive to inhibition by the APP herbicides [129]. The partial reaction at the carboxyltransferase site (malonyl-CoA—acetyl-CoA exchange) is much more sensitive to these herbicides than that of the reaction at the carboxylation site (ATP-Pi exchange).

$$\text{ACCase-biotin} + HCO_3^- + ATP \longrightarrow \text{ACCase-biotin-}CO_2^- + ADP + Pi$$

$$\text{ACCase-biotin-}CO_2^- + \text{Acetyl-CoA} \xrightarrow{\text{APP Herbicides}} \text{Malonyl-CoA} + \text{ACCase-biotin}$$

Fig. 7.17. Site of action of aryloxyphenoxypropionic (APP) herbicides.

Shimabukuro and Hoffer [142] reported that the sensitive sites in the mechanism of action of diclofop-methyl were the plasma membrane and ACCase. The inhibition of auxin-induced growth in susceptible oat coleoptiles was related to the inhibition of lipid biosynthesis. Diclofop-induced lipid profile changed within 1 h of treatment. They concluded that inhibition of coleoptile growth, reduction in cellular lipid biosynthesis and lipid profile changes are probably due to factors other than inhibition of ACCase *in vivo*.

N-Phenylphthalamides

Flumiclorac

The primary site of action of flumiclorac, like that of diphenyl ethers, appears to be protoporphyrinogen oxidase (Protox), an enzyme of chlorophyll and heme biosynthesis catalyzing the oxidation of protoporphyrinogen (PPG IX) to protoporphyrin IX (PP IX). Inhbition of Protox leads to uncontrolled accumulation of PP IX, the first light-absorbing precursor of chlorophyll, because the unprocessed PPG IX overflows the normal localization in the thylakoid membrane and is oxidized by molecular oxygen to PP IX. This oxidation is catalyzed by a plasmalemma enzyme that has Protox activity, but it is insensitive to diphenyl ethers. PP IX formed outside its native environment probably is separated from Mg chelatase and other pathway enzymes that normally prevent accumulation of PP IX. Light absorption by PP IX apparently produces triplet state PP IX which interacts with ground-state oxygen to form singlet oxygen ($^1O_2$). Both triplet PP IX and singlet oxygen can abstract hydrogen from unsaturated lipids, producing a lipid radical and initiating a chain reaction of lipid peroxidation. Lipids and proteins are attacked and oxidized, resulting in loss of chlorophyll and carotenoids, as well as in leaky membranes; consequently cells and cell organelles dry and disintegrate rapidly.

Phenylpyridazines

Pyridate

Pyridate, upon entering the plant, is hydrolyzed to 3-phenyl-4-hydroxy-6-chloropyridazine, which then inhibits PS II electron transport.

Phenyl Triazinones (Aryl Triazinones)

Sulfentrazone

Sulfentrazone, an experimental phenyl triazinone (also termed as aryl triazinone) herbicide, does not affect germination of susceptible species, but inhibits seedling development by causing chlorosis. Nandihalli and Duke [114] found that sulfentrazone inhibited protoporphyrinogen oxidase (Protox) in barley, the susceptible species. Most photobleaching herbicides, including sulfentrazone, are bicyclic compounds that mimic half of the tetrapyrrole ring of protoporphyrinogen (the substrate of Protox) and compete for the catalytic site on the enzyme [47, 115]. Like other photobleaching herbicides, sulfentrazone is believed to cause accumulation of photodynamic protoporphyrin IX (PP IX) which reacts with oxygen to form highly destructive oxygen radicals (vide discussion on Diphenylethers)

Phthalamates

Naptalam

Naptalam appears to have both herbicidal and plant growth regulator activity. It is an auxin-inhibitor. It appears to attach to a phytotropin binding site and inhibit auxin efflux from the basal end of cells, thereby blocking basipetal auxin transport.

Pyrazoliums

Difenzoquat

The mechanism of action of difenzoquat is not well understood. However, its effects include inhibition of nucleic acid biosynthesis, photosynthesis, ATP production, K^+ absorption and P incorporation into phospholipids and DNA.

Pyridazinones

Metflurazon (SAN 6706), Norflurazon, Pyrazon, SAN 9785

Pyridazinones inhibit the development of plant pigments, causing chlorosis. They destroy chloroplast morphology and block carotenoid biosynthesis. Anderson and Schaelling [5] found that preemergence and early postemergence applications of pyrazon caused rapid desiccation of the seedling, causing death. Leaves showed chlorosis at the margins, with abnormal changes in the form and arrangement of chloroplasts. Grana formation was inhibited and thylakoids became swollen and perforated, possibly due to the inhibition of thylakoid structural protein formation. Bartels and Watson [12] reported that norflurazon inhibited carotenoid biosynthesis in wheat seedlings and caused accumulation of colourless carotenoid precursors, phytoene and phytofluene. Their results suggested that dehydrogenation reactions following phytoene formation were inhibited by norflurazon. They concluded that the disruption of chloroplasts and loss of chlorophyll were due to the absence of carotenoids in treated plants, which would normally act to protect the chloroplasts from photoreduction.

Update: Research done so far indicates that substituted pyradazinone herbicides inhibited one or more of the following: a) photosynthesis, b) carotenoid biosynthesis and c) fatty acid desaturation.

Norflurazon causes bleaching in susceptible species [63, 72, 133]. It inhibits carotenoid biosynthesis, resulting in the photo-oxidation of chlorophyll. Current evidence suggests that norflurazon blocks carotenoid biosynthesis by inhibiting phytoene desaturase, located in chloroplast membranes, that catalyzes the destruction of phytoene (vide discussion on **Carotenoid Biosynthesis** in Pigment Biosynthesis and Fig. 7.6). This hypothesis is supported by the work of Mayer et al. [104] who found norflurazon to be a reversible, non-competitive inhibitor of phytoene desaturase.

Harwood et al. [63] reported that norflurazon and its dimethylamino analogue metflurazon (SAN 6706) inhibit the desaturation of phytoene to lycopene during caro-tene biosynthesis, thus causing bleaching of leaves. Metflurazon inhibits both carotenoid biosynthesis and fatty acid desaturation. At higher concentrations it inhibits desaturation of 18:2 in galactolipids and at lower concentrations actively inhibits desaturation of 16:0 to form trans-Δ^3-hexadecenoic acid in the sn-2 position of phosphotidylglycerol, the major phospholipid found in plastids [36, 84] and linolenate synthesis [34]. The inhibitory effect of metflurazon on fatty acid desaturation is attributed to its ability to inhibit desaturases that catalyze the above reactions. These enzymes which utilize complex galactolipids (located in the chloroplast envelope), **monogalacto-syldiacylglycerol (MGDG)** and **digalactosyldiacylglycerol (DGDG)** as substrates, are located in the chloroplast envelope. It is hypothesized that the chloroplast envelope contains multiple desaturases and the selective effect of certain substituted pyridazinones is due to their ability to selectively inhibit one or more of these enzymes [45, 72].

The herbicidal effect of pyrazon is largely due to its ability to inhibit photosynthetic electron transport. It binds D1, the quinone-binding protein on the reducing side of PS II [77]. Pyrazon, however, is not a potent inhibitor compared to triazines and ureas at this site. Unlike SAN 9785 (see next paragraph), pyrazon has little effect on the degree of unsaturation. Generally, pyradazinones, except pyrazon, alter the fatty acid composition of lipids of the MGDG fraction. They particularly increase the degree of saturation of certain lipids, primarily galactolipids by blocking conversion of linoleic acid to linolenic acid [45, 63, 72, 158]. The galactolipids, MGDG and DGDG, make up a large portion of chloroplast membranes and the fatty acids associated with these lipids exhibit a high degree of unsaturation [78].

Compared to other pyridazinones, BASF 13338 (SAN 9785) is rather selective in its mode of action. It inhibits desaturation of fatty acids with little or no effect on photo-synthesis and carotenoid biosynthesis [45, 63, 72, 77]. Other workers have shown that this herbicide causes an increase in the 18:2/18:3 ratio found in the galactolipid (MGDG) fraction of plant tissues [94, 113, 117] by inhibiting the Δ-15 desaturase that catalyzes the desaturation of 18:2 bound to MGDG. It causes reduction in the level of linolenic acid, with a concomitant increase in the level of linoleic acid.

Pyridinecarboxylic Acids

Clopyralid, Picloram, Triclopyr

The mechanism of action of pyridinecarboxylic acids is similar to that of exogenous auxin (IAA) and auxin-type herbicides The primary action appears to involve cell-wall plasticity and nucleic acid metabolism. Clopyralid, picloram and triclopyr are believed to acidify the cell wall by stimulating activity of the membrane-bound ATPase proton pump. The reduction in apoplasmic pH induces cell elongation by increasing the activ-ity of enzymes responsible for cell-wall loosening. Low concentrations of these herbi-cides also stimulate RNA polymerase, resulting in subsequent increases in DNA, RNA and protein biosynthesis. An abnormal increase in these processes would lead to un-controlled cell division and growth, causing destruction of vascular tissue. In contrast, high concentrations of pyridinecarboxylic acid and other auxinic herbicides inhibit cell division and growth, usually in meristematic regions that accumulate photosynthate assimilates and herbicide from the phloem. Clopyralid, picloram and triclopyr, like other auxinic herbicides, stimulate ethylene evolution which in some cases produces the characteristic epinastic symptoms associated with exposure to these herbicides.

Pyridines

Dithiopyr, Thiazopyr

Dithiopyr and thiazopyr disrupt cell division by inhibiting mitosis in late metaphase and causing multipolar mitosis. These do not bind to tubulin and to another protein, a **65-68 kDa protein**, which may be a **microtubule associated protein (MAP)**. MAPs function in the stabilization of growing microtubules. The pyridine herbicides shorten microtubules such that they cannot form spindle fibres normally responsible for separating chromosomes to the poles of the cell during mitosis. Branched or multiple phragmoplasts are found in the zone of cell division. Cortical microtubules, which normally prevent isodiametric cell expansion, are also essentially absent, resulting in club-shaped root tips.

Pyridinones

Furidone

Fluridone blocks carotenoid biosynthesis by inhibiting phytoene desaturase. Carotenoids are largely absent in fluridone-treated plants, allowing $^1O_2$ (singlet oxygen) and ^{3}chl (triplet state chlorophyll) to abstract a hydrogen from an unsaturated lipid (e.g., membrane fatty acid, chlorophyll), producing a lipid radical (vide Carotenoid Biosynthesis in this Chapter). Lipid radical interacts with O_2, yielding a peroxidized lipid and another lipid radical. Lipid peroxidation leads to destruction of chlorophyll and membrane lipids. Proteins are also destroyed by $^1O_2$. Destruction of integral membrane components leads to leaky membranes and rapid tissue desiccation.

Pyrimidinylthio-benzoates (Benzoates)

Pyrithiobac

Like sufonylureas and imidazolinones, pyrimidinylthio-benzoate herbicides, represented by pyrithiobac, inhibit acetolactate synthase (ALS), an enzyme in the synthesis of branched-chain amino acids valine, leucine and isoleucine. ALS inhibition causes plants to appear stunted and chlorotic. Death generally occurs slowly, in 1 to 2 wk, as protein synthesis, cell division and growth are inhibited and metabolic precursors accumulated.

Quinolinecarboxylic Acids

Quinclorac

Quinclorac, the quinolinecarboxylic acid, is a new auxin-type herbicide used for control of dicot and monocot weeds in rice. Its phytotoxicity on *Echinochloa crus-galli* (barnyardgrass) is characterized by inhibition of shoot growth with chlorosis and subsequent necrosis of the entire tissue and a simultaneous stimulation of ethylene production by activating the synthesis of an enzyme, 1-aminocyclopropane-1-carboxylic acid (ACC) synthetase, in the root tissue [60]. The ACC and its conjugate (N-malonyl-1-aminocyclopropane-1-carboxylic acid) (MACC) accumulate predominantly in the shoot, suggesting transfer of ACC into the roots. Although the principal site of biochemical action of quinclorac in barnyardgrass is localized in the root tissue, the shoot appears to be the predominant target of the phytotoxic effect. Ethylene is not, however, inherently lethal to plant cells even at unphysiologically high concentrations. Grossman and Kwiatkowski [60] found that quinclorac induced a threefold increase in endogenous cyanide levels of shoot tissue. The cyanide formed as a coproduct during quinclorac-stimulated ethylene biosynthesis could be the agent by which the phytotoxicity of the herbicide on shoots was induced.

Grossman and Kwiatkowski [61] reported that the increase in cyanide (HCN) levels in the quinclorac-treated shoot tissue of barnyardgrass was closely related to the reduction in shoot growth. They also reported that the increase in β-cyanoalanine synthase activity, the main HCN detoxifying enzyme, ethylene production and the exogenous levels of ACC preceded cyanide accumulation. They concluded that cyanide, derived ultimately from quinclorac-stimulated ACC synthesis, is a causative factor in the herbicidal effects.

The quinclorac-resistant rice plants showed no notable changes in ACC, ethylene and cyanide production [60]. The HCN production in root tissue was 240-fold lower than that in barnyardgrass. In addition, plants and detached shoots of rice were less susceptible to damage by HCN than the respective parts of barnyardgrass. Thus, quinclorac-induced cyanide production could be involved in the mechanism of herbicide selectivity between the two species.

Sulfonylureas

Bensulfuron, Chlorimuron, Chlorsulfuron, Halosulfuron, Metsulfuron, Nicosulfuron, Primisulfuron, Prosulfuron, Sulfometuron, Triasulfuron, Thifensulforon, Triflusulfuron, Tribenuron

Sulfonylurea herbicides first affect the meristematic tissues where growth ceases soon after treatment. Chlorosis and the necrosis of these tissues soon follow, with dieback to the mature parts of the plant taking a further 3-4 wk. Chlorsulfuron has an early effect on plant tissue with a decrease in mitotic frequency, accompanied by inhibition of (^{3}H)-thimidine incorporation into DNA (128). Protein and RNA synthesis were only slightly affected and there was no direct effect on either photosynthesis (the oxygen-evolution step) or respiration. Although inhibition of DNA synthesis was the major effect, chlorsulfuron does not appear to inhibit DNA·polymerase directly and neither could its effects be reversed by addition of the four nucelotide precursors. This suggests that inhibition of DNA synthesis is not a direct effect of chlorsulfuron. The work of LaRossa and Schloss [91] with sulfometuron provided the first evidence that sulfonylureas inhibit the biosynthesis of branched-chain amino acids. Scheel and Casida [136] found that only trace levels of chlorsulfuron were required to inhibit plant tissue growth by 50% and mixtures of isoleucine and valine (but neither amino acid alone) were effective in counteracting the herbicide.

Chlorsulfuron was shown to be an extremely potent inhibitor of acetolactate synthase (ALS) (also referred to as acetohydroxyacid synthase, AHAS), the first enzyme in the pathway in the biosynthesis of branched-chain amino acids, valine, leucine and isoleucine in chloroplasts (Fig. 7.8) [128]. Genetic data have provided additional evidence that ALS is the target of the sulfonylurea herbicides.

The metabolic fates of pyruvate may include branched-chain amino acids, fatty acids, or both. Inhibition of ALS causes an increase in biosynthesis of fatty acids via the pyruvate dehydrogenase complex. Increase in fatty acid biosynthesis may be due to an increased pool of pyruvate as a result of inhibition of ALS.

Tetrahydropyrimidinones

Tetrahydropyrimidinones (1,3-diaryl-2-pyrimidinones), also called cyclic ureas, is a new class of photobleaching herbicides. Herbicides of this family are yet to be commercialized. Chemically, these herbicides combine several structural elements of well-known chlorotic herbicides including fluometuron, fluridone and norflurazon. Fluridone and fluometuron are parent structures for tetrahydropyrimidinones [9].

This new class of herbicides is active on a broad spectrum of monocot and dicot weeds and sedges, including *Echinochloa crus-galli*, *Cyperus difformis*, *Scirpus juncoides*, *Monochoria vaginalis*, *Cyperus serotinus*, *Eleocharis acicularis*, *Sagittaria pygmaea*, etc. while not affecting rice.

Babczinski et al. [9] found that these cyclic urea herbicides caused bleaching of green leaf tissue. The cytological changes included cell deterioration, vacuole disorientation, mitochondrial accumulation around the chloroplasts; and the chloroplasts having no thylakoid stacks. They also found inhibition of carotene biosynthesis at the phytoene desaturation step. The seedlings of *Lepidium sativum* bleached out completely and accumulated large amounts of phytoene. Chlorophyll bleaching was demonstrated to be a secondary photo-oxidative event caused by the loss of carotenoids. These authors further showed that the phytoene desaturase step is the primary molecular target of the tetrahydropyrimidinone herbicides.

Thiocarbamates

Butylate, EPTC, Diallate, Pebulate, Triallate

Thiocarbamate herbicides, primarily applied to the soil, inhibit shoot and root growth of grass weeds depending on the depth of incorporation. Dawson [35] observed EPTC causing kinking (zigzag pattern) on the first internode. The emerging leaves did not unfurl but stayed within the coleoptile, which was the major site of EPTC injury. The leaf primordia on apical buds as well as young leaves were seriously injured. He concluded that EPTC injury was concentrated in the mesophyll of the unfolding leaves within the coleoptile.

Wilkinson and Smith [161] found that EPTC and diallate inhibited fatty acid biosynthesis by inhibiting the incorporation of acetate-2-^{14}C and malonate-2-^{14}C into lipids of spinach chloroplasts. They suggested that inhibition of lipid synthesis and unsaturated fatty acid synthesis is the major mode of action of diallate and EPTC. This was confirmed by the work of other scientists who found that EPTC inhibited polyunsaturated fatty acid formation in wheat roots [80], galactolipid polyunsaturated fatty acid content in wheat leaf [79] and galactolipid and phospholipid polyunsaturated fatty acid content in soybean leaf [162].

Update: The considerable evidence accumulated so far indicates that the inhibitory effect of thiocarbamates (EPTC, diallate, triallate, vernolate and pebulate) on the synthesis of surface lipids such as waxes, cutin and suberin is due to their ability to inhibit the biosynthesis of very long-chain fatty acids **(VLCFA)**. VLCFAs are defined as fatty acids of chain length greater than 18 dinucleotide phosphate (NADPH)-dependent elongation of selected acyl-CoA substrates using malonyl-CoA as the condensing agent (or greater than C18 fatty acids) [14, 97]. Prevention of acyl-CoA elongases, which catalyze the synthesis of VLCFA, or inhibition of their activity, is suggested to be affected by thiocarbamates. This inhibitory effect of acyl-CoA elongases could be attributed to the selective effect of thiocarbamate on the *in vivo* synthesis of VLCFA or the metabolite of the thiocarbamate, such as its sulphoxide derivative. It is commonly assumed that the sulphoxide derivatives of thiocarbamates alkylate (carbomylate) CoA and interfere with CoA metabolism. However, Gronwald [59] has suggested that the thiocarbamate molecule itself (as opposed to its sulphoxide derivative) inhibits key enzymes involved in the synthesis of acetyl-CoA.

Inhibition of VLCFAs would lead to reduction in epicuticular wax formation on the plant foliage which increases leaf wettability, allowing the leaf to be more effectively treated by the foliage-applied herbicides.

Thiobencarb

Thiobencarb, when applied at preemergence, causes temporary leaf deformities. It produced leaf dieback, leaf cohesion, separation of laminal joints of wings in a single leaf, an increase in tillering and decrease in adventitious roots, abnormal emergence of panicle with fewer spikelets and abnormal spikelets in *Echinochloa crus-galli* [2]. Although thiobencarb is a non-hormonal herbicide, these aberrations suggested that it had growth-regulating properties similar to hormones.

When applied at postemergence, thiobencarb caused adnation of leaves or longitudinal fusion of leaf margins, separation of laminal joints by two wings in a single leaf, shorter and narrower leaves and shorter panicles with malformed spikelets in *Echinochloa crus-galli* [3]. Irrespective of growth stage, the herbicide produced shorter plants and panicles, fewer primary branches and spikelets per panicle and more shoots and panicles per plant.

Thiobencarb affects protein and RNA metabolism in bringing out the morphological changes described above.

Triazines

Atrazine, Ametryn, Propazine, Prometon, Prometryn, Simazine, Terbutryn

Triazines are widely known to inhibit growth of emerged seedlings. Injury symptoms appear a few days after seedling emergence. The foliage shows chlorosis followed by necrosis and eventually death of the tissue and whole plant ensues. All the effects of triazine herbicides occur only in light, not in darkness, thus indicating that these inhibitory effects are associated with photosynthetic processes.

Atrazine causes degradation of the chloroplasts and also swelling followed by disruption of the granal discs [69]. The membranes of grana and chloroplast envelope are also ruptured. Liang et al. [99] reported that atrazine caused several abnormalities in sorghum including multinucleate cells, bridges and increased chromosome numbers and interference with meiotic stability.

Inhibition of photosynthesis is a characteristic effect of triazines and is widely documented. Their specific site of action is the oxygen evolution step or photolysis in photosynthesis. They also inhibit ATP formation via photophosphorylation in the chloroplasts when flavin mononucelotide (FMN) is used as an electron acceptor but not when N-methyl phenozonium (PMS) is used as an electron acceptor [58]. Bishop [16] found that simazine inhibited FMN-catalyzed photophosphorylation but not that catalyzed by vitamin K_3. Blockage of oxygen evolution and light reactions results in a lack of reducing power, which is required for CO_2 reduction to carbohydrates.

All triazine herbicides inhibited state 3 respiration most and cyclic photophosphorylation [149, 150]. This decrease in state 3 oxidation of malate and succinate was found to be overcome by 2,4-dinitrophenol, indicating that the effect of triazines was similar to that of oligomycin [105]. Copping and Davis [30] reported that s-triazine herbicides inhibited non-cyclic photophosphorylation and were toxic only in light.

Brewer et al. [20] found that atrazine inhibited electron transport, but did not uncouple photophosphorylation. Its primary site of action was on the reducing side of PS II. They suggested that the electron transfer step between the primary acceptor (Q, the primary fluorescence quencher associated with PS II) and the plastoquinone pool of the electron transport chain is affected by atrazine.

The mechanism tolerance to triazines between biotypes of weed species is related to their abilities to inhibit PS II. Machado et al. [100] reported that atrazine inhibited electron transport on the reducing side of PS II in chloroplasts of a susceptible biotype of (*Chenopodium album*) common lambsquarters while it had little effect on chloroplasts of a tolerant biotype. Radosevich et al. [123] also reported that triazine herbicides, atrazine, atraton, propazine, prometryn, prometon and terbutryn, inhibited photosynthetic electron transport on the reducing site of PS II in thylakoid membranes from the susceptible biotype of *Senecio vulgaris* (common groundsel). Thylakoids isolated from the resistant biotype were 60 to 3200 times more tolerant to the inhibitors than the thylakoids from the susceptible biotype.

Update: Triazine herbicides inhibit photosynthesis by binding to the PQ_B-binding niche on the D1 protein of the PS II complex in chloroplast thylakoid membranes, thus blocking electron transport from PQ_A to PQ_B. This stops CO_2 fixation and production of ATP and $NADPH^{++}$, all needed for plant growth, but plant death occurs in most cases by other processes. Inability to reoxidize PQ_A promotes the formation of triplet state chlorophyll (^{3}chl) which interacts with ground-state oxygen to form singlet oxygen ($^1O_2$). Both ^{3}chl and $^1O_2$ can abstract hydrogen from unsaturated lipids, producing a lipid radical and initiating a chain reaction of lipid peroxidation. Lipids and proteins are attacked and oxidized, resulting in loss of chlorophyll and carotenoids, as well as leaky membranes; consequently cells and cell organelles dry and disintegrate rapidly.

Triazoles

Amitrole, Amitrole-T

The development of albino leaves and shoots is a distinct feature of amitrole toxicity. Higher concentrations inhibit growth and cause chlorosis, while lower concentrations stimulate growth. Chlorosis leads to death. Amitrole causes destruction of chlorophyll and impairment of chlorophyll development in tissues. The inhibitory effect on chlorophyll appears to occur prior to the photochlorophyll stage. Thus, the developing leaves show yellowing and browning followed by death. Amitrole-induced albinism is due to the destruction of chloroplast pigments or an inhibition of their synthesis. Bartels [10] reported that amitrole blocked light-induced plastid development as evidenced by the absence of a normal lamellar system and membranal disorganization. Inhibition of chloroplast development exerts an effect on photosynthesis. Bartels and Hyde [11] suggested that amitrole brings about a metabolic change in the developing plastids by destroying the chloroplast DNA or blocking the synthesis of chloroplast enzymes or structural components necessary for chloroplast synthesis.

Update: Aminotriazole inhibits carotenoid biosynthesis by inhibiting the phytoene desaturation step, resulting in accumulation of phytoene, phytofluene and ζ-carotene (Fig. 7.6). It also causes accumulation of lycopene, inhibiting the cyclization process.

Triazolopyrimidine Sulfonanilides

Flumetsulam

Flumetsulam, a triazolopyrimidine sulfonanilide herbicide, inhibits acetolactate synthase (ALS), a key enzyme in the biosynthesis of branched-chain amino acids valine, leucine and isoleucine. Plant death results from events occurring in response to ALS inhibition, but the actual sequence of phytotoxic events is not clear.

Uracils

Bromacil, Terbacil, UCC-C4243

Bromacil and terbacil inhibit photosynthesis by binding to the PQ_B-binding niche on the D1 protein of the PS II complex in chloroplast thylakoid membranes, thereby blocking the electron transport from PQ_A to PQ_B. This inhibits CO_2 fixation and production of ATP and $NADP^{++}$. Inability to reoxidize PQ_A promotes the formation of ^{3}chl (triplet chlorophyll) which interacts with ground-state oxygen to form singlet oxygen ($^1O_2$). Triplet chlorophyll and $^1O_2$ then cause lipid peroxidation, resulting in loss of chlorophyll and carotenoids and in leaky membranes.

UCC-C4243 requires light for phytotoxicity and the toxicity symptoms are similar to inhibitors of porphyrin synthesis, such as diphenyl ethers. It potently inhibits protoporphyrinogen oxidase (Protox). It causes light-dependent chlorophyll bleaching. Like other uracil herbicides, bromacil and terbacil, UCC-C4243 also causes electron transport from PQ_A to PQ_B and the sequence of events described above, resulting in loss of chlorophyll and carotenoids, as well as leaky membranes.

Ureas

Diuron, Fluometuron, Isoproturon, Linuron, Methabenzthiazuron, Monuron, Siduron

As substituted herbicides are applied to the soil, they are translocated to the foliage and mature leaves turn light green in colour followed by burning at the tips. This develops into chlorosis, growth retardation and eventually death of the plant. Ashton [8] reported that chlorophyll was the principal absorbing pigment involved in monuron injury and that no injury occurred in darkness. Normally, the higher the light intensity, the greater the injury.

It is widely believed that substituted urea herbicides inhibit the oxygen evolution step or photolysis in photosynthesis. The primary site of action of urea herbicides, like that of triazine herbicides, is located in PS II involving the oxygen evolution step. They interfere with the reducing side of PS II [123]. Ureas also inhibit non-cyclic electron transport as well as cyclic electron transport. Giannpolitis and Ayers [57] reported that monuron markedly accelerated photo-oxidations (chlorophyll bleaching and lipid oxidation) normally occurring in isolated chloroplasts. Their results provided additional support to the hypothesis that depletion of the source of reducing potential (NADPH) is responsible for chloroplast oxidations and plant death following treatment with a photosynthesis-inhibiting herbicide monuron.

Update: Ureic herbicides inhibit photosynthesis by binding to the PQ_B-binding niche on the D1 protein of the PS II complex in chloroplast thylakoid membranes, thus blocking electron transport from PQ_A to PQ_B. This stops CO_2 fixation and production of ATP and $NADPH^{++}$ (all needed for plant growth), but plant death occurs in most cases by other processes. Inability to reoxidize PQ_A promotes the formation of triplet state chlorophyll (^{3}chl), which interacts with ground-state oxygen to form singlet oxygen ($^1O_2$). Both ^{3}chl and $^1O_2$ can abstract hydrogen from unsaturated lipids, producing a lipid radical and initiating a chain reaction of lipid peroxidation. Lipids and proteins are attacked and oxidized, resulting in loss of chlorophyll and carotenoids, as well as leaky membranes; consequently, cells and cell organelles dry and disintegrate rapidly.

Unclassified Herbicides

Bensulide

Bensulide, a preemergence herbicide, inhibits root growth. Cutter et al. [33] reported that the affected roots became curved, root hairs were often present at the root tip and epidermal cells elongated radially. It causes binucleated cells in epidermal and cortical tissues of the root, suggesting that it may directly inhibit mitosis, although mitosis is not completely inhibited. Ashton et al. [8] found that bensulide inhibited proteolytic activity in squash seedlings.

Ethofumesate

Ethofumesate, a preemergence as well as postmergence herbicide, inhibits epicuticular wax on the leaf surface, suggesting that it affects lipid biosynthesis. Levitt et al. [98] reported that ethofumesate almost totally eliminated the epicuticular wax on cabbage leaves and, as a result, epicuticular transpiration increased.

Glufosinate

Glufosinate, also referred to as phosphinothricin, is an analogue of glutamate (glutamic acid). It is a mixture of the D- and L-isomers, with only the L-isomer being active. Glufosinate inhibits glutamine synthetase (GS) [1] which converts glutamate to glutamine in plant chloroplasts (vide **Glutamine Biosynthesis**). Inhibition is a two-step process: a) a reversible inhibition, with glufosinate being competitive with glutamate and b) an irreversible inhibition of phosphorylation.

Kocher et al. [89] found that photosynthetic carbon assimilation of sorghum was severely reduced after spraying with glufosinate, with a corresponding increase in ammonia level. Build-up of 'toxic' ammonia level could cause both inhibition of photosynthesis and the death of plants. The increase in ammonia caused by the herbicide stimulated leakage of K^+ ions. Lea et al. [93] reported that glufosinate caused an increase in ammonia levels with a decrease in free glutamine in a range of higher plants. The herbicide enters the cell and inhibits GS activity.

Glufosinate can also inhibit the rate of photosynthetic CO_2 assimilation. This inhibition is not related to the build-up of ammonia, but due to a direct effect upon the operation of the Calvin cycle or photosynthetic electron transport. The more likely explanation is that under higher ammonia levels, sufficient amino donors are not available to convert glyoxalate to glycine in the photorespiratory nitrogen cycle, which would prevent the recycling of glycerate back into the chloroplast. Prevention of operation of the Calvin cycle is analogous to placing the leaves in light in the absence of CO_2. Under such conditions, energy cannot be dissipated in the formation of ATP and $NADPH^{++}$, resulting in photoinhibition. Extensive photoinhibition will lead to the formation of triplet state chlorophyll, singlet oxygen ($^1O_2$) and the hydroxy radical ($OH^\bullet$) (vide Fig. 7.3 and relevant text).

Higher levels of ammonia can cause the uncoupling of photosynthetic phosphorylation. It is probable that unprotonated NH_3 diffuses freely through the chloroplast and thylakoid membranes. NH_3 is then able to take up a H^+ to form $NH4^+$ and in order to maintain an ionic balance there must be an inward flux of anions. Following the influx of ions, there is an osmotic uptake of water that causes the chloroplast and thylakoids to swell dramatically, causing protrusions in the affected leaves. At higher concentrations, ammonia can also bind to the water-splitting, manganese-containing site in PS II and thus also inhibit O_2 evolution in electron transport.

The above-described metabolic processes affected by glufosinate are considered to be only secondary, following its primary inhibitory effect on glutamine synthetase.

Inhibition of GS leads to multiple deleterious effects, thus making it an exquisitely toxic site [1] for glufosinate.

Glyphosate

Glyphosate is a highly specific competitive inhibitor of the enzyme 5-enolpyruvyl-shikimate-3-phosphate (EPSP) synthase involved in the shikimate pathway of biosynthesis of aromatic compounds (amino acids) including tryptophan, tyrosine and phenylalanine, all needed for protein synthesis or for biosynthetic pathways leading to growth (Fig. 7.9). The EPSP synthase catalyzes the reaction between shikimate-3-phosphate and 5-enolpyruvyl-shikimate-3-phosphate. The inhibition is centred at the ternary (three elements) dead-end complex of EPSP enzyme-shikimate-3-phosphate-glyphosate. A build-up of shikimate-3-phosphate due to inhibition of EPSP synthase by glyphosate has a deleterious effect on the formation of chorismic acid, a precursor required for the biosynthesis of aromatic compounds. This indicates that EPSP synthase inhibition is the primary site of action of glyphosate.

Glyphosate is considered to be an extraneous site inhibitor because it binds either entirely or to a large extent to a site outside of, or extraneous to, the active site of action. Recent studies showed that glyphosate may be trapped on the EPSP synthase enzyme in the presence of EPSP [132]. The carboxyl group of EPSP does not prevent glyphosate binding, since glyphosate itself has very little affinity for the free enzyme. In addition, glyphosate is a mixed inhibitor with respect to phosphate [121]. Abell [1] suggested that glyphosate binds near, but not at, the active site of EPSP synthase.

The secondary effects of glyphosate, responsible for bringing about cell death in plants, are the arrest of both protein synthesis and diverse phenolic compound formation. These are directly attributable to a deficit of aromatic amino acids. Other potentially significant effects are the increased destruction of indolacetic acid (IAA) and inhibition of porphyrin synthesis.

REFERENCES

1. Abell, L.M. 1996. Biochemical approaches to herbicide discovery: Advances in enzyme target identification and inhibitor design. Weed Sci. **44**: 734-742.
2. Al-Mamun, A. and M. Shimizu. 1978a. Studies on herbicide, benthiocarb. I. Effects on grass morphology of barnyardgrass. Japanese J. Crop Sci. **47**: 563-572.
3. Al-Mamun, A. and M. Shimizu. 1978b. Studies on herbicide, benthiocarb. 2. Effects of different amounts of benthiocarb treated at different growth stages on the morphogenesis of barnyardgrass. Japanese J. Crop Sci. **47**: 573-580.
4. Amen, R.D. 1968. A model of seed dormancy. Bot. Rev. **34**: 1-31.
5. Anderson, J.L. and J.P. Schaelling. 1970. Effects of pyrazon on bean chloroplast ultrastructure. Weed Sci. **18**: 455-459.
6. Anderson, R.N., A.J. Linck and R. Behrens. 1962. Absorption, translocation and fate of dalapon in sugar beets and yellow foxtail. Weeds **10**: 1-3.
7. Arnold, W.F. and J.D. Nalewaja. 1971. Effect of dicamba on RNA and protein. Weed Sci. **19**: 301-305.
8. Ashton, F.M., D. Penner and S. Hoffman. 1968. Effect of several herbicides on proteolytic activity of squash cotyledons. Weeds Sci. **16**: 169-171.
9. Babczinski, P. R.R. Schmidt, K. Shiokawa and K. Yasui. 1995. Substituted Tetrahydropyrimidinones. A new herbicidal class of compounds inducing chlorosis by inhibition of phytoene desaturation. I. Biological and biochemical results. Pestic. Biochem. Physiol. **52**: 33-44

10. Bartels, P.G. 1965. Effect of amitrole on the ultra-structure of plastids in seedlings. Plant Cell Physiol. **6**: 227-230.

11. Bartels, P.G. and A. Hyde. 1970. Buoyant density studies of chloroplast and nuclear deoxyribonucleic acid from control and 3-amino-1,2,4-triazole-treated wheat seedlings. Plant Physiol. **45**: 825-830.

12. Bartels, P.G. and C.W. Watson. 1978. Inhibition of carotenoid synthesis by fluridone and norflurazon. Weed Sci. **26**: 198-203.

13. Baur, J.R. and J.J. Bowman. 1972. Effect of 4-amino-3,5,6-trichloropicolinic acid on protein synthesis. Physiol. Plant. **27**: 354-359.

14. Bessoule, J.J., R. Lessire and C. Cassagne. 1989. Partial purification of the acyl-CoA elongase of *Allium porrum* leaves. Arch. Biochem. Biophys. **268**: 475-484.

15. Bethlenfalvay, G. and R.N. Norris. 1975. Phytotoxic action of desmedipham: Influence of temperature and light intensity. Weed Sci. **23**: 499-503.

16. Bishop, N.T. 1962. Inhibition of oxygen evolving system of photosynthesis by amino-triaz-ines. Biochem. Biophys. Acta **57**: 186-189.

17. Bjelk, L.A. and T.J. Monaco. 1992. Effect of chlorimuron and quizalofop on fatty acid biosynthesis. Weed Sci. **40**: 1-6.

18. Boulware, M.A. and N.D. Camper. 1972. Effect of selected herbicides on plant protoplasts. Physiol. Plant. **26**: 313-317.

19. Bowler, C., M.V. Montagu and D. Inze. 1992. Superoxide dismutase and stress tolerance. Ann. Rev. Plant Physiol. Biol. **43**: 83-116.

20. Brewer, P.E., C.J. Arntzen and F.W. Slife. 1977. Effect of atrazine, cyanazine and procyanazine on the photochemical reactions of isolated chloroplasts. Weed Sci. **27**: 300-308.

21. Briggs, B.A. 1978. Manipulation of herbicides and effect of herbicides on rooting. In Combined Proc. Inter. Ph. Propagator's Soc. 1977, **27**: 426-467.

22. Britton, G., P. Barry and A.J. Young. 1989. Carotenoids and chlorophylls: Herbicidal inhibition of pigment biosynthesis. *In* A.D. Dodge (ed.). Herbicides and Plant Metabolism. Cambridge Univ. Press, Cambridge, pp. 51-72.

23. Burgstahler, R.J. and H.K. Lichtenthaler, 1984. Inhibition by sethoxydim of phospho- and galactolipid accumulation in maize seedlings. *In* P.A. Siegenthaler and W. Eichenberger (eds.). Structure, Function and Metabolism of Plant Lipids. Elsevier, Amsterdam, pp. 619-622.

24. Burton, J.D., J.W Gronwald, D.A. Somers, B.G. Gengenbach and D.L. Wyse. 1989. Inhibition of acetyl-CoA carboxylase by cyclohexanedione and aryloxyphenoxypropionate herbicides. Pestic. Biochem. Physiol. **34**: 76-85.

25. Cardenas, J., F.W. Slife, J.B. Hanson and J. Butler. 1967. Physiological changes accompanying the death of cocklebur plants treated with 2 4-D. Weed Sci. **16**: 96-100.

26. Chandler, I M., I I Croy and P.W. Santelmann. 1972. Alachlor effects on plant nitrogen metabolism and Hill reaction. J. Agr. Food Chem. **20**: 661-664.

27 Chen,L.G., C.M. Switzer and R.A. Fletcher. 1971. Nucleic acid and protein changes induced by auxin-like herbicides. Weed Sci. **20**: 53-55.

28. Cho, H-Y., J.M. Widholm and F.W. Slife. 1986. Effects of haloxyfop on corn (*Zea mays*) and soybean (*Glycine max*) cell suspension cultures. Weed Sci. **34**: 496-501.

29. Chrispeels, M.J. and J.B. Hanson. 1962. The increase in ribonucleic acid content of cytoplasmic particulates of soybean hypocotyl induced by 2,4-dichlorophenoxyacetic acid. Weeds **10**: 123-125.

30. Copping, L.G. and D.E. Davis. 1972. Effects of atrazine on chlorophyll retention in corn leaf discs. Weed Sci. **20**: 86-89.

31. Corio-Costet, M.F., J. Lherminier and R. Scalla. 1991. Effects of isoxaben on sensitive and tolerant cell cultures. II. Cellulose alterations and inhibition of synthesis of acid-insoluble cell wall material. Pestic. Biochem. Physiol. **40**: 255-265.

32. Cosio, E.G., G. Weissenbock and J.W. McClure. 1985. Acifluorfen-induced isoflavonoids and enzymes of their biosynthesis in mature soybean leaves—whole leaf and mesophyll responses. Plant Physiol. **78**: 14-19.

33. Cutter, E.G., F.M. Ashton and D. Huffstutter. 1968. The effects of bensulide on the growth, morphology and anatomy of oat roots. Weed Res. 8: 346-352.
34. Davies, A.O. and J.L. Harwood. 1983. Effect of substituted pyridazinones on chloroplast structure and lipid metabolism in greening barley leaves. J. Exp. Bot. 34: 1089-1100.
35. Dawson, J.H. 1963. Development of barnyardgrass seedlings and their response to EPTC. Weeds 11: 60-66.
36. Deal, L.M. and F.D. Hess. 1980. An analysis of the growth inhibitory characteristics of alachlor and metolachlor. Weed Sci. 28: 168-175.
37. Deal, L.M., J.T. Reeves, B.A. Larkins and F.D. Hess. 1980. Use of an in vitro protein synthesizing system to test the mode of action of chloroacetamides. Weed Sci. 28: 334-340.
38. Derrick, P.M. and A.H. Cobb. 1987. The effects of acifluorfen on membrane integrity in *Galium aparine* leaves and protoplasts. Proc. Br. Crop Protec. Conf. on Weeds, 3: 997-1004.
39. Derrick, P.M., A.H. Cobb and K.E. Pallett. 1988. Ultrastructural effects of the diphenyl ether herbicide acifluorfen and experimental herbicide M&B 39279. Pestic. Biochem. Physiol. 32: 153-163.
40. Devlin, R.M. and R.P. Cunningham. 1970. The inhibition of gibberellic acid induction of α-amylase activity in barley endosperm by certain herbicides. Weed Res. 10: 316-320.
41. Devlin, R.M. and I.E. Demoranville. 1974. Influence of dichlobenil and three experimental herbicides on bud break, terminal growth and root development of cranberry cuttings. Abstr. Weed Sci. Soc. Amer. pp. 14-15.
42. Dhillon, N.S. 1970. An investigation of the mode of action of N-isopropyl-chloroacetamide. Ph.D. thesis, Utah State Univ., Logan, Utah, USA, 88 pp.
43. Dhillon, N.S. and J.L. Anderson. 1972. Morphological, anatomical and biochemical effects of propachlor on seedling growth. Weed Res. 12: 182-189.
44. Duggan, J. and M. Gassman. 1974. Induction of porphyrin synthesis in etiolated bean leaves by chelators of iron. Plant Physiol. 53: 206-215.
45. Duke, S.O. 1985. Effects of herbicides on nonphotosynthetic biosynthetic processes. In S.O. Duke (ed.). Weed Physiology. Vol. II. Herbicide Physiology. CRC Press, Boca Raton, Fl., USA.
46. Duke, S.O., J. Lyndon and R.N. Paul. 1989. Oxadiazon activity is similar to that of p-nitrodiphenyl ether herbicides. Weed Sci. 37: 152-160.
47. Duke, S.O., U.B. Nandihalli and M.V. Duke. 1994. Protoporphyrinogen oxidase as the optimal herbicide site. In S.O. Duke and C.A. Rebeiz (eds.). Porphyric Pesticides: Chemistry, Toxicology and Pharmaceutical Applications. ACS Sym. Ser. 559.
48. Duke, S.O., J.M. Becerril, T.D. Sherman, J. Lyndon and H. Matsumoto. 1990. The role of protoporphyrin IX in the mechanism of action of diphenyl ether herbicides. Pestic. Sci. 30: 367-378.
49. Duke, S.O., J.M. Becerril, T.D. Sherman and H. Matsumoto. 1991. Photosensitizing porphyrins as herbicides. In P.A. Hedin (ed.). Naturally Occurring Pest Regulators. Am. Chem. Soc. Symp. Ser. No. 449.
50. Duke, W.B. 1967. An investigation of the mode of action of 2-chloro-N-isopropylacetanilide. Ph.D. thesis, Univ. of Illinois, Urbana, Illinois USA, 124 pp.
51. Duke, W.B., F.W. Slife and J.B. Hanson. 1967. Studies on mode of action of 2-chloro-N-isopropylacetanilide. Abstr. Weed Sci. Soc. Amer. p. 50.
52. Duke, W.B., F.W. Slife, J.B. Hanson and H.S. Butler. 1975. An investigation on the mechanism of action of propachlor. Weed Sci. 23: 142-147.
53. Fawcett, R.S. and F.W. Slife. 1978. Effects of 2,4-D and dalapon on weed seed production and dormancy. Weed Sci. 26: 543-547.
54. Foy, C.L. and D. Penner. 1965. Effect of inhibitors and herbicides on tricarboxylic acid cycle substrate oxidation by isolated cucumber mitochondria. Weeds 13: 226-231.
55. Furest, E.P. 1987. Understanding the mode of action of the chloroacetamide and thiocarbamate herbicides. Weed Technol. 1: 270-277.
56. Fuerst, E.P. and M.A. Norman. 1991. Interactions of herbicides with photosynthetic electron transport. Weed Sci. 39: 458-464.

57. Giannopolitis., C.N. and G.S. Ayers. 1978. Enhancement of chloroplast photooxidations with photosynthesis-inhibiting herbicides and protection with NADH and NADPH. Weed Sci. **26**: 440-443.

58. Good, N.E. 1961. Inhibitors of Hill reaction. Plant Physiol. **36**: 788-803.

59. Gronwald, J.W. 1991. Lipid biosynthesis inhibitors. Weed Sci. **39**: 435-439.

60. Grossman, K. and J. Kwiatkowski. 1993. Selective induction of ethylene and cyanide biosynthesis appears to be involved in the selectivity of the herbicide quinclorac between rice and barnyardgrass. J. Plant Physiol. **142**: 457.

61. Grossman, K. and J. Kwiatkowski. 1995. Evidence for a causative role of cyanide, derived from ethylene biosynthesis, in the herbicidal mode of action of quinclorac in barnyardgrass. Pestic. Bichem. Physiol. **51**: 150-160.

62. Gruenhagen, R.D. and D.E. Moreland. 1971. Effect of herbicide on ATP levels in excised soybean hypocotyls. Weed Sci. **19**: 319-323.

63. Harwood, J.L., S.M. Ridley and K.A. Walker. 1989. Herbicides inhibiting lipid synthesis. In A.D. Dodge (ed.). Herbicides and Plant Metabolism. Cambridege Univ. Press, Cambridge, 277 pp.

64. Hatzios, K.K. 1987. Comparative effects of oxadiazon and its metabolism on biochemical processes of enzymatically isolated leaf cells of soybean. Zizaniology **1**: 235-242.

65. Hatzios, K.K. and D. Penner. 1980. Some effects of buthidazole on corn (*Zea mays*) photosynthesis, respiration, anthocyanin formation and leaf ultrastructure. Weed Sci. **28**: 97-100.

66. Hawkes, T.R., J.L. Howard and S.E. Pontin. 1989. Herbicides that inhibit the biosynthesis of branched chain amino acids. In A.D. Dodge (ed.). Herbicides and Plant Metabolism. Cambridge Univ. Press, Cambridge, 277 pp.

67. Hess, F.D. 1987. Herbicide effects on the cell cycle of meristematic plant cells. Ann. Rev. Weed Sci. **31**: 183-203.

68. Hewitt, E.J. and B.A. Notten. 1966. Effect of substituted uracil derivatives on induction of nitrate reductase in plants. Biochem. J. **101**: 39-40.

69. Hill, E.R., E.C. Putala and J. Vengris. 1968. Atrazine-induced ultrastructural changes of barnyardgrass chloroplasts. Weed Sci. **16**: 377-380.

70. Hoagland, R.E. 1989. Acifluorfen action on growth and phenolic metabolism in soybean (*Glycine max*) seedlings. Weed Sci. **37**: 743-747.

71. Hofstra, G. and C.M. Switzer. 1968. The phytotoxicity of propanil. Weed Sci. **16**: 23-28.

72. Hoppe, H.H. 1989. Fatty acid biosynthesis — a target site of herbicide action. In P. Boger and G. Sandman (eds.). Target Sites of Herbicide Action. CRC Press, Inc., Boca Raton, Fl., USA.

73. Ishihara, K., H. Hosaka, M. Kubota, H. Kamimura, N. Takakusa and Y. Yasuda. 1987. Effects of sethoxydim on the metabolism of excised root tips of corn. In R. Greenhalgh and T.R. Roberts (eds.). Pesticide Chemistry and Technology. Blackwell Scientific, Palo Alto, CA, USA, pp. 187-190.

74. Jacobs, J.M., J.M. Wehner and N.J. Jacobs. 1994. Porphyrin stability in plant supernatant fractions: Implications for the action of porphyrinogenic herbicides. Pestic. Biochem. Physiol. **50**: 23-30.

75. Jain, M.L., E.B. Kurtz and K.C. Hamilton. 1966. Effect of dalapon on glucose utilization in the shoot and root of barley. Weeds **9**: 431-442.

76. Jaworski, E.G. 1969. Analysis of mode of action of herbicidal α-chloroacetamides. J. Agr. Food Chem. **17**: 165-170.

77. John, St. J.B. 1982. Effects of herbicides on the lipid composition of plant membranes. In D.E. Moreland, J.B. St. John and F.D. Hess (eds.). Biochemical Responses Induced by Herbicides. Amer. Chem. Soc., Washington, D.C., pp. 97-109.

78. Joyard, J. and R. Douce. 1987. Galactolipid synthesis. In P.K. Stumpf and E.E. Conn (eds.). Biochemistry of Plants. Vol. 9. Lipids: Structure and Function. Academic Press, New York, pp. 215-274.

79. Karunen, P. and L. Eronen. 1977. Influence of S-ethyl dipropylthiocarbamate (EPTC) on the fatty acid composition of wheat leaf galactolipids. Physiol. Plant. **40**: 101-104.

80. Karunen, P., N. Ualanne and R.E. Wilkinson. 1976. Influence of S-ethyl dipropylthiocarbamate on growth, chlorophyll and carotenoid production and chloroplast ultrastructure on germinating *Polytrichum commune* spores. Bryologist **79**: 332-338.

81. Keitt, G.W. and R.A. Baker. 1966. Auxin activity of substituted benzoic acids and their effect on polar auxin transport. Plant Physiol. **41**: 1561-1569.

82. Kerr, M.W. and R.L. Wain. 1964. The uncoupling oxidative phosphorylation in pea shoot mitochondria by 3,5-diido-4-hydroxy benzonitrile (ioxynil) and related compounds. Ann. Appl. Biol. **54**: 441-446.

83. Key, J.L., C.Y. Lin, E.M. Gifford, Jr. and R. Dengler. 1966. Relation of 2,4-D induced growth aberrations to changes in nucleic acid metabolism in soybean seedlings. Bot. Gaz. **127**: 87-94.

84. Khan, M-U., N.W. Lem, K.R. Chandorkar and J.P. Williams. 1979. Effect of substituted pyridazinones (SAN 6706, SAN 9774 and SAN 9785) on galactolipids and their associated fatty acids in the leaves of *Vicia faba* and *Hordeum vulgare*. Plant Physiol. **64**: 300-305.

85. Kleczkowski, L.A. 1994. Inhibitors of photosynthetic enzymes/carriers and metabolism. Ann. Rev. Plant Physiol. **45**: 339-367.

86. Klepper, L.A. 1975. Inhibition of nitrate reduction by photosynthetic inhibitors. Weed Sci. **23**: 188-190.

87. Kobek, K., M. Focke and H.K. Lichtenthaler. 1988. Fatty acid biosynthesis and acetyl-CoA carboxylase as a target of diclofop, fenoxaprop and other aryloxy-phenoxy-propionic acid herbicides. Z. Naturforsch. **43c**: 47-54.

88. Kocher, H. 1983. Influence of the light factor on physiological effects of the herbicide Hoe 39866. Aspects of Applied Biology **4**: 227-234.

89. Kocher, H., H.M. Kellner, K. Lotzsch and E. Dom. 1982. Mode of action and metabolic fate of the herbicide fenoxyprop-ethyl, Hoe 33171. Proc. Br. Crop Protec. Conf. On Weeds, pp. 341-347.

90. Komives, T. and J.E. Casida. 1983. Acifluorfen increases the leaf content of phytoalexins and stress metabolites in several crops. J. Agric. Food Chem. **31**: 751-755.

91. LaRossa, R.A. and J.V. Schloss. 1984. The sulfonylurea herbicide sulfometuron methyl is an extremely potent and selective inhibitor of acetolactate synthase in *Salmonella typhrimurium*. J. Biol. Chem. **259**: 8753-8757.

92. Lea, P.J. and S.M. Ridley. 1989. Glutamine synthetase and its inhibition. In A.D. Dodge (ed.). Herbicides and Plant Metabolism. Cambridge Univ. Press, Cambridge, 277 pp.

93. Lea, P.J., K.W. Joy, J.L. Ramos and M.G. Guerrero. 1984. The action of 2-amino-4-(methylphosphinyl)-butanoic acid (phosphinothricin) and its 2-oxo derivative on the metabolism of cyanobacteria and higher plants. Phytochemistry **23**: 1-6

94. Leech, R.M. C.A. Walton and N.R. Baker. 1985. Some effects of 4-chloro-5-(dimethylamino)-2-phenyl-3(2H)-pyridazinone (SAN 9785) on the development of chloroplast thylakoid membranes in *Hordeum vulgare* L. Planta **165**: 277-283.

95. Lehnen, L.P. T.D. Sherman, J.E. BeCerril and S.D. Duke. 1990. Tissue and cellular localization of acifluorfen-induced porphyrins in cucumber cotyledons. Pestic. Biochem. Physiol. **37**: 239-248.

96. Leis, J.P. and E.B. Keiler. 1970. Protein chain initiating methionine RNAs in chloroplasts and cytoplasm of wheat leaves. Proc. Nat. Acad. Sci. (USA) **67**: 1593-1599.

97. Lessire, R. J-J. Bessoule and C. Cassagne. 1989. Involvement of a β-ketoacyl-CoA intermediate into acyl-CoA elongation by an acyl-CoA elongase purified from leek epidermal cells. Biochem. Biophys. Acta **1006**: 35-40.

98. Levitt, J.R.C., D.N. Duncan, D. Penner and W.F. Meggitt. 1978. Inhibition of epicuticular wax deposition on cabbage ethofumesate. Plant Physiol. **61**: 1034-1036.

99. Liang, G.H.L., K.C. Feltner, Y.T.S. Liang and J.L. Morrill. 1967. Cytogenetic effects and responses of agronomic characters in grain sorghum (*Sorghum vulgare* Pens.) following atrazine application. Crop Sci. **7**: 245-248.

100. Machado, V.S., C.J. Arntzen, J.D. Bandeen and G.R. Stephenson. 1978. Comparative triazine effects upon photosystem II. Photochemistry in chloroplasts of two common lambsquarters (*Chenopidium album*) biotypes. Weed Sci. **26**: 318-322.

101. Magalhaes, A.C. and F.M. Ashton. 1969. Effect of dicamba on oxygen uptake and cell membrane permeability in leaf tissue of *Cyperus rotundus*. Weed Res. **9**: 48-52.

102. Matringe, M. and R. Scala. 1988. Effects of acifluorfen-methyl on cucumber cotyledons: Porphyrin accumulation. Pestic. Biochem. Physiol. **32**: 164-172.

103. Matsumoto, H., J.J. Lee and K. Ishizuka. 1993. A rapid and strong inhibition of protoporphyrinogen oxidase from several plant species by oxyfluorfen. Pestic. Biochem. Physiol. **47**: 113-118.

104. Mayer, M.P., D.L. Bartlett, P. Beyer and H. Kleinig. 1989. The *in vitro* mode of action of bleaching herbicides on the desaturation of 15-*cis*-phytoene and *cis-d* -carotene in isolated daffodil chloroplasts. Pestic. Biochem. Physiol. **34**: 111-117.

105. McDaniel, J.L. and R.E. Frans. 1969. Soybean mitochondrial response to prometryn and fluometuron. Weed Sci. **17**: 192-196.

106. Mercado, L.R. 1975. Some effects of amitrole on the respiratory activities of *Zea mays*. Weeds **8**: 29-38.

107. Morejohn, L.C., T.E. Bureau, J. Mole-Bajer, A.S. Bajer and D.E. Fosket. Oryzalin, a dinitroaniline herbicide, binds to plant tubulin and inhibits microtubule polymerization *in vitro*. Planta **172**: 252-264.

108. Moreland, D.E. and K.L. Hill. 1963. Inhibition of photochemical activity of isolated chloroplasts by acylanilides. Weeds **11**: 55-60.

109. Moreland, D.E. and S.C. Huber. 1979. Alterations to properties of the inner mitochondrial membrane induced by chlorpropham and other (phenyl carbamate and phenylamide) herbicides. Abstr. Weed Sci. Amer. p. 104.

110. Moreland, D.E., T.E. Corbin and J.E. McFarland. 1993. Oxidation of multiple substrates by corn shoot microsomes. Pestic. Biochem. Physiol. **47**: 206.

111. Moreland, D.E., S.S. Malhotra, R.D. Gruenhagen and E.H. Shokrahil. 1969. Effects of herbicides on RNA and protein synthesis. Weed Sci. **17**: 556-563.

112. Muhitch, M.J., D.L. Shaner and M.A. Stidham. 1987. Imidazolinones and acetohydroxyacid synthase from higher plants. Plant Physiol. **83**: 451-456.

113. Murphy, D.J. J.L. Harwood, K.A. Lee, F. Roberto, P.K. Stumpf and J.B. St. John. 1985. Differential responses of a range of photosynthetic tissues to a substituted pyridazinone, SAN 9785: Specific effects on fatty acid desaturation. Phytochemistry **24**: 1923-1929.

114. Nandihalli, U.B. and S.O. Duke. 1993. The porphyrin pathway as a herbicide target site. *In* S.O. Duke, J.J. Menn and J.R. Plimmer (eds.). Pest Control with Enhanced Environmental Safety. ACS Symp. Ser. 524, pp. 62-78.

115. Nandihalli, U.B. and S.O. Duke. 1994. Structure-activity relationships of 12 protoporphyrinogen oxidase-inhibiting herbicides. *In* Porphyric Pesticides: Chemistry, Toxicology and Pharmaceutical Applications. ACS Symp. Ser. 559, pp. 143-146.

116. Nishimura, M. and A. Takamiya. 1966. Energy and electron transfer systems in algal photosynthesis. I. Action of two photochemical systems in oxidation-reduction reactions of cytochrome in *Porphyra*. Biochem. Biophys. Acta **120**: 45-56.

117. Norman, H.A. and J.B. St. John. 1987. Differential effects of a substituted pyridazinone, BAS 13-338, on pathways of monogalactosyldiacylglycerol synthesis in *Arabidopsis*. Plant Physiol. **85**: 684-688.

118. O'Brien, J.J., B.C. Jarvis, J.H. Harvey and J.B. Hanson. 1968. Enhancement by 2,4-dichlorophenoxyacetic acid of chromatin RNA polymers in soybean hypocotyl tissue. Biochem. Biophys. Acta **169**: 35-43.

119. Paton, D. and J.E. Smith. 1965. The effect of 4-hydroxy-3,5-idobenzonitrile in CO_2 fixation, ATP formation and NADP reduction in chloroplasts of *Vicia faba* L. Weed Res. **5**: 75-77.

120. Pillai, C.G.P. and D.E. Davis. 1975. Mode of action of CGA-18762, CGA-17020 and CGA-24705. Proc. 28th Meeting South. Weed Sci. Soc. USA, pp. 308-313.

121. Pohlenz, H-D. and R. Hofgen. 1994. Antisense gene expression as tool for evaluating potential molecular herbicide targets. Eighth IUPAC Intnl. Congress of Pesticide Chemistry, July 4-9, 1994, Washington, D.C.

122. Quimby, P.C. 1967. Studies relating to the sensitivity of dicamba for wild buckwheat (*Polygonum convolvulus* L.) vs. Selkirk wheat (*Triticum aestivum* L.) and a possible mode of action. Ph.D. thesis. N. Dakota State Univ. , Fargo, USA, 87 pp.

123. Radosevich, S.R., K.E. Steinback and C.J. Arntzen. 1979. Effect of photosystem II inhibitors on thylakoid membranes of two common groundsel (*Senecio vulgaris*) biotypes. Weed Sci. **27**: 216-218.

124. Rao, V.S. 1974. Mechanisms of action of acetanilide herbicides. Ph.D. thesis, Cornell University, Ithaca, New York, USA, 116 pp.

125. Rao, V.S. and W.B. Duke. 1974. The effects of acetanilide herbicides on polysome and protein formation. Abstr. Weed Sci. Soc. Amer. pp. 87.

126. Rao, Sivaji V. and W.B. Duke. 1976. Effect of alachlor, propachlor and prynachlor on GA_3^- induced production of protease and *a*-amylase. Weed Sci. **24**: 616-618.

127. Rawlinson, C.J., G. Muthylu and R.H. Turner. 1978. Effect of herbicides on epicuticular wax of winter oilseed rape (*Brassica napus*) and infection by *Pyrenopeziza brassicae*. Trans. British Mycol. Soc. **71**: 441-451.

128. Ray, T.B. 1984. Site of action of chlorsulfuron. Plant Physiol. 75:827-831.

129. Redina, A.R., J.D. Beaudoin, A.C. Craig-Kennard and M.K. Breen. 1989. Kinetics of inhibition of acetyl-coenzyme A carboxylase by the aryloxyphenoxypropionate and cyclohexanedione graminicides. Proc. Brighton Crop Prot. Conf. Weeds **1**: 163-172.

130. Retzinger Jr., E.J. and C. Mallory-Smith. 1997. Classification of herbicides by site of action for weed resistance management strategies. Weed Technol. **11**: 384-393.

131. Sachs, R.M. and J.L. Michael. 1971. Comparative phytotoxicity among four arsenical herbicides. Weed Sci. **17**: 550-564.

132. Sammons, R.D., K.J. Gruys, K.S. Anderson, K.A. Johnson and J.A. Sikorski. 1995. Re-evaluating glyphosate as a transition-state inhibitor of EPSP synthase: Identification of an EPSP synthase•EPSP•glyphosate ternary complex. Biochemistry **34**: 6433-6440.

133. Sandman, G., I.E. Clark, P.M. Bramley and P. Boger. 1984. Inhibition of phytoene desaturase-mode of action of certain bleaching herbicides. Z. Naturforsch. **39c**: 443-449.

134. Sandman, G., A. Schmidt, H. Linden and P. Boger. 1991. Phytoene desaturase, the essential target for bleaching herbicides. Weed Sci. **39**: 474-479.

135. Scalla, R. and M. Matringe. 1994. Inhibitors of protoporphyrinogen oxidase as herbicides: diphenyl ethers and related photobleaching molecules. Rev. Weed. Sci. **6**: 103-132.

136. Scheel, D. and J.E. Casida. 1985. Sulfonylurea herbicides: Growth inhibition in soybean cell suspension cultures and in bacteria correlated with block in biosynthesis of valine, leucine and isoleucine. Pestic. Biochem. Physiol. **23**: 398-412.

137. Schweizer, E.E. and D.M. Weatherspoon. 1971. Response of sugarbeets and weeds to phenmedipham and two analogs. Weed Sci. **19**: 635-639.

138. Secor, J. and C. Cseke. 1988. Inhibition of acetyl-CoA carboxylase activity by haloxyfop and tralkoxydim. Plant Physiol. **86**: 10-12.

139. Shaner, D.L. and M.L Reider. 1986. Physiological responses of corn (*Zea mays*) to AC 243,997 in combination with valine, leucine and isoleucine. Pestic. Biochem. Physiol. **25**: 248-257.

140. Shaner, D.L. B.K. Singh and M.A. Stidham. 1990. Interaction of imidazolinones with plant acetohydroxyacid synthase: Evidence for *in vivo* binding and competition with sulfometuron methyl. J. Agric. Food Chem. **38**: 1279-1282.

141. Shaner, D.L., M.A. Stidham, M. Muhitch, M. Reider, P. Robson, and P. Anderson. 1985. Mode of action of the imidazolinones. Proc. British Crop Protec. Conf. **1**: 147-154.

142. Shimabukuro, R.H. and B.L. Hoffer. 1994. Effects on transmembrane proton gradient and lipid biosynthesis in the mode of action of diclofop-methyl. Pestic. Biochem. Physiol. **48**: 85-97.

143. Smith, J.E., D. Paton and M.M. Robertson. 1966. Herbicides on electron transport. Proc. 8th British. Weed Control Conf. **1**: 279-282.

144. Smith, L.W., R.L. Peterson and R.F. Worton. 1971. Effects of dimethylpropynyl benzamide herbicide on quackgrass rhizomes. Weed Sci. **19**: 174-177.

198

145. Stidham, M.A. 1991. Herbicides that inhibit acetohydroxyacid synthase. Weed Sci. **39**: 428-434.

146. Stidham, M.A. and D.L. Shaner. 1990. Imidazolinone inhibition of acetohydroxyacid synthase *in vitro* and *in vivo*. Pestic. Sci. **29**: 335-340.

147. Strachan, S.D. and F.D. Hess. 1983. The biochemical mechanism of action of the dinitroaniline herbicide oryzalin. Pestic. Biochem. Physiol. **20**: 141-150.

148. Sutton, D.L. and C.L. Foy. 1971. Effect of diquat and several surfactants on membrane permeability in red beet root tissue. Bot. Gaz. **132**: 299-304.

149. Thompson, O.C., B. Truelove and D.E. Davis. 1969. Effect of herbicide prometryne [2,4-bis-(isopropylamino)-6-(methylthio)-S-triazine] on mitochondria. J. Agr. Food Chem. **17**: 997-999.

150. Thompson, O.C., B. Truelove and D.E. Davis. 1974. Effects of triazines on energy relations of mitochondria and chloroplasts. Weed Sci. **22**: 164-166.

151. Truelove, B., A.M. Dinner, D.E. Davis and J.D. Weete. 1979. Metolachlor, membranes and permeability. Abstr. Meeting. Weed Sci. Soc. Amer. pp. 99-100.

152. Tsay, R. and F.M. Ashton. 1974. Influence of 2,4-D, dicamba and naptalam on hormonal control of dipeptidase. Weed Sci. **22**: 72-74.

153. Vaughn, K.C. and M.A. Vaughn. 1990. Structural and biochemical characterization of dinitroaniline-resistant *Eleusine*. *In* M.B. Green, H.M. LeBaron and W.K. Moberg (eds.). Managing Resistance to Agrochemicals. Amer. Chem. Soc., Washington, D.C. **421**: 364-375.

154. Vaughn, K.C. and L.P. Lehnen, Jr. 1991. Mitotic disrupter herbicides. Weed Sci. **39**: 450-457.

155. Veerasekaran, P., R.C. Kirkwood and W.W. Fletcher. 1977. Studies on the mode of action of asulam in bracken (*Pteridium aquilinum* L. Kuhn). II. Biochemical activity in the rhizome buds. Weed Res. **17**: 85-92.

156. Wain, R.L. 1964. Ioxynil — some considerations on its mode of action. Proc. 7th Brit. Weed Control Conf. **1**: 306-311.

157. Walker, K.A., S.M. Ridley, T. Lewis and J.L. Harwood. 1988. Fluazifop, a grass-selective herbicide which inhibits acetyl-CoA carboxylase in sensitive plant species. Biochem. J. **254**: 307-310.

158. Wang, X-M, D.F. Hildebrand, H.A. Norman, M.L. Dahmer, J.B. St. John and G.B. Collins. 1987. Reduction of linoleate content in soybean cotyledons by a substituted pyridazine. Phytochemistry. **26**: 955-960.

159. Warmund, M.R., H.D. Kerr and E.J. Peters. 1985. Lipid metabolism in grain sorghum (*Sorghum bicolor*) treated with alachlor plus flurazole. Weed Sci. **35**: 25-28.

160. Watson, M.C., P.G. Bartels and K.C. Hamilton. 1980. Action of selected herbicides and Tween 20 on oat (*Avena sativa*) membranes. Weed Sci. **28**: 122-127.

161. Wilkinson, R.E. and A.E. Smith. 1975. Reversal of EPTC induced fatty acid synthesis inhibition. Weed Sci. **23**: 90-92.

162. Wilkinson, R.E., B. Michel and A.E. Smith. 1977. Alteration of soybean complex lipid biosynthesis by S-ethyl dipropylthiocarbamate (EPTC). Plant Physiol. **60**: 86-88.

163. Yenne, S.P. and K.K. Hatzios. 1989. Influence of oxime ether safeners and metolachlor on acetate incorporation into lipids and on acetyl-CoA carboxylase of grain sorghum. Pestic. Biochem. Physiol. **35**: 146-154.

164. York, A.C. and C.J. Arntzen. 1979. Photosynthetic electron transport inhibition with buthidazole. Abstr. Weed Sci. Soc. Amer., p. 103.

165. York, A.C., C.J. Arntzen and F.W. Slife. 1978. Effect of buthidazole on the photochemical reactions of isolated pea (*Pisum sativum* L.) chloroplasts. Abstr. Weed Sci. Soc. Amer. pp. 76.

166. York, A.C., C.J. Arntzen and F.W. Slife. 1981. Photosynthetic electron transport inhibition by buthidazole. Weed Sci. **29**: 59-64.

167. Zweig, G., N. Shavit and M. Avron. 1965. Diquat (1,1-ethylene-2,2-dipyridilium dibromide) in photoreactions of isolated chloroplasts. Biochem. Biophys. Acta **109**: 332-346.

8

Herbicide Transformations in Plants

The ability of one plant species to detoxify a herbicide as opposed to the inability of another species to do so forms the basis for herbicide selectivity. Such differential detoxification ability among plant species is economically valuable for selective weed control.

Herbicides absorbed by the plant undergo structural and chemical modifications to form innocuous or less phytotoxic substances. These metabolic transformations, which include degradation and detoxification, occur both in tolerant and susceptible plant species, but in the tolerant species they take place at a rate much faster than herbicide accumulation and before the chemical can disrupt the plant metabolic processes. In susceptible plant species, the herbicide undergoes transformations in small amounts and more slowly. Some herbicides undergo metabolic transformations that may result in increased phytotoxicity. Thus, this Chapter covers the mechanisms related to both detoxification and activation of herbicides.

There is a growing apprehension from an ecological standpoint about the accumulation of herbicide residues in the plant either in the form of parent compounds or their metabolites. As a result, knowledge of the mechanisms of herbicide degradation and detoxification is becoming increasingly essential.

PATHWAYS OF HERBICIDE TRANSFORMATION

Plants accomplish herbicide transformation in different metabolic pathways. These include: oxidation, sulphoxidation, hydroxylation, hydrolysis, decarboxylation, N-dealkylation, dealkylthiolation, deamination, dethioation, dehalogenation, conjugation, ring cleavage, etc. Some of these important reactions are briefly discussed below.

Oxidation

There are three types of oxidation, that is, alpha (α), beta (β) and omega (ω), which involve oxidation at three different sites on the side chain. The ω-(2,4-dichlorophenoxy) alkane nitriles undergo α-oxidation [24] as shown in Fig. 8.1.

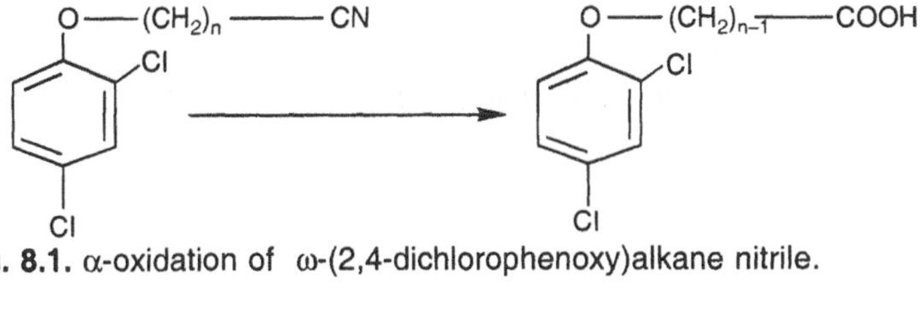

Fig. 8.1. α-oxidation of ω-(2,4-dichlorophenoxy)alkane nitrile.

β-oxidation of the side chain of 2,4-dichlorophenoxy acids with even–numbered carbons results in the formation of active 2,4-D, while the same with an odd number causes the formation of an unstable intermediate, which degrades to CO_2 and 2,4-dichlorophenol [90] as illustrated in Fig. 8.2.

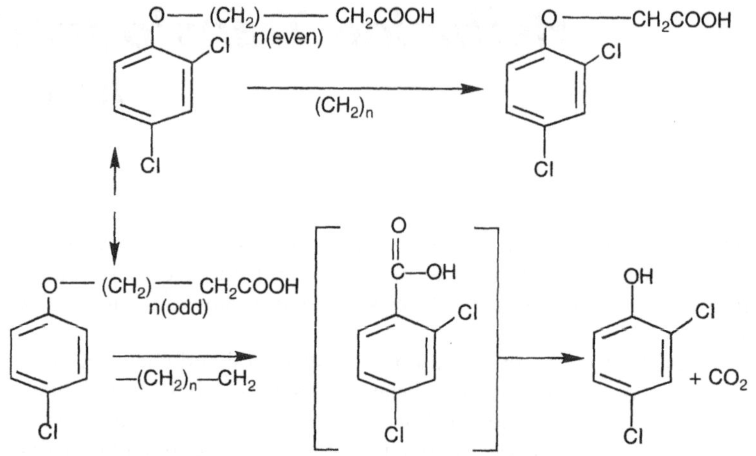

Fig. 8.2. β-oxidation of (2,4-dichlorophenoxy)alkane carboxylic acids with even-numbered and odd-numbered carbons on the side chain.

β-oxidation is an important route of transformation in that the non-herbicidal higher ω-phenoxyalkanoic acids are converted into herbicidal compounds. Similarly, tolerant plants are capable of degrading active herbicidal products into inactive products. This explains the tolerance of grasses to 2,4-D. 2,4-DB, the 4-carbon analogue of 2,4-D, undergoes β-oxidation in certain weed species to form 2,4-D. Hence, a plant species can be resistant to 2,4-DB either by virtue of resistance to 2,4-D or by virtue of the absence of β-oxidizing enzymes that alter 2,4-DB.

Fawcett et al. [24] reported that 10-phenoxy-*n*-deconoic acid produced large amounts of phenol during its breakdown through ω-oxidation. Other phenoxyalkanoic acids with an even number of carbons on the side chain produce little or no phenol. The ω-oxidation is shown in Fig. 8.3.

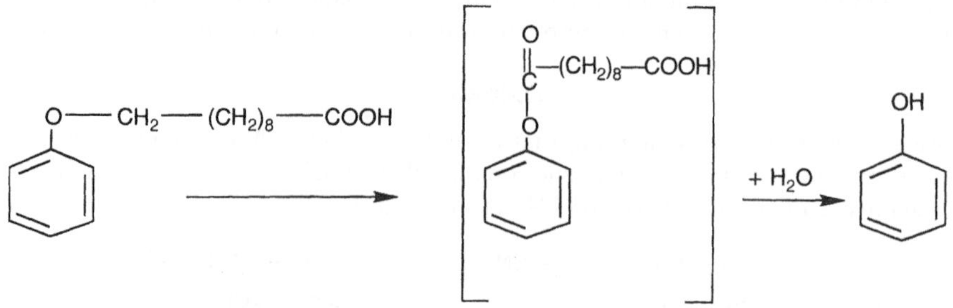

Fig. 8.3. ω-oxidation of 10-phenoxy-*n*-deconoic acid.

Sulphoxidation (S-Oxidation)

Oxidation of sulphide groups and other sulphur moieties is a common occurrence in certain herbicides like triazines (SCH$_3$-triazines) and thiocarbamates. Sulphoxidation is catalyzed primarily by microsomal cytochrome P-450-dependent mono-oxygenases. The products of sulphoxidation are more water-soluble and tend to be more chemically active and short-lived in the environment. In the case of detoxification, sulphoxidation of alkylsulphide groups is often an important initial step. The sulphoxides generated by the oxidation of the methylsulphide (SCH$_3$) ring substituents of terbutryn and metribuzin are substrates for glutathione-s-transferases, which catalyze their conversion to non-phytotoxic peptide conjugates. In the case of thiocarbamates, the sulphone derivative is the carbamoylating agent rather than the sulphoxide.

Hydroxylation

Hydroxylation is a common pathway of herbicide degradation in benzoic, phenoxy and triazine groups of herbicides. Ring hydroxylation occurs in the case of benzoics and phenoxyacids with or without a shift of chlorine atom as shown in Fig. 8.4.

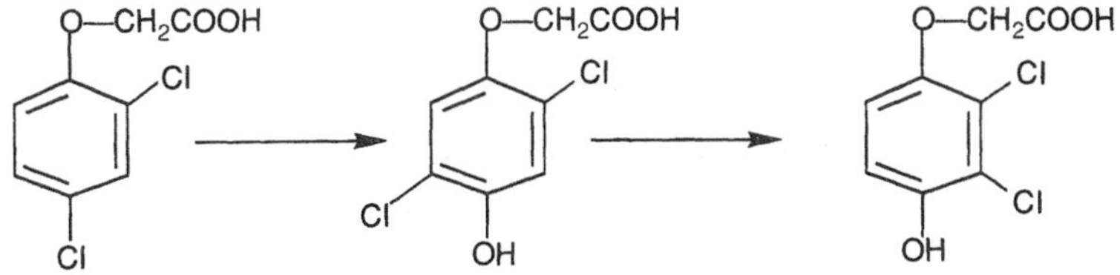

2,4-D(chlorine in 4-position)   Hydroxylated 2,4-D(chlorine in 5 or 3-position)

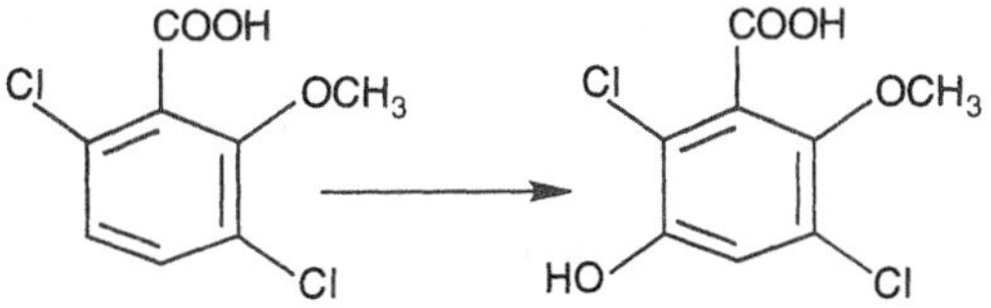

Dicamba   Hydroxylated dicamba
(no shift in chlorine position)

Fig. 8.4. Hydroxylation of 2,4-D and dicamba with or without change in chlorine position.

Ring hydroxylation may also involve dechlorination, demethylation or demethylthiolation as in the case of triazine herbicides. For example, in the case of simazine, hydroxylation takes place by replacing the chlorine molecule at the 2-position (Fig. 8.5).

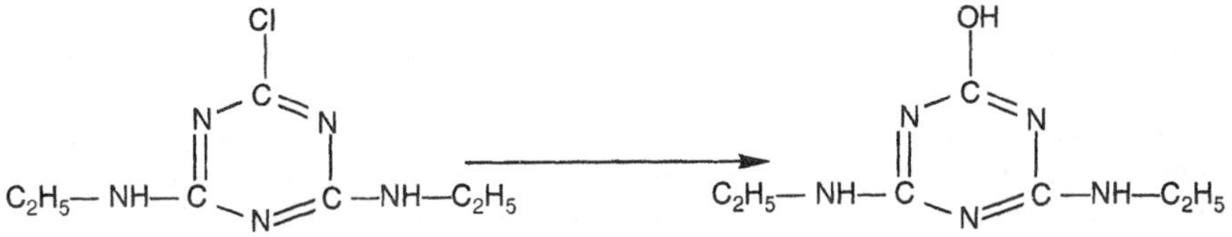

Fig. 8.5. Hydroxylation of simazine with replacement of chlorine at the 2-position

In the case of hydroxylation with demethoxylation and demethylthiolation, the methoxy molecule at the 2-position on the simeton ring and the methylthio molecule at the 2-position on the simetryn ring are replaced by OH.

Hydrolysis

Esters of carboxylic acids readily undergo hydrolysis. Hydrolysis can be expected if the herbicide is a carboxylic acid ester, a phosphate ester or an amide. The hydrolyzing enzymes can be accordingly classified as esterases, phosphatases or amidases. Hydrolysis usually occurs in the case of carbamates, thiocarbamates, triazines, pyridinecarboxylic acids, nitriles, phenoxyalkanoic acids, etc. Cleavage of carboxylic acids by carboxylesterases, while causing detoxification of several herbicides in these herbicide families, could also lead to bioactivation of certain herbicides as in the case of aryloxyphenoxypropionics. For example, diclofop methyl requires activation by ester hydrolysis to diclofop acid, a herbicidally active compound. Carboxylesterases could cleave the methylesters, ethylhexyl esters and ethoxybutyl esters of herbicides. The hydrolysis of amide linkage is somewhat similar to that in esters and is catalyzed by amidases. The hydrolyzed herbicides are made inactive by conjugation with sugars and amino acids.

Decarboxylation

In the presence of water, CO_2 is removed from the herbicide molecule as shown in Fig. 8.2. Benzoics, phenoxyacids and ureas undergo decarboxylation.

Deamination

In deamination, the aromatic nitro group is reduced to the corresponding aniline. Deamination represents an important detoxification mechanism of triazinone herbicide metribuzin in the tolerant soybean. The deaminase activity is apparently located in the peroxisome fraction.

N-Dealkylation

During this reaction, the alkyl groups attached to amide nitrogen are substituted. N-dealkylation occurs in the case of triazines, carbamates, thiocarbamates, ureas and dinitroanilines. In the case of atrazine, the ethyl group is removed by N-dealkylation as shown in Fig. 8.6.

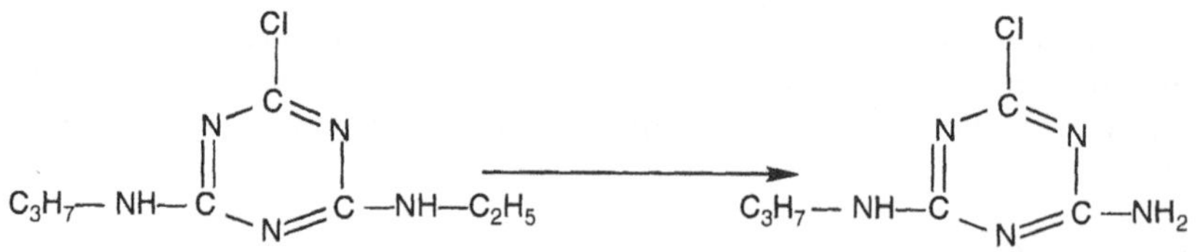

Fig. 8.6. Substitution of ethyl (C_2H_5) group at the 4-position of atrazine molecule.

N-dealkylation may be partial, occurring at either the 4-position to result in 2-chloro-4-amino-6-isopropylamino-s-triazine or at the 6-position to yield 2-chloro-4-ethylamino-6-amino-s-triazine [78, 79], or complete, occurring at both positions to form 2-chloro-4,5-amino-s-triazine [82].

Ring Cleavage

During this reaction, the aromatic and heterocyclic ring structure is split, resulting in formation of a less phytotoxic or non-phytotoxic compound in the plant. Triazines and phenoxyacetics are subject to ring cleavage. Triazine heterocyclic ring can be cleaved, resulting in liberation of CO_2. Similarly, cleavage of the 2,4-D ring results in formation of monochloroacetic acid. The cleaved products are further degraded readily.

Ring cleavage transforms intrinsically inactive herbicides to herbicidally active products. For example, methazole, an oxadiazolidine, is converted by the ring cleavage mechanism to the Photosystem II inhibitor DCMU [1-(3,4-dichlorophenyl)-3-methyl urea]. Similarly, buthidazole, with an imidazolinone ring, also undergoes ring cleavage to yield an active N-methyl urea.

Conjugation

Conjugation of a parent compound or a metabolite is a common pathway of herbicide transformation in plants. Conjugation involves endogenous plant products such as sugars, amino acids, peptides, lignins and lipids. The conjugated products can be esters, amides, ethers, thioethers or glycosides. One fraction of conjugates is usually soluble, and can be extracted by water or other solvents. Much of the herbicide may form an insoluble fraction that can be extracted by standard extraction procedures. The insoluble fraction may contain the herbicide and/or its conversion products chemically bound to insoluble matrix molecules such as lignin, hemicellulose or protein. Also, adsorption to cell-wall components may be very strong.

Glutathione Conjugation

Glutathione conjugation is an important pathway in the metabolism of several herbicides including triazines, acetanilides, diphenyl ethers and sulfonylureas. In corn and sorghum, chloroacetanilide herbicides are detoxified by conjugation with the tripeptide glutathione (GSH). **Glutathione-s-transferases (GSTs)** are dimeric, multifunctional proteins found in plants. A major function of these enzymes is to detoxify a wide variety of hydrophobic, electrophilic compounds by catalyzing their conjugation with GSH. Maize is known to have at least three GST isozymes [62] in addition to several constitutive and inducible mono-oxygenases. They are known to play a role in the detoxification of EPTC and atrazine. Not all cultivars are atrazine-tolerant and the susceptible ones contain a low GST activity. Similarly, GST activity is deficient in susceptible grass weed species.

The chemical mechanism of glutathione conjugation is a nucleophilic displacement in the herbicide molecule. The glutathione anion GS^- serves as a nucleophile, while chlorine, p-nitrophenol or an alkyl-sulphoxide is the group involved in conjugation. GSH-mediated conjugation is described in greater detail when transformation of individual herbicides is discussed.

Conjugation with Sugars (Glycosylation)

Transformations of herbicides to hydroxylated derivatives are almost invariably followed by conjugation to glucose and form O-β-D-glucosides. β-D-glucopyranoside is one of the most common glycoside conjugates formed in plants. O-glycosides and glucose esters are also formed. In addition to glucose, other sugars can form the glucoside linkage or can be added to the already formed glycoside, thereby causing growth of the sugar moiety with time.

In the example given in Fig. 8.7, a 2,4-D molecule is conjugated to form a glucose ester of 2,4-D.

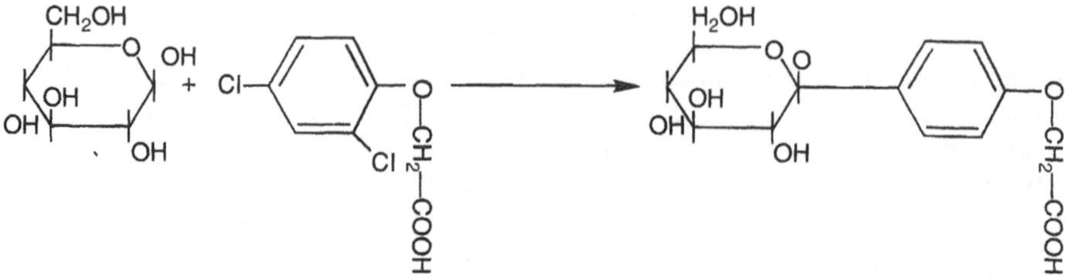

Fig. 8.7. Conjugation of 2,4-D with glucose to form a glucose ester of 2,4-D.

The glucosides can undergo hydrolysis enzymatically with *O*-glycosyl-transferase. In 2,4-D, a phenoxyacetic herbicide, the hydroxyl (OH) group may undergo esterification of glucose. A free hydroxyl group on almost any position of a herbicide or its metabolite is susceptible to conjugation with glucose, although some positions are more favoured than others. Glycosidation may contribute to detoxification by virtue of enhanced water solubility of the products.

Plants also form *N*-glycosides through the catalytical action of *N*-glucosyl transferase as in the case of metribuzin and chloramben. In the case of metribuzin, the *N*-glucoside is malonylated by the action of arylamine *N*-glucosyl transferase (vide Metribuzin, later in this Chapter).

Conjugation with Amino Acids

Several amino acids such as glutamate, asparagine, cysteine, alanine, leucine, proline, etc. are also involved in formation of conjugates with herbicides or their metabolites. This conjugation occurs in the case of auxinic herbicides. The glutathione and homoglutathione conjugates are broken down by peptide bond hydrolysis, yielding cysteine conjugates which are eventually malonylated (e.g. fluorodifen). The amino acid conjugates of auxin herbicides are bound to the cell wall.

Cytochrome P-450—Mediated Transformation

Cytochrome P-450 mono-oxygenases are membrane-bound, heme-containing proteins, which have been implicated in a variety of oxidative reactions in plant tissues. These are found in various subcellular locations including endoplasmic reticulum (ER), plasma membranes, glyoxysomes and mitochondria. While these enzymes are involved in biosynthetic pathways, leading to the synthesis of lignin phenolics, membrane sterols, phytoalexins and terpenoids, they have been found responsible for detoxification of herbicides and other xenobiotics. The enzymes mediate hydroxylation and demethylation reactions used by plants to detoxify a range of herbicides [21], including phenylphureas and sulphonylureas [26], acetanilides [59], imidazolinones and triazolopyrimidine sulphonanilides [27], and bentazon, a buthiadiazole herbicide [58].

In the cytochrome P-450 oxygenase reaction, hydroxylation, two electrons from cytochrome P-450 are donated to O_2, and then one oxygen atom combines (hydroxylates) with the substrate (RH) and the second oxygen atom forms water:

$$NADPH + H^+ + O_2 + R\text{-}H \longrightarrow NADP^+ + H_2O + R\text{-}OH$$

In oxidative demethylation (and dealkylation), the hydroxylation mechanism is followed by elimination of the hydroxylated methyl (alkyl) group as the aldehyde. Cytochrome P-450 is reduced with cytosolic NADPH by NADPH: Cyt P-450 reductase, a membrane-bound flavoprotein. Thus, the electron flow is from cytosolic reductant to O_2. The cytochrome P-450s involved in transformation reactions may metabolize a number of substrates. Many of the herbicide-metabolizing activities are enhanced by the treatment of plant tissue with herbicide safeners (vide Chapter 12).

The P-450-mediated metabolism of sulfonylurea herbicides results in both hydroxypyrimidine and hydroxy phenyl metabolites. More than one herbicide can be metabolized by the same P-450. Cytochrome P-450 is the electron donor in the ring hydroxylation of sulfonylurea herbicide metabolism.

TRANSFORMATION OF INDIVIDUAL HERBICIDES

Acetamides

Alachlor

Alachlor is metabolized rapidly by *Cyperus esculentus* to at least one water-soluble metabolite within 2 d of application [3]. Rice and Putnam [70] found alachlor and two polar metabolites in the roots, and alachlor plus one metabolite in the shoots. In a study with ^{14}C-alachlor, Rao [67] suggested that corn, a tolerant species, may have metabolized the herbicide faster than barley, a susceptible species, into inactive products.

Update: Alachlor is detoxified by conjugation with glutathione (GSH) or homoglutathione (hGSH). The involvement of glutathione-s-transferases is less certain. The GSH conjugate is subsequently metabolized to the malonylcysteine conjugate. Homoglutathione may be further metabolized by peptide hydrolysis, sulphoxidation and O-malonylation.

Butachlor

Butachlor is metabolized rapidly, primarily to polar water-soluble metabolites. Its detoxification may involve GSH or hGSH conjugation as in the case of alachlor.

Acetochlor

Acetochlor is rapidly metabolized by maize and soybean seedlings to GSH and hGSH conjugates respectively. The tolerance of a plant to acetochlor depends on the rate of metabolism by GSH or hGSH conjugation [5] and on endogenous levels of GSH and glutathione-s-transferases [43]. Metolachlor is detoxified by cleavage of the methyl ether followed by conjugation with glucose. It is also detoxified by conjugation of the chloroacetyl group with GSH or hGSH. Conjugation occurs with a half-life of a few hours. GSH conjugate (S-metolachlor-glutathione) is subsequently metabolized by the concerted action of carboxypeptidase and γ-glutamyltranspeptidase enzymes to the cysteine conjugate followed by oxidative deamination and reduction to thioacetic acid, and oxidation to sulphoxide derivatives.

Propachlor

Degradation of propachlor in plants is very rapid and no parental compound could be detected 5 d after treatment in maize and soybean roots [44, 45]. It conjugates through its chlorine to a natural plant product to form a glycosidic linkage and the resultant metabolite, 2-hydroxy-N-isopropylacetanilide, is quite stable and relatively non-phytotoxic. Propachlor is also degraded to form a hydrophilic water-soluble metabolite N-isopropylaniline. Besides maize and soybean, other resistant plant species also metabolize propachlor in a similar manner. Lamoureux et al. [49] isolated two water-soluble metabolites, glutathione and γ-glutamylcysteine conjugates of propachlor from maize, sorghum, sugarcane and barley.

Update: Propachlor is conjugated to homoglutathione by chlorine displacement. The homoglutathione is further metabolized by peptide hydrolysis, sulphur oxygenation (sulphoxidation) and/or O-malonylation to form thioacetic acid sulphoxide.

Propanil

Rice plants hydrolyze propanil to 3,4-chloroaniline [99] as shown in Fig. 8.8. There appears to be a transient intermediate 3,4-dichloroacetanilide between propanil and 3,4-dichloroaniline (3,4-DCA). This 3,4-DCA complexes with plant constituents such as cellulose, lignin, glucose, xylose, fructose and glucosylamine [99]. Frear and Still [27] found that hydrolysis of propanil to 3,4-DCA was catalyzed in rice by an enzyme acylamidase. They found that rice leaves contained 60 times more enzyme units than the leaves of *Echinochloa crus-galli* (barnyardgrass), a susceptible species. Thus, resistance of a rice plant to propanil was due to its higher acylamidase activity, leading to a faster rate of detoxification than in barnyardgrass. Metabolism of propanil in rice to form 3,4-DCA and *N*-(3,4-dichlorophenyl) glucosylamine was more rapid under high temperature and long-day conditions [41]. Propanil does not persist for more than 26 d following postemergence application in transplanted rice, with a half-life of 5 to 9 d.

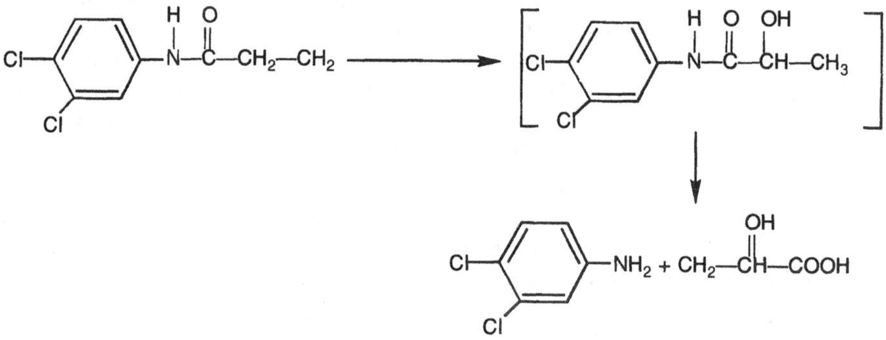

Fig. 8.8. Pathway of propanil degradation in plants (99).

Dimethenamid

Dimethenamid is rapidly and extensively metabolized in maize and soybean to about 30 metabolites. Residues in maize and soybean consist of numerous highly polar metabolites, each less than 0.1 ppm of the sample, and none representing more than 10% of the applied dimethenamid.

Napropamide

Napropamide undergoes *N*-dealkylation and ring hydroxylation (# 4 carbon) followed by hexose conjugation. This constitutes the primary detoxification mechanism in tolerant species such as tomato and fruit trees.

Aliphatics

Dalapon

Dalapon is generally considered to be non-degradable by plants. It remains essentially non-metabolized for long periods, especially in dormant and quiescent tissues. In cotton (a tolerant species), sorghum (a susceptible species) and wheat dalapon accumulates principally as the intact molecule or its dissociate salt. The herbicide is so stable and persistent that it is carried over to three generations of wheat via the seed.

Arsenicals

DSMA, MSMA

Organic arsenicals are not known to be degraded by plants. Instead, they conjugate with sugars, amino acids, organic acids and other molecules, and form complexes. Rice roots and shoots slowly metabolize DSMA to the demethylated inorganic arsenic and to the methylated dimethyl and trimethyl arsenic species. MSMA is not, however, demethylated to form inorganic arsenicals, and is not reduced to trivalent arsenic compounds in beans. MSMA binds rapidly to another molecule to form a ninhydrin-positive complex. It is not known whether this MSMA-complex in plants has the same or altered phytotoxicity compared to MSMA.

Benzamides

Isoxaben

Isoxaben is readily metabolized in cereals, primarily through hydroxylation of the alkyl side chain. It is hydroxylated on the 2-carbon of the propyl side chain as also at the propyl 3-carbon. More than 50% of these metabolites are then glucosylated. A minor metabolite, 2,6-dimethoxybenzamide, is also formed. Tolerant and susceptible species metabolize up to 50% of absorbed isoxaben within 4 d following root application. The quantitative or qualitative differences in metabolism of isoxaben are not sufficient enough to explain the tolerance or sensitivity of wheat and soybean cell cultures [15].

Benzoics

Chloramben

Chloramben complexes with endogenous plant compounds to form conjugates. This is found in several crop species such as soybean, barley, cucumber, sugarbeet, carrot, pea, tomato and others. One of these conjugates includes N-glycosyl chloramben [(N-3-carboxy-2,5-dichlorophenyl)-glycosylamine]. This conjugate is relatively non-phytotoxic and heat-labile. As chloramben is a soil-applied herbicide, this conjugation occurs more in the root tissue than in the leaf tissue. Chloramben could be released from the conjugates by alkaline hydrolysis. Chloramben also undergoes decarboxylation, but it is not a major degradation pathway.

Dicamba

Dicamba is completely metabolized by wheat plants in 18 d to form a major metabolite, 5-hydroxy-2-methoxy-3,6-dichlorobenzoic acid (Fig. 8.4), and a minor metabolite, 3,6-dichlorosalicylic acid, with the major metabolite accounting for 90% of the radioactivity and the minor metabolite for 5% [6]. Ray and Wilcox [68] found that maize roots also converted dicamba into 5-hydroxy derivative as the major metabolite and 3,6-dichlorosalicylic acid as the minor metabolite. Wheat and barley alter the dicamba molecule more extensively than buckwheat and wild mustard (*Sinapis arvensis*) [14]. Ray and Wilcox [69] detected no dicamba metabolism in *Cyperus rotundus* even after 10 d. Quimby and Nalewaja [66] observed more rapid metabolism of dicamba in wheat, a resistant species, than in wild buckwheat, a susceptible species. Thus, differential transformation of dicamba is apparently related to species selectivity.

Benzothiadiazoles

Bentazon

Bentazon forms in tolerant species forming glucosyl conjugates, which are then incorporated into the metabolic system.

Bipyridiliums

Diquat and Paraquat

Bipyridilium herbicides are not metabolized by higher plants. They, however, undergo photodecomposition when applied to the foliage. When paraquat-applied plants are exposed to sunlight, 4-carboxyl-1-methylpyridilium chloride and methylamine are formed. Since plants are killed rapidly in bright sunlight, significant quantities of the breakdown products are formed only on the surface of dead tissues.

Carbamates

Carbamate herbicides, which have amide bonds, are effectively broken down in plants, leaving little residue.

Chlorpropham

Chlorpropham is metabolized by plants to form water-soluble products. Treated soybean plants rapidly formed polar products and insoluble residues in roots, while only a small amount of insoluble residues were formed in shoots [84, 85, 86]. Polar metabolites are immobile and not translocated from the site of their anabolism. The phenyl ring of chlorpropham is hydroxylated by both root and shoot tissues. The predominant polar metabolite in roots is O-glucoside of 2-hydroxychlorpropham, which by the action of β-glucosidase gives the predominant aglycon 5-chloro-2-hydroxycarbanilate (Fig. 8.9).

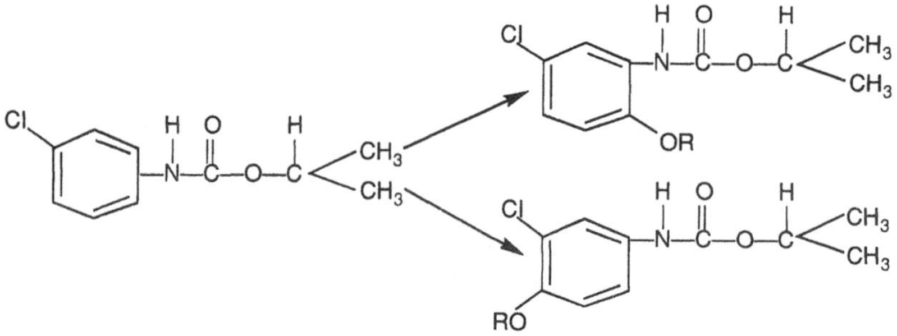

Fig. 8.9. Transformation of chlorpropham in soybean [85, 86].

In shoot tissue, the polar metabolites were the O-glucoside of 2-hydroxychlorpropham and isopropyl-3-chloro-4-hydroxycarbanilate (4-hydroxychlorpropham); they were found in the ratio of 1:1 (Fig. 8.9). The root and shoot tissues hydroxylate only the phenyl ring of chlorpropham, but this does not alter the isopropylmoiety or the carbamate bond.

In cucumber, Still and Mansager [87] found that chlorpropham was hydroxylated to yield 4-hydroxychlorpropham, which was subsequently conjugated to form unknown polar metabolites. The conjugate was not a β-glucoside or a common glucoside because it was not hydrolyzed either by β-glucosidase or the broad-spectrum enzyme hesperidinase. The authors concluded that the apparent absence of glucosides in cucumber and the formation of unknown conjugates could be the key to the sensitivity of cucumber to chlorpropham.

Desmedipham, Phenmedipham

Desmedipham and phenmedipham break down by hydrolysis and the decomposition products bind to the plant constituents. Desmedipham undergoes hydrolysis to form ethyl N-(3-hydroxyphenyl)carbamate and subsequently *m*-aminophenol. This appears to be one of the main pathways of transformation in sugarbeets. Similarly, phenmedipham hydroxylation and glycosylation followed by hydrolysis of the carbamate linkage appear to be the major factors in the tolerance of sugarbeets. Several metabolites including a sulphate conjugate have been identified, but metabolism varies with species. The bacterial enzyme phenmedipham hydrolase, coded by a plasmid gene (*pcd*), catalyzes hydrolysis of the carbamate linkage of desmedipham and phenmedipham [63].

Cineoles

Cinmethylin

Cinmethylin is metabolized by plants to form hydroxylation and oxidation products, most of which become conjugated.

Cyclohexanediones (Cyclohexenones)

Clethodim, Cycloxidim, Sethoxydim

Th primary detoxification mechanism of clethodim in tolerant soybean involves oxidative cleavage, yielding metabolites with greater polarity. Cycloxidim rapidly undergoes oxidation, followed by conjugation with plant constituents to yield cycloxidim-conjugate. Tolerant plants detoxify sethoxydim rapidly, metabolizing 98% of the applied herbicide to at least nine products within 24 h after application [10]. Two of these metabolites are photodegradation products and one was identified as desethoxy-sethoxydim.

Dinitroanilines

Benefin, Dinitramine, Ethalfluralin, Fluchloralin, Isopropalin, Nitralin, Trifluralin

Dinitroanilines and their degradation products observed in the soil do not accumulate in the edible portions of crops tolerant to these herbicides. Golab et al. [34] found only traces of benefin in the foliage of alfalfa (lucerne) and groundnut (*Arachis hypogea*) grown in benefin-treated soil. They suggested that plants absorb benefin and its degradation products from the soil and transport them to different plant tissues rather than metabolizing the herbicide.

The accumulation of dinitramine, isopropalin and nitralin, and their degradation products is believed to be negligible. Ethalfluralin is converted to an unidentified water-soluble metabolite(s) within 24 h of application. Besides, a methanol-soluble metabolite is also formed. However, significant amounts of ethalfluralin remain unmetabolized. Ethalfluralin is relatively more rapidly metabolized than trifluralin. The herbicide does not appear to be present in detectable levels in treated soybean, cotton, groundnut and drybean. Fluchloralin is metabolized rapidly by soybean roots to several chloroform-soluble and water-soluble metabolites and high levels of methanol-soluble residue [57]. No single metabolite represented more than 4% of the total fluchloralin in the roots [57]. Soybean roots metabolize fluchloralin at a higher rate than maize roots.

Trifluralin appears to be relatively slowly metabolized in many plants [33, 65], although sweet potato and groundnut (peanut) metabolized 83% and 99% respectively [4]. In groundnut, the degradation identified was α,α,α-trifluoro-2,6-dinitro-N-(*n*-propyl)-*p*-toluidine, but not in sweet potato [4]. Trifluralin metabolism primarily occurs by way of one- or two-step dealkylation, reduction of NO_2 to NH_2 and oxidation to produce phenol derivatives.

Dinitroaniline herbicides and their degradation products are not translocated into the edible portions of the plant, except in the case of root crops, e.g. carrot [64]. Root crops apparently accumulate them in minor amounts but not in sufficient quantities to be hazardous.

Diphenylethers

Acifluorfen, Fluorodifen, Fluoroglycofen, Fomesafen, Lactofen, Oxyfluorfen

Acifluorfen is metabolized by the tolerant soybean plants by rapid cleavage of the ether bond by homoglutathione, yielding S-(3-carboxy-4-nitrophenyl)homoglutathione conjugate and 2-chloro-4-trifluoromethylphenol, followed by further metabolism. The metabolism appears to be much slower in susceptible weed species than in tolerant soybean.

Fluorodifen is degraded by hydrolysis. Eastin [22] found p-nitrophenol and unknowns I and II as the major degradation products of fluorodifen in groundnut seedlings. He also detected some 2-amino-4-trifluoromethylphenol and traces of 2-amino-, p-amino-, and p-amino-2-aminofluorodifens, and several minor unknowns. He speculated that unknowns I and II might be conjugates of p-nitrophenol and 2-amino-4-trifluoromethylphenol respectively. He postulated that fluorodifen is reduced to 2-aminofluorodifen which is cleaved at the ether linkage to yield p-nitrophenol and 2-amino-4-fluoromethylphenol, which were subsequently conjugated with natural plant substances to form water-soluble conjugates. Rogers [72] proposed that fluorodifen metabolism in soybean plants primarily involved cleavage of the ether linkage and that reduction of the two nitro substituents on the rings was of minor importance. This was at variance with Eastin's observation, which showed reduction of the nitro group at the 2-position of the 2-nitro-4-trifluorophenyl ring to an amino group yielding 2-aminofluorodifen before cleavage of the ether linkage.

Fluoroglycofen is rapidly detoxified in tolerant species by de-esterification to 4-trifluoromethyl-2-chloro-3'-carboxy(methoxy)carbonyl-4'-nitro-diphenyl ether and 4-trifluoromethyl-2-chloro-3'-carboxy-4'-nitro-diphenyl ether followed by conjugation with glutathione or homoglutathione.

Fomesafen is rapidly cleaved in tolerant plants at the diphenyl ether bond, yielding inactive metabolites. Lactofen is readily metabolized by plants, with no measurable residues appearing longer than 24 d after treatment. Oxyfluorfen appears to be metabolized very slowly in plants.

Imidazolidinones

Buthidazole

Buthidazole was reported to be concentrated at 87% in sugarcane leaves but contributed residues neither to sugar nor to molasses [2]. After 12 wk, only 15% of the label was buthidazole and the main metabolite was the desmethyl derivative. Hatzios and Penner [39] found buthidazole and 6 metabolites in resistant alfalfa leaf extracts 6 d after treatment. One unidentified metabolite accounted for 40% of the detected radioactivity 6 d after root application. In the case of susceptible *Elytrigia repens* (quackgrass), unmetabolized buthidazole accounted for 87% of the total radioactivity after 16 d. They concluded that differential rate and type of metabolism appeared to contribute to buthidazole selectivity between the species.

Imidazolinones

Imazapyr, Imazaquin, Imazethapyr, Imazamethabenz

Imazaquin and imazethapyr are selective herbicides in several crops including soybean and groundnut. Soybean has the ability to rapidly metabolize these herbicides to inactive compounds. These two herbicides have different routes of metabolism.

In the case of imazaquin, soybean first hydroxylates the imidazolinone ring at the imine carbon and then cyclize the imidazolinyl nitrogen with the pyridyl carboxyl group (Fig. 8.10) [93]. This may be followed by dehydration to yield a cyclic intermediate (1), which is then ring-opened to give metabolite (2) which when hydrolyzed yields a water-soluble metabolite (3). Metabolite (4), which arises from a non-enzymatic hydrolysis of metabolite (2), is a minor metabolite. Metabolite (5), on the other hand, may arise from the hydroxylation of intermediate (1). Tecle et al. [93] reported that the polar metabolites resulted from metabolic processes in soybean tissue.

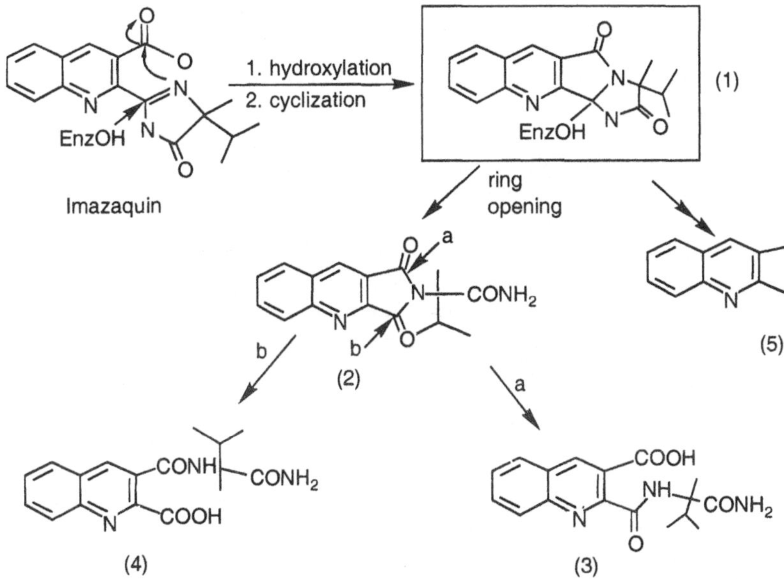

Fig. 8.10. Proposed pathway of imazaquin metabolism in soybean seedlings [93].

Imazethapyr undergoes a detoxification process different from imazaquin in soybean. The ethyl group on the pyridine ring is first hydroxylated at the vinylic carbon to form metabolite (a) followed by conjugation of this compound to yield an O-glucoside (b) (Fig. 8.11) [93]. The hydroxylated metabolite of imazethapyr is herbicidally active, although at a greatly reduced level compared to that of the parent herbicide. However, the glucose conjugate of this hydroxylated metabolite is completely inactive. Conjugation of the hydroxylated metabolite to glucose is the true detoxification step. Differential tolerance to imazethapyr is associated with the relative capacity of the plant to conjugate the hydroxylated metabolite with glucose.

In the case of imazamethabenz, the tolerant wheat causes oxidation of the benzene methyl group to the corresponding hydroxymethyl derivative, followed by glucoside conjugation. The susceptible *Avena* spp. (wild oats) de-esterifies imazamethabenz methyl ester, producing the phytotoxic imazamethabenz acid.

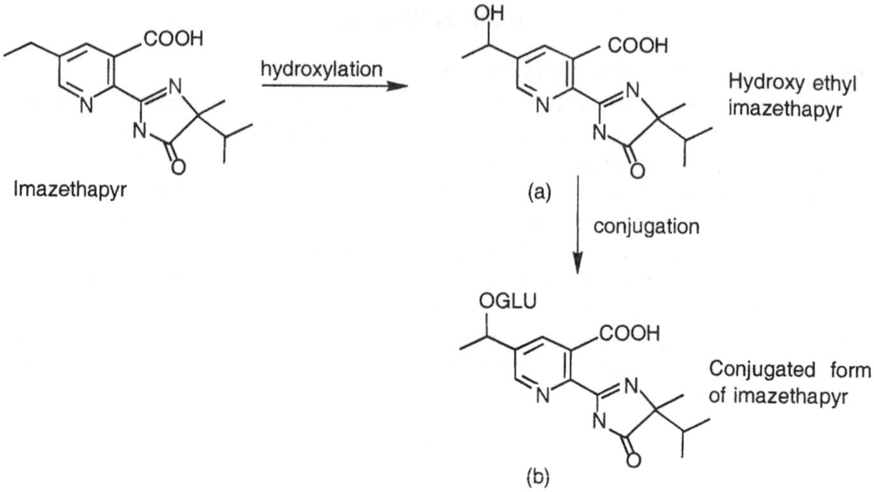

Fig. 8.11. Proposed pathway of imazethapyr metabolism in soybean seedlings [93]. (GLU: glucose)

The tolerant plants metabolize imazapyr by initiating hydroxylation of the imidazolinone ring to form 2-carbomylnicotinic acid. In this hydrolysis-mediated metabolism, an imidazopyrrolopyridine derivative also is formed. Susceptible weed species metabolize imazapyr slowly or not at all. In tolerant wheat plants, imazamathabenz undergoes oxidation at the benzene methyl group, yielding the corresponding hydroxymethyl derivative. This is followed by glucoside conjugation. The susceptible *Avena* spp. primarily de-esterify imazamethabenz methyl ester, producing the phytotoxic imazamethabenz acid.

Isoxazolidinones

Clomazone

Clomazone undergoes oxidative cleavage, yielding metabolites with greater polarity. This appears to be the primary mechanism of detoxification in soybean and *Abutilon theophrasti* (velvetleaf). Differences in metabolism among species do not correlate with relative susceptibility.

Nitriles

Bromoxynil, Ioxynil

Bromoxynil is degraded in both susceptible *Amsinckia intermedia* (coast fiddleneck) and resistant winter wheat (*Triticum aestivum*) [77]. It is also degraded to CO_2 in both susceptible and resistant species. Ioxynil is rapidly degraded by resistant barley after foliar treatment [16]. Four days after treatment, unchanged ioxynil, traces of benzamide and benzoic acid derivatives, and at least two other unknown compounds were identified in resistant barley [16]. The unhindered nitrile groups of bromoxynil and ioxynil were slowly hydrolyzed in the plant to the corresponding benzamide and benzoic acid derivatives, with possible further degradation through decarboxylation, dehalogenation or conjugation of the benzyl moiety or the aromatic ring. Decarboxylation and/or possible conjugation reactions may be much faster than hydrolysis. This is supported by the detection of trace amounts of trace amounts of hydrolysis products [16, 77] and the rapid accumulation of insoluble residues [77] with (cyano-^{14}C) ioxynil and (cyano-^{14}C) bromoxynil. The expected benzamide and benzoic acid analogues may be only transitory intermediates. Ioxynil selectivity is partly attributed to differential metabolism but may involve other factors. Liberation of iodide ions from ioxynil followed by oxidation of iodide to iodine via plant peroxidases varies with species; this may be important in selectivity.

Oxadiazolidines

Methazole

Methazole is decarboxylated and N-demethylated to 1-(3,4-dichlorophenyl) urea (DCPU) in cotton, beans, wheat, onions, corn and alfalfa. The herbicide is first rapidly metabolized to 1-(3,4-dichlorophenyl)-3-methylurea (DCPMU) which in turn is demethylated to DCPU. Methazole and DCPMU are 20 times as toxic as DCPU for susceptible species.

Phenols

Dinoseb

Phenols rapidly destroy the foliage through contact action, thus precluding investigations on their metabolism. Reduction of the nitro group with further conjugation of the amine groups appears to be the degradative mechanism of dinoseb [47]. The nitro groups may be removed and converted to ammonium or amine ions, which could also inhibit cellular processes.

Phenoxyalkanoic Acids

Phenoxyacetics – 2,4-D, MCPA and Phenoxybutyrics – 2,4-DB, MCPB

Phenoxyacetics are metabolized by plants by three mechanisms, which include degradation of the acetic acid side chain, hydroxylation of the aromatic ring and conjugation with a plant constituent.

Degradation of Side Chain: In this mechanism, CO_2 is released from 2,4-D. Side chain degradation is slow and hence this pathway is not considered to be of any significance in 2,4-D metabolism. This degradation takes place by oxidation at carboxyl and methylene carbons of the side chain. Resistant species are better able to metabolize 2,4-D via oxidation and decarboxylation pathways than the susceptible species.

Ring Hydroxylation: This mechanism involves hydroxylation of the ring and oxidation of hydroxyl to carboxyl, with a split in the ring. 2,4-D also undergoes β-oxidation followed by ring hydroxylation forming 4-hydroxyphenoxyacetic acid; this hydroxylation pathway is considered to be the mechanism of detoxification in resistant species. The hydroxylated-2,4-D may conjugate with glucose to form 4-O-β-D-glucoside. The higher phenoxyalkanoic acids, with even-numbered carbons on the side chain, may also be first α-oxidized to phenoxyacetic acid, which is then hydroxylated to form 4-hydroxyphenoxyacetic acid, which in turn may conjugate with glucose to form 4-O-β-D-glucoside. Hydroxylation serves a significant detoxification mechanism. The resistance of graminaceous species to phenoxyacetics is apparently due to their ability to hydroxylate these compounds.

Conjugation with Plant Constituents: Phenoxyacetics form complexes with plant constituents such as sugars, proteins and amino acids. The phenoxyacetics with no hydroxyl group also conjugate with glucose during esterification to form the glucose ester of

2,4-D. 2,4-D-protein complex is a product of the detoxification process and probably an intermediate in the metabolism of 2,4-D. The amino acid-2,4-D conjugates possess herbicidal properties [25]. The most active compounds were the less polar amino acid conjugates of leucine, isoleucine, valine and methionine. In general, the aromatic and polar amino acid conjugates exhibited poor herbicidal properties. Like hydroxylation, conjugation also serves as a major detoxification mechanism and resistance is perhaps determined by the rate of conjugate formation. The tolerant graminaceous species form 2,4-D conjugates much more rapidly than 2,4-D-susceptible species such as beans, tomatoes and cotton (9). The conjugates do not have growth regulating activity.

β-Oxidation: Plants metabolize ω-phenoxyalkanoic acids, which contain a side chain with more than two carbon atoms, by β-oxidation of the side chain as shown in Fig. 8.2. The role of β-oxidation in selective weed control is significant. A long-chain substituted phenoxyalkanoic acid could be used for weed control in a crop if the weeds, but not the crop plants, can metabolize it by β-oxidation to the corresponding phenoxyacetic acid. The phenoxybutyrics, which undergo β-oxidation are 2,4-DB and MCPB.

In the final analysis, differential rate in 2,4-D metabolism can account for varying degrees of susceptibility among plant species to the herbicide. The rapid detoxification of 2,4-D side chain forms the basis of selectivity. 2,4-D metabolism reactions are several and can be divided into two phases [17]. In phase I, the reactions include hydroxylation (where no. 4 Cl is displaced by a hydroxyl group, and moved to the no. 5 or 3 carbon), decarboxylation, dealkylation and dechlorination. Phase II reactions include conjugation of amino acids, particularly glutamate and aspartate, to the side chain, and conjugation with glucose at the hydroxylated positions.

Aryloxyphenoxypropionics (APPs) — Diclofop-methyl, Fenoxaprop, Fluazifop-P, Haloxyfop-P, Mecoprop, Quizalofop-P

Diclofop is commercially available as a methyl ester formulation. Upon entering the plant, the methyl ester of diclofop is hydrolyzed, causing demethylation in both tolerant and susceptible plants, and forming the herbicidally active diclofop acid (Fig. 8.12). Subsequent to ester hydrolysis, the diclofop acid is arylhydroxylated by the tolerant wheat plant and inactivated. This is followed by glycosidation, i.e., glucoside conjugation, to yield O-glucoside of diclofop. Hydroxylation occurs at all three available positions on the phenyl ring, forming at least three isomeric arylhydroxylated metabolites, which then undergo glycosidation to form respective O-glycosides of diclofop. Cytochrome P-450-dependent mono-oxygenases catalyze the hydroxylation reactions leading to diclofop detoxification.

In susceptible wild oat plants, aryl hydroxylation of diclofop acid is of minor importance. Instead, diclofop is primarily conjugated at the carboxyl group to form a neutral 'glucose ester' (glycosyl ester), which is non-phytotoxic. The glucose ester is readily hydrolyzed, resulting in liberation of the active diclofop acid. This implies that glucose ester is not a true detoxification metabolite and its production does not protect wild oats from diclofop phytotoxicity.

Fenoxaprop is the hydrolyzed form of fenoxaprop ethyl. The first step in the plant metabolism of fenoxaprop ethyl is its de-esterification to form fenoxaprop, followed by cleavage at the ether bond by nucleophilic displacement of the phenyl group of gluthathione (GSH) and/or cysteine, resulting in three possible metabolites (Fig. 8.13): the GSH-conjugate, cysteine-conjugate and 4-hydroxyphenoxy-propionic acid [91]. Subsequently, the GSH may be further conjugated with glucose to yield an N-glucoside conjugate (Metabolite I). The 4-hydroxyphenoxy-propionic acid residue may also

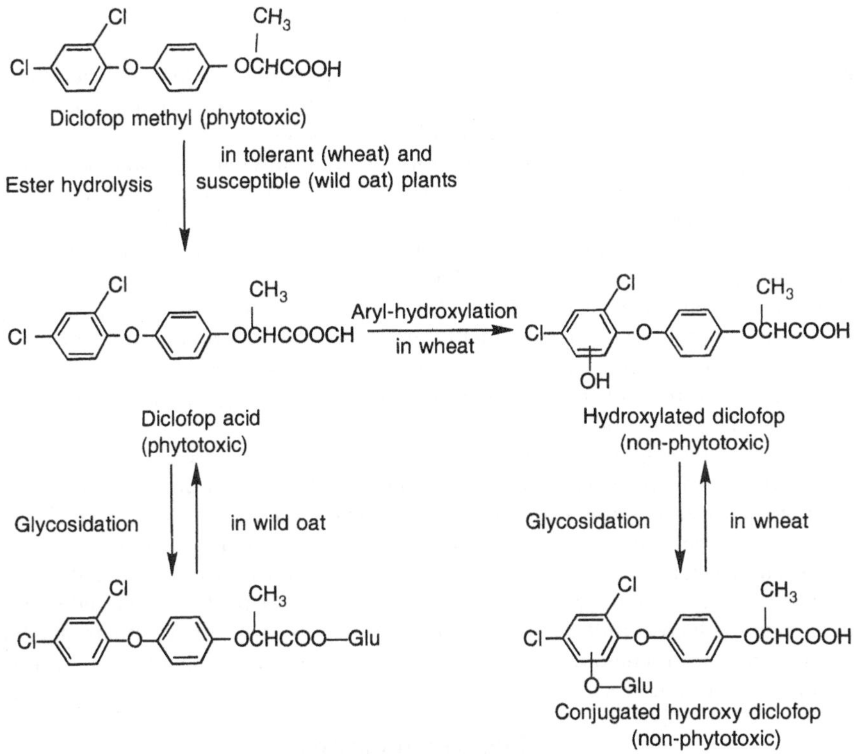

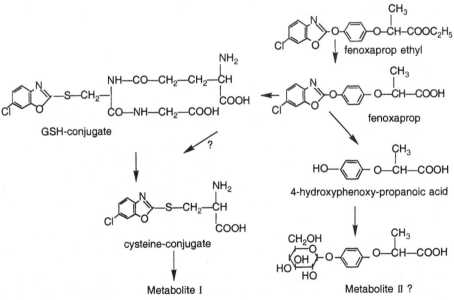

Fig. 8.12. The metabolic pathway of diclofop methyl.

Fig. 8.13. Proposed pathway of fenoxaprop ethyl metabolism in grass species [91].

conjugate with glucose to produce an β-*O*-glucoside conjugate (Metabolite II). In susceptible species such as *Avena* spp. (wild oats) and *Digitaria ischaemum* (crabgrass), only trace amounts of the conjugates are found, apparently due to insufficient levels of substrates and/or enzymes to detoxify fenoxaprop [91]. Fenoxaprop tolerance in wheat and barley is associated with high rates of detoxification.

Fluazifop-P butyl ester is hydrolyzed rapidly to the phytotoxic fluazifop-P acid. The susceptible *Elytrigia repens* (quackgrass) retains most of the applied fluazifop as the acid after 48 h, while a small fraction is metabolized to polar and non-polar conjugates.

Haloxyfop methyl ester is hydrolyzed rapidly in tolerant and susceptible plants to· the herbicidally active haloxyfop acid. Five hours after application of haloxyfop methyl ester was applied to corn leaves, 50% of that applied was discovered as haloxyfop acid, 31% as unmetabolized haloxyfop methyl ester, and 19% as unidentified metabolites, which could be the conjugated forms of the herbicide [38]. Rates of de-esterification and subsequent metabolism to polar products vary among grass species and account for differential susceptibility to the herbicide [8].

Mecoprop is metabolized readily by the tolerant perennial broadleaf weeds, including *Stellaria media* (chickweed). Most (63 to 99%) of the applied mecoprop is converted to conjugated (with amino acids, proteins or sugars, attached at the carboxyl moiety) forms of the herbicide between 3 d and 3 wk after application in tolerant species.

The ethyl ester of quizalofop-P is hydrolyzed rapidly to quizalofop-P acid. The tolerant soybean and maize convert quizalofop acid to phenol metabolites or it is conjugated to glucose. In the susceptible *Elytrigia repens* (quackgrass), polar metabolites of quizalofop represented 26% to 48% of the absorbed herbicide after 1 d and 5 d of treatment respectively, while the remainder was the ester forms of quizalofop [92].

Phenylpyridazines

Pyridate

Pyridate is rapidly hydrolyzed, within a few minutes to days, to 3-phenyl-4-hydroxy-6-chloropyridazine, followed by conjugation to form non-phytotoxic *O*- and *N*-glycosides.

Phenyl Triazinones (Aryl Triazinones)

Sulfentrazone

The primary detoxification of sulfentrazone appears to be oxidation of the methyl group of the triazolinone ring, resulting in formation of the more polar hydroxymethyl derivative [18]. The methyl group on position 3 of the triazolinone ring is necessary for maximum biological activity [94].

Dayan et al. [18] found that little sulfentrazone was metabolized in the roots of *Senna obtusifolia* (sicklepod), the relatively more tolerant weed species, and *Cassia occidentalis* (coffee senna), the relatively more susceptible weed species, indicating that most of the parent chemical reached the foliage in both species. However, 3 h after exposure to ^{14}C-sulfentrazone, only 49.2% of the radioactivity taken up by the roots and translocated to the laves of *S. obtusifolia* remained as the parent compound, while 83.3% of the radioactivity in *C. occidentalis* was associated with the herbicide. The metabolites, hydroxymethyl derivative (II), carboxylic acid derivative (III) and conjugated derivative (IV) accounted for 22%, 7.3%, and 19.5% of the total ^{14}C-sulfentrazone activity. After 9 h exposure, only 8.4% of the parent herbicide remained in *S. obtusifolia* whereas *C. occidentalis* contained 83%. The difference in sensitivity between *S. obtusifolia* and *C. occidentalis* to sulfentrazone may be primarily attributed to differential rate of metabolism of the herbicide.

Pyridazinones

Metflurazon (SAN 6706), Norflurazon (9789), Pyrazon

Norflurazon is readily degraded by several species. Degradation occurred more rapidly in soybean and maize, the susceptible species, than in cotton, the tolerant species [88]. The major degradation product in all species was desmethyl derivative, with other unidentified compound(s) also becoming apparent. The degradation mechanism involved N-demethylation. The rapid degradation of metflurazon (SAN 6706) by maize did not represent a detoxification mechanism since the primary degradation product formed (norflurazon), was more phytotoxic than metflurazon itself. Strang and Rogers [88] suggested the possibility that the susceptibility of a plant species to SAN 6706 (metflurazon) was probably dependent upon its ability to metabolize it to the phytotoxic metabolite SAN 9789 (norflurazon). In such a case, the tolerant cotton was probably lacking an active enzyme system capable of degrading SAN 6706 (metflurazon). The N-demethylation mechanism of metflurazon to norflurazon is shown in Fig. 8.14.

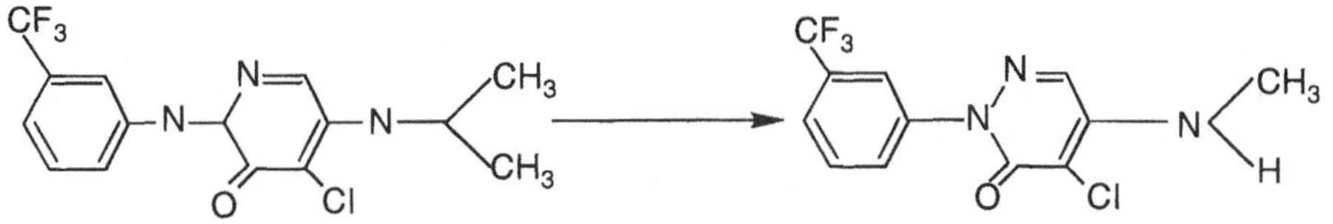

Fig. 8.14. Metabolism of metflurazon to norflurazon [88].

Pyrazon breaks down to 5-amino-4-chloro-3(2H)-pyridazinone as a result of the split of the phenyl group. This product is inactive. Ries et al. [71] reported that pyrazon conjugated with glucose to form N-glucosyl metabolite of pyrazon in red beet. They found that this metabolite N-2-chloro-4-phenyl-3(2H)-pyridazinone glucosamine was formed only in shoots and not in roots. They postulated that the tolerance of beets to pyrazon results from the metabolic conversion of the herbicide to this metabolite. Stephenson et al. [83] observed that pyrazon conjugation with glucose was dependent upon the carbohydrate status of the leaf tissue and hence formation of N-glucosyl pyrazon was greater in light than in darkness.

Pyridinecarboxylic Acids

Clopyralid, Picloram, Triclopyr

Clopyralid is metabolized slowly in most plants. The susceptible *Cirsium arvense* (Canada thistle) requires 6 to 9 d after application to form water-soluble metabolites while the tolerant rapeseed plant forms these metabolites within 1 to 6 h after treatment.

Picloram, like clopyralid, is metabolized slowly in susceptible species but more rapidly in tolerant ones. Most metabolites are water-soluble and suspected to be glucose conjugates. In sunflower, picloram conjugates with glucose to form N-glucoside, while in *Euphorbia esula* (leafy spurge) it is metabolized to isomeric glucose esters, gentibiose esters and N-glucosides. The water-soluble glucoside conjugates, upon ammonolysis, yield the carboxylic acid amide of picloram.

Triclopyr esters are rapidly hydrolyzed to triclopyr acid. Metabolism of the acid may be slow in susceptible plants, while it appears to be rapid in tolerant ones. Wheat and barley, the tolerant plants, metabolized at least 85% of the acid and ester forms of triclopyr within 3 d of application [55]. The metabolites included a mixture of unidentified water-soluble sugar conjugates as well as glutamate and aspartate conjugate and a methyl ester. In *Stellaria media* (chickweed), the major metabolite was triclopyr aspartate. A glutamate conjugate was also formed.

Pyridines

Thiazopyr

Thiazopyr is extensively and rapidly degraded to a large number of polar metabolites, each comprising less than 10% of the total applied thiazopyr residues. Major degradation reactions include sulphur oxidation, thiazoline ring opening and methyl ester hydrolysis, and transformation of the isobutyl side chain. The 2-difluoromethyl-4-(2-methylpropyl)-6-trifluoromethyl-3-pyridinecarboxylate moiety is found in most of the metabolites as well as in the parent thiazopyr.

Quinolinecarboxylic Acids

Quinclorac

Quinclorac metabolized at a moderate rate in young *Euphorbia esula* (leafy spurge) leaves and apex at a lower rate than in other organs [48]. It metabolized first to a C-1 glucose ester that was subsequently converted to a pentosylglucose ester (major product) or a malonylglucose ester (minor product). The metabolites were labile and readily hydrolyzed by β-glucosidases. It is uncertain whether these metabolites (glucose esters) are sufficiently stable to result in effective detoxification Therefore, the role of metabolism in quinclorac detoxification in *E. esula* is not definitively known.

Sulfonylureas

A. Detoxification

Bensulfuron, Chlorimuron, Chlorsulfuron, Metsulfuron, Nicosulfuron, Primisulfuron, Triasulfuron, Triflusulfuron

Bensulfuron is metabolized by tolerant species to herbicidally inactive compounds. Conjugation with homoglutathione at the 4-chloro group on the pyrimidine ring is a major metabolic pathway of chlorimuron in tolerant soybean. Homoglutathione conjugate may undergo further degradation to a cysteinyl conjugate. Chlorimuron methyl also undergoes esterification, followed by hydrolysis of the carboxylic acid analogue to sulfonamide and aminopyrimidine. A minor metabolite following esterification in soybean is chlorimuron acid.

In the case of chlorsulfuron, the tolerant wheat metabolizes 97% of the herbicide to a phenyl-*o*-glycosylated metabolite in 24 h [89]. This metabolite has no herbicidal activity. Metabolism of chlorsulfuron is by way of hydroxylation at the 5-carbon position of the phenyl ring, apparently via cytochrome P-450 mono-oxygenases. The hydroxylated chlorsulfuron is then rapidly conjugated with glucose. Thus, like other sulfonylurea herbicides, the metabolism of chlorsulfuron involves a two-step process, i.e., ring hydroxylation followed by glycosylation at the site of ring hydroxylation. Ring hydroxylation is the ultimate step in its metabolism. Tolerant broadleaf weeds such as *Linum catharticum* (flax) and *Solanum nigrum* (black nightshade) hydroxylate the methyl group on the triazine ring, followed by rapid glucose conjugation. The sensitive soybean and sugarbeet (*Beta vulgaris*) metabolize slowly. Nicosulfuron is metabolized in maize with half-lives ranging from 1.5 to 4.5 h. The sensitive species of sorghum and *Setaria* spp. metabolize the herbicides much more slowly. Maize metabolizes nicosulfuron on the 5-position of the pyrimidine ring (Fig. 8.15) [7] and this

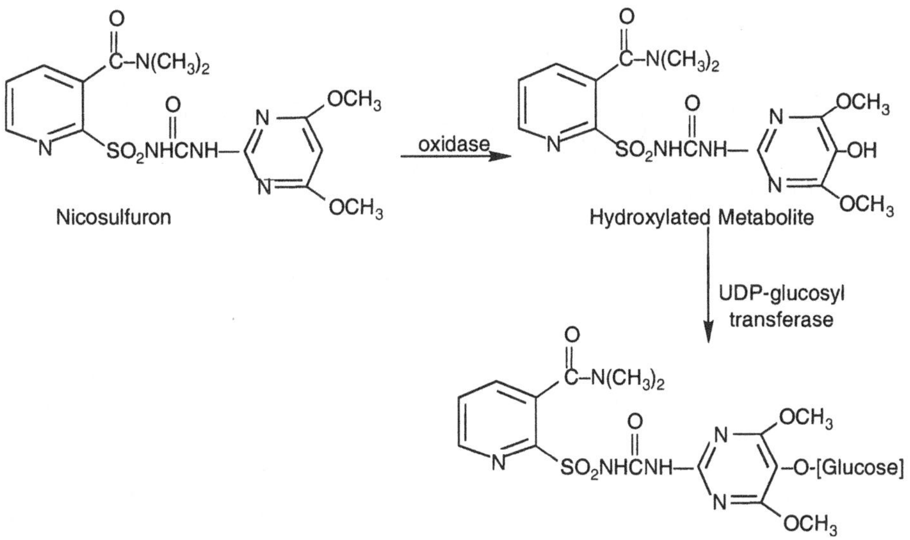

Fig. 8.15. Metabolism of nicosulfuron in tolerant maize. (Reproduced from [7] with permission of the publisher).

hydroxylated metabolite is inactive against acetolactate synthase, which is inhibited by the herbicide. This hydroxylated metabolite is subsequently rapidly conjugated to glucose. The selectivity of maize is strongly affected by substituents on the pyridine half of the molecule, a site relatively distant from the actual site of metabolism [7].

Primisulfuron is metabolized rapidly by tolerant maize, with a half-life of 3 to 5 h, via hydroxylation of the phenyl and the 5-carbon position of the pyrimidine ring, with a subsequent conjugation to glucose. The reactions are mediated by cytochrome P-450 mono-oxygenases. The tolerant annual grass weed *Echinochloa crus-galli* (barnyardgrass) also metabolizes primisulfuron more rapidly (half-life 1.5 h) than maize via hydroxylation and conjugation pathway. In addition, *E. crus-galli* also produces a non-glycosylated metabolite. The hydroxylated primisulfuron and both the conjugated forms are herbicidally inactive. Besides, a relatively minor reaction involves hydrolytic cleavage of the sulfonylurea bridge yielding CO_2, a pyrimidine amine metabolite and a benzene sulfonamide metabolite. Hinz and Owen [40] reported that maize metabolized primisulfuron very rapidly, with a half-life of less than 4 h, while the sensitive *Sorghum bicolor* (shattercane) metabolized it very slowly, with a half-life of 36 h. The plants are tolerant if they can metabolize 50% of the herbicide within 5 h.

Triasulfuron is metabolized by tolerant wheat more rapidly than by the sensitive grass weed *Lolium perenne* (perennial ryegrass). Wheat foliage retains only 14% of the applied triasulfuron intact compared to 70% by the foliage of *L. perenne*. The metabolic pathway of triasulfuron is similar to other sulfonylureas, beginning with rapid phenyl hydroxylation followed by conjugation to glucose (Fig. 8.16) [7]. Wheat microsomes contain several different cytochrome P-450-linked mono-oxygenases having different properties and substrate specificities that are responsible for herbicide metabolism.

Triflusulfuron is rapidly metabolized by the tolerant sugar beet, with a half-life of less than 1 h, while the moderately tolerant *Chenopodium album* has an intermediate rate (half-life of 7 h). The sensitive weeds (*Brassica capus*), *Matricaria inodora* (scentless

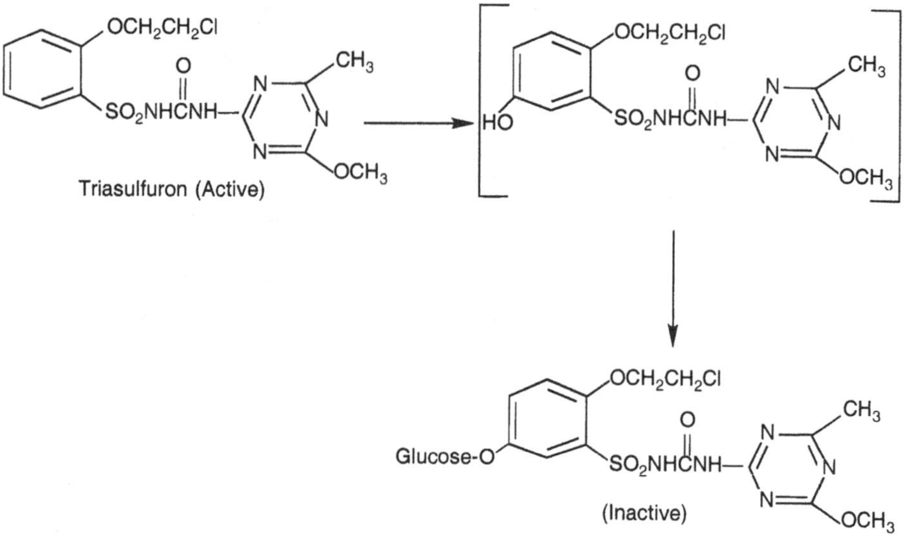

Fig. 8.16. Metabolism of triasulfuron in tolerant wheat. (Reproduced from [7] with permission of the publisher)

chamomile), and *Veronica persica* (Persian speedwell) have slow rates (half-lives of more than 35 h) of metabolism [98], The initial metabolism of triflusulfuron in sugar beets involves nucleophile attack by glutathione at the urea carbonyl group, producing the *S*-carbonyl glutathione conjugate plus 7-methylsaccharin and its free acid.

B. BIOACTIVATION

DPX-H6564

While metabolic inactivation by tolerant crops is the common mechanism of selectivity to sulfonylurea herbicides, a few compounds are activated by weed species to become herbicidally active. DPX-H6564, an experimental herbicide, is selective for soybean and active against grass and broadleaf weeds at 16-70 g ha^{-1}. It is inactive against acetolactate synthase (ALS), the site of action of sulfonylurea herbicides. The sensitive species of maize and *Sorghum halepense* (Johnsongrass) metabolize DPX-H6564 relatively rapidly, with half-lives of 7 to 9 h, while the tolerant soybean does not metabolize measurably, with more than 95% of the herbicide remaining intact after 24 h. The metabolized product 4-hydroxymethyl derivative (Fig. 8.17) [7] in the sensitive species is highly active against ALS. Thus, the herbicidal activity of DPX-H6564 results

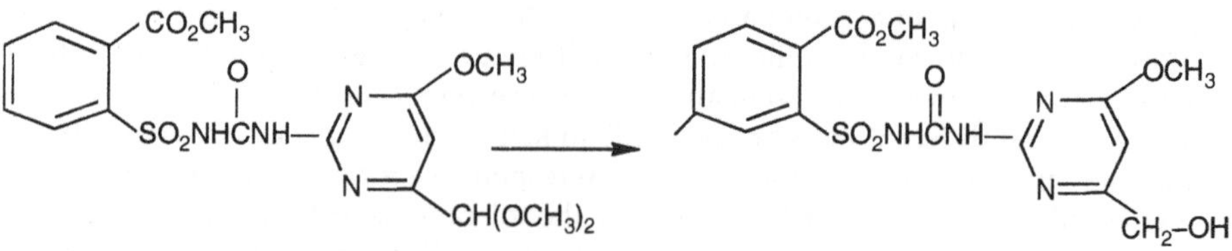

Fig. 8.17. Metabolic activation of sulfonylurea herbicide DPX-H6564 in sensitive maize and *Sorghum halepense* (Johnsongrass). (Reproduced from [7] with permission of the publisher).

from metabolic inactivation by sensitive plant species, and its tolerance for soybean results from the inability of soybean to similarly activate this herbicide [7]

C. SELECTIVITY FOR SULFONULUREA HERBICIDES

The primary basis for selectivity for sulfonylurea herbicides is the differential rate of herbicide metabolism. Tolerant species rapidly detoxify sulfonylureas to herbicidally inactive products, while metabolism is much lower and less sensitive in susceptible species.

Of the various sulfonylurea herbicides, nicosulfuron and primisulfuron were the first selective postemergence herbicides that effectively control perennial and annual grasses as well as some broadleaf weeds while being selective for maize. Nicosulfuron is rapidly metabolized within 20 h in the tolerant maize, while there is no perceptible metabolism in the sensitive *Sorghum halepense* (Johnsongrass) even after 24 h [61]. Primisulfuron is rapidly metabolized by the tolerant *Echinochloa crus-galli* (barnyardgrass) into metabolites that do not inhibit acetolactate synthase (ALS) [63]. ALS from tolerant and sensitive plants is generally similar in sensitivity to nicosulfuron and primisulfuron [20, 35, 60], suggesting that metabolism is the primary basis for differential response.

Nicosulfuron and primisulfuron display differential selectivity between species, between varieties within a species and between herbicides against a species. Maize is tolerant to both herbicides, while seedling *S. halepense* is sensitive to both herbicides, *E. crus-galli* sensitive to nicosulfuron and tolerant to primisulfuron, *Setaria faberi* (giant foxtail) sensitive to both nicosulfuron and primisulfuron, and *Solanum nigrum* (black nightshade) tolerant to nicosulfuron and sensitive to primisulfuron [12]. Both herbicides are metabolized by tolerant species more rapidly and extensively than by sensitive species. However, tolerance of *S. nigrum* to nicosulfuron and sensitivity to primisulfuron is not due to differential uptake, translocation, or metabolism but rather to the lower sensitivity of ALS to nicosulfuron than to primisulfuron [12]. The work done so far indicates that plant tolerance to sulfonylurea herbicides is not always a function of herbicide metabolism, but due to the interaction of several physiological and biochemical factors.

Thicarbamates

Butylate, Diallate, EPTC, Molinate, Pebulate, Triallate

Butylate is apparently oxidized in tolerant crops to butylate sulphoxide and then detoxified by glutathione (homoglutathione). Subsequent cleavage of amino acids from the glutathione tripeptide and conjugation with malonate yields the malonyl-3-thiolactic acid conjugate of butylate. The half-life of butylate in tolerant crops is a few hours or less. Butylate residues disappear from stems and leaves of tolerant plants 7-14 d after treatment.

Diallate and triallate are oxidized initially to their sulphoxides, followed by conjugation with glutathione. Sulphoxides of diallate are also further metabolized to yield 2-chloroacrolein and 2,3-chloro-2-propane-1-sulphonic acid. The herbicidal activities of diallate and triallate probably result from the metabolically formed sulphoxides acting as carbamoylating agents for critical enzyme of thiol groups or liberating chloroallyl-containing toxicants such as 2-chloroacrolein. Tolerant species apparently detoxify diallate and triallate by the sulphoxidation-conjugation mechanism.

EPTC is hydrolyzed to ethyl mercaptan, CO_2 and dipropylamine in various crops. EPTC is initially oxidized to EPTC-sulphoxide, which in turn is cleaved by a glutathione S-transferase enzyme system to form glutathione conjugate, S-(N,N dipropylcarbamyl) glutathione (Fig. 8.18.) [13, 53, 54]. The glutathione conjugate of EPTC is further metabolized, through malonylation, to yield N-malonylcysteine conjugates.

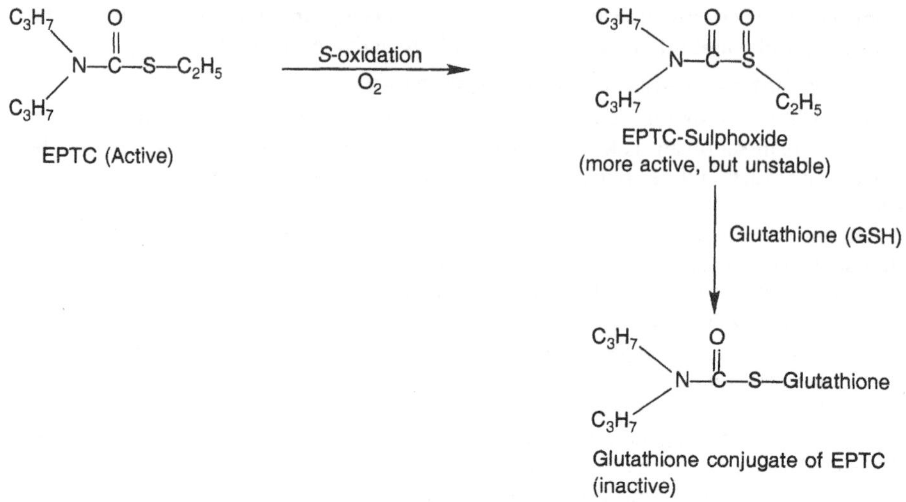

Fig. 8.18. Metabolic pathway of EPTC.

In rice, molinate undergoes hydrolysis and decarboxylation reactions on the alkyl and carbonyl groups, yielding CO_2 which is incorporated into amino acids asparagine, glycine, threonine, alanine, tryptophan, phenylalanine and isoleucine, and organic acids, lactic and glycolic [36].

Pebulate is rapidly metabolized by resistant mung bean (*Phaseolus aureus*), with metabolism virtually completing within 8 h of treatment [23]. Wheat, the susceptible species, lacked an active pebulate detoxification system. In treated tobacco seedlings, pebulate was so rapidly metabolized that radioactivity could not be detected, possibly due to its immediate translocation to shoots as metabolic constituents [56]. The pattern of labelling suggested that these metabolites might have been quickly incorporated into the acetyl CoA pool. Hydrolysis and decarboxylation reactions on the alkyl and carbonyl groups of pebulate yield CO_2. Pebulate may also be oxidized to pebulate sulphoxide.

Triazines

Atrazine, Propazine, Simazine

Generally triazine herbicides are metabolized by higher plants by the following mechanisms: a) 2-hydroxylation, b) N-dealkylation, c) glutathione conjugation, and d) ring cleavage.

2-Hydroxylation: In this mechanism atrazine (a chlorotriazine) is converted to 2-hydroxyatrazine, and simazine to 2-hydroxysimazine. This hydroxylation (Fig. 8.5) occurs in resistant plant species such as maize, which contains a resistant factor 2,4-dihydroxy-3-keto-7-methoxy-1,4-benzoxazine [37]. This benzoxazine factor does not appear to be present in other tolerant plants or active on the methylthiotriazines. Chlorotriazines also transform into corresponding hydroxy derivatives by some other mechanism. The methoxy and methylthio triazines are more stable to hydroxylation than chlorotriazines. However, they do yield small amounts of hydroxytriazine and other metabolites.

N-Dealkylation: In this process, the alkyl group, i.e. the ethylamino group and/or isopropylamino group in the case of atrazine is removed, leaving the rest of the molecule intact. In peas, atrazine is dealkylated (Fig. 8.6) by removal of the ethyl and isopropyl groups to form 2-chloro-4-amino-6-isopropylamino-s-triazine and 2-chloro-4-ethylamino-6-amino-s-triazine respectively. These dealkylated products are formed in soybean, wheat, maize and sorghum. Complete N-dealkylation of atrazine results in the formation of 2,chloro-4,6-amino-s-triazine. 2-hydroxyatrazine also undergoes dealkylation in maize. The delakylated-hydroxy-atrazine in maize is 2-hydroxy-4-ethylamino-6-amino-s-triazine. Simazine was completely N-dealkylated by the perennial grass *Imperata cylindrica* (thatchgrass, cogongrass or alang-alang) to form 2-hydroxy-4,6-*bis*(amino)-s-triazine [42]. The N-dealkylated atrazine metabolites in sorghum were partially phytotoxic while the 2-hydroxyatrazine formed in maize was non-phytotoxic [74]. Thompson et al. [97] reported that the order of tolerance of monocot species, maize, *Panicum dichotomiflorum* (fall panicum), *Digitaria sanguinalis* (large crabgrass), *Setraia faberi* (giant foxtail) and oats, was identical to their ability to metabolize atrazine. The completely N-dealkylated atrazine showed no inhibition of the Hill reaction and non-cyclic photophosphorylation in pea chloroplasts [82]. However, the relatively low turnover rate of N-dealkylated atrazine and the less active N-dealkylation reaction suggested that N-dealkylation was not a major degradation pathway in plants [82].

Conjugation: Maize completely transforms atrazine to peptide conjugates [80, 83, 96]. Similar peptide conjugates are also formed in the case of propazine and [96]. In sorghum, another tolerant species, the first step in atrazine metabolism is conjugation with glutathione by displacing the 2-chloro substituent [50, 51]. This conjugation with glutathione is catalyzed by glutathione-s-transferase [80, 81], present in the leaves of sorghum, maize and other triazine-resistant species [28]. The metabolic pathway of atrazine in sorghum [52] is illustrated in Fig. 8.19. The first four reactions proceed at a faster rate than the last two.

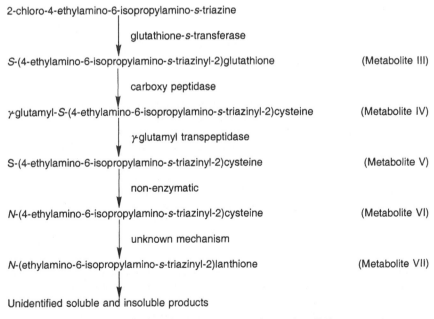

Fig. 8.19. Glutathione conjugation of atrazine [52].

Lamoureux et al. [52] reported that about 40 to 87% of the atrazine entering sorghum through the roots was metabolized via this pathway. They also suggested that this metabolic pathway could also occur after first undergoing N-dealkylation.

A grass weed, Sorghum bicolor (shattercane), also forms 2-hydroxy derivatives as well as peptide (glutathione and L-cysteine) conjugates of atrazine as well as simazine [28, 95]. Formation of glutathione conjugates by glutathione-s-transferase was much faster in the case of atrazine than simazine, and this caused greater accumulation of simazine in the roots of S. bicolor [95].

Ring Cleavage: Triazines are also metabolized via the ring cleavage pathway. During ring cleavage CO_2 is formed, which can be utilized by the plant. However, degradation via the ring cleavage pathway appears to be so small as to preclude its importance in the metabolism of chloro-s-triazine derivatives.

Cyanazine

Like other s-triazines, cyanazine also undergoes 2-hydroxylation, hydrolysis and N-dealkylation in plants. The chlorine atom is removed during hydroxylation, the cyano group is hydrated during hydrolysis and the ethyl group is removed during N-dealkylation.

Hexazinone

Hexazinone is converted to several hydroxylated and/or demethylated trazinone metabolites. Tolerance to hexazinone appeared to be associated with higher levels of the mono-demethylated metabolite 3-cyclohexyl-6-methylamino-1-methyl-1,3,5-triazine-2,4-dione. In sugarcane, the metabolites include a hydroxylated derivative, a hydroxylated-demethylated derivative and a hydroxylated-deaminated derivative.

Triazinones (*as*-Triazines)

Metribuzin

The most important metribuzin detoxification reactions in soybean are N-deamination, N-glucoside conjugation and sulphoxidation, followed by conjugation with homoglutathione. N-deamination is possibly mediated by mixed function oxidase systems and results in deaminated (DA) metribuzin. The rate of N-deamination seems to be in correlation with tolerance in soybean cultivars. N-glucoside conjugation is mediated by metribuzin-N-glucosyl transferase and the resultant glucoside is conjugated rapidly with malonic acid to form the malonyl β-D-(N-glucoside) conjugate. This two-step conjugation is the major detoxification pathway in tomato [29]. Sulphoxidation forms an unstable sulphoxide that is most commonly conjugated with homoglutathione. The sulphoxide may also undergo demethylation to form diketo (DK) metribuzin. The DK metabolite can be conjugated with malonate, deaminated to form the deaminated diketo (DADK) metabolite, or can be incorporated into soluble residues. Winter wheat metabolizes metribuzin to the DA, DADK and DK derivatives [19].

Triazoles

Amitrole, Amitrole-T

Conjugation with plant constituents is the major transformational pathway of amitrole in plants. The resultant conjugates are considered to be less phytotoxic than the parent amitrole compound. The major conjugate or metabolite is 3-amino-1,2,4-triazolylalanine (3-ATL), a condensation product of amitrole and serine. Of the various conjugated products, 3-ATL is the most predominant one. As the triazole molecule is highly stable, it is resistant to ring cleavage and other normal transformation pathways other than conjugation.

Triazolopyrimidine Sulfonanilides

Flumetsulam

Flumetsulam is metabolized by tolerant species faster than by susceptible species. The half-life of flumetsulam is 2 h in tolerant corn, 18 h in soybean and 104 h in susceptible *Amaranthus retroflexus* (redroot pigweed). No metabolism occurs after 144 h in highly susceptible *Abutilon theophrasti* (velvetleaf).

Uracils

Bromacil, Terbacil

Bromacil is rapidly degraded in tolerant plants to 5-bromo-3-*sec*-hydroxymethyluracil as the major metabolite and 5-bromo-3-(3-hydroxyl-1-methylpropyl)-6-methyluracil as a minor metabolite [46]. No 5-bromocil is found as a metabolite.

The major metabolite of terbacil is the glucosidic conjugate of terbacil. Genez and Monaco [31] reported that greater than 80% of terbacil was converted to glucosyl conjugate after 48 h in cucumber (*Cucumis sativus*) and *Solidago fistulosa* (goldenrod). Anderson et al. [1] also found that the major metabolite of terbacil in alfalfa (lucerne) is a glucosidic conjugate indicating the presence of a very active terbacil glucosyltransferase enzyme in alfalfa.

Ureas

Chlortoluron, Diuron, Fluometuron, Linuron, Monuron

Substituted urea herbicides are metabolized via the N-dealkylation (demethylation) pathway (Fig. 8.20). Following demethylation, the metabolites possibly undergo hydrolysis involving deamination and decarboxylation, yielding the corresponding aniline (R-NH$_2$) which will undergo further oxidation and conjugation. Removal of one methyl group reduces the phytotoxicity of the urea herbicide while removal of the second group would make the herbicide non-phytotoxic. Diuron, monuron, and fluometuron undergo demethylation. Demethylation occurred in cotton and cucumber plants, with the shoots of cotton more active in demethylation than roots [73].

Linuron, which contains a methoxy group, undergoes demethoxylation as thoroughly as demethylation in the case of monuron. Demethoxylation is a two-step reaction involving hydroxylation with removal of the methyl group, and subsequent dehydroxylation.

Wheat can detoxify chlortoluron via two routes of oxidative metabolism: aryl (ring-methyl) hydroxylation and N-dealkylation. N-dealkylation, a minor route towards plant tolerance to chlortoluron, involves the removal of both methyl groups. Demethylation of one methyl group yields a partially phytotoxic mono-demethylated metabolite, while removal of two methyl groups results in a phytotoxic di-demethylated metabolite. Metabolism via ring-methyl hydroxylation is, however, primarily responsible for chlortoluron resistance in wheat. This is mediated by cytochrome P-450. The non-phytotoxic hydroxymethyl metabolite derived from ring-methyl hydroxylation undergoes conjugation with glucose.

Urea herbicides also undergo conjugation with proteins. Camper and Moreland [11] found that the free amino groups of bovine serum albumin (BSA) were involved in the binding of diuron to proteins.

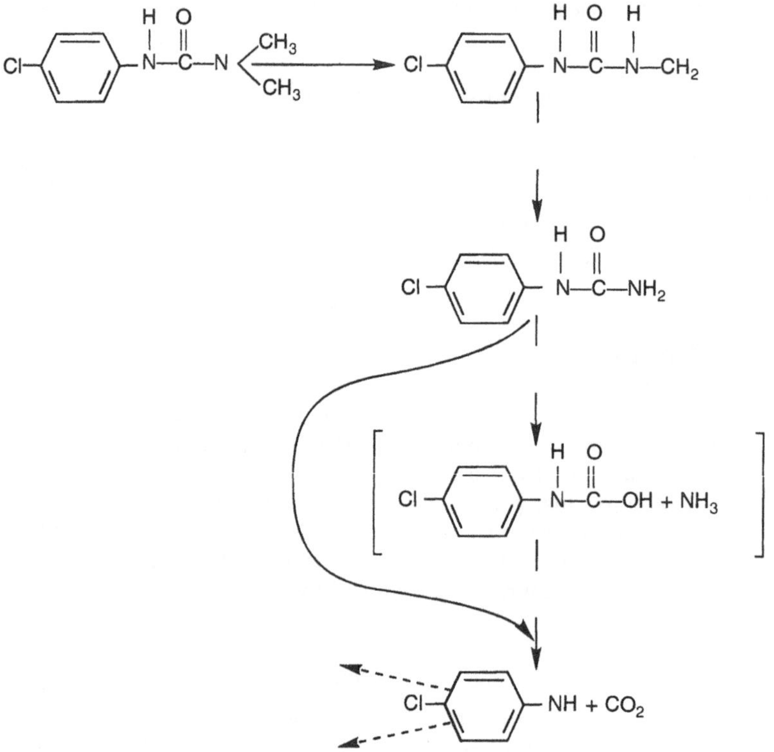

Fig. 8.20. Degradation of monuron by *N*-dealkylation by plants.

Unclassified Herbicides

Fosamine

The primary metabolite of fosamine in tolerant pasture and turf grasses is carbamoylphosphonic acid (CPA) which reached a maximum concentration after 2-4 wk. Besides, a minor metabolite carboxyphosphonic acid is also formed.

Glufosinate

D-glufosinate appears to be stable in plants, but L-glufosinate gets metabolized to 4-methylphosphinico-2-oxo-butanoic acid, 3-methylphosphinico-propanoic acid, and 4-methylphosphinico-2-hydroxybutanoic acid in tobacco, alfalfa and carrots.

Glyphosate

Glyphosate is metabolized to aminomethylphosphonic acid (AMP) and glyoxalate in maize, soybean, cotton and wheat [75]. Metabolism occurs very slowly, with metabolites, being transient, rapidly broken down to CO_2. Sandberg et al. [76] noted very little, if any, metabolism of glyphosate to AMP in *Convolvulus arvensis* (field bindweed) and *Ipomaea purpurea* (tall morningglory).

Melfluidide

No metabolites of melfluidide could be detected even after 3 to 14 d after treatment, indicating that metabolites are apparently degraded or are incorporated into non-extractable residues. Melfluidide was metabolized 14% in corn coleoptiles and 54% in soybean coleoptiles after 6 h, apparently to a conjugate releasing parent melfluidide upon acid treatment [32].

REFERENCES

1. Anderson, M.P. C. Bensch, J.F. Strikze and J.L. Caddel. 1995. Uptake, translocation, and metabolism in alfalfa (*Medicago sativa*) selected for enhanced tolerance to terbacil. Weed Sci. 43: 365-369.
2. Anonymous. 1977. Ann. Rep. USA, Hawaiian Sugar Planters' Association Expt. Sta., 56 pp.
3. Armstrong, T.F., W.F. Meggitt and D. Penner. 1973. Absorption, translocation, and metabolism of alachlor by yellow nutsedge. Weed Sci. **21**: 357-359.
4. Biswas, P.K. and K.C. Hamilton. 1969. Metabolism of trifluralin in peanuts and sweet potatoes. Weed Sci. **17**: 206-211.
5. Breux, E.J. 1987. Initial metabolism of acetochlor in tolerant and susceptible seedlings. Weed Sci. **35**: 463-468.
6. Broadhurst, N.A., M.L. Montgomery and V.H. Freed. 1966. Metabolism of 2-methoxy-3,6-dichlorobenzoic acid (dicamba) by wheat and bluegrass plants. J. Agr. Food Chem. **14**: 585-588.
7. Brown, H.M. and J.C. Cotterman. 1994. Recent advances in sulfonylurea herbicides. *In* Chemistry of Plant Protection, vol. 10. Springer-Verlag, Berlin Heidelberg, Germany, pp. 47-80.
8. Buhler, D.D., B.A. Swisher and O.C. Burnside. 1985. Behaviour of ^{14}C-haloxyfop-methyl in intact plants and cell cultures. Weed Sci. **33**: 291-299.
9. Butts, J.S. and S.C. Fang. 1956. *In* a Conference on Radioactive Isotopes in Agriculture. A.E.C. Rep. No. TID-7512, pp. 209-214.
10. Campbell, J.R. and D. Penner. 1985. Sethoxydim metabolism in monocotyledonous and dicotyledonous plants. Weed Sci. **33**: 771-773.
11. Camper, N.D. and D.E. Moreland. 1971. Sorption of substituted phenylamides into Bovine Serum Albumin. Weed Sci. **19**: 269-273.
12. Carey, J.B. D. Penner, and J.J. Kells. 1997. Physiological basis for nicosulfuron and primisulfuron selectivity in five plant species. Weed Sci. **45**: 22-30.
13. Carringer, R.D., C.E. Rieck and L.P. Bush. 1978. Metabolism of EPTC in corn (Zea mays) Weed Sci. **26**: 157-160.
14. Chang, F.Y. and W.H. VandenBorn. 1971. Dicamba uptake, translocation and metabolism in Canada thistle. Weed Sci. 176-181.
15. Corio-Costet, M.-F., M.D. Agnese and R. Scalla. 1991. Effects of isoxaben on sensitive and tolerant plant cell cultures. I. Metabolic fate of isoxaben. Pestic. Biochem. Physiol. **40**: 246-254.
16. Davies, P.J., D.S.H. Drennan, J.D. Fryer and K. Holly. 1968. The basis of differential phytotoxicity of 4-hydroxy-3,5-di-iodobenzonitrile. II. Metabolism. Weed Res. **8**: 241-248.
17. Davis, C. and D.L. Linscott. 1986. Tolerance of birdsfoot trefoil (*Lotus corniculatus*) to 2,4-D. Weed Sci. **34**: 373-376.
18. Dayan, F.E., J.D. Wheet and H.G. Hancock. 1996. Physiological basis for differential sensitivity to sulfentrazone by sicklepod (*Senna obtusifolia*) and coffee senna (*Cassia occidentalis*). Weed Sci. **44**: 12-17.
19. Devlin, D.L., D.R. Grealy and L.A. Morrow. 1987. Differential metabolism of metribuzin by downy brome (*Bromus tectorum*). Weed Sci. **35**: 741-745.
20. Diehl, K.E., H. Mukaida, R.A. Liebl and E.W. Stoller. 1993. Sensitivity mechanism in an ALS-susceptible corn hybrid. Abstr. Weed Sci. Soc. Amer. **33**: 191.
21. Durst, F. and I. Benvenist. 1993. Plant 450. *In* J.B. Schenkman and H. Grein (eds.). Hand Book of Experimental Pharmacology: Cytochrome P450. Springer-Verlag, New York, USA, pp. 293-310.
22. Eastin, E.F. 1971. Fate of fluorodifen in resistant peanut seedlings. Weed Sci. **19**: 261-265.
23. Fang, S.C. and M. George. 1962. Metabolism of propyl-1-^{14}C-(N,N-ethyl, n-butyl) thiocarbamate (Tillam) in plants. Plant Physiol. Suppl. **37**: 26.

24. Fawcett, C.H., R.C. Seeley, H.F. Taylor, R.L. Wain and F. Wightman. 1955. Alpha oxidation of omega-(2,4-dichlorophenoxy)alkane nitriles and 3-indolylacetonitrile within plant tissues. Nature **176**: 1026-1028.

25. Feung, C.S., R.H. Hamilton and R.O. Mumma. 1977. Metabolism of 2,4-dichlorophenoxyacetic acid. II. Herbicidal properties of amino acid conjugates. J. Agr. Food Chem. **25**: 898-900.

26. Fonne-Pfister, R. and K. Kruez. 1990. Ring-methyl hydroxylation of chlortoluron by an inducible cytochrome P450-dependent enzyme from maize. Phytochemistry **29**: 2793.

27. Frear, D.S. and G.G. Still. 1968. The metabolism of 3,4-dichloropropionanilide in plants. Partial purification and properties of an aryl acylamidase from rice. Phytochemistry **7**: 913-920.

28. Frear, D.S. and H.R. Swanson. 1970. The biosynthesis of s-(4-ethylamino-6-isopropylamino-2-s-triazine) glutathione: partial purification and properties of a glutathione-s-transferase from corn. Phytochemistry. **9**: 2123-2132.

29. Frear, D.S., E.R. Mansager, H.R. Swanson and F.S. Tanaka. 1983. Metribuzin metabolism in tomato: isolation and identification of N-glucoside conjugates. Pestic. Biochem. Physiol. **19**: 270.

30. Frear, D.S., H.R. Swanson and F.S. Tanaka. 1993. Metabolism of flumetsulam (DE-498) in wheat, corn, and barley. Pestic. Biochem. Physiol. **37**: 165.

31. Genez, A.L. and T.J. Monaco. 1983. Metabolism of terbacil in strawberry (*Fragaria ananassa*) and goldenrod (*Solidago fistulosa*). Weed Sci. **31**: 221-225.

32. Glenn, S. and C.E. Rieck. 1985. Auxin-like activity and metabolism of melfluidide in corn (*Zea mays*) and soybean (*Glycine max*) tissue. Weed Sci. **33**: 452-456.

33. Golab, T., R.J. Herberg, S.J. Parka and J.B. Tepe. 1967. Metabolism of carbon-14 tirfluralin in carrots. J. Agr. Food Chem. **15**: 638-641.

34. Golab, T., R.J. Herberg, J.V. Gramlich, A.P. Raun and G.W. Probst. 1970. Fate of benefin in soils, plants, artificial rumen fluid and the ruminant animal. J. Agr. Food Chem. **18**: 838.

35. Green, J.M. and J.F. Ulrich. 1993. Response of corn (*Zea mays* L.) inbreds and hybrids to sulfonylurea herbicides. Weed Sci. **41**: 508-516.

36. Gray, R.A. 1969. Abstr. Weed Sci. Soc. Amer., no. 174.

37. Hamilton, R.H. and D.E. Moreland. 1962. Simazine: degradation by corn seedlings. Science **135**: 373-374.

38. Harrison, S.K. and L.M. Wax. 1986. Adjuvant effects on absorption, translocation, and metabolism of haloxyfop-methyl in corn (*Zea mays*). Weed Sci. **34**: 185-195.

39. Hatzios, K.K. and D. Penner. 1980. Absorption, translocation, and metabolism of 14C-buthidazole in alfalfa (*Medicago sativa*) and quackgrass (*Agropyron repens*). Weed Sci. **28**: 635-639.

40. Hinz, J.R. and M.D.K. Owen. 1996. Nicosulfuron and primisulfuron selectivity in corn (*Zea mays*) and two annual grass weeds. Weed Sci. **44**: 219-223.

41. Hodgson, R.H. 1971. Influence of environment on metabolism of propanil in rice. Weed Sci. **19**: 501-507.

42. Hurter, J. 1967. The inactivation of simazine in resistant grasses. 6th Intl. Cong. Pl. Prot., Vienna, Austria, pp. 398-399.

43. Jablonkai, I. and K.K. Hatzios. 1991. Role of glutathione and glutathione-S-transferase in the selectivity of alachlor in maize and wheat. Pestic. Biochem. Physiol. **41**: 221-231.

44. Jaworski, E.G. 1969. Chloroacetamides. In P.C. Kearney and D.D. Kaufman (eds.). Degradation of Herbicides. Mercel Dekker, New York, USA, pp. 165-185.

45. Jaworski, E.G. and C.A. Porter. Uptake, and metabolism of 2-chloro-n-isopropylacetanilide in plants. Abstr. 149th Meeting Am. Chem. Soc. 21A.

46. Jordan. L.S. and W.A. Clerx. 1981. Accumulation and metabolism of bromacil in pineapple, sweet orange (*Citrus sinensis*) and Cleopatra mandarin (*Citrus reticulata*). Weed Sci. **29**: 1-5.

47. Kaufman, D.D. 1975. Phenols. In P.C. Kearney and D.D. Kaufman (eds.). Herbicides: Chemistry, Degradation, and Mode of Action. Mercel Dekker, New York, USA, pp. 665-707.

48. Lamoureux, G.L. and D.G. Rusness. 1995. Quinclorac absorption, translocation, metabolism, and toxicity of leafy spurge (*Euphorbia esula*). Pestic. Biochem. Physiol. **53**: 210-226.

49. Lamoureux, G.L., L.E. Stafford and F.S. Tanaka. 1971. Metabolism of 2-chloro-N-isopropyl acetanilide (propachlor) in the leaves of corn, sorghum, sugarcane, and barley. J. Agric. Food. Chem. **19:** 346-350.

50. Lamoureux, G.L., L.E. Stafford and R.H. Shimabukuro. 1972. Conjugation of 2-chloro-4,6-bis(alylamino)-s-triazines in higher plants. J. Agr. Food Chem. **20:** 1004-1010.

51. Lamoureux, G.L., R.H. Shimabukuro, H.R. Swanson and D.S. Frear. 1970. Metabolism of 3-chloro-4-ethylamino-6-isopropylamino-s-triazine (Atrazine) in excised sorghum leaf sections. J. Agr. Food Chem. **18:** 81-86.

52. Lamoureux, G.L., L.E. Stafford, R.H. Shimabukuro and R.G. Zaylskie. 1973. Atrazine metabolism in sorghum: Catabolism of the glutathione conjugate of atrazine. J. Agr. Food Chem. **21:** 1020-1030.

53. Lay, M.M. and J.E. Casida. 1976. Dichloroacetamide antidotes enhance thiocarbamate sulfoxide detoxification by elevating content and glutathione-s-transferase activity. Pestic. Biochem. Physiol. **6:** 442-456.

54. Lay, M.M., J.P. Hubbell and J.E. Casida. 1975. Dichloroacetamide antidotes for thiocarbamate herbicides: mode of action. Science **189:** 287-289.

55. Lewer, P. and W.J. Owen. 1990. Selective action of herbicide triclopyr. Pestic. Biochem. Physiol. **36:** 187-200.

56. Long, J.W., L. Thompson, Jr. and C.E. Rieck. 1974. Metabolism of ^{14}C-pebulate in seedling tobacco. Weed Sci. **22:** 91-94.

57. Marquis, L.Y., R.H. Shimabukuro, G.E. Stolzenberg, V.J. Feil and R.G. Zayalskie. 1979. Metabolism and selectivity of fluchloralin in soybean roots. J. Agr. Food Chem. **27:** 1148-1156.

58. McFadden, J.J., J.W. Gronwald and C.V. Eberlein. 1990. *In vitro* hydroxylation of bentazon by microsomes from napththalic anhydride-treated corn shoots. Biochem. Biophys. Res. Commun. **168:** 206.

59. Moreland, D.E., F.T. Corbin and J.E. McFarland. 1993. Oxidation of multiple substrates by corn shoot microsomes. Pestic. Biochem. Physiol. **47:** 206-210.

60. Neighbors, S. and L.S. Privalle. 1990. Metabolism of primisulfuron by barnyardgrass. Pestic. Biochem. Physiol. **37:** 145-153.

61. Obrigawitch, T.T., W.H. Kenyon and H. Kurtale. 1990. Effect of application timing on rhizome johnsongrass (Sorghum halepense) control with DPX-V9360. Weed Sci. **38:** 45-49.

62. O'Connell, K.M., E.J. Breux and R.T. Fraley. 1988 Different rates of metabolism of two chloroacetanilide herbicides in pioneer 3320 corn. Plant Physiol. **86:** 359-363.

63. Pohlenz, H.D., W. Boidol, I. Schuttke and W.R. Streber. 1992. Purification of *Arthrobactor oxydans* carbamate hydrolase specific for the herbicide phenmedipham and nucleotide sequence of the corresponding gene. J. Bacteriol. **174:** 6600-6607.

64. Probst, G.W., T. Golab and W.L. Wright. 1975. Dinitroanilines. *In* P.C. Kearney and D.D. Kaufman (eds.). Herbicides:Chemistry, Degradation, and Mode of Action. Marcel Dekker, New York, USA, pp. 454-500.

65. Probst, G.W., T. Golab, R.J. Herberg, F.J. Hazler, S.J. Parka, C. Vander Sachans and J.B. Tepe. 1967. Fate of trifluralin in soils and plants. J. Agr. Food Chem. **15:** 592-599.

66. Quimby, P.C., Jr. and J.D. Nalewaja. 1971. Selectivity of dicamba in wheat and wild buckwheat. Weed Sci. **19:** 598-601.

67. Rao, V.S. 1974. Mechanisms of action of acetanilide herbicides. Ph.D. Thesis. Cornell University, Ithaca, New York, USA. 116 pp.

68. Ray, B. and Wilcox. 1967. Metabolism of dicamba in *Zea* and *Hordeum*. Abstr. 154th Meeting of Amer. Chem. Soc. p. A28.

69. Ray, B. and M. Wilcox. 1969. Translocation of the herbicide dicamba in purple nutsedge, *Cyperus rotundus*. Physiol. Plant. **22:** 503-505.

70. Rice, R.P. and A.R. Putnam. 1980. Temperature influences on uptake, translocation, and metabolism of alachlor in snap beans (*Phaseolus vulgaris*). Weed Sci. **28:** 131-134.

71. Ries, S.K., M.J. Zabik, G.R. Stephenson and T.M. Chen. 1968. N-glucosyl metabolite of pyrazon in red beets. Weed Sci. **164:** 1-40.

72. Rogers, R.L. 1971. J. Agr. Food Chem. **19:** 32-36.

73. Rogers, R.L. and H.H. Funderburk. 1968. Physiological aspects of fluometuron in cotton and cucumber. J. Agr. Food Chem. **16:** 434440.

74. Roth, F.W. and T.L. Lavy. 1971. Atrazine translocation and metabolism in sudangrass, sorghum, and corn. Weed Sci. **19:** 98-101.

75. Rueppel, M.L., J.T. Marvel and L.A. Suba. 1975. The metabolism of N-phosphonomethylglycine in corn, cotton, soybeans, and wheat. 170th Meeting Abstr. Am. Chem. Soc., Division of Pestic. Chem. No. 26.

76. Sandberg, C.L., W.F. Meggitt and D. Penner. 1980. Absorption, translocation, and metabolism of ^{14}C-glyphosate in several weed species. Weed Res. **20:** 195-200.

77. Schafer, D.E. and D.O. Chilcote. 1970. Weed Sci. **18:** 725-730.

78. Shimabukuro, R.H. 1967. Significance of atrazine dealkylation in root and shoot of pea plants. J. Agr. Food Chem. **15:** 557-562.

79. Shimabukuro, R.H., R.E. Kundance and D.S. Frear. 1966. Dealkylation of atrazine in mature pea plants. J. Agr. Food Chem. **14:** 392-395.

80. Shimabukuro, R.H., H.R. Swanson and W.C. Walsh. 1970. Glutathione conjugation: atrazine detoxification mechanism in corn. Plant Physiol. **46:** 103-107.

81. Shimabukuro, R.H., D.S. Frear, H.R. Swanson and W.C. Walsh. 1971. Glutathione conjugation: an enzymatic basis for atrazine resistance in corn. Plant Physiol. **47:** 10-14.

82. Shimabukuro, R.H., W.C. Walsh, G.L. Lamoureux and L.E. Stafford. 1973. Atrazine metabolism in sorghum: chloroform-soluble intermediates in the N-dealkylation and glutathione conjugation pathways. J. Agr. Food Chem. **21:** 1031-1036.

83. Stephenson, G.R., D.R. Dilley and S.K. Ries. 1971. Influence of light and sucrose on N-glucosyl pyrazon formation in red beet. Weed Sci. **19:** 406-409.

84. Still, G.G. and G.R. Mansager. 1971. Metabolism of isopropyl-3-chlorocarbanilate by soybean plants. J. Agr. Food Chem. **19:** 879-884.

85. Still, G.G. and E.R. Mansager. 1972. Aryl hydroxylation of isopropyl-3-chlorocarbanilate by soybean plants. Phytochemistry. **11:** 515-520.

86. Still, G.G. and E.R. Mansager. 1973a. Metabolism of isopropyl carbanilate by soybean plants. Pest. Biochem. Physiol. **3:** 289-299.

87. Still, G.G. and E.R. Mansager. 1973b. Metabolism of isopropyl-3-chlorocarbanilate by cucumber plants. J. Agr. Food Chem. **21:** 787-791.

88. Strang, R.H. and R.L. Rogers. 1974. Behaviour and fate of two phenylpyridazinone herbicides in cotton, corn, and soybean. J. Agr. Food Chem. **22:** 1119-1125.

89. Sweetser, P.B., G.S. Schow and J.M. Hutchinson. 1982. Metabolism of chlorsulfuron by plants: biological basis for selectivity of a new herbicide for cereals. Pestic. Biochem. Physiol. **17:** 18-23.

90. Synerholm, M.E. and P.W. Zimmerman. 1947. Preparation of a series of (-(2,4-dichlorophenoxy) aliphatic acids and some related compounds with consideration of their biochemical role as plant-growth regulators. Contrib. Boyce Thompson Inst. **14:** 369-382.

91. Tal, A., M.L. Romano, G.R. Stephenson, A.L. Schwan and J.C. Hall. 1993. Glutathione conjugation: A detoxification pathway for fenoxaprop-ethyl in barley, crabgrass, oat, and wheat. Pestic. Biochem. Physiol. **46:** 190-199.

92. Tardif and Leroux. 1991. Weed Technol. **5:** 525.

93. Tecle, B., A.D. Cunha, and D.L. Shaner. 1993. Differential routes of metabolism of imidazolinones: Basis for soybean (*Glycine max*) selectivity. Pestic. Biochem. Physiol. **46:** 120-130.

94. Theodoridis, G., J.S. Braun, F.W. Hotzamn, M.C. Manfredi, L.L. Maravetz, J.W. Lyga, J.M. Tymonko, K.R. Wilson, K.M. Poss and M.J. Wile. 1992. Synthesis and herbicidal properties of aryltriazolinones. A new class of pre- and postemergence herbicides. In D.R: Baker, J.G. Fenyes and J.J. Steffens (eds.). Synthesis and Chemistry of Agrochemicals. III. ACS Symposium Series 504, pp. 135-136.

95. Thompson, L., Jr. 1972a. Metabolism of simazine and atrazine by wild cane. Weed Sci. **20:** 153-155.

96. Thompson, L., Jr. 1972b. Metabolism of chloro-s-triazine herbicides by *Panicum* and *Setaria*. Weed Sci. **20:** 584-587.

97. Thompson, L., Jr., J.M. Houghton, F.W. Slife and H.J. Butler. 1971. Metabolism of atrazine by fall panicum and large crabgrass. Weed Sci. **19:** 409-412.

98. Wittenbach, V.A., M.K. Koeppe, F.T. Lichtner, W.T. Zimmerman and R.K. Reiser. 1994. Basis of selectivity of triasulfuron methyl in sugar beets (*Beta vulgaris*). Pestic. Biochem. Physiol. **49:** 72-81.

99. Yih, R.Y., D.M. McRae and H.P. Wilson. 1968. Metabolism of 3,4-dichloropropionanilide: 3,4-dichloroaniline-lignin complex in rice plants. Science **161:** 376-378.

9

Persistence and Behaviour of Herbicides in Soil and Environment

When a herbicide reaches the soil, it is subjected to various reactions with soil and environmental factors. Herbicides reach the soil through preplanting (PPL) and preemergence (PRE) applications, as foliage run-off from postemergence (POST) applications, and through the return of crop residues to the soil. Interactions of a herbicide with soil and environmental factors determine its immediate phytotoxicity and subsequent degradation, and hence its persistence of activity and behaviour in the soil.

Persistence in soils is an important feature of a herbicide as it determines its suitability or otherwise in a particular soil and cropping situation. Herbicides that decompose too readily are less desirable in some situations as they cannot be very effective on the weeds emerging later. In other situations, herbicides that have longer persistence of activity are unsuitable as their toxic residues can injure the sensitive crops grown in rotation. Thus, a herbicide which becomes more desirable in some situations may not be useful in other situations. Knowledge of the persistence and residual effects of herbicides is essential to use them safely and effectively and to programme non-hazardous chemical weed management schedules. Besides, it also helps in ascertaining their long-term effects on the soil environment, which includes soil microorganisms, soil fertility, etc.

Herbicides reaching the soil become dissipated or removed through 1) uptake and metabolism by plants, 2) volatilization, 3) photodecomposition, 4) adsorption, 5) leaching, 6) microbial degradation and 7) chemical degradation (Fig. 9.1).

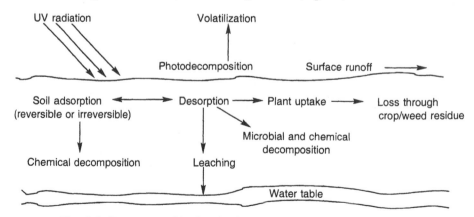

Fig. 9.1. Processes of herbicide dissipation after application to the soil

UPTAKE AND METABOLISM BY PLANTS

Herbicides are absorbed and metabolized by plants as discussed in Chapters 6 and 8. They return to the soil, either in the original form or as metabolites, through crops and weed residues.

Most of the soil-active herbicides are applied at rates of 0.5 to 4.0 kg ha^{-1}, equivalent to 0.5 to 4 ppm in soil of one hectare (up to 15 cm depth). Relatively, a small proportion of the applied herbicide is taken up and retained by plants and the remainder is subjected to dissipation in the soil. The portion of the herbicide absorbed by germinating weed seeds and emerging weed seedlings returns to the soil when they are killed and decomposed. A part of the herbicide absorbed by crop plants is retained in the edible portions of the plant, i.e., grain, fruits, leaves, stems, etc., and the rest is returned to the soil through crop residues.

In the case of POST herbicides, a rate of 1 kg ha^{-1} applied on the crop would theoretically amount to 100 ppm on the basis of raw agricultural product of 10 tons ha^{-1} at harvest, assuming that all produce was taken into account. However, in practice only a relatively small percentage (0.1 to 1.0%) of the applied herbicide reaches the crop plants, mostly through spray drift, and the remainder falls on the weeds or soil. At 1% level (under direct spraying conditions), the crop material at harvest accumulates herbicide residues of 1.0 ppm. All the herbicide absorbed by the crop, except that retained in the edible parts, will eventually return to the soil through plant residues. Residues of herbicides in harvested produce would normally range from undetectable to below 1.0 ppm [26].

The foregoing brief discussion indicates that uptake and metabolism by plants do not, over a long time, cause substantial loss of herbicides and their residues from the soil.

VOLATILIZATION

Volatilization is a process wherein a condensed phase, such as liquid or solid, is transformed into vapour by elevation of temperature or reduction of external pressure. The term **vapour** describes a substance in the gaseous stage below its critical temperature. The tendency of a substance to volatilize is expressed by its vapour pressure. Vapour pressure is usually expressed in millimetres of mercury (mmHg) or occasionally in microns of mercury. Herbicides with vapour pressure of more than 1×10^{-5} mmHg, at 20°C, are generally considered volatile.

If a liquid or solid contained in a closed vessel is in equilibrium with its own vapour in the space above it, then the pressure exerted by that vapour is known as the **vapour pressure** of the substance. Vapour pressure is solely a function of the temperature for any substance. Evaporation of the substance will continue until equilibrium is achieved at a given temperature. At equilibrium, the number of molecules evaporating is equal to those condensing, and this, too, is a function of temperature.

The volatilization of a herbicide from soil or foliar surfaces depends mainly on the vapour pressure of the compound, its concentration, its adsorption by soil and its solubility in water. It is also affected by air temperature, wind velocity above the soil surface, relative humidity, soil temperature and soil moisture. Drift during spraying also provides an opportunity for volatilization. It is also influenced by the chemical nature of the compound. Some herbicides are metabolized or degraded relatively rapidly to more polar products, which may be strongly adsorbed by soil. Volatility can be a factor in reducing the effectiveness of a herbicide. In such cases, incorporation into the soil or improved formulation reduces the loss by volatilization.

Codistillation with water is sometimes considered responsible for substantial loss of some herbicides from soil surfaces. Codistillation with water involves physical or chemical reaction between water and the chemical, with the product being more volatile than the parent chemical itself. However, codistillation is generally referred to as the creation, by water, of an interface at which the chemical is concentrated and/or at which the chemical is held with reduced energy. If the net result of the presence of water increases partition into the air phase, it is commonly said that the volatility of the chemical has been enhanced by codistillation with water.

Many volatile herbicides, such as carbamates, dinitroanilines and thiocarbamates, show a direct relationship between volatilization and soil moisture. Their vapour loss is greater in moist soil than in dry soil. At low soil moisture, the vapour of the herbicide is considerably reduced and it is adsorbed by the soil although the binding energy is weak. In high soil moisture situation, water competes with herbicide for sites of adsorption, leading to greater volatilization. Gray and Weierich [32] found that during the first 15 min after spraying on the soil surface, 20%, 27% and 44% of the applied EPTC was lost by volatilization from dry, moist and wet soils respectively. They also found that increasing the air temperature from 0 to 15.5° C increased the rate of EPTC vaporization from moist soil more than from dry soil. Parochetti and Hein [64] showed increased vapour losses of dinitroaniline herbicides, trifluralin and benefin, as temperature was increased from 30° to 50° C and soil moisture from air dryness to field capacity, but failed to increase at moisture levels above field capacity at temperatures of 40° or 50° C. Harper et al. [37] reported that under field conditions, trifluralin flux increased during the day when soil surface water content was low, even though air turbulence, soil temperature and evaporative demand were high. During the night, when evaporative demand subsided and soil water content increased, the trifluralin flux also increased. They also reported that trifluralin adsorption by soil particles upon drying was apparently a reversible process since efflux of the herbicide was rapid when the soil was re-wetted by dew or rainfall to above the equivalent of three molecular layers of adsorbed soil water. Hollingsworth [41] found trifluralin vapours accumulating rapidly during a period in which several centimetres of rain fell, but with continued precipitation the vapour density declined. The influence of rainfall and resultant soil moisture upon vaporization became less apparent as the soil concentration of the herbicides declined.

The vapours of herbicides emitted during volatilization may be phytotoxic. Swann and Behrens [85] observed severe shoot growth inhibition from emission of trifluralin (vapour pressure 1.99×10^{-4} mmHg at 30°C) vapours even 16 to 22 d after application to soil. Jaques and Harvey [44] found that all the 8 dinitroaniline herbicides except oryzalin inhibited primary root growth of oats through vapour activity. Dinitramine vapours were most inhibitory. Inhibition due to herbicide vapours increased with increasing temperature. Benefin, which has a vapour pressure of 3.89×10^{-5} mmHg at 30°C, caused severe injury to tobacco seedlings through its vapours [94]. Gray and Weierich [32] recommended soil incorporation to a depth of 5 to 7.5 cm to prevent large vapour losses of EPTC, followed by light rain or sprinkler irrigation.

Adsorption of herbicides by soil reduces volatilization. Vapour loss has a direct relationship with the percentage of sand and inverse relationship with the percentage of clay and organic matter. Volatile herbicides that are strongly adsorbed may remain on the surface of clay soils or soils of high organic matter. Bardsley et al. [6]] found that increasing organic matter from 1.5 to 6.0% resulted in proportionally greater retention of active trifluralin in the soil. They suggested that the increased adsorptive capacity of the organic materials was instrumental in retaining trifluralin vapours.

Soil pH affects vapourization of herbicides from soils by influencing their adsorption and through differential ionization properties of the compounds. The degree of dissociation of a herbicide in the soil solution may determine its rate of vaporization. A decrease in pH favours the undissociated form of anionic herbicides such as 2,4-D and increases their potential for vapour loss. Conversely, the dissociation of s-triazines or anilines may increase with a resultant vapour pressure decrease.

The chemical structure of the herbicide also has a great influence on its volatility consequent to adsorption by soil. The esters of phenoxyacetic acids are more volatile than free acids and their salts. The isopropyl ester of 2,4-D is much more volatile than the propylene glycolbutyl ether ester. Similarly, among the triazines, prometon, with a vapour pressure of 7.9×10^{-6} mmHg at 30°C is more volatile than its analogue prometryn (4.0×10^{-6} mmHg), which in turn is more volatile than chlorpropham. Generally, carbamates, dinitroanilines, phenols, phenoxyacid esters, thiocarbamates and certain triazines have high vapour losses. Soil incorporation immediately after application will minimize these losses. On the other hand, uracils have low vapour pressure. At 100°C, bromail has a vapour pressure of 0.8×10^{-3} mmHg. Hence, volatilization losses are considered to be a minor factor in the dissipation of substituted uracil herbicides in the soil.

PHOTODECOMPOSITION

Photodecomposition, i.e., decomposition by light, is an important mechanism of detoxification of herbicides in the soil under field conditions especially where surface applications are made without receiving subsequent incorporation, rainfall or overhead irrigation [46]. Light is electromagnetic in nature. The effect of light involves the transfer of energy. The energy transferred depends on the wavelength and intensity of the light. The shorter the wavelength, the greater the energy of radiation [67].

When exposed to light, organic molecules absorb the radiation, resulting in an increase in molecular energy. The energy thus absorbed may destabilize the molecules by increasing their transpirational, rotational, vibrational or electronic energy. If this energy interacts with the electrons of the molecules, electronically excited molecules result. These excited molecules may lose energy in a number of ways, one of which is chemical reaction.

Following light absorption, the organic molecules undergo photochemical reaction. The majority of aromatic compounds commonly used as herbicides exhibit a rather intense absorption in the ultraviolet (UV) region of the spectrum. Therefore, UV sources are commonly used to study photochemical reactions of herbicides. Most of the herbicides exhibit their principal absorption maxima in the region of 220 to 440 nm. The products of photochemical reactions are often wavelength-dependent. Failure of a chemical to react to light may result from its failure to absorb light.

The products of photochemical reaction depend on the sequence of events following the excitation process. The common modes of decomposition include aromatic substitution, reduction, elimination, cyclization and polymerization. However, the reaction of a particular herbicide is influenced by light intensity, concentration, pH and other factors. In many instances, photolysis provides the same products as metabolism by plants and microorganisms do. Hydrolysis of esters and amides, oxidative dealkylation of amines and amides, and reduction of nitro groups are some of the pathways of photodecomposition [18]. The source and intensity of light as well as the physical state, sensitization and, of course, the intrinsic chemical and physical properties of the compounds regulate the nature of the products and the rate of decomposition [18].

PHOTODECOMPOSITION OF INDIVIDUAL HERBICIDES

Acetamides

Acetamides are very slow to undergo photolyzed reactions, particularly under soil conditions. **Alachlor** and **propachlor,** the acetanilides, form the corresponding anilinoazobenzenes. Photolysis of **propanil** results in hydrolysis, forming 3-chloroaniline and 3,4-dichloroaniline, followed by oxidation and coupling to form 3,3',4,4'-tetrachlorobenzene. Photodecomposition of **butachlor** involves debutoxymethylation, dechlorination followed by hydroxylation, O-dealkylation and polymerization [14]. The photodecomposed products include 2-chloro-2',6'diethyl-acetanilide, 2-hydroxy-2',6'-diethyl-N-(butoxymethyl)acetanilide, and N-2',6'-diethylphenyl-2,3-dihydro-oxazole-4-one [14]. The decomposition follows first order kinetics.

Photodecomposition is a major contributor of **metolachlor** dissipation in the field, particularly when rainfall or irrigation is delayed after herbicide application to the soil surface. The half-life of metolachlor is 8 d on sandy soil at 15° to 52° C under natural light conditions. There is negligible photodecomposition loss of **acetochlor** and **pronamide. Napropamide** is lost up to 50% by photodegradation during summer, after 4 d of exposure on the soil surface. In winter, the loss is around 30% after 8 d.

Aliphatics

Aliphatics are very strong organic acids. **Dalapon,** on the other hand, undergoes hydrolysis to pyruvic acid. It may also undergo dehydrochlorination to 2-chloroacrylic acid or reduction to form propionic acid [25].

Arsenicals

Arsenicals are not subject to photodecomposition losses. The half-life of **MSMA** is 990 d when irradiated on soil. In water, the photolysis half-life is less than 30 d.

Benzamides

Photodecomposition loss of **isoxaben** is negligible.

Benzoics

Dicamba is fairly resistant to photolysis under field conditions. Under a xenon lamp, it is slowly photodegraded, with a half-life of 269 d.

Benzothiadiazoles

Bentazon loss due to photodegradation is negligible under field conditions. Its field half-life in the soil is 4 mon. About 30% of applied bentazon is degraded when exposed to light of 200-400 nm.

Bipyridiliums

Sunlight and UV light degrade **paraquat** dichloride (1,1-dimethyl-4,4-bipyridilium dichloride), first by demethylation followed by ring cleavage of one of the heterocyclic rings to form 4-carboxy-1-methylpyridinium ion and methylamine. Paraquat can be photodegraded from desiccated leaf surfaces and possibly to the extent of 25-50% in 3 wk under bright sunlight.

Carbamates

Propham is photolyzed by at least two routes, one leading to formation of phenylisocyanate and 2-propanol, and the other producing propylene and carbamic acid, which later undergo decarboxylation to form aniline and CO_2. **Chlropropham** also undergoes similar photodecomposition.

Cineoles

Cinmethylin is rapidly photodegraded under field conditions in sunlight.

Cyclohexanediones (Cyclohexenones)

Photodegradation of cyclohexanedione herbicides is significant. During degradation, they produce the corresponding sulphoxides. **Clethodim** has a half-life of 1.7 d on a sandy loam soil in natural sunlight, while **sethoxydim** has a half-life of less than 4 h.

Dinitroanilines

Trifluralin is readily decomposed at sunlight wavelengths to form numerous products [55]. In addition to minor dealkylated intermediates, the principal product under acidic conditions was 2-amino-6-nitro-α,α,α-trifluoro-*p*-toluidine [55]. At alkaline pH, 2-ethyl-7-nitro-5-trifluoromethyl benzimidazole represented about 80% of photolysis products within 24 h. The highly polar products 2,3-hydroxy-2-ethyl-7-nitro-1-propyl-5-trifluoromethyl benzimidazoline and 2-ethyl-7-nitro-5-trifluoro benzimidazoline were present in significant amounts but were degraded by heat of further irradiation. The photochemical formation of benzimidazolines and benzimidazole oxides conforms to a general mechanism which should apply to many dinitroaniline herbicides. Sullivan et al. [84] reported formation of azoxybenzene derivatives which were chemically reduced to yield azobenzenes during photolysis of trifluralin.

Significant photodecomposition losses of **benefin, ethalfluralin, isopropalin** and **oryzalin** occur when they are allowed to remain on the soil surface. The field half-life of pendemethalin is 7 d in water at 25° C in full summer sunlight. Less than 5% of **pendimethalin** is degraded by 30 d after application on sandy loam with 11% clay, 1.6% organic matter, pH 6.4 and at 10.2% moisture. The contribution of photolysis to field dissipation of pendemethalin is minor.

Diphenylethers

The photodegradation losses of **bifenox** are minimal. **Acifluorfen** readily photodegrades, with a half-life of 2 to 2.5 d in water and 4.5 d on soil. **Oxyfluorfen** has a half-life of 20 to 30 d on dry soil and when rainfall or irrigation is delayed photolysis may play an important role in its degradation in the environment. The half-life of oxyfluorfen in water is as short as 5 d. **Lactofen** and **fomesafen** readily photodecompose even under relatively low sunlight conditions.

Imidazolinones

Photodegradation plays a negligible role in the dissipation of imidazolinone herbicides on the soil surface. **Imazapyr** produces, under xenon arc light, quinolinic acid, quinilinimide and furo[3,4-*b*]pyridine-5(7*H*)-one and 7-hydroxy-furo[3,4-*b*)pyridine-5(7*H*)-one. However, photolysis on the soil surface is limited and is a minor contributor to degradation of imazapyr in the soil.

Photolysis is a major means of **imazaquin** degradation in aquatic systems, with a half-life of less than one day at 18-19° C. The metabolites include 3-quinolinecarboxylic acid, 2,3-quinolinedicarboxylic acid, 2-carboxamido-3-quinolinecarboxylic acid, and 2,3-dihydro-3-imino-1*H*-pyrrolo[3,4-*b*]quinoline-1-one. Imazaquin degrades more slowly on a dry soil surface than in an aqueous solution. Although photolysis contributes to imazaquin degradation, this may vary with environmental conditions.

Imazethapyr is photodegraded in water to 5-ethyl-2,3-pyridinedicarboxylic acid and 5-ethyl-2,3-pyridinecarboxylic acid. However, its photolysis is limited. Imazethapyr is relatively less susceptible to photodecomposition than imazaquin. The half-life of **imazamethabenz** on soil is 30-60 d.

Imines

CGA-248757 is rapidly photodegraded on soil and in water, with half-lives of < 0.5 d and 4.9 d respectively.

Nitriles

UV light decomposes **ioxynil** to 3,5-diphenyl-4-cyanophenol and **bromoxynil** to polyphenols and coloured polymeric materials.

Phenols

Dinoseb undergoes side-chain hydroxylation through photodecomposition to form polar materials and resin.

Phenoxyalkanoic Acids

Phenoxyacetics, Phenoxybutyrics

The effect of sunlight on **2,4-D** degradation is similar to UV irradiation. Photodecomposition of 2,4-D takes place in the presence of riboflavin, a sensitizer. The major products of photolysis include 2,4-dichlorophenol and 4-chlorocatechol. **2,4-DB** also forms corresponding phenols during photodecomposition.

Aryloxyphenoxypropionics

Sunlight has a negligible effect on the dissipation of aryloxyphenoxy propionic herbicides on dry soil surface. They may be degraded slowly in the presence of moisture, by cleaving of the aryl ether linkage and opening of the phenyl ring as in the case of **haloxyfop**. The half-lives of these herbicides on sandy loam soils is 40-50 d. Photolysis may not be an important mechanism of dissipation in field conditions when these herbicides are soil incorporated. Photodecomposition may, however, be a problem with foliar applications. Haloxyfop is more photolabile to UV light when applied to foliage than 2,4-D; this photolability increases with the addition of adjuvants.

N-Phenylphthalimide

The half-life of **flumiclorac** is 1 to 2 d on soil and 3 to 4 d in water in the presence of sunlight. It is degraded by light, releasing CO_2.

Phenylpyridazines

Photodecomposition of **pyridate** is negligible, but the primary metabolite, CL 9673 (3-phenyl-4-hydroxy-6-chloropyridazine), has a half-life of 6 to 7 d.

Pyridazinones

Photodecomposition contributes significantly to field dissipation of **norflurazon** on the soil surface. It is degraded on the soil surface to desmethyl norflurazon, CO_2, and small amounts of other products, with a half-life of 20 d. In water, it is rapidly degraded (half-life 1.0–1.5 h), forming deschloro norflurazon and small amounts of other products. **Pyrazon** degradation by sunlight is, however, negligible.

Pyridinecarboxylic Acids

Photodegradation of **picloram** is rapid in water and on soil and plant surfaces. Its photolysis half-life in water is 2.6 d. The effect of sunlight involves cleavage of the pyridine ring of picloram. **Triclopyr** has a half-life of 10 h in water in the presence of sunlight, producing trichloropyridinol as the major metabolite. The effect of sunlight on **clopyralid** is negligible.

Pyridines

Photolysis of **dithiopyr** is insignificant, with only 5% of the herbicide being degraded to its monoacid form after exposure to sunlight for 33 d at 25° C. Its dissipation is faster, with a half-life of 17 d, in the presence of moisture. The primary metabolites include two monoacids and one diacid. **Thiazopyr** undergoes rapid photolysis in water, with a half-life of 15 d, while its dissipation on the soil surface is negligible.

Sulfonylureas

Photodegradation plays a negligible role in the dissipation of sulfonylurea herbicides. Their half-lives on the soil surface in the presence of sunlight range from 25 to 90 d. At 25° C and natural light conditions, **primisulfuron** is stable at pH 9, but half-life is 22 d at pH 5, producing saccharin (10.1% of applied) and [benzoic acid, 2-(amino sulfonyl) methylester] (53.3%). Although photolysis of **triflusulfuron** is minimal under field conditions, initial metabolites are primarily triazine amine and methylsaccharin, with small amounts of triazine urea, N-demethylated products and CO_2.

Thiocarbamates

EPTC undergoes photodecomposition by oxidation. **Thiobencarb** is degraded by sunlight to form 4-chlorobenzylmercaptan, 4-chlorobenzylalcohol, and 4-chlorobenzoic acid as the metabolites. The volatilized EPTC and **pebulate** may be further degraded by UV light to produce formides, dialkylamines, mercaptans and disulphides [56]. The products of photolysis become more toxic than the parent herbicide.

Triazines

Triazines are relatively stable in light. However, they do undergo photodegradation by methylation and dehalogenation. The half-life of **simazine** on a sandy loam soil at 25° C is 21 d, producing low levels of the mono-N-de-ethylated metabolite (6-chloro-N-ethyl-1,3,5-trazine-2,4-diamine), the di-N-de-ethylated metabolite (6-chloro-1,3,5-triazine-2,4-diamine) and hydroxy simazine [4,6-*bis*(ethylamino)-1,3,5-triazin-2(1H)-one] 207 d after application. Sunlight causes moderate dissipation of simazine in the field, particularly when the surface-applied simazine is exposed to prolonged drought conditions.

The half-life of **atrazine** is 45 d on a sandy loam soil at 25° C and pH 7.5 under artificial light. The major products after 30 d are de-ethylated atrazine (13.3%) and N-de-ethyl-N-demethylethyl atrazine (11.9%). Under prolonged drought conditions when atrazine remains on the soil surface, it undergoes moderate photodecomposition. It has a half-life of 335 d at 12-45° C and pH 7 (under natural light conditions), with major degradation products after 15 d being hydroxy atrazine {4-(ethylamino)-6-[(1-methylethyl)amino]-1,3,5-triazine-2(1H)-one} (2.6% applied) and N-de-ethylated atrazine [6-chloro-N-(1-methylethyl)-1,3,5-triazine-2,4-diamine] (2.89%).

Triazoles

Amitrole is rapidly photolyzed in the presence of riboflavin and oxygen. It is principally converted to CO_2, urea, and cyanamide by ring cleavage.

Uracils

Substituted uracil herbicides are not adversely affected by sunlight. However, **bromacil** can be decomposed by UV light. By exposing to solar radiation bromacil can be transformed to a major intermediate 5-acetyl-5-hydroxy-3-s-butyl-6-hydantoin and a minor intermediate, the 5,5-dimer photoproduct of 3-s-butyl-6-methyluracil [1]. Under optimal conditions, photo-oxidation may be completed after about 1 h and further irradiation could result in the decomposition of intermediate products and formation of unidentified polar products.

Ureas

Photodecomposition plays a significant role in the loss of substituted urea herbicides from the soil surface in dry areas. Considerable amounts of **diuron**, **monuron** and **fenuron** may be lost after exposure to light. One of the products of monuron hydrolysis includes 3-(p-hydroxyphenyl)-1,1-dimethylurea. The structures of diuron, fenuron and monuron are altered by far UV light in the range of 240-260 nm, with peak emission at 153.7 nm [45].

Fluometuron is photolyzed to a mono-N-demethylated metabolite N-methyl-N'-[3-(trifluoromethyl)phenyl] urea. Photodegradation losses are substantial when no rainfall or irrigation follows fluometuron application. Its losses are moderately low under adequate rainfall and irrigation conditions. Photodegradation loss is negligible in the case of **siduron**.

Unclassified Herbicides

Photodegradation has a negligible role in the field dissipation of **ethofumesate**, **fosamine**, **glufosinate** and **glyphosate**.

From the foregoing discussion, it is clear that some organic herbicides undergo photodecomposition. During this process, dehalogenation and hydroxylation appear to be the predominant reactions. Other reactions, such as hydrolysis, oxidation and reduction also take place. The metabolites are similar to those formed by plant metabolism of herbicides. Although photodecomposition may not be of great practical importance in the temperate regions, it could become a significant pathway of herbicide dissipation in the tropics where the intensity and duration of sunlight are far greater.

HERBICIDE ADSORPTION BY SOIL

Adsorption refers to the attraction, adhesion and accumulation of molecules at the soil-water or soil-air interface, resulting in one or more ionic or molecular layers on the surface of soil particles. The term **sorption** is also used generally while referring to adsorption. Sorption refers to surface-induced removal from solution by adsorption, absorption or precipitation. The term **adsorption** is widely used in this book in preference to the term sorption, as adsorption controls the availability of herbicide molecules to soil solution, affects the movement of herbicides in the soil, and regulates their availability to plants and rate of decomposition in the soil (Fig. 9.2). Strongly adsorbed molecules decompose very slowly. Decomposition is rapid in the case of herbicides which are freely available in soil solution.

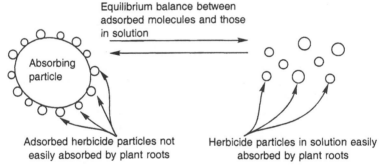

Fig. 9.2. Adsorption of herbicide particles and their availability in soil solution for plant absorption.

Desorption is the reverse of adsorption, i.e., movement of the herbicide particles from the soil surface into the soil solution. Desorption replenishes herbicide as it dissipates from the soil solution. Desorption is stimulated by herbicide removal from the solution, by plant roots or by other organic materials that extract herbicides from the soil solution.

Adsorption and desorption are the key processes controlling herbicide efficacy, dissipation and behaviour in soil as also the contamination of ground and surface waters. An understanding of the interactions between adsorption and desorption mechanisms and the factors affecting these processes will ensure more efficient use of herbicides.

Adsorption Mechanisms

Adsorption of herbicides by soil particles occurs through a number of mechanisms involving varying bond strengths. It is dependent on the herbicide characteristics, soil surface characteristics and competing solutes. The physicochemical mechanisms responsible for adsorption of herbicides by soil include hydrophobic partitioning, London-van der Waals forces, hydrogen bonding, cation and water bridging, anion exchange, ligand exchange, protonation, cation exchange and covalent bonding [38].

Hydrophobic partitioning involves partitioning of an organic chemical from the aqueous phase into the more hydrophobic organic matter. Hydrophobic partitioning, though not widely applicable for herbicides, is the proposed mechanism of adsorption for organic chemicals in the soil. **London-van der Waals forces** are short-range bonds, resulting from a correlation in electron movement between two molecules to produce a small net electrostatic attraction. This mechanism is characterized by low heats of adsorption (20 kcal mol^{-1} or less) or low binding strength and short residence time on the adsorptive surface (1×10^{-2} s or less). This indicates that the strength of these bonds is weak and decreases with distance. The interaction is additive, however, and increases with area of contact. As a result, interactions caused by London-van der Waals forces are particularly important for neutral high molecular weight compounds.

Hydrogen bonding involves dipole-dipole interactions involving an electrostatic attraction between an electropositive hydrogen nucleus or functional groups such as –OH and –NH and exposed electron pairs on electronegative atoms such as O^- and N^-. Bonding occurs between two highly electronegative atoms through the medium of an electropositive atom. These bonds are stronger than London-van der Waals forces and can occur both intra- and intermolecularly. Hydrogen bonding is considered to be most prevalent in the bonding of herbicides to organic surfaces in the soil. Hydrogen bonding is a significant binding mechanism for triazines, sulfonylureas and aryloxyphenoxypropionics.

Cation and water bridgings involve complex formation between an exchangeable cation and an anionic or polar functional group on the herbicide. In cation bonding, the organic functional group displaces hydrating water molecules around the cation to form the complex. Water bridging occurs when the organic functional group is unable to displace hydrating water molecules from the cation. Water bridging is more likely to occur with the larger, high valency cations such as Fe^{+++}, Al^{+++} and Mg^{++}. Cation and water bridging are the proposed mechanisms of adsorption for fluazifop-butyl, glyphosate, and picloram.

Anion bonding is a stronger adsorption mechanism caused by an electrostatic attraction of an anion to a positively charged site on the soil surface. This mechanism becomes significant only in acidic soils containing a pH-dependent charge from kaolinite or amorphous aluminosilicates and iron oxides. Anion bonding is a predominant adsorption mechanism in herbicides having dissociated functional groups such as carboxylates.

Ligand exchange mechanism involves displacement of an inorganic hydroxyl or water molecule from a metal ion at a hydrous oxide surface by a carboxylate or hydroxyl on an organic molecule. This is the mechanism proposed for adsorption of chlorsulfuron and possibly other sulfonylurea herbicides on iron oxides. In **protonation**, a charge-transfer complex is formed when a functional group like an amino or carboxyl group forms a complex with a surface proton. For herbicides with basic functional groups, protonation is a significant adsorption mechanism at acidic mineral surfaces at low pH and water content. These include fluazifop, chlorsulfuron and certain triazines.

Cation exchange involves exchange of cations through an electrostatic attraction on the soil surface. Cation exchange is predominant in herbicides such as paraquat and diquat, which are ·predominantly cationic. Weakly basic herbicides with functional groups such as amines and heterocyclic N compounds may protonate to form the cationic form. Bipyridiliums, triazines and fluridone are adsorbed by the cation exchange mechanism.

Covalent bonding is an irreversible binding of herbicides and other metabolites to soil organic matter. This binding is formed by oxidative coupling and enzymatic polymerization. Covalent bonding reduces the availability and extractability of absorbed herbicides.

Several factors affect the adsorption of herbicides by soils. These include: a) type of clay colloid, b) soil organic matter, c) soil pH, d) soil moisture content, e) chemical nature of the herbicide and f) leaching.

Clay Colloid

Clay colloid, which has high adsorptive capacities, refers to the microscopic (0.001 to 1.0 μ in diameter) inorganic and organic particles in the soil. These particles have an extremely large surface in proportion to a given volume. Clay, particles have negative charges and hence can attract to their surface positive ions (cations). Adsorption is closely associated with inorganic and organic colloids of the soil. Inorganic colloids are principally clay.

There are three major groups of clays: **montmorillonite, illite** and **kaolinite**. Montmorillonite is an expanding lattice clay providing both external and internal adsorptive surfaces. It has three layers, with one layer of aluminium oxide lying between two layers of silicon oxides. Illite is also a three-layered clay, but it lacks the expanding lattice character which makes it less adsorptive of herbicide molecules than montmorillonite clay. On the other hand, kaolinite is only a two-layered clay, with alternate layers of aluminium and silicon oxides. There are few residual charges in kaolinite clay, making it the least adsorbent of the three clays. However, kaolinite does have one hydroxyl surface, which makes it adsorb some organic chemicals more strongly than the other clays.

Montmorillonite clay has a high cation exchange capacity (CEC) (80 to 150 meq 100 g^{-1}) and a high surface area (600 to 800 $m^2 g^{-1}$). It has a high capacity for adsorption due to **Coulombic** and London-Van der Waals forces. Vermiculite also has a high CEC (100 to 150 meq 10 g^{-1}) and a high surface area (600 to 800 $m^2 g^{-1}$). Illite and kaolinite clays, which have low cation exchange capacity (10 to 40 and 3 to 15 meq 10 g^{-1}) and low surface area (65 to 100 and 7 to 30 $m^2 g^{-1}$) do not have as large an adsorptive capacity as montmorillonite. The strength of adsorption follows the order of montmorillonite > illite > kaolinite. Montmorillonite adsorbs considerably more of various herbicides than do illite and kaolinite. Adsorption of monuron is highest in the soils in which montmorillonite is the major clay constituent [95]. Paraquat and diquat are adsorbed by montmorillonite to its exchange capacity. Paraquat adsorbed by montmorillonite has little toxicity, while that adsorbed by kaolinite and vermiculite clays becomes available with time [91]. Scott and Weber [77] reported that phytotoxicity of paraquat and prometon for cucumber plants was reduced by montmorillonite and kaolinite clays, and this reduction in phytotoxicity was related to an increase in the adsorption of herbicides. Glyphosate is inactivated rapidly by clay loam and muck soil [82]. In the case of urea herbicides, adsorption was greater on bentonite than on montmorillonite clay largely because the lattice charge in the former originated from tetrahedral and octahedral layers, whereas in the latter only from octahedral layers [88].

The persistence of EPTC in dry soils is due to the ability of soil particles to adsorb this chemical. Mortland and Meggitt [58] proposed three mechanisms of adsorption of EPTC by montmorillonite: a) coordination to exchangeable metal cations through the carboxyl group, b) coordination to metal cations through the nitrogen and c) hydrogen bonding to methylene hydrogens through surface oxygen atoms on the clay surface.

Soil Organic Matter

Soil organic matter is the most significant factor affecting adsorption and hence, the behaviour of herbicides in soils. The organic matter (OM) consists of humic materials, plant and animal residues, and soil microbes. Soil humic materials consist of three components: a) humic acid, the alkaline soluble and acid insoluble fraction, b) fulvic acid, the alkaline soluble and acid soluble fraction, and c) humin, the alkaline insoluble and acid insoluble fraction [38]. The humic acids are responsible for stable bonding during herbicide adsorption. Plant residues, when decomposed in soil, have a much greater adsorption capacity than the soil itself. Dao [20] found that decaying wheat straw adsorbed metribuzin better than the undecayed wheat straw. This increase in sorptive capacity was associated with a decline in cellulose and its accompanying proportional increase in lignin. The humic material has a primary influence in the adsorption of several herbicides including 2,4-D, chlorsulfuron, picloram, linuron and metribuzin. Boyd et al. [11] estimated that the sorption coefficients of corn plant cuticles were 5 to 15 times greater than corn residues, indicating that the cuticle material is the primary sorbing component. Other organic materials such as activated carbon, sewage sludge and animal manure have large adsorption capacities.

The adsorption of herbicides by organic materials involves H-bonding, London-Van der Waals forces and cation exchange [78]. Organic matter has a variable effect on the phytotoxicity of herbicides. It has little effect on the phytotoxicity of CDAA. It reduces the phytotoxicity of simazine grass weeds more than for cotton, of diuron for cotton more than grass weeds and of chlorpropham about equally for cotton and grass weeds but more than for soybean [87]. This suggests that the influence of OM on the phytotoxicity of herbicides in soil varies according to crop species. An increase in OM content may reduce the activity of certain herbicides, warranting an increase in rate of application. Parochetti [63] found that an increase in OM content resulted in an increase in GR_{50} values (rate of herbicide required for 50% growth reduction) for alachlor, propachlor, prynachlor, CDAA and atrazine. Rahman and Matthews [69] observed a similar correlation between OM and 13 triazine herbicides. Warren [90] suggested that for soils containing 1, 2, 3, 4, 6, 8 and 10% OM, trifluralin should be applied at 0.25, 0.50, 0.75, 1.0, 2.0 and 4.0 lb acre^{-1} (0.28, 0.56, .84, 1.12, 2.24 and 4.48 kg ha^{-1}) rates respectively.

Addition of organic matter to soils low in OM content increases the activity of herbicides. Warren [90] reported that on a soil with 0.7% OM, propachlor leached out so rapidly that it failed to control *Digitaria sanguinalis* (large crabgrass). However, it performed well on a soil of 3% OM, indicating that OM enhanced the adsorptive capacity of the soil and hence reduced the leaching losses.

Soil pH

Soil pH affects the detoxification of herbicides by affecting the ionic or molecular character of the chemical, the ionic character and the CEC of soil colloids, and the inherent capacity of soil microorganisms to react with the herbicides.

The term soil pH refers to the pH of the soil solution, the water and other elements that exist in a free state around soil particles. Basically, the pH number is related to the number of hydrogen (H^+) ions in the water solution; the more the hydrogen ions, the more acidic the solution becomes. As hydrogen ions decrease, the hydroxyl (OH^-) ions increase, making the solution more basic. A pH value of 7 has an equal number of hydrogen and hydroxyl ions and it is considered neutral.

Hydrogen ions have a positive electrical charge, indicating that they can be bound to the negatively charged soil and organic matter particles. The more free sites clay and organic matter particles have on them, the more hydrogen and other ions that can be bound to these particles. These binding sites, also called exchange sites, indicate the CEC of the soil. Thus, soils with greater CEC have more exchange sites. Also, the more exchange sites a soil has, the more hydrogen ions that can be held to the soil for eventual release into the soil solution. This is referred to as reserve acidity. This explains why an acid soil with high CEC needs more time for neutralization than an acid soil with low CEC.

Many herbicides are ionic which enables them, when in soil solution, to give off or attract hydrogen ions depending on the pH of the soil solution. For example, 2,4-D, MCPA, dicamba, chloramben, picloram, etc., which are acidic in character, can release hydrogen ions in a neutral or basic solution, while herbicides such as s-triazines, substituted ureas, phenyl carbamates, amides, etc., which are chemically basic in nature, can accept hydrogen ions in an acidic solution. Other herbicides, such as diquat and paraquat, are so basic that they are positively charged in virtually all soil pH values. Glyphosate is acidic in soils, but has both negative and positive charge sites on it and, as a result, its pH charge interactions are complex. Herbicides which are non-ionic in character do not react with water and do not carry electrical charge. These include alachlor, chlorpropham, diuron, metolachlor, monuron, etc. Even though they are not ionic, many of them are polar in nature and can be affected by soil pH, but the effect is generally smaller than with ionic herbicides.

Differences in the pH of the soil affect its ability to adsorb and retain herbicide molecules. Different herbicides respond differently to changes in soil pH. As soil pH is lowered, more hydrogen ions are associated with triazine molecules to give them more cationic characteristics, leading to more adsorption [39]. Soil pH enhanced adsorption of chloramben through its effect on the number of soil anion exchange sites and sites for polyvalent cation bridging and hydrogen bonding [92]. Schnappinger et al. [75] reported that atrazine exhibited better control where soil pH was highest, and poorest where soils were more acidic. Degradation of atrazine occured more rapidly when surface pH was less than 5.0 compared with a pH greater than 6.5. The disappearance of dinitroaniline herbicides such as trifluralin, dinitramine and profluralin was greatest from soils at pH 6.4 [22].

Soil pH also influences the phytotoxicity of herbicides by affecting their adsorption by soil and availability for plant uptake. Grover [35] found an increase in GR_{50} values of picloram as the soil pH was lowered or raised from 6.5. The increase in GR_{50} values in the acidic range was attributed to adsorption of the unionized molecules of picloram on the organic matter in the soil. The increase in these values in the alkaline range was attributed to the reduction in uptake of the ionized acid by plant roots. Similar results were also reported by Corbin et al. [16] who found an increase in phytotoxicity as the soil pH increased and reached a maximum of 6.5 for weak aromatic acids such as dicamba and 2,4-D, and for weak bases as prometon and amitrole. They also found an increase in phytotoxicity as soil pH decreased and reached a maximum of 4.3 for the weak aliphatic acid dalapon, for the cationic herbicides diquat and paraquat, and for the non-ionic herbicide vernolate. They observed that soil pH levels between 4.3 and 6.5 had no effect on the phytotoxicity of the weak aromatic acids chloramben and picloram, and non-ionic herbicides dichlobenil, isocil, diuron and nitralin. A change in one pH unit decreased the phytotoxicity of 2,4-D, dicamba, dalapon, prometon, amitrole, paraquat and vernolate by a factor of two to four depending on the herbicide and the pH values considered.

Soil Moisture

The moisture content of a soil system has a considerable effect on both the degree of adsorption and the phytotoxicity of herbicides present in the aqueous phase. When a herbicide is applied to the soil, it is partitioned into adsorption and solution phases.

Herbicides move through the soil either by **molecular diffusion** or by **mass flow** with the movement of water. The amount of herbicide present in solution depends on the solubility of the herbicide in water and the amount adsorbed by the soil colloids. The availability of a herbicide for plant uptake is related to its desorption into water solution. Thus, if a herbicide moves with the water, it may be distributed through the soil profile or it may be completely removed from the soil profile and leached down to groundwater and streams.

Green and Obien [34] found that the effect of a change in soil water content on herbicide concentration in soil solution was dependent on the magnitude of adsorption. Their results suggested that herbicide phytotoxicity would increase with increasing soil water content under most circumstances. Baumann and Merkle [8] observed increased phytotoxicity of diuron and fluridone as soil moisture increased from 35 to 95% of field capacity. They found a positive correlation between herbicide phytotoxicity and soil moisture.

Most of the herbicides have lower phytotoxicity at lower soil moisture contents. This is attributed to the degree of competition of the organic compound for adsorption sites at different moisture levels. Water is a very polar molecule and is strongly adsorbed by mineral colloids. At low moisture levels, the number of water molecules present to compete for adsorption sites is relatively small and fewer polar organic molecules may be able to compete more favourably for the available sites to be adsorbed. As the moisture content increases, the number of water molecules increases, resulting in reduced adsorption of the organic molecules. If the organic molecules have been adsorbed under conditions of low moisture and then the moisture level is increased, the adsorbed organic molecules may be displaced by water molecules and made available in soil solution for plant absorption. Competition is important only when not enough water molecules are present to cover all colloidal surfaces. If there is enough water in the system for the plants to take it up, all surfaces will be covered with water, and so the competition phenomenon will not be observed.

Chemical Nature of Herbicides

Herbicides are separated into groups according to the base chemical structure of the herbicide molecule. Herbicides within the principal group can be loosely categorized as **permanently ionized** (i.e., quaternary ammonium compounds), **ionizable** (i.e. triazines) or **neutral** (carbamothioates) [38]. Different functional groups on the base structure lead to a range in polarity and ionizability within a herbicide group. Substitution of functional groups on the base chemical structure also brings about changes in water solubility, volatility, adsorption strength and adsorption mechanisms as also changes in herbicidal activity.

The charge on a herbicide molecule may be strong, resulting in dissociation, or weak arising from an unequal distribution of electrons producing polarity in the molecule [2]. Generally, soil and organic matter particles have negative electrical charges. Herbicides that have positive charges are attracted and bound to them. During this process, they either displace the positively charged ions of the soil and organic matter particles or react directly with the hydrogen ions.

Most organic molecules ionize only under certain pH conditions. The pH of ionization may range from −0.5 to 11.2 depending on the functional group in question [2]. Compounds that ionize at these extremes would be unlikely to occur as ions in soils. Within the normal pH range of soils, 4.0 to 9.0, dissociation usually takes the form of H^+ loss by acids and H^+ gain by bases.

The weak base herbicides such as triazines and triazoles, which are less effective in soils of low pH, adsorb hydrogen ions in an acidic solution and become cationic. More atrazine is adsorbed by a muck soil at pH 3.2 than at 5.3, as little atrazine (ionization constant, pKa, 1.85) would exist as cation at pH 5.3. Adsorption is generally more pronounced when the pH of the soil is near the pKa of the herbicide. In high pH soils, triazines are desorbed into the soil solution, which results in greater availability of the chemical for uptake by plants and possible risk of injury even at rates considered safe. The strongly basic herbicides such as paraquat and diquat are so rapidly and tightly bound to montmorillonite clay and organic matter that they are virtually inactivated as soon as they come into contact with the soil.

The strongly acidic herbicide glyphosate is adsorbed more at low pH. The Fe^{+++}– and Al^{+++}– saturated clays and organic matter adsorb more organic matter than Na^+–or Ca^+–saturated clays or organic matter [82]. Glyphosate is readily bound to kaolinite, illite and bentonite clays, and to charcoal and muck. The strongly acidic herbicides such as benzoic acids, phenols, aliphatics and nitriles possess carboxyl, phenolic or phosphonic functional groups and ionize in soil solution to become anions. The weak acidic herbicides such as 2,4-D, dicamba and dinoseb are less active at a soil pH 5.0 or below. They tend to be repelled by, rather than attracted to, the negatively charged soil and organic matter surfaces. As the percentage of negatively charged herbicide molecules decreases at low pH, adsorption increases, and hence their low activity at pH below 5.0. Generally, acidic herbicides are adsorbed more strongly by charcoal, organic matter and anion-exchange resins, and the basic herbicides are adsorbed more strongly by cation-exchange resins and soils.

The non-ionic herbicides such as diuron and other urea herbicides, and trifluralin and other dinitroanilines, which do not ionize significantly in soil solution can also be affected by soil pH, but to a much lesser degree than the basic and acidic herbicides. These non-ionic herbicides are adsorbed through physical adsorption forces. Hance [36] reported the following order of increasing adsorption tendency among urea herbicides: fenuron, neburon, monuron, monolinuron, diuron, linuron and chloroxuron. He found that N-alkyl and N-aryl substituents played a part in the adsorption of substituted ureas by soils. Increasing chain length in the alkyl substituents increased adsorption. The presence and amount of such functional groups as carboxyl, amino, phenolic hydroxyl and alcoholic hydroxyl have a considerable effect on the cation and anion adsorption of herbicides [5].

Leaching

The movement of herbicides with water within the soil profile is referred to as **leaching**. This is influenced by the chemical nature of a herbicide, the adsorptive capacity of the soil and the amount of water available for downward movement through the soil. These aspects have already been discussed.

A particular adsorption-desorption equilibrium is necessary for the availability of herbicides in the soil solution for plant uptake. A change in this equilibrium in favour of desorption results in leaching. This reduces the persistence of the herbicides in the soil. Leaching is more of a problem in lighter soils where the adsorptive forces of the soil colloids are not strong enough to hold the herbicidal molecules tightly. Metribuzin leached to a depth of 21 cm in a loamy soil at 60 mm simulated rainfall while it leached only to 9 cm in clay soil [65].

Leaching is influenced by the chemical nature of a herbicide. Triazines have the following decreasing order of adsorption: propazine, atrazine, simazine and prometryn [72]. The order of leaching is, however, exactly the opposite [86]. Thus, leaching and soil adsorption have an inverse relationship. A similar inverse relationship between herbicide leaching and soil organic matter also exists. Coultas and Harvey [17] reported that leaching of buthidazole and metribuzin was most highly correlated with soil organic matter, bulk density and field moisture capacity. Maximum movement was observed in sandy soil, which contained 0.8% OM, and least movement occurred in a sandy loam with 11.7% OM.

Rodgers [72] placed seven s-triazines in the following order of leaching in a fine sandy soil irrigated at different periods after herbicide application: atraton, propazine, atrazine, simazine, ipazine, ametryn and prometryn, and this showed no strong relationship with their solubilities in water. Gray and Weierich [33] reported direct correlation between the depths of leaching of thiocarbamate herbicides and their water solubilities. The order of leaching in decreasing order was EPTC, vernolate and pebulate. They also found that the depth of leaching decreased as the clay content of the soil is decreased. In peat soil with 35% OM, no movement out of the treated zone could be detected with any of the thiocarbamate herbicides tested when leached with 20 cm of water solubility [51]. These reports suggest that the relationship between leaching and solubility varies with herbicides and that besides solubility, other factors also affect leaching.

Leaching affects selectivity of herbicides. Excessive leaching to the deeper soil layers may render the herbicide less effective on shallow-rooted weed species, but could make it effective on deep-rooted ones. In this situation, a shallow-rooted crop species may show tolerance while the deep-rooted crop plants become susceptible. Irrigation or rainfall following herbicide application has a profound effect on leaching and crop and weed tolerance to a herbicide.

HERBICIDE TRANSPORT IN SOIL

Herbicide absorption by plants occurs primarily from free herbicide content available in soil water. The processes that control the concentration of herbicide in soil water are: a) solubility of the herbicide, b) adsorptive capacity of the soil for the herbicide and c) water content of the soil.

There is no general correlation between the water solubility of herbicides and the concentration of herbicides that remain free in equilibrium soil solution, because the adsorption of herbicides by soil is the main factor controlling the concentration of the solution. Soil water content affects the rate of transpiration, mass flow and molecular diffusion in the liquid phase, thus controlling the rate at which the herbicide is transported in the liquid phase to the site of action in plants (Fig. 9.3) [34]. Soil water content also determines the pore space diffusion in the vapour phase, which affects the rate of herbicide uptake by roots.

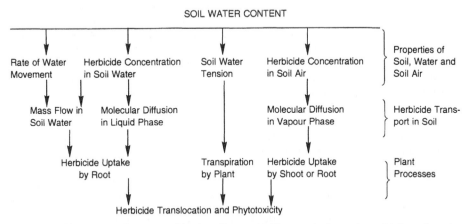

Fig. 9.3. Schematic diagram showing the probable mechnisms by which soil water content influences the phytotoxicity of a soil-applied herbicide at water contents above the wilting point [34].

Diffusion Transport

Herbicides are distributed in soil among solid, air and liquid phases. Molecules in the liquid or gas phase are in a state of random motion [60]. More molecules move out of the high concentration region to the dilute region along the concentration gradient. The diffusion rate in a very dry soil increases rapidly as soil moisture content increases, and the diffusion rate in moist soils increases as the temperature increases. For example, trifluralin diffuses primarily in soil air. The diffusion coefficient of trifluralin increases rapidly from zero in very dry soils to a maximum level as the soil moisture content increases. At low moisture levels, most of the trifluralin is adsorbed on the soil surface. At high soil moisture levels, the change in water content does not affect the degree of adsorption, but the air-filled polarity of the soil and diffusion coefficients are reduced.

Herbicides which diffuse primarily in soil air are not active in extremely dry soil. Hence, incorporation of herbicides such as trifluarlin and triallate may not lead to satisfactory weed control in dry soils, with soil moisture content below the permanent wilting point, unless soil moisture content is increased by way of rainfall or irrigation. A herbicide like simazine diffuses primarily in soil water and its diffusion coefficient increases as the soil water content decreases. The diffusion rate in the vapour phase increases rapidly to a maximum level at moisture levels just above those where complete adsorption occurs, while in the liquid phase there is a gradual increase in the diffusion rate as the soil moisture content increases. Thus at moisture levels near or at the permanent wilting point, diffusion in the vapour phase may be important even for herbicides with low vapour pressures. The slight activity of herbicides at low vapour pressure in dry soils may be due to diffusion in the vapour phase. The efficacy of herbicides with low vapour pressures increases as the soil moisture content increases. The lack of their efficacy in dry soil may be due to insufficient water content to move the herbicide into the root zone or due to reduced diffusion of some herbicides in soil water to the emerging shoots.

Mass Flow Transport

The mass flow of dissolved herbicides to plant roots results from a bulk flow of soil solution to plant roots [60]. The amount of herbicide that reaches the roots by mass flow is calculated by the volume of water transpired from plant leaves. The water flow through the soil-plant system occurs along a potential gradient, which must decrease continuously from the soil through the plant to the atmosphere for transpiration to occur. The soil moisture level or potential does not have a large influence on the herbicide concentration in soil water. However, the mass flow of water and dissolved herbicide to the plants varies greatly depending on soil and atmospheric conditions [34].

The rate of water transfer in the soil-root-leaf part of the pathway to the air is described by the following equation [62]:

$$\text{Rate of transpiration (Q)} = \frac{\text{Water potential (Ws) } - \text{ Water Potential in Leaf (Wl)}}{\text{Resistance to water flow in plant (Rp) } + \text{ Resistance to water flow in soil (Rs)}}$$

The resistance in plant and soil (Rp + Rs) increases greatly as the soil water content falls and as the relative transport rate also decreases. Generally, water will be transported through the soil-plant system fast enough to meet the potential transpiration demands only when the soil water potential is at a very small absolute value. The potential transpiration rate will depend on meteorological parameters such as incident radiation, temperature and wind speed [60]. When plant resistance is constant, transpiration increases linearly with: a) an increase in radiation, b) a decrease in relative humidity and c) an increase in air temperature when both radiation and relative humidity are constant; the variation with wind speed is more complex [48]. Thus, the flow of water and dissolved herbicide to the roots depends on atmospheric conditions and soil water potential.

Normally, a good correlation exists between the uptake of herbicides by plants and the rate of transpiration at varying soil moisture contents. The concentration of a herbicide in plant leaves is related to total uptake and rate of plant growth.

Effect of Rainfall on Herbicide Transport

Rainfall is required for the activation of soil-applied herbicides. The downward flow of water transports herbicides to the root zone. Besides, some downward movement of herbicides occurs for uptake in the shoot zone. As the soil surface is constantly exposed to extreme wetting and drying cycles, the herbicides will need to be distributed in the top 0.5 cm or more of soil for continuous uptake to occur.

In soils that are uniformly moist with depth, water will move downwards under the force of gravity. The gravitational force and the matric potential gradient determine the movement of water in the soil. As the soil surface dries up, the matric potential gradient near the soil surface overcomes gravitational force and enables the water to move from lower layers of soil towards the soil surface. The placement or position of the herbicide moving along the water flow is dependent on the amount of rainfall, amount of moisture in the soil, soil composition, adsorption of the herbicide by soil, speed of adsorption and desorption, permeability and amount of dispersion. Herbicides diffuse into small pores with time, while the mass flow of water following a rain will take place primarily in the large soil pores.

Movement of herbicides following rainfall depends on their relative mobilities in soil. The herbicides range between those which move freely (moving with a wetting point) to those which are almost immobile. The herbicides with lowest adsorption coefficients move into the 5 to 10 cm soil layer, while those with higher adsorption coefficients will remain above the 5 to 10 cm depth. However, herbicides with lower adsorption coefficients move readily into the root zone, in which case herbicide selectivity is based on depth protection. Herbicides with higher adsorption coefficients remain in the surface layer, although these may flow through the large soil pores and cracks to the root zone even with a moderate rainfall.

For most herbicides, rainfall shortly after their application to the soil is important for good herbicidal activity. Soil incorporation tends to improve weed control under all climatic conditions for herbicides that are highly volatile and strongly adsorbed by soil or absorbed into plants primarily by roots.

MICROBIAL DEGRADATION

Microbial degradation plays a major role in the persistence and behaviour of herbicides in soil. Soil microorganisms have the capacity to detoxify and inactivate the herbicides present in the soil. Some groups of herbicides are more easily degradable by microbes than others. The difference lies in the molecular configuration of the herbicide. Besides, microbial degradation is also governed by many factors, the most critical ones being soil temperature, soil moisture content and soil acidity. Generally a temperature range of 25° C to 35° C and a soil moisture range of 50% to 100% of field capacity are considered optimum for the action of microbes on herbicides. The temperature and moisture regimes coincide with those required for crop growth and with the normal herbicide application time. The effect of soil pH on microbial degradation of herbicides is, however, highly specific depending on the herbicides and microorganisms involved. Generally, bacteria predominate in alkaline soils, while fungi are more active in acidic soils. The greatest total microbial life is in soils where the pH is neutral. Generally, soil-applied herbicides tend to lose activity fastest in a neutral soil where the microbial activity is greatest.

The microorganisms involved in herbicide detoxification include bacteria, fungi, algae, moulds, etc. Of these, bacteria predominate. Bacteria include representatives of the genera *Agrobacterium, Achromobacterium, Alcalaginese, Arthrobacter, Bacillus, Flavobacterium, Pseudomonas, Nocardia, Streptomyces, Rhizobium*, etc. The fungi include those of the genera *Fusarium, Penicillium*, etc. The moulds (of the order Mucorales) include the species of *Aspergillus, Rhizobacteria, Sclerotium*, etc.

Microbes degrade herbicides predominantly through ester or amide hydrolysis, alkylation, dealkylation, dehalogenation, reduction, oxidation, aromatic ring hydroxylation, ring cleavage and conjugation pathways.

During microbial degradation, herbicides exhibit varying degrees of inhibitory effects on soil microorganisms, which in turn interact with them and show interesting adaptive mechanisms. These interactions are as under:

a) In one interaction, herbicides inhibit none of the segments of soil microbes and are rapidly degraded by them.

b) In another interaction, herbicides do not affect the soil microflora but are degraded only after a lag period.

c) In a different interaction, herbicides are non-degradable but they are inhibitory to some of the microorganisms, thus reducing their population. The resistant ones survive on the dead organic matter of the killed flora. In the end, the entire microbial population is composed of organisms tolerant to this non-degradable herbicide.

The adaptive mechanisms enable the microbes to develop the capacity to degrade chemicals in the soil. Soil microbes develop this capacity through the formation of **adaptive enzymes** or through **mutation** to form new strains. During this **adaptive phase**, also called the **lag phase**, the population of microbes is built up to the levels required for detoxification (Fig. 9.4). Duration of the lag phase is dependent on the herbicide under attack and the organisms present as briefly discussed earlier. For example, the lag phase of MCPA is considerably longer than that of 2,4-D. Once the organisms develop the potential to degrade a herbicide, subsequent applications of the same chemical are rapidly degraded.

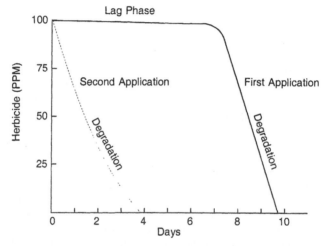

Fig. 9.4. Herbicide degradation by soil microorganisms showing the lag phase (solid line) following an initial application to the soil and rapid detoxification (broken line) after the second application of the same chemical.

The ability of soil organisms to degrade a herbicide results from the possession of enzyme systems that normally metabolize structurally related herbicides. These enzymes may have wider specificity so that they can also degrade different herbicides and their breakdown products. If the organisms lack specific enzymes to metabolize them, mutations give the enzymes the necessary specificity.

Soil microbes degrade herbicides both inside (**intracellular**) and outside (**extracellular**) their bodies. Intracellular types form their own enzymes for degradation and extracellular microbes utilize the enzymes excreted by other specific organisms. During degradation, microorganisms use carbon atoms of the herbicide particle for their energy requirements. This results in the release of CO_2 during microbial degradation of herbicides.

Enhanced Herbicide Degradation

Enhanced herbicide degradation refers to the phenomenon of an increase in the rate of herbicide decomposition induced by prior treatment. This phenomenon occurs in soils previously treated with the same or different herbicide but not in soils left untreated in the past. The net result of enhanced herbicide degradation is unsatisfactory weed control.

Enhanced degradation could refer to '**self-enhancement**' or '**cross-enhancement**'. Self-enhancement refers to degradation of the same herbicide in the soil both initially and subsequently. Cross-enhancement refers to degradation of a herbicide in the soil previously treated with a different herbicide. In the self-enhancement phenomenon, the subsequent applications of biodegradable herbicides are subjected to more rapid degradation without a lag phase, while in cross-enhancement degradation of a herbicide applied subsequently is preceded by an initial lag phase as though it were applied for the first time (Fig. 9.4).

Cross-enhancement degradation requires fresh adaptation by soil microbes already adapted to the herbicide(s) applied in the past, to the new herbicide. As a result, the degradation rate usually becomes slow. The slower degradation is attributed to the following mechanisms: 1) In the first, each herbicide may induce its own enrichment flora with different efficiencies for utilization of structurally similar chemicals. 2) In the second mechanism, the microorganisms may be the same, but may possess different inducible enzymes that a have broad recognition of similar substrates but with different affinities for each herbicide. 3) In the third, the same microbes and inducible enzymes are active in degradation, but the genetic operon may possess different affinities for each herbicide. In this case, the operon itself would undergo an adaptive change. Whatever may be the adaptive mechanism, which could differ with various herbicides, it must allow for a pathway different from the one used for the earlier herbicide. Cross-enhancement occurs not only between structurally dissimilar herbicides, but between structurally similar herbicides as well, but the adaptation caused by a prior herbicide treatment has considerable specificity, resulting in a lower rate of enhanced degradation.

The ability of microbes to degrade a herbicide (i.e., microbial enrichment) is lost in the absence of the enhancing herbicide in about a year in soil. The survival ability of the microbes is aided by a) resistant spore-like structures, b) trace amounts of the original enriching substrate, c) an ability to survive alternate substrates and d) the existence of a quiescent state free of competition in soil caused by winter or unfavourable soil temperature conditions.

Management of Enhanced Herbicide Degradation

As farming becomes more intensive with mixed croppings and crop rotations, some closely related, or even different herbicides are to be applied annually to the same field. This is in addition to various insecticides, fungicides and other xenobiotics applied to the same field. This may eventually lead to a steady selection pressure which may result in weed resistance, weed shifts or enhanced herbicide degradation. Enhanced degradation may occur much sooner than weed resistance shifts because microbial generation time is shorter. Selection pressure is strongest when the herbicide is applied repeatedly to a site. Thus, crop and herbicide rotations are important to reduce selection pressure and maintain herbicide effectiveness. Crop rotations provide the opportunity to use unrelated herbicides and allow for application of alternative control strategies such as employing cultural and mechanical methods, using foliage-applied herbicides to reduce reliance on soil-applied herbicides, etc. Crop rotation also favours different rhizosphere populations and may facilitate an alteration of the microbial balance in a particular soil.

Enhanced degradation may also be managed by changing the herbicide formulation. Application of granular formulations may slow the rate of enhanced degradation. Formulating short residual herbicides into microcapsules or granules of clay, starch, or other polymers to gain slow and steady release may improve weed control in a non-enhanced environment.

Herbicide extenders, which prolong the duration of herbicide effectiveness, represent feasible short-term aids for coping with enhanced degradation situations in which an enhancing herbicide must be used for a better weed control. Extenders, however, slow the enhanced herbicide degradation to a normal rate but do not cure enhancement, i.e., cause reversion to normality. Dietholate is a good extender for EPTC. It causes doubling of the half-life of EPTC in soils, thus slowing down EPTC degradation. Dietholate is also known to enhance the performance of other thiocarbamate herbicides, butylate and vernolate.

Extrachromosomal Elements

Extrachromosomal elements (ECEs) such as **plasmids, transposons** and **insertion sequences** are being increasingly implicated as providing the mechanisms for structural change and adaptation of bacteria to degrade herbicides in soil. ECEs are small units of DNA that can readily alter the metabolic characteristics of the host. Plasmids are ECEs that are often transferable between bacteria.

Plasmids act as transfer agents of degradative genes between bacteria in soil and account for the conservation of the degradative ability in the absence of the herbicide substrate. The lag phase in herbicide degradation represents the period of plasmid exchange and recombination (adaptation) followed by proliferation of those microbes having the essential enzymes for the task. Plasmid degradative enzymes are inducible and have broader specificity than chromosomally coded enzymes. Degradative plasmids usually have a broad host range that allows rapid transmission between microbes and provide microbes with an access to a pool of degradative genes.

The difference between non-persistent and persistent herbicides appears to be the ability of non-persistent types to elicit the rapid evolution and spread of plasmids encoding their degradation. The herbicides for which plasmid-coded degradation takes place include 2,4-D, MCPA and 2,4,5-T.

MICROBIAL DEGRADATION OF INDIVIDUAL HERBICIDES

Acetamides

Acetochlor, Alachlor, Butachlor, Dimethenamid, Metolachlor, Propachlor, Propanil, Pronamide

Acetochlor is degraded by soil microbes, yielding under aerobic conditions, the metabolites N-ethoxymethyl-2′ethoxy-6′-methyl-oxanilic acid; [N-ethoxymethyl-N-(2′-ethyl-6′-methyl)phenyl]-2-amino-2-oxoethyl]sulphinylacetic acid; and N-ethoxymethyl-2′-ethyl-6′-methyl-2-sulphoacetamide.

Microbial degradation is a major means of degradation of alachlor. Although several microbes are capable of transforming alachlor, no microorganism is capable of mineralizing it as a sole source of carbon. Major metabolites of microbial degradation of alachlor include 2,6-diethyl-N-methoxy-methyloxanilic acid; 2',6'-diethyl-N-methoxymethyl-2-sulphoacetanilide; 2,6-diethyloxanilic acid; 2',6'-diethyl-2-sulphoacetanilide; 2',6'-diethyl-2-hydroxy-N-methoxymethyl-acetanilide; and 2',6'-1-diethyl-N-methoxymethyl-2-methylsulphinyl acetanilide. Half-life of alachlor in aerobic soils is 6.1-15.8 d, but only 4.5 d in anaerobic (aquatic) conditions. Butachlor and dimethenamid are readily degraded by soil microbes. Under aerobic conditions, the half-lives of butachlor and dimethenamid are 12 and 38 d respectively.

Microbial degradation is a major pathway of field dissipation of metolachlor, producing [(2-ethyl-6-methylphenyl)(2-methoxy-1-methylethyl)amino]oxo-acetic acid as the major metabolite. In a sandy loam soil, its half-life under aerobic conditions is 67 d, while under anaerobic conditions it is 81 d. Propachlor is readily degraded under aerobic conditions in all soils into water-soluble metabolites. The N-chloroacetyl group is degraded to form metabolites containing an oxanilic, oxoacetic, or oxoethanesulfonic acid group.

Propanil is rapidly metabolized to propionic acid and 3,4-dichloroaniline (DCA) in aerobic rice soils, followed (7 d after application) by DCA degradation to volatiles and residues strongly bound to soil sediment. Metabolism in anaerobic rice field conditions is similar except that degradation of DCA occurs in 14 d after application. Most DCA becomes covalently bound to soil OM and degrades very slowly, with a half-life of several years. Some DCA is converted by microbial peroxidases to 3,4-dichlorophenylhydroxylamine, which condenses with DCA to form 3,3',4,4'-tetrachlorobenzene. Soil microbes convert pronamide to the cyclized metabolite and to subsequent hydrolysis products. Pronamide is inactive against common soil microbes.

Aliphatics

Dalapon

Dalapon is dehalogenated to yield 2-chloro-2-hydroxy-propionate from which pyruvic acid is formed. Pyruvic acid is used by microbes to liberate CO_2. Decomposition of dalapon is complete with CO_2 liberation. The bacteria responsible for dalapon degradation include those belonging to the genera *Agrobacterium, Alcalaginese, Bacillus, Pseudomonas, Nocardia* and *Streptomyces*, and the moulds *Penicillium* and *Trichoderma*. The fastest degradation of dalapon occurs in plant and clay loam soils, while the slowest decomposition occurs in sandy or silt loams.

Arsenicals

DSMA, MSMA

Organoarsenical herbicides in the soil are metabolized by two pathways: oxidation to form inorganic orthoarsenic acid and CO_2, and reduction and methylation to yield volatile methylarsines, mostly dimethyl arsine [70]. The average half-live of DSMA and MSMA is 180 d.

Benzoics

Dicamba

Microbial degradation is the most important dissipation pathway of dicamba in soil. It is decarboxylated and demethylated. In aerobic soils, dicamba is metabolized to CO_2, with 3,6-dichlorosalicylic acid as the only major metabolite and low levels of 2,5-dihydroxy-3,6-dichlorosalicylic acid. Over 50% of the applied dicamba is lost within 4 weeks. Non-biological degradation is negligible. Dicamba degrades very slowly in anaerobic soils.

Benzothiadiazoles

Bentazon

Biodegradation of bentazon proceeds with the formation of bound residue. The metabolites formed include 6-hydroxy bentazon, 8-hydroxy bentazon, 2-amino-N-isopropyl benzamide (AIBA), and 2-aminobenzoic acid. AIBA is produced by hydroly-sis of the sulphamide function, while 2-aminobenzoic acid is a product of AIBA. The N-methyl derivative of bentazon is also found in bentazon-treated soils. The metabolites AIBA, N-methyl bentazon, and 2-aminobenzoic acid are strongly adsorbed by soils of high organic carbon and lower pH [28]. Among these metabolites, adsorption of 2-aminobenzoic acid is highest. Sulphamide hydrolysis is the pathway of bentazon degradation that leads to bound residues similar to binding of hydroxy bentazons.

Bipyridiliums

Diquat, Paraquat

Diquat and paraquat are strongly adsorbed in soils, preventing microbial degradation. In soils with high OM and montmorillonite clay, bipyridilium herbicides are so rapidly adsorbed that they are not accessible for microbial attack and degradation. There is, however, abundant evidence for the microbial degradation of diquat and paraquat in solution. An unidentified bacterium could reduce ^{14}C-methyl- and ^{14}C-ring-lebelled paraquat to the free radical in culture solutions [27]. The identified metabolites included 1-methyl-4,4(-bipyridilium ion and 4-carboxy-1-methylpyridinium ion.

Carbamates

Asulam, Desmedipham, Phenmedipham

Carbamates readily undergo hydrolysis by the action of soil microbes. Actinomycetes use asulam as a source of CO_2. Phenmedipham and desmedipham are hydrolyzed microbially to yield 3-aminophenols which form complexes with soil. Most of the phenmedipham and desmedipham remains in the top 5 cm of soil, with little leaching. Asulam has medium to high mobility. However, its leaching potential is very low because of rapid degradation.

Dinitroanilines

Benefin, Dinitramine, Pendimethalin, Trifluralin

Degradation of benefin by soil organisms accounts for only a small fraction of their total dissipation in the soil. Dinitramine is degraded by demethylation into a monoalkylated product, N^3-ethyl-2,4-dinitro-6-trifluoromethyl-m-phenylenediamine and the completely dealkylated product, 2,4-dinitro-6-trifluoromethyl-m-phenylenediamine [53]. The microorganisms responsible for this degradation include *Aspergillus fumigatus, Fusarium oxysporum* and *Pacilomyces* spp. The other products include 6-amino-1-ethyl-2-methyl-7-nitro-5-(trifluoro)-m-benzimidazole, and 6-amino-2-methyl-7-nitro-5-(trifluoromethyl)benzimidazole.

Pendimethalin undergoes slow degradation under aerobic conditions. Trifluralin is degradad microbially, more rapidly in flooded anaerobic than in moist aerobic soils.

Initial detoxification reactions under aerobic conditions primarily are N-dealkylation, hydrolytic cleavage of the dipropylamine group to produce the phenolic derivative and oxidation of an N-propyl group followed by cyclization to produce the benzimidazole derivative. Intitial detoxification reactions under anaerobic conditions are nitro reduction and N-dealkylation. Non-biological degradation rates appear to be negligible.

Diphenylethers

Acifluorfen, Bifenox, Fomesafen, Fluoroglycofen, Lactofen, Oxyfluorfen

Acifluorfen, bifenox, fomesafen, fluoroglycofen and lactofen are rapidly degraded biologically. Aerobic microbial degradation rate of oxyfluorfen is slow. It gets dissipated to CO_2. Under anaerobic conditions its degradation is insignificant.

Imidazolinones

Imazapyr, Imazaquin, Imazethapyr, Imazamethabenhz

Imidazolinone herbicides are primarily degraded by microbes. Studies with ^{14}C-imazaquin and ^{14}C-imazethapyr showed that [14]CO_2 evolution was greatest from carboxyl-labelled imazaquin and imazethapyr compared to ring-labelled imazaquin and imazethapyr [24]. This indicated that microbial degradation did not involve cleavage of the ring or carboxyl group. The rate of degradation (CO_2 evolution) was higher when soil temperature increased from 15° to 30° C and soil moisture from 15% to 75% of field capacity. Non-microbial degradation was insignificant. Anaerobic microbial degradation occurred at much slower rates. Imidazolinone herbicides have a low potential for forming bound residues.

Imines

Cinmethylin

Cinmethylin is readily degraded in soil microbially, with CO_2 as the major terminal degradation product. Its degradation is reduced under anaerobic conditions (irrigated rice) because of slower microbial degradation. Cinmethylin has limited mobility in soil. Its field half-life ranges from 25 to 78 d, increasing with higher OM content.

Isoxazolidinones

Clomazone

Clomazone readily undergoes microbial degradation under aerobic conditions. Aerobic degradation undergoes oxidation, resulting in CO_2 evolution and formation of bound residues. Degradation is faster under flooded (anaerobic) than aerobic conditions. Anaerobic degradation involves reductive ring opening. Field half-life varies from 5 to 29 d depending on the type of soil.

Nitriles

Bromoxynil, Ioxynil

The nitrile herbicides are extremely effective inhibitors of nitrification in soils. However, soil microorganisms are believed to have an ability to adapt to these herbicides and degrade them. They degrade bromoxynil and ioxynil to the respective benzamide and benzoic acid derivatives.

Phenols

Dinoseb

Dinoseb is readily broken down by soil microorganisms in two pathways: reduction of a nitro group to an amine and oxidative elimination of the nitro group with subsequent formation of dihydric phenol.

Phenoxyalkanoic Acids

Phenoxyacetics – 2,4-D, MCPA; Phenoxybutyrics – 2,4-DB, MCPB

Phenoxy acids are metabolized by various organisms in the soil. Many phenoxyacetic acids are degraded completely or almost completely, with the loss of their aromatic structure and release of their chlorine as a chloride ion.

The important pathway of degradation of phenoxyacetics, 2,4-D and MCPA, is the removal of the acetic acid side chain by *Arthrobacterium* and *Pseudomonas* to yield the corresponding phenol, which is hydroxylated at the *ortho* position to produce a catechol. Catechols are subjected to *ortho* cleavage of the aromatic ring and converted to muconic acid (Fig. 9.5). Another pathway of 2,4-D and MCPA breakdown is dehalogenation by *Pseudomonas* at the 4-position on the aromatic ring and replacement by OH to yield 2-chloro-4-hydroxyphenoxyacetic acid (Fig. 9.6).

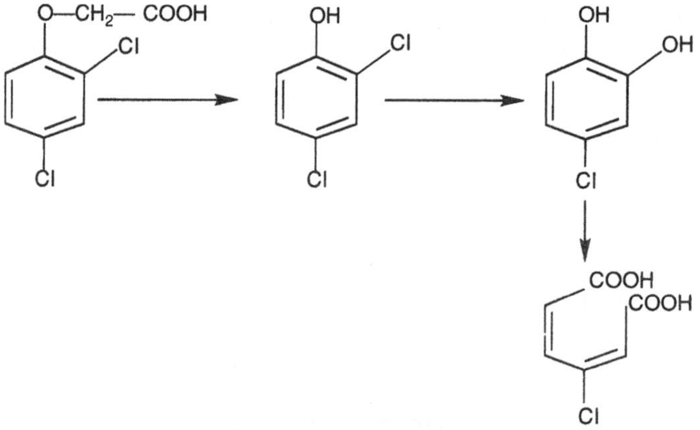

Fig. 9.5. Pathway of 2,4-D degradation by *Arthrobacterium* and *Pseudomonas* resulting in a phenol, catechol and muconate.

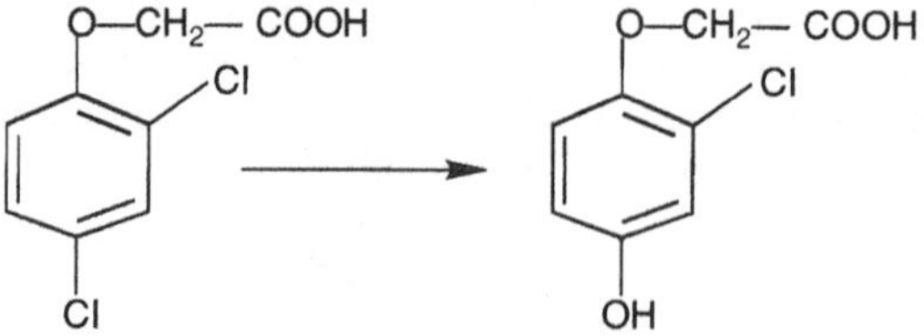

Fig. 9.6. Degradation of 2,4-D by *Pseudomonas* bacteria through the dehalogenation pathway.

In another pathway, the side chain of 2,4-D and MCPA is removed by *Arthrobacter* bacteria as a glyoxalate (CHO-COOH), yielding 2,4-dichlorophenol. The released glyoxalate undergoes condensation with decarboxylation and simultaneous incorporation of ammonia to form alanine.

Fungi also degrade phenoxyacetic acid compounds. For example, *Aspergillus niger* hydroxylates 2,4-D at the 2-, 4- or 2- and 4- positions to yield the corresponding hydroxyphenoxyacetic acids, It is also metabolized by 4-hydroxylation followed by a shift in chlorine position to form 2,5-dichloro-4-hydroxyphenoxyacetic acid.

2,4-DB is metabolized by *Flavobacterium* by cleavage at the ether linkage to yield 2,4-dichlorophenol, 4-chlorocatechol, butyric acid and crotonic acid. The fungus *Aspergillus niger* degrades 2,4-DB through β-oxidation to form 4-hydroxyphenoxyacetic acid with small amounts of 2-hydroxyphenoxyacetic acid. However, degradation by ring cleavage of the ether linkage leads to immediate detoxification of 2,4-DB than by β-oxidation.

Aryloxyphenoxypropanoics — Diclofop methyl, Fenoxaprop, Haloxyfop-P

Diclofop methyl is hydroxylated by microbes within a few days under aerobic conditions to the herbicidally active diclofop acid which is then metabolized. It has a half-life of 10 d in sandy soils and 30 d in sandy loam soils. Small amounts of 4-(2,4-dichlorophenoxy)phenol are also produced. Under anaerobic conditions, ester hydrolysis, yielding diclofop acid, is extremely rapid (within 1 h); after 2 d up to 86% of applied diclofop had metabolized to various metabolites of diclofop acid while up to 3.7% was found as phenol metabolites. Diclofop has no bactericidal or fungicidal effects under aerobic or anaerobic conditions. There is also no effect of diclofop on nitrogen fixation and nitrification, nor utilization of cellulose, starch and protein by microbes.

Fenoxaprop (the ethyl ester form), with a half-life of less than 1 d, is converted to the phytotoxic fenoxaprop acid, under both aerobic and anaerobic conditions. The acid is degraded primarily to 6-chloro-2,3-dihydrobenzoxazole-2-one and 4-(6-chloro-2-benzoxazolyloxy)phenol.

Haloxyfop methyl ester is rapidly hydrolyzed to acid form. Further degradation produces pyridinol and phenol under aerobic conditions. No degradation occurs under anaerobic conditions.

Phenylpyridazines

Pyridate

Pyridate undergoes microbial degradation only after it is chemically degraded by hydrolysis to 3-phenyl-4-hydroxy-6-chloropyridazine.

Pyridazinones

Norflurazon, Pyrazon

Norflurazon is microbially degraded, under aerobic conditions, yielding CO_2 and desmethyl norflurazon as major metabolites. Microbial degradation of pyrazon is moderately rapid depending on soil temperature and moisture. The primary degradation product is dephenylated pyrazon, which is non-phytotoxic.

Pyridinecarboxylic Acids

Clopyralid, Picloram, Triclopyr

Clopyralid and triclopyr are rapidly degraded by soil microbes. Microbial degradation of picloram, which proceeds somewhat slowly, involves no complex intermediates, but yields a spectrum of simpler molecules which may be easily assimilated by living systems [71]. CO_2 is an end-product metabolite. Subsequent degradation of primary metabolites is rapid, leaving only trace amounts of primary products.

Pyridines

Dithiopyr, Thiazopyr

Microbial degradation is the primary dissipation pathway of pyridine herbicides. Dithiopyr is microbially hydrolyzed to monoacids and a diacid, each at > 6% of applied herbicide. The monoacids include 2-(difluoromethyl)-4-(2-methylpropyl)-5-[(methylthio)carbonyl]-6-(trifluoromethyl)-3-pyridinecarboxylic acid and 6-(difluoromethyl)-4-(2-methylpropyl)-5-[(methylthiocarbonyl]-2-(trifluoromethyl)-3-pyridinedicarboxylic acid. The diacid includes 2-(difluoromethyl)-4-(2-methylpropyl)-6-(trifluoromethyl)-3,5-pyridinedicarboxylic acid. Non-biological degradation is negligible.

In the case of thiazopyr, the monoacid degradation product is formed at low levels 20 d after application. Its half-life ranges from 12 to 14 d, indicating minimal potential for carryover injury to sensitive rotational crops.

Pyridinones

Fluridone

In aquatic environments, fluridone is degraded principally by photolysis. Photolysis, however, is affected by geographic location, application rate, water depth, turbidity, weather and weed cover, all of which may affect sunlight penetration in the water.

Sulfonylureas

Chlorsulfuron, Halosulfuron, Metsulfuron

Microorganisms play a major role in the degradation of sulfonylurea herbicides in the soil. They account for 50% to 90% of the total dissipation, depending on the soil type. The microbes include actinomycetes, fungi and bacteria.

The soil actinomycete *Streptomyces griseolus* rapidly metabolizes chlorsulfuron via conversion of the methoxy group to a hydroxyl group and hydroxylation of the methyl group [9] (Fig. 9.7). Metsulfuron, which contains the same heterocycle as chlorsulfuron, also undergoes the same microbial degradation. In addition, metsulfuron undergoes

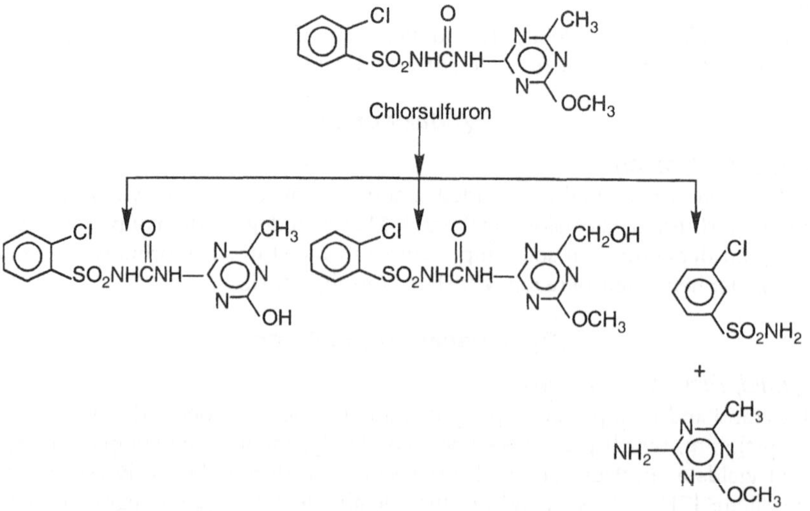

Fig. 9.7. Metabolic pathways of chlorsulfuron by soil microorganisms.

de-esterification of the *ortho*-carboxymethoxy group on the aryl portion, resulting in the formation of free acid [9] (Fig. 9.8).

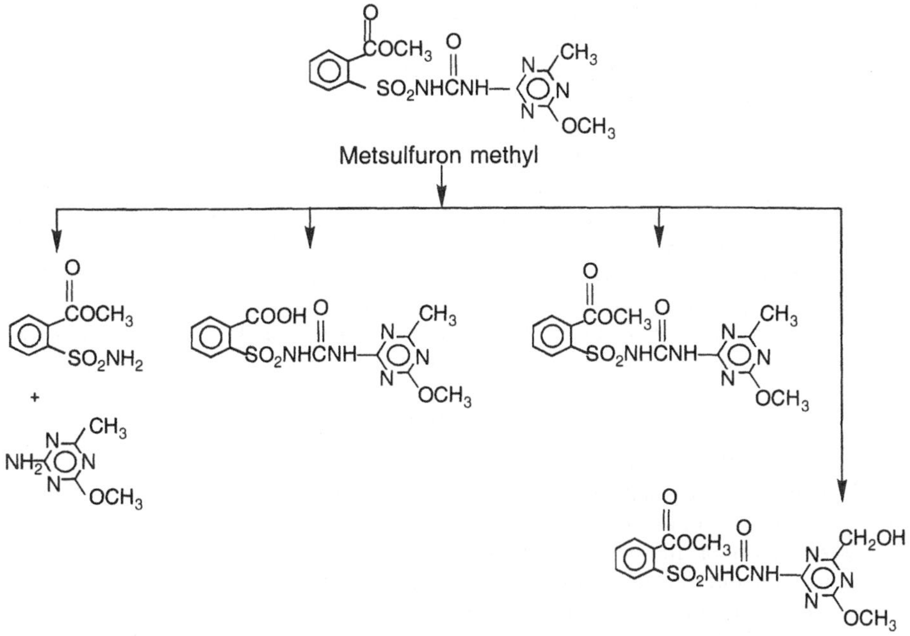

Fig. 9.8. Metabolic pathway of metsulfuron methyl by soil microorganisms.

Two soil fungi, *Aspergillus niger* and *Penicillium* spp., catalyze hydrolysis of the sulfonylurea bridge, yielding the corresponding sulfonamide and heterocycle. These fungi may also convert the methoxy group on the heterocycle to a hydroxyl group, but this is a relatively minor pathway.

Thiocarbamates

Butylate, EPTC, Diallate, Molinate, Pebulate, Thiobencarb, Triallate, Vernolate

Soil microorganisms make a significant contribution to the disappearance of thiocarbamate herbicides when incorporated into the soil. About 50% of the applied butylate, EPTC, diallate, pebulate and vernolate is lost within 2 to 4 wk, while the loss of triallate is much slower.

Microbial degradation of thiocarbamate herbicides involves hydrolysis of the ester linkage, with the formation of a mercaptan, CO_2 and a secondary amine (Fig. 9.9). The mercaptan is converted into an alcohol by transthiolation and further oxidized to an acid prior to entering the metabolic pool. The amines are oxidized to ammonia, CO_2 and water.

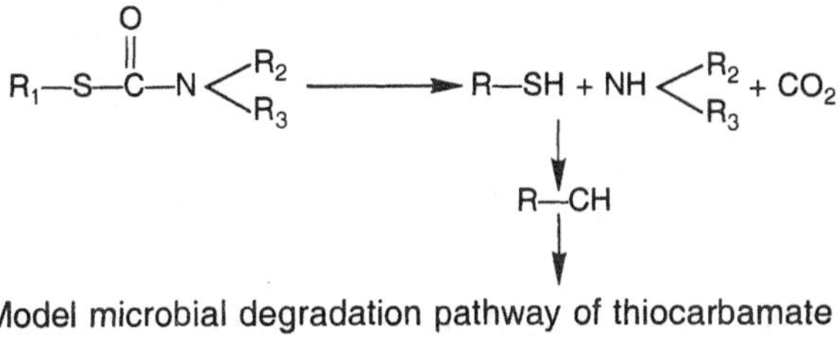

Fig. 9.9. Model microbial degradation pathway of thiocarbamate herbicides.

Thiobencarb undergoes more rapid microbial degradation under oxidative conditions than under reductive conditions, liberating considerable amounts of CO_2. Bacteria are capable of utilizing molinate as the sole source of carbon or nitrogen. Thiobencarb undergoes hydrolysis, forming a mercaptan, CO_2, and a dialkylamine.

The microorganisms adapted to one thiocarbamate herbicide are cross-adapted to other thiocarbamate herbicides. The butylate-adapted microorganisms are cross-adapted to EPTC, pebulate and vernolate. Similarly, enhanced biodegradation of butylate, EPTC and vernolate occurs in soils previously treated with EPTC. Lawrence et al. [54] found that prior applications of butylate resulted in cross-adapted degradation of EPTC, but not of vernolate, pebulate or cycloate. This indicates that even though cross-adapted enhanced biodegradation occurs among thiocarbamate herbicides, it is not a generalized phenomenon. It depends on the type of soils, climatic conditions and field management practices—all of which affect microbial reproduction. This is especially true since the degradative ability of soil microorganisms may be plasmid-encoded and transferred to conjugal mating. The plasmids may be lost in extreme temperatures.

Triazines

Atrazine, Ametryn, Simazine, Terbutryn

Triazines are microbially degraded by hydrolysis at the no. 2 carbon, N-dealkylation of side chains, and deamination. They are degraded by ring cleavage.

Degradation of atrazine to hydroxyatrazine is dependent on soil type, atrazine concentration and moisture content [80]. The isopropyl and ring constituents of atrazine are subject to minimal attack. The hydroxyatrazine ring is attacked more readily than the atrazine ring. The content of hydroxyatrazine varies in different soils. In one soil, hydrolysis was the dominant pathway of detoxification, but in other soils detoxification was a combination of chemical hydrolysis and slow microbial degradation by N-dealkylation of the ethyl side chain constituent [80].

The atrazine molecule is also metabolized by soil fungi (*Aspergillus* spp.) through N-dealkylation (Fig. 9.10). Dealkylation is easier at the ethylamino moiety than at the isopropylamino moiety. As observed in plants, microbial degradation by dealkylation of the ethyl group results in the formation of 2-chloro-4-amino-6-isopropylamino-*s*-triazine. Similar monodealkylated derivatives are also formed after the application of simazine and terbutryn.

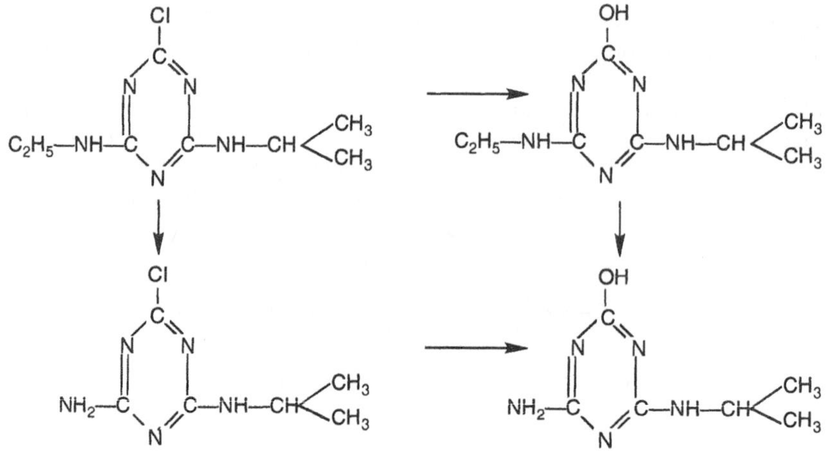

Fig. 9.10. Mircobial degradation of atrazine by hydroxylation and dealkylation.

The hydroxylated triazine compounds are also dealkylated to form 2-hydroxy-4-amino-6-alkylamino-s-triazines (Fig. 9.10). These metabolites are formed in the case of simazine ametryn and terbutryn.

Triazines also undergo deamination during decomposition by *Aspergillus fumigatus*. Deamination of dealkylated and hydroxylated products of simazine and atrazine, 2-hydroxy-4-amino-6-ethylamino-s-triazine and 2-hydroxy-4-amino-6-isopropylamino-s-triazine, yield the corresponding 2,4-dihydroxy derivatives. The ammelines are degraded readily to CO_2, cyanuric acid and traces of ammelide.

The bacteria *Nocardia* spp., which utilize triazine as the sole source of carbon and nitrogen, degrade atrazine, forming 2-amino-2-chloro-1,3,4-triazine [30]. This degradation occurs through dealkylation as well as deamination. That this metabolite was not phytotoxic to oats suggest that deamination ensures detoxification [30].

Triazoles

Amitrole, Amitrole-T

Microbial degradation is not considered to be important in the case of triazoles. However, approximately 69% of the 6-carbon from amitrole was released as $^{14}CO_2$ in 20 d by non-sterilized soil, while autoclaved soil released only 2% [49]. Soil moisture, temperature and pH markedly affect amitrole degradation, indicating possible microbial involvement.

Triazolopyrimidine Sulfonanilides

Flumetsulam

PRE application of flumetsulam is primarily limited to the upper 8 cm of the soil, regardless of soil type and days after treatment [61]. No significant concentrations were detected below 15 cm. The half-life ranges from 20 to 46 d, depending on rainfall, soil OM content and soil pH. Microbial degradation is considered to be the primary route of flumetsulam dissipation in soils.

Uracils

Bromacil, Terbacil

Soil microbes play a very important role in the disappearence of substituted uracil herbicides from soils. *Pseudomonas* and *Arthrobacter* bacteria are active in the dissipation of bromacil and terbacil. They use uracils as a source of energy. Biological activity is required for breakdown of bormacil and terbacil and their presence has no influence on microbial activity in the soil.

Ureas

Chlorbromuron, Chloroxuron, Diuron, Fluometuron, Monuron

The substituted urea herbicides are subject to active microbial degradation. N-dealkylation is the major initial pathway of breakdown by soil microbes. Soil fungi (*Penicillium, Aspergillus*, etc.) degrade chloroxuron to yield two metabolites, 1-methyl-3-(p-chlorophenoxy)phenylurea and 3-(p-chlorophenoxy)phenylurea [29]. Subsequent degradation (oxidation) of these metabolites releases CO_2 and ammonia.

Monuron, fluometuron and monolinuron are demethylated by *Rhizopus japonicus* into the metabolites 3-(p-chlorophenyl)-1-methylurea, 3-(m-trifluoromethyl-phenyl)-1-methyl urea, and 3-(p-chlorophenyl)-1-methylurea respectively [89]. Fluometuron degradation involves a 2-step demethylation followed by hydrolysis of the urea linkage to form the aniline derivative (Fig. 9.11) [12].

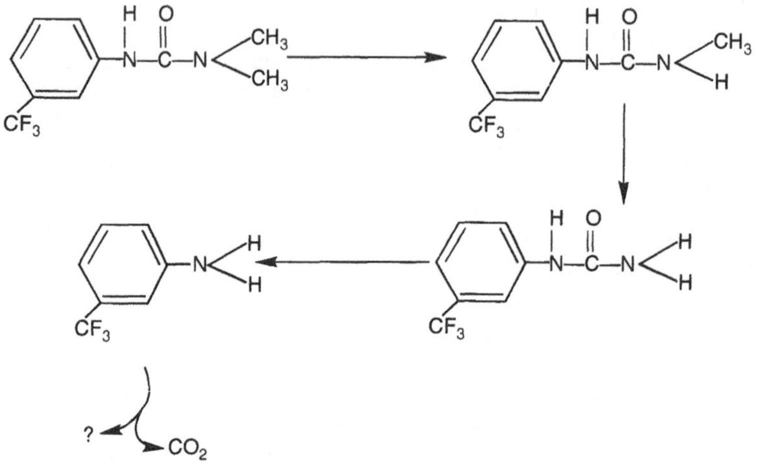

Fig. 9.11. Microbial degradation of fluometuron [12].

Chlorbromuron is broken down to monomethyl-chlorbromuron and monomethoxy-chlorbormuron, which are rapidly degraded [74]. As in the case of fluometuron, the chlorbromuron degradation pathway involves a stepwise dealkylation followed by hydrolysis of the urea linkage to form the aniline derivative, which may be further degraded or may become strongly bound to soil humic substances.

Diuron is extensively degraded to 3,4-dichloroaniline [23]. It is also demethylated to form 3-(3,4-dichlorophenyl)-1-methylurea and 3-(3,4-dichlorophenyl)urea [23]. During microbial degradation CO_2 is evolved. Diuron has a half-life of 62 d in red ferrallitic soil, while 3,4-dichloroaniline, the toxic metabolite, has a half-life of 35 d [3].

Unclassified Herbicides

Fosamine

Fosamine is rapidly degraded by soil microbes, yielding CO_2.

Glyphosate

Glyphosate is primarily biodegraded to CO_2. Factors influencing microbial activity such as pH, moisture, temperature, organic substrate availability and nutrient levels affect glyphosate degradation in soil. Moshier and Penner [59] reported that 40%, 9.5% and 3% of glyphosate was metabolized to CO_2 in 32 d in sandy loam, silty loam and loamy sand soils respectively. The addition of phosphate stimulated glyphosate degradation to a limited extent in silt loam but not in sandy loam soil. Additions of Fe^{+++} and Al^{+++} ions reduced degradation in sandy loam soil. The authors postulated that formation of colloidal Fe and Al precipitates in modified soils with concomitant adsorption of glyphosate was responsible for the decreased availability of glyphosate to microorganisms.

Rueppel et al. [73] found that 90% of glyphosate was dissipated in less than 12 wk. Microbial degradation resulted in the formation of aminomethylphosphonic acid, a significant soil metabolite of glyphosate, and this also underwent rapid degradation.

Decomposition of glyphosate by microorganisms increases the CO_2, nitrate and phosphoric acid contents of the soil [52].

Glufosinate

Glufosinate is rapidly degraded by microbes in soil or surface water to 3-methyl phosphinicopropionic acid and ultimately to CO_2.

CHEMICAL DEGRADATION AND BEHAVIOUR OF INDIVIDUAL HERBICIDES

Generally, chemical degradation is not considered to be of major importance in herbicide decomposition compared to microbial degradation. However, for certain herbicides such as dinitroanilines, diphenyl ethers, sulfonylureas, triazines, etc. chemical degradation constitutes a major pathway of decomposition. Chemical degradation may involve hydrolysis, oxidation, reduction, dealkylation, etc., and it follows first order kinetics, i.e., the rate of degradation is proportional to its concentration. It usually begins as soon as a herbicide is applied to soil without any lag phase, and the process continues at a steady state or declining rate.

Acetamides

Chemical degradation is negligible in the case of **acetochlor, dimethenamid,** and **metolachlor**. These herbicides have low leaching potential in moist soils and thus do not pose an acceptable risk to groundwater. **Napropamide** is rapidly leached in coarse- or fine-textured soils. Heavy rains immediately after napropamide application can cause crop injury and reduced weed control. Non-microbial degradation may contribute significantly to **pronamide** dissipation. It is readily cyclized to 2-(3,5-dichlorophenyl)-4,4-dimethyl-5-methyleneoxazoline, which is subsequently hydrolyzed to N-(1,1-dimethylacetonyl)-3,5-dichlorobenzamide. **Propanil** is chemically hydrolyzed to form 3,4-dichloroaniline (the major product) and 3,3',4,4'-tetrachlorobenzene. Plimmer et al. [67] reported that 1,3-bis(3,4-dichlorophenyl)triazene was a major product isolated from ^{14}C-ring-labelled propanil. They observed that this triazene was probably formed of an intermediate diazonium cation, which subsequently coupled with 3,4-dichloroaniline. They proposed that soil nitrite initially reacts with 3,4-dichloroaniline to form the intermediate diazonium cation.

Aliphatics

Chemical degradation is of minor importance in the dissipation of chlorinated aliphatics as they are readily subject to soil microbial degradation. As a result, no problems of residues occur in soils.

Arsenicals

Chemical degradation of arsenical herbicides **(DSMA, MSMA)** is mostly through transformations to form insoluble salts reacting with Ca, Fe, and Al present in soils. These insoluble salts are considerably less phytotoxic to plants. Woolson and Kearney [93] observed that arsenicals have a natural methylation-demethylation cycle during which alkylarsines are volatilized from the soil. They proposed that 12% of the arsenicals applied and present in the soil is lost through volatilization of alkylarsines each year. Hiltbold et al. [40] reported that 67%, 57% and 39% of MSMA were recovered after application over a 6-yr period to fine sandy loam, silt loam and loamy sand soils respectively. All of the arsenic recovered in the soils occurred in the plough layer, and the rate of MSMA movement through the surface horizon was faster in loamy sand and silt loam soils.

Benzothiadiazoles

Bentazon does not undergo chemical degradation. It does not leach below the plough layer. Bentazon residues decline to undetectable levels within 6 wk.

Benzoics

Chemical degradation is not considered to be the major pathway of metabolism of benzoic herbicides in soils. **Dicamba** is highly mobile in soils. It is readily leached down to deeper soil layers. It is adsorbed very little in most soils. Adsorption is directly related to soil pH. Generally, adsorption is strongest at pH 4.0 to 6.0 and minimum at pH higher than 6.0. Corbin and Upchurch [15] reported that herbicidal injury to plants was not observed at pH 5.3 and 6.5 at the end of an 8-week incubation period but was evident at pH 4.3 and 7.5. It was detoxified at pH 6.5 and not at 7.5. Its persistence in soils was intermediate compared with other herbicides.

Bipyridiliums

Bipyridilium herbicides (**paraquat, diquat**) do not undergo chemical degradation as they are tightly bound to various soil fractions. Most of the applied-paraquat is found in the upper 5 cm layer, but significant amounts reach the lower layers. The build-up of paraquat does not lead to any phytotoxic symptoms. Paraquat adsorption in a sandy loam soil is not significantly affected by soil pH, but is influenced by porosity, moisture content, residence time and adsorption capacity of the soil [81]. Staiff et al. [83] simulated decontamination treatment of paraquat and detected N,N'-dimethyl-1,1',2,2',3,3',6,6'-octahydro-4,4'-bipyridyl as the degradation product, with no further degradation.

Carbamates

Carbamate herbicides are not particularly persistent in soils. Besides microbial degradation, volatilization is an important factor in the dissipation of carbamate herbicides. Chemical degradation is considered to be of little importance. **Asulam** has medium to high mobility in soils, but leaching potential is very low because of rapid degradation. No appreciable leaching occurs in the case of **desmedipham** and **phenmedipham**. Their residues remain in the top 5 cm of soil.

Dinitroanilines

Chemical degradation of dinitroaniline herbicides under aerobic and anaerobic conditions is the major pathway of dissipation after their incorporation in the soil. The transformation products do not build up in the soil. **Benefin** is degraded to a major reduction product α,α,α-trifluoro-N-butyl-N-ethyl-5-nitrotoluene-3,4-diamine, with one nitro group replaced by an amino group. Most of the residues are accumulated in the top 15 cm soil layer. There is no accumulation of benefin even with repetitive applications to the soil. The persistence of **nitralin** is greater in acid than in neutral soil, possibly due to a physiochemical change in the adsorption-desorption complex of nitralin in soil rather than actual herbicidal degradation. The disappearance of nitralin is due more to microbial activity than to chemical decomposition.

Pendimethalin is rapidly degraded under anaerobic conditions. Aerobic biological degradation is slow. It is strongly bound to OM and clay. It is immobile. Most pendimethalin washed into surface water via sediment will remain bound to sediment and thus be unavailable to aquatic organisms.

Chemical degradation of **trifluralin** appears to be negligible. It, however, proceeds by reduction and N-dealkylation pathways. One nitro group of trifluralin is reduced to an amino group to form a major product α,α,α-N^4,N^4-dipropyl-5-nitrotoluene. With the gradual decline of this compound, there is a simultaneous rise in α,α,α-trifluoro-N^4,N^4-dipropyltoluidine-3,4,5-triamine and some extractable polar products [68]. Trifluralin and its residues are mostly confined to the upper 30 cm of soil, with about 80% being in the upper 15 cm [57]. Trifluralin persists as the depth of incorporation is increased. The retention of trifluralin in toxic form is directly related to the level of OM in the soil. Under high OM levels, vapour losses are apparently reduced and adsorption of the herbicide increased.

Diphenylethers

Diphenylether herbicides are generally strongly adsorbed by soils. They undergo no significant chemical degradation. They are immobile in soils but slightly mobile in extremely sandy soils.

Imidazolinones

The dissipation rates for **imazaquin** and **imazethapyr** are similar regardless of application method. Curran et al. [19] found that PPI application of imazaquin and imazethapyr had higher residue concentrations in the upper 10 cm than PRE treatments. Four months after application, 56% of imazaquin was found in PPI treatments while 35% of the herbicide was found in PRE treatments. Their residues below 10 cm soil depth were low regardless of application method and tillage. Volatilization of imidazolinone herbicides is minimal and not an important dissipation process. Goetz et al. [31] found that volatilization of imazethapyr from soil was less than 2%. Under normal conditions, evaporation from the soil surface may result in considerable movement of imazaquin [7].

Isoxazolidinones

Clomazone does not undergo chemical degradation. It has low mobility in moist soils but is moderately mobile in fine sandy soils. Most of the herbicide remains in the top 30 cm in sandy loam soils.

Phenols

Phenols are volatile. **Dinoseb** is apparently lost from soil by codistillation. An increase in temperature enhances volatilization losses of dinoseb. It is water soluble and gets readily leached in soil, but more rapidly in alkaline soils. Adsorption of dinoseb is pH-dependent. Greatest adsorption of dinoseb is on illite and montmorillonite and least adsorption on kaolinite at pH 4.6. The major pathway of dinoseb degradation is through soil microbes, as discussed earlier.

Phenoxyalkanoic Acids

Phenoxyacetics (2,4-D, MCPA) and Phenoxybutyrics (2,4-DB, MCPB) undergo very little chemical degradation in soils. 2,4-D is potentially mobile, but rapid degradation in soil and removal from soil by plant uptake minimizes leaching. About 95% of the soil-applied 2,4-D moves within 15 cm in clay loam soils, while it moves to deeper layers in sandy soils. Its field half-life is 10 d. In the case of 2,4-DB, the field half-life is 5 d for the acid, 7 d for the butoxyethyl ester and 10 d for the dimethylamine acid.

Chemical degradation plays a minor role in the dissipation of **aryloxyphenoxy-propionic** (APP) herbicides. However, the ester of **diclofop methyl** undergoes chemical hydrolysis at soil moisture levels above the wilting point for plants, yielding 4-(2,4-dichlorophenoxy)phenol as a metabolite. Hydrolysis proceeds more slowly at lower pH (5 %) than at higher pH levels. The pH range of 5.5-7.5 has no effect on subsequent degradation rate or metabolites formed. **Haloxyfop methyl** also undergoes hydrolysis at the ester bond, but it is much slower in air-dried soil than in moist soil.

The relatively inactive S-enantiomer of APPs may be activated by racemization reactions in the soil. After hydrolysis of **fluazifop butyl**, the S-enantiomer of the herbicide undergoes inversion with a half-life of 1 to 2 d. After 2 d of application of racemic mixture (R:S 1:1), the R:S ratio becomes 3:1, indicating that racemization reactions in soil favour the R form.

Volatilization has little effect on the dissipation of APPs. Of these herbicides, haloxyfop is more volatile. Generally, very little residues of these herbicides are found below a soil depth of 5 cm. The degradative products bind to the soil with limited metabolism to CO_2. Soil type may strongly influence the degradation rates of APPs. The half-lives of haloxyfop in clay loam, heavy clay and sandy loam soils are 92, 38 and 28 d respectively.

APPs are strongly adsorbed by soil and have very low solubility in water. As a result, very little of these herbicides enter groundwater, streams and other aquatic environments through leaching or run-off of soil-applied APPs, or APPs washed off foliage. However, diclofop, which is relatively more soluble in water, is an exception. Nevertheless, even with diclofop, less than 10% of a soil-applied herbicide is leached to depths greater than 10 cm by application of 10 cm water.

N-Phenylphthalimides

Flumiclorac undergoes rapid chemical hydrolysis, releasing CO_2. It does not leach. No residues are observed below 7.5 cm. Run-off of flumiclorac is very little due to rapid breakdown and strong adsorption by the soil. Field half-life is 1 to 5 d in moist soils.

Phenylpyridazines

Pyridate is rapidly hydrolyzed chemically to 3-phenyl-4-hydroxy-6-chloropyridazine, which is then biologically degraded. It is relatively immobile in soils.

Phthalamates

Naptalam is poorly adsorbed by clay and OM and hence leaches rapidly in coarse- and fine-textured soils. Heavy rains immediately after application can cause crop injury and reduced weed control. About 1 to 1.5 cm rainfall is sufficient to inactivate its preemergence activity. It is slowly degraded by soil microbes.

Pyridazinones

Leaching does not contribute greatly to the dissipation of pyridazinone herbicides. Schroeder and Banks [76] found that more than 60% of **norflurazon** remained within

8 cm of the soil surface, with less than 12% deeper than 15 cm, 100 d after its application at 3.4 k ha^{-1}. The greatest movement of norflurazon was found in a loamy sand soil that had the lowest cation exchange capacity. Hubbs and Lavy [42] reported that norflurazon adsorption increased and mobility decreased as soil OM and clay content increased. Pyrazon is dephenylated chemically to 5-amino-4-chloro-3-pyridazinone in soil.

Pyridinecarboxylic Acids

Chemical degradation is not considered a significant pathway of **picloram** dissipation. It has very little volatility losses. It is readily leached down to lower layers of the soil by heavy rainfall. Adsorption of picloram increases with decreasing pH and increasing OM in the soil. Movement and dissipation are negligible under low soil temperature conditions, but under high rainfall conditions picloram was leached into the 30-60 cm depth [43]. Dissipation increased with an increase in temperature over 27° C and decreased with increasing depth in the soil profile. The leaching potential of **clopyralid** and **triclopyr** is low to moderate.

Pyridines

Chemical degradation of **dithiopyr** and **thiazopyr** is negligible. Neither of these herbicides and their metabolites (of microbial degradation) leach below 61 cm. Potential for movement in run-off water is unlikely due to low water solubility and strong adsorption by soil.

Pyridinones

Fluridone is strongly adsorbed by soil organic matter. It is more strongly adsorbed by sediments, but gradually desorbs into water, and then is subjected to photodegradation.

Sulfonylureas

Chemical hydrolysis plays an important role in the degradation of sulfonylurea herbicides, particularly in soils with lower pH. The major hydrolysis is via ring cleavage of the sulfonylurea bridge to give the corresponding sulfonamide and heterocyclic amine. This cleavage at the sulfonylurea bridge of **triasulfuron** yields 2-(2-chloroethoxy) benzenesulfonamide as the major metabolite. This cleavage is pH-sensitive.`With an increase in pH, the rate of hydrolysis of **chlorsulfuron** is six times faster at a soil pH of 5.7 than at a pH of 7.5 [9]. The half-life of chlorsulfuron in sterilized soil at 30° C increases from 3 to 15 wk (i.e., 15-fold) as the soil pH increases from 5.9 to 8.0 [47]. Degradation of chlorsulfuron proceeded slowly as the pH of soil increased from 6.0 to 7.2.

Soil moisture and temperature have considerable influence on the dissipation of sulfonylurea herbicides. The persistence of chlorsulfuron decreased when soil temperature was increased from 20° C to 40° C. Increasing soil water content increased degradation of chlorsulfuron in a sandy loam soil, with little or no effect in a loam soil. Zimdahl et al. [96] reported that decreasing the soil moisture at a constant temperature resulted in a slower rate of chlorsulfuron degradation. When the soil moisture was held constant, decreasing the temperature had a similar effect. The half-life of chlorsulfuron increased by 3.7-fold (from 8.9 to 32.7 weeks) as the temperature was reduced from 40° C to 10° C.

Generally, the rate of degradation of sulfonylurea herbicides in soil is fastest in warm, moist, light-textured and low pH soils, and slowest in cold, dry, heavy and high pH soils. Under normal field conditions, their degradation proceeds at rates that are similar to, and often faster than conventional soil-active herbicides. The half-life of sulfonylurea herbicides is usually between 1 and 2 mon. Despite their relatively rapid dissipation, crop injury could result after the first season of application with a herbicide like chlorsulfuron. This injury, however, depends on the sensitivity of crops even to the extremely low herbicide residues in soil and the herbicide used.

The mobility of sulfonylurea herbicides in soil increases with increasing soil pH and decreasing soil OM. Mobility under field conditions is a function of soil characteristics such as OM content, pH, soil type, porosity, etc. Sulfonylurea herbicides can move vertically upwards with rising capillary water under semi-dry and dry soil conditions. Thus, during periods of net upward flow of soil water, the herbicide may enter the root zone from deeper in the soil profile where it had penetrated during earlier periods of net downward water flow. In spite of their relatively high mobilities, sulfonylurea herbicides are not expected to pose groundwater contamination problems because of their exceptionally low use rates, low toxicities and relatively rapid soil degradation/ dissipation characteristics.

Thiocarbamates

Volatilization is an important mechanism for the loss of thiocarbamate herbicides from soils. **Pebulate** and **vernolate** are less volatile than **EPTC**. As microbial degradation is more predominant in soils, chemical decomposition of thiocarbamate herbicides is almost non-existent. **Thiobencarb** is rapidly adsorbed by soil colloids and is classified as having little mobility. Most of the applied thiobencarb remains in the top 1 cm of a clay loam soil. Its adsorption by soil correlates with soil organic carbon. The half-life of thiobencarb is 2 to 3 wk, indicating that degradation is an important factor in thiobencarb activity. Thiobencarb degradation at 25° C or 35° C is more rapid under moist conditions than dry conditions. The degradation products include 4-chlorobenzoic acid, 4-chloro-benzylmethyl sulphone and desethyl thiobencarb, and the percentages of these metabolites vary with organic matter content and temperature [13]. Braverman et al. [13] reported that during the degradation and mineralization of ^{14}C-thiobencarb, characteristics of chemical structure of metabolites may have caused differences in adsorption and extraction efficiencies from the muck soils containing high OM content. The degradation of thiobencarb to metabolites without evolution of CO_2 indicated that thiobencarb may be cometabolized. It is proposed that following ring hydroxylation of thiobencab, conjugate products may form rather than being mineralized to CO_2.

Triazines

The formation of hydroxy derivatives of triazines as major degradation products in sterilized soil indicated the participation of non-biological mechanism of triazine decomposition. Chemical hydrolysis, the important pathway of atrazine degradation of atrazine degradation to hydroxyatrazine, followed first-order kinetics in soil-free, sterilized soil and perfusion systems [4]. An increased rate of **atrazine** hydrolysis was consistent with the effect of pH on hydrolysis. Soil pH and OM content largely controlled the rate of atrazine hydrolysis; in soils of similar pH, atrazine degradation rate increased with increased atrazine adsorption. The concentration of H ions around soil particles would create a zone of lower pH, resulting in more rapid hydrolysis. Acid hydrolysis may result from protonation of a ring or chain N atom, followed by cleavage of the C-Cl bond by H_2O. Protonation of N would increase the electron efficiency of C bonded to Cl and increase the tendency for nucleophilic displacement of Cl by H_2O. Hydroxyatrazine, like atrazine, was resistant to microbial degradation [4].

Similar hyrdoxy derivatives are also formed in the case of **simazine, propazine, ametryn** and **cyanazine** in soils. Atrazine and simazine were degraded in soils at a slower rate than cyanazine [10]. Atrazine and simazine were hydrolyzed at the chlorine atom to form hydroxyl derivatives, while in the case of cyanazine hydrolysis of the nitrile group occurred more rapidly than that of chlorine atoms [10]. Sirons et al. [79] reported that hydrolysis of the nitrile group in the case of cyanazine preceded microbial degradation. The cyanomethylethyl group with its hydrolysis products in the 4-position became the primary target for microbial degradation in cyanazine. These authors found that cyanazine was changed to deisopropylated atrazine (2-chloro-4-amino-6-ethylamino-s-triazine), a major metabolite. They proposed that in the case of cyanazine, chemical hydrolysis precedes microbiological degradation to the deisopropylated atrazine.

These reports suggest that hydrolysis of triazines to form hydroxytriazines is a chemical process, and it precedes dealkylation and other subsequent reactions catalyzed by microbes, as discussed earlier.

Triazines are relatively more persistent in soils. Chloro-s-triazines either complex with some soil constituents or form compounds that are more difficult to extract. Higher content of adsorptive clays in the soil protects triazines from degradation and prolongs their persistence. Khan and Saidak [50] showed the presence of residues of the parent and its monodealkylated hydroxy derivative following application of atrazine for 20 consecutive years at rates ranging from 1.40 to 2.24 kg ha^{-1}.

Triazinones (*AS*-Triazines)

Metribuzin undergoes chemical hydrolysis. The primary degradation product of metribuzin is 6-(1,1-dimethylethyl)-1,2,4-triazine-3,5-(2*H*,4*H*)-dione. Metribuzin is readily leached in sandy soils of low OM content. It has intermediate leaching potential on medium-texture soils and is immobile in heavy soils of high OM. Higher pH increases its mobility. Metribuzin can move with soil run-off and is phytotoxic to algae and other phytotoxic plants. However, rapid hydrolysis in water greatly limits the impact of metribuzin in aquatic environments.

Triazoles

Triazoles disappear readily from soils but this does not appear to be due to microbial degradation. Most of the **amitrole** degradation occurring in soils proceeds by non-biological reactions [49]. It is degraded by an oxidative mechanism involving an attack on the triazole nucleus by (OH) or other free radicals. That amitrole degradation follows first-order kinetics suggests the involvement of chemical reaction. Plimmer et al. [66] reported that degradation of amitrole proceeded through oxidation, resulting in ring cleavage, loss of CO_2 and production of urea, cyanamid and possibly molecular nitrogen.

Uracils

The importance of chemical degradation of substituted uracils in soils has not been established. Microbial degradation is an important pathway in the dissipation of uracils.

Mobility of **bromacil** and **terbacil** is considered to be moderate. They are less tightly adsorbed by soil particles than other herbicides.

Ureas

Chemical degradation is of relatively lower importance compared to microbial degradation. Chemical degradation, however, involves N-demethylation, N-demethoxylation or oxidation of ring substituents, followed by a second phase where the first-phase metabolites are further degraded to anilines.

Ureas are very persistent in soils. Their persistence is dependent on OM content, clay content and cation exchange capacity (CEC) of the soil. Of these, OM appears to contribute more to toxicity reduction than clay or CEC. The average field half-life of diuron is 90 d. Most of the **diuron** and **monuron** residues are located near the soil surface. Of the total residue present in the surface 30 cm, 62 to 89% was found in the upper 5 cm of the soil, and 86 to 100% in the upper 10 cm [21]. The higher the rate of each herbicide, the greater the penetration to deeper layers. Inactivation of **fluometuron** is more rapid in silty clay loam/silty loam soil with 7.2% OM.

Unclassified Herbicides

Fosamine is so rapidly degraded by soil microbes that there is little downward movement of the parent compound or its degradation products in the soil.

Chemical degradation of **glufosinate** is negligible. It is highly mobile in soils. Despite high leaching potential, glufosinate may be found no deeper than 15 cm in field, presumably because of rapid microbial degradation.

Chemical degradation of **glyphosate** is negligible in soils but it could undergo decomposition to ultimately yield CO_2 and ammonia. During this process, a CH_2CO_2 fragment is lost to form an intermediate product, aminomethylphosphonic acid, which then reacts further to lose ammonia and CO_2. The rate of degradation varies considerably in different soils due to differences in binding capacity of the soil for the herbicide. Binding is greatest in soils with the highest phosphate fixation, the lowest pH and the highest OM content. Glyphosate is rapidly inactivated in organic and mineral soils but not by washed quartz sand [82]. Fe^{+++}-and Al^{+++}-saturated clays and OM adsorb more glyphosate than Na^+-and Ca^+-saturated clays and OM. It was proposed that glyphosate may be binding to the soil similar to phosphate, possibly through the phosphonic acid moiety [82].

REFERENCES

1. Acher, A.J. and S. Saltzman. 1980. Dye-sensitized photooxidation of bromacil in water. J. Environ. Quality **9**: 190-194.
2. Adams, R.S., Jr. 1973. Factors influencing soil adsorption and bioactivity of herbicides. *In* F.A. Gunther (ed.). Residue Reviews, Springer-Verlag, New York, 198 pp.
3. Alfonso, H.M.M. 1980. Persistence of diuron and its metabolite, 3,4-dichloroaniline, in two types of Cuban soil. Ciencias de la Agriculture **5**: 125-133.
4. Armstrong, D.E., G. Chesters and R.F. Harris. 1967. Atrazine hydrolysis in soil. Proc. Soil Soc. Amer. **31**: 61-66.
5. Bailey, G.W. and J.L. White. 1964. Review of adsorption and desorption of organic pesticides by soil colloids with implications concerning pesticide bioactivity. J. Agr. Food Chem. **12**: 324-332.
6. Bardsley, C.E., K.E. Savage and V.O. Childers. 1967. Trifluralin behaviour in soil. I. Toxicity and persistence as related to organic matter. Agron. J. **59**: 159-160.

7. Basham, G.W., T.L. Lavy, L.R. Oliver and H.D. Scott. 1987. Imazaquin persistence and mobility in three Arkansas soils. Weed Sci. **35**: 576-582.

8. Baumann, P.A. and M.G. Merkle. 1979. The effects of soil moisture on the phytotoxicity of diuron, fluridone, and trifluralin. Proc. Meeting South Weed Sci. Soc. Amer., pp. 315.

9. Beyer, E.M., Jr., M.J. Duffy, J.V. Hay and D.D. Schlueter. 1988. Sulfonylureas. In P.C. Kearney and D.D. Kaufman (eds.). Herbicides: Chemistry, Degradation, and Mode of Action. Merecl Dekker, New York, pp. 117-183.

10. Beynon, K.I., G. Stoydin and A.N. Wright. 1972. A comparison of the breakdown of the triazine herbicides cyanazine, atrazine, and simazine in soils and in maize. Pest. Biochem. Physiol. **2**: 153-161.

11. Boyd, S.A., J. Xiangcan and J.F. Lee. 1990. Sorption of nonionic organic compounds by corn residues from a no-tillage field. J. Environ. Qual. **19**: 734-738.

12. Bozarth, G.A. and H.H. Fundrburk, Jr. 1971. Degradation of fluometuron in sandy loam soil. Weed Sci. **19**: 691-695.

13. Braverman, M.P., J.A. Dusky, S.J. Locasto and A.G. Hornsby. 1990. Sorption and degradation of thiobencarb in three Florida soils. Weed Sci. **38**:583-588.

14. Chen, Y.L. and C.C. Chen. 1977. Photodecomposition and some behaviour of herbicide butachlor in soils. Proc. 6th Conf. Asian-Pacific Weed Sci. Soc., pp. 570-578.

15. Corbin, F.T. and R.P. Upchurch. 1967. Influence of pH on detoxification of herbicides in soil., Weeds **15**: 370-377.

16. Corbin, F.T., R.P. Upchurch and F.L. Selmen. 1971. Influence of pH on the phytotoxicity of herbicides in soil. Weed Sci. **19**: 233-239.

17. Coultas, J.S. and R.G. Harvey. 1979. Adsorption and leaching of buthidazole and metribuzin in Wisconsin soils. Abstr. Weed Sci. Soc. Amer., pp. 124-125.

18. Crosby, D.G. 1975. Herbicide photodecomposition. In P.C. Kearney and D.D. Kaufman (eds.). Herbicides: Chemistry, Degradation, and Mode of Action. Mercel Dekker, New York, pp. 835-890.

19. Curran, W.S., M.M. Loux, R.A. Liebl and F. William Simmons. 1992. Photolysis of imidazolinone herbicides in aqueous solution and on soil. Weed Sci. **40**: 143-148.

20. Dao, T.H. 1991. Field decay of wheat straw and its effects on metribuzin and S-ethyl metribuzin sorption and elution from crop residues. J. Environ. Sci. Technol. **20**: 203-208.

21. Dawson, J.W., V.F. Bruns and W.J. Clore. 1968. Residual monuron, diuron, and simazine in a vineyard soil. Weed Sci. **16**: 63-65.

22. Duseja, D.R. 1981. Soil dissipation of three herbicides. Proc. 8th Asian-Pacific Weed Sci. Soc. Conf., pp. 487-490.

23. Ellis, P.A., N.D. Camper and J.M. Shively. 1979. Degradation of the herbicide diuron by pond water microorganisms. Proc. Meeting South. Weed Sci. Soc. Amer., p. 318.

24. Flint, J.L. and W.W. Witt. 1997. Mircobial degradation of imazaquin and imazethapyr. Weed Sci. **45**: 586-591.

25. Foy, C.L. 1975. The chlorinated aliphatic acids. In P.C. Kearney and D.D. Kaufman (eds.). Herbicides: Chemistry, Degradation, and Mode of Action. Mercel Dekker, New York, pp. 399-452.

26. Fryer, J.D. and S.A. Evans (eds.). 1968. Weed Control Handbook. Vol. I. Principles. The British Crop Protection Council. Blackwell Sci. Public., Oxford and Edinburg, 494 pp.

27. Funderburk, H.H., Jr., and G.A. Bozarth. 1967. J. Agr. Food Chem. **15**: 563-568.

28. Gaston, L.A., M.A. Locke, S.C. Wagner, R.M. Zablotowicz and K.N. Reddy. 1996. Sorption of bentazon and degradation products in two Mississippi soils. Weed Sci. **44**: 678-682.

29. Geissbuhler, H., C. Haselbach, H. Aebi and L. Ebner. 1963. The fate of N-(chlorophenoxy)-phenyl-N,N-dimethylurea (C-1983) in soils. 3. Breakdown in soils and plants. Weed Res. **3**: 277-297.

30. Giardina, M.C., M.T. Giardi and G. Filacchioni. 1980. 4-amino-2-chloro-1,3,5-triazine: a new metabolite of atrazine by a soil bacterium. Agri. Biol. Chem. **44**: 2067-2072.

31. Goetz, A.J., T.L. Lavy and E.E. Gbur, Jr. 1990. Degradation and field persistence of imazethapyr. Weed Sci. **38**: 421-428.

32. Gray, R.A. and A.J. Weierich. 1965. Factors affecting the vapor loss of EPTC from soils. Weeds **13**: 141-147.

33. Gray, R.A. and A.J. Weierich. 1968. Leaching of five thiocarbamate herbicides in soils. Weed Sci. **16**: 77-79.

34. Green, R.E. and S.R. Obien. 1969. Herbicide equilibrium in relation to soil water content. Weed Sci. **18**: 514-519.

35. Grover, R. 1968. Influence of soil properties on the phytotoxicity of 4-amino-3,5,6-trichloro picolinic acid (picloram). Weed Res. **8**: 226-232.

36. Hance, R.J. 1965. The adsorption of urea and some of its derivatives by a variety of soils. Weed Res. **5**: 98-107.

37. Harper, L.SA., A.W. White, Jr., R.R. Bruce, A.W. Thomas and R.A. Leonard. 1976. Soil and microclimate effects on trifluralin volatilization. J. Environ. Qual. **5**: 236-242.

38. Harper, S.S. 1994. Sorption-desorption and herbicide behavior in soil. Rev. Weed Sci. **6**: 207-225.

39. Harris, C.I. and G.F. Warren. 1964. Adsorption and desorption of herbicides by soil. Weeds **12**: 120-126.

40. Hiltbold, A.E., B.F. Hajck and G.A. Buchanan. 1974. Distribution of arsenic in soil profiles after repeated application of MSMA. Weed Sci. **22**: 272-275.

41. Hollingsworth, E.B. 1980. Volatility of trifluralin from field soil. Weed Sci. **28**: 224-228.

42. Hubbs, C.W. and T.L. Lavy. 1990. Dissipation of norflurazon and other persistent herbicides in soil. Weed Sci. **38**: 81-88.

43. Hunter, J.H. and E.H. Stobbe. 1972. Movement and persistence of picloram in soil. Weed Sci. **20**: 486-489.

44. Jaques, G.L. and R.G. Harvey. 1979. Dinitroaniline herbicide phytotoxicity as influenced by soil moisture and herbicide vaporization. Weed Sci. **27**: 536-539.

45. Jordan, L.S., J.D. Mann and B.E. Day. 1965. Effects of ultraviolet light on herbicides. Weeds **13**: 43-46.

46. Jordan, L.S., C.W. Coggins, Jr., B.E. Day and W.A. Clerx. 1964. Photodecomposition of substituted phenylureas. Weeds **12**: 1-6.

47. Joshi, M.M., H.M. Brown and J.A. Romesser. 1985. Degradation of chlorsulfuron by soil microorganisms. Weed Sci. **33**: 888-893.

48. Karamanos, A.J. 1981. The development of water deficits in plants. *In* G.M. Simpson (ed.). Water Stress in Plants. Praeger Publishers, Eastbourne, England, pp. 34-88.

49. Kaufman, D.D., J.R. Plimmer, P.C. Kearney, J. Blake and F.S. Guardia. 1968. Weed Sci. **16**: 226-230.

50. Khan, S.U. and W.J. Saidak. 1981. Residues of atrazine and its metabolites after prolonged usage. Weed Res. **21**: 9-12.

51. Koren, E., C.L. Foe and F.M. Ashton. 1967. Adsorption, leaching, and lateral diffusion of four thiocarbamate herbicides in soils. Abstr. Weed Sci. Soc. Amer., pp. 72.

52. Kruglov, Yu.V., N.B. Gersh and M. Shtal'Berg, M.V. 1980. The influence of glyphosate on the soil microflora. Khimiya v Sel'skom Khozyaistve **18(10)**: 52-54.

53. Laanio, T.L., P.C. Kearney and D.D. Kaufman. 1973. Pest. Biochem. Physiol. **3**: 271.

54. Lawrence, E.G, H.D. Skipper, D.T. Gooden, J.P. Zublena and J.E. Strubble. 1990. Persistence of carbamothioate herbicides in soils pretreated with butylate. Weed Sci. **38**: 194-197.

55. Letis, E. and D.G. Crosby. 1974. Photodecomposition of trifluralin. J. Agr. Food Chem. **22**: 842-848.

56. Marco, A.C.De and E.R. Hayes. 1979. Photodegradation of thiocarbamate herbicides. Chemosphere **8**: 321-326.

57. Miller, J.H., P.C. Keeley, C.H. Carter and R.J. Thullen. 1975. Soil persistence of trifluralin, benefin, and nitralin. Weed Sci. **23**: 211-214.

58. Mortland, D.M. and W.F. Meggitt. 1966. Interaction of ethyl-N,N-di-n-propylthiocarbamate (EPTC) with montmorillonite. J. Agr. Food Chem. **14**: 126-129.

59. Moshier, L.J. and D. Penner. 1978. Factors influencin$_g$ microbial degradation of ^{14}C-glyphosate to ^{14}CO$_2$ in soil. Weed Sci. **26**: 686-691.

60. Moyer, J.R. 1987. Effect of soil moisture on the efficiency and selectivity of soil-applied herbicides. Rev. Weed Sci. **3**: 19-34.

61. Murphy, G.P. and D.R. Shaw. 1997. Field mobility of flumetsulam in three Mississippi soils. Weed Sci. **45**: 564-567.

62. Nye, P.H. and P.B. Tinker. 1977. Studies in ecology. Vol. 4. Solute movement in the soil-root system. Univ. Calif. Press, Berkeley and Los Angeles, USA, 342 pp.

63. Parochetti, J.V. 1973. Soil organic matter effect on activity of acetanilides, CDAA and atrazine. Weed Sci. **21**: 157-160.

64. Parochetti, J.V. and E.R. Hein. 1973. Volatility and photodecomposition of trifluralin, benefin, and nitralin. Weed Sci. **21**: 469-473.

65. Paulo, E.M., L.H. Signori and R. Deuber. 1979. Leaching of metribuzin, oxadiazon, and bromacil in two soil types. Planta Daninha **2**: 111-115.

66. Plimmer, J.R. 1967. Photolysis of amiben. Abstr. Weed Sci. Soc. Amer., p. 76.

67. Plimmer, J.R. 1970. The photochemistry of halogenated herbicides. In F.A. Gunther (ed.). Residue Reviews, Springer-Verlag, New York, pp. 47-74.

68. Probst, G.W., T. Golab, R.J. Herberg, F.J. Holzer, S.J. Parka, C. Van der Scans and J.B. Tepe. 1967. Fate of trifluralin in soils and plants. J. Agr. Food Chem. 15:592-599.

69. Rahman, A. and L.J. Matthews. 1979. Effect of soil organic matter on the phototoxicity of thirteen s-triazine herbicides. Weed Sci. **27**: 158-161.

70. Ray, B. 1975. Fate of organic chemicals in soils and plants. Interntl. Pest Control. Jan/Feb. 1975, pp. 9-14.

71. Redemann, R.T., R.W. Meikle, P. Hamilton, V.S. Banks and C.R. Youngson. 1968. The fate of 4-amino-3,5,6-trichloro-picolinic acid in spring wheat and soil. Bull. Environ. Contamin. Toxicol. **3**: 80-96.

72. Rodgers, E.G. '968. Leaching of seven triazines. Weed Sci. **16**: 117-120

73. Rueppel, M.L., B.B. Brightwell, J. Schaefer and J.T. Marvel. 1977. Metabolism and degradation of glyphosate in soil and water. J. Agr. Food Chem. **25**: 517-527.

74. Savage, K.E. 1973. Nitralin and trifluralin persistence in soil. Weed Sci. **21**: 285-288.

75. Schnappinger, M.G., C.P. Trap, J.M. Boyd and S.W. Pruss. 1977. Soil pH and atrazine activity in no-tillage corn as affected by nitrogen and lime applications. Proc. Northeast Weed Sci. Soc. Amer. **13**: 116.

76. Schroeder, J. and P.A. Banks. 1986. Persistence of norflurazon in five Georgia soils. Weed Sci. **34**: 595-599.

77. Scott, D.C. and J.B. Weber. 1967. Herbicide phytotoxicity as influenced by adsorption. Soil Sci. **104**: 151-158.

78. Senesi, N. and C. Testini. 1984. Theoretical aspects and experimental evidence of the capacity of humic substances to bind herbicides by charge-transfer mechanisms (electron donor-acceptor process). Chemosphere. **13**: 461-468.

79. Sirons, G.J., R. Frank and T. Sawyer. 1973. Residues of atrazine, cyanazine, and their phytotoxic metabolites in a clay loam soil. J. Agr. Food Chem. **21**: 1016-1020.

80. Skipper, H.D. and V.V. Volk. 1972. Biological and chemical degradation of atrazine in three organic soils. Weed Sci. **20**: 344-347.

81. Smith, A.E. and C.I. Mayfield. 1978. Paraquat: determination, degradation, and mobility in soil. Water, Air, and Soil Pollution **9**: 439-452.

82. Sprankle, P. , W.F. Meggitt and D. Penner. 1975. Adsorption, mobility, and microbial degradation of glyphosate in the soil. Weed Sci. **23**: 229-234.

83. Staiff, D.C., L.C. Butler and J.E. Davis. 1981. A field study of the chemical degradation of paraquat dichloride following simulated spillage on soil. Bull. Environ. Contam. Toxicol. **26**: 16-21.

84. Sullivan, R.G., H.W. Knoche and J.C. Merkle. 1980. Photolysis of trifluralin: characterization of azobenzene photodegradation products. J. Agr. Food Chem. **28**: 746-755.

85. Swann, C.W. and R. Behrens. 1972. Phytotoxicity of trifluralin vapors from soil. Weed Sci. **20**: 143-146.

86. Talbert, R.E. and O.H. Fletchall. 1965. The adsorption of some s-triazines in soils. Weeds **13**: 46-52.

87. Upchurch, R.P. and D.D. Mason. 1962. The influence of soil organic matter on the phytotoxicity of herbicides. Weeds 10: 9-14.
88. Van Bladel, R. and A. Moreale. 1974. Adsorption of fenuron and monuron (substituted ureas) by two montmorillonite clays. Proc. Soil Sci. Soc. Amer. 38: 244-249.
89. Wallonofer, P.R., S. Safe and O. Hutzinger. 1973. Microbial demethylation and debutynylation of four phenylurea herbicides. Pest. Biochem. Physiol. 3: 253-258.
90. Warren, G.F. 1973. Action of herbicides in soilà. affected by organic matter. Weeds Today 4(2): 10-11.
91. Weber, J.B. and D.C. Scott. 1966. Availability of a cationic herbicide adsorbed on clay minerals to cucumber seedlings. Science 152: 1400.
92. ·Wildung, R.E., G. Chesters and D.E. Armstrong. 1968. Chloramben (amiben) degradation in soil. Weed Res. 8: 213-215.
93. Woolson, E.A. and P.C. Kearney. 1973. Behaviour of arsenic herbicides in soil. Abstr. Weed Sci. Soc. Amer. pp. 78.
94. Yamasue, Y. and A.D. Worsham. 1980. Effects of benefin vaporizing from soils on tobacco (*Nicotiana tabacum*) foliage. Weed Sci. 28: 306-311.
95. Yuen, Q.H. and H.W. Hilton. 1962. The adsorption of monuron and diuron by Hawaiian sugarcane soils. J. Agr. Food Chem. 10: 386-392.
96. Zimdahl, R.L., K. Thirunarayanan and D.E. Smika. 1985. Chlorsulfuron adsorption and degradation in soil. Weed Sci. 33: 558-563.

10

Herbicide Resistance and Genetic Engineering

RESISTANCE

The intensive and continuous use of herbicides over the last five decades has resulted in the development and evolution of weeds resistant to the normally phytotoxic chemicals. Since the first reports of triazine resistance in the mid-1960s, over 150 weed species have been identified as having developed biotypes resistant to at least one or more herbicide classes. Resistance to at least 15 classes of herbicides have been noted [82, 153]. Moreover, the area of land infested with resistant weeds is increasing rapidly. Reports of evolution of newly resistant biotypes continue to emanate from all over the world, including developing countries, e.g., India [90], where extensive use of herbicides in grain crops began only in the late 1970s. These reports indicate that herbicide resistance problems are accelerating and, consequently, management of weeds is becoming increasingly more difficult and complex.

The phenomenon of herbicide resistance in plants, especially in weeds, is analogous to the evolution of changes that arise in insects and pathogens in response to the continuous use of insecticides and fungicides. Herbicide resistance is the result of selection for traits that allow weed species to survive specific herbicide usage, which would otherwise cause mortality. Herbicide resistance contributes a trait of resistance within the plant species to a herbicide and resistant weeds are those plant species that express the genetic variation required to evolve mechanisms to escape control. Like other crop pests (insects, pathogens, etc.) with reference to pesticides (insecticides, fungicides, etc.), herbicide-resistant weeds are the result of intensive selection pressure in weed populations.

Generally, **resistance** is defined as the ability of a plant species to withstand the phytotoxicity of a chemical. This is the trait normally found in crop plants, as against weeds, thus forming the basis for herbicide selectivity. Crop resistance to a herbicide is often different from the one that exists in the phenomenon of herbicide resistance in weeds.

Herbicide resistance in the context of weeds is defined as a characteristic of the weed species to withstand a herbicide dosage substantially higher than the wild type of the same plant species can withstand. It is the inherent or acquired ability to withstand the dosage of herbicide normally used for satisfactory weed control. Resistant weed populations are unharmed by a concentration of herbicide that completely kills all unselected populations. This herbicide concentration is the one that in field practice produces an effective kill.

Herbicide susceptibility represents the other end of the scale, encompassing natural variability in sensitivity to herbicides within the unselected populations. **Tolerance**, on the other hand, indicates reduced susceptibility (or conversely enhanced resistance) which may sometimes result from selection by herbicides [128]. It is a response rather less than 100% survival accompanied by zero reduction in growth at herbicide concentrations, which normally give 100% kill of susceptible biotypes. Tolerance may also be considered the natural or normal variability of response to herbicides that exists within a species and can easily and quickly evolve.

There is no clear distinction between **highly tolerant** and **resistant biotypes**, and low levels of tolerance usually merge into susceptibility. The word **resistance** used in this book encompasses both tolerance and resistance to herbicides in plant species and biotypes.

DEVELOPMENT OF HERBICIDE RESISTANCE

Development of herbicide resistance in weeds is an evolutionary process. There are two pre-requisites for the evolution of herbicide resistance in plant populations: a) the occurrence of heritable variation in genetic composition for herbicide resistance and b) natural selection for increased resistance to herbicides.

In response to repeated treatment with a particular herbicide or class (family) of herbicides, weed populations change in genetic composition such that the frequency of resistance alleles and resistant individuals increase [73]. In this way, weed populations become adapted to the intense selection pressure imposed by herbicides. The evolution of resistance under continuous application of a herbicide may be considered as an example of recurrent selection in which there is a progressive and sometimes rapid shift in average fitness of populations of weeds exposed to a herbicide. This shift in fitness, a genetic trait, is directly related to an increase in frequency of the resistance trait (phenotype) in the population.

The selection pressure for herbicide resistance is contributed by three factors: a) efficiency of the herbicide, b) frequency of herbicide use and c) duration of herbicide effect. Selection intensity in response to herbicide application is a measure of the relative mortality in target weed populations and/or the relative reduction in seed production of survivors; this will be proportional, in some manner, to herbicide dose [100]. The duration of selection is a measure of the period of time over which phytotoxicity is imposed by a herbicide. Both intensity and duration will interact to give seasonal variation in the process of selection, which will in turn depend upon the phenology and growth of a weed species. For example, with preemergence application of a herbicide that inhibits seedling emergence over a time period, the intensity of selection may be much higher on weed seedlings emerging early in the life of a crop than those seedlings emerging later.

The occurrence and speed of evolution of herbicide resistance are determined by several factors. These include:
   a) Number of alleles involved in the expression of functional resistance.
   b) Frequency of resistance alleles in natural (unselected) populations of the weed species.
   c) Mode of inheritance of the resistant alleles.
   d) Reproductive and breeding characters of the weed species.
   e) Longevity of weed seeds in the soil.
   f) Intensity of selection which differentiates resistant biotypes from susceptible ones.
   g) Relative fitness of resistance and susceptible genotypes.

Genetic Variation and Mutation

The mechanisms which interrupt the transport of herbicides to biochemical sites of action, reduce the sensitivity of target sites and detoxify the chemical or enhance repair can potentially confer resistance. The specific biochemical mechanisms include:

a) sequestration of the herbicide in the apoplast,
b) modification of cell membrane function and structure,
c) changes in sensitivity of the key target enzyme,
d) enhanced production of the herbicide target,
e) enhanced metabolic breakdown and conjugation of the herbicide and
f) enhanced degradation of herbicide-generated toxic products.

These mechanisms and consequently the expression of resistance, are controlled by genetic loci. Development of herbicide resistance is dependent on the extent of genetic variation and frequency of occurrence of mutation.

Genetic variation for resistance must be present in a susceptible population for the evolution of herbicide resistance to occur. There are two ways in which resistance traits may arise within a weed population. A major resistant gene, or genes, may be present at low frequency so that selection acts to change a population, which is initially susceptible [100]. Alternately, recurrent selection may occur continuously to achieve a progressive increase in average resistance from generation to generation, with changes in gene frequency at many loci conferring resistance. Thus, herbicide resistance is developed by gene mutations or conferred by pre-existing genes.

In general, gene mutations conferring resistance to a herbicide class are not induced by application of the herbicide, but rather occur spontaneously. Spontaneous mutations at gene loci recur with characteristic frequency such that new mutations are continuously generated in natural populations of weeds. Mutations at some loci, particularly those encoding herbicide site of action, may confer resistance. Typical spontaneous mutation rates in biological organisms are often cited as 1×10^{-5} or 1×10^{-6} gametes per locus per generation [104]. These values are for single, nuclear gene inheritance of resistance related to herbicide resistance evolution. At a rate of mutation to a single, dominant resistant allele of 1×10^{-6}, the probability of occurrence of at least one resistant plant in a 30 ha field with a weed density equal to, or exceeding, five plants m^2 is greater than 0.95 for a random mating species [73]. Thus, at least one resistant plant is almost certain to occur in a weed population of this size despite the low rate of mutation to resistance. Factors such as seed inviability and seedling mortality may cause the plant's mortality prior to reproduction [73]. However, should it survive and reproduce and if the corresponding herbicide or herbicide class is applied for several generations, the single resistant plant could give rise to a predominantly resistant population of weeds [73].

If the mutation rates are lower than 1×10^{-6}, i.e., 1×10^{-8} to 1×10^{-10}, the probability of occurrence of a resistant mutant is markedly reduced. It requires densities greater than 50 plants per m^2 for a resistant mutant to occur if the mutation rate is 1×10^{-8} [73]. The positive correlation between the size of a susceptible weed population and probability of occurrence of resistant mutant plants may partly explain why herbicide-resistance has evolved in some species but not in others. The probability of herbicide-resistant plants occurring in a population is greater for weeds with high densities than for those which occur at low densities.

The frequency of mutation may be influenced by environment and dosage of a herbicide. For example, atrazine applied at sublethal doses to certain susceptible genotypes of *Chenopodium album* resulted in a progeny with triazine-resistant characteristics similar to highly resistant plants [11]. This indicates that herbicide resistance could be induced by low doses of the chemical in certain genotypes.

Most herbicides are extensively screened for efficacy in the field before release for commercial use. Herbicides are not accepted for market without at least 85-90% control of the target weeds. Therefore, it is very likely that resistance traits are present, if undetectable, in weed populations before large-scale selection with herbicides.

Inheritance of Herbicide Resistance

There are two modes of inheritance of herbicide resistance: **nuclear inheritance** and **cytoplasmic inheritance**. Resistance to most classes of herbicides is caused by nuclear inheritance. These include bipyridiliums, benzoics, aryloxyphenoxypropionics, sulfonylureas, substituted ureas, etc. In nuclear inheritance, the resistant alleles are transmitted through pollen and ovules. Adaptive evolution is achieved by the selection of phenotypes encoded by many genes (i.e., polygenes), each with a small additive effect. Generally, herbicide resistance is conferred by major genes present in weeds. In the majority of cases in which the number of genes has been determined, resistance is controlled by a single, major gene [73]. The predominance of major gene inheritance is attributed to the two following factors:

1. The most recently developed herbicides are highly target site-specific and interfere with single enzymes in major metabolic pathways. Mutation of the gene encoding for the enzyme may alter a plant's sensitivity to the herbicide and result in resistance.
2. Repeated application of these herbicides imposes strong selection, often causing 95-99% mortality, against susceptible phenotypes in weed populations.

Adaptation to the herbicide is possible only if resistant genes are present in a population and have a significantly large phenotypic effect to allow the survival of a few individuals in a single generation [102]. With polygenic inheritance, recombination among individuals for many generations is required to bring together a sufficient number of favourable alleles to produce a highly resistant phenotype. Polygenic inheritance of resistance is thus more likely under conditions of weak selection as would occur with sublethal herbicide application [73].

Cytoplasmic inheritance of resistance occurs with triazine herbicides in several weed species. The gene conferring triazine resistance is located in the chloroplast genome[66]. Transmission of the chloroplast resistant gene mostly occurs by pollen, the paternal parent. The mutation that confers maternally inherited triazine resistance involves a single base substitution in the *psbA* chloroplast gene which codes for a Photosystem II membrane protein to which triazine herbicides bind. The expected frequency with which a mutation occurs in the chloroplast genome and gives rise to gamete-transmissible triazine resistance is very low [73]. The probability ranges from 1×10^{-9} to 1×10^{-12} mutations per gene locus [5].

Selection by Herbicides

The evolution of herbicide resistance in weeds is dependent on the intensity of selection imposed by herbicides. Most herbicides are applied at rates that result in the mortality of 90-99% of the susceptible weeds (weed seedlings). If genetic variation for resistance is present due to mutation or gene flow, even at very low frequencies, repeated herbicide application will normally result in a rapid increase in the frequency of resistant individuals until they dominate the population [73]. The selection pressure imposed by a herbicide is a primary factor that determines the rate of enrichment of herbicide resistance in a weed population. The higher the intensity of selection imposed by a herbicide against susceptible species, the faster the expected rate of evolution and spread of resistance. Several herbicide characteristics and patterns of use result in higher mortality than others do and these impose a more intense selection pressure for the development of resistance. These include a single target site and highly specific mechanism of action, long-term soil residual activity and frequent applications [83].

Seed production is an essential component to assess selection pressure. Effective kill by a herbicide is measured as the percentage reduction in seed yield at the end of the growing season. Values obtained for both resistant and susceptible plants are then used to estimate the selection pressure. The selection pressure of a herbicide is calculated as the ratio of the fraction of resistant plants that survive herbicide application to the corresponding fraction of susceptible plants [52a]. Gressel and Segel [52a] defined the selection pressure of a herbicide on a particular weed species as:

$$\frac{1 - \textbf{Effective kill of resistant plants}}{1 - \textbf{Effective kill of susceptible plants}}$$

Measurement of selection intensity exerted by herbicides is done theoretically and phenotypically. Population geneticists measure selection frequency after the action of the selection agent. However, weed scientists measure selection frequency at the phenotype level. Selection coefficients may be variously defined at the gametic (gene) or zygotic (genotypic) level. The coefficient of selection, S, may be defined as the proportionate contribution of a particular genotype to future generations compared with a standard genotype, whose contribution is usually taken to be unity [100].

For a detailed discussion of this subject, the reader may refer to a textbook on population genetics.

Fitness

In the theoretical plant population model constructed to predict herbicide resistance, two sets of biological processes serve as major factors. These are **ecological fitness** and **gene flow**. Knowledge about both factors is necessary to develop effective strategies for management of herbicide-resistant weeds.

An individual plant is a unique genotype with variation at many loci affecting fitness. The fitness of a group of plants having a certain genotype is assessed in relation to the fitness of other genotypes lacking key traits of interest. During breeding of two types of plants in a homogeneous environment allelic changes occur at a single locus, causing genotypic changes. When genotypic response is measured against a constant environment and genetic background, the expression **'genotypic fitness'** is used. It is possible to measure and calculate 'genotypic fitness' in the field and laboratory for a given genotype (homozygote and heterozygote).

Fitness is a measure of survival and ability to produce viable offspring. Under conditions of natural selection, genotypes with greater fitness produce, on the average, more offspring than less fit genotypes. It is measured over the whole life cycle of a plant, encompassing the effects of selection on mortality and seed production of survivors. For an annual species, the seed produced by a genotype per generation constitutes a fitness estimate for a given environment only at one point in the evolutionary time. While determining the fitness of a weed with a persistent seedbank, the rate of loss of seed from the soil needs to be measured. Seed carryover from previous generations plus the seed produced in the current generation contribute to total seed production. In order to understand the rate of evolution of resistance or management of resistance, measurements of fitness need be conducted only under field conditions with the crop and with and without herbicide application. For vegetatively reproducing weed species, fitness is assessed from the measurement of total biomass or plant parts over a time period appropriate for the species concerned.

Fitness is a dynamic integration of processes that can change with the environment and promote natural selection for more fit genotypes within the resistance phenotype. Fitness is a relative term whereby genotypes are compared among themselves relative to the most successful one. For example, susceptible biotypes produce up to 64% more biomass than their triazine-resistant counterparts under competitive and non-competitive conditions in a selective environment [24, 70, 154]. As many triazine-resistant biotypes are less fit than their susceptible counterpart, it is widely believed that plants resistant to other classes of herbicides also would be less fit than susceptible biotypes.

Certain new genes introduced through mutation to a population can result in a less well-adapted (fit) phenotype than is associated with the original gene if there is no change in the environment [57]. The new genotype would then be displaced by the wild biotype, or increase in frequency to some equilibrium in population. The rate of decrease or increase and the new equilibrium frequency would depend on the relative fitness advantage or disadvantage of the genotype and the rate at which the genotype can acquire compensatory traits by crossing with the wild type. If changes occur in the environment or in the wild-type genotype, then the new genotype may become more frequent in the population within a few generations. In the absence of herbicides, a new herbicide-resistant genotype would become less fit, to a certain degree, than the susceptible genotype and its success will be determined by the rate at which it can acquire compensatory traits. If a herbicide is introduced into the system, the resistant genotype would then immediately become more fit than the susceptible genotype and rapidly account for a greater proportion of the population.

Gene Flow and Spread of Herbicide Resistance

Gene flow has a significant impact on the rate of evolution of herbicide resistance. If a single plant or set of plants survives a herbicide application, the allele for resistance is passed to other plants in the field. Gene flow among plants can occur through: a) pollen dispersal and resultant fertilization and b) seed dispersal.

Rates of gene flow are generally believed to be higher than rates of mutation and hence would result in a higher frequency of plants resistant to a particular herbicide prior to its initial application. Therefore, gene flow and the resultant increase in initial frequency of resistant genes would reduce the time required to reach a specific level of resistance within a field once a herbicide is applied [73].

Gene flow could occur from herbicide-treated fields to adjacent unsprayed areas. Similar flow could also occur from resistant plants within the crop to susceptible plants at the field edge. Devlin and Ellstrand [33] showed that a significant proportion of a population's seed crop can be fathered by plants from outside the population. Such gene flow could lead to a rapid spread of a herbicide resistance gene among weed populations in a particular area.

Pollen movement of resistance alleles within populations occurs in several weed species. For example, outcrossing of resistant pollen to susceptible plants was 1.4% at a distance of 28.9 m in *Kochia scoparia*, resistant to sulfonylurea herbicide and 1% at > 6.84 m in *Lolium multiflorum*, resistant to diclofop-methyl [99, 143]. Mulugeta et al. [107] recovered *Kochia scoparia* pollen as far as 50 m from its source, indicating that pollen dispersal could lead to long-distance spread of herbicide-resistant genes. Although pollen dispersal is generally assumed to be the major mechanism of interpopulation gene flow in plants, seed dispersal may play a far greater role in weed populations. In self-fertilizing plants, pollen flow is minimal.

Cross and Multiple Resistance

Cross resistance is an expression of a mechanism that confers on plants the ability to withstand herbicides from different classes [61]. It may be conferred either by a single gene or, as in the case of quantitative inheritance, by two or more genes.

The most common form of cross resistance occurs when a change at the site of action of one herbicide also confers resistance to herbicides from a different class that inhibits the same site of action [61]. For example, weed biotypes resistant to acetolactate synthase (ALS)-inhibiting herbicides (sulfonylureas) become resistant to other ALS-inhibiting herbicides (imidazolinones and triazolopyrimidines). A biotype of resistant *Lolium rigidum* (annual ryegrass) selected with chlorsulfuron, metsulfuron and eight other sulfonylurea herbicides was also selected for resistance to imazapyr and imazethapyr [91, 92]. Similarly a *Kochia scoparia* (kochia) biotype resistant to five sulfonylurea herbicides is also resistant to one imidazolinone herbicide, imazapyr [125]. While being resistant to ALS-inhibitor herbicides, these weed biotypes are not resistant to herbicides with alternate modes of action.

Weed biotypes selected for resistance to aryloxyphenoxypropionic acid herbicides (APPs) which act on acetyl-CoA carboxylase (ACCase) are also less sensitive to cyclohexanedione herbicides (CHDs) and vice versa [119, 146]. There are exceptions for cross resistance in weeds. For example, the R biotype of *Xanthium strumarium* is sensitive to imazaquin and other imidazolinone herbicides, but not to chlorimuron, a sulfonylurea herbicide used in soybean. As this biotype has a significantly different cross resistance pattern, it probably contains a different ALS mutation than the one selected for resistance to sulphonylurea herbicides. This biotype is apparently controlled by herbicides with alternate mode(s) of action.

Multiple resistance in a weed biotype is an expression of more than one mechanism of resistance and hence is endowed with the ability to withstand herbicides from different chemical classes. Multiple resistant plants possess two or more distinct resistance mechanisms. For example, Gill [49] screened 242 populations of herbicide-resistant *Lolium rigidum* in Australia and found that 32% of them were resistant to sulfonylureas and APPs, 12% to sulfonylureas, APPs and CHDs, 6% to APPs and CHDs, 39% only to sulfonylureas and 11% only to APPs. Resistance to many herbicides makes weed biotypes similar to *Lolium rigidum* difficult to manage at the field level.

Target Site Resistance

Target site resistance involves a decrease of the herbicide target site to inhibition by the herbicides. The first target site resistance to ALS-inhibiting herbicides was found in *Kochia scoparia* [132]. Similar decreased ALS sensitivity to ALS-inhibiting herbicides has been reported in several other weed species. In the cases of *Kochia scoparia* [132], *Kochia iberica* and *Lolium perenne* [134], the resistance mechanism was due to an insensitive ALS and not to differential uptake and translocation. Saari et al. [133] reported that all the weeds selected for resistance to one sulfonylurea herbicide are also cross sensitive at the enzyme level to all other sulfonylurea herbicides. Triazolopyrimidine herbicides are also similarly affected by the mutation conferring resistance to the sulfonylureas.

There is also a lower but consistent cross insensitivity to imidazolinone herbicides, especially imazapyr, for weeds selected for sulfonylureas.

The target site resistance also exists in the case of ACCase-inhibiting herbicides. In the case of *Lolium rigidum*, one biotype (WLR 96) developed resistance after 10 yr of exposure to diclofop methyl (an APP herbicide), while the other biotype (SLR 3) developed tolerance after only three consecutive years of exposure to sethoxydim (a CHD herbicide) [69, 146]. Both types exhibit target site cross resistance to the APP and CHD herbicides. However, despite dissimilar periods of exposure, resistance to APPs is higher than resistance to CHDs. Tardif et al. [146] reported that the resistant *L. rigidum* biotypes with resistant ACCase exhibited biotype-specific patterns of resistance both at the whole plant level and in ACCase-assay level, in a manner similar to maize. In all the herbicide-resistant plant ACCases, the inheritance is apparently controlled by single, nuclear-coded, dominant or partially dominant genes [9, 146].

Non-Target Site Cross Resistance

Certain weed biotypes exhibit enhanced rates of herbicide metabolism. In such cases the degree of resistance at the whole plant level, while being sufficient to provide resistance at the recommended rates, is much less than conferred by the target site cross resistance mechanism. Uptake of herbicides and/or its movement to the site of action are the other mechanisms that confer non-target site cross resistance.

Mathews and Powles [97] reported that selection with an ACCase-inhibiting herbicide could lead to a resistant population that displays non-target cross resistance to ALS-inhibiting herbicides without exposure to these herbicides.

Wheat and some biotypes of *L. rigidum* have a similar resistance mechanism. Wheat has a sensitive ALS but can rapidly metabolize some ALS-inhibitors by aryl-hydroxylation, catalyzed by cytochrome P450 (Cyt 450) mono-oxygenases, followed by glycosylation forming non-toxic substances. Similarly, two chlorsulfuron-resistant biotypes of *L. rigidum* which have sensitive ALA, like wheat, can oxidatively metabolize chlorsulfuron more rapidly than the susceptible biotypes, yielding the same metabolites as produced by wheat [22]. The enhanced metabolism of ALS-inhibiting herbicides in *L. rigidum* is catalyzed by Cyt 450 mono-oxygenase. However, chlorsulfuron resistance can be reversed if the herbicide is combined with another chemical.

Similarly, selection with atrazine in conjunction with amitrole, a triazole herbicide, resulted in triazine resistance and cross resistance to a substituted urea herbicide in a *L. rigidum* biotype, WLR 2 [16]. In the case of chlortoluron, a substituted urea , herbicide-resistant plants exhibited an enhanced capacity for N-demethylation as an initial metabolic step [17]. An enhanced capacity for N-demethylation may explain the broad cross resistance to many substituted ureas which have N-alkyl groups in common but differ in phenyl substituents [61].

As another example, two maize inbreds with different sensitivities to thifensulfuron-methyl, more of the absorbed ^{14}C-herbicide was translocated out of the treated leaf in the susceptible inbred than in the resistant inbred [37]. As the observed differences in translocation were too small to account for the large differences in sensitivity, the greater and more rapid metabolism of thifensulfuron-methyl in the resistant inbred appeared to be the major mechanism conferring resistance [37]. In a similar case, Hall et al. [59] reported that uptake and distribution of foliage-applied ^{14}C-thifensulfuron-methyl was nearly identical in the resistant commercial mustard (*Brassica juncea*) and the susceptible wild mustard (*Sinapis arvensis* L.). They attributed resistance to more rapid metabolism of the herbicide, yielding two major metabolites, in the commercial mustard than in wild mustard.

These examples indicate that the primary mechanism of non-target cross resistance is metabolic inactivation or target site insensitivity while the differences in uptake and translocation play only a secondary role.

MECHANISMS OF RESISTANCE DEVELOPMENT IN WEEDS

It is important to understand the mechanism of resistance to herbicides in order to develop methods to increase herbicide resistance in crop plants and strategies to overcome development of herbicide resistance in weed species. The crops are resistant because they can detoxify herbicides. The resistance mechanism in weeds is, however, different from that found in crops. The herbicide resistance in weeds can be due to a number of mechanisms: a) modified target site, b) enhanced detoxification (metabolism), c) reduced absorption and/or translocation, d) sequestration or compartmentation and e) repair of the toxic effects of herbicides.

Although the mechanisms of action of various herbicide classes were discussed in Chapter 7, their primary targets of action are briefly mentioned here to bring relevance to the discussion on mechanisms of resistance to various herbicide classes and herbicides. The following presentatiom pertains only to herbicide classes and herbicides that are known to inhibit major biochemical processes in plants.

Resistance to Photosystem I Inhibitors

Photosystem I (PS I) inhibitors include bipyridilium herbicides (paraquat and diquat) and arsenical herbicides (DSMA and MSMA).

Target Site

Paraquat enters the chloroplast to reach the site of action at PS I. Being a divalent cation, paraquat short-circuits the light-driven photosynthetic electron transport in PS I by accepting electrons that normally flow to ferredoxin. The paraquat radical thus generated reduces oxygen to form superoxide, which is catalyzed by superoxide dismutase yielding a highly toxic hydroxy radical. This toxic form of oxygen, superoxide, causes lipid peroxidation, rapid loss of membrane integrity, photooxidation, desiccation and plant death. Sunlight, which enables photosynthetiç electron transport, aids in the toxicity of paraquat.

Weed Resistance

Weed resistance to paraquat was first reported in the late 1970s and early 1980s, after about two decades of use, when resistance was observed in *Conyza bonariensis* (hairy fleabane) from vineyards and citrus plantations in Egypt [47]. Similar resistance was reported in *Erigeron philadelphicus* (Philadelphia fleabane), *Erigeron canadensis* (horseweed), *Hordeum glaucum* (wall barley) and *Poa annua* (annual bluegrass) [75, 127, 153, 156]. The other species which showed resistance to paraquat include *Amaranthus retroflexus* (redroot pigweed), *Parthenium hysterophorus* (ragweed parthenium), *Solanum nigrum* (black nightshade), *Epilobium ciliatum* (willowweed), *Crassocephalum crepidioides*, etc.

The resistant (R) biotypes of these weeds showed evidence of reduced uptake into the leaf and restricted translocation compared to susceptible (S) biotypes. Preston et al. [123] applied paraquat at field doses to the intact plants of R and S biotypes of *Hordeum glaucum* and *H. leporinum* under conditions close to field applications and found reduced paraquat translocation in a basipetal direction. The herbicide moved more rapidly in the S biotypes. Preston et al. [123] reported that there was no difference in gross translocation of paraquat in the R biotype of *Arctotheca calendula*, but movement to the target site was reduced, probably as a result of reduced penetration of herbicide into the leaf cells. These and other studies suggest that paraquat is rapidly detoxified or removed from the chloroplasts of the R biotype [71, 138].

Although herbicide metabolism is a major selectivity and resistance mechanism in many plant species, metabolism is unlikely to be responsible for resistance to paraquat [68]. The rapid action of paraquat in plants is unlikely to allow metabolic detoxification to non-herbicidal metabolites such as *N*-methyl isonicotinic acid, monopyridone and methylamine found when detoxified by bacteria and fungi. Furthermore, no metabolism of paraquat or diquat has yet been detected for any other higher plants. It has been proposed that resistance to paraquat is possibly due to the detoxification of paraquat-generated toxic oxygen species (superoxide) catalyzed by superoxide dismutase (SOD) and other chloroplast enzymes, ascorbate peroxidase and glutathione reductase. Elevated amounts of these enzymes have been proposed as the mechanism of resistance to paraquat. However, no difference in the amounts of these enzymes has been observed in isolated chloroplasts from R and S biotypes of *Ceratopteris richardii* [19], in crude extracts from R and S biotypes of *Hordeum glaucum* [121], or from separate accessions of R and S biotypes of *Erigeron canadensis* (also referred to as *Conyza canadensis*) [120]. In contrast, other studies showed that a resistant biotype of *Conyza bonariensis* had increased amounts of SOD, ascorbate peroxidase and glutathione reductase [113, 137, 161]. However, increased amounts of these enzymes in themselves will not confer much resistance to paraquat as the reaction-producing superoxide regenerates the paraquat cation (Chapter 7, Fig. 7.5).

Other Photosystem I Inhibiting herbicides

DSMA and MSMA affect chlorophyll synthesis, amino acid content and respiration in sensitive plants. These arsenic herbicides are reduced by PS I to form sulphydryl groups of enzymes involved in CO_2 fixation and its regulation, thereby inhibiting CO_2 fixation. Uptake, translocation and metabolism are not involved in the mechanism of resistance of the R biotype of *Xanthium strumarium* (common cocklebur) to MSMA [76, 111]. The R biotype is not cross resistant to other herbicides. Inhibition of carbon assimilation is more rapid and of greater magnitude in the S biotype than in the R biotype of *X. strumarium* [112]. MSMA has no direct effect on photosynthetic electron transport. The mechanism of resistance to MSMA and DSMA remains a mystery.

Resistance to Photosystem II Inhibitors

The photosystem II (PS II) inhibitors include triazines and substituted ureas as also herbicide belonging to such structurally diverse groups as nitriles (bromoxynil and ioxynil), carbamates (desmedipham and phenmedipham), dinitroanilines, acetamides (propanil), pyridazinones (pyrazon), benzothiadiazole (bentazon), oxadiazolidines (methazole), phenylpyridazines (pyridate), imidazolidinones (buthidazole), etc. Some of these herbicides/herbicide classes such as nitriles, pyridazinones, dinitroanilines, etc. also have other sites of action.

Herbicide resistance was predicted in the 1950s, even before triazines and ureas were made available for commerical use [62]. That prediction came true in 1968 when the first herbicide-resistant weed, *Senecio vulgaris* (common groundsel) was discovered, as its resistant biotype could no longer be controlled by the PSII inhibitor simazine [131]. In the 30 years since this discovery, there have been reports of scores of weed biotypes exhibiting resistance to PS II-inhibiting herbicides. Of these weeds, dicots, which are more susceptible to triazines, outnumbered monocots, which are partially resistant. Millions of hectares of cropped land under maize monoculture and receiving triazines repeatedly for several years, particularly in developed countries, are now infested with triazine-resistant weeds.

Target Site

As discussed in Chapter 7, the PS II complex catalyzes the light-dependent oxidation of water with the reduction of plastoquinone (PQ). Light energy absorbed by the light-harvesting chlorophyll (LHC) antenna pigment (or proteins) is transferred to the reaction centre chlorophyll, P680, causing a charge separation. An electron is then passed to pheophytin and then to PQ_A on the D_2 protein (a polypeptide). Plastoquinone can be reduced by two sequential one-electron acceptors. The bound PQ_A transfers two electrons, one at a time, to the second plastoquinone molecule, PQ_B, which is reversibly bound to the D1 protein (a polypeptide). Upon transfer of the second electron and the acceptance of two protons from the stroma, fully reduced plastoquinone, plastohydroquinone (PQH_2), is formed. This molecule leaves the PQ_B site on the D1 protein, diffuses across the thylakoid membrane and donates two electrons to the cytochrome **b/f** complex with the release of two protons into the thylakoid lumen. Electrons from the cytochrome **b/f** complex are then transferred to PS I via plastocyanin.

The D1 polypeptide, also known as the 32-kDa protein or the PQ_B protein, is encoded by the chloroplast *psbA* gene. In the light, the D1 protein undergoes a continuous cycle of degradation and re-synthesis, with a half-life ranging from 30 min to a few hours [1]. Continuous turnover of the D1 protein is necessary for electron transport because photoinhibition (photodamage) occurs during the normal operation of the PS II reaction centre [1]. D1 protein incurs greater photodamage under conditions that promote photoinhibition (i.e. strong illuminations), thus requiring a faster turnover rate. The photoinhibition is due to formation of radicals or singlet oxygen by redox components associated with the D1 protein [1].

The PS II-inhibiting herbicides displace plastoquinone at the PQ_B binding site on the D1 protein and thereby block electron flow from PQA to PQ_B. [Triazine and urea herbicides act as non-reducible analogues of plastoquinone while nitriles and phenols act as non-reducible analogues of semiquinone anion of plastoquinone]. Binding of the herbicide at the PQB niche not only causes photodamage to the D1 protein by preventing electron transfer from PQ_A to PQ_B, but also compounds herbicidal activity by preventing replacement (re-synthesis) of the damaged D1 protein.

Target Site Modification

LeBaron and McFarland [83] reported that of the 57 weed species (40 dicots and 17 monocots) exhibiting biotypes resistant to PS II inhibitors (primarily triazines), all but a few were resistant due to a modification at the target site. The target site modified in these resistant biotypes was the amino acid residues in the PQ_B-binding niche on the D1 protein [45, 148]. This modification reduces the affinity of PS II herbicides at this site so that they can no longer effectively compete for the exchangeable PQ_B.

In all cases of target site resistance, resistance is due to a point mutation in the chloroplast *psbA* gene, resulting in the substitution of **glycine** for **serine 264** [45, 148]. This modification greatly reduces the affinity of atrazine at the PQ_B-binding site, since binding affinity is strongly dependent on H-bonding with the OH-side chain of serine 264. As a result of this modification, there is a 1000-fold reduction affinity at the PQB-binding site and greater than 100-fold increase in atrazine resistance at the whole plant level [46]. Triazine-resistant weeds with ser **264-to-glycine mutation** exhibit a modified galactolipid composition, an increase in unsaturation of fatty acids and altered chloroplast structure. The chloroplasts in resistant weeds are similar to 'shade chloroplasts' which develop under low light conditions. In relation to triazine-susceptible chloroplasts, the chloroplasts of triazine-resistant biotypes exhibit increased grana stacking and a reduced **chlorophyll a/b** ratio.

Aside from reducing the affinity of atrazine at the PQ_B-binding site in a resistant biotype, the **ser 264-to-glycine mutation** of the *psbA* gene also reduces the rate of electron transfer between PQ_A and PQ_B by 3-fold to 10-fold. Modification of amino acid residues in the PQ_B-binding niche on the D1 protein is both herbicide-specific and biotype-specific. The **ser 264-to-glycine mutation** is specific to the R biotypes of *Amaranthus* spp. The R biotypes of other weeds may have a different amino acid residue modification. Each mutation confers varying levels of resistance, each to a different spectrum of herbicides, particularly triazines, uracils and ureas. Biotypes with this mutation exhibit a reduction in CO_2 fixation, quantum yield and seed, and biomass production.

Inheritance of PS II-Inhibitor Resistance

As the D1 protein is a chloroplast gene product, triazine resistance is maternally inherited. Certain weed species exhibit greater than expected incidence of mutation rate of the *psbA* gene, resulting in higher triazine resistance. This is hypothesized as due to the presence of a **plastone mutator**, a nuclear gene, that increases the frequency of chloroplast DNA mutation [5]. The triazine R biotypes of some weed species (e.g., *Chenopodium album*) rapidly develop resistance to other PS II-inhibiting herbicides since the presence of the mutator gene would increase the likelihood of other mutations in the PQ_B-binding pocket [142].

In certain populations of *C. album*, triazine resistance involves a two-step process [11, 28]. It requires the presence of `R precursor' genotypes that exhibit a high random mutation rate of the *psbA* gene. This precursor genotype mutates in the absence of herbicides to form an `intermediate biotype' which has the **ser 264-to-glycine mutation** in the *psbA* gene. This intermediate biotype exhibits a high level of atrazine resistance (1000-fold) at the chloroplast level. When treated with other herbicides, e.g., carbamates, ureas, etc. the intermediate biotype produces progeny that are highly resistant to atrazine. This induced expression of a high level of resistance in the intermediate biotype involves an extrachloroplastidic genetic mechanism [11].

The triazine resistance in the R biotypes of *Abutilon theophrasti* is not maternally inherited. It is controlled by a single, partially dominant nuclear gene that is not cytoplasmically inherited [3]. Unlike most triazine R weed biotypes, the resistant *A. theophrasti* biotypes are only 10 times as resistant as susceptible biotypes. Resistance in this weed species is due to enhanced **glutathione s-transferase** activity, which confers enhanced capacity to detoxify atrazine via glutathione conjugation [4].

In the case of *Alopecurus myosuroides*, the inheritance of resistance to chlortoluron and isoproturon is polygenic and under nuclear control. The rapid degradation of these herbicides in the R biotypes is catalyzed by **cytochrome P450 mono-oxygenases**.

Lolium rigidum, a ubiquitous weed widely found in Southern Australia exhibits multiple resistance to several herbicide classes. The R biotype, WLR 2, is cross resistant to several triazines (atrazine, simazine, propazine, cyanazine, prometryn, ametryn, etc.), ureas (chlortoluron, diuron and isoproturon), asymmetrical triazines (metribuzin) and amitrole. The resistance to these herbicides varies from 3- to 9-fold depending on the herbicide. VLR 69, the R biotype of *Lolium perenne* (perennial ryegrass), is not only resistant to triazines, ureas and metribuzin, but also to aryloxyphenoxypropionics (diclofop, fluazifop and haloxyfop), cyclohexhanediones (tralkoxydim), imidazolinones (imazaquin) and sulfonylureas (chlorsulfuron and triasulfuron) [15]. Both WLR 2 and VLR 69 metabolize simazine and chlortoluron twice as fast as the S biotype [16, 17]. The inheritance of resistance in these R biotypes is a polygenic trait under nuclear control.

A resistant (R) biotype of the dicot weed *Polygonum lapathifolium* (pale smartweed) was 35.5 times more resistant to atrazine than an S biotype [116]. Electron transport in chloroplast thylakoids isolated from the leaves of the R biotype of *Polygonum* was 781 times less sensitive to atrazine than in S biotypes. The atrazine-resistance mechanism in *P. lapathifolium* was due to reduced affinity for atrazine to its target site.

Resistance to Acetolactate Synthase Inhibitors

Acetolactate synthase (ALS) is the first enzyme common to the biosynthesis of the branched-chain amino acids, valine, leucine and isoleucine (vide Chapter 7). Currently, three chemically dissimilar classes of herbicides, sulfonylureas, imidazolinones and triazolopyrimidines share a common target site, ALS inhibition. These herbicides are commercially used for selective weed control in such diverse crops as wheat, maize, soybean and rice. The list of ALS inhibitors may grow in future as more are discovered.

Resistance in weeds to ALS-inhibiting herbicides was first observed in *Lactuca serriola* (prickly lettuce) in 1987 [91], only five years after the commercial introduction of chlorsulfuron. Later, *Kochia scoparia* (kochia) was identified to have developed resistance to sulfonylurea and imidazolinone herbicides [125]. Since then, numerous weed species, most of them dicots, have been reported worldwide to have developed resistance to ALS-inhibiting herbicides. These weeds include *Amaranthus retroflexus* (redroot pigweed), *Alopecurus myosuroides* (blackgrass), *Cyperus difformis* (umbrella sedge), *Lolium perenne* (perennial ryegrass), *Lolium rigidum* (rigid ryegrass), *Saggitaria montevidensis* (arrowhead), *Sonchus oleraceus* (annual sowthistle), *Stellaria media* (common chickweed), *Salsola iberica* (Russian thistle), *Xanthium strumarium* (common cocklebur), *Eleusine indica* (goosegrass), etc.

Resistance to ALS-inhibitors evolved relatively quickly—after only four to seven continuous applications of the herbicide in monoculture crop areas and non-crop areas. The long residual activity of many of these herbicides has also contributed to the rapid development of resistance by increasing the effective kill and hence the selection pressure.

Target Site and Its Modification

Resistance to target site involves a reduction in the sensitivity of the herbicide target site to inhibition by the herbicides. The first target site resistance to ALS-inhibiting herbicides was identified in *Kochia scoparia* [132]. In the R biotypes of *K. scoparia* and other weed species, the resistant mechanism identified was an insensitive ALS. The other possibilities such as increased metabolism and differential uptake and translocation have had no role in the resistance mechanism exhibited by the R biotypes.

Saari et al. [132] reported that the insensitive ALS in the R biotypes of *K. scoparia* was due to an altered form of ALS. The R/S ratios of I_{50} values for LAS inhibition ranged from 28 for metsulfuron to 2 for imazethapyr. The chlorsulfuron-resistant isolate demonstrated both intra- and interfamily cross resistance with sulfonylurea, imidazolinone and triazolopyrimidine sulfonanilide herbicides. They suggested that differences in ALS sensitivity to different classes of herbicides might be due to slightly different binding domains within a common binding site on the protein. Wiersma et al.

[159] found that two protein domains, **Domain A** and **Domain B**, appeared to influence resistance. In Domain A, consisting of 13 amino acids, substitution of serine, glutamine, or alanine for proline residue at amino acid 173 confers resistance to ALS inhibitors. Domain B, with four amino acids, enhanced the level of resistance conferred by mutation in Domain A in tobacco [84]. Wiersma et al. [159] reported that proline residue in Domain A was pivotal for developing resistance to chlorsulfuron.

Seeds from sulfonylurea-resistant *K. scoparia* accessions contained 2-fold higher levels of branched-chain amino acids than seeds from susceptible accessions [36]. Mutations conferring resistance may concomitantly reduce or abolish ALS sensitivity to normal feedback inhibition patterns, resulting in elevated levels of branched-chain amino acids ([36].

Guttieri et al. [56] observed that the Domain A DNA sequence from the chlorsulfuron-R biotype of *Lactuca serriola* (prickly lettuce) differed from that of the S biotype by a single point mutation, which substituted histidine from a proline. The Domain A sequence from a resistant *K. scoparia* biotype also differed from the S biotype by a single point mutation in the same proline codon (**Pro$_{173}$**), substituting threonine from proline. Further work by Guttieri et al. [55] found point mutation in the codon for the proline residue at position 173 in Domain A of the ALS protein in seven of the R biotypes. Among these seven R biotypes, mutation to threonine, serine, arginine, leucine, glutamine and alanine was observed. All the R biotypes were resistant to modified ALS, indicating that at least one non-Domain A mutation site for resistance exists in kochia. Their results indicated the existence of multiple mutations for resistance controlled by multiple resistant alleles.

Resistance to ALS-inhibiting herbicides, usually inherited as a semi-dominant trait, is conferred by single nuclear genes coding for less susceptible ALS. The ALS gene codes for an enzyme of 670 amino acids with a molecular weight of about 73 kDa, an enzyme highly conserved in various organisms, including plants and microbes. Mutant ALS genes contain a single nucleotide change, resulting in a single amino acid substitution for proline, as discussed earlier.

Cross Resistance

All of the weed biotypes resistant to one class of ALS-inhibiting herbicides are also resistant to other classes of ALS-inhibiting herbicides. A biotype of resistant *Lactuca serriola* selected with chlorsulfuron and metsulfuron-methyl is also resistant not only to eight other sulfonylurea herbicides but also to imidazolinone herbicides such as imazapyr and imazethapyr [91]. Similarly, a *Kochia scoparia* biotype is resistant to five sulfonylurea herbicides and one imidazolinone herbicide, imazapyr [125]. The R biotype of *Stellaria media* (common chickweed), selected with and resistant to chlorsulfuron, is also resistant to an ALS-inhibiting triazolopyrimidine herbicide, DE-489. The R biotype of *Datura innoxia* (sacred datura) is resistant to both sulfonylurea and imidazolinone herbicides [135]. The resistant biotype of *Amaranthus rudis* (common waterhemp) which exhibited 130-fold resistance to imazethapyr was found cross resistant to sulfonylureas, chlorimuron and thifensulfuron, both at the whole plant and enzyme levels [87]. These R biotypes, while being resistant to ALS-inhibiting herbicides, are not resistant to herbicides with alternate mode of action.

A notable exception to cross resistance is *Xanthium strumarium* (common cocklebur) resistant to imazaquin. The R biotype is resistant to several imidazolinone herbicides but not to chlorimuron-ethyl, a sulfonylurea herbicide used in soybean. This R biotype, which probably contains a different ALS mutation than those selected with other sulfonylurea herbicides, is controlled by herbicides with alternate modes of action.

The biotypes resistant to one class of herbicides may vary in their magnitudes of resistance to other ALS-inhibiting herbicides. Furthermore, some ALS-inhibiting herbicides may still be very effective in controlling weed biotypes resistant to other ALS-inhibiting herbicides. For example, Kwon and Penner [80] found that the R biotypes of *Kochia scoparia* which showed 22, 18 and 16 times more resistance to primisulfuron, chlorsulfuron and thifensulfuron respectively, exhibited only 1, 2 and 3 times more resistance to imazethapyr, nicosulfuron and flumetsulam (a triazolopyrimidine) than the susceptible biotype. The altered ALS enzyme system of the R biotype has a differential response for the ALS-inhibiting herbicides. A specific alteration in ALS does not confer automatic cross resistance across the different ALS-inhibiting herbicide classes. Given the structural dissimilarities between and within the three ALS-inhibiting classes that share a common ALS binding site, it is likely that individual herbicides have overlapping binding niches [68].

Inheritance of Resistance to ALS-Inhibiting Herbicides

Inheritance of sulfonylurea resistance in weeds has been investigated in *Lactuca serriola* [92], *Kochia scoparia* [108] and *Salsola iberica* [56]. All these weed species are diploid. However, one R biotype of *Salsola iberica* is a polyploid [56]. Sulfonylurea resistance in *K. scoparia* is inherited as a dominant, nuclear trait. However, the resistance of *Lactuca* spp. is controlled by a single gene with incomplete dominance [92]. The best fit for chi- square analysis of the F_2 generation of both the susceptible × resistant *Lactuca serriola,* and *L. sativa* × resistant *L. serriola* crosses was a 1:2:1 ratio, indicating that the trait was controlled by a single nuclear gene with incomplete dominance. Evaluation of F_3 plants grown from seed collected from intermediate and resistant F_2 plants confirmed the F_2 generation findings. Resistant F_3 plants did not segregate, while intermediate F_3 *L. sativa* and *L. serriola* plants segregated 1:2:1 and 3:1 respectively.

In most cases the ALS-inhibiting herbicide resistance trait is controlled by a single nuclear gene with incomplete dominance, which results in survival of both homozygous-resistant and the heterozygous intermediate plants treated with herbicide applied at normal doses. Migration of resistant pollen or seed from fields of non-crop areas infested with ALS-inhibiting-resistant weeds into fields previously infested with susceptible weed population could increase the proportion of resistant plants, even in the absence of selection pressure. The subsequent use of ALS-inhibiting herbicides would accelerate selection of R biotypes. ALS-inhibitor resistance in resistant mutant populations of *L. rigidum* is associated with a modified target site (ALS) [133]. Molecular genetic studies conducted by Saari et al. [133] indicated that a number of mutations at various locations along the DNS sequence encoding for ALS enzyme result in resistance. Murray et al. [109] reported that resistance in both the R populations of *Avena fatua* was encoded at the same gene locus.

Resistance to Acetyl Coenzyme A Carboxylase Inhibitors

Although aryloxyphenoxypropionic (APP) and cyclohexanedione (CHD) herbicides are structurally dissimilar, they exhibit similarities in their activity at the whole plant and physiological/biochemical levels. They share some common structural element(s) responsible for their herbicidal activity. These postemergence herbicides control annual and perennial grass weeds in cereal and dicot crops, with little activity against dicots or non-graminaceous monocots.

APP herbicides are characterized by two aryl groups, one of which is substituted with an isopropyl group on the 1 position. Variation in the second aryl group gives rise to herbicides with differing biological activity and selectivity. These herbicides are usually formulated as esters of the parent acid (e.g., diclofop-methyl, fenoxaprop-ethyl, fluazifop-butyl, etc.). Once inside the plant, the ester is hydrolyzed to release the parent acid. As APPs are lipophilic, they do not show high phloem mobility. CHD herbicides vary considerably in the substitution at the 2 and 5 positions of the cyclohexane ring.

The synthesis of malonyl-CoA, catalyzed by acetyl-CoA carboxylase (ACCase), is the first step in fatty acid synthesis (vide Chapter 7). Malonyl-CoA is a key intermediate in the synthesis and elongation of fatty acids and for the synthesis of flavonoids. Inhibition of ACCase would deprive the plant of this key intermediate.

Target Site and Its Modification

APPs and CHDs are effective and potent inhibitors of ACCase from susceptible grass species but not from resistant dicot species. Selectivity between grass and dicot species is due to the relative insensitivity of dicot ACCase to these herbicides. Both classes of herbicides are reversible, linear and noncompetitive inhibitors of ACCase [18, 129]. The herbicides interact with an acetyl-CoA binding site, with the release of malonyl-CoA. Both APP and CHD herbicides act at the same site on ACCase.

Resistance to ACCase-inhibiting herbicides in R biotypes is controlled by a single, nuclear, partially dominant gene. The R biotypes possess altered forms of ACCase that are much less sensitive to these herbicides. The R and S biotypes do not differ in herbicide uptake, translocation or metabolism, suggesting that resistance in R biotypes is due solely to resistant ACCase forms.

Elecetrophysiological evidence suggests the existence of subtle differences in membrane function between the R and S biotypes. Diclofop and haloxyfop appear to interact with specific membrane sites to cause increased permeability to protons. This clearly establishes that resistance to APP herbicides in the R biotypes may be due to unknown structural or functional changes not found in the S biotypes.

It is speculated that APP and CHD herbicides are sequestered or compartmentalized in such a way as to prevent them from entering the plastids [31]. Rapid removal of herbicides from the available pool by such a mechanism (for example, by binding to the cell wall or accumulation in the vacuole) would effectively limit the herbicide concentration in the plastid, allowing lipid synthesis to proceed normally. Besides, the removal of herbicide in R biotypes may allow the membrane potential in these R biotypes to recover to its pretreatment value [31]. Devine et al. [32] found reduced uptake in isolated protoplasts and plasma membrane vesicles from a R *Avena fatua* biotype compared to an S biotype.

Inheritance of Resistance to ACCase-Inhibiting Herbicides

As mentioned earlier, the target enzyme-based resistance to ACCase inhibitors is controlled by a single dominant or partially dominant nuclear gene. This suggests that resistance can be conferred by a number of different point mutations, each of which confers a different pattern of resistance and cross resistance to APP and CHD herbicides.

The R biotype of *Eleusine indica* is resistant even at 100 times the application rate of fluazifop butyl that was lethal to the S biotype [95]. Leach et al. [81] found that the R biotype ACCase was less sensitive to the APP (fluazifop and fenoxaprop) and CHD (clethodim and sethoxydim) herbicides than the S biotype ACCase. They concluded that resistance to ACCase inhibitors in the R biotype of *E. indica* is conferred by a mutant target enzyme with reduced herbicide sensitivity. Besides the R biotype of *E. indica*, the R biotypes of *Setaria viridis*, *Lolium rigidum*, *L. multiflorum* and *Avena sterilis* also possess altered target enzyme sensitivities to APP and CHD herbicides [54, 93, 94, 146]. However, the R biotype ACCase is not equally resistant to all of the compounds, suggesting that the mutation does not affect the binding of all herbicides equally [81].

While the resistance mechanism in *A. fatua* is controlled by a single, partially dominant nuclear gene [106, 109], the inheritance of resistance in *A. sativa* (cultivated oats) conforms most closely to a two-gene model, with the genes controlling resistance being recessive [151].

Resistance to Auxinic Herbicides

The prominent auxinic herbicides 2,4-D and MCPA, the phenoxy acids, belong to the group of phenoxyalkanoics. These herbicides mimic the action of the natural plant growth substance IAA (indol-3-yl acetic acid), often simply referred to as auxin. Phenoxyalkanoics are a diverse group of herbicides; the following discussion concerns only the auxin-analogue herbicides.

Although auxinic herbicides have been in use for over 50 years, the precise mode of action of these compounds remains unclear. As these herbicides mimic auxins, their phytotoxic effects are believed to be manifested by initiating a plethora of auxin-induced changes, which eventually lead to plant death. The changes include cell elongation, epinasty, hypertrophy (excessive growth), root initiation and ethylene biosynthesis. These morphological and physiological changes are attributed to several biochemical responses, viz. a rapid Ca^{++} influx, activation of plasma membrane ATPase and enhancement in protein and nucleic acid biosynthesis. These biochemical responses are believed to be elicited by a specific receptor, which when it binds auxin or an auxinic herbicide initiates this biochemical cascade. The auxin-binding proteins (ABPs) are known to possess receptor functions. ABPs are located in the plasma membrane and within the cell, primarily associated with the endoplasmic reticulum (ER), but they may also be localized in the nucleus.

Sensitivity or resistance to auxinic herbicides has been correlated with their ABP binding activity and this relationship explains the mechanism of auxinic herbicide resistance in the R biotype of *Sinapis arvensis* (wild mustard) [156].

Target Site

Despite several decades of study, the mechanisms of resistance to phenoxyacetic acid herbicides and hence their target site(s), have not been clearly established. While there is overwhelming evidence that the ABPs possess receptor function, their exact location in relation to the ER is not clear. They are believed to be associated with both the plasmalemma and internal membranes. One possibility for this bilocation is that ABPs are synthesized in the ER, but they are secreted from the ER lumen for subsequent migration to the outer face of the plasma membrane. Another possibility is that two types of ABPs are needed, one type at the plasma membrane, ultimately regulating the H^+ pumping and the other type sensing changes in auxin level within the cell and affecting changes in gene transcription.

Irrespective of where the receptor (ABP) is located, after auxin has become bound, the reception signal is transmitted and amplified to affect the multitude of biochemical events that precede growth stimulation. Amplification of receptor is required due to the relatively small number of available receptors. Phospholipids and Ca^{++} are considered to play a key role in the signal transmission pathway.

It has a long been widely known that auxin and auxinic herbicides affect nucleic acid and protein synthesis due to their effect on transcription and translation. Auxin and auxinic herbicides rapidly and specifically regulate mRNA biosynthesis. An increase in specific mRNA sequences could result from increased transcription rates, or post-transcriptional events such as increased transport of mRNA from the nucleus to the cytoplasm. The available evidence clearly shows that auxin and auxin-analogue herbicides have a direct (not dependent on protein synthesis), selective and rapid effect on nucleic acid biosynthesis and that the primary site of action is at the level of transcription [26]. This becomes the target site for development of plant resistance to auxinic herbicides.

Development of Resistance

Despite extensive and persistent usage of 2,4-D and related herbicides throughout the world for over 50 years, auxinic herbicides are relatively free of resistance problems. Development of resistance has been unequivocally found in only six weed species [68]. There have been reports of differential effects of auxin-type herbicides between biotypes of particular species, but clear evidence of resistance is lacking in most cases [7, 52].

In *Stellaria media*, the R and S biotypes have equal rates of herbicide uptake and translocation [8], with little differences in plasma membrane ATPase activity [27]. But the R biotype exhibited greater rate of metabolism of mecoprop to conjugated metabolites, indicating that increased metabolism (mediated by cytochrome P450 mono-oxygenases) may be the basis of resistance [8]. However, the resistance in the R biotype of *Sinapis arvensis* to 2,4-D, dicamba and picloram cannot be explained on the basis of differences in rates of herbicide uptake, translocation or metabolism [119]. This suggests that each weed species exhibits a resistance mechanism different from another. Jasieniuk et al. [72] reported that the inheritance of dicamba resistance in this wild mustard biotype was determined by a single, completely dominant nuclear allele. The susceptible biotype of wild mustard produces significantly more ethylene than the resistant biotype following picloram treatment [60]. Although both biotypes possess functional ethylene biosynthesis, increased ethylene in the R biotype is due to differences in pathway regulation. Deshpande and Hall [30] found picloram affecting the ATP-dependent activity of susceptible wild mustard protoplasts to a greater extent than resistant protoplasts. They attributed the differences in ATP-dependent activity to differences in the ability of biotypes to modulate calcium ion channels. Webb and Hall [156] reported that differences between R and S biotypes of wild mustard were due to the differential response of ABPs to auxinic herbicides. Their results also showed that differences in biotype response to auxinic herbicides were due to a link between ABP activity and calcium levels and ATP activity in protoplasts and whole plants. The endogenous levels of calcium were two to three times higher in the R than in the S biotype.

Estelle and Somerville [40] found that a single gene mutation in *Arabidopsis thaliana* resulted in several auxin-resistant lines which display pleiotropic (multiple phenotypic expressions) morphological effects. The isolated mutant lines of *A. thaliana* were 50 and 8 times more resistant to 2,4-D and IAA respectively. The resistance was due to a recessive mutation at the *axrl* locus. The mutant of *A. thaliana* produced bushy and short plants, with small, thin roots and smaller leaves. These authors hypothesized that *axrl* gene (in *A. thaliana*) may encode for a different auxin receptor and resistance may be due to an alteration that has a greater effect on the affinity of this receptor for 2,4-D and IAA. Later, Lincoln et al. [85] reported that all the 20 *axrl* mutants of *A. thalinana*, with at least five different *axrl* alleles, had similar phenotypes and the extent of auxin resistance of each mutant line could be directly correlated with severity of the morphological alterations. The distinct morphological phenotypes of the mutant *A. thaliana* described by them were similar to those described for the R biotype of

Sinapis arvensis [58]. Hall and Romano [58] suggested that in the presence of high levels of calcium and cytokinin, such as those found in the R biotype of *S. arvensis* as a result of mutation, the classic cytokinin responses (such as delayed senescence and morphological alterations) may be involved in the resistance mechanism of the R biotype.

Resistance to Mitotic Inhibitors

The mitotic inhibitors include dinitroaniline herbicides. Dinitroanilines are typically applied preplanting and preemergence and are incorporated to avoid loss through volatilization and photodecomposition. In general, these herbicides remain in the upper layer of soil due to low water solubility and a tendency to bind to organic matter.

Target Site

The most striking symptoms of dinitroaniline injury is a reduction in root length and swelling of seedling roots into a characteristic club shape. In the affected roots, cells undergo prometaphase but not the later stages of mitosis, resulting in a number of cells being arrested at prometaphase. Eventually, a nuclear membrane reforms around the chromosomes, but because of the doubling in chromosome number and the spreading of the chromosomes throughout the cytoplasm, the nucleus takes on a highly lobed morphology. The cells eventually turn isodiametric (square-shaped) in shape, rather than become elongate (rectangular), contributing to the swollen appearance of the root tips.

Dinitroanilines inhibit polymerization of tubulin, a major protein constituent of microtubules, into microtubules. These herbicides are believed to bind to tubulin heterodimers in the cytoplasm and when attached to the growing end of the microtubule, prevent further polymerization of tubulin. By blocking polymerization, the depolymerization process continuously shortens the microtubules until they eventually become undetectable.

Dinitroaniline herbicides are effective against several annual grasses and certain annual dicots. They exhibit poor activity on perennial weeds. Resistance to dinitroanilines may be due to the paucity of detoxification mechanisms in target plants.

Resistance has, however, been reported in *Amaranthus palmeri, Eleusine indica, Setaria viridis, Lolium rigidum, Alopecurus myosuroides*, etc. While *E. indica* biotypes exhibited high and moderate resistance, the R biotypes of *S. viridis* [105] and *L. rigidum* [64] showed only moderate resistance. Mitosis in root tips of the highly R biotype of *E. indica* was 1000 to 10000 times more resistant to dinitroanilines than in the S biotype [150]. Mitosis in the R biotype of *S. viridis* was only 2 to 10 times as resistant as in the S biotype [141]. The highly R biotypes of *E. indica* contained higher amounts of tubulin and higher-molecular-weight form of β-tubulin than S biotypes [149, 150]. In the presence of oryzalin, greater polymerization of tubulin takes place in the R biotypes. The resistance trait in the R biotype of *E. indica* is dominant, controlled by several genes (multigenetic effect). Inheritance of resistance in *S. viridis*, however, appears to be under the control of a single, nuclear recessive gene. This indicates that different mechanisms of resistance operate in *E. indica* and *S. viridis*.

Resistance to Inhibitors of Aromatic Compound Biosynthesis

GLYPHOSATE: EPSP synthase (5-enolpyruvylshikimate-3-phosphate synthase) is the penultimate enzyme in the aromatic amino acid biosynthetic (shikimate) pathway. The enzyme catalyzes formation of EPSP and inorganic phosphate from phosphoenolpyruvate (PEP) and shikimate-3-shikimate in an unusual carboxyvinyl transfer reaction (vide Chapter 7). Glyphosate is a specific competitive inhibitor of EPSP synthase, involved in the shikimate pathway of biosynthesis of aromatic compounds.

Target Site and resistance in Plants

Glyphosate interacts with the EPSP synthase:shikimate-3-phosphate complex and inhibits this reaction, resulting in the build-up of shikimate-3-phosphate. This leads to a deleterious effect on the formation of chorismic acid, a precursor required for the biosynthesis of aromatic compounds. EPSP synthase inhibition is considered to be the primary site of action of glyphosate. Glyphosate binds either entirely or to a large extent to the site outside the site of action. It is trapped on the EPSP synthase enzyme in the presence of EPSP. The carboxyvinyl group of EPSP strongly facilitates the binding of glyphosate to the enzyme. Mutations in this domain confer resistance to glyphosate.

Glyphosate-Resistant Transgenic Plants

Resistance in R biotypes of various plant species such as carrot, tobacco, petunia, chickory, pea, etc. is due to elevated levels of EPSP synthase activity, although the enzyme is still sensitive to glyphosate. Overproduction of EPSP synthase effectively increases the number of molecules that must be inhibited in order to block carbon flow through the shikimate pathway. Enzyme overproduction is due to amplification of the gene(s) encoding EPSP synthase. EPSP synthase activity, gene copy number and glyphosate resistance level correlated positively.

Research on whole plant germplasm screening has demonstrated that variability in glyphosate exists among various plant species [35, 63, 74, 86]. Perennial ryegrass (*Lolium perenne*), soybean and *Convolvulus arvensis* displayed varying degrees of resistance to glyphosate. The resistant lines showed 2- to 3-fold resistance to the herbicide. When *Lotus corniculatus* (bird's-foot trefoil) was subjected to recurrent selection for two cycles, the selected populations showed an increase in resistance levels to glyphosate treatment [14]. Resistance in three clones picked from selected populations correlated positively with elevated EPSP synthase-specific activities.

Although glyphosate has been extensively used worldwide since its commercial production in 1974, there are no confirmed reports of glyphosate-resistant weeds developing in field situations. Although multiple treatments have been applied each year for several years in coffee and banana and in non-agricultural systems, no verified cases of glyphosate resistance have resulted. The lack of control of certain weeds cannot be attributed to the presence of resistant biotypes selected by the herbicide. For example, the low susceptibility of a naturally resistant biotype of *Convolvulus arvensis* (field bindweed) is not due to inherent, physiological, genetic and/or behaviouristic properties of the plants. The resistant biotype has not evolved due to selection pressure exerted by repeated field applications of glyphosate.

GLUFOSINATE: Phosphinothricin (PPT), also known as glufosinate, is an analogue of glutamine that inhibits the amino acid biosynthetic enzyme glutamine synthetase (GS) of bacteria and plants. Bialaphos is a tripeptide precursor of PPT produced by some strains of *Streptomyces*, in which two alanine residues are linked to the PPT moiety. The active PPT moiety is released intracellularly by peptidase activity. Both PPT (glufosinate) and bialaphos are marketed as broad-spectrum contact herbicides.

Target Site

Glutamine synthetase, the first enzyme involved in the assimilation of ammonia, catalyzes the conversion of glutamate (glutamic acid) to glutamine (vide Chapter 7). Glutamine thus formed serves as an amino donor for a variety of biosynthetic pathways including amino acid and nucleotide biosynthesis. Glufosinate inhibits GS, resulting in rapid accumulation of ammonia, which leads to death of the plant. GS is the target site for glufosinate resistance. However, no naturally occurring resistance mechanisms are known in plants.

DEVELOPMENT OF HERBICIDE-RESISTANT CROPS

Herbicides are an integral part of modern agriculture because they provide cost-effective increase in agricultural productivity. These have traditionally been discovered by screening novel compounds in a series of tests, including greenhouse and field testing, toxicological testing, etc., spreading over several years. These tests are done to meet various biological, environmental, toxicological and economic criteria, so that the new compounds can compete with the many excellent existing compounds. In the 1950s, about 1 in 2000 screened compounds were commercialized. By the 1970s, the rate had dropped to approximately 1 in 7000 compounds, while in the 1980s hardly 1 in 20,000 compounds emerged from these tests [101]. Thus, the herbicide industry has been finding it increasingly harder and far more expensive to develop new selective herbicides.

Selectivity is a function of the physicochemical properties of a compound and of the biochemical interactions of the compound with the crop and weeds. In developing selective herbicides, the chemical companies resort to a biorational approach in which the pathways of enzymatic degradation of herbicides, specific to the crop, are studied. The herbicidal moiety is then modified by adding a chemical group that will be recognized by those enzymes, thus aiding in the herbicide degradation by the crop. A number of important classes of herbicides such as triazines, sulfonylureas and imidazolinones have been developed by this approach. Another biorational approach is to make current herbicides more effective by adding synergists (safeners or another herbicide) that either prevent their breakdown or prevent detoxification of toxic products in some weeds.

In the past 15 years, biotechnologists in industry and the public sector have been resorting to another approach by making susceptible crops genetically and biochemically amenable to herbicides by conferring resistance. This genetic approach could remove a major factor in determining the choice of herbicides available for use by farmers. It could allow for the wider use of more effective herbicides with broader weed-control spectrums. In addition, compounds that are effective at low application rates have shorter persistence in the soil and also more favourable toxicological properties that might become more generally useful if the constraints of crop selectivity were removed. These possibilities would extend beyond major crops to minor crops that have not yet been benefited from effective herbicides. The introduction of herbicide resistance into crops could give the farmer greater flexibility in choosing crops for rotational or multiple cropping systems. Thus, the combination of chemical approach and genetic approach could provide the farmer a broad range of crop management options for more effective weed control.

Biotechnological Approaches

Herbicide-resistant crops may be developed by using the following biotechnological approaches.

Plant Breeding and Selection

The conventional plant breeding and selection approach is very hard to implement because of the low numbers of plants involved, low selection pressure and longer regeneration time. Even under high selection pressures, there may be many more susceptible plants that have escaped than plants that are genetically tolerant. If resistance is dominantly inherited on an allele of a single gene, one plant in 10^{-5} to 10^{-7} should be resistant and if it is recessively inherited on a single gene, the frequency is calculated at 10^{-10} to 10^{-14} [51]. For success of this approach, very high seeding rates would be required, many progenies would be needed in the first year and many years of field selection necessitated. Furthermore, there is always the possibility that there will not be a single gene for resistance in the genome and thus the many years of selection wasted [51].

However, Stannard [144] has been successful in obtaining resistance using field screening for chlorsulfuron resistance in alfalfa; he used high rates of chlorsulfuron and applied a very high selection pressure. After sowing 20 million alfalfa seeds in a field pre-incorporated with 16 g ha^{-1} chlorsulfuron, he found 15 putatively resistant plants. Four of these retained resistance after postemergence and laboratory treatments and have been cloned for breeding purposes. None of the resistant plants, however, showed resistance at the level of ALS, the target enzyme.

The plant breeding approach involves crossing, backcrossing and field testing before resistance is transferred into the varieties of choice. It requires two generations per year and at least five years of work before a herbicide-resistant crop variety can be field tested and released. Metribuzin resistance in soybean is inherited by a single dominant gene [39] and was transferred to an important but sensitive soybean variety. Similarly, paraquat tolerance in *Lolium perenne* (perennial ryegrass), polygenically inherited, has been used to breed resistance in perennial ryegrass [41]. The resultant material allows the use of low rates of paraquat in establishing stands of perennial ryegrass in pastures in England.

Many weed species show maternally inherited triazine resistance. Some of these weeds are sufficiently related to crops to allow transfer of the trait into related crop varieties. For example, transferring the trait from *Setaria viridis* (green foxtail) into the crop *Setaria italica* (foxtail millet) was moderately successful. The yields of the resultant crop varieties were considerably lower than the original variety [130]. This loss could be alleviated by further hybridization but so far the resultant hybrids have always yielded less than the reciprocal hybrids whose offspring were triazine-sensitive [12]. The loss of yield is probably due to the low fitness that is usually inherent in the chloroplast genome mutation conferring triazine resistance. The triazines are good candidates for developing crop resistance as they are the least expensive herbicides. However, there is the possibility of evolution of more triazine-resistant weeds if triazine herbicides are used continuously without rotating with non-photosystem II-inhibiting herbicides.

Resistance can be transferred by selection at the whole plant level. Seeds may be mutagenised to isolate resistant genes for genetic engineering, as in the case of *Arabidopsis thaliana*. Another plant in which the whole plant selection approach is used is the fern *Ceratopteris* spp. Selection pressure is applied on the photoautotrophic haploid gametophytes, allowing easier selection of recessive mutants for resistance to herbicides such as paraquat and acifluorfen. Paraquat resistance is monogenically inherited [65]. These mutants can be used, after analysis of fitness, for selecting the resistant genes for genetic engineering. Using similar methods, large numbers of mutation-treated soybean seeds were selected for chlorsulfuron tolerance [136]. However, none of the mutants had target site resistance at the level of ALS.

Tissue culture

Cell cultures have short generation times (doubling in 3-5 d), 10^5 to 10^6 cells ml^{-1}, with each cell theoretically representing a whole plant. This allows the possibility of very uniform treatment with herbicides at high selection pressures. Thus, cell cultures have many advantages over field cultures for selection of resistance. However, many species cannot be regenerated from tissue cultures and only some varieties of some species can be regenerated, although many recalcitrant crop species including maize, wheat and soybean have been regenerated from cell cultures in the recent past.

In tissue culture, selection can be done using non-photosynthetic cells, green tissues and protoplast fusion transfer.

Non-Photosynthetic Cell Culture: Non-photosynthetic cell tissue culture technique is used to select herbicide resistance to herbicides that do not specifically inhibit photosynthesis. Initial selection can be in suspension cultures, large callus species on microcalli derived from plating suspension cultures on a medium and then overlaying the herbicide [103]. The technique involves: a) preparation of cell suspension culture, b) spreading and growing the culture on agar medium containing the herbicide, c) identification and isolation of tolerant callus, d) testing of tolerant isolates for a second passage on herbicide-containing agar medium and e) development of shoots, which finally lead to fully grown plants. The plants are then used for genetic analysis via sexual crosses.

This technique has been used to develop tobacco resistant to picloram [20], paraquat [48] and sulphonylureas [21].

Selection has been achieved with a modicum of success with maize for resistance to imidazolinone herbicides. The resistant maize strains have been transferred to a seed producer for introducing the gene into commercial inbreds for eventual release [2]. Better resistance is achieved when both resistant alleles are present. The resistance is due to a site modification in acetolactate synthase (ALS). The imidazolinone-resistant maize has cross resistance to sulfonylurea herbicides, which act by inhibiting the same enzyme.

Green Tissue Cell Culture: Cell tissue culture technique involving green tissues is used to develop resistance to herbicides that affect photosynthesis. In this method, the seedlings and seeds are mutagenized by irradiation or chemical treatment. The young leaves of the mutagenized plant are then treated with herbicide and allowed to grow. As the leaves show chlorotic appearance, they develop small green islands of cells. These green cells are excised and propagated on a culture medium. The plants that grow from these propagated and regenerated plantlets are tested for herbicide tolerance and then used for genetic analysis via sexual crosses. Using this technique, nuclearly inherited resistance was introduced into tobacco for resistance to the photosynthetic inhibitors bentazon and phenmedipham. Later, tobacco having chloroplast-inherited streptomycin resistance was obtained after using a specific DNA [43]. This technique, using seedlings and isolating 'green islands', is useful in developing resistance to PS II-inhibiting herbicides that interact with the *psbA* gene product.

Protoplast Fusion Transfer: Protoplast fusion transfer technique is used to transfer resistance to a related but non-cross-breeding species. This interspecific protoplast fusion achieves the transfer of genetic material, but the resultant plants are usually infertile. This procedure is useful only when the genes for resistance have unknown products and thus are hard to isolate by molecular biological procedures for genetic engineering. It is also useful when there is no easy way to introduce the known gene(s), as in the case of chloroplast inherited genes.

In the protoplast fusion (hybridization) transfer procedure [157], the protoplasts of triazine-resistant donor are irradiated (X-ray) to cause functional enucleation (inactive nucleus) as the nuclear genome is undesirable in the product. The irradiated protoplasts are then fused with protoplasts of the susceptible crop. The subsequent cell division gives calli and plants with one or the other plastid type. This protocol is the modified technique used to transfer cytoplasmic male sterility. Triazine resistance has been transferred from resistant *Solanum nigrum* to sensitive potato [53, 157] and tobacco using the protoplast fusion procedure.

GENETIC ENGINEERING OF HERBICIDE RESISTANCE IN CROPS

Genetic engineering in crops for herbicide resistance has become a major area of weed research for over a decade. The process of genetic engineering involves isolating the gene from resistant plants or microorganisms (fungi), inserting it into plasmid and then into a vector system, introducing it into plant cultures and regenerating the cultures as a newly resistant crop plant. The main advantage of the genetic engineering approach is that once the gene is successfully engineered into a plasmid vector system, it can be used to transform many different plant species. Besides *Solanaceous* species, which have been used as models, the *Agrobacterium* infection systems (with model marker genes) have been used to successfully transform soybean, cotton, sunflower and maize [51]. Many species have been transformed by incubating protoplasts with plasmid DNA and pushing it through the cell membrane with polyethyleneglycol, or electrophorating it through the membrane with a high voltage pulse or even pelting cells with DNA or microprojectiles [79].

The progress of genetic engineering with respect to development of resistance of major classes of herbicides shall now be discussed.

Engineering for Acetolactate Synthase Resistance

As mentioned earlier in this chapter, the sulfonylurea and imidazolinone herbicides inhibit acetolactate synthase (ALS). In tolerant crops, resistance to ALS inhibitors is based on enhanced capacity for metabolic inactivation rather than differential herbicide uptake, translocation, or altered susceptibility of the ALS target site. Wheat is resistant to chlorsulfuron by way of metabolic inactivation involving aryl hydroxylation followed by conjugation to herbicidally inactive glucose conjugates. Hydroxylation is catalyzed by microsomal cytochrome P450. In susceptible crop species, the same metabolic inactivation can occur but with insufficient activity to confer resistance. Gene transfer and tissue culture technology have been used to produce transgenic plants with less susceptible ALS [110, 114].

Earlier efforts were directed to isolate microbial (yeast and bacteria) genes with a resistant ALS. A single amino acid difference between the resistant and susceptible ALS was found in bacteria. The ALS genes from microbes have been isolated, cloned and sequenced. However, there have been problems transforming plants using the bacterial genes, possibly because of promoters. The bacterial ALS is composed of two different subunits, while plant ALS is localized in the chloroplast. Thus, the generation of herbicide-resistant plants using bacterial ALS genes requires not only an expression of these two protein subunits of the enzyme, but also their translocation into and assembly in plant chloroplasts. For this reason, the isolation and use of plant ALS genes appears to be a more attractive route for engineering sulfonylurea-resistant transgenic crop plants.

Plant species such as *Nicotiana tabacum* and *Arabidopsis thaliana* possess ALS genes. Many crop species carry multiple ALS genes. Efforts have been made to isolate genes carrying ALS mutations from herbicide-resistant plants. In tobacco, the **Hra** and **C₃** lines, which carry mutations at the *SURB* locus and *SURA* locus respectively, were used as sources of mutant ALS genes. Four genes representing all of the ALS loci from these two mutant lines were isolated and sequenced. The *SURB-Hra* gene contained two mutations, which resulted in **proline 196-to-alanine** and **tryptophan 573-to-leucine** substitutions. The *SURA-C₃* gene carried a single mutation that resulted in a **proline 196-to-glutamine** substitution. The double tobacco mutant that carried the proline 196-to-alanine and tryptophan 573-to-leucine susbstitutions were cross resistant to imidazolinone herbicides.

The tobacco *SURB-Hra* gene was used to transform a number of heterologous species to sulfonylurea herbicide resistance at the cellular and in some cases, whole plant level. These crop species include tomato, lucerne, lettuce, melon, rapeseed and sugarbeet [10]. In some of these species (e.g., tomato), expression of the resistant tobacco ALS gene was efficient while in others (rapeseed) only a low level of resistance was observed.

The *SURB-Hra* gene was introduced into a number of commerical lines of tobacco and the transformed plants showed high levels of resistance in the field. The transformed plants showed no damage at 32 g ha⁻¹, four times that of a typical field application rate. Thus, expression of the *SURB-Hra* gene provides an effective means of producing sulfonylurea herbicide-resistant crops.

As metabolic detoxification is the basis of selectivity of certain crops to sulfonyl ureas, isolation of genes responsible for the detoxification pathways from resistant crops and transferring them to sensitive ones could provide another approach for genetically engineering sulfonylurea resistant crops. This method requires a microbial model. In *Streptomyces griseolus*, chlorimuron ethyl, chlorsulfuron and sulfometuron methyl are metabolized by a three-step process, requiring an NADP-reductase, an iron-sulphur protein and cytochrome P450 enzymes. The genes encoding these proteins could be isolated from this bacterium and introduced into plants by transformation to confer herbicide resistance.

As most of the agronomically important crops are not amenable to transformation, other alternative methods were used to generate resistance to the sulfonylureas and other ALS-inhibiting herbicides. Using the genetic selection of cultured maize cells, imidazolinone-resistant mutants of maize were isolated. Plants exhibiting more than 100-fold increase in resistance to imidazolinones and cross resistance to the sulfonylureas were regenerated from one line. The homozygous progeny showed more than 300-fold increase in resistance to the herbicides. Resistance was inherited in a single dominant nuclear gene. The resistant mutant, which did not show growth or yield reduction, is being developed by Pioneer Hi-Bred International, Inc. [101]. This maize mutant will be the first ALS-targeted herbicide-resistant crop when commercialized.

In another effort, mutagenized soybean seeds were used to obtain herbicide-tolerant soybean mutants. While some of the mutants showed only a 5 to 10-fold increase in tolerance, with little alteration in ALS, others recorded higher levels of tolerance with an altered ALS.

Engineering for Photosystem II Resistance

Such structurally diverse herbicides and classes of herbicides as triazines, ureas, uracils, propanil (acetamide), pyrazon (pyridazinone), desmedipham and phenmedipham (phenylcarbamates) and bromoxynil and ioxynil (benzonitriles) inhibit electron transport in PS II. Triazine herbicides are metabolized in a naturally resistant crop like maize by several pathways, including 2-hydroxylation, conjugation with glutathione and, to a lesser extent, N-hydroxylation of the side chains with subsequent oxidation. Selectivity to substituted urea herbicides is due to limited absorption and translocation, rapid cytochrome P450 mono-oxygenase-catalyzed degradation, or oxidation. Thus, most PS II inhibitors are selective by virtue of degradation by crops.

Triazine-resistant weeds have been used to generate resistant crops. The resistance trait has been transferred by sexual crosses from the weed *Brassica campestris* (bird's-rape mustard) into several related cole crops, leading to release of atrazine-resistant *Canola* (rapeseed) cultivar 'Triton' [13]. Despite a yield penalty of about 20% due to resistant cytoplasm, this R cultivar was planted on about 250,000 acres in 1986 and proved economical in areas where weed densities were high and triazines were used (44). In another case, resistance from the weed *Setaria viridis* (green foxtail) was crossed into the sensitive crop *Setaria italica*. Such interspecific and intergeneric sexual crosses were hampered because only a few resistant weeds are sexually compatible with crop plants. Protoplast fusion has also been used to transfer atrazine resistance between *Solanum* lines but with only limited success.

Even with the availability of a number of resistant genes, genetic engineering of triazine-resistant crop plants via transformation has been difficult. This was because of difficulties in introducing genes into chloroplasts. To bypass this technical problem, transgenic tobacco plants were produced by following nuclear transformation with a mutant *psbA* gene, which encodes for the D1 protein (the PQ_B protein). To accomplish this, a chimeric gene was constructed in which a nuclear promoter and a chloroplast transit peptide-encoding sequence were attached to a mutant *psbA* structural gene. The transgenic plants showed higher tolerance for atrazine.

In other studies, attempts were made to engineer atrazine resistance through the introduction of genes encoding detoxifying enzymes, glutathione-s-transferases (GSTs) from maize resistant to atrazine and alachlor. The transgenic tobacco plants expressed an atrazine-resistant GST enzyme, resulting in increased tolerance to atrazine.

Engineering for ACCase Resistance

Dicot crops and non-graminaceous monocot crops exhibit intrinsic resistance to aryloxyphenoxypropionic (APP) and cyclohexanedione (CHD) herbicides. Dicot crops possess resistant ACCase and many of them are capable of detoxifying APPs and CHDs. Wheat and barley are resistant only to diclofop and tralkoxydim, while most other major graminaceous crops are sensitive to all APPs and CHDs.

In laboratory studies, resistance to APPs and CHDs has been introduced into crops. Resistance to diclofop and sethoxydim was selected in maize tissue cultures exposed to sethoxydim [117, 118]. In one study, resistance was due to increased expression of herbicide-susceptible ACCase, while in the other, resistance in regenerated plants was associated with herbicide-resistant ACCase. The trait of resistance inheritance was controlled by a partially dominant nuclear mutation. A subsequent study [96] suggested that three and possibly five, alleles of the maize ACCase structural gene were present in ACCase mutants which exhibited different levels of herbicide resistance.

Engineering for Auxinic Herbicide Resistance

Many monocot crops are naturally resistant to auxinic herbicides. It is believed that resistance of cereal crops to 2,4-D is due to differences in rates of herbicide translocation, metabolism and anatomical characteristics. In a resistant crop like wheat, 2.4–D is arylhydroxylated through the catalytic role of cytochrome P450 mono-oxygenases and glucosylated. This was confirmed by recombinant studies.

Soil-borne microbes, in particular *Alcaligenes* spp., contain plasmids that confer ability to degrade auxinic herbicides. A gene encoding for a mono-oxygenase capable of degrading 2,4–D has been identified, cloned and sequenced from a plasmid of *Alcaligenes*. A single, dominant gene from *Alcaligenes* coding for a 2,4–D-degrading mono-oxygenase has been introduced into tobacco, resulting in transgenic plants resistant to high levels of 2,4–D [89]. Commercial cultivars of otherwise susceptible cotton have been transformed with this 2,4-D-resistance gene, producing transgenic plants resistant to 2,4–D [88]. These studies show that increased cytochrome P450-catalyzed metabolism is an important mechanism that can confer crop selectivity to auxinic herbicides.

Engineering for Resistance to Other Herbicides

BROMOXYNIL: Bromoxynil, a benzonitrile herbicide, inhibits photosynthetic electron transport. It is used for dicot weed control in cereal crops. A gene encoding for nitrilase enzyme that detoxifies bromoxynil has been isolated from a strain of soil bacterium *Klebsiella ozaenae*, which uses the herbicide as its sole source of nitrogen. This strain produces a bromoxynil-specific nitrilase protein that degrades the herbicide to 3,5-dibromo 4-hydroxybenzoic acid in a single step. The **bxn** gene that encodes this enzyme was plasmid-borne and it was later cloned. The gene has been placed under the control of plant promoters and has been transferred to tobacco and tomato plants to confer resistance to bromoxynil in these plants.

GLYPHOSATE: In view of the efficacy and desirable environmental and toxicological properties of glyphosate, considerable efforts have been made in the past 10 years to increase the utility of this herbicide by producing resistant crops.

The first report of introduced glyphosate resistance in transgenic plants, using commercial applications of plant biotechnology, was published in 1985 by Comai et al. [23]. A mutant *Salmonella typhimurium aroA* gene, encoding a resistant EPSP synthase, was introduced into tobacco cells under the control of octopine synthase or mannopine synthase promoter. The transformed regenerated plants expressed chimeric enzyme levels in the leaves and were two or three times more resistant to glyphosate, although their growth rate was reduced compared to untreated controls. The same gene was also introduced into tomato plants subsequently, with essentially similar results [42].

Glyphosate resistance has also been achieved in crop plants by overproduction of EPSP synthase. A wild petunia cDNA encoding EPSP synthase was linked to the constitutively expressed cauliflower mosaic virus 35S promoter and reintroduced into petunia [78, 139]. The resultant transgenic plants were about 4-fold more resistant to glyphosate, with growth rates lower than those in untreated controls.

The relatively low levels of resistance and poor field performance of transgenic plants, described above, led to efforts to achieve higher levels of resistance. As part of these efforts, the **petunia gene** was subjected to site-specific mutagenesis and a number of mutant enzymes were recovered and tested [77]. Some of these genes overexpressed in transgenic tobacco, enabling the plants to develop 4-fold more resistance to glyphosate, with growth rates and field performance being equal to plants in untreated control. Similar constructions, including recently isolated mutant genes, with very little or no loss of catalytic efficiency, were later introduced into several crop species. These mutant genes, some under the control of meristem-specific promoters, provide high levels of glyphosate resistance. These efforts may lead to the evolution of glyphosate-resistant soybean, tomato and canola cultivars.

Certain bacteria belonging to genera *Pseudomonas* and *Arthrobactor* are capable of growing on glyphosate as a sole source of carbon. These strains metabolize glyphosate to phosphate, glycine and a one-carbon unit. This is in contrast to soil microbial metabolism of glyphosate being primarily degraded to aminomethylphosphonic acid. The bacterial genes that carry out these metabolic degradations could theoretically be isolated and expressed in plants to produce glyphosate-tolerant plants. No glyphosate-resistant plants have yet been reported in which expression of closed genes enables metabolism of glyphosate to non-toxic products.

GLUFOSINATE: Several approaches have been used to produce plants tolerant to glufosinate, also known as phosphinothricin. In the first one, an alfalfa (lucerne) cell line tolerant to phosphinothricin (PPT), producing elevated levels of GS, was isolated following selection. The amino acid sequences of two internal GS peptides were isolated from this line and used to confirm the identification of an alfalfa cDNA clone. The corresponding genomic DNA clone for alfalfa GS was 4 kb long and contained 11 introns [34]. The encoded protein was 356 amino acids long. In order to produce herbicide-resistant transgenic plants, the alfalfa GS cDNA clone was linked to the cauliflower mosaic promoter and introduced into tobacco plants via transformation. The resultant high level of expression of the alfalfa GS gene in the transgenic tobacco plants conferred a low level of tolerance to PPT [38].

In another approach, a bacterial expression assay system was used. An alfalfa GS cDNA clone was linked to a bacterial promoter and transferred to a strain of *Escherichia. coli* devoid of GS activity [29]. Mutations that allowed bacterial growth in the presence of PPT were then selected. A mutant alfalfa GS gene was then transferred to tobacco cells and plants were generated. Although the mutant, GS gene protected plants when herbicide was taken up by the roots, it did not protect against foliar application of PPT. This indicated the existence of multiple nuclear GS genes in plants [147]. In pea, one GS gene encodes an isozyme found in nodules, a second encodes a GS enzyme found in chloroplasts and two others encode cytoplasmic forms of GS. This suggests that resistant mutations may need to be introduced into several GS genes to confer whole-plant resistance to PPT.

An alternative approach of introducing resistance into plants via a gene that produces a detoxifying enzyme has been more successful. The gene, designated *bar*, that encodes for PPT-acetyl transferase activity is found in strains of *Streptomyces hygroscopicus* that produce bialaphos, the tripeptide precursor of PPT. This enzyme, PPT-acetyl transferase, converts PPT to a non-phytotoxic acelyated metabolite by transferring the acetyl group from acetylcoenzyme A onto the amino group of PPT. The *bar* gene isolated from *S. hygroscopicus* was placed under the control of the cauliflower mosaic virus 35S promoter and introduced into tobacco, potato and tomato. The genetically engineered crops were thus able to rapidly detoxify PPT (glufosinate). The trait is inherited as a simple dominant allele. The transgenic crops produced in this manner are in various stages of field testing for possible commercialization.

MANAGEMENT OF HERBICIDE RESISTANCE IN WEEDS

Development of herbicide resistance in weeds is a worldwide phenomenon and the number and frequency of resistant biotypes have increased at an alarming pace in recent years. During the past 10 years, herbicide resistance has increased at a rate equivalent to that of insecticide and fungicide resistance. The number of insecticide-resistant biotypes increased over 500-fold between 1948 and the 1980s [67]. A similar phenomenon has occurred in plant pathogens and microorganisms in developing resistance to fungicides and antibiotics [140]. The rapid spread of resistance to various pesticides is due to many reasons, including the difficulty in predicting the occurrence and spread of resistance and the lack of viable and effective alternatives to pesticides.

Much before weed resistance to triazines was reported in the mid-1960s, Whitehead and Switzer [158] found that control of wild carrot (*Datura carota*) by 2,4-D was diminished after several years of use in Ontario, Canada. Despite frequent reports of weed resistance to triazines, dinitroanilines, bipyridiliums and pyridazinones in the 1970s and early 1980s, only a few weed species were identified as resistant and the area infested by resistant biotypes was only limited. This relatively slow development of resistance to other classes of herbicides coupled with the much lower fitness (of particularly the triazine-resistant biotypes), the ability of herbicide-thinned weed populations to produce high amounts of seed and the large reservoir of susceptible weed seeds in the soil, did not warrant considering herbicide resistance a major problem [128]. However, beginning late 1980s, there has been a rapid and widespread increase in the number of biotypes that developed resistance to a wide range of herbicides, including the ACCase inhibitors (aryloxyphenoxypropionics, cyclohexanediones, etc.), ALS inhibitors (sulfonylureas, imidazolinones and triazolopyrimidines), acetamides, arsenicals, nitriles, carbamates, substituted ureas and uracils [82]. The problem assumed alarming proportions when the R biotypes of *Lolium rigidum* and *Alopecurus myosuroides* exhibited cross resistance to two or more herbicides. Some of these biotypes have the ability to rapidly metabolize a large number of herbicides to non-phytotoxic forms via the same mechanism that allows these compounds to be used successfully in wheat [25, 98]. If this trend continues, the number of effective herbicides for use in crops could be severely reduced. This necessitates development of alternative weed management programmes before resistance becomes a greater problem. Before dealing with weed management strategies, the factors responsible for development of herbicide resistance in weeds need to be examined.

Factors Responsible for Development of Herbicide Resistance

Continuous Use of Herbicides

The continuous use of a herbicide or herbicides sharing the same mode of action as the primary method of weed control is a major factor in the development of herbicide resistance in weeds. LeBaron [82] reported that the time required for weeds to develop resistance ranged from 18 yr (from 1945, the year first introduced, to 1963, the year first reported) for 2,4-D, to 10 yr for atrazine (1958 to 1968) and trifluralin (1963 to 1973) and 5 yr for diclofop (1977 to 1982) and chlorsulfuron (1982 to 1987). Resistance largely depends on the herbicide and its use pattern. If the same herbicide is used in each crop, crop rotation does little to arrest resistance development, as in the case of selection for resistance to ACCase inhibitors (APPs and CHDs). Malik and Singh [90] found that control of *Phalaris minor* in wheat by isoproturon, in continuous use in India since the early 1980s, reduced from 78% to 21% from 1990 to 1993.

Seedbank in the Soil

Development of herbicide resistance development is more rapid whenever weed seed turnover in the soil is rapid. This turnover is characteristic of weed species (e.g., *Kochia scoparia*) and is aided by cultural practices [140]. Changes in tillage practices may result in a more rapid turnover of the seedbank population. Ball [6] found that the number of seeds in the upper 2.5 cm of soil increased after chisel ploughing compared to mouldboard ploughing. Application of farmyard manure from one field to another is another means of resistance development.

Farming Practices

Ever since the discovery of herbicides, weed scientists and farmers have exploited the efficacy and cost effectiveness of herbicides to maximize crop productivity and economic returns from the land. As a part of their efforts, however, the traditional multicrop systems were discarded in favour of monocrop systems and the normal tillage practices that eliminated persistent weeds more effectively were replaced by no-till or minimal practices, to fit in the existing herbicides and reduce the time needed for weed control. The consequence of this excessive reliance on herbicides is the selection of resistant weed species and biotypes.

Farmer Ignorance

For most farmers, both in developed and developing countries, herbicide resistance is not a high priority issue. Stoller et al. [145] reported that soybean farmers ranked development of strategies for resistant weeds only 7th in 14 categories related to weed management, with the first rank going to development of more economical weed control. This suggested that farmers generally are more concerned with short term economics of weed management than with herbicide resistance management. Most farmers rely on herbicides with the same mode of action and show reluctance to change their weed control programmes.

Strategies to Manage Herbicide Resistance

The selection and evolution of herbicide-resistant weed population, if left unattended, will have serious consequences on farmers' ability to control weeds safely and efficiently. Hence, it is essential for weed scientists to design strategies to manage problems related to herbicide resistance in weeds effectively and economically. These strategies are briefly discussed below.

Containment of Resistant Alleles

Containment of resistant alleles conferring herbicide resistance on weeds is one way of minimizing the spread of herbicide resistance. This strategy requires sound knowledge of the ecology, genetics and biochemistry of weed species undergoing selection. Models of dynamics of resistant and susceptible weed populations serve as a useful tool for understanding the factors that determine rates of evolution of resistance in the field, particularly fitness and gene flow. Screening of unselected weed populations for resistance alleles is valuable in identifying weed species that are likely to become resistant.

Weed scientists and plant geneticists will find it more challenging to understand the genesis of variation in weed populations than in crop plants. For example, in *Avena* spp., there is more genetic diversity in populations found in the habitat of origin than in a population found on agricultural land. This increased diversity may be due to increased niche differentiation driven by higher species diversity in natural plant communities. The species that have evolved resistance to herbicides would naturally be from those with the most genetic diversity. Powles and Matthews [122] reported that there were a number of different resistant mechanisms that contribute to multiple herbicide resistance in *Lolium rigidum* and they attributed the rapid evolution of multiple resistance to general genetic diversity within the species. Guttieri et al. [55] found that target site resistance of *Kochia scoparia* to ALS inhibitors was due to several differ-ent amino acid substitutions, indicating that genetic variation exists within a single mechanism of resistance.

Introduction of Herbicide-Resistant Crops

Introduction of herbicide-resistant crops may help in mitigating the problem of herbicide resistance in weeds. While certain benefits may be realized from using the herbicide-resistant crops, at least initially, resistant-weed problems may be exacerbated when appropriate resistance management is not practiced [68].

Seedbank and its Exhaustion

The simplest option in preventing development of herbicide resistance is to use techniques to reduce weed populations such that the expected weed numbers remain below the economic threshold. A successful adoption of these techniques requires knowledge of seedbank dynamics because the life of the seed propagules in the soil dictates the length of time that weed control measures are needed to be maintained in order to reduce the reservoir of resistant seeds in the seedbank to a low level.

Generally, annual grass weeds such as *Lolium rigidum, Lolium multiflorum, Lactuca serriola, Kochia scoparia, Setaria viridis*, etc. have a seedbank life of 3 to 5 yr, with an annual rate of decline of about 50% to 80%. The R and S biotypes of a weed species, as in the case of *K. scoparia* [36], exhibit different germination patterns. The viable population of most of the weed species in the seedbank can be substantially reduced by preventing seed set for 3 to 5 yr.

Methods to reduce the number of viable seeds in the soil and thereby exhaust the seed bank, include:

  a) application of low rates of nonselective herbicides such as paraquat and glyphosate to kill the existing weed growth and prevent the development of fertile seed;
  b) soil disturbance during crop planting to stimulate germination of some weed species; and
  c) adopting chemical or plough fallow method before a crop is planted, so the germinating weed seedlings can be controlled.

Modified Herbicide Usage

The modified herbicide usage approach includes rotation of herbicides with different modes of action and use of combinations of herbicides with different modes of action.

Herbicide Rotation: The logic behind herbicide rotation is that exposure of weeds to different classes of herbicides will decrease the frequency of resistance alleles. This decrease is expected as long as weed species with resistance alleles are less fit than weed species without these alleles. When one herbicide is very useful but prone to resistance problems, it may be useful to rotate it with another herbicide that may be less effective but more reliable. However, before a certain herbicide rotation is recommended, the following factors need to be considered [50]. 1. It is important to understand that rotation of herbicides must involve alternate generations of the weed species. If the same field is treated twice within a single weed generation with two compounds, there is no period during which significant reversion is expected. 2. The cost of resistance is almost always measured in the absence of any herbicide. For a rotation to be successful, it is important for the 'cost' to be expressed when the rotational compound is used. If under the stress of the alternate compound the weed genotypes with resistance to the first compound are as fit as the susceptible genotypes, then the utility of the rotation is compromised. Besides, if there is any cross resistance between the two compounds, a rotation is not generally expected to show the overall resistance build-up. Gould [50] suggested that from a practical perspective, it may be difficult to implement rotations unless two or more chemicals are of approximately equal cost to the farmers, or farmers are acutely concerned with a resistance problem. In order to provide the farmer a viable and useful rotation programme, it would be necessary for weed scientists to examine, over a period of time, the changes in the frequency of resistance alleles in single fields where rotations are used, so that the empirical basis for judging the rotations can be strengthened.

Herbicide Mixtures: It has been suggested that the use of herbicide mixtures combining herbicides of different modes of action would substantially delay or preclude evolution of resistance to the more vulnerable or at-risk herbicides. This is because weeds resistant to the vulnerable herbicide would be destroyed by the mixing partner, or at least be rendered relatively unfit compared to the wild biotype.

Herbicide mixtures are of two types. In the first type, the **broad spectrum mixture (BSM)**, each member of the mixture affects different weed spectra as in the case of mixtures of grass and broadleaf herbicides (Table 10.1). These BSMs increase weed control benefits but have no influence in delaying evolution of resistance in weeds they do not affect. They might even exacerbate resistance by controlling competing weeds, creating a more open niche for resistant biotypes [160]. This could be true even when the mixing partner has a moderate effect on a species excellently controlled by the vulnerable herbicide. The second type, the **target weed mixtures (TWM)**, both the vulnerable herbicide and the mixing partner, aim at complete kill of the same weed species. In order for the TWM to be effective in preventing resistance, the mixing partner should have the following traits as compared to the vulnerable herbicide: a) control the same spectra of weeds, b) have the same persistence, c) have a different target site, d) be degraded in a different manner and e) preferably exert negative cross resistance [160]. These are elaborated as below.

Table 10.1. Herbicide mixtures and their effects on different weed species.

Type of mixture	Herbicide	Weed Species	
		A	B
Broad Spectrum Mixture (BSM)	Vulnerable herbicide	++	0
	Mixing partner	0	++ (or +)
Target Weed Mixture-I (TWM)	Vulnerable herbicide	++	0
	Mixing partner	++	0
-II (TWM)	Vulnerable herbicide	0	++
	Mixing partner	0	++

With respect to control of the same target weed species, the mixing partner should effectively kill or severely weaken the weed most sensitive to the vulnerable herbicide because the target weed is most likely to evolve resistance. It will be of little use if the vulnerable herbicide kills only 95% of the target weed population and the mixing partner kills only 75%, unless the remaining 20% of the population are severely inhibited to incapacitate its reproduction. Otherwise, resistance could quickly evolve in the remaining 20% of the weed population. As seeds of many weed species have many flushes of germination during a cropping season, both components of mixture should have similar persistence to prevent weed germination throughout the cropping season. If the mixing partner has a shorter period of activity than the vulnerable herbicide, then the latter chemical will select for plants resistant only to it after the former has dissipated.This, however, may be a problem if all seeds of the target weed germinate in a single flush without further germination and if both herbicides outlast the flush with equal efficacy.

In a TWM, the ideal mixing partner should have a target site of action different from that of the vulnerable herbicide. Two Photosystem II inhibitors (triazines and substituted ureas or uracils) or two ALS inhibitors (sulfonylureas and imidazolinones) cannot be considered ideal target weed mixtures for resistance management. Furthermore, the two herbicides should not be degraded in an identical manner. For example, if the vulnerable herbicide is degraded by a glutathione-s-transferase, the mixing partner will then have no chemical site that can be attacked by that enzyme. Possessing a negative cross resistance is a useful attribute for the mixing partner. In this, weeds resistant to the vulnerable herbicide become more susceptible than the wild type to the mixing partner.

Alternative Weed Control Methods

Excessive dependence on herbicides for weed control is the primary reason for the evolution of herbicide resistance in weeds. Integrating herbicide usage with manual, mechanical and biological methods could help reduce the herbicide selection pressure on the weed populations while maintaining crop productivity. This approach would lead to delaying and hence mitigating the herbicide resistance problems.

Implementation of Herbicide Resistance Management Programmes

Successful implementation of programmes related to herbicide resistance management requires an integrated and cooperative approach between research/academic institutions, herbicide industry, government and farmers. Each of these parties has a specific role and responsibility (Fig. 10.1) [140].

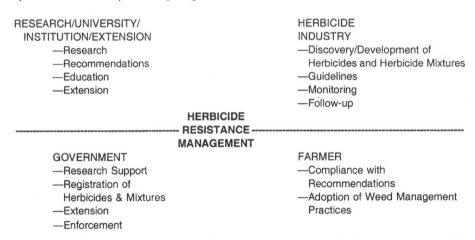

Fig. 10.1. Role of various organizations (research/academic institutions, herbicide industry, government and farmers) in herbicide resistance management

Research/Academic Institutions

Research and academic (universities) institutions, the primary source of information for weed management for farmers, are required to intensify research on herbicide resistance in weeds and develop strategies for effective resistance management programmes. There is a need to integrate chemical weed control with other methods of weed control, monitor the development and spread of herbicide-resistant biotypes and provide the farmers cost-effective, ecologically sound and resistance-mitigating weed management programmes. They should also develop different models to predict yield losses due to weed pressure and the rate of selection of herbicide-resistant biotypes [52a, 100]. These models are useful in determining which factors are important in selecting for herbicide resistance and in assessing the effectiveness of different management strategies. The models should be thoroughly field-tested.

The extension personnel should be educated and trained in various aspects of weed management and herbicide resistance management so they can form a vital link between the researcher and farmer.

Herbicide Industry

Herbicides will continue to remain as the principal method of effective and economical weed control. Hence, the herbicide industry will continue to be the source of new herbicides and herbicide mixtures. They, however, need to monitor development of resistance to herbicides and take steps to minimize the selection and spread of resistance by developing target weed mixtures consisting of two or more herbicides with different modes of action and those that meet the criteria listed earlier (vide Herbicide Mixtures). The herbicide mixtures should also control an overlapping spectrum of weeds that are high risk for developing resistance. The industry may also need to work with farmers to bring out weed management programmes that do not depend on a single herbicide or mode of action for control. It should also monitor changes in weed spectra and resistance management problems at the field level and develop alternative management programmes. The companies need to educate and train their staff, farmers and herbicide distributors and retailers and certain government agencies on effective herbicide resistance management. They may also work with research personnel of research and academic institutions and government in the implementation of resistance management programmes.

Government

Government agencies, both at the central (federal) and state level and their policies have a great impact on the implementation of effective herbicide resistance management programmes. As herbicide usage has created a multitude of secondary problems (environmental, weed resistance, etc.), several countries began taking steps, in the recent past, to reduce herbicide use, eliminate registration of older herbicides, scale down the number of registered products and shift from dependence on broad-spectrum herbicides to narrow-spectrum herbicides. As a consequence, herbicide resistance problems became more aggravated and widespread than ever.

Government should encourage and support research programmes that offer more options for farmers to tackle herbicide resistance in weeds. Herbicide usage may need to be integrated with other means of weed control. As herbicides will continue to be used extensively and be the vital part of weed management programmes, more herbicides and herbicide mixtures (as premixes) may need to be made available to farmers. Government is required to fund research on integrated weed management (IWM) programmes, incorporating herbicides as one of the management components. It should also educate and train its personnel in various aspects of weed management, particularly in herbicide-resistant management.

Farmer

Currently, herbicide resistance is not a high priority issue for most farmers. Hence, the success or failure of a herbicide resistance management programme depends on the farmer. If farmers do not change weed control practices that lead to weed selection for herbicide resistance or cannot afford to change their programmes because of financial and logistic constraints, then all the efforts of government, research and academic institutions and the herbicide industry will be fruitless [140]. Farmers should be educated in the principles of weed management and development programmes that they can adapt in their farming practices. Reducing reliance on herbicides as the primary means of weed control and moving towards integrated weed management (IWM) programmes will require the cooperation and compliance of farmers. The IWM systems should be flexible enough to allow adjustments to changes in environmental, technological and social factors.

REFERENCES

1. Anderson, B. and S. Styring. 1991. Photosystem II: Molecular organization, function and acclimation. Curr. Topics Bioenergetics. **16**: 1-81.
2. Anderson, P.C. 1986. Cell culture of herbicide tolerant corn and its ramifications. Proc. 41st Annual corn and Sorghum Industry Res. Conf. Public, no. 41, Amer. Seed Trade Assoc., Washington, D.C.
3. Anderson, R.N. and J.W. Gronwald, 1987. Noncytoplasmic inheritance of a triazine tolerance in velveltleaf (*Abutilon theophrasti*). Weed Sci. **35**: 496-498.
4. Anderson, R.N. and J.W. Gronwald. 1991. Atrazine resistance in velvetleaf (*Abutilon theophrasti*) biotype due to enhanced glutathione-s-transferase activity. Plant Physiol. **96**: 104-109.
5. Arntzen, C.J. and J.H. Duesing. 1983. Chroplast-encoded herbicide resistance. *In* K. Downey, R.W. Voellmy, F. Ahmad and J. Schultz (eds.). Molecular Genetics of Plants and Animals. Academic Press, New York, pp. 273-294.
6. Ball, D.A. 1992. Weed seed bank response to tillage, herbicides and crop rotation sequences. Weed Sci. **40**: 654-659.
7. Bandeen, J.D., G.R. Stephenson and E.R. Cowett. 1982. Discovery and distribution of herbicide resistant weeds in North America. *In* H. LeBaron and J. Gressel (eds.). Herbicide Resistance in Plants. John Wiley & Sons, Inc., New York, pp. 9-30.
8. Barnwell, P. and A.H. Cobb. 1989. Physiological studies of mecoprop-resistance in chickweed (*Stellaria media* L.). Weed Res. **29**: 135-140.
9. Barr, A.R., A.M. Mansooji, J.A.M. Holtum and S.B. Powles. 1992. The inheritance of herbicide resistance in *Avena sterilis* ssp. *ludoviciana*, Biotype SAS 1. Proc. First Intl. Weed Control Congress, Melbourne: Weed Sci. Soc., Victoria, pp. 70-72.
10. Bedbrook, J., R.S. Chaleff, S.C. Falco, B.J. Mazur and N. Yadav. 1988. Nucleic acid fragment encoding herbicide resistant plant acetolactate synthase. Eur. Patent Appl. 0257993.
11. Bettini, P., S. McNally, M. Sevingnac, H. Darmency, J. Gasquez and M. Dron. 1987. Atrazine resistance in *Chenopodium album*: low and high levels of resistance to the herbicide related to the same chloroplast *psbA* gene mutation. Plant Physiol. **84**: 1442-1446.
12. Beversdorf, W.D., D.J. Hume and M.J. Donnely-Vanderloo. 1988. Agronomic performance of triazine-resistant and susceptible reciprocal spring Canola biotypes. Crop Sci. **28**: 932-934.
13. Beversdorf, W.D., J. Weiss-Lerman, L.R. Erickson and V. Souza Machado. 1980. Transfer of inherited triazine resistance from bird's rape to cultivated *Brassica campestris* and *Brassica napus*. Can J. Genet. Cytol. **22**: 167-172.
14. Boerboom, C.M., D.L. Wyse and D.A. Somers. 1990. Mechanism of glyphosate tolerance in birdsfoot trefoil (*Lotus corniculatus*). Weed Sci. **38**: 463-467.
15. Burnet, M.W.M. 1992. Mechanisms of herbicide resistance in *Lolium rigidum*. Ph.D. thesis. Univ. of Adelaide, Australia.

16. Burnet, M.W.M., B.R. Loveys, J.A.M. Holtum and S.B. Powles. 1993a. Increased detoxification is a mechanism of simazine resistance in *Lolium rigidum*. Pestic. Biochem. Physiol. **46**: 207-218.

17. Burnet, M.W.M., B.R. Loveys, J.A.M. Holtum and S.B. Powles. 1993b. A mechanism of chlortoluron resistance in *Lolium rigidum*. Planta **190**: 182-189.

18. Burton, J.D., J.W. Gronwald, R.A. Keith, D.A. Somers, B.G. Gengenbach and D.L. Wyse. 1991. Kinetics of inhibition of acetyl-coenzymeA carboxylase by sethoxydim and diclofop. Pestic. Biochem. Physiol. **39**: 100-109.

19. Carroll, E.W., O.J. Schwarz and L.G. Hickok. 1988. Biochemical studies of paraquat-tolerant mutants of the fern *Ceratopteris richardii*. Plant Physiol. **87**: 651-654.

20. Chaleff, R.S. and M.F. Parson. 1978. Direct selection *in vitro* for herbicide resistant mutants of *Nicotiana tabacum*. Proc. Natl. Acad. Sci., USA, **75**: 5104-5107.

21. Chaleff, R.S. and T.B. Ray. 1984. Herbicide-resistant mutants from tobacco cell cultures. Science **223**: 1148-1151.

22. Christopher, J.T., S.B. Powles, D.R. Liljegren and J.A.M. Holtum. 1991. Cross resistance to herbicides in annual ryegrass (*Lolium rigidum*). II. Chlorsulfuron resistance involves a wheat-like detoxification system. Plant Physiol. **95**: 1036-1043.

23. Comai L., D. Faccioti, W.R. Hiatt, G. Thompson, R.E. Rose and D.M. Stalker. 1985. Expression in plants of a mutant *aroA* gene from *Salmonella typhimurium* confers tolerance to glyphosate. Nature **317**: 741-744.

24. Conrad, S.G. and S.R. Radosevich. 1979. Ecological fitness of *Senecio vulgaris* and *Amaranthus retroflexus* biotypes susceptible or resistant to atrazine. J. Appl. Ecol. **16**: 171-177.

25. Cotterman, J.C. and L.L. Saari. 1992. Rapid metabolic inactivation is the basis for cross-resistance to chlorsulfuron in diclofop-methyl-resistant rigid ryegrass (*Lolium rigidum*) biotype SR2/84. Pestic. Biochem. Physiol. **43**: 182-192.

26. Coupland, D. 1994. Resistance to the auxin analog herbicides: *In* S.B. Powles. (ed.). Herbicide Resistance in Plants., CRC Press, Inc., Boca Raton, Florida, USA, pp. 171-214.

27. Coupland, D., D.T. Cooke and C.S. James. 1991. Effects of 4-chloro-2-methylphenoxypropionate (an auxin analogue) on plasma membrane ATPase activity in herbicide-resistant and herbicide-susceptible biotypes of *Stellaria media* L. J. Exp. Bot. **42**: 1065-1071.

28. Darmency, H. and J. Gasquez. 1990. Appearance and spread of triazine resistance in common lambsquarters (*Chenopodium album*). Weed Technol. **4**: 173-177.

29. DasSarma, S., E. Tischer and H.M. Goodman. 1986. Plant glutamine synthetase complements a *glnA* mutation in *Escherichia coli*. Science **232**: 1242-1244.

30. Deshpande, S. and J.C. hall. 1995. Comparison of flash-induced light scattering transients and proton efflux from auxin-herbicide resistant and susceptible wild mustard protoplasts: A possible role for calcium mediating auxinic herbicide resistance. Biochem. Biophys. Acta. **1244(1)**: 69-78.

31. Devine, M.D. and R.H. Shimabukuro. 1994. Resistance to acetyl coenzymeA carboxylase inhibiting herbicides. *In* S.B. Powles (ed.). Herbicide Resistance in Plants. CRC Press, Inc., Boca Raton, Florida, USA, pp. 141-169.

32. Devine, M.D., S. Renault and X. Wang. 1993. Alternative mechanisms of resistance to acetyl-coA carboxylase inhibitors in grass weeds. Proc. Brit. Crop Prot. Conf. û Weeds. Farnham, U.K., pp. 541-548.

33. Devlin, B. and N.C. Ellstrand. 1990. The development and application of a refined method for estimating gene flow from angiosperm paternity analysis. Evolution **44**: 248-259.

34. Donn, G., E. Tisher, J.A. Smith and H.M. Goodman. 1987. Herbicide-resistant alfalfa cells: an example of gene amplification in plants. J. Mol. Appl. Genet. **2**: 621-635.

35. Duncan, C.N. and S.C. Weller. 1987. Heritability of glyphosate susceptibility among biotypes of field bindweed. J. Hered. **78**: 257-260.

36. Dyer, W.E., P.W. Chee and P.K. Fay. 1993. Rapid germination of sulfonylurea-resistant *Kochia scoparia* L. accessions is associated with elevated seed levels of branched chain amino acids. Weed Sci. **41**: 18-22.

37. Eberlein, C.V., K.M. Rosow, J.L. Geadelmann and S.J. Openshaw. 1989. Differential tolerance of corn genotypes to DPX-M6316. Weed Sci. 37 :651-657.

38. Eckes, P. and F. Wengenmayer. 1987. Overproduction of glutamine synthetase in transgenic plants. *In* Regulation of Plant Gene Expression. 29th Harden Conf. Prog. Abstr., Wye College, Ashford, U.K.

39. Edwards, Jr., C.J., W.L. Barrentine and T.C. Klein. 1976. Inheritance of sensitivity to metribuzin in soybeans. Crop Sci. **16**: 119-120.

40. Estelle, M.A. and C. Somerville. 1987. Auxin-resistant mutants of *Arabidopsis thaliana* with an altered morphology. Mol. J. Genet. **206**: 200-206.

41. Faulkner, J. 1982. Breeding herbicide-tolerant crop cultivars by conventional methods. *In* H.M. LeBaron and J. Gressel (eds.). Herbicide Resistance in Plants. J. Wiley & Sons, New York, pp. 235-256.

42. Fillatti, J.J., J. Kiser, R. Rose and L. Comai. 1987. Efficient transfer of a glyphosate tolerance gene into tomato using a binary *Agrobacterium tumefaciens* vector. Bio/Technology **5**: 726-730.

43. Flur R., D. Aviv, E. Galun and M. Edelman. 1985. Efficient induction and selection of chloroplast-coded antibiotic-resistant mutants in *Nicotiana*. Proc. Natl. Acad. Sci. (USA). **82**: 1485-1489.

44. Forcella, F. 1987. Herbicide-resistant crops: yield penalties and weed thresholds for oilseed rape (*Brassica napus* L.) Weed Res. 27: 31-34.

45. Fuerst, E.P. and M.A. Norman. 1991. Interactions of herbicides with photosynthetic electron transport. Weed Sci. **39**: 458-464.

46. Fuerst, E.P., C.J. Arntzen, K. Pfister and D. Penner. 1986. Herbicide cross-resistance in triazine-resistant biotypes of four species. Weed Sci. **34**: 344-353.

47. Fuerst, E.P., H.Y. Nakatani, A.D. Dodge, D. Penner and C.J. Arntzen. 1985. Paraquat resistance in *Conyza*. Plant Physiol. **77**: 984-989.

48. Furu Sawa, I., K. Tanaka, P. Thanutong, A. Mizuguchi, M. Yazaki and K. Sada. 1984. Paraquat resistant tobacco calluses with enhanced superoxide dismutase activity. Plant & Cell Physiol. **25**: 1247-1254.

49. Gill, G.S. 1992. Herbicide resistance in annual ryegrass in Western Australia. Proc. Natl. Herb.Resistance Ext. Workshop, Glen Osmond, S.A., Australia, pp. 8-11.

50. Gould, F. 1995. Comparison between resistance management strategies for insects and weeds. Weed Technol. **9**: 830-839.

51. Gressel, J. 1994. Conferring herbicide resistance on susceptible crops. *In* S.P. Powles (ed.). Herbicide Resistance in Plants. CRC Press, Inc., Boca Raton, Florida, USA, pp. 237-259.

52. Gressel, J. and L.A. Segel. 1982. Interrelating factors controlling the rate of appearance of resistance: the outlook for the future. *In* H.M. LeBaron and J. Gressel (eds.). Herbicide Resistance in Plants. John Wiley & Sons, New York, pp. 325-347.

52a Gressel, J. and L.A. Segel. 1990. Modeling the effectiveness of herbicide rotation and mixtures as strategies to delay or preclude resistance. Weed Technol. 4:186-198.

53. Gressel, J., D. Aviv and A. Perl. 1985. Methods of producing herbicide resistant plant varieties and Plants produced thereby. US Patent Application **707**: 416.

54. Gronwald, J.W., C.V. Eberlein, K.J. Betts, R.J. Baerg, N.J. Ehlke and D.L. Wyse. 1992. Mechanism of diclofop resistance in an Italian ryegrass (*Lolium multiflorum*) biotype. Pestic. Biochem. Physiol. **44**: 126-132.

55. Guttieri, M.J., C.V. Eberlein and D.C. Thill. 1995. Diverse mutations in the acetolactate synthase gene confer chlorsulfuron resistance in kochia (*Kochia scoparia*) biotypes. Weed Sci. **43**: 175-178.

56. Guttieri, M.J., C.V. Eberlein, C.A. Mallory-Smith, D.C. Thill and D.L. Hoffman. 1992. DNA sequence variation in Domain A of the acetolactate synthase genes of herbicide resistant and susceptible weed biotypes. Weed Sci. **40**: 670-676.

57. Haldane, J.B.S. 1960. More precise expressions for the cost of natural selection. J. Genet. **57**: 351-360.

58. Hall, J.C. and M.L. Romano. 1995. Morphological and physiological differences between the auxinic herbicide-susceptible (S) and –resistant (R) wild mustard (*Sinapis arvensis* L.) biotypes. Pestic. Biochem. Physiol. **42**: 149-155.

59. Hall, J.C., C.J. Swanton and M.D. Devine. 1992. Physiological and biochemical investigation of the selectivity of ethametsulfuron in commercial brown mustard and wild mustard. Pestic. Biochem. Physiol. **42**: 188-195.

60. Hall, J.C., S.M.M. Alam and D.P. Murr. 1993. Ethylene biosynthesis in resistant and susceptible wild mustard (*Sinapis arvensis* L.) biotypes following foliar application of picloram. Pestic. Biochem. Physiol. **47**: 36-41.

61. Hall, L.M., J.A.M. Holtum and S.B. Powles. 1994. Mechanisms responsible for cross resistance and multiple cross resistance. *In* S.B. Powles (ed.). Herbicide Resistance in Plants. CRC Press, Inc., Boca Raton, Florida, USA, pp. 243-261.

62. Harper, J.L. 1956. The evolution of weeds in relation to resistance to herbicides. Proc. Brighton Crop Protec. Conf. – Weeds (Farnham, U.K.: British Crop Protection Council), pp. 179-188.

63. Hartwig, E.E. 1987. Identification and utilization of variation in herbicide tolerance in soybean (*Glycine max*) breeding. Weed Sci. 35 (Suppl. 1): 4-8.

64. Heap, I.M. and I.N. Morrison. 1992. Resistance to auxin-type herbicides in wild mustard (*Sinapis arvensis*) populations in Western Canada. Abstr. Weed Sci. Soc. Amer., p. 163.

65. Hickok, L.G., T.R. Warne and M.K. Slocum. 1987. *Ceratopteris richardii*. Application for experimental plant biology. Amer. J. Bot. **74**: 1304-1316.

66. Hirschberg, J. and L. McIntosh. 1983. Molecular basis for herbicide resistance in *Amaranthus hybridus*. Science **22**: 1346-1349.

67. Holt, J.S. and H.M. LeBaron. 1990. Significance and distribution of herbicide resistance. Weed Technol. 4: 141-149.

68. Holt, J.S., S.B. Powles and J.A.M. Holtum. 1993. Mechanisms and agronomic aspects of herbicide resistance. Ann. Rev. Plant Physiol. Plant Mol. Biol. **44**: 203-209.

69. Holtum, J.A.M. and S.B. Powles. 1991. Annual Ryegrass: an abundance of resistance, a plethora of mechanisms. Brighton Crop Protec. Conf. – Weeds (Farnham, U.K.: British crop Protection Council), pp. 1071-1078.

70. Jacobs, B.F., J.H. Duesing, J. Antonovics and D.T. Patterson. 1988. Growth performance of triazine-resistant and ûsusceptible biotypes of *Solanum nigrum* over a range of temperatures. Can J. Bot. **66**: 847-850.

71. Jansen, M.A.K., C. Malan, Y. Shaalitiel and J. Gressel. 1990. Mode of evolved photooxidant resistance to herbicides and xenobiotics. Z. Naturforsch. **45c**: 463-469.

72. Jasieniuk, M., I.N. Morrison and A.L. Brule-Babel. 1995. Inheritance of dicamba resistance in wild mustard (*Brassica kaber*). Weed Sci. **43**:192-195.

73. Jasieniuk, M., A.L. Brule-Babel and I.N. Morrison. 1996. The evolution and genetics of herbicide resistance in weeds. Weed Sci. **44**: 176-193.

74. Johnston, D.T. and J.S. Faulkner. 1991. Herbicide resistance in Graminaceae—a plant breeder's view. *In* J.C. Caseley, G.W. Cussans and R.K. Atkin. (eds.). Herbicide Resistant Weeds and Crops. Buttersworth-Heinemann Ltd., Oxford, U.K., pp. 319-330.

75. Kato, A. and Y. Okuda. 1993. Paraquat resistance in *Erigeron canadensis* L. Weed Res. (Japan) **28**: 54-56.

76. Keese, R.J. and N.D. Camper. 1994. Uptake and translocation of [^{14}C]MSMA in cotton and MSMA-resistant and -susceptible cocklebur. Pestic. Biochem. Physiol. **49**: 138-143.

77. Kishore, G.M. and D. Shah. 1988. Amino acid biosynthesis inhibitors as herbicides. Ann. Rev. Biochem. **57**: 627-663.

78. Kishore, G.M., D. Shah, S.R. Padgette, G. della-Cioppa, C. Gasser, D. Re, C. Hironaka, M. Taylor, J. Wibbenmeyer, D. Eichholtz, M. Hayford, N. Horrman, X. Delannay, R. Horsch, H. Klee, S. Rogers, D. Rochester, L. Brundage, P. Sanders and R.T. Fraley. 1988. 5-enolpuruvylshikimate 3-phosphate synthase: from biochemistry to genetic engineering of glyphosate tolerance. Amer. Chem. Soc. Symp. Ser. 379: 37-48.

79. Klein, T.M., E.D. Wolf, R. Wu and J.L. Sanford. 1987. High-velocity microprojectiles for delivering nucleic acids into living cells. Nature **327**: 70-73.

80. Kwon, C.S. and D. Penner. 1995. Response of chlorsulfuron-resistant biotype of *Kochia scoparia* to ALS inhibiting herbicides and piperonyl butoxide. Weed Sci. **43**: 561-565.

81. Leach, G.E., M.D. Devine, R.C. Kirkwood and G. Marshall. 1995. Target enzyme-based resistance to acetyl-coenzymeA carboxylase inhibitors in *Eleusine indica*. Pestic. Biochem. Physiol. **51**: 129-136.

82. LeBaron, H.M. 1991. Distribution and seriousness of herbicide-resistant weed infestations worldwide. *In* J.C. Caseley, G.W. Cussans and R.K. Atkin (eds.). Herbicide Resistance in Weeds and Crops. Butterworth–Heinemann Ltd., Oxford, U.K., pp. 27-43.

83. LeBaron, H.M. and M. McFarland. 1990. Herbicide resistance in weeds and crops: an overview and prognosis. *In* M.B. Green, H.M. LeBaron and W.K. Moberg. (eds.). Managing Resistance to Agrochemicals: From Fundamental Research to Practical Strategies. Amer. Chem. Soc., Washington, D.C., pp. 336-352.

84. Lee, K.Y., J. Townsend, J. Tepperman, M. Black, C.F. Chui et al. 1988. The molecular basis of sulfonylurea herbicide resistance in tobacco. EMBO J. **7**: 1241-1248.

85. Lincoln, C., J.H. Britton and M. Estelle. 1990. Growth and development of the axrl mutants of *Arabidopsis*. Plant cell **2**: 1071-1076.

86. Loux, M.M. R.A. Liebl and T. Hymowitz. 1987. Examination of wild perennial Glycine species for glyphosate tolerance. Soybean Genet. Newsl. **14**: 268-272.

87. Lovell, S.T., L.M. Wax, M.J. Horak and D.E. Peterson. 1996. Imidazolinone and sulfonylurea resistance in a biotype of common waterhemp (*Amaranthus rudis*). Weed Sci. **44**: 789-794.

88. Lyon, B.R. Y.L. Cousins, D.J. Llewellyn and E.S. Dennis. 1993. Cotton plants transformed with a bacterial degradation gene are protected from accidental spray drift damage by the herbicide 2,4-dichlorophenoxyacetic acid. J. Transgenic Res. **2(3)**: 162-169.

89. Lyon, B.R., D.J. Llewellyn, J.L. Huppatz E.S. Dennis and W.J. Peacock. 1989. Expression of a bacterial gene in transgenic tobacco plants confers resistance to the herbicide 2,4-dichlorophenoxy-acetic acid. Plant Mol. Biol. **13**: 533-540.

90. Malik, R.K. and S. Singh. 1995. Littleseed canarygrass (*Phalaris minor*) resistance to isoproturon in India. Weed Technol. **9**: 419-425.

91. Mallory-Smith, C.A., D.C. Thill and M.J. Dial. 1990a. Identification of sulfonylurea herbicide-resistant prickly lettuce (*Lactuca serriola*). Weed Technol. **4**: 163-168.

92. Mallory-Smith, C.A., D.C. Thill, M.J. Dial and R.S. Zemetra. 1990b. Inheritance of sulfonylurea herbicide resistance in *Lactuca* spp. Weed Technol. **4**: 787-790.

93. Maneechote, C., J.A.M. Holtum, C. Preston and S.B. Powles. 1994. Resistant acetyl-coA carboxylase is a mechanism of herbicide resistance in a biotype of *Avena sterilis* spp. *ludoviciana*. Plant Cell Physiol. **35**: 627-632.

94. Marles, M.A.S., M.d. Devine and J.C. Hall. 1993. Herbicide resistance in *Setaria viridis* conferred by a less sensitive form of acetyl coenzymeA carboxylase. Pestic. Biochem. Physiol. **46**: 7-13.

95. Marshall, G., R.C. Kirkwood and G.E. Leach. 1994. Comparative studies on graminicide resistant and susceptible biotypes of *Eleusine indica*. Weed Res. **34**: 177-184.

96. Marshall, L.C., D.A. Somers, P.D. Dotray, B.G. Gengenbach, D.L. Wyse et al. 1992. Allelic mutations in acetyl-coenzymeA carboxylase confer herbicide tolerance in maize. Theor. Appl. Gen. **83**: 435-442.

97. Matthews, J.M. and S.B. Powles. 1992. Aspects of population dynamics of selection for herbicide resistance in *Lolium rigidum* (Gaud.). Proc. First Intl. Weed Control Cong., Melbourne: Weed Sci. Soc. Victoria, Australia, pp. 318-320.

98. Matthews, J.M., J.A.M. Holtum, D.R. Liljegren, B. Furness and S.B. Powles. 1990. Cross-resistance to herbicide in annual ryegrass (*Lolium rigidum*). I. Properties of the herbicide target enzyme acetyl-coenzymeA carboxylase and acetolactate synthase. Plant Physiol. **94**: 1180-1186.

99. Maxwell, B.D. 1992. Predicting gene flow from herbicide resistant weeds in annual agriculture systems. Bull. Ecol. Soc. Am. Abstr. **73**: 264.

100. Maxwell, B.D. and A.M. Mortimer. 1994. Selection for herbicide resistance. *In* S.B. Powles and J.A.M. Holtum (eds.). Herbicide Resistance in Plants. Lewis Publishers., Ann harbor, Michigan, USA, pp. 1-25.

101. Mazur, B.J. and S.C. Falco. 1989. The development of herbicide resistant crops. Ann. Rev. Plant Physiol. Plant Mol. Biol. **40**: 441-470.

102. Mcnair, M.R. 1991. Why the evolution of resistance to anthropogenic toxins normally involves major gene changes: the limits to natural selection. Genetica **84**: 213-219.

103. Meredith, C.P. and P.S. Carlson. 1982. Herbicide resistance in plant cell cultures. *In* H.M. LeBaron and J. Gressel (eds.). Herbicide Resistance in Plants. John Wiley & Sons, New York, pp. 275-292.

104. Merrell, D.J. 1981. Ecological Genetics. Univ. of Minnesota Press, Minneapolis, Minnesota, USA, 500 pp.

105. Morrison, I.N., H. Beckie and K.M. Nawlosky. 1991. The occurrence of trifluralin-resistant green foxtail (*Setaria viridis*) in Western Canada. *In* J.C. Caseley, G.W. Cussans and R.K. Atkin. (eds.). Butterworth–Heinemann Ltd. Oxford, U.K., pp. 67-75.

106. Morrison, I.N., I.M. Heap and B. Murray. 1992. Herbicide resistance in wild oat—the Canadian perspective. Proc. Fourth Intl. Oat Conf., Adelaide, Australia, pp. 36-40.

107. Mulugeta, D., P.K. Fay and W.E. Dyer. 1992. The role of pollen in the spread of sulfonylurea resistant *Kochia scoparia* L. (Schrad.). Weeds Sci. Soc. Amer. Abstr. **32**: 16.

108. Mulugeta, D., P.K. Fay, W.C. Dyer and L.E. Talbert. 1991. Inheritance of resistance to the sulfonylurea herbicides in *Kochia scoparia* L. (Schrad.) Proc. Western Weed Sci. Soc., pp. 81-82.

109. Murray, B.G., A.L. Brule-Babel and I.N. Morrison. 1996. Two distinct alleles encode for acetyl-coA carboxylase inhibitor resistance in wild oat (*Avena fatua*). Weed Sci. **44**: 476-481.

110. Newhouse, K., B. Singh, D. Shaner and D. Stidham. 1991. Mutations in corn (*Zea mays* L.) conferring resistance to imidazolinone herbicides. Theor. Appl. Genet. **83**: 65-70.

111. Nimbal, C.I. , J.J. Hetholt, D.R. Shaw and S.O. Duke. 1995a. Photosynthetic performance of MSMA-resistant and -susceptible Mississippi biotypes of common cocklebur. Pestic. Biochem. Physiol. **53** :129-137.

112. Nimbal, C.I., G.D. Wills, S.O. Duke and D.R. Shaw. 1995b. Uptake, translocation and metabolism of ^{14}C-MSMA in organic arsenical-resistant and ûsusceptible Mississippi biotypes of common cocklebur (*Xanthium strumarium*). Weed Sci. **43**: 549-554.

113. Norman, M.A., E.P. Fuerst, R.J. Smeda and K.C. Vaughn. 1993. Evaluation of paraquat resistance mechanisms in *Conyza*. Pestic. Biochem. Physiol. **46**: 236-249.

114. Padgette, S.R., G. Della-cioppa, D.M. Shah, R.T. Fraley and G.M. Kishore. 1988. Selective herbicide resistance through protein engineering. *In* J. Schell and I. Vasil (eds.). Cell Culture and Somatic Cell Genetics. Vol. 6. Academic Press, New York, p. 490 .

116. Padro, R.D., E. Romera and J. Menendaz. 1995. Atrazine detoxification in *Panicum dichotomiflorum* and target site in *Polygonum lapathifolium*. Pestic. Biochem. Physiol. **52**: 1-11.

117. Parker, W.B., L.C. Marshall, J.D. Burton, D.A. Somers D.L. Wyse et al. 1990a. Dominant mutations causing alterations in acetyl-coenzymeA carboxylase confer tolerance to cyclohexanedione and aryloxyphenoxypropionate herbicides in maize. Proc. Natl. Acad. Sci. USA **87**: 7175-7179.

118. Parker, W.B., D.A. Somers, D.L. Wyse, R.A. Keith, J.D. Burton et al. 1990b. Selection and characterization of sethoxydim-tolerant maize tissue cultures. Plant Physiol. **92**: 1220-1225.

119. Peniuk, M.G., M.L. Romano and J.C. Hall. 1993. Physiological investigations into the resistance of a wild mustard (*Sinapis arvensis* L.) biotype to auxinic herbicides. Weed Res. **33**: 431-440.

120. Polos, E., J. Mikulas, Z. Szigeti, B. Malkovics, D.Q. Hai, A. Parducz and Lehoczki. 1988. Paraquat and atrazine co-resistance in *Conyza canadensis* (L.) Cronq. Pestic. Biochem. Physiol. **30**: 142-154.

121. Powles, S.B. and G. Cornic. 1987. Mechanism of paraquat resistance in *Hordeum glaucum*. I. Studies with isolated organelles and enzymes. Aust. J. Plant Physiol. **14**: 81-89.

122. Powles, S.P. and J.M. Matthews. 1991. Multiple herbicide resistance in annual ryegrass (*Lolium rigidum*): A driving force for the adoption of integrated weed management. *In* I. Denholm, A.L. Devonshire and D.W. Holloman (eds.). Resistance 91:Achievements and Developments in Combating Pesticide Resistance. Elsevier Applied Science, London, pp. 75-87.

123. Preston, C., J.A.M. Holtum and S.B. Powles. 1992. On the mechanism of resistance to paraquat in *Hordeum glaucum* and *H. laporium*: Delayed inhibition of photosynthetic CO_2 evolution after paraquat application. Plant Physiol. **100**: 630-636.

124. Preston, C., S. Balachandran and S.B. Powles. 1994. Mechanisms of resistance to bipyridyl herbicides in *Arctotheca calendula* (L.) Levyns. Plant Cell Environ. **17**: 1113-1123.

125. Primiani, M.M. J.C. Holtum and L.L. Saari. 1990. Resistance of Kochia (*Kochia scoparia*) to sulfonylurea and imidazolinone herbicides. Weed Technol. **4**: 169-172.

126. Putwain, P.D. 1982. Herbicide resistance in weeds—an inevitable consequence of herbicide use? British Crop Prot. Conf.—Weeds, pp. 719-728.

127. Putwain, P.D. and H.A. Collin. 1994. Mechanisms involved in the evolution of herbicide resistance in weeds. *In* S.B. Powles and J.A.M. Holtum (eds.). Herbicide Resistance in Plants. Lewis Publishers, Ann Harbor, Michigan, USA, pp. 211-235.

128. Radosevich, R.S. 1983. Herbicide resistance in higher plants. *In* G.P. Georghiou and T. Saito (eds.). Pest Resistance to Pesticides. Plenum Press, New York, pp. 453-479.

129. Redina, A.R., A.C. Craig-Kennard, J.D. Bandeen and M.K. Breen. 1990. Inhibition of acetyl coenzymeA carboxylase by two classes of selective herbicides. J. Agric. Food Chem. **38**: 1282-1287.

130. Ricroch, A, M. Mousseau, H. Darmency and J. Pernes. 1987. Comparison of triazine-resistant and-susceptible cultivated *Setaria italica* growth and development and photosynthetic capacity. Plant Physiol. Biochem. **25**: 29-34.

131. Ryan, G.F. 1970. Resistance of common groundsel to simazine and atrazine. Weed Sci. **18**: 614-616.

132. Saari, L.L., J.C. Cotterman and M.M. Primiani. 1990. Mechanism of sulfonylurea herbicide resistance in the broadleaf weed, *Kochia scoparia*. Plant Physiol. **93**: 55-61.

133. Saari, L.L., J.C. Cotterman and D.C. Thill. 1994. Resistance to acetolactate synthase inhibiting herbicides. *In* S.B. Powles (ed.). Herbicide Resistance in Plants. CRC Press, Inc. Boca Raton, Florida, USA. pp. 83-139.

134. Saari, L.L., J.C. Cotterman, W.F. Smith and M.M. Primiani. 1992. Sulfonylurea herbicide resistance in common chickweed, perennial ryegrass and Russian thistle. Pestic. Biochem. Physiol. **42**: 110-118.

135. Saxena, P.K. and J. King. 1988. Herbicide resistance in *Datura innoxia*. Cross resistance of sulfonylurea-resistant cell lines to imidazolinones. Plant Physiol. **86**: 863-867.

136. Sebastian, S.A. and R.S. Chaleff. 1987. Soybean mutants with increased tolerance for sulfonylurea herbicides. Crop Sci. **27**: 948-952.

137. Shaaltiel, Y. and J. Gressel. 1986. Multienzyme oxygen radical destroying system correlated with paraquat resistance in *Conyza bonariensis*. Pestic. Biochem. Physiol. **26**: 22-28.

138. Shaaltiel, Y. and J. Gressel. 1987. Kinetic analysis of resistance to paraquat in *Conyza*. Evidence that paraquat transiently inhibits leaf chloroplast reactions in resistant plants. Plant Physiol. **85**: 869-871.

139. Shah, D.M., R.B. Horsch, H.J. Klee, G.M. Kishore, J.A. Winter, N.E. Tumer, C.M. Hironaka, P.R. Sanders, C.S. Gasser, S.A. Aykent, N.R. Siegel, S.G. Rogers and R.T. Fraley. 1986. Engineering herbicide tolerance in transgenic plants. Science 233: 478-481.

140. Shaner, D. 1995. Herbicide resistance: Where are we? How did we get here? Where are we going? Weed Technol. **9**: 850-856.

141. Smeda, R.J., K.C. Vaughn and I.N. Morrison. 1992. A novel pattern of herbicide cross resistance in a trifluralin-resistant biotype of green foxtail (*Setaria viridis* (L.) Beauv.). Pestic. Biochem. Physiol. **42**: 227-241.

142. Solymosi, P. and E. Lehoczki. 1989. Characterization of a triple (atrazine-pyrazon-pyridate) resistant biotype of common lambsquarters (*Chenopodium album* L.) J. Plant Physiol. **134**: 685-690.

143. Stallings, G.P., D.C. Thill and C.A. Mallory-Smith. 1993. Pollen-mediated gene flow of sulfonylurea-resistant kochia (*Kochia scoparia* (L.) Schrad.). Weed Sci. Soc. Amer. Abstr. **33**: 60.

144. Stannard, M.E. 1987. Weed control in alfalfa (*Medicago sativa* L.) grown for seed. M.S. thesis, Montana State University, Bozman, MT, USA.

145. Stoller, E.W., L.M. Wax and D.M. Alm. 1993. Survey results on environmental issues and weed science research priorities within the corn belt. Weed Technol. **7**: 763-770.

318

146. Tardif, F.J., J.A.M. Holtum and S.B. Powles. 1993. Occurrence of a herbicide resistant acetyl-coenzymeA carboxylase mutant in annual ryegrass (*Lolium rigidum*) selected by sethoxydim. Planta **190**: 176-181.

147. Tingey, S.V. and G.M. Coruzzi. 1987. Glutamine synthetase in *Nicotiana plumbaginifolia* cloning and *in vivo* expression. Plant Physiol. **84**: 366-373.

148. Trebst. A. 1991. The molecular basis of resistance of photosystem II herbicides. *In* J.C. Caseley, G.W. Cusans and R.K. Atkin. Herbicide Resistance in Weeds and Crops. Butterworth—Heineman, Oxford, U.K. pp. 145-164.

149. Vaughn, K.C. and M.A. Vaughn. 1991. Dinitroaniline resistance in *Eleusine* may be due to hyperstabilized microtubules. *In* J.C. Caseley, G.W. Cussans and R.K. Atkin. Herbicide Resistance in Weeds and Crops. Butterworth—Heinemann Ltd., Oxford. pp. 177-186.

150. Vaughn, K.C., D.M. Marks and D.P. Weeks. 1987. A dinitroaniline-resistant mutant of *Eleusine indica* exhibits cross-reistsance and susceptibility to antimicrotubule herbicides and drugs. Plant Physiol. **83**: 956-964.

151. Warkentin, T.D., G. Marshall, R.I.H. McKenzie and I.N. Morrison. 1988. Diclofop-methyl tolerance in cultivated oats (*Avena sativa* L.). Weed Res. **28**: 27-35.

152. Warner, P.B. and W.B.C. Mackie. 1983. A barley grass (*Hordeum leporinum* sp. *glaucum* Steud). population tolerant to paraquat (Gramoxone). Austr. Weed Res. Newsl. **31**: 16-21.

153. Warwick, S.I. 1991. Herbicide resistance in weedy plants: physiology and population biology. Ann. Rev. Ecol. Syst. **22**: 95-114.

154. Warwick, S.I. and L. Black. 1981. The relative competitiveness of atrazine susceptible and resistant populations of *Chenopodium album* and *C. strictum*. Can J. Bot. **59**: 689-693.

155. Watanabe, Y., T. Hanna, K. Ito and M. Miyahara. 1982. Paraquat resistance in *Erigon philadelphicus*. Weed Res. (Japan) **7**: 49-54.

156. Webb, S.R. and J.C. Hall. 1995. Auxinic herbicide-resistant and ûsusceptible wild mustard (*Sinapis arvensis* L.) biotypes: Effect of auxinic herbicides on seedling growth and auxin-binding activity. Pestic. Biochem. Physiol. **52**: 137-148.

157. Weller, S.C., J.B. Masiunas and J. Gressel. 1987. Biotechnologies of obtaining herbicide tolerance in potato. *In* Y.P.S. Bajaj (ed.). Biotechnology of Plant Improvement. vol. 3. Springer-Verlag, Berlin, pp. 281-297.

158. Whitehead, C.W. and C.M. Switzer. 1963. The differential response of strains of wild carrot to 2,4-D and related herbicides. Can J. Plant Sci. **43**: 255-262.

159. Wiersma, P.A., M.G. Schmiemann, J.A. Condie, W.L. Crosby and M.M. Moloney. 1989. Isolation, expression and phylogenetic inheritance of an acetolactate synthase gene from *Brassica napus*. Mol. Gen. Genet. **219**: 413-420.

160. Wrubel, R.P. and J. Gressel. 1994. Are herbicide mixtures useful for delaying the rapid evolution of resistance? A case study. Weed Technol. **8**: 635-648.

161. Youngman, R.J. and A.D. Dodge. 1981. On the mechanism of paraquat resistance in *Conyza* spp. In G. Akoyunoglou (ed). Photosynthesis. VI. Photosynthesis and productivity, photosynthesis and environment. Balaban Intl. Science Services, Philadelphia, PA, USA, pp. 537-544.

11

Biological Approaches in Weed Management

Currently, herbicides account for 55% of the global market of all pesticides used in crop production. Although the organic herbicides have been playing a major role in providing the food and fibre required by the world population, there has been increasing concern, both real and perceived, about their safety for food products, their adverse impact on environment and widespread weed resistance to herbicides. These factors, coupled with rising prohibitive costs of developing, testing and registering synthetic materials as herbicides, have provided the impetus to develop alternative weed management strategies. In this context, biological control as an alternative or supplemental weed management method appears to play a major role in agricultural production systems.

Biological approach to weed management involves various biological weed control agents such as pathogens, insects, nematodes, etc. In fact, insects and fungi have long been used in biological control of noxious weeds. The fact that weeds serve as alternative hosts to crop pathogens, insects and nematodes which, if properly used may eliminate certain noxious weeds, led to extensive research to develop a 'bioherbicide approach' for controlling weeds. Significant progress has been made in this area of weed science during the past 20 years.

The biological approach includes the conventional biocontrol agents such as insects, nematodes, fungi and bacteria as well as plant-based chemicals that exhibit herbicidal properties. This Chapter encompasses these and other related topics.

INSECT-BASED BIOCONTROL

Mode of Action

Insects kill plants by exhausting the plant food reserves, sometimes through destruction of photosynthetic parts. The damage depends on the intensity of attack, the size and condition of the plant and environment. The soil and climatic conditions and the role of competing plants determine whether the plants can survive insect attack. The weed usually competes with neighbouring plants for essential requisites and the attacking insect may reduce or nullify this competitive advantage. This kind of indirect effect is also important in the biological control of weeds. The number of insects required to kill a plant is also a function of the competitive advantage of the weed. An aggressive weed in a favourable environment requires a large number of insects. The number of insects required for control is influenced by the size of the insect, its mode of attack and whether it transmits pathogens.

Kinds of Insects and Host Specificity

Of the wide range of insects used for biological control of weeds, those belonging to Lepidoptera, Hemiptera, Coleoptera, Diptera, Hymenoptera, Thysanoptera, etc. are more widely used.

The success of biological control of weeds by insects depends on host specificity. The host-specific insect does not attack a plant outside its host range. Plants generally contain the necessary nutrients to support many kinds of insects. But the host-specific insects are highly discriminatory in their choice of food. The presence or absence of certain chemical and physical features in the host determines acceptance or rejection. Host-specificity is determined by starvation tests, which make sure that the insect will not eat representative plants other than those within the restricted host range, including the weed and it is not likely to attack economic species.

The insect-host relationship involves two phases of insect behaviour: the insect can disperse and search, or it can stay and feed. A hungry insect continues to wander or disperse until it encounters a plant that it recognizes as palatable. This recognition is primarily based on chemical stimuli, although visual and tactile stimuli may also play a role. Many of the stimulating chemicals are common nutrients, such as sugar and ascorbic acid. The preferences of polyphagous insects may be largely determined by the abundance of such compounds in the plants. Stimulants are detected by sensillae, which in lepidopterous larvae are partially grouped on the maxillary lobe. The insect may also respond to repugnant substances with the same or other sensillae.

Factors Affecting Insect-based Biological Weed Control

The various factors that affect the success of biological control by insects are discussed below.

Choice of Enemy Agents

The major factor in successful biological control is the introduction of a good enemy agent. To be an effective agent in the control of a weed, an insect should possess the following qualities:

a) Ability to kill the weed or prevent its reproduction in a direct or indirect way; feeding on young leaves and growing points is much more damaging than feeding on old, mature leaves.
b) High ability to disperse successfully and to locate its host plant.
c) Good adaptation of the weed host and environmental conditions in which the weed is infesting; a species found in all situations is to be preferred to one restricted in distribution.
d) Reproductive capacity at a rate sufficient to maintain control of host. The species with greater reproductive potential due to high fecundity and for longevity and/or multivoltinism are to be preferred. The potential control agent should be able to feed on old, mature leaves and maintain itself on a low population density of the weed and, for this, mobility is a desirable trait.

It is said that greater success of biological control of weeds as compared to biological control of insect pests is largely due to the fact that the insects used against weeds have high host specificity as compared to those used against insect pests.

Aggressiveness of the Weed

Competitive ability or aggressiveness is affected by favourable climatic and soil conditions, reduced competition from other plant species and absence of enemies that attack it. Resistance of the host weed to an insect attack is largely determined by its inherent ability to withstand or recover from tissue destruction by insects and pathogens, ability to multiply faster than the attacking agent even under the agent's heaviest attack and greater ability to disperse than that of the enemy.

Climatic Conditions

Climate affects the adaptability of the insect and competitive ability of the weed host. An insect has to adapt to extreme fluctuations in the climate and also to the special way in which the plant itself has met those extremities. Ability to survive extreme conditions and to rebound in numbers before the weed can recover its vigour is an important asset in an enemy agent.

Discontinuity is an essential host condition, which may reduce an enemy's ability to survive during critical periods and as a result the enemy may not quickly regain its controlling status when favourable conditions return. Such discontinuity may arise seasonally, as with winter dieback of aerial parts, defoliation of deciduous perennials or the death of annuals. Except for extreme cases, polyphagous insects are efficient in meeting evolutionary challenges presented by such discontinuity. Many of them are remarkably adapted to the physical conditions and specific living relations of their hosts.

Competition of Plant Species

The effects of other plant species in the environment on the enemy-weakened weed species determine the degree of biological control attained and also the nature of replacement vegetation. If some weed species have competitive superiority over others, biological control may result in the replacement of the target species by another species, perhaps as bad or worse. This possibility, however, exists with any kind of control.

Examples of Insect-based Weed Control

The first attempt at using insects to control weeds was made in the early 1900s to control *Lantana camara*, a prickly shrub introduced into the Hawaiian Islands around 1860. Tests conducted in 1902 and later showed that some of the insects were very effective in controlling *Lantana*. These included: a) larvae of *Crocidosema lantana*, the tortricid moth, which bcre into the flower stems, fall on réceptacles of flower clusters and eat flowers and fruits, b) larvae of *Agromyza lantanae*, the seed fly, which eat many berries and cause others to dry so that birds do not eat and carry them and c) larvae of *Thecla echion* and *Thecla bazochi*, the lycaenid butterflies, which destroy many flowers. These insects were so effective in controlling *Lantana* that the weed has not taken possession again even when the land was abandoned. *Lantana* has also been controlled by insects in India, Australia and Fiji [40]. Of the insects introduced into Australia via Hawaii, *Telenemia scrupulosa*, the lace bug, was the most effective. Similar successes were also recorded in controlling *Lantana* in East Africa.

Striking successes have also been obtained in the control of prickly pears, *Opuntia* spp., in Australia. The seven species of *Opuntia* found in Australia were: *O. inermis, O. stricta, O. aurantiaca, O. monocantha, O. streptacantha, O. imbricata* and *O. tomentosa*. Of these, *O. inermis* and *O. stricta* were the most serious ones. In 1925, the spread of *Opuntia* was around 60 million acres. Of the many insects tested for control of *O. inermis* and *O. stricta, Cactoblastis cactorum*, the moth borer from Argentina and *Dactylopius opuntiae*, the cochinial insect from the United States of America, were found very effective.

The moth borers of *Cactoblastis cactorum* were introduced in 1926 after importing 2750 eggs of the insect from Argentina in 1925. Following their introduction, these insects multiplied so rapidly that their numbers in the span of 15 mon had increased to several millions. By 1935, about 95% of the prickly pear infestation had been destroyed. The rapidity and completeness of destruction stands out as the finest example of biological control of weeds by insects. The larvae of *C. cactorum* tunnel through the plants and eventually destroy all aboveground portions. They also penetrate into the underground bulbs and roots. They eat out the inside of young cladodes but leave the fibrous vascular tissues in old cladodes. This facilitates the entry of bacteria and fungi, which also aid in destruction. This insect was also introduced against *Opuntia* in Hawaii and it met with success there too. In India, the infestation of *Opuntia* was controlled by the cochinial insects *Dactylopus indicus* and *Dactylopus tomentosus*.

Another success was the control of *Hypericum perforatum* (common St. Johnswort), a common roadside plant, with heavy infestations in Europe, USA and Australia. In Australia and New Zealand, this weed was controlled by *Crysolins hyperici*, the leaf-eating beetle. In the USA, it was controlled by *C. hyperici*, *Agrilus hyperici* (a root borer) and *C. gemellata* between 1945 and 1950. These insects helped to free thousands of acres of excellent rangeland of common St. Johnswort.

Another weed that was found very susceptible to insect predators is *Cuscuta* spp., (dodder), a parasitic weed occurring on a variety of plants almost throughout the world in very different climates [4]. The insects found effective against *Cuscuta* were *Melangromyza cuscutae*, *Smicronyx cuscutae* and *Acro-clita* spp.

Biological control was also found successful on other weeds, such as *Eupatorium adenophorum* in Hawaii and Australia by a gall fly *Procecidochares utilis*; *Leptospermum scoparium* in New Zealand by a mealy bug, *Ericoccus orariensis*; and *Acaena sanguisorbae* in Australia by a chrysomelid beetle *Haltica pagana*. An interesting success of control was seen in the case of *Clidemia hirta*. Its infestation was reduced by the combined effect of a thrips, *Liothrips urichi* and competition from other plants. This insect does not kill the weed outright, but so inhibits its growth that other plants can successfully compete with it. Mangoendihardjo and Soerjani [53] identified a gall midge *Orseoliella javanica* as a potential natural enemy of *Imperata cylindrica* (thatchgrass) and a plant feeder *Bactra vermosana* of *Cyperus rotundus* (purple nutsedge or nut grass).

Biological control of aquatic weeds was also achieved in the destruction of vegetation in ditches by herbivorous fish. A member of the sunfish family *Tilapia mossambica* destroys the roots of aquatic weeds and control their growth by eating the vegetation and burrowing in the bottom of ditches to form spawning nests.

In the recent past, aphids were found useful as biological agents to control weeds in rice fields. Waterlily aphid (*Rhopalosiphum nymphaceae*) destroys much of the above-water vegetation of *Heteranthera limosa* (ducksalad) in irrigated rice in Arkansas, USA. The aphid population is quite resilient. This aphid has a broad host range, which provides a constant food source, thereby helping to maintain population stability.

Good progress has been made on the biological control of weeds in tropical countries. For example,. In the mid-nineteenth century, *Opuntia vulgaris* was controlled in Central and North India by the cochinial mealy bug *Dactylopius indicus* obtained from Brazil in 1975. This bug was, however, a failure on *Opuntia dellenii* found in South India. Another species of mealy bug *Dactylopius opuntiae*, introduced from Ceylon in 1926 was very effective and eradicated *O. dellinii* within 5 or 6 y [65].

For control of *Lantana camara*, the lace bug *Teleonemia scrupulosa* was found very successful in the 1940s. Currently, the other insect enemies found successful include the leaf-eating moths or the leaf webbers *Hypena striatalis, Syngamia haemorrhoidalis, Diastema tigris* and *Catabaena esula* and wood-boring beetle, *Plagiohammus spinipennis* [34].

Eupatorium adenophorum a predominant broadleaf weed in the higher elevations of North-east India and South India. This was successfully controlled by the Mexican gall fly (*Procecidochares utilis*) imported from New Zealand in 1963 and released in the Nilgiris, Darjeeling and Kalimpong areas of India [34]. Gupta [34] reported that gall-forming weevil (*Smicronyx albovariegatus* Fst.) produced galls on the root, stem, branches of *Striga* spp. and the larvae of the Noctuid, *Eulocastra argentisparsa* Hmps., feed on the ripening seeds in the fruit pods of witchweed.

Vaidya and Vartak [84] found the larvae of *Diacrisia obliqua* (Wlk.) feeding voraciously on *Parthenium hysterophorus*, the older larvae destroying the whole plant. They observed that since this insect species also attacks some of the crop species it cannot be used to control *Parthenium* unless a biotype can be isolated by genetic selection that feeds exclusively on the weed.

NEMATODE-BASED BIOCONTROL

There are several plant parasitic nematodes which, on the basis of their life cycle, offer potential as biological agents for weed control. Two prominent parasitic nematodes *Orrina phyllobia* (Thorne) Brzeski and *Subanguina picridis* (Kirjanova) Brzeski offer a natural potential for biological control of two economically important weeds, *Solanum elaeagnifolium* (silverleaf nightshade) and *Centaurea repens* (Russian knapweed) respectively.

Silverleaf nightshade is a serious perennial weed in cotton. The perennial root system of the weed has the ability to regenerate itself from root pieces, making cultivation practices ineffective to control the weed. Chemical herbicides are also largely ineffective, besides being uneconomical. The nematode *O. phyllobia* and silverleaf nightshade have a natural host-parasite relationship. Robinson et al. [71] postulated that augmentative release of the nematode could result in a substantial reduction in weed numbers and vigour.

The nematode *O. phyllobia* produces galls on the affected weed foliage. During a natural infection cycle, nematodes remain in a dormant state (anhydrobiosis) within silverleaf nightshade leaf debris on top of or within the upper soil surface [63]. When dormancy is broken during extended periods of rainfall, the nematodes swim up the stems in a film of water and move towards meristematic growing portions of the plant including young leaf tissue, stem and floral parts. Nematode penetration and infection are possible only if plant surfaces remain moist for several hours. Visual symptoms of infection become evident 10 to 14 d following initiation of infection. Galls first appear as small pimple-like protrusions on the leaf surface, then increase in size and engulf the entire leaf surface. During gall formation, the nematodes draw upon the food reserves of the plant, placing a severe stress upon the weed and subsequently causing defoliation of the plant. When the galled leaves dry, the nematodes become dormant and remain so until moisture conditions are correct for the cycle to start over [63].

Commercialization of the nematode requires mass production and appropriate application technology. For mass production, the weed seeds may be germinated and the seedlings transplanted. The transplanted field may be inoculated by broadcasting the inoculum (dried galled leaves) on the soil @ 3.4 to 11.2 kg ha^{-1} and the soil kept moist with a modified sprinkler irrigation system. Galls will appear for several weeks following inoculation. Prior to leaf fall, galls are collected, dried and stored in a freezer at −17° C [63]. Successful inoculation of the nematode depends on formulation and application carrier.

The major problems using nematodes for biological weed control include the host range of the nematode, the economics of nematode inoculum production and the degree of control that can be achieved under field conditions. These can be circumvented with adequate research at field and laboratory levels.

MICROBIAL HERBICIDES/AGENTS

The microbial bioherbicide or bioagent approach involves application of indigenous plant pathogens, such as fungi, bacteria, microplasms and viruses to control weed species. Plant pathologists and weed scientists have identified over 100 microorganisms that are candidates for development as commercial microbial herbicide agents. Of these, four fungal pathogens have been registered so far for use as commercial herbicides. These include those with commercial names, DeVine®, COLLEGO®, BIOMAL® and Dr.Biosedge.

Fungi-Based Herbicides (Mycoherbicides)

Fungi are nucleated, usually filamentous, spore-bearing microorganisms, reproducing both sexually and asexually. They presumably evolved from algae by loss of ability to produce chlorophyll. They are heterotrophs that live on organic nutrients, either as saprophytes on dead plants or animal matter, or symbionts or parasites that assimilate nutrients from living plants or animals in aquatic, marine, terrestrial and subaerial habitats. The fungi that incite plant diseases are a diverse assemblage of species that vary markedly in morphology and physiology [1]. They range from single cell, intracellular, obligate parasites with simple life cycles to forms that are multicellular that produce asexual spore forms in their life cycle. Fungal pathogens are termed **incitants** of disease rather than the cause of disease.

The type of parasitism exhibited by a fungus affects its potential as a mycoherbicide [79]. Obligate parasites, nearly symbionts on the scale of parasitism, are less damaging to their hosts than facultative parasites or facultative saprophytes. Hence, obligate parasites have less potential for use as mycoherbicides if rapid, complete kill of the weed is required. If, however, slowing weed growth or suppressing seed production is an adequate requirement of control, obligate parasites have good potential to form mycoherbicides. Fungi that are either facultative parasites or facultative saprophytes are usually the best candidates for development as mycoherbicides.

DeVine®, developed by Abbott Laboratories, USA, was the first mycoherbicide derived from fungi. The fungal organism *Phytophthora palmivora* (Butl.) is a facultative parasite that produces lethal root and collar rot of its host plant *Morrenia odorata* (stranglervine) and persists in soil saprophytically for extended periods of residual control. It was the first product to be fully registered as a mycoherbicide. DeVine® infects and kills stranglervine, a problem weed in citrus plantations of Florida.

When a sexual spore is produced by a fungal pathogen, it is usually a dormant structure for the preservation of the organism during winter or other adverse environments. The DeVine® pathogen produces a sexual spore that may provide a mechanism for surviving during the absence of a host or favourable environment for saprophytic growth. The DeVine® fungus belongs to a group of Oomycetes, to which belong pathogens causing late blight of potato and tomato, damping off of seedlings, downy mildews, white rusts of crucifers and other plants and stem and leaf galls of various hosts. Each of these pathogens has a unique disease cycle during which the fungus continues its life cycle on the host. In the DeVine® fungus, the infective units are chlamydospores, but they are not stable. The formulated material has a shelf-life of only six weeks and must be handled much like fresh milk. This is a manageable problem if the marketing area is small enough for practical distribution of a refrigerated product based on custom-order sales [70]. However, DeVine® is considered to be too active. Kenny [50] reported that citrus plantations treated only once between 1978 and 1980 were still showing in 1986, with 95% to 100% control of stranglervine weed. This total control was occurring in spite of a continuous infestation of new seedlings from the wind-blown seeds. The new flush of seedlings caused an increase in the *Phytophthora palmivora* levels in the soil, which killed the weeds before they became established.

COLLEGO® was developed from the fungus *Colletotrichum gloeosporioides* (Penz.) Penz. Sacc. f.sp. *aeschynomene* (CGA), a facultative saprophyte that generally produces innocuous stem cankers in natural environments. However, it causes a lethal stem and foliage blight of its host weed *Aeschynomene virginica* (northern jointvetch) when inundatively inoculated with spores. It was the first fungus to be evaluated as a mycoherbicide [22]. It persists saprophytically on dead host tissue above the ground and colonizes host tissue in the soil. In nature, it persists from season to season in infected seeds.

The endemic fungus pathogen CGA was developed and registered in 1982 and 1983 for control of northern jointvetch in rice and soybean. The disease of northern joint-vetch was discovered in 1969 at the University of Arkansas Rice Research and Extension Center. The pathogen was then isolated in pure culture and was found to sporulate abundantly. It was found to be indigenous in the rice-growing areas of Arkansas, USA, where the weed grows in natural habitats. COLLEGO® performs better when weeds have just emerged through the crop canopy than when they are smaller or larger. The developmental work was done by the U.S. Department of Agriculture and University of Arkansas and the Upjohn Company developed CGA as a mycoherbicide for control of northern jointvetch.

Fungal pathogens usually grow vegetatively as microscopic filaments, known individually as **hyphae** and collectively as **mycelium**. Asexual spores are produced from vegetative mycelium and they serve as dissemination and conservation propagules in the absence of suitable growing environment or as inoculum for infection of the host. Fungi that produce no sexual spores lose the sexual cycle and utilize asexual spores to maintain the spores [80]. These are classified as **imperfect fungi** and many pathogens belong to this group. COLLEGO® pathogen belongs to this group and can be mass-produced.

The disease cycle of a pathogen must be understood before assessing its potential as a mycoherbicide. The important factors to be considered are: a) the source of inoculum, b) the method of dissemination, c) the cyclic parameters that favour rapid infection and disease development, d) the age and physiology of the host that favours or suppresses plant infection, e) variation in the genetic resistance of the host or virulence of the pathogen, f) the method and rapidity of secondary spread of the pathogen and g) the means of overwintering or survivability in hostile environment [80]. In addition to the aforesaid, knowledge of the climate in the geographic region where the target weed grows and the growth stage during which the weed must be controlled is required in order to assess the mycoherbicidal potential of a particular fungus. The biological successes of DeVine® and COLLEGO® are directly attributable to the disease cycles that permit these fungi to grow over a wide range of temperatures and moisture levels [80].

The successful development of COLLEGO, led to the discovery of another *Colletotrichum*-based mycoherbicide BIOMAL® by Philom Bios Inc., Sarkatoon, Canada. BIOMAL® contains spores of *Colletotrichum gloeosporioides* (Penz.) Sacc. f. sp. *malvae*. It is used to control *Malva pusilla* (round-leaved mallow) in Canada and the USA.

The rust fungus *Puccinia canalicuta* (Schw.) Lagerh. is commercialized under the name **Dr.Biosedge** for control of *Cyperus esculentus* L. (yellow nutsedge). It parasitizes yellow nutsedge.

Another fungal pathogen, which showed potential as a source of mycoherbicide, is *Cercospora rodmanii* (Conway) for controlling *Eichhornia crassipes* (waterhyacinth). The experimental formulation, ABG-5003 (developed by Abbott Laboratories) consists of mycelial fragments and spores applied as a wettable powder [21]. Although biological control of waterhyacinth was achieved, the efficacy was less than desirable for commercialization because of the restrictive environmental requirements by the fungus [30]. Later, it was found that weed control by the fungus could be enhanced by using it in combination with sublethal doses of various chemical herbicides or with certain insects [15].

Rhizobacteria-Based Biocontrol Agents

Deleterious Rhizobacteria

The presence of plant roots exerts profound influence on the species composition and activities of microorganisms. As roots grow through the soil, large quantities of carbon compounds are excreted or diffused into the adjacent soil. The soil environment around the root contains a large number of bacteria and other microorganisms than soil just a few millimetres away from the root. This zone of influence is called the **rhizosphere** which is described as a very intense, narrow zone of bacterial activity around legume roots.

Microbial populations in the rhizosphere are dynamic, represent a number of taxonomic genera and can affect plant nutrients and plant health. The dominant bacteria normally found in the rhizosphere belong to the genera **Pseudomonas, Xanthomonas, Phyllobacterium, Streptomyces, Bacillus** etc.

Bacteria that elicit a negative effect on plant growth, but do not parasitize the plant are considered exopathogens and these are termed **Deleterious Rhizobacteria (DRB)**. Thus, DRB are characterized as non-parasitic bacteria which colonize plant root surfaces and suppress plant growth. Many DRB are plant-specific and thus their existence on weeds and their potential as biological control agents have been recently investigated.

The mode of action of DRB is primarily through production of phytotoxins, which are absorbed by the seedling roots [82]. They regulate development of weeds before or coincident with emergence of crop plants. Thus, DRB do not necessarily eradicate the problem weed, but significantly suppress early growth of the weed and allow the developing crop plants to effectively compete with the weakened weed seedlings. DRB are most effective when weed growth coincides with environmental factors conducive to bacterial growth and plant-suppressive activity [45]. Application of DRB to soil for controlling weeds through toxigenicity is similar to the use of necrogenic fungi applied to soil and aquatic habitats where phytotoxins are subsequently produced during fungal proliferation in the weed root zone.

The efficacy of DRB agents is a function of several characteristics of DRB including root- and seed-colonizing ability, extent and rate of suppression and adaptability to pesticide application methods. DRB as biological control agents are considered less efficacious than herbicides since DRB are effective primarily through root suppression. Kennedy et al [47] found that crop yields in plants where weed growth was suppressed by DRB were significantly higher than for crop yields in plots with healthy weeds. They also observed weed control efficacy of DRB agents not only as reduction in competitive ability but also inhibition of biomass accumulation, reduced density and reduced seed production by the affected weed *Bromus tectorum* (downy brome)

Weed Control by DRB

The potential of DRB as biological control agents was first reported on *Bromus tectorum*, a predominant weed in winter wheat fields [17]. The plant-suppressive bacterium *Pseudomonas fluorescens* strain D proliferates on downy brome roots.

In field experiments, Kennedy et al. [49] found that the application of *P. fluorescens* strain D7 (D7) to the soil surface reduced the above-ground growth of downy brome up to 50% and seed production up to 64% and increased wheat yields up to 31%. This yield response was equivalent to the increase expected from chemical control of a moderate infestation of downy brome. This bacterial strain suppresses rather than kills downy brome and this gives wheat a competitive advantage. However, in other field trials, the efficacy of D7 in the suppression of downy brome was not always consistent [60]. The poor efficacy in certain trials was attributed to the following: a) the bacteria either did not survive, or, if they did survive, their numbers were too low to suppress weeds and b) the level of toxin production by surviving bacteria was too low to inhibit downy brome. This showed the need to improve the survival and colonization of roots by D7 by changing the application methodology.

Skipper et al. [74] studied the suitability of wheat roots as carrier of rhizobacteria. They monitored the presence of D7 in the soil and rhizosphere of winter wheat and downy brome roots by using a spontaneous antibiotic-resistant mutant of D7, *Pseudomonas fluorescens* strain **D7rif**, which is resistant to rifampicin (at 100 mg L^{-1}) [45]. The bacteria were applied in 0.1 M $CaSO_4$ at 5 ml 30 cm^{-1} row and approximately 108 cfu ml^{-1} in the seed furrow and onto winter wheat seeds. The results demonstrated the presence of D7rif on the seminal, internode and crown roots of winter wheat. Seminal roots are the first to emerge from wheat seed and the introduced rhizobacteria colonized these roots prior to migrating up the subcrown internode, and colonizing the crown (adventitious) roots as they emerged (Fig. 11.1). Thus, the wheat root system appeared to be a suitable host and carrier for biocontrol agents. They postulated that D7rif would be transferred from wheat roots to downy brome roots upon contact and subsequently suppress the weeds as shown in Fig. 11.1. Based on these results, they concluded that host plant roots may serve as carriers for biological control agents against diseases, nematodes or insects, but the transfer process and perhaps lack of toxin production, upon transfer, appear to be too restrictive for acceptable weed management.

The efficacy of rhizobacteria is influenced by soil conditions (water content, temperature, clay content, pH, etc.), the application techniques, storage conditions and target or non-target pathogen interferences. The work done so far indicated that rhizobacteria have a good potential to become viable biological control agents in the near future. In order for them to play a significant role in sustainable agriculture, Skipper et al (74) suggested that research needs to be done in the following areas of rhizobacteria technology: a) microbial ecology as influenced by biotic and abiotic

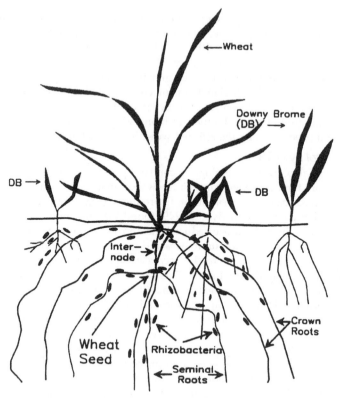

Fig. 11.1. Graphic depiction of rhizobacteria applied in the wheat seed furrow, being carried by developing wheat roots, and transferred to nearby roots of *Bromus tectorum* (downy brome) and suppressing the weed's growth. [74]. (Reproduced with permission of WSSA)

parameters in the soil environment, including the host plant, b) survival of introduced microorganisms under field conditions and adequate colonization of seeds and roots, c) isolation of herbicidal (toxin) genes and insertion into a 'field-proven microbial isolate', d) regulation of genes that produce toxins and e) more holistic approaches to understand this complex area.

The process to obtain bacteria suppressive to downy brome and *Aegilops cylindrica* (jointed goatgrass) has been patented and is under consideration for commercial development [27, 48]. Since downy brome and jointed goatgrass are difficult to control with available herbicides, biological control based on DRB appears to be a very promising weed management strategy.

Recently, 18 weed species were identified in the USA and Canada as targets for weed control by DRB [51, 52]. These include *Bromus tectorum* (downy brome), *Bromus secalinus* (cheat weed), *Bromus japonicus* (Japanese brome), *Aegilops cylindrica* Host. (jointed goatgrass), *Sisymbrium altissimum* (tumble mustard), *Amaranthus retroflexus* (redroot pigweed), *Xanthium strumarium* (common cocklebur), *Poa annua* (annual bluegrass), *Setaria viridis* (green foxtail) and *Abutilon theophrasti* (velvetleaf) in cereal and row crops and *Euphorbia esula* (leaf spurge), *Acroptilon repens* (Russian knapweed) and *Avena fatua* (wild oat), *Rubus* spp. (blackberry) and *Calamagrostis canadensis*

(Canada redgrass) of forestry, pastures and rangelands. Nearly all the weed species have been targeted by *Pseudomonas* spp. for biological control. Besides, *Agrobacterium*, *Xanthomonas* spp., *Enterobacter* spp., *Erwinia herbicola* and *Flavobacterium* spp. have also been identified as potential agents of biological control. These bacterial groups are typical rhizosphere bacteria with high root-colonizing abilities.

Integration of Rhizobacteria into Weed Management

Despite the high potential of DRB as a viable weed management tool, the success of rhizobacteria is limited by their narrow spectrum of activity, moderate efficacy and less reliability. The limited spectrum of activity and susceptibility to adverse environmental conditions contributes to the perception of DRB as less efficacious than herbicides. This necessitates consideration of integrating DRB with currently available weed management methods.

DRB and Herbicides: Integration of DRB with chemical herbicides has been used successfully to improve the activity of rhizobacteria. For example, Greaves and Sargent [33] found that colonization of wheat roots by *Pseudomonas* spp. was greatly enhanced and resulted in extensive cellular and tissue damage when plants were treated with mecoprop. They suggested that exploitation of plant-herbicide-microorganism interactions could potentially be an effective strategy for biological control of weeds. Rhizobacteria inhibitory to *Bromus tectorum* (downy brome) and *Aegilops cylindrica* (jointed goatgrass) exhibited higher growth-suppressive activity in soil when combined with herbicides at reduced rates [37, 78]. Growth suppression by some DRB combined with metribuzin was additive. Diclofop plus DRB increased root growth suppression of downy brome by 12% over diclofop alone [78]. Further research is required to determine the mechanisms of rhizobacteria-herbicide interactions to develop strategies wherein DRB selected to control a particular weed could be paired with a specific herbicide that increases susceptibility of the weed to the DRB. A successful integration strategy would not only enhance the efficacy of DRB agents, but also reduce herbicide volume required for weed control and decrease potential environmental problems.

DRB and Cultural Practices: Cultural and tillage practices can influence the frequency of inhibitory bacteria occurring in soil and their growth-suppressive activity. Downy brome and jointed goatgrass were suppressed by rhizobacteria under either conventional or minimum tillage suggesting that application of selected DRB during tillage may be effective in integrated weed management [47]. Crop residues at or near the soil surface could serve as substrates for production of weed-suppressive agents by DRB applied directly to the residues. Stroo et al. [77] reported that the population of wheat-inhibitory DRB increased dramatically when applied on crop residues, which promoted production of toxins inhibitory to wheat. This suggests the possibility of an approach in which DRB are applied on surface plant residues to produce phytotoxins that suppress weed growth prior to planting the crop, similar to the preemergence herbicide approach. The feasibility of this approach is possible as demonstrated in the case of disease control wherein application of biocontrol agents to wheat straw in the field led to reduction in population of the tan spot pathogen [*Pyrenophora tritici-repentis* (Died.) Drechs.] of winter wheat [64].

Crop rotation could be manipulated to encourage development of specific inhibitory bacteria on weed roots. Certain rhizobacteria are associated with corn roots, suggesting the potential for using DRB to achieve suppression of weeds in crop rotation systems [83].

DRB and Other Biocontrol Agents: Combining different DRB strains or integrating them with other types of biological agents (fungi, insects, etc.) may enhance efficacy of control over that exhibited by either agent alone [51]. Souissi [76] reported that a combination of equivalent amounts of inocula of strains *Pseudomonas fluorescens* LS102 and *Flavobacterium balustinum* LS105 applied to soil in the greenhouse resulted in a synergistic increase in inhibition of seedling growth of *Euphorbia esula* (leafy spurge). Bacteria co-inoculated with mycoherbicide agents enhanced diseases caused by the fungal pathogens on *Abutilon theophrasti* (velvetleaf) [28] and *Sesbania exaltata* (hemp sesbania) [72]. The crown galls caused by *Agrobacterium* spp. on *Euphorbia esula* (leafy spurge) and *Acropilum repens* (Russian knapweed) were often also infected by other soil-borne bacteria and fungi [11]. Similarly, the inhibitory effects caused by insects on growth of weeds could be enhanced by fungi. In this case, the insect agent feeds on roots or crowns of target weeds. For example, control of leafy spurge caused from feeding by root-boring larvae of flea beetles is enhanced by secondary invasion by plant pathogens naturally present in soils [69]. Exploitation of flea beetle larvae as vectors of DRB selective for suppression of leafy spurge could contribute an additional strategy for control of this noxious weed and serve as a model for integration of root insect-DRB combination on other weeds.

Allelochemicals released from plant residues and growing plants could stimulate phytotoxic activity of endemic microorganisms. This would become a viable strategy in minimum tillage systems and in situations where crop residues are turned into the soil. The accumulated plant residues are the readily available sources of allelochemicals.

DRB Delivery System: Formulation of a delivery system that promotes survival and colonization of weed seeds and roots by DRB in the field is critical in attaining a high level of efficacy [51]. Alginate encapsulation of *Pseudomonas fluorescens* causes better survival and efficient colonization of wheat roots in soil. Similarly, clay encapsulation of *P. fluorescens* D7 enhanced survival and increased biocontrol of downy brome by 20% to 40% [51]. This suggests that preparation of inocula based on clay and alginate formulations are likely to maintain a high level of efficacy of DRB under field conditions.

Phytopathogenic Bacteria-Based Biocontrol

In recent years, phytopathogenic bacteria have shown great potential as biocontrol agents because they can be applied directly to the foliage of target weeds. These bacterial bioherbicides are similar to the phytopathogenic fungi-based mycoherbicides. Bacteria grow quickly in liquid culture, can be stabilized in frozen or dried formulations and are amenable to mutant selection and genetic manipulation.

The bacterium *Pseudomonas syringae* pv. *tagetis* (*Pst*) causes chlorosis on such different plant species as *Ambrosia artemisiifolia* (common ragweed), *Helianthus tuberosus* (Jerusalem artichoke), *Cirsium arvense* (Canada thistle) and *Tagetes erecta* L. (marigold). In all these natural weed hosts, *Pst* causes plant vigour reduction, flowering inhibition and plant mortality.

Despite the promise shown by bacterial bioherbicides, the major impediment in the successful use of phytopathogenic bacteria is their inability to penetrate intact plants. The spread of bacteria on the plant foliage is usually limited and localized. The phytopathogenic bacteria require a film of water for dispersal on the phylloplane (leaf surface), but the mere presence of water on the foliage is normally inadequate for entry through stomata and other openings [46]. The penetration of aqueous bacterial suspensions is precluded by the high surface tension. Addition of surfactants to bacterial suspensions or wounding of leaf surface would greatly improve foliage penetration by the bacteria. Once inside, the bacteria would be protected and, hence, free to move into intracellular spaces. Zidack et al. [90] demonstrated that the organosilicone surfactant Silwet L-77 (polyalkyleneoxide-modified polydimethylsiloxane) facilitated entry of bacteria into stomata and hydathodes without phytotoxicity.

Thus, use of appropriate surfactants with *Pst* suspension is crucial. Johnson et al. [46] found that spray application of aqueous *Pst* bacterial suspension (5×10^8 cells ml^{-1}) was ineffective on natural hosts (disease incidence being low to zero), while the addition of non-ionic organosilicone surfactants Silwet L-77 (0.1% v/v) or Silwet 408 (0.2% v/v) to the bacterial suspension caused 100% decrease in incidence and greater severity than that observed in natural infections. The surfactant-added suspension rapidly penetrated the leaf through stomata, hydathodes and natural wounds. The leaf tissue became heavily soaked within minutes of application, including the susceptible apical and axillary meristems.

Johnson et al. [46] studied bioefficacy of surfactant-added *Pst* bacterial suspensions (buffered in 0.01 M phosphate, pH 7.0) on various weed species infesting a commercial maize field. They found that the populations of *Xanthium strumarium* (common cocklebur), *Ambrosia artemisiifolia* (common ragweed), *Conyza canadensis* (horseweed) and *Lactuca serriola* (prickly lettuce) were virtually eliminated, while the populations of *Cirsium arvense* (Canada thistle) were significantly reduced compared to controls. In Canada thistle plants, which showed only apical chlorosis but not death, seed production was inhibited. The affected plants showed chlorotic symptoms within 3 to 4 d of application. The tissue that developed after treatment was yellow or white and sometimes became brown and necrotic, while the tissue that formed before treatment remained unaffected. The weed species *Setaria viridis* (green foxtail), *Polygonum persicaria* (ladysthumb) and *Abutilon theophrasti* (velvetleaf) were infected inconsistently, while *Chenopodium album* (common lambsquarters), *Amaranthus retroflexus* (redroot pigweed) and *Elytrigia repens* (quackgrass) showed no signs of infection or injury. Maize remained unaffected.

Wounding the foliage considerably improves foliage-penetration and consequently the efficacy of bacterial herbicides. The bacterial pathovar *Xanthomonas campestris* pv. *poannua (Xcp)* causes vascular wilt in *Poa annua* ssp. *annua* under field conditions. Johnson et al. [46] found that when aqueous suspensions of *Xcp* were applied during mowing of *Poa annua* ssp. *annua* in recalcitrant turf fields, the weed showed lethal and systemic wilt symptoms, while applications made to this annual grass that had not been recently mowed showed no effect. The affected plants showed yellowing symptoms within 14 d of application and death occurred within 4 wk. The safety of *Xcp* on grass mixtures and its efficacy relative to alternatives makes *Xcp* a useful and promising product in high value turf management.

The preceding discussion has clearly shown that phytopathogenic bacteria such as *Pst* and *Xcp* offer viable, alternative weed management strategies. Suitable formulation and application technology will greatly enhance their suitability as bacterial herbicides. Under conditions that favour and facilitate rapid foliage penetration, the injury of susceptible plants from bacterial disease may exceed that caused by chemical herbicides. However, caution may be exercised to prevent unwanted pathogen spread and host infection.

PHYTOCHEMICAL-BASED HERBICIDES

Plants synthesize numerous chemicals in a natural environment. These naturally occurring compounds exhibit plant growth regulating and herbicidal properties on other plants. Most of these phytochemicals are easily and rapidly degraded or detoxified in the environment. Study of these compounds may lead to the discovery of new classes of herbicides that affect sites of action different from those exhibited by currently used herbicides. As the traditional methods of searching for new herbicides are yielding diminishing returns, natural plant compounds offer a viable tool to discover herbicides with newer chemistries and phytotoxic properties. In fact, many herbicide companies have begun or strengthened research in this direction with this end in mind.

Many of these compounds are not involved in the primary metabolism of the plants. These are termed secondary products, which are believed to be involved in interactions of the plant with other organisms such as insects, pathogens, nematodes and other plants [25]. Allelochemicals involved in allelopathy exhibit phytotoxic or herbicidal properties to other plant species, or even to the species producing them. The various natural compounds produced by plants are briefly reviewed here.

Aromatic Compounds

A large number of secondary plant products have aromatic chemistries. These include phenolic acids, coumarins, flavonoids, tannins, alkaloids, quinones, etc. Most of them are derivatives of aromatic amino acids or their precursors (vide **Shikimate pathway**, Chapter 7). Many of the allelochemicals belong to this group.

Phenolic acids are derivatives of benzoic acid or cinnamic acid. But these are not considered to have herbicidal properties. Coumarins, the lactones of O-hydroxycinnamic acid, inhibit wheat root growth by about 50% within 4 d [58]. Coumarin inhibits both oxidative phosphorylation and photophosphorylation. It acts more like an uncoupler of electron transport. Scopoletin, a coumarin product, is a widely known allelochemical that inhibits seed germination. Scopoletin is a potent IAA-oxidase inhibitor. Scopolin and esulin are more inhibitory of glucose-6-phosphate dehydrogenase than their aglycones scopoletin and esculetin. Another coumarin, xanthotoxin, is a good germination inhibitor.

Flavonoids (flavone, quercitin, phloridzin, phloretin, kaemferol, melitin, etc.) occur in plants both as glycones (conjugation with sugars) and aglycones (without sugar to the basic structure). There are several compounds in this group, but only aglycones exhibit phytotoxicity. Tannins are water-soluble polyphenolic compounds. They do exhibit seed germination and plant growth-inhibiting properties. However, their nonspecific binding to proteins and relatively low efficacy preclude tannins from becoming suitable herbicides.

Alkaloids are mostly alkaline and heterocyclic. All of them contain nitrogen as a part of the ring structure. They are believed to be involved in plant defence against pathogens. Alkaloids include caffeine, nicotine, quinine, strychnine, colchicine, sparteine, trigonelline, theophylline etc. Many of them, widely known for their pharmacological properties, exhibit phytotoxic properties in the form of inhibition of cell division and seed germination. Although several phytotoxic quinones from microorganisms are known, juglone (5-hydroxy-napthoquinone) is the only phytotoxic quinone produced by higher plants. It is one of the most phytotoxic allelochemicals known. Hydroquinones and benzoquinones are also phytotoxic.

Other aromatic compounds such as catechol, phenylheptatriyne, benzaldehyde, and salicylaldehyde and substituted phenols such as O-cresol have shown seed germination and plant growth inhibition properties. Another phenolic compound, α-terthienyl, has a site of action near photosystem II. It has been patented as a herbicide [81]. Catechol (1,2-dihydroxybenzene) is a plant growth inhibitor. The hydroxamic acid (2,4-dihydroxy-7-methoxy-1,4-benzoxain-3-one), a component of wheat, maize and rye, inhibits photophosphorylation by affecting the energy transfer in chloroplast [66].

Terpenoids and Steroids

Terpenoids are classified as hemi, mono, sesqui, di, tri and tetra (with 5, 10, 15, 20, 30 and 40 carbons respectively) terpnoides. They are made up of isoprene (also termed isopentane) units. The more prominent terpenoids are camphor and 1,8-cineole. The herbicide cinmethylin is closely related to 1,8-cineole, which shows well-established phytotoxic properties. The monoterpenoid p-methane-3,8-diol is a potent inhibitor of germination and growth inhibitor of several weed species at concentrations that have little effect on rice [59]. Steroids are tetracyclic triterpenoids and sterols are steroid alcohols. There is a large variation in the structure and chemistry of these compounds. Some of the essential oils, hormones like gibberellic and abscisic acids, and carotenoids belong to this group.

Other Compounds

Plants synthesize over 200 known non-protein amino acids. The more prominent amino acid δ-aminolevulinic acid (ALA) has strong herbicidal properties. The combination of ALA and 2,2^1-dipyridyl (DPy), a non-natural chemical causes massive accumulation of tetrapyrrole intermediates of chlorophyll, primarily protochlorophyllide [68]. These intermediates are photodynamic sensitizers that result in production of lethal levels of toxic radicals and toxic oxygen species (vide Chapter 7) when the treated plant is exposed to sunlight. This principal mechanism of action is similar to the one exhibited by bipyridilium and diphenyl ether herbicides. Another photodynamic compound showing herbicidal properties is hypercin.

The aquatic plant *Eleocharis* spp. produces several allelochemicals, including trihydroxycyclopentylcarboxylic acid, hydroxycyclopentenone, chrycorin etc. Benzyl isothiocyanate, isolated from papaya, almost completely inhibits germination and growth of *Abutilon theophrasti* (velvetleaf), while it is less phytotoxic to maize and soybean. When formulated in granules, it is also highly toxic to *Cassia obtusifolia* (sicklepod) and *Sorghum bicolor* (sorghum). Similarly, caprolactum, a component of sunflower, also has herbicidal properties.

MICROBIAL CHEMICAL-BASED HERBICIDES

Although chemicals produced by microorganisms have revolutionized modern medicine, little attention has been paid to exploring these chemicals for development of herbicides. Gibberellins were the first group of products derived from microorganisms over sixty years ago. These compounds cause dramatic effects on higher plants. Many fungal and bacterial compounds with phytotoxic properties have been discovered since the development of gibberellins. Most of these phytotoxins are toxic only for crops and have an extremely narrow range of activity [25]. However, the unique relationship between plants and microorganisms may serve as a better source of herbicides.

Anisomycin, a product of *Streptomyces* spp. 638 is very selective, with good bioefficacy on *Echinochloa crus-galli* (barnyardgrass) and *Digitaria* spp. (crabgrass) and no effect on tomato and turnip.

Bialaphos {L-amino-4[(hydroxy) (methyl) phosphinoyl]butyryl-L-alanyl-L-alanine}, a tripeptide, produced by the saprophytic actinomycete *Streptomyces hygroscopicus* (Jensen) Waksman and Henrici and *S. viridochromogenes* (Krainsky) Waksman and Henrici [57] controls a broad spectrum of dicot and monocots. It has non-selective herbicidal properties. Bialaphos is rapidly metabolized to phosphinothricin [L-amino-4(hydroxymethylphosphinyl)butanoic acid] in higher plants. Phosphinothricin is the active compound of bialaphos. Meiji Seika Kaisha Ltd. of Japan produced bialaphos as a product of conventional microbial fermentation and developed and marketed it under the trade name **Herbiac**. Hoechst AG of Germany synthesized the active ingredient phosphinothricin, developed with a common name **glufosinate** and marketed under the trade name BASTA. Phosphinothricin is a competitive inhibitor of **glutamine** synthetase, an enzyme important in ammonia assimilation in higher plants (vide Chapter 7).

Phosalacine (an analogue of bialaphos that also contains phosphinothricin) is produced by *Kitasatosporia phosalacinea* [62]. Another analogue of phosphinothricin, tabtoxinine-β-lactam, is a product of *Pseudomonas tabaci*. It is also a glutamine synthetase inhibitor.

Tentoxin is a cyclic tetrapeptide produced by *Alternaria alternata*. It causes extreme chlorosis in several monocot and dicot weed species but not in soybean and maize. It kills most of the weeds in soybean. It can also kill *Sorghum halepense* (Johnsongrass). It can be used as a preemergence herbicide at very low rates. Tentoxin inhibits energy transfer in chloroplasts, affecting photophosphorylation. It also inhibits transport of nuclear-coded plastid proteins to plastid, causing chlorosis.

Rhizobitoxine, a product of certain strains of *Rhizobium japonicum*, causes chlorosis and other symptoms of phytotoxicity in many weed species in soybean which is the host of this bacterium. Rhizobitoxine, an analogue of cystathione, severely inhibits β-cystathionase, an enzyme required in the synthesis of methionine. It also inhibits enzymatic production of ethylene from methionine (vide Chapter 7).

IMPROVEMENT OF EFFICACY OF BIOHERBICIDES/BIOAGENTS

The commercial success of a bioherbicide or bioagent depends on overcoming difficulties in production, formulation and application of living microbial control agents. It is important to remember that for any bioherbicide to be successful, the host-pathogen balance must be tipped to favour the bioagent. Innovative and economic production, formulation and application systems can help overcome the constraints that limit the development of bioherbicides.

The production and formulation of bioherbicides is a three-step process involving development of a production medium, formulation of the propagules produced and stabilization of the formulation.

Production of Biopropagules

Production technology for bioherbicidal agents must be low-cost and yield-high concentrations of viable, highly effective propagules. These propagules must be amenable to dry preparations for long-term storage. A bioherbicide can be produced by three methods: 1) use of living host plants, 2) solid substrate fermentation and 3) liquid culture fermentation. The first two methods are too costly for commercial use. Liquid culture fermentation is the most economical method for producing biocontrol agents.

Development of a defined medium, which supports good growth and propagule formation of the biocontrol agent, is the first step in the production of bioherbicides. The biopropagule can be bacterial cells, bacterial or fungal spores, fungal sclerotia, or mycelial fragments. Once a basal medium is developed, nutritional components can be varied in terms of propagule growth, propagule fitness as a bioherbicidal agent and propagule stability as a dry preparation [43]. Nutritional components such as carbon sources, nitrogen sources, trace metals, vitamins, carbon loading and carbon-nitrogen (CN) ratio have an influence on growth, propagule formation and biocontrol efficacy. Once an optimized defined medium has been developed, a production medium can be formulated by replacing the nutritional components of the defined medium with low-cost, complex substrates [43]. Use of this directed optimization strategy not only helps in the development of production media for specific bioherbicides but also provides nutritional information, which is useful in developing production media for other microbial bioherbicides.

Jackson et al. [43] made extensive studies with submerged conidial cultures of *Colletotrichum truncatum* NRRL 18434, a specific fungal pathogen of the weed *Sesbania exaltata* (hemp sesbania) to understand the nutritional regulation on propagule yield, formation and bioefficacy. Carbon concentration and carbon-to-nitrogen (CN) ratio are the two nutritional factors that have a dramatic effect on propagule formation. Carbon concentration in the range of 4 to 16 g L^{-1} produced high concentrations of conidia. Greater than 25 g L^{-1} inhibited conidiation and promoted microsclerotia, the highly melanized hyphal aggregates. Conidia produced in a medium with a CN ratio were longer, thinner and more efficacious. The '10:1' conidia also germinated more rapidly, formed appressoria more frequently and incited more disease in hemp sesbania seedlings. The rapidly germinating spores have a significant advantage in causing infection under field conditions where limited free-moisture represents a significant constraint to biocontrol efficacy. The conidia of 10:1 CN ratio contained more protein and less lipid than the conidia of 30:1 and 80:1 ratios [42]. Conidia yields may be further increased and fermentation time reduced if the media contain low concentrations of methionine, cysteine and tryptophan.

The microsclerotia formed under higher carbon concentrations (greater than 25 g^{-1}) may also serve as useful bioherbicidal propagules as they are amenable to drying and storage. They retain over 80% viability after 8 mon storage and over 50% viability after 12 mon storage. Soil incorporation of microsclerotia (150 microsclerotia/cc potting soil) killed over 95% of the emerging hemp sesbania seedlings [42].

Formulation

Formulation is the blending of active ingredient, the fungal and bacterial propagules, with inert carriers and diluents in order to alter physical characteristics of the bioherbicide to a more desirable form. This includes diluting the final product to a common potency, enhancing stability and biological activity, improving mixing and spraying, integrating the bioherbicide into a pest management system and possibly altering the weed control spectrum of the organism [10]. Bioherbicides can be made into liquid-based formulations and solid-based formulations.

Liquid-based Formulations

Liquid-based formulations are best suited for postemergence control of the target weeds. They are used to incite leaf and stem diseases. Most weeds are covered by a waxy cuticle that prevents a water-based product from spreading evenly, resulting in unequal distribution of active ingredient, the propagules. Surfactants help to wet the plants and aid in dispersing the fungal spores throughout the spray mix. Certain surfactants e.g. Tween-20 and Tween-80, inhibit germination or growth of fungi. However, a non-ionic non-oxynol surfactant enhances spray coverage without affecting spore germination.

The formulated material of a selective postemergence herbicide COLLEGO® is a two-component product, with one component consisting of water-suspendible dried CGA spores and the other containing a water-soluble spore rehydrating agent. Both components are packaged in an 18.9 L plastic mixing container. A mixture containing 0.95 L of rehydrating agent and one package of CGA spores will treat 4 ha. The dried spores are packaged on the basis of number of viable spores, then by weight, with each package containing 75.7×10^{10} viable spores. Components of the two packages are added to the desired volume of water just prior to application.

DeVine®, the first mycoherbicide to be registered, is based on chlamydospores of the fungus produced upon fermentation in a vegetable juice medium. Chlamydospores, the infective units, are not stable and the formulated material, as already mentioned, has a shelf-life of only 6 wk and must be handled like fresh milk. Another mycoherbicide, BioMal® is formulated by using silica gel as a carrier. It provides over 90% control of the target weed. The wettable powder formulation disperses easily in water and is applied as a spray to the weed [55, 56]. The experimental formulation of the herbicide ABG-5003 consists of mycelial fragments and spores applied as a wettable powder.

Solid-based Formulations

Pathogens that infect at below the soil surface are usually delivered in a solid or granular formulation. Granular formulations are better used as preplant and pre-emergence mycoherbicides than spray formulations because the granules: a) provide a buffer from environmental extremes, b) serve as a food-base for the fungus, resulting in longer period of persistence and c) are less likely than spores to be washed away from the treated areas [10].

A formulation based on cornmeal and sand was successfully used to produce mycelia and a mixture of microconidia, macroconidia and chlamydospores of *Fusarium solani* f.sp. *cucurbitae* (Snyd.) Snyad. and Hans. (FSC), a pathogen that controls *Cucurbita texana* (Texas gourd) [8]. This granular formulation achieved 96% control of the weed when applied at preplant and preemergence.

Sodium alginate is another good medium to produce fungal granular formulation [85, 86]. In this method, fungal mycelia are mixed with sodium alginate and a filler like kaolin clay and dripped into a 0.25M solution of calcium chloride. The Ca^{++} ions react immediately with sodium alginate to form gel beads. The beads, after allowing them to harden in the calcium chloride solution for a few minutes, can be collected, rinsed and subsequently air-dried. The granules of fairly uniform size and shape may be used in a manner similar to preplant and preemergence granular herbicides. The granules may also be rehydrated and exposed to UV light to induce fungal spore production. Spores may then be collected and used for postemergence application.

In another granular formulation, the conidia, microsclerotia and mycelia from a liquid culture of *Colletotrichum truncatum* (Schw.) Andrus and Moore, is mixed with wheat flour and kaolin [19, 20]. The product, called 'PESTA' can be stored at room temperature for several months with little decline in pathogen viability or virulence. Preemergence application and soil incorporation gives over 90% control of hemp sesbania (*Sesbania exaltata*) in soybean. PESTA granules made of micro- and macroconidia of *Fusarium oxysporum* Schlect. emend. Snyad. and Hans., isolated from *Cassia obtusifolia* (sicklepod), provide simultaneous control of sicklepod, *Cassia occidentalis* (coffee senna) and hemp sesbania [9].

Stabilization and Improvement of Formulation

For a successful commercial use of a bioherbicide or bioagent, the formulation must be stabilized by the addition of adjuvants and amendments. The stabilized formulation should preferably have a shelf-life of 6 to 18 mon at ambient temperatures [43].

In a process developed by Quimby et al. [67], the microorganism-incorporated alginate granules are coated with an inverting oil. An oil absorbent is then added until the granules become free flowing. The oil component slows the drying process of the living bioagents, required for successful formulation. Using this process, they extended the shelf-life of *Sclerotinia sclerotiorum* (Libert) deBarry, a potential mycoherbicide for control of several broadleaf weeds, for at least 2 yr and that of *C. truncatum* for hemp sesbania for 1 yr. In studies with *Trichoderma hariazum*, a fungal biocontrol agent effective against various plant pathogens, the nutritional environment during submerged culture sporulation was found to significantly influence the ability of the spores to withstand drying and retain viability during storage [36, 44]. Osmotic stress induced through the addition of polyethylene glycol (PEG) to the culture medium enhanced *T. hariazum* spore stability. Osmotic stress appears to trigger accumulation or synthesis of trehalose in fungi and bacteria [32, 44]. Trehalose is an osmoprotective compound.

Adjuvants and amendments also improve or modify spore germination, pathogen stability and virulence, environmental requirements and host preference, all of which greatly influence the potential of a bioherbicide [10].

The addition of sucrose to aqueous suspensions of *Alternaria macrospora* results in increased disease severity on spurred anoda. Similarly, spore germination and disease severity also result on *Desmodium tortuosum* (Florida beggarweed) when small quantities of sucrose and gum xanthan are added to aqueous spore suspensions of a host-specific strain of *C. truncatum* [13]. Disease severity on *Sorghum halepense* (Johnsongrass) was significantly increased with the addition of 1% Soy-Dex (a commercial adjuvant) to the spray mix of the fungus *Bipolaris sorghicola* (Lef. and Sherw.) [87]. Addition of sorbitol to the spray mixture of *C. coccodes* (Wallr.) Hughes. caused a 20-fold increase in viable spores re-isolated from inoculated velvetleaf [88].

Most pathogens from which bioherbicides are made require free moisture (usually from dew) in order to germinate, penetrate, infect and kill the target weed. They require 6 to 24 h free moisture period, depending upon the pathogen and the host weed [10]. Invert emulsions provide a potential method to retard evaporation and trap water in the spray mixture, thereby decreasing the amount of free moisture required for spore germination and infection. Diagle et al. [23] used lecithin as an emulsifying agent and paraffin oil and wax to further retard evaporation and help retain droplet size. Delivery of this viscous material, however, requires special delivery equipment. This fungus-invert formulation system helped to enhance control of hemp sesbania (*Sesbonia exaltata*) by *C. truncatum* to 95% which was comparable with that (96%) in an acifluorfen-treated field.

Similar effects can be achieved with vegetable oil emulsions. The vegetable oil-based formulations are less viscous and can be applied with conventional spraying equipment. Auld [3] found that a vegetable oil emulsion containing *C. orbiculare* Berk. and Mont. effectively controlled *Xanthium spinosum* (spiny cocklebur) under moisture-limiting conditions. Boyette et al. [10] reported that unrefined corn oil and *C. truncatum* emulsion formulation reduced the minimum dew period requirement for infection and weed mortality from 12 h to 2 h and reduced spray volume 100-fold, from 500 L ha^{-1} to 5 L ha^{-1}. Unrefined corn oil stimulates conidial germination and appressorial formation, resulting in increased infection of hemp sesbania [6, 26].

Invert emulsions, besides minimizing free moisture requirement, have been shown to expand the host selectivities of some mycoherbicides. Amsellen et al. (2) reported that the host selectivities of *Alternaria cassiae* and *A. crassa* were eliminated when spores were formulated in an invert emulsion. Host ranges of *Colletotrichum truncatum* and CGA were also expanded when spores were formulated in an invert emulsion. Hemp sesbania (*Sesbania exaltata)* was immune to both inundative and wound-inoculation of CGA by conidia in water only. However, when conidia were formulated in the invert emulsion, hemp sesbania became highly susceptible to the fungus. Similarly, water- or wound-inoculation by *C. truncatum* had no effect on *Aeschynomene virginica* (northern jointvetch), but this weed became susceptible to the fungus when it was formulated in an invert emulsion. Yang et al. [89] found that virulence of *A. alternata* (Fr:Fr) Keissl. and *A. angustivoidea* E. Simmons on leafy spurge (*Euphorbia esula*) were improved significantly when spores of pathogens were applied in an invert emulsion.

Host specificity can be expanded by alternating the formulation of mycoherbicides. Host selectivity of *A. crassa* (Sacc.) Rands, a mycoherbicide agent for *Datura stramonium* (jimsonweed) can be altered by the addition of water-soluble filtrates of jimsonweed or dilute fruit pectin to spore *suspension formulation* [10]. Several resistant plant species, including hemp sesbania, *Solanum ptycanthum* (eastern blacknightshade), *Xanthium strumarium* (common cocklebur) and *Crotolaria specatbilis* (showy crotolaria), exhibited higher levels of susceptibility following these amendments. Several Solanaceae crop species, including tomato, potato, tobacco and eggplant (brinjal) were also susceptible to infection by *A. crassa* following formulation amendments.

As most mycoherbicides have a limited host range, which prohibits their use in a mixed weed spectrum, mixtures of pathogens may be applied to widen the host specificity and control more weeds in a single application. For example, the rice weeds *Aeschynomene virginica* (northern jointvetch) and *Ludwigia decurrens* (winged waterprimrose) can be simultaneously controlled by a single application of CGA and *C. gleosporiodes* (Penz.) Sacc. f.sp. *jussiaea* [7]. A mixture of these two pathogens along with *C. malvarum* (A. Braun. and Casp.) Southworth also effectively controlled northern jointvetch, *Ludwigia decurrens* (winged waterprimrose) and *Sida spinosa* (prickly sida).

Genetic Manipulation

Many of the potential bioherbicidal agents are insufficiently virulent to cause acceptable levels of control of their target weeds. Besides, a majority of pathogens are extremely host-specific, thus restricting the usefulness of these agents only to control single weed species. This limits the economic feasibility of developing them as bioherbicides. **Pathogenicity** (the ability to cause disease), **virulence** (the capacity to cause a certain level of disease damage) and **host range** (the ability to cause disease on different hosts) of the phytopathogenic microbes can be genetically altered by mutagenesis, or by cloning and expressing some innate genes controlling these traits [16]. Genes or cultures of genes are involved in pathogenicity, virulence, hypersensitivity and host range, and these genes could be altered to produce a change in a pathogenicity trait. For example, in *Erwinia chrysanthemi* Burkh. et al., a soft-root causing bacterium, the extracellular pectic enzyme complex, comprising pectin lyase and exo-poly-α-D-galacturonosidase, is a key factor in virulence [5, 18]. The deletion/inactivation of these genes involved in the biosynthesis of these enzymes, the *pel* genes, is correlated with the loss of ability of the bacterium to injure plant tissues. Cloning and characterization of genes involved in the biosynthesis of host-specific and host-non-specific toxins and the roles of these toxins as virulence pathogenicity determinants have been well documented.

Gene manipulation or modification in a pathogen can be done by using a gene donor from among other pathogenic microbes or from non-pathogenic microbes. In most cases of gene manipulation, the gene encoding production of a phytotoxin is transferred from a pathogen into the pathogen in question that is already equipped with genes for pathogenicity. Such genes could even be derived from an organism that is phylogenetically unrelated to the intended transformant.

Transfer of Bialaphos Genes

Bialaphos, as mentioned earlier in this Chapter, is a tripeptide produced by the saprophytic actinomycete *Streptomyces hygroscopicus* (Jensen) Waksman and Henrici and *S. viridochromogenes* (Krainsky) Waksman and Henrici. The herbicidally active moiety in bialaphos is phosphinothricin (the L-glutamic acid analogue) containing the unique C-P-C bond. The biosynthesis of bialaphos involves the *bap* genes (genes that catalyze 13 biosynthetic steps starting from phosphoenol pvruvate to bialaphos), the *brp*A gene (a regulatory gene) and the *bar* gene (a resistant gene). The *bar* gene, which encodes phosphinothricin N-acetyl-transferase (PAT) in *S. hygroscopicus* plays a role in auto-resistance to bialaphos as well as in the synthesis of this metabolite. The *bar* gene is also linked to resistance of several crop plants and microorganisms to bialaphos. Nagaoka [57] subcloned the *bap* and *bar* genes to a 16-kb region of the *S. hygroscopicus* genome. The regulatory gene, *brp*A, which activates transcription of the *bar* gene and at least six *bap* genes, appears to play a role in the expression of the bialaphos biosynthetic gene cluster [57]. As bialaphos is a broad-spectrum phytotoxin, active at very low doses and is easily degraded in soil, it can serve well to increase both the virulence and host range of a phytopathogen if 'bialaphos genes' are transformed by genetic manipulation.

In a classical study, Charudattan et al. [16] isolated a cluster of 'bialaphos genes' from several biosynthetically blocked mutants of *S. hygroscopicus* and constructed into a plasmid vector, **pBG9**. They also transferred a fragment of the gene cluster into **pLAFR3**, a plasmid that functions in both *Escherichia coli* and *Xanthomonas campestris* pv. *campestris* (**XCC**) and contains a tetracycline resistance marker. The resultant plasmid, **pIL-1**, was used to transform *E. coli* and was incorporated into XCC by conjugation. The genes were maintained in XCC for about 47 generations in the absence of selection for tetracycline and no changes in cultural phenotypes were observed in the transformed XCC (**XCC/pIL-1**). They found that XCC/pIL-1 cells caused pathogenicity to their natural hosts, cabbage and broccoli and induced an altered hypersensitive response in the non-hosts, bean, pepper, sunflower and tobacco. The pathogenic symptom caused by the parent XCC, XCC/pLAFR3 and XCC/pIL-1, was a typical black rot disease in the inoculated leaves of the two hosts. The reaction induced by XCC and XCC/pLAFR3 on the non-host leaves was necrotic hypersensitivity, while that induced by XCC/pIL-1 was hypersensitivity accompanied only by chlorosis. They hypothesized that the altered hypersensitivity phenotype may be due to transformed XCC becoming more compatible with non-host plants, a step towards acquiring virulence by non-hosts, or due to disruption of the normal expression of the hypersensitivity and pathogenicity genes in the transformed XCC. They further suggested that more work was needed to confirm their findings.

The foregoing brief review clearly shows that the virulence of a pathogen can be enhanced through the introduction of a genetically encoded virulence factor and the host range of bioherbicide agents can be enlarged within generic and familial bounds. This may be one avenue for improving weed control of pathogenic bacteria.

ENHANCEMENT OF BIOHERBICIDE EFFICACY BY HERBICIDES

Chemical herbicides can enhance or reduce weed control when applied in combination with microbial and other bioherbicides or bioagents. The enhancement effect through synergistic and additive interaction is possible if weed defences can be lowered using herbicides, so that the weeds become more susceptible to pathogen attack. In such an event, the quantity and concentration of a bioherbicide and concentration of herbicide could be lowered and the host range of a pathogenic agent expanded.

Plants have inherent ability to protect themselves against attack from essentially all microorganisms. The pathogenicity of fungi, bacteria and viruses depends on their ability to either evade or break down the defence mechanisms that plants possess. Thus, enhancement of the efficacy of bioherbicides is possible only if a particular herbicide in a combination is capable of evading or breaking down the plant defences. The mechanisms of plant defence are discussed briefly below.

Mechanisms of Plant Defence

Although many pathogens with bioherbicidal potential have been discovered in the recent past, many of them lack sufficient aggressiveness to overcome weed defences to achieve adequate control [39]. Plants use various physical and biochemical mechanisms in defence against pathogen inactivity, including callose deposition, hydroxyproline-rich glycoprotein accumulation, epicuticular wax formulation, pathogen-related protein (PR protein) synthesis, phytoalexin production, lignin and phenolic formation, free radical generation, etc.

Callose is composed mostly of β-1-3 glucans situated at surfaces of intact plant cells. Its production is induced by cellular damage via pathogen attack or other stress. The active site of callose synthase, the enzyme responsible for synthesis of callose, is on the cytoplasmic side of the plasma membrane [29]. Chemicals that regulate and decrease callose production will weaken this defence mechanism to facilitate greater efficacy of bioherbicides. The accumulation of hydroxyproline-rich glycoproteins, also called extensins, is a defence reaction of several dicot plants to several pathogens and stress [54]. These proteins are believed to be involved in pathogen recognition and their defensive action appeared to be via physical agglutination of negatively charged fungal and/or bacterial pathogens. Ethylene causes accumulation of extensins in plant species.

Epicuticular waxes formed on the plant foliage serve not only as physical barriers to pathogen attack, but also reduce the spore germination of pathogens. Various herbicides affect wax biosynthesis and deposition by inhibiting the formation of long-chain (C_{28} to C_{32}) alcohols and fatty acids and increasing short-chain species. Adjuvants, surfactants and other non-herbicidal additives also alter wax structure or its formation on plant surfaces and thereby change protection properties.

Pathogen-related proteins (PR proteins) are a group of proteins whose synthesis is induced by infection, specific environmental stress or developmental stage [14]. These are generally of low molecular weight (14,000 to 30,000) and differ from heat shock proteins. Each plant produces a characteristic set of proteins. The PR-proteins, also called **stress proteins**, are associated with resistance to fungal, bacterial and viral infections in host plants. During pathogen attack or wounding, these polypeptides accumulate. Although PR-proteins are induced in stressed plants, their function and role in the defence mechanism of weeds are little known. Herbicides which inhibit protein synthesis might inhibit PR-protein synthesis and lower weed defence responses, but are likely to inhibit protein synthesis in biocontrol agents (bioherbicides) as well. More work is needed to identify chemicals that inhibit PR-proteins in weeds without affecting protein synthesis in pathogens.

Phytoalexins are secondary plant compounds, produced post-infection, that exhibit anti-microbial activity [31]. These include phenolics, isoprenes (mono-, di- and sesquiterpenes) from the acetate-mevalonic acid (mavalonate) pathway and polyacetylenes from the condensation of acetate. Bacteria, fungi and plants have enzymes that can degrade phytoalexins. Since many phytoalexins are phytotoxic, the combinations of phytoalexins and bioherbicides may bring out synergistic responses. Weed control using combinations of bioherbicides with phytoalexins is theoretically possible to elicit additive or synergistic mechanisms by reducing phytoalexins in plants, thereby allowing increased pathogenicity, or by increasing the levels of phytoalexins (by supplemental phytoalexins) to phytotoxic levels while maintaining pathogen viability [39]. These possibilities need to be examined to find synergistic interactions of herbicides and bioherbicides.

Elicitors are glucans, glycoproteins and other pathogen cell wall components that induce phytoalexin production and accumulation. Many of them possess carbohydrate moieties. Microbial elicitors (of fungal genera *Phytophthora*, *Colletotrichum* and *Fusarium*) are high molecular weight cell wall components (glucan-cellulose or glucan-chitin) of pathogenic fungi. Besides inducing accumulation of phenolics and phytoalexins, elicitors cause development of necrotic symptoms in plants. If incorporated into a bioherbicide formulation, supplemental elicitors may enter weeds during the infection process, be recognized and cause necrosis [39]. Thus, the combination of an elicitor and pathogen could provide additive or synergistic responses. In some plant systems, **oxygen radicals** are generated in response to pathogenic attack. These radicals are highly reactive and attack the double bonds in unsaturated membrane fatty acids [35]. This reaction leads to autocatalytic lipid peroxidation, with subsequent loss of membrane function leading to cell death. In some plants, rapid cell death may be caused by disruption of the tonoplast due to lipid peroxidation which releases toxic phenolic compounds previously sequestered. Some plant tissues possess an O_2^--generating NADH oxidase, which is activated by digitonin, leading to enhanced resistance to pathogens. The use of free radical scavengers, such as glutathione, ascorbate, α-tocopherol and Tiron (4,5-dihydroxy-1,3-benzenedisulphonic acid), might be useful in combination with bioherbicides to lower free radical defence systems, if operable in weeds [39]. Some of these chemicals have been used to lower defence responses of potato [24].

Suppressors are non-phytotoxic compounds secreted by pathogens that can delay or prevent elicitation of host defence responses [73]. These compounds render host cells susceptible even without avirulent and non-pathogenic organisms. They are termed 'determinants' for pathogenicity without apparent phytotoxicity [61]. Little, however, is known about their identities, characterization and sites of action. More research is needed to identify suppressors and understand their abilities to reduce weed defences.

Theoretically, suppressors provide synergistic responses on weeds when mixed with bioherbicide formulations. The most useful suppressor compounds could be those that are of low molecular weight and stable and that can enter the foliage and translocate to active sites in effective bioherbicide concentrations [39]. More work is needed to identify these compounds to elicit additive and synergistic responses when combined with bioherbicides.

Phenylalanine ammonia lyase (PAL) is involved in the shikimate pathway of carbohydrate metabolism and formation of products of secondary plant metabolism. The level of PAL in plants increases in response to stimuli such as light, invasion, mechanical injury, action of some chemicals such as elicitors and pathogenic attack [12]. Increased enzyme activity can lead to increased production of phenolic compounds. The activity of PAL is rapidly increased in plant foliage within hours of pathogenic infection. The fungal pathogen *Alternaria cassiae* Jurair and Kahn, a bioherbicidal agent against sicklepod (*Cassia obtusifolia* L.), alters phenylpropanoid metabolism in this weed [38]. The activity of PAL was increased 3-fold prior to appearance of visual symptoms after the spores of *A. cassiae* were applied to sicklepod seedlings. Thus herbicides which inhibit PAL and other enzymes (peroxidase and polyphenol oxidase) associated with secondary plant metabolism have the potential to lower weed resistance and, hence, interact synergistically with bioherbicides.

Interaction of Bioherbicides/Bioagents and Herbicides

Some herbicides can enhance weed control when applied in combination with microbial control agents. This synergistic interaction has been found for a diverse group of herbicides, bioherbicides and weed hosts. The herbicides acifluorfen and bentazon are good effective synergists in providing increased control of *Aeschynomene virginica* (northern jointvetch) by *Alternaria cassiae*, *Sesbania exaltata* (hemp sesbania) by *Colletotrichum truncatum* and *Desmodium tortuosum* (Florida beggarweed) by *Fusarium lateritium* [39]. Similar synergistic response occurs with atrazine on *Echinochloa crus-galli* (barnyardgrass) by *Colletotrichum coccodes*, trifluralin on *Cucurbita texana* (Texas gourd) by *Fusarium solani* f.sp. *cucurbitae*, glyphosate on *Cassia obtusifolia* (sicklepod) by *Alternaria cassiae* and diquat on *Eichhornia crassipes* (waterhyacinth) by *Cercospora rodmanii*. This indicates that many herbicides with vastly different chemistries and molecular modes of action can act synergistically with microbial agents to improve weed control.

Mixing bioherbicides with certain herbicides could cause detrimental effect on the living microbes. COLLEGO® mixed with propanil and molinate resulted in a loss of bioherbicide efficacy [75]. Bromacil is detrimental to the pathogen *Phytophthora palmivora* when chlamydospores were applied with the herbicide. This problem can be mitigated if the herbicide and bioagent are applied sequentially.

The combinations of living organisms and herbicides provide additive, synergistic and antagonistic effects on weed control. As two living and dynamic systems (weeds and microbes) are involved in this interaction, the resultant effect is influenced by several factors [39]. These include:

a) Toxicity of the herbicide to the pathogen spores, mycelia or cells.
b) The concentrations of the herbicide and inoculum: the concentrations of each component should be adequate enough to weaken weed defences and allow pathogen growth and infectivity to proceed normally.
c) Timing of herbicide and pathogen application in relation to each other and weed stage: if a herbicide is too deleterious to the pathogen, its application could occur before or after pathogen application rather than in a combination; and as

younger weeds are generally easier to control, the herbicide-pathogen application should be made during the early stages of weed growth.

d) Environmental conditions: UV light can cause chemical degradation and death of pathogen propagules; dew or free moisture condition can influence the germination and infection processes; light enhances the activity of some enzymes that are involved in plant defences (physical/chemical barriers and biochemical responses) that attempt to protect plants against attack from essentially all microorganisms.

e) Weed resistance and susceptibility: weed resistance and susceptibility to a herbicide are determined by differences in absorption, translocation and metabolism of the compound. These differences could alter the microbial effects on target weeds. Application of a herbicide at sublethal rates is a strategy that could cause synergistic chemical-pathogen interaction.

f) Effect of herbicides on metabolism in weeds: the herbicide used as a synergist should not alter plant metabolism to cause an increase in: i) production of plant-defence chemicals such as phenols and phytoalexins, ii) enzymatic degradation of pathogen cells and iii) metabolism of pathogen-produced phytotoxin.

The preceding discussion indicates that chemical herbicides have a certain role in improving the potential of bioherbicides which generally lack sufficient aggression to overcome weed defences, discussed earlier, to achieve adequate control. In view of the availability of numerous herbicides in the market, several pathogen-herbicide combinations may be tested for additive and synergistic effects. Such effects could result in reduced rates of both the chemicals, i.e., herbicides and bioherbicides, for effective and economic weed control. When combined applications of herbicides and bioherbicides are made, formulation materials will need to facilitate or promote microbial germination and infection [39]. Research in this area will lead to a better understanding of the additive and synergistic interactions and development of desirable combinations of bioherbicides and herbicides.

REFERENCES

1. Alexopoulos, C.J. and C.W. Mims. 1979. Introductory Mycology. John Wiley & Sons, New York, 3rd ed., 632 pp.
2. Amsellen, Z., A. Sharon and J. Gressel. 1991. Abolition of selectivity of two mycoherbicidal Organisms and enhanced virulence of avirulent fungi by an invert emulsion. Phytopathology 81: 925-929.
3. Auld, B.A. 1993. Vegetable oil suspension emulsions reduce dew dependence of a mycoherbicide. Crop Prot. 12: 477-479.
4. Baloch, G.M. 1968. Possibilities for biological control of some species of Cuscuta (Convolvulaceae) PANS (C) 14: 27-33.
5. Barras, F. and A.K. Chatterjee. 1987. Genetic analysis of the pe/A-pe/E cluster encoding the acidic and basic pectate lyases in Erwinia chrysanthemi EC 16. Mol. Gen. Genet. 209: 615.
6. Boyette, C.D. 1994. Unrefined corn oil improves the mycoherbicidal activity of Colletotrichum truncatum for hemp sesbania (Sesbania exaltata) control. Weed Technol. 8: 526-529.
7. Boyette, C.D., G.E. Templeton and R.J. Smith, Jr. 1979. Control of winged waterprimrose (Jussiaea decurrens) and northern jointvetch (Aeschynomene virginica) with fungal pathogens. Weed Sci. 27: 497-501.
8. Boyette, C.D., G.E. Templeton and L.R. Oliver. 1985. Texas gourd (Cucurbita texana) control with Fusarium solani f.sp. cucurbitae. Weed Sci. 32: 649-654.
9. Boyette, C.D., H.K. Abbas and W.J. Connick, Jr. 1993. Evaluation of Fusarium oxysporum as a potential bioherbicide for sicklepod (Cassia obtusifolia), coffee senna (C. occidentalis) and hemp sesbania (Sesbania exaltata). Weed Sci. 41: 678-681.

344

10. Boyette, C.D., P.C. Quimby, Jr., A.J. Caesar, J.L. Birdsall, W.J. Connick, Jr., D.A. Daigle, M.A. Jackson, G.H. Egley and H.K. Abbas. 1996. Adjuvants, formulations and spraying systems for improvement of mycoherbicides. Weed Technol. **10:** 637-644.

11. Caesar, A.J. 1994. Pathogenicity of Agrobacterium species from the noxious rangeland weeds *Euphorbia esula* and *Centaurea repens*. Plant Dis. **78:** 796-800.

12. Camm, E.L. and G.H.N. Towers. 1973. Phenylalanine ammonialyase. Phytochemistry **12:** 961-973.

13. Cardina, J., R.H. Littrell and R. Hanlin. 1988. Anthracnose of Florida beggarweed (*Desmodium Carr, tortuosum*) caused by *Colletotrichum truncatum*. Weed Sci. **36:** 329-334.

14. Clessing, J.P. and O.F. 1990. The pathogenesis-related proteins of plants. *In* J.K. Setlow (ed.). Genetic Engineering, Principles and Methods, Vol. II. Plenum Press, New York, pp. 65-109.

15. Charudattan, R. 1986. Integrated control of waterhyacinth (*Eichhornia crassipes*) with a pathogen, insects and herbicides. Weed Sci. **34 (Suppl. 1):** 26-30.

16. Charudattan, R., V.J. Prange and J.T. DeValerio. 1996. Exploration of the use of the "bialaphos genes" for improving herbicide efficacy. Weed Technol. **10:** 625-636.

17. Cherrington, C.A. and L.F. Elliott. 1987. Incidence of inhibitory pseudomonads in the Pacific Northwest. Plant Soil **101:** 159-165.

18. Collmer, A., J.L. Ried, G.L. Cleveland, S.Y. He and A.D. Brooks. 1989. Mutational analysis of the role of pectic enzymes in the virulence of *Erwinia chrysanthemi*. *In* B. Staskowicz, P. Ablquist and O. Yoder (eds.). Molecular Biology of Plant-Pathogen Interactions. Alan R. Liss, New York, pp. 35-48.

19. Connick, W.J., Jr., C.D. Boyette and J.R. McAlpine. 1991. Formulation of mycoherbicides using a pasta-like process. Biol. Control **1:** 281-287.

20. Connick, W.J., Jr., W.R. Nickle and C.D. Boyette. 1993. Wheat flour granules containing mycoherbicides and entomogenous nematodes. *In* R.D. Lumsden and J.L. Vaughn, (eds.). Pest Management: Biologically Based Technologies. Amer. Chem. Soc., Washington, D.C., USA, pp. 238-240.

21. Conway, K.E. 1976. Evaluation of *Cercospora rodmanii* as a biological control of waterhyacinth. Phytopathology **66:** 914-917.

22. Daniel, J.T., G.E. Templeton, R.J. Smith, Jr. and W.T. Fox. 1973. Biological control of northern jointvetch in rice with an endemic fungal disease. Weed Sci. **21:** 303-307.

23. Diagle, D J., W.J. Connick, Jr., P.C. Quimby, Jr., J.P. Evans, B. Trask-Merrell and F.E. Falgham. 1989. Invert emulsions: Delivery system and water source for the mycoherbicide, *Alternaria cassiae*. Weed Technol. **3:** 442-444.

24. Doke, N. Y. Miura, L. Sanchez and K. Kawakita. 1994. Involvement of superoxide in signal transduction: Responses to attack by pathogens, physical and chemical shocks and UV irradiation. *In* C.H. Foyer and P.M. Mullineaux (eds.). Causes of Photooxidative Stress and Amelioration of Defence Systems in Plants. CRC Press, Boca Raton, Fl., USA, pp. 177-197.

25. Duke, S.O. 1986. Naturally occurring chemical compounds as herbicides. Rev. Weed Sci. **2:** 17-44.

26. Eagley, G.H. and C.D. Boyette. 1994. Water-corn oil emulsion enhances conidia germination and mycoherbicidal activity of *Colletotrichum truncatum*. Weed Sci. **43:** 312-317.

27. Elliott, L.F. and A.C. Kennedy. 1991. Method for screening bacteria and application thereof for field control of the weed downy brome. U.S. Patent No. 5,030,562.

28. Fernando, W.G.D., A.K. Watson and T.C. Paulitz. 1994. Phylloplane *Pseudomonas* spp. enhance disease caused by *Colletotrichum coccodes* on velvetleaf. Biol. Control **4:** 125-131.

29. Fredrickson, K. and C. Larsson. 1989. Activation of 1,3-β-glucan synthase by Ca^{++} spermine and cellobiose: localization of activator sites using inside-out plasma membrane vesicles. Physiol. Plant. **77:** 96-201.

30. Freeman, T.E. and R. Charudattan. 1984. *Cercospora rodmanii* Conway-a biocontrol agent for waterhyacinth. Tech. Bull. 843, Agric. Exp. Stn., Inst. Food and Agric. Sci., Univ. of Florida, Gainesville, FL, USA.

31. Friend, J. 1981. Plant phenolics, lignification and plant disease. Prog. Phytochem. **7:** 197-261.

32. Graham, J.E. and 'B.J. Wilkinson. 1992. *Staphylococcus aureus osmoregulation*: roles for choline, glycine betaine, proline and taurine. J. Bacteriol. **174:** 2711-2716.

33. Greaves, M.P. and J.A. Sargent. 1986. Herbicide-induced microbial invasion of plant roots. Weed Sci. **34 (Suppl. 1):** 50-53.
34. Gupta, Meera. 1977. Biological control of weeds in India. Indian J. Weed Sci. **9:** 113-119.
35. Halliwell, B. 1978. Biochemical mechanisms accounting for the toxic oxygen in living organisms: The key role of superoxide dismutase. Cell Biol. Int. Rep. **2:** 113-128.
36. Harman, G.E., X. Jin, T.E. Stasz, G. Peruzzotti, A.C. Leopold and A.G. Taylor. 1991. Production of conidial biomass of *Trichoderma harzianum* for biological control. Biol Control **1:** 23-28.
37. Harris, P.A. and P.W. Stahlman. 1992. Soil bacteria combined with various herbicides suppress winter and annual grass weeds. Agron. Abstr. **56:** 258.
38. Hoagland, R.E. 1990. *Alternaria cassiae* alters phenylpropanoid metabolism in sickelpod (*Cassia obtusifolia*). J. Phytopathol. **130:** 177-187.
39. Hoagland, R.E. 1996. Chemical interactions with bioherbicides to improve efficacy. Weed Technol. **10:** 651-674.
40. Huffaker, C.D. 1957. Biological control of weeds with insects. Ann. Rev. of Ent. **4:** 251-276.
41. Jackson, M.A. and D.A. Schisler. 1992. The composition and attributes of *Colletotrichum truncatum* spores are altered by the nutritional environment. Appl. Environ. Microbiol. **58:** 2260-2265.
42. Jackson, M.A. and D.A. Schisler. 1995. Liquid culture production of microsclerotia of *Colletotrichum truncatum* for use as bioherbicidal propagules. Mycol. Res. **99:** 879-884.
43. Jackson, M.A., D.A. Schisler, P.J. Slininger, C.D. Boyette, R.W. Silman and R.J. Bothast. 1996. Fermentation strategies for improving the fitness of a bioherbicide. Weed Technol. **10:** 645-650.
44. Jin, X., G.E. Harman and A.G. Taylor. 1991. Conidial biomass and desiccation tolerance of *Trichoderma harzianum* produced at different medium water potentials. Biol. Control **1:** 237-243.
45. Johnson, B.N., A.C. Kennedy and A.G. Ogg, jr. 1993. Suppression of downy brome growth by a rhizobacterium in controlled environments. Soil Sci. Soc. Am. J. **57:** 73-77.
46. Johnson, D.R., D.L. Wyse and K.J. Jones. 1996. Controlling weeds with phytopathogenic bacteria. Weed Technol. **10:** 621-624.
47. Kennedy, A.C., T.L. Stubbs and F.L. Young. 1989. Rhizobacterial colonization of winter wheat and grass weeds. Agron. Abstr. **53:** 220.
48. Kennedy, A.C., A.G. Ogg, Jr. and F.L. Young. 1992. Biocontrol of jointed goatgrass. Patent No. 5,163,991.
49. Kennedy, A.C., L.F. Elliott, F.L. Young and C.L. Douglas. 1991. Rhizobacteria suppressive to the weed downy brome. Soil Sci. Soc. Am. J. **55:** 722-727.
50. Kenny, D.S. 1986. DeVine® -the way it was developed in an industrialist's review. Weed Sci. **34 (Suppl. 1):** 15-16.
51. Kremer, R.J., M.F.T Begonia, L. Stanley and E.T. Lanham. 1990. Characterization of rhizobacteria associated with weed seedlings. Appl. Environ. Microbiol. **56:** 1649-1655.
52. Kremer, R.J. and A.C. Kennedy. 1996. Rhizobacteria as biocontrol agents of weeds. Weed Technol. **10:** 601-609.
53. Mangoendihardjo, S. and M. Soerjani. 1977. Weed management through biological control in Indonesia. Proc. Plant Prot. Conf. Malaysia, pp. 323-337.
54. Mazau, D., D. Rumeau, M.T. Esquerre-Tugaye. 1987. Molecular approaches to understanding cell surface interactions between plants and fungal pathogens. Plant Physiol. Biochem. **25:** 337-343.
55. Mortensen, K. 1988. The potential of an endemic fungus, *Colletotrichum gloeosporioides* f.sp. *malvae*, as a bioherbicide for round-leaved mallow (*Malva pusilla*) and velvetleaf (*Abutilon theophrasti*). Weed Sci. **36:** 473-478.
56. Mortensen, K. and R.M.D. Makowski. 1989. Field efficacy at different levels of *Colletotrichum gloeosporioides* f.sp. *malvae* as a bioherbicide for round-leaved mallow (*Malva pusilla*). Proc. VII Intl. Symp. Biol. Contr. Weeds, 6-March 1988. Rome, Italy. Ist Sper. Pathol. Veg. (MAF), pp. 523-530.

346

57. Nagaoka, K. 1987. Cloning, expression and application of Streptomyces genes. *In* R.F. Cape, M.L. Goldberg, T. Hata and K. Maeda (eds.). Antibiotic Research and Biochemistry. Proc. US-Japan Colloq. on Antibiotic Res. and Biotechnol. J. Antibiot. Res. Assn., Tokyo, pp. 45-60.

58. Nicollier, G.F. and A.C. Thompson. 1982. Phytotoxic compounds from *Melilotus alba* (white sweet clover) and isolation and identification of two new flavonoids. J. Agric. Food Chem. **30**: 760-764.

59. Nishimura, H., T. Nakomura and J. Mizutani. 1984. Allelopathic effects of *p*-methane-3,8-diols in *Eucalyptus citriodora*. Phytochemistry **23**: 2777-2779.

60. Ogg, A.G., Jr., F.L. Young, H.D. Skipper and A.C. Kennedy. 1991. Integrating rhizobacteria into management strategies for downy brome control in wheat. *Rhizobacteria* Prospects for Weed Management, National Workshop sponsored by ARD and Washington State Univ., Pullman, WA, pp. 12-13.

61. Oku, H. T. Shiraishi, S. Ouchi and M. Ishiura. 1980. A new determinant of pathogenicity in fungal plant disease. Naturwissenschaften **67**: 829-835.

62. Omura, S., M. Murata, H. Hanaki, K. Hinotozawa, R. Oiwa and H. Tanaka. 1984. Phosalacine, new herbicidal antibiotic containing phosphinothricin. Fermentation, isolation, biological activity and mechanism of action. J. Antibiot. **37**: 829-835.

63. Parker, P.E. 1986. Nematode control of silverleaf nightshade (*Solanum elaeagnifolium*); a biological control pilot project. Weed Sci. **34 (Sppl. 1)**: 33-34.

64. Pfender, W.F., W. Zhang and A. Nus. 1993. Biological control to reduce inoculum of the tan spot pathogen *Pyrenophora tritici-repentis* in surface-borne residues of wheat fields. Phytopathology **83**: 371-375.

65. Pruthi, H.S. 1969. The textbook of agricultural entomology. Indian Council of Agri. Res. New Delhi. 977 pp.

66. Queirolo, C.B., C.S. Andreo, H.M. Niemeyer and L.J. Corcuera. 1983. Inhibition of ATPase from chloroplasts by a hydroxamic acid from Gramineae. Phytochemistry **22**: 2455-2458.

67. Quimby, P.C., Jr., J.L. Birdsall, A.J. Caesar, C. Hertoghe, W.J. Connick, Jr., C.D. Boyette, T.C. Caesar and D.C. Sands. Formulation process for stabilizing living microbial biocontrol agents for extended storage. Weed Sci. Sco. Amer. Abstr. **33**: 172.

68. Rebeiz, C.A., A. Montazer-Zouhoor, H.J. Hopen and S.M. Wu. 1984. Photodynamic herbicides: I. Concept and Phenomenology. Enzyme Microb. Technol. **5**: 390-401.

69. Rees, N.E. and N.R. Spencer. 1991. Biological control of leafy spurge. *In* L.F. James (ed.). Noxious Range Weeds. Westview Press, Boulder, CO, USA, pp. 182-192.

70. Ridings, W.H., D.J. Mitchell, C.L. Schoultics and N.E. El-gholl. 1976. Biological control of milkweed vine in Florida citrus groves with a pathotype of *Phytophthora citropthora*. *In* T.E. Freeman (ed.). Proc. 4th Int. Symp. Biol. Control Weeds. University of Florida, Gainesville, FL, USA, pp. 224-240.

71. Robinson, A.F., C.C. Orr and J.R. Abernathy. 1978. Distribution of *Nothanguina phyllobia* and its potential as a biological control agent for silver-leaf nightshade. J. Nematol. **10**: 362-366.

72. Schisler, D.A., K.M. Howard and R.J. Bothast. 1991. Enhancement of disease caused by *Colletotrichum truncatum* in *Sesbania exaltata* by co-inoculating with epiphytic bacteria. Biol. Control **1**: 261-268.

73. Shiraishi, T., K. Saitoh, H.M. Kim, T. Kato, M. Tahara, H. Oku, Y. Yamada and Y. Ichinose. 1992. Two suppressors, suppressins A and B, secreted by a pea pathogen *Mycosphoerella pinodes*. Plant Cell Physiol. **33**: 663-667.

74. Skipper, H.D., A.G. Ogg, Jr. and A.C. Kennedy. 1996. Root biology of grasses and ecology of *Rhizobacteria* for biological control. Weed Technol. **10**: 610-620.

75. Smith, R.J., Jr. 1982. Integration of microbial herbicides with existing pest management programs. *In* R. Charudattan and H.L. Walker (eds.). Biological Control of Weeds with plant pathogens. John Wiley & Sons, New York, pp. 189-203.

76. Souissi, T. 1994. *Rhizobacteria* in weed management: biological control of leafy spurge (*Euphorbia esula*). Ph.D. diss. University of Missouri, Columbia, MO, USA, 130 pp.

77. Stroo, H.F., L.F. Elliott and R.I. Papendick. 1988. Growth, survival and toxin production of root- inhibitory pseudomonads on crop residues. Soil Biol. Biochem. **20**: 201-207.

78. Stubbs, T.L. and A.C. Kennedy. 1993. Effect of bacterial and chemical stresses in biological weed control systems. Agron. Abstr. **57**: 261.

79. Templeton, G.E. and R.J. Smith, Jr. 1977. Managing weeds with pathogens. *In* Horsefall and E.B. Cowling (eds.). Plant Diseases: An Advanced Treatise, Vol. 1. Academic Press, New York, pp. 167-176.

80. Templeton, G.E., R.J. Smith, Jr. and D.O. Tebeest. 1986. Progress and potential of weed control with mycoherbicides. Rev. Weed Sci. 2:1-14.

81. Towers, G.H.N., J.T. Arnason, C.K. Wat and D.H. Lambert. 1984. Controlling weeds using a naturally occurring conjugated polyacetylene. Canadian patent, CA 1,172,460.

82. Tranel, P.J., D.R. Gealy and A.C. Kennedy. 1993. Inhibition of downy brome (*Bromus tectorum*) root growth by a phytotoxin from *Pseudomonas fluorescens* strain D7. Weed Technol. 7:134-139.

83. Turco, R.F., M. Bischoff, D.P. Breakwell and D.R. Griffith. 1990. Contribution of soil-borne bacteria to the rotation effect in corn. Plant Soil **122**: 115-120.

84. Vaidya, V.G. and V.D. Vartak. 1977. Larvae of *Diacrisia obliqua* Wlk. (Arctiidae: Lepidoptera) feeding on the weed *Parthenium hysterophorus* L. Science and Culture **43**: 394-395.

85. Walker, H.L. and W.J. Connick, Jr. 1983. Sodium alginate for production and formulation of mycoherbicides. Weed Sci. 33:333-338.

86. Weidemann, G.J. and G.E. Templeton. 1988. Efficacy and soil persistence of *Fusarium solani* f.sp. *cucurbitae* for control of Texas gourd (*Cucurbita texana*). Plant Dis. **72**: 36-38.

87. Winder, R.S. and C.G. Van Dyke. 1990. The pathogenicity, virulence and biocontrol potential of two *Bipolaris* species of Johnsongrass (*Sorghum halepense*). Weed Sci. **38**: 89-94.

88. Wymore, L.A. and A.K. Watson. 1986. An adjuvant increases survival and efficacy of *Colletotrichum coccodes*, a mycoherbicide for velvetleaf (*Abutilon theophrasti*). Phytopathology. **76**: 1115-1116.

89. Yang. W.W., D.R. Johnson, W.M. Dowler and W.J. Connick, Jr. 1993. Infection of leafy spurge by *Alternaria alternata* and *A. angustivoidea* in the absence of dew. Phytopathology **83**: 953-958.

90. Zidack, N.K., P.K. Backman and J.J. Shaw. 1992. Promotion of bacterial infection of leaves by an organosilicone surfactant: implications for biological weed control. Biol. Control **2**: 111-117.

12

Herbicide Interactions with Herbicides, Safeners and Other Agrochemicals

Under the current agricultural production practices, crops receive, in a single season, multiple seed treatments, preemergence and postemergence herbicides, insecticides, fungicides, nematicides, fertilizers, and sometimes plant growth regulators. Most of these chemicals, whether applied as mixtures, simultaneously or sequentially, may undergo a change in physical and chemical characters which could lead to enhancement or reduction in the effect of one or more compounds. A normally safe herbicide may become toxic to a crop or an effective product may show reduced activity on a weed species. These interactions between chemicals are of common occurrence in today's agriculture. Their significance is more evident when the interaction effects are seen much later in the growing season or the following season and year due to build up of persistent chemicals or their residues in the soil. Application of a preplant insecticide to soil may prove to be toxic when a postemergence herbicide is applied several weeks later. Knowledge of the interaction of herbicides with various chemicals will help in formulating and adopting a sound and effective weed management programme. It can also help to exploit the synergistic and antagonistic interactions between various herbicides to evolve an effective and economical eradication of weeds.

INTERACTION RESPONSES

Definition of Interaction Terms

Interaction is the term used to express the relationship of one agent against the other in a combination. When two or more chemicals accumulate in the plant, they may interact and bring out responses different from those obtained when they are used alone. These responses are generally described as **additive response, synergistic response (synergism)** and **antagonistic response (antagonism)**.

Additive Response

Additive response is the total effect of a combination, which is equal to the sum of the effects of the two components taken independently [113]. It is also known as the cooperative action of two agrochemicals such that the observed response of a test organism to their joint application is equal to the response predicted to occur by an appropriate reference model [54].

Synergism

In synergistic response, the total effect of a combination is greater or more pro-longed than the sum of the effects of the two taken independently [113]. It is also defined as the cooperative action of two agrochemicals such that the observed response of a test organism to their joint application appears to be greater than the response predicted to occur by an appropriate reference model [54].

Antagonism

In antagonistic response, the total effect of a combination is smaller than the most active component applied alone [113]. This definition, however, differs from the one suggested by Hatzios and Penner [54] who defined it as a type of joint action of two agrochemicals such that the observed response of a test organism to their combined application appears to be less than the response predicted to occur by an appropriate reference model.

Evaluation of Interaction Response

Proper evaluation of interaction responses of herbicide mixtures or mixtures of herbicides with other agrochemicals requires careful consideration of: a) proper experimental design, b) an appropriate parameter for measuring plant growth responses and c) a suitable statistical method for analyzing and interpreting the data.

Experimental Design

In designing and conducting an experiment, the researcher should clearly state the following: a) questions that the experiment will answer, b) the factors or variables to be controlled or kept constant during the experiment, c) the levels of factors to be used in the study and d) the number of observations to be taken [54].

The experimental design that is suitable is the factorial experiment in which treatments of all possible combinations of the levels of two or more factors are undertaken. The factorial experiment could be conducted in a completely randomized, randomized block or split-plot design. The minimum number of five levels of each factor may be included because the description of a curve with less than five points is not valid [84]. The levels of each factor must be in an appropriate range to test for potential interactions. In practice, it is desirable to select those levels of each factor that cause inhibition of plant growth in the range of 20% or 30% to 70% or 80% [54]. The interaction response may need to be evaluated only on one plant (crop or weed) species at a time. The mixtures found on one plant species may subsequently be tested on several other plant species of different families, as responses found synergistic may show only additive or antagonistic responses on other species treated with the same mixture.

Plant Responses

In field evaluation of interaction responses, the plant parameters that could be used include visual injury or growth rating, dry or fresh weight of root and/or shoot, stand reduction, shoot height and survival. Of these, visual injury and growth rating are subjective and open to bias and should be excluded. Quantitative measurements such as dry and fresh weights of root and shoot and qualitative measurements involving growth and development may be used. In laboratory studies, interaction may be evaluated by measuring the specific plant metabolic processes affected by herbicides. These processes include pigment biosynthesis (chlorophylls, carotenoids etc.), CO_2 uptake or evolution, oxygen consumption, synthesis of proteins, nucleic acids and lipids as well as specific enzyme systems [54].

Methods of Evaluation

The purpose of evaluation is to determine: a) whether the observed plant growth responses to treatment combinations of an agrochemical mixture differ from predicted or expected values calculated for additive responses and b) whether differences between observed and calculated responses are statistically significant [54]. Several statistical procedures are available for this purpose. These include the following:

1. Pair-wise multiple comparison tests (e.g., Duncan's Multiple Range Test etc.).
2. Orthogonal comparisons of planned sets of contents among means or groups of means.
3. Algebraic methods calculating an expected response (e.g., Colby's method, Analysis of Variance procedure, etc.).
4. Regression techniques for fitting surface response functions (e.g., isobole method, regression estimation analysis, logit analysis, calculus method, etc.).

Each of these procedures is briefly described below. For more detailed information on these methods, the reader should see standard textbooks on statistics and experimental methods.

Multiple Comparison Tests: The multiple comparison tests (**MRCs**), including Duncan's Multiple Range Test, are used to make comparisons of each mean with every other mean, and their purpose is to detect possible groups among a set of unstructured treatments. In spite of their popularity among weed scientists, MRCs are very often used inappropriately for the detection of significant differences among treatment combinations of two or more agrochemicals. The use of MRCs is considered statistically incorrect in the following situations: a) experiments in which treatments are factorial in nature, b) experiments in which treatments are graded levels of a quantitative factor (rates of agrochemicals) and c) experiments in which meaningful orthogonal comparisons among certain treatments can be formulated in advance (before the experiment is conducted) [54]. Varietal trials or agrochemical screening trials represent situations wherein the use of MRCs is statistically proper. MRCs are, however, inappropriate in evaluating agrochemical interactions, ant their use should be avoided in such experiments.

Orthogonal Comparison: In orthogonal comparison, the degrees of freedom (d.f.) for treatments are partitioned either completely into single groups of degrees of freedom or partially into groups of degrees of freedom, before performing the experiment. This partitioning of treatment degrees of freedom is also known as 'functional analysis of variance' [76]. Although orthogonal comparisons correctly identify significant interactions, they do not characterize the type of interaction as synergistic or antagonistic. This disadvantage is, however, minimized when the use of orthogonal comparisons is combined with the use of another procedure (e.g., Colby's method or Analysis of Variance procedure) which can correctly identify the character of an interaction as synergism or antagonism [54].

Algebraic methods: The Algebraic methods are based on mathematical principles. These include Analysis of Variance (ANOVA) procedure and Colby's method. In the ANOVA procedure, one can calculate a theoretical expected response for each treatment combination of two separate levels of factors A and B. The calculated expected responses are then compared to those actually obtained (observed responses) in the experiment [54]. The types of interactions that can occur between factors A and B are characterized as synergistic when the observed responses are greater than expected or as antagonistic when the observed and expected responses are less than expected. When observed and expected responses are equal, the two factors act independently and the effects are described as additive. The statistical significance of the differences between observed and expected responses can be determined by an analysis of variance of the experimentally obtained set of data. If the interaction is statistically significant, then the departures of observed responses from the model (additive expected responses) are synergistic or antagonistic. If the interaction is not significant, the treatment combinations may be indicated as additive.

Colby's Method: Gowing [45, 46] was the first scientist to question direct arithmetic addition for analysis of herbicide action. He suggested that different modes of action would result in different types of response curves and make it difficult to predict herbicide activity. Gowing advocated probit analysis for herbicide mixture work and stressed comparison of activity at 50% inhibition, where the response curve is steepest. His formula is,

$$E = X + Y \frac{(100 - X)}{100}$$

where, E is the expected per cent inhibition of growth by the mixture of herbicides A and B, X is percent inhibition of growth induced by herbicide A, and Y is the per cent inhibition of growth induced by herbicide B.

Gowing's formula was subjected to algebraic manipulation first by Limpel et al. [77] and later by Colby [22]. The modified formula suggested by Limpel et al. [77] is as under:

$$E = X + Y - \frac{(XY)}{100}$$

Colby [22] simplified the number of arithmetic operations required by initially converting the growth inhibition to percent of control values and the expected response is calculated as shown below:

$$E_1 = \frac{X_1 Y_1}{100}$$

where, E_1 represents the expected growth as per cent of control, and X_1 and Y_1 are growth inhibitions, expressed also as per cent control, induced by herbicides A and B respectively.

Colby's method can also be used for calculation of expected responses based on the original experimental data and not necessarily only from data that have been calculated to per cent of control values.

The methods suggested by Gowing [46], Limpel et al. [77] and Colby [22] are compared below, based on using the date presented in Table 12.1.

Table 12.1. Effect of herbicides A and B applied alone and in combination on weed control

Herbicide	Rate (kg ha^{-1})	Dry Wt (g plant^{-1})	Growth Inhibition (%)	Per cent of control
None (control)	—	50	0	100
A	2.0	35	30 (X)	70 (X1)
B	2.0	25	50 (Y)	50 (Y1)
A + B	1.0 + 1.0	10	80	20

Gowing (1960): $E = X + Y \dfrac{(100-X)}{100}$; $E = 30 + 50\dfrac{(100-30)}{100} = 30 + 35 = 65\%$ inhibition

Limpel et al. (1962): $E = X + Y - \dfrac{(XY)}{100}$; $E = 30 + 50 - \dfrac{(30 \times 50)}{100} = 30 + 35$

$$= 65\% \text{ inhibition}$$

Colby (1967): $E_1 = \dfrac{X_1 Y_1}{100}$; $E_1 = \dfrac{70 \times 50}{100} = 35\%$ of control $= 65\%$ growth inhibition

As the observed growth inhibition (80%) by the herbicide mixture (A+B) is more than the expected value (65%) obtained by the three methods, the interaction response is considered synergistic.

Colby [23] extended the formula $E_1 = \dfrac{X_1 Y_1}{100}$ for two-way combinations to three-way combinations, in which case, $E_1 = \dfrac{X_1 Y_1 Z_1}{10,000}$.

He suggested that these formulae are most accurate when values of X, Y and Z (of herbicides A, B and C) are near the 50% level since the dose-response curves deviate least from linearity at the 50% level.

Although Colby's method has become very popular among weed scientists because of its simplicity, it main disadvantage has been the difficulty of statistically testing individual departures of observed responses from those predicted by Colby's formula. To overcome this limitation, Colby [23] suggested transformation of the data to logarithms and analysis of the new set of data for variance. In most analysis of cases, on a logarithmic scale, Colby's multiplicative model becomes additive. However, analysis of variance of logarithmically transformed data can be used for characterization of the differences between observed and predicted responses for each treatment combination of two or more factors as synergistic, antagonistic or additive.

Regression Techniques: Regression techniques include the calculus method, regression estimation analysis, logit analysis and isobole method.

The **Calculus method** suggested by Drury [31] uses multiple regression polynomial models for fitting data from a biological function. It determines the first partial derivative of the multiple regression equation with respect to each factor and the second partial derivative of the multiple regression equation with respect to both factors as their interaction. Then, the numerical values of the actions and interactions at points of interest are calculated and graphed. Finally, comparisons of the sign of the interaction with the signs of the action of each factor are used to determine the promotion (enhancement or synergism) or the reversal (antagonism) of its action by the other factor at a specific point. Drury suggested that comparison of the interaction with a reference model be considered a descriptive device rather than a quantitative measurement of synergism or antagonism.

The **Regression Estimate Analysis** suggested by Nash and Jensen [85] and Nash [84] represents a modification of Colby's method in an effort to overcome the difficulty of statistically testing individual departures of observed responses from those predicted by Colby's formula. It involves the first, second or higher order polynomial equations to the data for various levels of each herbicide (or agrochemical) in the absence of the other. This approach gives adequate fit to the data, but often produces a cumbersome array of model terms and a contour plot that is often very sensitive and difficult without the use of computers and statistical analysis systems.

The **Logarithmic or Logit Analysis** involves [9, 38] plotting the logarithmic concentration of a chemical against the percentage response on a probability scale, and fitting a weighted regression line to the data according to the equation: $E(y) = \beta_0 + \beta_1 \log_x$. The logit analysis converts a non-linear regression model into a linear one. This method, also called **probit analysis**, is widely used in pharmacology, entomology and plant pathology almost exclusively with quantal (all-or-nothing) responses. Gowing [46] proposed that probit analysis was applicable for evaluation of the joint action of herbicide mixtures.

The **Isobole Method** developed by Tammes [113] uses logarithmically transformed data to form curves of isoboles, the lines representing various herbicide mixture compositions giving the same effect. He presented two types of isoboles; one with one-sided effects, obtained when a non-toxic component is added to a toxic one, and the other with two-sided effects when two toxic chemicals are mixed. The isobole method has not gained much acceptance, however because it is time consuming and often not very revealing for phytotoxic interaction studies [85]. It also requires a large number of test plants to obtain accurate probit graphs for the necessary interpolations [95].

In an effort to eliminate this drawback, Akobundu et al. [2] proposed a **modified isobole method** in which they avoided the probit transformation of the percentages by using the dosage-response curve at the 50% level (ID_{50}) and the reproducibility of measurements found around this region. From each dosage-response curve, the level of the herbicide at which a 50% decrease in plant growth occurs is interpolated. This modified isobole method, in effect, is the study of plant responses at the ID_{50} of each herbicide and their various mixtures. It eliminates the tedious probit analysis and substitutes the dosage-response curve at the 50% level as the criterion for measuring plant response to herbicide mixtures as additive, synergistic and antagonistic, and enhancement effects (Fig. 12.1).

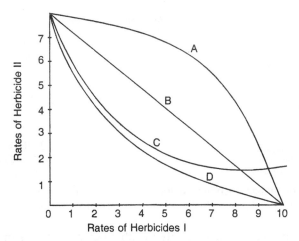

Fig. 12.1. The ID50 isoboles showing antagonistic (A), additive (B), enhancement (C), and synergistic (D) effects involving two herbicides [2].

In order to obtain data suitable for evaluating and classifying plant responses to herbicide mixtures, Akobundu et al. [2] suggested that in addition to the usual good techniques of herbicide research, the following points need to be observed: a) choosing a non-infinite criterion for plant response, e.g., fresh or dry weight, height, leaf area,

CO_2 evolution, etc. in preference to dead plants, stand counts etc.; b) selecting data primarily from ID_{50} range rather than from the near zero or 100% ranges; c) interpreting data primarily on the basis of trends from many single and combination dosages rather than from a few dosages or combinations; and d) restricting conclusions as to classification of plant responses to those individual plant species for which data are available, rather than for weeds or crops in general.

PHYSIOLOGICAL BASIS OF HERBICIDE INTERACTION

Enhancement or reduction of the activity of a given herbicide on a target plant brought about by the prior or simultaneous application of another chemical is the result of changes in the amount of herbicide reaching, or after reaching, the site of action in the active form. In a mixture, one component influences the toxicity of another component by interfering with the patterns of its penetration, translocation and biotransformation in plants. The physiological mechanisms explaining the synergistic, antagonistic and other interactions of herbicides with other agrochemicals in higher plants are briefly discussed below.

Mechanisms of Synergistic Interaction

Several classes of agrochemicals including adjuvants, fertilizers, herbicides, insecticides, growth regulators and other additives are known to synergize the activity of several herbicides on certain plant species. The mechanism by which these additives synergize herbicide activity is shown in Fig. 12.2.

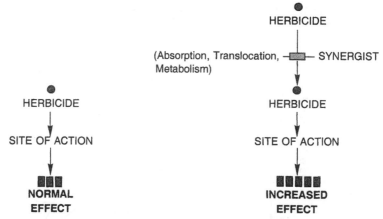

Fig. 12.2. Mechanism responsible for synergism of herbicides by synergists (other agrochemicals) in higher plants [54]. (Reproduced with permission of WSSA).

The synergism of herbicide activity by other agrochemicals is the result of an enhanced penetration and/or translocation or an altered metabolism (biotransformation) of a given herbicide consequent to the prior or simultaneous application of another chemical.

Mechanisms of Antagonistic Interaction

The mechanisms by which chemicals would reduce the activity of a given herbicide on a target plant species can be generally classified into four groups. They are **biochemical** antagonism, competitive antagonism, physiological antagonism and chemical antagonism [54]. These mechanisms are analogous to those used to describe pharmacological drug interactions.

Biochemical Antagonism

This is the opposite of the synergism presented in Fig. 12.2. Biochemical antagonism occurs when one agrochemical (the antagonist) decreases the amount of a given herbicide that would otherwise be available to its site of action in the absence of the antagonist (Fig. 12.3A). The antagonist reduces the rate of herbicide penetration into the plant or its transport to the site of its action within the plant. It could also increase the rate of biotransformation in certain plants.

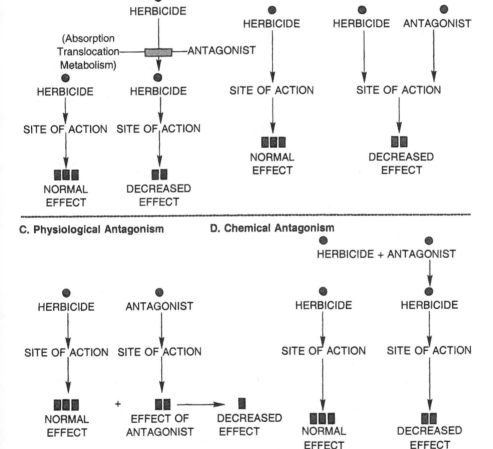

Fig. 12.3. Mechanisms Responsible for the antagonism of herbicides by other agrochemicals in higher plants. A. Biochemical Antagonism, B. Competitive Antagonism, C. Physiological Antagonism and D. Chemical Antagonism [54]. (Reproduced with permission of WSSA).

Competitive Antagonism

In competitive antagonism, the antagonist acts reversibly at the same site as the herbicide (Fig. 12.3B). In some cases, the antagonist, although capable of reacting with a receptor site within the cell, may lack intrinsic activity. However, by reacting with the receptor site the antagonist prevents the binding of a given herbicide to its receptor site and the formation of a herbicide-receptor complex required for the exhibition of herbicidal activity.

Physiological Antagonism

In this type of antagonism, two components of a mixture, acting at different sites, counteract each other by producing opposite effects on the same physiological process (Fig. 12.3C). In this situation, each component has the ability to react with its receptor site and bring out a characteristic response, but when combined one opposes the effect of the other on the same physiological process. Thus, physiological antagonism is quite different from competitive antagonism where the components act at the same site and one of them actually prevents the other from exerting its effect.

Chemical Antagonism

Chemical antagonism occurs when an antagonist reacts chemically with a herbicide to form an inactive complex (Fig. 12.3D). The inactivation of the herbicide by a given antagonist is directly proportional to the extent of the components in a mixture.

HERBICIDE-HERBICIDE INTERACTIONS

Use of two- or three-way mixtures is quite common in modern weed management programmes. The synergistic herbicide mixtures offer the following immediate and long-term benefits.

1. Provide more effective weed control.
2. Increase the spectrum of weeds controlled. The most successful combinations are those based on the pairing of a grass-effective herbicide with a broadleaf weed-effective herbicide in order to complement the phytotoxicity of each other at the physiological level.
3. Reduce the environmental impact of each herbicide in a mixture. The soil receives lower amounts of herbicides, resulting in far less amounts of herbicides reaching groundwater.
4. Improve crop safety (selectivity) as a result of application of reduced rates. Allow crops to be planted sooner after herbicide application; because of lower initial rates, herbicides will be more rapidly dissipated to non-toxic levels.
5. Cause delay in the evolution of weed resistance to herbicides.
6. Allow zero- or minimum-tillage and chemical fallow practices achieve effective and economical weed control.
7. Reduce danger of human toxicity directly and through crop residues because of reduced rates of application.

There are certain limitations in the use of synergistic herbicide mixtures. These include: a) additional costs of having them registered (premixes) and b) loss of selectivity of a given herbicide, as will be described later in this section. These limitations are not insurmountable, however. These problems need to be studied in greater detail to evolve and recommend only those herbicide mixtures which are more effective, useful and economical. Certain physical, chemical and physiological changes may cause herbicides applied in a mixture to interact adversely, resulting in antagonistic interaction.

Although synergism is a more desirable response, herbicide mixtures with antagonistic interaction also offer some benefits in certain situations.

There are two types of herbicide mixtures: a) premixed formulations, made by manufacturers, and b) tank mixes made by farmers before application with desired and recommended herbicides and rates.

The success of a herbicide mixture can be undermined by physical, chemical and biological incompatibility. This incompatibility leads to reduced performance of the mixture. Physical incompatibility results in the formation of agglomerates, crystals, phase separation, or thickening in the herbicide mixture. Thus, incompatibility is a frequent contributor to the variability of herbicide performance, particularly in the case of tank mixtures. Compatibility charts serve as an excellent source of information to help avoid these mixing problems. When making tank mixtures of herbicides, their compatibility needs to be tested. This may be done by premixing the materials by making a slurry of small quantities (equal proportion) before adding to the tank .If required, a compatibility agent, which is non-ionic in nature, may be used.

Herbicide-herbicide mixtures have been widely used for over 30 years. Scores of factory-made premixes have been developed and farm-level tank mixtures recommended during this period to control some of the most problematical weeds in agricultural and non-agricultural systems. Some of the premixes formulated and marketed in the USA are listed in Table 12.2.

Table 12.2. List of premix combinations available in the USA and their trade names

Herbicide Combination	Product Name
Acetochlor + Dichlormid (S)	SURPASS
Acetochlor + MON 4660	HARNESS PLUS
Alachlor + Glyphosate	BRONCO
Alachlor + Trifluralin	FREDDOM
Amitrole + Simazine	LIQUID AMIZINE
Atrazine + Acetochlor	SURPASS
Atrazine + Alachlor	LARIAT, BULLET, BOXER
Atrazine + Bentazon Na salt	LADDOCK, PROMPT
Atrazine + Butylate + Dichlormid (S)	SUTAZINE
Atrazine + Cyanazine	EXTRAZINE II4L, EXTRAZINE IIDF
Atrazine + Dicamba	MARKSMAN
Atrazine + Dimethanamid	GUARDSMAN
Atrazine + Metolachlor + Benoxacor (S)	BICEP
Atrazine + Simazine	SIMAZAT
Benefin + Oryalin	XL2G
Bentazon + Acifluorfen	GALAXY, STORM, TACKLE, CONCLUDE B
Bentazon + Sethoxydim	REZULT
Bifenox + Mecoprop + Ioxynil	FOXPRO D+
Bromacil + Diuron	KROVAR, WEED BLAST
Bromacil + Diuron + Sodium metaborate + Sodium chlorate	BAREGROUND BD, TOTAL
Bromacil + Sodium metaborate + Sodium chlorate	BARESPOT, UREABOR
Bromoxynil + Atrazine	BUCTRIL+ATRAZINE
Bromoxynil + Ioxynil	OXYTRIL
Bromoxynil (ester) + MACPA (ester)	BRONATE
Butylate + R-29148 (S)	SUTAN

(S) = Safener   (Contd.)

Table 12.2. *(Contd.)*

Herbicide Combination	Product Name
Chlorimuron + Linuron	GEMINI 60DF, NEW LOROX PLUS, LOROX PLUS 60DF
Chlorimuron + Metribuzin	CLASSIC, CANOPY, PREVIEW
Chlorimuron + Thifensulfuron	CONCERT, SYNCHRONY
Chlorsulfuron + Metsulfuron	FINESSE, GLEAN, TELAR
Clomazone + Trifluralin	CURTAIL
Clopyralid + Flumetsulam	BROADSTRIKE PLUS CORN
Clopyralid + Triclopyr	CONFRONT
2,4-D + Dicamba	FOUR POWER PLUS, VETERAN, WEED MASTER
2,4-D + Dichlorprop	BRUSH OUT, ENVERT 171, WEEDONE
2,4-D + Dichlorprop + Mecoprop	DISSOLVE, TRIAMINO, TRI-ESTER
2,4-D + DSMA	CRABICIDE, SUMMER
2,4-D + Glyphosate	BRONCO, CAMPAIGN, LANDMASTER BW
2,4-D + Ioxynil	ACTRIL DS
2,4-D + Mecoprop	2 PLUS 2
2,4-D + Mecoprop + Dicamba	ACME, HIDEP, TRIPLET, TRIMEC SOUTHERN
2,4-D + Picloram	PATHWAY, TORDON RTU
2,4-D + Prometon	VEGEME
2,4-D + Triclopyr	CHASER, CROSSBOW, TURFLON
2,4-D isooctyl ester + Mecoprop isooctyl ester + Dicamba	ACME SUPER BRUSH KILLER, SUPER TRIMEC
Dichlobenil + Diuron	CORSAGE
Diuron + MSMA	DIUMATE
Diuron + Paraquat dichloride salt	SUREFIRE
Diuron + Tebuthiuron	SPRAKIL, SPRAKIL SK-26
Diuron + Thidiazuron	GINSTAR
EPTC + Dichlormid (S)	SURPASS
EPTC + Dietholate (S)	ERADICANE EXTRA
Fenoxaprop-P ethyl ester + 2,4-D + MCPA	TILLER
Fenoxaprop-P ethyl ester + Fluazifop-P butyl ester	HORIZON 2000, FUSION
Fenoxaprop-P ethyl ester + MCPA	CHEYENNE, DAKOTA
Flumetsulam + Clopyralid	BROADSTRIKE PLUS CORN PRE/PPI
Flumetsulam + Metolachlor	BROADSTRIKE + DUAL
Flumetsulam + Trifluralin	BROADSTRIKE + TREFLAN
Flumeturon + MSMA	CROAK, COTORAN + MSMA
Fomesafen + Fluazifop-P butyl ester	TORNADO
Glyphosate + Dicamba	FALLOW MASTER
Halosulfuron methyl + MON 13900 (S)	BATALLION
Imazaquin + Acifluorfen	SCEPTOR O.T.
Imazapyr + Diuron	TOPSITE
Imazaquin + Pendemethalin	OFFICE, SQUADRON
Imazaquin + Trfluralin	TRI-SCEPT
Imazethapyr + Atrazine	CONTOUR
Imazethapyr + Pendemethalin	PURSUIT PLUS
Imazethapyr + Trifluralin	PASSPORT
Ioxynil + Bromoxynil + Mecoprop	OXYTRIL M

(S) = Safener *(Contd.)*

Table 12.2. *(Contd.)*

Herbicide Combination	Product Name
Isoxaben + Oryzalin	SNAPSHOT
Metolachlor + Atrazine	BICEP DF
Metolachlor + Cyanazine	CYCLE
Metribuzin + Metolachlor	TURBO 8EC
Metolachlor + Simazine	DERBY
Metribuzin + Trifluralin	SALUTE
Molinate + Propanil	ARROSOLO
MCPA + Mecoprop + Dicamba	TRI-POWER
Oxyfluorfen + Oryzalin	ROUT
Oxyfluorfen + Pendimethalin	ORNAMENTAL HERBICIDE II
Phenmedipham + Desmedipham	BETAMIX, BETAMIX PROGRESS
Phenmedipham + Desmedipham + Ethofumesate	BETAMIX PROGRESS
Picloram + Triclopyr butoxyethyl ester	ACCESS
Prometryn + MSMA	CAPAROL+MSMA
Propachlor + Atrazine	RAMROD/ATRAZINE
Prosulfuron + Primisulfuron	RING, EXCEED
Prosulfuron + Terbuthylazine	BRONS
Quinclorac + Propanil	BAS 52701H
Simazine + Prometon + Sodium metaborate + Sodium chlorate	PROMETON 5PS
Thifensulfuron + Tribenuron	HARMONY EXTRA, CHEYENNE X-TRA
Trifluralin + Benefin	TEAM 2G
Trifluralin + Tebuthiuron	SPIKE TREFLAN 6G
Trifluralin + Triallate	BUCKLE

(S) = Safener

Synergistic and Additive Responses

Many reports of synergistic responses in herbicide combinations are related to 2,4-D and other phenoxyacid derivatives. The mixture of 2,4-D and chlorpropham was found synergistic on monocot species generally resistant to 2,4-D [82], while it showed antagonism on *Amaranthus retroflexus*, a broadleaf weed [67]. Pretreatment of chlorpropham followed by 2,4-D spray 9 to 17 d later resulted in complete kill of shoot growth of *Cyperus esculentus* in the greenhouse and good control in the field for at least 2 mon [82]. Agbakoba and Goodin [1] found synergistic response on *Convolvulus arvensis* at low rates of 2,4-D and picloram. They observed that picloram uptake was antagonized by 2,4-D while 2,4-D uptake was unaffected by picloram. However, more 2,4-D was translocated from the site of application in the presence of picloram. Synergistic reduction in the weight of bean plants was observed when 2,4-D or mecoprop was applied with picloram [52]. In contrast, picloram and 2,4-DB were antagonistic. In this combination, picloram prevented the movement of 2,4-DB, while picloram transport and distribution were enhanced by phenoxy compounds. Bovey et al. [10] found synergism with 2,4,5-T and picloram. The three-way combination of 2,4-D, mecoprop and dicamba also showed synergism on several broadleaf weed species [109].

The combinations of 2,4-D and triazine herbicides also show synergism. Dickerson and Sweet [28] observed synergism with the addition of three ounces (85 g) of 2,4-D to one pound (452 g) of atrazine and one gallon (3.785 L) of crop oil per acre (0.4 ha) on

Cyperus esculentus. This combination was much more effective than four times the rate of atrazine applied with oil. Lynch et al. [78] found synergism with low rates of 2,4-D and atrazine on bean leaves. Diem and Davis [29] reported that non-toxic concentrations of 2,4-D increased the toxicity of atrazine when applied to the roots of water hyacinth (*Eichhornia crassipes*) and to a lesser extent when applied to the roots of soybean. Non-toxic concentrations of 2,4-D frequently also increased the absorption of ametryn, water and calcium by corn, soybean and water hyacinth. The mixtures of toxic concentrations of 2,4-D and ametryn were not synergistic and they were less toxic than ametryn alone.

One of the better known examples of a synergistic combination is amitrole and ammonium thiocyanate. In this mixture, ammonium thiocyanate increases the translocation of amitrole and consequently reduces the contact action of amitrole, resulting in a better kill of the entire plant. This occurs only when ammonium thiocyanate is applied along with amitrole or one day after amitrole application. This mixture is marketed as 'Amitrole-T'. Amitrole and amitrole-T produce additive and synergistic effects when applied in combination with other herbicides. Sheets and Leonard [108] found that mixtures of amitrole and dalapon were additive while mixtures of amitrole and monuron were more than additive on kidney bean and barley. The characteristic chlorotic symptoms of amitrole were not observed when monuron or dalapon was present. Similar increased activity was obtained when amitrole-diuron mixture was applied on several weed species [93]. Putnam and Ries [94] observed better control of quackgrass (*Agropyron repens*, currently known as *Elytrigia repens*) when low rates of amitrole were added to simazine or diuron.

Paraquat, a contact herbicide, was found to act synergistically when applied with certain triazines and urea against *Elytrigia repens* [93, 94]. Horowitz [62] reported that when paraquat was combined with triazines and phenylureas, the control of *Portulaca oleracea* (common purslane) was much better. Headford [56] found application of amitrole-T followed 2 wk later by paraquat gave more effective control of *Paspalum conjugatum* (sour paspalum), while the mixture showed antagonistic response. In a sequential application, paraquat destroyed the foliage by contact action causing depletion of carbohydrate reserves, and amitrole-T, already accumulated in the shoot meristems, exerted its toxic action on the new growth. When applied together, paraquat destroyed the foliage rapidly and impaired the action of amitrole-T. Rao et al. [103] reported better control of *Imperata cylindrica* (thatchgrass, cogongrass or alang-alang) when paraquat was applied one week after the application of dalapon.

Atrazine and alachlor combination, which shows synergism, is widely used for an effective weed control in maize. Akobundu et al. [3] investigated the physiological basis for the synergistic response of this mixture. They reported that alachlor which showed no effect of the Hill reaction activity of isolated chloroplasts of *Echinochloa crus-galli* (baenyardgrass) had no influence on the inhibitory effect of atrazine on the Hill reaction. However, atrazine and alachlor combinations reduced chloroplast protein and severely inhibited chloroplast protein synthesis relative to protein synthesis by other particulate fractions. This inhibition appeared to be the basis for the synergistic effect of this herbicide mixture on *Echinochloa.*

In certain combinations, one herbicide makes plants more sensitive to another herbicide, resulting in the synergistic response. This **predisposition** phenomenon may occur when herbicides are applied in mixtures or in a sequence. Gentner [44] reported that EPTC and other thiocarbamates inhibited the deposition of foliar waxes on cabbage, and application of dinoseb on cabbage previously treated with EPTC increased toxicity. Reduction in the deposition of waxes enhances spray retention and absorption. Davis and Dusabek [24] reported that exposure of pea plants to diallate vapours resulted in enhancement of subsequent foliar uptake of 2,4-D, atrazine, TCA and diquat; the increase was due to a reduction of epicuticular wax on the leaves of treated plants. Johnson [68] observed increased toxicity for soybean when chloroxuron was applied as a semi-directed postemergence treatment following preplant treatment with nitralin. Similarly, Johnson [69] also found more toxicity for soybean when postemergence applications of prometryn or chloroxuron followed preplant application of vernolate. Rao et al. [104] reported that in a combination, 2,4-D predisposed *Setaria palmifolia* (palmgrass) for increased activity by glyphosate. This predisposition phenomenon is also observed when in the presence of one herbicide, the sublethal residue of another herbicide applied (at normal rates) in the previous season (or year) becomes toxic for plants normally tolerant. Duke et al. [33] and Rao and Duke [98] investigated this and found that when EPTC was applied to alfalfa (lucerne) grown on soil with small amounts of atrazine residue, toxicity increased. The injury symptoms resembled those caused by higher concentrations of atrazine. The enhanced toxicity was attributed to the prevention by EPTC of the formation of an enzyme responsible for the detoxification of atrazine in alfalfa.

The activity of glutathione-s-transferases (GSTs), enzymes that catalyze the conjugation of atrazine with glutathione, could be inhibited by propachlor, propachlor, and barban [39]. Similar inhibition of plant GSTs that catalyze fluorodifen with glutathione by diuron, propachlor, trifluralin and nitralin has been demonstrated [40]. Tridiphane (DOWCO 356), a grass herbicide, synergizes atrazine toxicity on certain normally tolerant grass weed species (*Setaria faberii, S. italica, Panicum miliaceum* and *Digitaria sanguinalis*) which catalyze the conjugation of atrazine with glutathione [37, 75, 80]. Tridiphane has no effect on the uptake of atrazine but inhibits the activity of GST enzymes of these grass weeds. Tridiphane does not, however, affect GST enzyme activity in maize. Thus, tridiphane synergizes atrazine, killing grass weeds in maize. Tridiphane also synergizes EPTC and alachlor by a similar mechanism [37].

Antagonistic Responses

With respect to **biochemical antagonism**, picloram has been reported to antagonize the toxicity of bromacil on oats [111] and glyphosate on *Cirsium arvense* (Canada thistle) [89]. Picloram also prevents translocation of 2,4-DB or its metabolites from the treated leaf of *Phaseolus vulgaris* [52]. On the other hand, the uptake of picloram in *Convolvulus arvensis* (field bindweed) and its transport in *Prosopis juliflora* (honey mesquite), *Acacia farnesiana* (huisache) and bean were antagonized significantly by 2,4-D [1] and paraquat [25] respectively. Similarly, MCPA reduced the uptake and translocation of asulam in wild oats, while the translocation of MCPA was significantly inhibited by the herbicides ioxynil and bromoxynil [42]. The tank mixture application of paraquat and glyphosate or dalapon resulted in poor control of *Imperata cylindrica* [4]. In this, the contact herbicide rapidly destroyed the leaf tissue, thus preventing the uptake and movement of systemic herbicides.

The uptake and translocation of diclofop-methyl was reduced in wild oats in the presence of 2,4-D and MCPA [87, 96, 114]. Although the rate of de-esterification of diclofop-methyl to diclofop, the free acid responsible for the inhibition of meristematic activity, was found to be much reduced by MCA, de-esterification appeared to be a secondary effect [59]. The primary effect of the antagonistic activity of diclofop-methyl was due to the reduction of diclofop acid that moves basipetally to meristematic areas of wild oats. Marked reduction in the uptake of difenzoquat by propanil was also the basis of the observed antagonism of these two herbicides on wild oats [106].

Dinitroaniline or chloroacetanilide herbicides inhibit uptake of photosynthesis-inhibiting herbicides such as atrazine, cyanazine, metribuzin, diuron and methazole by oat, soybean and tomato, and induced antagonistic responses when the two classes of herbicides were applied together in the soil.

The mixture of diclofop-methyl ester and 2,4-D exhibits **competitive antagonism**. Hall et al. [49] reported that the effects of diclofop-methyl and 2,4-D combinations on oat coleoptiles involved auxin vs. anti-auxin competition for a common receptor site. ·

Physiological antagonism is the likely mechanism involved in the observed reduction of glyphosate activity on certain plants by 2,4-D and growth regulator compounds like auxins. EPTC inhibited growth of soybean and corn tissue, and 2,4-D in combination with EPTC caused an increase in growth by enhancing RNA synthesis compared to EPTC alone [8]. Similarly, inhibition of soybean tissue growth by butylate was also antagonized by 2,4-D. The antagonism of the phytotoxic effects of benzoylprop-ethyl and flamprop-methyl on wild oats by 2,4-D results also from opposing actions of these herbicides on the same physiological function. Benzoylprop-ethyl and flamprop-methyl retarded growth and development of wild oats by interfering with both cell division and elongation of stem internodes, while 2,4-D enhanced these functions [88]. The contact action of paraquat is antagonized by PS II inhibitors such as monuron and linuron. Antagonism of the paraquat effect by monuron or linuron results from the effect of the PS II inhibitors on electron transport that prevents reduction of paraquat.

The majority of herbicides are organic molecules which have functional groups such as aliphatic amino (R-NH$_2$), aromatic amino (N:), carboxylate (-COO$^-$), phosphonate (CH$_2$ –P—O$^-$) in their chemical structures [54]. These functional groups serve as electron donors in the binding of metal ions, leading to metal chelation of organic herbicides. Picloram is capable of forming complexes with iron (Fe^{+++}) or copper (Cu^{++}), leading to reduction of herbicidal activity. This kind of chemical antagonism has also been demonstrated for glyphosate. Metal ions such as Cu^{++}, Fe^{+++}, Zn^{++}, and Al^{+++} form complexes with glyphosate and reduce the latter's activity. The formation of glyphosate-metal complexes has been proposed as the likely mechanism for the observed antagonism of glyphosate by metal cations [32, 53, 107, 115]. The antagonism of paraquat phytotoxicity to barley by MCPA is the result of paraquat and the dimethylamine salt of MCPA leading to the production of two compounds with reduced biological activity [86].

HERBICIDE-SAFENER INTERACTIONS

Safeners

Herbicide safeners (also referred to as antidotes, protectants, antagonists, or modifiers), as briefly mentioned in Chapter 5, increase the tolerance level of crop plants to herbicides. As a result, the herbicide–safener mixture allows for improved control of weeds. Certain herbicides, when applied alone, would either not be sufficiently active on the weeds at lower rates, or would cause injury at the higher rates required for satisfactory weed control. Safeners help in mitigating the problem by allowing herbicide application at higher rates.

One of the earliest reports on the effect of safeners came from Hoffman [60] when he found 75% of the inhibitory effect of a foliar application of barban on wheat was eliminated by dusting the seed with 4-chloro-hydroxyminoacetanilide. Other highly effective antidotes found in this study were 2,4-dichloro-9-xanthenone and N-methyl-3,4-dichlorobenzene sulphonamide. In later studies, 1,8-napthalic anhydride (NA) applied as seed treatment was found to protect corn seedlings from injury to EPTC [14, 19, 58, 61]. NA was found effective in preventing alachlor injury to sorghum [58, 97]. Chang et al. [20] reported that coating oat seeds with NA at rates of 0.5% to 1.0% by seed weight significantly reduced the phytotoxicity of barban. NA also slightly reduced oat injury from diallate and triallate. The protective nature of NA on alachlor was physiological in nature and not one of physical deactivation of the herbicide [57].

NA protected sorghum from injury by EPTC, alachlor and diallate [70]. Chang et al. [19] established that the antidotal action of NA was not very specific in that it reduced EPTC injury to other crop species as well as corn. They suggested that that the non-selective nature of NA could make it useful as a seed-applied herbicide antidote for a wide range of crop species. NA also protects corn against injury caused by sulfonylurea herbicides such as chlorsufuron, metsulfuron, primisulfuron and thifensulfuron. Another safener found to exhibit a safening effect in the 1970s was R-25788, later renamed as dichlormid. Dichlormid, when tank-mixed or sprayed with EPTC, protected corn from herbicide injury without affecting weed control [17, 97]. Dichlormid was an effective antidote when applied to seed, soil, or nutrient culture [19]. Rao and Khan [99, 101] reported that dichlormid reduced EPTC and alachlor injury in a moderately sensitive variety of barley better than in a highly sensitive variety. Dichlormid is used to protect corn against EPTC, alachlor and butylate injury.

Furilazole (MON 13900), a dichloroacetamide safener, has a very good safening effect on sulfonylureas, particularly halosulfuron (MON-12037). R-29148 provides excellent protection to corn against tribenuron (72). MG-191 is a highly active, non-dichloroacetamide safener against thiocarbamate and chloroacetanilide injury to corn. In MG-191, the dichloroacetyl ($Cl_2CH-CO-$) moiety rather than dichloroacetamide ($Cl_2CH-N\equiv$) moiety may be responsible for safening activity [34]. Thus MG-191 is the most useful member of a new class of safeners. OTC is structurally similar to thioproline (L-thiazolidine-4-carboxylic acid) and flurazole. OTC increases the thiol content in roots and acts as a safener for herbicides tridiphane and alachlor in sorghum. Dietholate, a microbial inhibitor (an organophosphate), is a herbicide extender mixed with EPTC, butylate and vernolate. It inhibits soil microbes that degrade these herbicides and extends their soil residual life, thereby increasing the duration of weed control efficacy. It is used in commercial formulations, as it is highly effective in suppressing mineralization of these herbicides in soils.

Currently, several safeners are available or have the potential to be available for commercial use. They are categorized into three groups:

Group I. Dichloroacetamides or structurallly related compounds, dichlormid (R-25788), benoxacor (CGA-154281), furilazole (MON 13900), MG 191, R-29148, BAS-145138, dietholate (R-33865).

Group II. Oximes or structurally related compounds, cyabetrinil, oxabetrinil (CGA 92194), fluxofenim (CGA 133205), flurazole (MON 4606), naphthalic anhydride (NA).

Group III Fenclorin (CGA 123,407), fenchlorazol (HOE-70,542), CGA-185072, OTC.

Group I and II safeners are generally used in corn and sorghum, but do show a safening effect in wheat. Group II safeners are also used in rice and wheat. Group I safeners are applied with herbicides at a dosage range of 1:10 (1:30 (safener: herbicide) at PPI, PRE, or POST depending on the herbicide. Group II safeners are used as seed treatment at 0.40 (1.25 kg ha^{-1} of seed. Group III safeners are used with herbicides at 1:2 –1:4 safener:herbicide mixture.

A broad range of safener-herbicide interactions are summarized in Table 12.3. This Table includes some of the safeners currently available. Generally, NA and oxabetrinil are fairly unspecific and they exert safening interactions in a wide range of crop plants. Dichlormid and other structurally related dichloroacetamides and α-chloropropionamides show strong safening interactions with thiocarbamate and chloroacetanilide herbicides. NA, oxabetrinil and dichlormid safen against injury of sulfonylureas. Furilazole and R-29148 are specific for sulfonylurea herbicides.

Table 12.3. Interactions of major safeners with various herbicides in different crops

Safeners	Herbicides	Crops
Dichlormid	Carbamates, Imidazolinones, Sulfonylureas, Thiocarbamates (EPTC, butylate), Cyclohexanediones, Chloroacetamides (alachlor, acetochlor)	Maize, Sorghum, Oat, Wheat
Cyometrinil	Thiocarbamates, Cyclohexanediones	Maize, Sorghum
Flurazole	Imidazolinones, Chloroacetamides (alachlor, acetochlor)	Maize, Sorghum
Fenclorin	Chloroacetamides (pertilachlor)	Rice
MG-191	Chloroacetamides, Thiocarbamates	Maize, Sorghum
NA	Carbamates, Imidazolinones, Sulfonylureas, Thiocarbamates, Aryloxyphenoxypropionics Cyclohexanediones	Maize, Sorghum, Rice, Oat, Wheat
Oxabetrinil	Imidazolinones, Sulfonylureas, Chloroacetamides, (metolachlor), Aryloxyphenoxypropionics, Cyclohexanediones	Maize, Sorghum Wheat, Oat, Rice
Furilazole	Sulfonylureas	Maize
R-29148	Sulfonylureas	Maize

Mode of Action of Safeners

The primary mode of action of safeners in reducing the injury caused by chloroacetanilide and thiocarbamate herbicides, is stimulation of herbicide detoxification via glutathione conjugation. Elevation of glutathione (GSH) content and glutathione-related enzyme activities, such as cytosolic and microsomal glutathione-s-transferases (GSTs), as well as glutathione reductase (GR) are of great importance in thiocarbamate and chloroacetanilide herbicide detoxification in plants [35]. Safeners also cause elevation of such other metabolic activities as: a) herbicide uptake, b) uptake of sulphate, c) soluble protein content (in relation to fresh weight), d) herbicide mono-oxygenation, e) lipid synthesis (from acetate), f) glucosylation, g) acetolactate synthase, h) sulphate activation (ATP sulphurylase), etc. The increase in GSH content after application of dichlormid or structurally similar safeners in maize ranges between 200% and 400%. The consequent increase in GSH activity explains the greater herbicide detoxification. Jablonkai [64] showed that MG-191, a highly active non-chloroacetamide safener, safened maize against EPTC by enhancing the early rate of conjugation with GSH after initial oxidative metabolism of EPTC. MG 191 also enhances acetochlor conjugation via GSH conjugation. An oxime safener, oxabetrinil (CGA-92194) enhances detoxification of metolachlor in sorghum [41]. In most of the safeners, the proportional increases of the GSTs and the rate of detoxification correlated positively with the safening activities.

During herbicide detoxification several GST isozymes may be present in the plant tissue. Dean et al. [26] reported that treating sorghum seedlings with various safeners, flurazole, oxabetrinil, fluxofenim, NA and dichlormid resulted in the appearance of 4 to 5 additional GSTs (depending on the particular safener) which exhibited activity on metolachlor. There is increasing evidence that safeners are involved in the transcription of gene coding. Wiegand et al. [117] found that flurazole treatment of maize caused a 4-fold increase in mRNA coding for a subunit of GST I. This indicated that the safener either activated the transcription of gene coding for the subunit or increased the half-life of mRNA. They also found that at least three GSTs are different dimers of the monomeric polypeptide subunits GST-A, GST-B, GST-C and GST-D. While GST-I conjugates with both atrazine and alachlor, GST-III is more specific and preferentially conjugates with the chloroacetamide herbicides alachlor and metolachlor. Jablonkai and Hatzios [65] reported that pretreatment of maize shoots and roots with MG-191 enhanced MG-191-metabolizing activity as well as acetochlor-metabolizing activity, but depressed EPTC-metabolizing activity.

Herbicide safeners were also found to protect crops from herbicide injury by enhancing the activity of cytochrome P-450 mono-oxygenases [81]. As discussed in Chapter 8, herbicides, during detoxification, undergo oxidation by the addition of a single oxygen atom into the herbicide molecule (with the other atom of O_2 forming water). This reaction is called mono-oxygenation, which is catalyzed by a mono-oxygenase. As the mono-oxygenase is an NADPH-dependent cytochrome-P450 mixed function oxidase (Cyt. P-450), it is known as P450-mono-oxygenase. Dichlormid and cyometrinil stimulate mono-oxygenation of chlortoluron [15]. Kreuz et al. [73] reported that safeners increase tolerance towards sulphonylurea and aryloxyphenoxypropanoic acid herbicides by inducing increased rates of herbicide mono-oxygenation. NA is known to have marked effects on the P-450 mono-oxygenase-catalyzed metabolism of sulphonylurea herbicides. The enhanced activity of P-450 mono-oxygenases following safener treatments is due to increased levels of P-450 enzymes [81]. Mono-oxygenases are most important during early seedling life, which is also the time period that is relevant for safener action.

Safeners are also known to act at the molecular level. Rao and Khan [99, 101] reported that the inhibition of polyribosome formation by propachlor was completely reversed by dichlormid [R-25788] (Fig. 12.4). They proposed that alachlor and dichlormid compete for the site(s) of action. This is consistent with the competitive antagonism that exists between herbicide and safener.

Information on the uptake, translocation and metabolism of safeners is limited. Flurazole, a thiazolecarboxylate safener, is rapidly absorbed and metabolized to a glutathione conjugate by 3- and 5-d old etiolated shoots of maize and grain sorghum [13]. The uptake of some safeners is faster than that of others. Yenne et al. [119] studied the uptake of oxime ether safeners, oxabetrinil and fluxofenim, and found that uptake of oxabetrinil by sorghum seedlings was almost two times greater than fluxofenim uptake. Sorghum metabolized both safeners mainly to unknown water-soluble products, and one of the metabolites was postulated as glutathione conjugate of oxabetrinil. Jablonkai and Hatzios [65] reported that microsomes in maize have the ability to convert MG-191 to a single, unknown metabolite. Jablonkai and Dutka [66] found that only a small portion of root-applied MG-191 was absorbed by maize seedlings. A concentration-dependent, rapid translocation of MG-191 and its labelled metabolites over time occurred in concert with rapid metabolism of the parent molecule. The relatively greater mobility of safeners within the plant appears to be related to their protective efficacy [64].

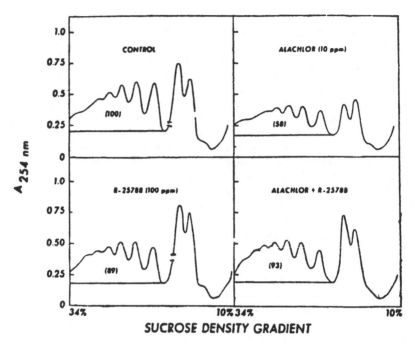

Fig.12.4. Effect of alachlor and dichlormid (R-25788) alone and in combination on polyribosome formation. The numbers in parentheses (inside polyribosome profiles) indicate the percentages of polyribosome content compared to control [99, 101].

Interaction Responses

Safeners reduce herbicide activity, and hence the interaction between the two is an antagonistic response. Safener-induced reduction in herbicide activity is due to an increase in herbicide detoxification, as discussed earlier. As the safener decreases the amount of a herbicide that would otherwise be available to its site of action in the absence of safener, the antagonist, the interaction response may be termed as biochemical antagonism. Most of the safeners act as biochemical antagonists when mixed with herbicides.

Certain safeners, however, act as competitive antagonists by acting at the same site as herbicides. In this competitive antagonism, both herbicide and safener have a common site of action. Dichlormid shows competitive antagonism while interacting with EPTC and alachlor.

Safeners and herbicides also show physiological antagonism. The effects of EPTC may be reversed or counterbalanced by the simultaneous addition of dichlormid. The inhibition of lipid, terpenoid and gibberellin biosynthesis by the thiocarbamate and acetanilide herbicides is believed to be reversed by safeners, and this antagonism appears to be physiological. Pretreatment of sorghum seeds with the safener cyometrinil alters the pattern of alachlor-induced effects on terpenoid and gibberellin biosynthesis in sorghum seedlings. Cyometrinil also prevents metolachlor-induced inhibition of cuticular integrity of sorghum plants. Thus, depending on the safener and herbicide used, three antagonistic responses, viz., biochemical antagonism, competitive antagonism and physiological antagonism, are believed to be in play during herbicide-safener interaction.

HERBICIDE-INSECTICIDE INTERACTIONS

Herbicides and insecticides are often applied simultaneously or serially to crops within a short period. These chemicals are usually not harmful when used as per recommended practices. The tolerance of plants to a herbicide may be altered in the presence of an insecticide and vice versa.

In one of the earlier reports on herbicide-insecticide interactions, Hacskaylo et al. [48] observed that phytotoxicity of monuron and diuron on cotton increased when applied with phorate or disulfoton. Nash [83] found similar synergistic phytotoxicities in oats with combinations of diuron and disulfoton or phorate. The toxicity of diuron-phorate combination was more initially and persisted longer than the diuron-disulfoton combination. Arle [5], however, noted that phorate or disulfoton interacted antagonistically with trifluralin to increase cotton yield, possibly by stimulating secondary roots in the zone of pesticide incorporation.

In peanut, disulfoton and phorate had little influence on the phytotoxicity of trifluralin and chloramben [16]. Parks et al. [90] reported that absorption of prometryn by beans (*Phaseolus vulgaris*) plants was increased in the presence of phorate. Both phorate and ametryn inhibited the state 3 respiration of mitochondria isolated from 6-d old bean hypocotyls. Penner [91] showed that preemergence application of disulfoton increased whereas postemergence applications of disulfoton and diazinon decreased atrazine injury. Combinations of organophosphate insecticides and atrazine on phytotoxicity appeared to involve an effect of the insecticides on herbicide absorption and translocation. The combination of metribuzin with either phorate or disulfoton interacted synergistically to reduce both soybean grain yield or plant population significantly even at recommended rates of both chemicals [55].

Del Rosario and Putnam [27] reported that foliar injury by linuron of carrot was enhanced when carbaryl was combined with herbicide and sprayed. Synergism occurred when the two compounds were applied as tank mixes or applied with a one-day interval. Carbaryl enhanced the penetration of linuron into intact leaves. The rate of linuron metabolism was also decreased in the presence of carbaryl. Thus, their results suggested that increased absorption and reduced rate of metabolism of linuron in the presence of carbaryl formed the basis for synergistic interaction between the two pesticides. Simultaneous treatment with carbaryl inhibited degradation of monuron in cotton leaf discs [112]. Besides, the carbamate insecticides did not prevent the metabolism of monuron to monomethyl-monuron, but they inhibited the subsequent degradation of monomethyl-monuron. This suggested that the carbamate insecticides act primarily to inhibit the second demethylation of monuron.

In combination, alachlor and carbofuron interact synergistically to reduce barley but not corn growth [50]. Similar synergistic interaction and physiological basis for the response was observed in corn and barley shoots when chlorbromuron was combined with seed treatment of carbofuron [51].

Various reports showed that propanil, used for weed control in rice, is involved in interaction responses with various insecticides. Bowling and Flinchum [11] found that propanil interacted with certain carbamate and phosphate insecticides used as seed treatments on rice. They also found that chlorinated insecticides used as seed treatments did not interact with propanil. They concluded that the interaction may be associated only with certain carbamate and phosphate insecticides. Matsunaka [79] reported that paraoxon (O,O-diethyl O-p-nitrophenyl phosphate) or sumioxon [O,O-dimethyl O-(4-nitro-m-tolyl) phosphate] to be more active than parathion or sumithion in inhibiting propanil hydrolysis in rice. He also reported that BHC showed no synergistic effect when mixed with propanil. He concluded that a compound having no inhibitory activity to acetylcholinesterase has no joint action with propanil in rice.

The metabolism of linuron, chlorpropham, dicamba and pronamide in plantain, bean, wheat and lettuce is inhibited respectively by fonofos (Dyfonate), chlorfenvinphos, malathion, fensulfothion and disulfoton [17]. The carbamate insecticides carbaryl and carbofuran have little effect on the metabolism of these herbicides. Carbaryl and diazinon affect the hydrolytic enzyme activity in rice shoots at all rates of application [36]. When propanil was applied in the presence of diazinon, no additive phytotoxicity occurred in rice plants over that incurred from propanil alone. Synergistic phytotoxicity on rice was apparent when propanil was applied one day after carbaryl treatment and most plants killed.

Propanil interacts with carbofuran when applied in rice. Smith and Tugwell [110] found synergistic response when propanil was applied at intervals ranging from 7 to 56 d after carbofuran treatment, resulting in greater injury to rice vegetatively. This interaction was of sufficient magnitude to prohibit the use of propanil after carbofuran treatment.

These reports indicate that risk of crop injury, stand reduction and decreased yields could result when combinations or sequential applications of herbicides and insecticides are used. Increased crop injury may be due to greater persistence of the herbicide or insecticide or both, resulting in greater residue problems.

The herbicide-insecticide interactions are predominantly synergistic. Of the 121 herbicide-insecticide interactions listed by Hatzios and Penner [54], more than 93% were synergistic. Most synergistic activity by insecticides is attributed to the insecticide's inhibition of the plant's metabolic pathway to detoxify herbicides [47].

HERBICIDE INTERACTIONS WITH FUNGICIDES AND PATHOGENS

Unlike herbicide-insecticide interactions, herbicide-fungicide interactions tend to be antagonistic. Of the 23 herbicide-fungicide mixtures listed by Hatzios and Penner [54], 21 showed antagonism.

Herbicide-fungicide interactions are not as widespread as herbicide-insecticide interactions. Herbicides interact with fungicides as well as the disease-causing organisms. Dinoseb was shown to reduce the severity of stem rot (white mould) in groundnut [21, 43].

Bagga [6] found interaction between chloroxuron and the disease organism *Rhizoctonia solani* in peas. In sterilized soil, chloroxuron did not cause any apparent injury to pea plants, while in the presence of *Rhizoctonia* in unsterilized soil it did. Beckman et al. [7] reported that oxadiazon significantly reduced the incidence of stem rot caused by the soil-borne pathogen *Sclerotium rolfsii* in field-grown groundnut. Oxadiazon was non-toxic to the pathogen, indicating that disease control must be indirect, perhaps relating to the reduced spreading habit of plants observed in the treated plots. They proposed that the reduced ground contact was reducing the probability of pathogen inoculum contact. Dinoseb, which reduced stem rot in the field, was found toxic to *S. rolfsii* in laboratory tests.

Diuron and atrazine, which inhibit photosynthesis, may make plants more susceptible to the tobacco mosaic virus. On the other hand, diuron may decrease the incidence
of root rot in wheat. Atrazine exhibits antagonistic interaction with the fungicide Dexon
on many plant species. The physiological basis for this interaction appeared to be
reduced uptake of atrazine by corn, cucumber and soybean seedlings in the presence
of Dexon [116]. Kappelman et al. [71] reported that there was no interaction when 1-
chloro-2-nitropropane or PCNB plus 5-ethoxy-3-trichloromethyl-1,1,1,2,4-thiadazole was
mixed with trifluralin, prometryn or fluometuron.

HERBICIDE-FERTILIZER INTERACTIONS

There is growing evidence of herbicide-fertilizer/plant nutrient interactions occurring
in agriculture. The addition of ammonium nitrate greatly increases the absorption of
solutions of isooctyl ester of 2,4,5-T by tree leaves [12]. Ammonium ion speeded up
translocation of the herbicide in post oak (*Quercus stellata*) but not in other species. It
was suggested that in ammonium nitrate solution (0.1%) 2,4,5-T was hydrolyzed to
acid while some remained in the ester form, thus facilitating entry into the treated
plants by both polar and lipoidal routes. Rao and Rahman [100] and Rao and Kotoky
[102] found that addition of ammonium sulphate at 0.5% (w/v) concentration to
glyphosate solution enhanced control of *Imperata cylindrica*. Similar increase in the
activity of 2,4-D was also found when ammonium sulphate was tank mixed [103].
Turner and Loader [115] also obtained increased phytotoxicity for *Elytrigia repens* of
sprays containing 0.2 to 0.5 kg ha^{-1} glyphosate with the addition of 1 to 10% (w/v)
ammonium sulphate. However, higher ammonium sulphate concentrations were sometimes antagonistic. A mixture of ammonium sulphate and a lipophilic surfactant showed
superior effect on glyphosate activity.

Doll et al. [30] observed that in the presence of high but non-toxic levels of phosphorus, the suppression of corn and squash (*Cucurbita maxima* Duchesne) seedling
growth in darkness by chloramben or atrazine was enhanced. This increased suppression of growth was, however, not accompanied by an increase in root uptake of herbicides. Selman and Upchurch [105] reported that amitrole was phytotoxic to peanuts in
the presence of high phosphorus levels. A linear relationship between amitrole
phytotoxicity and phosphorus concentration existed for wheat, soybean, sorghum and
rye. They also reported diuron-phosphorus interaction in cotton and soybean. Wilson
and Stewart [118] observed that phosphorus and trifluralin interacted in their effects
on root growth of tomato.

Fertilizers have a significant effect on herbicide absorption. For example, the complete fertilizer involving N, P and K reduced atrazine absorption by plants, thus reducing phytotoxicity [92]. Atrazine was more toxic in the presence of P and K than in the
presence of N and P or N and K due to increased absorption of the herbicide by plants.
Conversely, herbicides also affect the absorption of plant nutrients. Ladonin et al. [74]
observed 2,4-D and dicamba decreasing N uptake by the sensitive pea plants by
55-64% after 6 d and reduced N incorporation into non-protein N compounds. Similarly, lenacil sharply reduced N uptake in the sensitive French bean plants and reduced
15N incorporation into proteins.

HERBICIDE-ADJUVANT INTERACTIONS

An adjuvant refers to any substance included in the formulation of a herbicide or added to the spray tank to enhance herbicide characteristics or application characteristics (vide Chapter 6). Surfactants and phytobland oils are the two types of chemicals commonly used as adjuvants in herbicide formulations. Once a herbicide is commercialized, it may not be necessary to use additional adjuvants. However, addition of extra surfactants may be needed to improve the herbicide efficacy at low rates of several water-soluble herbicides applied to the foliage. The applied adjuvant enhances herbicide activity either favourably or unfavourably depending on the interaction of herbicide-adjuvant-plant combination. Adjuvants enhance herbicide uptake by the plant and its translocation to the site of action. As more herbicide is likely to be available for its effect, herbicide-adjuvant interactions are generally synergistic. The effects of adjuvants on herbicide activity are the opposite of safeners.

HERBICIDE-BIOHERBICIDE INTERACTIONS

The mixtures of bioherbicides (bioagents) and organic herbicides interact and bring out synergistic responses. Synergism has been found for a diverse group of herbicides bioherbicides and weed hosts. At the same time, mixing herbicides with certain bioherbicides may have a detrimental effect on the living microbes. This problem can be mitigated if the herbicide and bioagent are applied sequentially. The combinations of herbicides and living organisms normally provide additive, synergistic and antagonistic effects on weed control. This subject was discussed in greater detail in Chapter 11.

REFERENCES

1. Agbakoba, C.S. and J.F. Goodin. 1970. Picloram enhances 2,4-D movement in field bindweed. Weed Sci. **18:** 19-22.
2. Akobundu, I.O., R.D. Sweet and W.B. Duke. 1975a. A method of calculating herbicide combinations and determining herbicide synergism. Weed Sci. **23:** 20-25.
3. Akobundu, I.O., R.D. Sweet, and W.B. Duke and P.L. Minotti. 1975b. Basis for synergism of atrazine and alachlor combinations on Japanese millet. Weed Sci. **23:** 43-48.
4. Anonymous (1978-79). Annual Scientific Report. Tocklai Experimental Station, Tea Research Association, Jorhat, Assam, India.
5. Arle, H.F. 1968. Trifluralin-systemic insecticide interactions on seedling cotton. Weed Sci. **16:** 430-432.
6. Bagga, D.K. 1970. Interaction of chloroxuron with *Rhizoctonia solani* on seedling southern peas. Abstr. Weed Sci. Soc. Amer. no. 44.
7. Beckman, P.A., R. Rodriguez-Kabana and G.A. Buchanan. 1977. Interactions of oxadiazon and dinoseb with stem rot in peanuts. Weed Sci. **25:** 260-263.
8. Beste, C.E. and M.M. Schreiber. 1972. Interaction of EPTC and 2,4-D on excised tissue growth. Weed Sci. **20:** 4-7.
9. Bliss, C.I. 1939. The toxicity of poisons applied jointly. Ann. Appl. Biol. **26:** 585-615.
10. Bovey, R.W., F.S. Davis and H.L. Morton. 1968. Herbicide combinations for woody plant control. Weed Sci. **16:** 332-335.
11. Bowling, C.C. and W.T. Flinchum. 1968. Interaction of propanil with insecticides applied as seed treatments on rice. J. Econ. Entomol. **61:** 67-69.
12. Brady, H.A. 1970. Ammonium nitrate and phosphoric acid increase on 2,4,5-T absorption by tree leaves. Weed Sci. **18:** 204-206.
13. Breaux, E.J., M.A. Hoobler, J.E. Patanella and G.A. Leyes. 1983. Mechanism of action of thiazole safeners. *In* K.K. Hatzios and R.E. Hoagland (eds.). Crop Safeners for Herbicides: Development, Uses, and Mechanism of Action. Academic Press, San Diego, CA, USA, pp. 163-165.

14. Burnside, O.C., G.A. Wicks and C.R. Fenster. 1971. Protecting corn from herbicide injury by seed treatment. Weed Sci. **19**: 565-568.

15. Canivenc, M.C., B. Cagnac, F. Cabenne and R. Scalla. 1968. Manipulation of chlortoluron fate in wheat cells. *In* Proc. EWRS Symp. Factors Affecting Herbicidal Activity and Selectivity. Ponsen & Looijen, Wageningen, Netherlands, pp. 115-120.

16. Cargill, R.L. and P.W. Santelman. 1971. Response of peanuts to combinations of herbicides with other herbicides. Weed Sci. **19**: 24-27.

17. Chang, F.Y., L.W. Smith and G.R. Stephenson. 1971. Insecticide inhibition of herbicide metabolism in leaf tissues. J. Agr. Food Chem. **19**: 1183-1186.

18. Chang, F.Y., J.D. Bandeen and G.R. Stephenson. 1972. A selective antidote for prevention of EPTC injury in corn. Can. J. Plant Sci. **25**: 707-714.

19. Chang, F.Y., G.R. Stephenson and J.D. Bandeen. 1973. Comparative effects of three EPTC antidotes. Weed Sci. **21**: 292-295.

20. Chang, F.Y., G.R. Stephenson, G.W. Anderson and J.D. Bandeen. 1974. Control of wild oats in oats with barban plus antidote. Weed Sci. **22**: 546-548.

21. Chappel, W.E. and L.I. Miller. 1956. The effect of certain herbicides on plant pathogens. Plant Dis. Rep. **40**: 52-56.

22. Colby, S.R. 1967. Calculating synergistic and antagonistic responses of herbicide combinations. Weeds **15**: 20-22.

23. Colby, S.R. 1970. Comparison of two methods for testing herbicidal antagonism. Abstr. Weed Sci. Soc. Amer, p. 13.

24. Davis, D.G. and K.E. Dusabek. 1973. Effect of diallate on foliar uptake and translocation of herbicides in pea. Weed Sci. **21**: 16-20.

25. Davis, R.S., R.W. Bovey and M.G. Merkle. 1968. Effect of paraquat and 2,4,5-T on the uptake and transport of picloram in woody plants. Weed Sci. **26**: 336-339.

26. Dean, J.V., J.W. Gronwald and C.V. Eberlein. 1990. Induction of glutathione-S-transferase isozymes in sorghum by herbicide antidotes. Plant Physiol. **92**: 467-473.

27. Del Rosario, D.A. and A.R. Putnam. 1973. Enhancement of foliar activity of linuron with carbaryl. Weed Sci. **21**: 465-468.

28. Dickerson, C.T., Jr. and R.D. Sweet. 1968. Atrazine, oil, 2,4-D for postemergence weed control. Proc. Northeast Weed Control Conf. (USA). **22**: 64.

29. Diem, J.R. and D.E. Davis. 1974. Effect of 2,4-D on ametryne toxicity. Weed Sci. **22**: 285-292.

30. Doll, J.D., D. Penner and W.F. Meggitt. 1970. Herbicide and phosphorus influence on root absorption of amiben and atrazine. Weed Sci. **18**: 357-359.

31. Drury, R.E. 1980. Physiological interaction, its mathematical expression. Weed Sci. **28**: 573-579.

32. Duke, S.O., R.D. Wauchope and R.E. Hoagland. 1982. Glyphosate effects on calcium uptake in soybean seedlings. Abstr. Weed Sci. Amer., pp. 83-84.

33. Duke, W.B., V.S. Rao and J.F. Hunt. 1972. EPTC-atrazine residue interaction effect on seedling alfalfa varieties. Proc. Northeastern Weed Control Conf. 26:258-262.

34. Dutka, F. 1991. Bioactive chemical bond systems. Z. Naturforsch. **46c**: 805-809.

35. Dutka, F. and T. Komives. 1987. MG-191: a new selective herbicide antidote. *In* R. Greenhalgh and T.R. Roberts (eds.). Pesticide Science and Biotechnology. Blackwell, Oxford, UK, pp. 201-204.

36. El-Refai, A.R. and M. Mowafy. 1973. Interaction of propanil with insecticides absorbed from soil and translocated into rice plants. Weed Sci. **18**: 357-359.

37. Ezra, G., J.H. Dekker and G.R. Stephenson. 1985. Tridiphane as a synergist for herbicides in corn (*Zea mays*) and proso millet (*Panicum miliaceum*). Weed Sci. **33**: 287-290.

38. Finney, D.J. 1979. Bioassay and the practice of statistical interference. Internat. Statist. Rev. **47**: 1-12.

39. Frear, D.S. and H.R. Swanson. 1970. Biosynthesis of S-(4-ethylamino-6-isopropylamino-2-s-triazino) glutathione: partial purification and properties of a glutathione S-transferase from corn. Phytochemistry **9**: 2123-2132.

40. Frear, D.S and H.R. Swanson. 1973. Metabolism of substituted diphenyl-ether herbicides in plants. I. Enzymatic cleavage of fluorodifen in peas (*Pisum sativum* L.) Pestic. Biochem. Physiol. **3**: 473-482.

41. Fuerst, E.P and J.W. Gronwald. 1986. Induction of rapid metabolism of metolachlor in sorghum (*Sorghum bicolor*) shoots by CGA-92194 and other antidotes. Weed Sci. **34**: 354-361.

42. Fyske, H. 1975. Untersuchungen über *Sonchus arvensis* L. II. Translocation von ^{14}C-MCPA unter verschiedenen Bedingungen. Weed Res. **15**: 165-170.

43. Garren, K.H. and G.B. Duke. 1957. The peanut stem rot problem and a preliminary report on interrelations of 'non-dirting' weed control and other practices to stem rot and yield of peanuts. Plant Dis. Rep. **41**: 424-431.

44. Gentner, W.A. 1966. The influence of EPTC on external foliage wax deposition. Weeds **14**: 27-30.

45. Gowing, D.P. 1959. A method of comparing herbicides and assessing herbicide mixtures at the screening level. Weeds **7**: 66-76.

46. Gowing, D.P. 1960. Comments on tests of herbicide mixtures. Weeds **8**: 379-391.

47. Green, J.M. and S.P. Bailey. 1988. Herbicide interactions with herbicides and other agricultural chemicals. *In* C.G. McWhorter and M.R. Gebhardt (eds.). Methods of Applying Herbicides. Weed Sci. Soc. Amer., Champaign, IL, USA, pp. 37-61.

48. Hacskaylo, J.J. K. Walker and E.G. Pires. 1964. Response of cotton seedling to combinations of pre-emergence herbicides and systemic insecticides. Weeds **12**: 288-291.

49. Hall, C., L.V. Eddington and C.M. Switzer. 1982. Effects of chlorsulfuron or 2,4-D upon diclofop-methyl efficacy in oat (*Avena sativa*). Weed Sci. **30**: 670-672.

50. Hamill, A.S. and D. Penner. 1973a. Interaction of alachlor and carbofuran. Weed Sci. **21**: 330-335.

51. Hamill, A.S. and D. Penner. 1973b. Chlorbromuron-carbofuran interaction in corn and barley. Weed Sci. **21**: 335-339.

52. Hamill, A.S., L.W. Smith and C.M. Switzer. 1972. Influence of phenoxy herbicides on picloram uptake and phytotoxicity. Weed Sci. **20**: 226-229.

53. Hanson, C.L. and C.E. Rieck. 1976. The effect of iron and aluminum on glyphosate toxicity. Proc. South. Weed Sci. Soc. (USA), **29**: 49.

54. Hatzios, K.K. and D. Penner. 1985. Interactions of herbicides with other agrochemicals in higher plants. Rev. Weed Sci. **1**: 1-63.

55. 1,01,250.00Hayes, R.M., K.V. Yeargan, W.W. Witt and H.G. Raney. 1979. Interaction of selected insecticide-herbicide combinations on soybeans (*Glycine max*). Weed Sci. **27**: 51-54.

56. Headford, D.W.R. 1966. An improved method of control of *Paspalum conjugatum* with amitrole-T and paraquat. Weed Res. **6**: 304-308.

57. Hickey, J.S. and W.A. Krueger. 1974a. Alachlor and 1,8-naphthalic anhydride effects on sorghum seedling development. Weed Sci. **22**: 86-90.

58. Hickey, J.S. and W.A. Krueger. 1974b. Alachlor and 1,8-naphthalic anhydride effects on corn coleoptiles. Weed Sci. **22**: 250-252.

59. Hill, B.D., B.G. Todd and E.H. Stobbe. 1980. Effect of 2,4-D on the hydrolysis of diclofop-methyl in wild oat (*Avena fatua*). Weed Sci. **28**: 725-729.

60. Hoffman, O.L. 1962. Chemical seed treatments as herbicide antidotes. Weeds **10**: 322-323.

61. Hoffman, O.L. 1969. Chemical antidotes for EPTC on corn. Abstr. Weed Sci. Soc. Amer. 12.

62. Horowitz, M. 1971. Control of established *Portulaca oleracea* L. Weed Res. **11**: 302-306.

63. Horowitz, M. and Y. Klefeld. 1968. Winter application of residual herbicides before cotton. Proc. 9th Brit. Weed Cotrnol Conf., p. 707.

64. Jablonkai, I. 1991. Basis for differential chemical selectivity of MG-191 safener against acetochlor and EPTC injury to maize. Z. Naturforsch. **46c**: 836-845.

65. Jablonkai, I. and K.K. Hatzios. 1994. Microsomal oxidation of the herbicides EPTC and acetochlor and of the safener MG-191 in maize. Petsic. Biochem. Physiol. **48**: 98-109.

66. Jablonkai, I. And F. Dutka. 1995. Uptake, translocation, and metabolism of MG-191 safener in corn (*Zea mays* L.). Weed Sci. **43**: 169-174.

67. James, C.S., G.N. Prendeville, G.F. Warren and M.M. Schreiber. 1970. Interactions between herbicidal carbamates and growth regulators. Weeds Sci. **18**: 137-139.

68. Johnson, B.J. 1970. Combinations of herbicides and other pesticides on soybeans. Weed Sci. **18**: 128-130.

69. Johnson, B.J. 1971. Effects of sequential herbicide treatments on weeds and soybeans. Weed Sci. **19**: 695-699.

70. Jordan, L.S. and V.A. Jolliffe. 1971. Protection of plants from herbicides with 1,8-napthalic anhydride as illustrated with sorghum. Bull. Environ. Contam. Toxicol. **6**: 417-421.

71. Kappelman, A.J., Jr., G.A. Buchanan and L.F. Lund. 1971. Effect of fungicides, herbicides and combinations on root growth of cotton. Agron. J. **63**: 3-5.

72. Kotula-Syka, E., K.K. Hatzios and S.A. Meredith. Interactions between SAN 582H and selected safeners on grain sorghum (*Sorghum bicolor*) and corn (*Zea mays*). Weed Technol. **10**: 299-304.

73. Kreuz, K., J. Gaudin, J. Stingelin, and E. Ebert. 1991. Metabolism of the arylphenoxypropanoate herbicide CGA 184927 in wheat, barley and maize: different effects of the safener, CGA 185072. Z. Naturforsch. **46c**: 901-905.

74. Ladonin, V.F., G.A. Chesalin, L.N. Samoilov, L.G. Spesivtsev and V.I. Taova. 1980. The application of ^{15}N to investigating the influence of herbicides on crop and weed plants. Report 2. The action characteristics of herbicides on the nitrogenous metabolism of resistant and susceptible plants. Agrokhimiya **17**(9): 116-122.

75. Lamoureux, G.L. and D.G. Russness. 1983. Glutathione-S-transferase inhibition as the basis of Dowco 356 synergism of atrazine. Abstr. 186th ACS National Meeting, Pest, p.115.

76. LeClerg, E.L. 1957. Mean separation by the functional analysis of variance and multiple comparisons. Agr. Res. Serv., U.S. Dept. Agr. ARS, pp. 20-23.

77. Limpel, L.E., P.H. Schuldt and D. Lamont. 1962. Weed control by dimethyl tetrachloroterephthalate alone and in certain combinations. Proc. Northeast. Weed Sci. Soc (USA), **16**: 48-53.

78. Lynch, M.R., R.D. Sweet and C.T. Dickerson, Jr. 1970. Synergistic response of atrazine in combination with other herbicides—A preliminary report. Proc. Northeast. Weed Control Conf. (USA) **24**: 33.

79. Matsunaka, S. 1968. Propanil hydrolysis: inhibition in rice plants by insecticides. Science **160**: 1360-1361.

80. McCall, P.J., L.E. Stafford, P.S. Zorner and P.D. Gavit. 1986. Modeling the foliar behavior of atrazine with and without crop oil concentrate on giant foxtail and the effect of tridiphane on the model rate constants. J. Agric. Food Chem. **34**: 235-238.

81. McFadden, J.J., J.W. Gronwald and C.V. Eberlein. 1990. In vitro hydroxylation of bentazon by microsomes from naphthalic anhydride-treated corn shoots. Biochem. Biophys. Res. Comm. **168**: 206-213.

82. Morre, D.J. and R.D. Cheetham. 1970. Response of 2,4-D resistant monocot species to CIPC-2,4-D combinations. Proc. N. Central Weed Control Conf. (USA) **25**: 103.

83. Nash, R.G. 1968. Synergistic phytotoxicities of herbicide-insecticide combinations in soil. Weed Sci. **16**: 74-78.

84. Nash, R.G. 1981. Phytotoxic interaction studies—techniques for evaluation and presentation of results. Weed Sci. **29**: 147-155.

85. Nash, R.G. and L.L. Jensen. 1973. Determining phytotoxic pesticide interactions in soil. J. Environ. Qual. **2**: 503-510.

86. O'Donovan, J.T., P.A. O'Sullivan and D.C. Caldwell. 1983. Basis for antagonism of paraquat phytotoxicity to barley by MCPA dimethylamine. Weed Res. **23**: 165-172.

87. Olson, W. and J.D. Nalewaja. 1982. Effect of MCPA on ^{14}C-diclofop uptake and translocation. Weed Sci. **30**: 59-63.

88. O'Sullivan, P.A. and W.H. VandenBorn. 1980. Interaction between benzoylprop ethyl, flamprop methyl or flamprop isopropyl and herbicides used for broadleaved weed control. Weed Res. **20**: 53-57.

89. O'Sullivan, P.A. and V.C. Kossatz. 1982. Influence of picloram on *Cirsium arvense* (L.) Scop. Control with glyphosate. Weed Res. **22**: 251-256.

90. Parks, J.P., B. Truelove and G.A. Buchanan. 1972. Interaction of prometryne and phorate on bean. Weed Sci. **20**: 89-92.

374

91. Penner, D. 1974. Effect of disulfoton, diazinon and fenusulfothion on atrazine absorption by soybean. Agron. J. **66**: 107-109.

92. Popova, D. 1980. The translocation and phytotoxicity of ^{14}C-atrazine in French beans in relation to the level of mineral fertilizers. Pochvoznanie I Agrokhimiya **15(2)**: 77-84.

93. Putnam, A.R. and S.K. Ries. 1965. The effects of adjuvants on the activity of herbicides for the control of quackgrass (*Agropyron repens* L. Beauv.). Proc. Northeast. Weed Control Conf. (USA). **19**: 300-304.

94. Putnam, A.R. and S.K. Ries. 1967. The synergistic action of herbicide combinations containing paraquat on *Agropyron repens* L. Beauv. Weed Res. **7**: 191-194.

95. Putnam, A.R. and D. Penner. 1974. Pesticide interactions in higher plants. *In* F.A. Gunther. (ed.). Residue Reviews, Vol. 50. Springer-Verlag, New York, pp. 73-110.

96. Quereshi, F.A. and W.H. VandenBorn. 1979. Interaction of diclofop methyl and MCPA on wild oats (*Avena fatua*). Weed Sci. **27**: 202-205.

97. Rains, L.J. and O.H. Fletchall. 1971. The use of chemicals to protect crops from herbicide injury. Proc. North Central Weed Control Conf. (USA) **26**: 42.

98. Rao, V.S. and W.B. Duke. 1975. Effect of EPTC-atrazine residue on seedling alfalfa. Indian J. Weed Sci. **5**: 92-97.

99. Rao, V.S. and A.A. Khan. 1975. Protection of barley from alachlor injury. Abstr. Weed Sci. Soc. Amer. no. 175.

100. Rao, V.S and F. Rahman. 1978. Weed control in tea with glyphosate. Two and A Bud. **25**: 71-73.

101. Rao, V.S. and A.A. Khan. 1979. Antidotal action of R-25788 on EPTC and alachlor injury to corn and barley. Indian. J. Weed Sci. **11**: 12-19.

102. Rao, V.S. and B. Kotoky. 1979. Enhancement of glyphosate activity by solubilization and fertilizer additives. Abstr. Meetings of Indian Soc. Weed Sci. p. 11.

103. Rao, V.S., F. Rahman, H.S. Singh, A.K. Dutta and M.C. Saikia. 1977a. Advances in weed research in tea of Northeast India. Abstr. Weed Sci. and Workshop in India, p. 57.

104. Rao, V.S., F. Rahman, H.S. Singh, A.K. Dutta, M.C. Saikia, S.N. Sarmah and B.C. Phukan. 1977b. Effective weed control in tea with glyphosate. Abstr. Weed Sci. Soc. and Workshop in India, pp. 58.

105. Selman, F.L. and R.P. Upchurch. 1970. Regulation of amitrole and diuron toxicity by phosphorus. Weed Sci. **18**: 619-623.

106. Sharma, M.P. and W.H. VandenBorn. 1982. Interaction between difenzoquat and propanil/MCPA combinations for wild oat and green foxtail control in barley. Can J. Plant Sci. **62**: 453-459.

107. Shea, P.J. and D.R. Tupy. 1983. Reversing divalent cation-induced reductions in glyphosate activity with chelating agents. Abstr. Weed Sci. Soc. Amer., p. 28.

108. Sheets, T.J. and O.A. Leonard. 1958. An evaluation of the herbicidal efficiency of combinations of ATA with dalapon, monuron, and several other chemicals. Weeds **6**: 143-146.

109. Skaptason, J.S. 1969. Synergism from phenoxy, propionic, and benzoic herbicide mixtures. Proc. N. Central Weed Control Conf. (USA) **24**: 58.

110. Smith, R.J., Jr. and N.P. Tugwell. 1975. Propanil-carbofuran interactions in rice. Weed Sci. **23**: 176-178.

111. Sterrett, J.P., J.T. Davis and W. Hurt. 1972. Antagonistic effects between picloram and bromacil with oats. Weed Sci. **20**: 440-444.

112. Swanson, C.R. and H.R. Swanson. 1968. Inhibition of degradation of monuron in cotton leaf tissue by carbamate insecticides. Weed Sci. **16**: 481-484.

113. Tames, P.M.L. 1964. Isoboles, a graphic representation of synergism in pesticides. Neth. J. Plant Pathol. **70**: 73-80.

114. Todd, B.G. and E.H. Stobbe. 1980. The basis of the antagonistic effect of 2,4-D on diclofop-methyl toxicity to wild oats (*Avena fatua*). Weed Sci. **28**: 371-377.

115. Turner, D.J. and M.C.P. Loader. 1978. Complexing agents as herbicide additives. Weed Res. **18**: 199-207.

116. Webster, H.L. and T.J. Sheets. 1969. A physiological explanation of the atrazine-dexon interaction. Abstr. Weed Sci. Soc. Amer. no. 349.

117. Wiegand, R.C., D.M. Shah, T.J. Mozer, E.I. Harding, J.Diaz-Collier, C. Saunders, E.G. Jaworski, and D.C. Tiemeier. 1986. Messenger RNA encoding a glutathione-*S*-transferase responsible for herbicide tolerance in maize is induced in response to safener treatment. Plant Mol. Biol. 7: 235-243.

118. Wilson, H.P. and F.B. Stewart. 1973. Relationship between trifluralin and phosphorus on transplanted tomatoes. Weed Sci. 21: 150-153.

119. Yenne, S.P., K.K. Hatzios and S.A. Meredith. 1990. Uptake, translocation, and metabolism of oxabetrinil and CGA-133205 in grain sorghum (*Sorghum bicolor*) and their influence on metolachlor metabolism. J. Agr. Food Chem. 38: 1957-1961.

13

Herbicide Discovery and Development

HERBICIDE DISCOVERY

Herbicide discovery and development is a continuing process because there is always a need for newer herbicides to meet the changing weed situations in agricultural and non-agricultural systems, to achieve greater efficacy and economy in chemical weed control and to minimize risks to the environment through toxicity and residues. This requires massive and expensive technical support by chemical industry, research institutions, universities and public agencies. The initial burden is, however, borne by the industry. In order to discover one herbicide, a chemical company may need to evaluate a number of chemicals, ranging between 1000 and 20,000. Normally, the success rate for an herbicide is about 1 in 25,000, but this is dynamic as the weed situation and economic viability keep changing.

During the 1970s the herbicide market became more competitive due to the discovery of a greater number of herbicides than in the 1960s and the price of oil and petroleum products registered a several-fold increase. As a result, the time required to develop a successful product ranged from 8 to 10 yr or even longer and the cost of inputs for developing it varied from US $10 to 15 million. In the 1980s and early 1990s, the cost of producing a successful herbicide ranged from US $20-25 million. The governmental clearance requirements in view of the strict crop, human and environmental safety regulations have also increased the time period and developmental costs. In India, where the cost of manpower is relatively lower, the developmental cost per successful compound would be far lower than that in the USA, Europe and other developed countries where most of the herbicides are discovered and developed at present.

Thus, the discovery and development of a new herbicide is a prolonged and expensive process with a low success rate. Hence, a company which embarks upon discovery and development of new herbicides takes into account the market potential in various crops in different countries of the world, the global weed problems and economic justifications. It also considers the potential economic and effective life of the product in the patent form, which varies from 12 to 20 yr and the commercial utility of the herbicide in the world market.

Approaches for Herbicide Discovery

There are four major approaches for discovering new herbicides. These include random screening, imitative chemistry, biorational approach and natural product approach [5].

Random Screening

In random screening, also called the empirical method, a large range of chemicals, derived from several sources, are tested against a biological screen. Occasionally, a compound of good activity appears and this can serve as a lead for the chemist. This speculative method has the attraction of a high chance of patentability, counterbalanced, unfortunately, by a low chance of success rate. Currently, 20,000 to 25,000 chemicals are needed to identify one useful product. In random screening, chemicals that show activity on one screen, either herbicidal, plant growth regulation (PGR) or pharmaceutical, can often act on another. For example, imidazolinones, sulfonylureas and glyphosate initially exhibited some PGR activity before being discovered for herbicidal properties. Several first members of the current herbicide families or groups have resulted from this approach. In fact, random screening has contributed about 50% of the current herbicides.

The random screening approach has certain disadvantages. Besides a low chance of success rate, an increasingly large number of chemicals are required to discover a lead chemical and the time taken to discover a new molecule is progressively longer. At the current requirement of 20,000 chemicals to produce one product and at a testing rate of 8,000 chemicals per year, it might take 2-3 y to find a new product [5]. Furthermore, there may be several false leads which require a greater effort to sort them out.

Imitative Chemistry

In this method, compounds related to known active herbicide products are synthesized in the hope of improving upon what has already been developed. The new compounds are structural variants of a proven active chemical family. The chances for success by this approach are very high. About 50% of the current herbicides have been discovered by this method. These include a number of aryloxyphenoxypropionics, sulfonylureas and diphenyl ethers. This method offers a better chance of discovering compounds of high herbicidal activity, but not necessarily novelty products. The principal disadvantage of imitative chemistry is that if many companies are working along parallel lines, there is a high risk of losing the invention as a result of competing patent claims [5].

Natural Products

The development of natural products to discover new herbicides is a relatively new approach. Phosphinothricin (glufosinate) was discovered by this method. A variant of this approach is discovery of mycoherbicides such as DeVine®, COLLEGO®, BIOMAL®, etc. The principal advantages of the natural product method are that their registration is relatively easier than conventional herbicides. The disadvantage is that the products are highly specific to one or fewer weed species.

Biorational Approach

In the biorational approach, also known as the biochemical or target-site directed approach, lethal targets are identified by mode of action studies and the potent inhibitors are designed and synthesized. For example, a number of herbicides inhibit enzymes in amino acid biosynthesis, suggesting that the enzymes in these pathways may be attractive sites. The enzymes include acetolactate synthase (ALS), the target of the sulfonylurea, imidazolinone and triazolopyrimidine sulfonanilide herbicides and 5-enolpyruvyl-shikimate-3-phosphate synthase (EPSPS), the target of glyphosate. After identification of a lethal target, detailed knowledge of the chemical and kinetic mechanism of the enzymes may be used to design potent inhibitors [1]. Various types of potent inhibitors may be designed for a given enzyme. From a practical standpoint, if the enzyme is to guide the herbicide discovery process, it should be readily available for high throughput (output or production) screening and inhibitor design based either on the mechanism or protein structural data or both [1]. In addition, enzyme targets using complex, highly charged or highly lipophilic substrates should be avoided, because inhibitors based on these substrates are not taken up efficiently. Enzymes that fulfil the above criteria offer the best opportunity for successful discovery [1].

The target-sites of other herbicides include lipid biosynthesis, nucleoside synthesis, photosynthetic electron transport, pigment biosynthesis, etc. In the case of PS II electron transport-inhibiting herbicides (ureas, triazines, uracils, etc.), the D1 protein has been identified as the herbicide binding site and hence the target site. Such knowledge is useful in designing new molecules to maximize (perhaps by irreversibly binding to particular amino acids) the interactions and to build a new generation of potentially useful herbicides [4].

The biorational approach has the potential to produce extremely low rate compounds that are less toxic to non-target organisms and are environmentally friendly. The main disadvantage of this approach is the difficulty of identifying truly lethal sites. However, recent advances in molecular biology (e.g., antisense RNA technique, i.e., the production of an RNA transcript from the non-coding, antisense, DNA strand of the gene encoding target protein) now offer new opportunities for the identification of lethal target enzymes [1].

HERBICIDE DEVELOPMENT

Once a product is discovered, its promise as a candidate herbicide is investigated through a lengthy and time-consuming developmental programme. During this stage, bioefficacy of the new compound is tested in different formulations. The herbicide, if found successful, will then be manufactured in a formulation that best exhibits its biological activity and made available for commercial use.

Product Evaluation

A new product, discovered and synthesized by the above-mentioned approaches is subjected to an elaborate evaluation programme.

In the initial stage of a primary screening programme, the new compounds are evaluated on representative weed and crop species planted in containers of soil. The chemicals are sprayed on the soil surface before planting or immediately after planting and on emerged plants. At this stage, the objectives are to determine whether the new compounds have herbicidal activity and if they do, whether the activity is through the soil (preemergence activity) or through the plant foliage (postemergence activity). Visual responses are scored and stored in conventional books or on computer discs. This programme is confined mostly to the greenhouse and laboratory. Usually, the failure rate at the end of this stage is high and no more than 10 to 15% of the compounds examined expected to progress further.

In the second stage of this screening programme, which includes the promising chemicals, the magnitude of their herbicidal activity, range of rates to be used, spectrum of weeds killed, crop tolerance etc. are determined. In the same testing programme or a different one, the activity of these chemicals under different formulations is determined. This testing is done mostly in the greenhouse, laboratory and possibly in the field. At this stage, it is necessary to include, for direct comparison, appropriate standard products, which are not necessary in the initial stage.

The most promising of these chemicals is put under further field-testing before they can be made commercially available. At this secondary screening level, the chemical companies seek and obtain the help of research institutions, universities, testing laboratories and other public agencies to determine efficacy, dosage and time of applications on major target weeds and crops under varying soil and climatic conditions. Alongside, the companies also undertake safety and toxicity tests by generating toxicological data for possible registration with the governmental agency. They also take into account economic considerations to justify development and ultimate manufacture, since a potential herbicide must promise a satisfactory return on investment and effort for the companies and the user (farmer) compared to alternative herbicides or other methods of weed control. As more accurate data on herbicide bioefficacy, cost performance, toxicological tests etc. become available, consideration is given to commercial manufacture.

While generating bioefficacy data, extensive residue and toxicological studies need to be conducted to determine the residues of intact as well as breakdown products of a herbicide in and on edible crop parts as well as in the food chain and toxicological effects on human beings and animals.

Development of the chemical process is an expensive and time-consuming effort. Chemical synthesis to produce thousands of kilograms of a product may be vastly different from methods used to produce a few grams of a herbicide or 5 to 50 kg in a pilot-plant facility. Economic considerations at this stage of herbicidal development are very important for the success or failure in the commercialization of the product. Throughout this herbicidal discovery, research and development programme, formulation improvements are continuously explored, since selection of the optimum formulation can enhance the success of a product.

Formulation

The biological efficacy of a herbicide depends, to a large degree, on the type of formulation it is made into. Formulation is the process by which the technical ingredients are made ready to be used by mixing with liquid and dry diluents by grinding and /or by the addition of emulsifiers, stabilizers and other formulation adjuvants. Formulation of a herbicide involves conversion of a particular technical active ingredient to a commercial product. Development of a formulation depends on the physicochemical properties of the active ingredient to be formulated and of the inert materials to be used in the formulation. Generally, development of a formulation composition precedes development of a herbicide product for commercial use. The formulation technology of one active ingredient cannot be automatically applied to another active ingredient because of differences in physicochemical properties. These differences affect the manufacturing process of a formulated product as well as the physical properties, biological (phytotoxic) properties and its storage stability.

Active Ingredient

Active ingredient (ai) is that part of a formulated product that is principally responsible for herbicidal effects. An active ingredient may be a solid, a liquid or gas. Most technical ingredients cannot be used for control of weeds without further processing into a suitable formulation. It is normally advantageous to make technical active ingredients as pure as possible.

During the synthesis of an active ingredient, very small quantities of by-products derived from the intermediate products used in synthesis or which are formed during synthesis contaminate the active ingredient. In addition, water, metal traces, salts, etc. may also contaminate the technical active ingredient. If these by-product contaminants are allowed with the active ingredients during storage, several undesirable decomposition products may arise, which if present even in small quantities, can cause:

a) reduction of the chemical stability of the active ingredient of the formulation;
b) changes in physical properties of the active ingredient, e.g., reduction of suspendability of wettable powders by flocculation, sedimentation or creaming;
c) impairment of storage stability of emulsifiable concentrates by separation of contaminants present in the colloidal system; and
d) reduction in the grindability of the technical active ingredient because of depression in the melting point, insufficient crystallization and change in the adhesiveness of particles.

Chemical Stability

Chemical stability of the active ingredient in the formulated product is the most important factor considered during the development of a herbicide formulation. Before suitable stabilizers are found, the active ingredient is tested alone and mixed with possible carrier materials for stability. For this, it is important to know the hydrolysis and thermostability of the active ingredient. Suitable packaging can greatly improve the storage stability of a product which is susceptible to high humidity and high temperature.

The chemical stability of an active ingredient is often influenced by reactions between active ingredient and inert materials. In products of powder form or liquids with more than one phase, e.g. aqueous suspension concentrates, such reactions can, under certain circumstances, lead to changes in the physical properties of the formulated product. In a powder formulation, the particles of active ingredient can become hydrophobic when there is a high content of active ingredient, resulting in impairment of the wettability of the product. In the case of liquid formulations like solution concentrates, changes in the physical properties such as suspendability, viscosity, hard-cake formation, etc. may occur due to chemical reactions on the particle surface, particularly during prolonged storage at high temperatures.

Generally, formulations containing low concentrations of active ingredient tend to lose the active ingredient more rapidly during storage than the more concentrated herbicide forms. Some liquid herbicide products tend to develop gas upon deterioration, which may make containers very hazardous to open or lead to explosive rupture of containers.

Particle Size

The particle size of a herbicide formulation affects the biological activity as well as such physical properties as flowability, adhesiveness, deposition on plant surface, suspendability, density, etc. Size of the particle is determined by such factors as biology of the weeds, the chemical nature and the physical properties of the active ingredient (solubility, vapour pressure, systemic behaviour, etc.), the nature of the target (leaf surface, moisture content and soil pH), the quantity and quality of the deposit, the mode of treatment, etc.

Distribution of particle size in a formulation can be determined by several ways.

a) *Microscopic method:* This measures the actual size, irrespective of the shape of particles.
b) *Sieving:* Sieving is done by using a wire cloth, with a calibrated square opening.
c) *Elutriation:* Elutriation is done by using a special kind of centrifuge called the Bacho apparatus. In this method, the particle is subjected to two forces—an air

current and a centrifugal force. Light particles which are carried away by the current are grouped as small. Heavy particles taken away by the centrifugal force are classified as large. By controlling the flow of air current, the size of particle carried away by the air current can be increased.

d) *Sedimentation method:* In this method, the pipette and beaker decantation technique is used.

Types of Formulation

Herbicides are manufactured and supplied in different formulations. Of these, only the most efficient and economical formulations will find wider use than the others. The most important types of formulations are wettable powder, soluble powder, emulsifiable concentrate, solution concentrate, granules and fumigants.

Wettable Powder: A wettable powder (WP) contains finely divided solid particles (particle size less than 3 microns) that can be readily suspended in water to form a suspension. Wettable powders are used if a solid concentrate is preferable to a liquid, or if the solubility of a herbicide is so limited that it is impossible to formulate an economical solution concentrate or emulsifiable concentrate. Suspensions of wettable powders provide a high degree of dispension of the active ingredient of the preparation.

Wettable powders can be divided into three groups: formulations with a high content of the active ingredient 60 to 90%, with a medium content 30-60% and with a low content below 30%. The last group is usually prepared from liquid or waxy compounds, since a high content of such compounds in a wettable powder yields a preparation that cakes easily.

In addition to the active ingredient, a wettable powder usually contains a diluent, a surface-active-agent and an auxiliary material. Sometimes a wetting agent to increase the retention of the preparation on plants or other surfaces is also included.

The diluent will generally have a low bulk density and relatively large sorptive capacity. Silica gel with a bulk density of less than 0.15, hydrated aluminium oxide, synthetic calcium silicate, etc. are used as diluents. Detergents act as surface-active agents and include sulphonates of the alkali metals such as sodium dibutoxydiphenyl sulphonate, sodium dodecylbenzene sulphonate, sodium alkanesulphonate, etc., alkylaryl ethers of polyethylene glycol and polypropyleneglycol and many others. Sodium salts of sulphoacids obtained by sulphonation of petroleum products and also the sodium salts of lignin sulphoacids are used as auxiliary materials. As wetting agents or stickers, carboxy methylcellulose, methylcellulose, polymers of unsaturated alcohols, gelatin, animal glues, caseinates, salts of resin acids, etc. are used.

Wettable powders containing 50 to 80% active ingredient are the most widely used. Talc, kaolinite, attapulgite, bentonite and diatomaceous earth are the natural products often found in wettable powder formulations. Most suspensions of wettable powder formulations require constant agitation in a spray tank to prevent settling of the solid particles. If the solid particle and liquid have identical densities, their rate of separation will be slower.

Wettable powders need to be protected against moisture. Therefore, paper bags are usually constructed of several plys including a polyethylene or aluminium foil liner. Larger quantities are packaged directly into drums containing an inner liner of polyethylene or equivalent material as a moisture barrier.

Soluble Powder: Soluble powders (SP) are similar to wettable powders except that the technical ingredient as well as the diluent(s) and formulating adjuvants used, completely dissolve in water or another liquid for which the wettable powder is formulated. Packaging of soluble powder is basically the same as that for wettable powders.

Emulsifiable Concentrates (EC): An emulsifiable concentrate consists of a herbicide dissolved in an organic solvent, with sufficient emulsifier added to create an oil-in-water (O/W) emulsion. An emulsion is a mixture in which one liquid is suspended in another liquid, like fat globules in milk. In a herbicide emulsion, water (the carrier) is the continuous phase while oil globules (consisting of solvent plus herbicide technical) are dispersed in it. This is termed an O/W emulsion. An emulsifying agent binds the two phases—the water phase and oil phase—together. As there is no direct contact between the two phases, adverse reactions between the chemicals are unlikely to occur. The ECs, upon dilution with water give stable emulsions suitable for spraying plant surfaces. The ECs are made when the active ingredient may not enter the waxy plant foliage due to high water surface tension or evaporation that leaves herbicide salts on the plant foliage. At equal concentrations of the active ingredient, emulsifiable concentrates are usually more effective than the corresponding suspension concentrates. Generally, two types of emulsifiable concentrate—**concentrated emulsions** and **miscible oils**—are made.

Concentrated emulsions are prepared by (mechanical) dispersing an aqueous solution of the herbicide in a water-miscible solvent by means of a colloid mill. Only herbicides which are stable to the action of water are suitable for preparation as concentrated emulsions. The homogenized concentrated emulsions withstand low temperatures and prolonged storage. They may sometimes thicken in storage but a slight stirring will be enough to bring them to normal condition. **Miscible oils** consist of herbicide, solvent and emulsifier. When mixed with water, they yield stable emulsions.

Hydrocarbons and their halogenated derivatives, esters, various petroleum products, creolin, coal-tar oils and many other compounds may be used as solvents. Emulsifiers act as surface-active agents, which readily emulsify the concentrate in the final spray liquid. Calcium sulphonates, ethers of polyethylene and polypropylene glycols, monoesters of sorbitol and mannitol with the higher fatty acids, various soaps, salts of naphthenic acids, etc. can be used as emulsifiers. Sometimes a mixture of two or more emulsifiers is used to get better results.

Most emulsifiable concentrates consist of 60 to 65% (W/W) herbicide dissolved in 30-35% organic solvent, with 3-7% of an appropriate emulsifying agent to create an O/W emulsion when the concentrate is added to water. ECs, when added to water, form opaque or milky solutions. These can be applied in hard water without adverse reactions and are less liable to be washed off the foliage by rain or overhead irrigation. Herbicide solutions made from ECs evaporate more slowly from plant surfaces. The solvent used in EC may sometimes be phytotoxic, aiding in herbicide activity.

Emulsifiable concentrate formulations are usually packaged in small (1 to 5 litres) metal cans or glass bottles or in light-head steel pails or steel drums from 10 to 200 L, capacity.

Solution Concentrate: A solution concentrate (SC), also known as soluble concentrate, consists a herbicide dissolved in a solvent system to provide a concentrate that is soluble in the carrier. The most common formulations of this type are water concentrates designed for use in water carriers. The salts of the following herbicides in water: amine salts of 2,4-D and MCPA, amine salts of DNBP, isopropylamine salt of glyphosate etc. These can be dissolved in water and sprayed. Some types of solution concentrates are oil-soluble. Herbicides such as dinoseb (DNBP), ester formulations of 2,4-D etc. are soluble in oil. These oil-soluble solution concentrates are preparations of various organic solvents.

The basic requirement for development of a solution concentrate is solubility-dependence. The herbicides must be soluble in a small enough quantity of concentrate solvent to make packaging and shipment of the formulation economical. The concentrate must withstand winter storage temperatures without freezing. It must also be completely and rapidly soluble in the carrier at all temperatures and concentrate-carrier ratios that might occur in the field.

Flowable Concentrate: Flowable concentrates (FC) are the liquid extension of wettable powders (WP). They are concentrated aqueous dispersions of herbicides that are insoluble or nearly so in water [10]. They contain little or no organic solvent but do include clays similar to those used in wettable powders for bulkiness, some oil, water, an emulsifying agent and a suspending agent. The FCs are easily dispersed in water and easy to measure. However, they are more difficult to make but can be used in lieu of WP formulations. Some times the FCs can gel during storage and become unusable or the system can solidify and a portion of the solid material rise to the surface as an oil.

Invert Emulsion: An invert emulsion (IE) is a water-in-oil emulsion, with oil constituting a continuous phase and water a dispersed phase. The primary advantage of IEs is reduction in drift, because emulsion is more viscous and produces large drops. This formulation is used in phenoxyalkanoic herbicides for application in industrial sites and rangelands and also in the case of bioherbicides.

Granules: Granules (GR) contain 1 to 20% concentrations of herbicide in particulate solid carriers suitable for direct field application. In a granular formulation, the technical active ingredient is mixed with, or coated on, inert carrier material of the approximate particle size of granulated sugar.

Many materials are available in granular form as carriers. These carriers are grouped in terms of their liquid-retention capacity. Pearlite, expanded vermiculite and diatomaceous earth are in the high range, possessing a liquid:solid ratio of 1:1. Clay minerals such as bentonite, illite, kaolinite and attapulgite have a liquid:solid ratio of 1:2. Expanded shales, pyrophillite and finely ground plant parts have liquid-retention capabilities ranging from 1:10 to 1:20. Limestone, gypsum, sand and fertilizer materials are also used as carriers, although their liquid loading capacities are less than 5%. Most of the carriers disperse into water after various periods of exposure. Dispersion may be prevented by calcining to temperatures of 30°C and higher. Calcining may produce harder granules that are more resistant to dust formation.

Uniformity of particle size is necessary to assure proper application and distribution of granular herbicides. The granular formulation of herbicides is applied directly to soil and the granules are dispensed close to the soil surface. The granules tend to remain where they are deposited. There is no drift problem. The granular formulation is advantageous, for it does not require a liquid carrier and causes little foliage injury, as the granules tend to fall off the leaves of crop plants without injuring them during application. A granular formulation, however, is more expensive to make, package and ship than spray concentrates or wettable powders. Besides, some herbicides have low mobility in soil and do not perform as well in granular formulation as in liquid concentrate and powder formulation.

Herbicide movement from granules in soil is affected by soil and environmental conditions. Erbach et al. [3] reported that environmental conditions, weed/herbicide combination, depth of granule placement, depth of germinating weed seed and interactions of these factors affected the size of the area in which weeds were adequately controlled by an individual herbicide granule. The radius of the area (area-of-influence) controlled by a granule ranged from 0 to 1.5 cm for atrazine with *Abutilon theophrasti* (velvetleaf) and from 1.5 to 4.0 cm for alachlor with *Setaria italica* (foxtail millet). They concluded that the number of granules per unit area and, therefore, the herbicide rate needed depended on the area each granule controlled.

A granular formulation is more appropriate in annual crops grown under irrigation or high rainfall conditions where liquid formulations are liable to greater surface run-off losses. In transplanted irrigated rice, butachlor (5% granules or 5G) and 2,4-D ethyl ester (4G) are widely used, either alone or in combination, in India [7] and other rice-growing countries. Other herbicides used as granular formulation in rice include anilophos and thiobencarb.

Encapsulated Formulation: Encapsulated formulation, also called controlled-release formulation, encloses the herbicide in microscopic, porous polymer particles that release the active ingredient slowly. They offer the advantage of timed release through a longer period of the crop season. Herbicide release from microcapsules is dependent on the encapsulating material, microcapsule environment and herbicide properties. The microencapsulated (ME) materials include water-soluble resins, waxes, lipids, gelatin, albumin, the starch and polyurea-polyamide matrices. Of these, the starch matrix is widely used.

The basic starch-encapsulation process includes: a) dispersion of the starch in water, b) gelatinization of the starch with heat, c) addition of the herbicide and d) cross-linkage of the starch or natural retrogradation [2]. The resultant encapsulated material is dried, ground and sieved to produce various sized particles (2 to 15 M). Following application, the herbicide is released into the soil matrix by diffusion after wetting and/or microbial and physical degradation of the starch [8, 9]. In the case of soil-applied herbicides such as alachlor, metolachlor and atrazine, starch-encapsulated (SE) formulation provides weed control similar to commercial formulations over a wide range of weed species and densities and soil types [2].

The herbicide is released from microcapsules by dissolution of the encapsulating material, swelling and bursting, drying and cracking, or diffusion. Diffusion was reported to be the primary mode of alachlor release, with diffusion rate controlled by herbicide and salt concentration gradients between the microcapsules and the surrounding aqueous solution [6].

Encapsulated formulations have the potential to reduce herbicide leaching by maintaining the threshold concentration of the active ingredient in the soil for weed control, while the unreleased, encapsulated herbicide remains unavailable for leaching and degradation [2]. Due to controlled release of active ingredient, the period of weed control is extended and the potential for residue carryover is reduced.

Fumigants: Fumigants are volatile chemicals applied in confined spaces or in the soil to produce a gas that will destroy weed seeds and act as a soil sterilant. They are often injected as pressurized liquids a few inches below the soil surface. The vapours penetrate the soil zone and act as a soil sterilant. Gradually, the fumigant decomposes or dissipates from the soil, making it possible to grow crops. Fumigants are good for killing deep-rooted perennial weeds and for eliminating weed seeds from the soil. Their effect is highly dependent on soil moisture, soil temperature, soil compaction and soil texture as these affect the movement and ultimate distribution of vapours.

The herbicides commonly available as fumigants are methyl bromide, metham, allyl alcohol, carbon disulphide, chloropicrin and tetrachloroethane. Fumigants, which develop pressure at ambient temperatures, have to be packaged in special, pressure-resistant containers.

HERBICIDE PACKAGING

All herbicide products are packaged into suitable and clean containers which do not affect the product nor affected by the product contained. Normally, the container should adequately protect the product from exterior influences including moisture, temperature, vaporization and contamination. The containers should be lined with a suitable material to prevent corrosion or deterioration of contents and container.

Dry herbicide formulations are packaged in a suitable bag, box, or fibre or metal drum, adequate to protect the product from external influences including compaction, moisture, oxidation, vaporization and contamination. Containers provide an adequate moisture barrier by an inner bag or liner of polyethylene of sufficient thickness. The inner liner or bag is carefully sealed after filling.

The herbicide product, which consists of both container and contents, must be sufficiently stable to withstand the rigours of handling, loading, stacking, unloading and meet all product specifications for a period of at least two years.

REFERENCES

1. Abell, L.M. 1996. Biochemical approaches to herbicide discovery: Advances in enzyme target identification and inhibitor design. Weed Sci. 44: 734-742.
2. Buhler, D.D., M.M. Schreiber and W.C. Koskinen. 1994. Weed control with starch-encapsulated alachlor, metolachlor and atrazine. Weed Technol. 8: 277-284.
3. Erbach, D.C., W.G. Lovely and C.W. Bokhop. 1976. Area of influence of herbicide granules. Weed Sci. 24: 170-174.
4. Huppatz, J.L. 1996. Quantifying the inhibitor-target site interactions of photosystem II herbicides. Weed Sci. 44: 743-748.
5. Parry, K.P. 1989. Herbicide use and invention. In A.D. Dodge, (ed.). Herbicides and Plant Metabolism. Cambridge University Press, Cambridge, U.K., pp. 1-20.
6. Peterson, B.B. and P.J. Shea. 1989. Microencapsulated alachlor and its behaviour on wheat (*Triticum aestivum*) straw. Weed Sci. 37: 719-723.
7. Rao, V.S. 1995. Broad spectrum weed control by herbicide mixtures in rice and wheat in India. Weed Sci. Soc. Amer. Meeting, Seattle, Washington, Feb. 1995.
8. Schreiber, M.M., B.S. Shasha, D. Trimnell and M.D. White. 1987. Controlled release herbicides. In C.G. McWhorter and M.R. Gebhardt, (eds.). Methods of Applying Herbicides. Weed Sci. Soc. Amer. Monograph 4, Champaign, IL, USA, pp. 171-191.
9. Wienhold, B.J. and T.J. Gish. 1992. Effect of water potential , temperature and soil microbial activity on release of starch encapsulated atrazine and alachlor. J. Environ. Qual. 21: 382-386.
10. Zimdahl, R.L. 1993. Fundamentals of Weed Science. Academic Press, Inc., New York.

14

Herbicide Application

Generally, herbicides are applied in the form of solution or granules. Of the two, spraying solutions of herbicides formulated as wettable powders, soluble powders and emulsifiable and solution concentrates is more common. When herbicides are formulated as granular or encapsulated materials, they are applied by hand or with the help of a granular applicator.

HERBICIDE SPRAYING

A spray is defined as a liquid discharged into particles and scattered as dispersed droplets. Herbicides are mixed with a suitable carrier to facilitate distribution and even coverage over the area to be treated. Water is the most common carrier for herbicide spraying. In a spray solution, the herbicide is dissolved, emulsified or suspended in water. Herbicide application is directed to the soil in the case of (PPL) preplanting and preemergenc (PRE) herbicides and to the foliage in the case of postemergence (POST) herbicides.

Spraying Equipment

The basic components of the sprayer include the tank to hold the spray solution, the nozzle through which the spray solution is discharged, the pump for generating the necessary pressure, a filter or strainer, a pressure regulator and a hose.

Tanks

Tanks are constructed of brass, galvanized iron, stainless steel, fibreglass or glass-reinforced plastic (FRP). Of theses, the stainless steel and FRP tanks are resistant to corrosion by all types of agricultural chemicals. Tanks with resin-bound glass fibres or lined with inert plastic materials resist corrosion. Tanks vary in capacity, ranging from one litre, or even less in the case of a hand-atomizing sprayer, to several hundred litres in the case of power-driven tractor-mounted sprayers. The hand-operated backpack sprayers have a tank capacity ranging from 5 to 15 L sufficient to discharge in 15 to 30 min depending on the nozzle type. All tanks should have a large inlet to allow easy loading and periodic inspection of the inside.

Sprayers have agitators to prevent separation and settling of emulsions or suspensions. There are two types of agitation: mechanical and hydraulic. Mechanical agitation is achieved by a series of paddles on a shaft that runs horizontally through the tank or by means of a propeller at one end of the tank. The paddles or propeller are power-driven at a low speed. Mechanical agitation is generally used for emulsions with a high percentage of oil and for wettable powders. Hydraulic agitation is achieved by routing some of the pressurized spray liquid back into the spray tank. Hydraulic agitation is most often used for readily soluble or self-emulsifying materials. Generally, mechanical agitation is more efficient in terms of power requirement and uniformity in dispersion.

Nozzles

The nozzle is the most important component of the sprayer, as it influences uniformity in spraying, rate of herbicide application, spray drift, spray pattern and droplet size. It breaks the pressurized spray liquid into droplets or application to the target. The function of the nozzle is to accelerate and disintegrate the flow of fluid spray passing through it into droplets to form a spray [19].

A nozzle consists of a body, cap, lip or disc, and core [18]. In some cases the nozzle body is incorporated with a lance or boom. A nozzle filter is sometimes used with the nozzle to prevent blocking of the nozzle orifice by dust and other foreign particles.

Different types of nozzles (Fig. 14.1) are available to produce varying spray patterns and droplet sizes. In the cone type nozzle, the liquid is forced through one or more tangential or helical passages into a swirl chamber through which the liquid passes to a circular orifice at a high rotatory velocity to form an air core within the orifice and swirl chamber [18]. The liquid emerges from the orifice as a hollow cone. In the case of a solid cone nozzle, the liquid also passes centrally through the nozzle to fill the air core. Different discharge rates and cone angles can be obtained by changing the disc and core [18]. The hollow cone nozzles produce the heaviest droplet distribution on the edges of the pattern. Cone nozzles are widely used for insecticide and fungicide applications as also for spot application of POST herbicides.

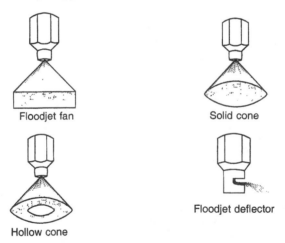

Floodjet fan   Solid cone

Hollow cone   Floodjet deflector

Fig. 14.1. Different types of nozzles for herbicide spraying.

Herbicide spraying is mostly done with fan and flooding or deflector type nozzles as they produce a more even distribution of spray and uniform coverage than the cone-spray types. In the case of a fan nozzle, the tip has a lenticular or rectangular orifice behind which two streams of liquid meet because of the shape of the bore [18]. Fan nozzles are ideal for spraying flat surfaces and for band spraying in row crops. In the flooding nozzle, also known as deflector, anvil or impact nozzle, a jet of liquid passes through a relatively large orifice and strikes a smooth surface at a high angle of incidence to form a fan-shaped spray pattern [18]. The flooding-type nozzles deliver coarse droplets of fluid spray under pressure, thus minimizing drift. They also produce a wide-angle fan pattern of spray. As the nozzle orifices are large, clogging is considerably reduced.

The rate of spray delivery through a nozzle is determined by the diameter of the orifice and the delivery pressure (pounds per square inch, **psi**, or grams per square centimetre). The delivery rates are usually based on water as the liquid, and they differ from those of the actual formulations of the herbicides to be sprayed in the field. The delivery pressure also changes the spray pattern (Fig. 14.2). When nozzles become worn, they deliver excessive amounts of spray and larger droplets.

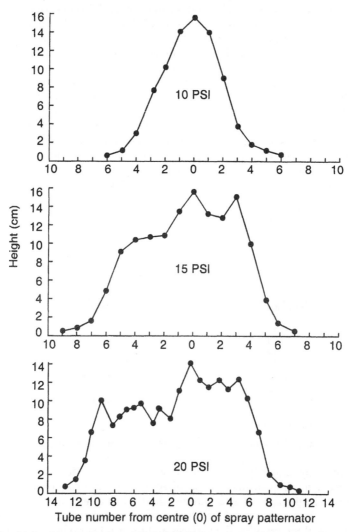

Fig. 14.2. Spray patterns of floodjet fan nozzle at different delivery pressures.

When the delivery pressure is low, the liquid escapes from the nozzle's tip as a liquid film which, on expanding, forms droplets at the outer edge. These droplets will be smaller in size if the liquid has low surface tension. At high pressure, the droplets escape in the form or a mist or fog, creating a drift hazard. For proper spraying using floodjet nozzles, the hand-operated backpack sprayer should maintain a pressure of 10 to 15 psi (700 to 1050 grams per square centimetre).

Nozzles are made of brass, aluminium, stainless steel or plastic, and nozzle orifices or tips are made of plastic and hard materials such as brass, stainless steel or ceramics. Because of rapid wear, materials harder then brass should be used for application of chemicals which are corrosive or abrasive. Repeated cleaning of the nozzle orifice with a wire, knife or other hard objects may widen the pore and damage the nozzle. Thorough rinsing of the nozzle with water or use of a bristle brush or wooden match may loosen and remove the obstruction.

Sprayers may have more than one nozzle fitted onto the lance. For treating the soil and small plants in narrow-spaced rows, the nozzle should be centered over the row pointing straight down. For larger plants and in widely spaced rows, two nozzles per row, one for each side of the row, may be used. Normally, the nozzles are best spaced from 30 to 60 cm apart, the exact distance depending on whether single or double boom coverage is desired and on how high the boom is to be carried. If carried high, the wind is likely to interfere with the spray pattern and also increase the drift. If carried low, the end of the boom may be very close to the ground on rough land and the tops of all weeds may not be covered.

Pumps

Pumps are devices that deliver fluids under pressure. The pressure provides the energy that permits the spray liquid to be broken into particles by the nozzles and determines the rate of flow through the nozzle orifice. An increase or decrease in pressure affects the size of spray droplets; increasing the pressure produces a finer spray with potential drift hazard.

Pumps are operated by gas or air pressure and liquid pressure. In gas pressure, the entire system is under pressure. The spray tank is pressurized to force the liquid from the tank through the nozzles. In this system, liquid air, liquid CO_2 or gases such as ammonia can be released through regulating valves to provide gas pressure for the sprayer. In liquid pressure systems, the spray tank is not pressurized. Part of the liquid is by-passed back to the tank for agitation and for pressure control.

The principal types of pumps used in spraying machines are piston pump, centrifugal or turbine pump, gear pump, impeller pump and diaphragm pump.

Piston pumps: Piston pumps are the most versatile of all pumps and were used prior to 1945 in most sprayers. They deliver a wide range of pressures from 40 to 1000 psi making them adaptable for chemicals other than herbicides. They are normally reliable and long-life pumps. They are easy to repair and resistant to wear from abrasive materials such as wettable powders. Each stroke of the pump delivers a given amount of spray liquid. Therefore, the delivery of the pump is directly proportional to the speed. Piston pumps are disadvantageous because of high initial cost and low delivery rate.

Centrifugal or turbine pumps: These develop pressure through the centrifugal force of rapidly rotating blades, fins or discs. The velocity of the liquid, combined with its weight, gives it pressure. The liquid is released at the outer edge of the pump housing. Centrifugal pumps are inexpensive and handle abrasive materials well. They provide a high delivery rate with an adequate supply for hydraulic agitation in the spray tank.

To maintain adequate pressure, centrifugal pumps must be operated at higher speed than other pumps. They are used mainly for low pressure sprayers.

Gear pumps: Gear pumps develop both vacuum and pressure through a meshing of gear teeth. Their volume varies with size but is usually adequate for hydraulic agitation. They are used with advantage for oil emulsions and other non-abrasive materials but have a limited life under adverse usage. However, use of these pumps has dwindled except in the case of low-volume sprayers.

Impeller pumps: These consist of a rotor set on one side within the pump housing. The rotor maintains contact with the outer pump housing through flexible rubber vanes or rollers that adjust to the outer wall of the housing. The space between the rotor and pump housing expands for every half revolution and contracts for the other half, thus creating alternate vacuum and pressure. The roller type of impeller pump is extremely popular. Rollers are made of nylon or hard rubber and work best with oil emulsions and non-abrasive materials. Hard rubber rollers are most satisfactory with abrasive materials. The rollers are held to the outer wall by centrifugal force.

Diaphragm pumps: These pumps produce moderately high pressures and have the advantage over a piston pump in that the spray fluid does not come into contact with the reciprocating parts. The chemical touches only the valves, the diaphragm and pump housing. Their performance is limited by the volume which can be displaced by the diaphragm and the quality of the component. Diaphragm pumps handle sprays which are both chemically and abrasively corrosive. They give long trouble-free service with minimum upkeep.

Pump Capacity

Pump capacity can be determined as follows:

$$\frac{\text{Pump capacity}}{(\text{L min}^{-1})} = \frac{\begin{array}{ccc} \text{Spray vol.} \times \text{swath width} \times \text{walking speed} \times \text{metres per km} \\ (\text{L ha}^{-1}) \quad\quad (\text{metres}) \quad\quad\quad (\text{km h}^{-1}) \end{array}}{\text{metres ha}^{-1} \times \text{min h}^{-1}}$$

If spraying is done at a spray volume of 500 litres per hectare while covering a swath of 1 metre wide and maintaining a walking speed of 1 km per hour, the pump capacity will be 0.83 litre per minute as shown below:

$$\frac{500 \times 1.0 \times 1 \times 1000}{10000 \times 60} = 0.83 \text{ litre per minute.}$$

When walking speed is increased to 2 and 3 km h^{-1} the pump capacity is 1.67 and 2.50 L ha^{-1} respectively. The required discharge is also proportionate to spray volume and spray width.

Filters and strainers

Generally filters or strainers are built as a part of the nozzle, placed on the inlet of the spray tank or installed as a line filter. In filling the tank, a coarse strainer also should be used to filter the spray fluid. This is essential if the water contains trash or dirt. Wettable powders rarely pass through screens finer than 50 mesh. Screens for emulsions should be about 100 mesh.

Pressure Regulators

A pressure regulator or relief valve is used to maintain a relatively constant desired pressure. The valves generally consist of a ball pressurized by a spring, which is screw-adjusted to vary the pressure. Variations include a double-spring type and a combination of spring and diaphragm. Excess liquid is released back into the spray tank through a bypass port. A pressure gauge is usually placed on the outlet side of the regulator to indicate nozzle pressure.

Spray Boom

A boom is a horizontal pipe to which the nozzles are fitted at the required spacing. Its length varies from 1 to 10 m, depending on the sprayer used. For hand-operated sprayers, a boom is normally fitted with 2 or 3 nozzles. More area is covered with a boom in one trip because of its wider swath.

Hoses

Hoses should be of a material that will withstand the chemicals used, particularly the petroleum solvent. Synthetic rubber and plastic are the materials most commonly used in spray hoses.

Spray Lance

Made of brass with a minimum diameter of 6 mm, The normal length of the lance of a hand-operated knapsack sprayer is 1 m. One end of the spray lance is fitted with a hose and the other with a nozzle. The nozzle end is normally bent to form a goose neck. A spray boom containing more than one nozzle can be fitted to the lance either horizontally or vertically. At the joint of the hose and lance, a trigger mechanism is fitted to shut off flow of the liquid. When necessary, a spray shield can be fitted at the delivery end of the lance.

Spray Calibration

Accurate calibration of spray equipment is essential if herbicides are to be used safely and effectively. A sublethal does of spray will fail to give satisfactory control of weeds. An overdose will increase costs besides killing a crop or resulting in accumulation of toxic residues in the soil. Achievement of a desired application rate is dependent on selection of the right combination of nozzle type, size of nozzle orifice, spraying pressure and spraying speed. The best way to calibrate is to spray an area of known size and measure the volume of spray delivered from the tank. This may be done using different nozzles at different spray pressure and walking speeds. Care must be taken to see that speed and pressure do not vary significantly from the calibration test to actual field spraying.

The area covered per hour can be calculated as shown under:

$$\text{Area (hectare)/h} = \frac{\text{Walking speed (kmh}^{-1}) \times \text{m km}^{-1} \times \text{spray width (m)}}{\text{m ha}^{-1}}$$

For example, if a person is walking at 1 km h^{-1} covering a swath of 0.6 m wide, the area covered ha is

$$\frac{1 \times 1000 \times 0.6}{10000} = 0.06 \text{ ha h}^{-1}$$

At this rate, it will take 16 h 40 min to cover an area of one hectare. At a spray discharge rate of 30 L ha^{-1}, it would require a spray volume of 500 L ha^{-1}.

The rate of spray applied can be determined by the following equation.

$$\text{L ha}^{-1} = \frac{(\text{L min}^{-1}) \times 6{,}000}{(\text{km ha}^{-1}) \times \text{W}}, \text{ where}$$

L min^{-1} : discharge per nozzle, in litres per minute

km h^{-1} : walking or ground speed in kilometres per hour
W   : spray width in centimetres
6,000   : a constant to convert litres per minute, kilometres per hour, and centimetres to litres per hectare

If more than one nozzle is used, spraying is done in bands or in crop rows. Then, spray width (W) can be calibrated as shown below.

$$W = \frac{\text{Row spacing (or band width)}}{\text{No. of nozzles per row (or band)}}$$

Sprayers

Sprayers are of two types: hand-operated sprayers and power-driven sprayers. Depending on the spray volume required to cover a unit area, sprayers are classified as high volume, low volume and ultralow volume (ULV).

Hand-operated Sprayers

Knapsack Sprayers

Of the hand-operated types of sprayers, the hydraulic knapsack sprayer with a capacity of 5 to 15 L is the most useful and widely used sprayer for herbicide application in developing countries. It can be carried on the back with the help of straps. It is provided with a double action lever-operated pump mounted either inside or outside the container. The operator operates the lever with his left hand and pumps the spray liquid into the lance held by his right hand. The spray liquid is drawn from the container by a suction stroke through a non-return valve and then forced through a delivery valve into the pressure chamber by a return stroke [18]. The air in the pressure chamber is then compressed due to the liquid charged by the pump. When the cut-off valve or trigger valve is operated the air which was pressurized in the pressure chamber forces the liquid out through the discharge hose and nozzle. The liquid gets delivered through the nozzles as fine spray. pumping needs to be done almost continuously to maintain a fairly uniform pressure. The sprayer can be fitted with different types of nozzles of varying delivery rates. The hydraulic knapsack sprayer is very simple, convenient and efficient.

The pneumatic or compressed system knapsack sprayer does not require pumping during spraying. The tank is pressurized after filling it with liquid with a built-in pump, a separate charge pump or CO_2 cylinders. The main disadvantage of this sprayer is that uniform delivery pressure cannot be maintained, resulting in an uneven delivery. It is more useful to spray on hilly terrain and irrigated fields where walking is difficult.

Logarithmic Sprayer

A logarithmic sprayer begins spraying at a high rate and decreases the concentration logarithmically. The sprayer has two tanks which are connected through tubing and valves. One tank is filled with herbicide and the other with water. The water flows from the diluter tank into the herbicide tank and form this tank through the nozzles. Initially, the liquid passes through the nozzles at maximum concentration. As the liquid in the herbicide tank gets diluted logarithmically with water from the diluter tank, lower and lower concentrations are sprayed.

The logarithmic sprayer is very useful for experimental work in herbicide evaluation to establish crop and weed tolerance to a chemical at varying rates.

Power-driven Sprayers

The most common type of power equipment for spraying herbicides is that which uses a hydraulic pump of plunger, rotary or centrifugal type. The fluid is forced through nozzles which break it up into droplets and direct the droplets at the surface to be covered.

Power-driven sprayers are usually mounted on a tractor on the three-point linkage. Tractor sprayers are equipped with a spray boom although sometimes clusters of nozzles are used. The spray from a boom is less affected by wind than that from the nozzles cluster, but unevenness of the ground can be more serious as the movement is magnified along the length of the boom. This can be minimized by a good design and by using full cone jets rather than hollow cone or fan jets.

Spray Droplet Size and Density

Spray Droplet

The spraying equipment used for applying herbicides atomizes the spray liquid into droplets of unequal size. This droplet spectrum can be measured to determine the amount of chemical used. The most widely used parameters of droplet size are (i) the volume median diameter (VMD) and (ii) the number median diameter (NMD) [19]. The unit of measurement of diameter of droplet size is the micron (μ).

The volume median diameter is the diameter of the droplet which divides the volume of the spray into two equal halves. A representative samples of droplets of a spectrum is divided into two equal parts by volume so that one-half of the volume contains droplets smaller than a droplet whose diameter is the VMD and the other half of the volume contains larger droplets. Figure 14.3A shows that 50% of the volume of

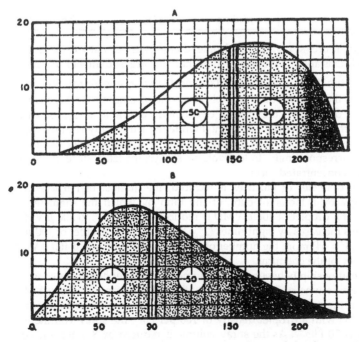

Fig. 14.3A. Volume median diameter (VMD) showing volume of droplets against droplet range in droplet spectrum and (B): Number median diameter (NMD) showing number of droplets against droplet range in droplet spectrum [19].

the spray is through particles smaller than 150 μ, while the rest is larger than 150 μ. Hence the VMD is 150 μ. In this, larger droplet can account for a large proportion of the spray which may result in higher VMD value.

The number median diameter is the diameter of a droplet which divides the number of droplets into two equal halves. Thus NMD is the average diameter of droplets without reference to their volume. In Fig. 14.3B, 50% of the total droplets are less than 90 μ in size, while the remaining 50% of the droplets are larger than 90 μ. Hence the NMD is 90 μ. In this method, the value of NMD depends on the number of small droplets.

Thus the values of VMD and NMD are affected by the extent of large and small droplets in a spray droplet spectrum respectively. The ratio between these parameters can give the extent of variation in droplet size. If the droplet size is uniform, the ratio will be nearer to 1 (Table 14.1). The slope of the curves presented in Fig. 14.3A and B gives an indication of the relative droplet spectrum. The steeper the curve, the narrower the droplet spectrum and hence the greater the uniformity in droplet size. A vertical curve could mean that all the droplets are of about the same size.

Table 14.1. VMD and NMD measurements of several hand-operated sprayers (adapted from Coffee, 1981 [3])

Sprayer type	Spray liquid	VMD	NMD	$\frac{VMD}{NMD}$ ratio
Knapsack (155 psi)				
(hydraulic)	Water	310.0	60.0	5.63
Herbi (1.0 ml s^{-1})	Water	230.0	195.0	1.18
Electrodyn				
(commercial)	Cypermethrin	44.0	37.6	1.17

In agricultural spraying, the droplet diameter is normally expressed as VMD. Generally, VMD is about 2.2 times smaller than the largest diameter in a droplet spectrum.

Relationship between Droplet size and Density, and Spray volume

Besides the rate and time of application, the bioefficacy of a herbicide is determined by droplet density, i.e. the number of droplets per unit area (e.g., sq cm) on the target. If the bioefficacy has to be maintained while reducing spray volume, the droplet density must be maintained as in the higher volume but the droplet size (VMD) should be reduced correspondingly. In low volume sprays, the lethal dose in the droplets will be in a more concentrated form.

As the diameter of particles is reduced, the number of particles increases correspondingly by a factor of cube of the ratio by which the diameter is reduce. For example, if a droplet of 400 μ VMD is reduced by half, i.e., to 200 μ VMD, the number of droplets increases by eight times ($2^3 = 8$). Similarly, a droplet of 100 μ VMD requires 64 times ($4^3 = 64$) more the number of particles for a droplets of 400 μ VMD to maintain the efficacy of the chemical.

The relationship between spray volume and droplet density is shown in Table 14.2. At a spray volume of 333.56 L ha^{-1} with spray droplets of 400 μ VMD, the droplet density in the spray would be 100 per cm^2. If one would like to maintain droplets of 200 μ VMD with a droplet density of 100 per cm^2, the spray volume is reduced by 8 times to 41.70 Lha^{-1}. As the spray volume is reduced by 8 times, the concentration of the spray liquid should be increased in the same proportion to maintain the efficacy of the chemical. This means the droplets of 200 μ should be concentrated 8 times more than those of 400 μ.

Table 14.2. Relationship between droplet density, droplet size, and spray volume on a horizontal surface (assuming complete recovery of spray)

Droplet size (microns)	No. of droplets per sq cm at applic. rate 1 L ha^{-1}	Spray volume L required to spray 1 ha at different droplet densities (no./sq cm)					
		25	40	50	80	100	200
40	298.57	0.083	0.134	0.167	0.267	0.334	0.667
50	152.88	0.163	0.261	0.326	0.521	0.651	1.303
75	45.44	0.555	0.887	1.109	1.775	2.219	4.437
100	19.11	1.303	2.085	2.606	4.170	5.212	10.424
150	5.70	4.437	7.100	8.875	14.200	17.750	35.500
200	2.39	10.425	16.680	20.850	33.360	41.700	83.400
250	1.22	20.362	32.580	40.725	65.160	81.450	162.900
300	0.71	35.504	56.806	71.008	113.602	142.015	284.030
400	0.30	83.390	133.440	166.780	266.850	333.560	667.120

Concentrated herbicides are used in the case of low volume and ultralow volume sprayers which atomize the spray liquid into a greater number of smaller size particles than the conventional high-volume knapsack sprayers.

Spray Precaution

Spray Drift

Spray drift usually occurs when the droplets are blown by the wind during spraying. This can be minimized by (a) spraying when no wind is blowing, (b) using suitable sprayer, (c) lowering the spray height from the ground, (d) using proper nozzles and (e) reducing the delivery pressure.

At high delivery pressure, the droplet size becomes smaller and the smaller droplets are readily carried away by wind. Suitable pressure-regulating or pressure-control valves are useful in maintaining a constant pressure at the outlet regardless of the inlet pressure. Pressure valves need frequent checking. Low-drift jets producing very few small droplets help in reducing the drift.

In order to prevent drift completely, spray shields are very useful. The shield must match the nozzle to ensure that the entire shielded area is sprayed. The shield is, however, disadvantageous because the operator cannot see the spray jet to direct it properly, and also dripping from the edges causes wastage of spray solution.

Spray Mixing

The effect of herbicide at a particular rate of application depends largely upon mixing the correct quantities of herbicide and water. Therefore, one must determine the quantity of spray volume required per unit area. This can be done by spraying a measured quantity of water on a known area and calibrating the actual quantity required for 100 metres, 1000 metres or one hectare. This is done as shown under.

$$\text{Volume applied per hectare} = \frac{\text{Water used (litres)} \times 10,000}{\text{Distance covered} \times \text{width of spray swath}}$$
$$\text{(metres)} \qquad \text{(metres)}$$

For example, if 80 meters were covered while spraying solution of 3.6 litres with a swath width of 1 metre, the volume applied per hectare would be:

$$\frac{3.6 \times 10,000}{80 \times 1} = \frac{36,000}{80} = 450 \text{ litres}$$

Once the spray volume rate is determined, one must calculate the exact amount of commercial formulation of a herbicide. The formulations contain less than 100% of the active ingredient (ai) and as the rates are usually quoted in terms of active ingredient the formulated product required is calculated thus:

$$\frac{\text{Rate of ai/ha to be applied} \times 100}{\text{Active ingredient} (\%)}$$

For example, if one needs to apply diuron, at 2 kg ai/ha and the commercial formulation contains 80% active ingredient, the total product required per hectare would be:

$$\frac{2 \times 100}{80} = 2.5 \text{ kg}$$

For some compounds, the activity is expressed on the basis of per cent concentration of the parent acid in the formulated product. In the case of Gramoxone, a commercial formulation of paraquat dichloride salt, the active ingredient in the form of paraquat dichloride is 24% and in the form of parquat ion 20%.

The wettable powders and emulsifiable concentrates should be initially mixed with an equal amount of water to ensure complete wetting and then brought to the required concentration by adding the remaining quantity of water. Perfectly clean water should be used for herbicide spraying.

Spraying Procedure

Hand-operated sprayers

In the case of large-scale field spraying with hand-operated knapsack sprayers, spray mixing should be done in drums or troughs of 50, 100 or 200 L capacity located at a convenient or central place in the area to be sprayed. The standard spray tank has 10 to 15 L capacity and it should be filled with one litre less than the capacity. Before transferring the spray solution to the spray tank, the spray mix should be stirred with a paddle to prevent settling at the bottom of the mixing tank.

After filling the spray tank, the sprayer is fixed on the back of the operator with the help of straps. The operator pumps the liquid continuously into the lance by stroking the handle with his left hand. He holds the lance with his right hand and releases the solution at the required pressure indicated on the pressure regulator fixed on the lance. The spray delivery height is usually maintained at 25 to 40 cm from the ground depending on the nozzle, delivery pressure, height of weed growth and row spacing. The spray swath should be narrower in crops with narrower row spacing. This can be achieved by using a low volume nozzle and/or by reducing the delivery height. The operator should walk forward in the centre of the row and cover the entire row space as far as possible with one sweep itself. The walking speed should be determined beforehand by spraying water on a measured area at the volume rate required.

When the spray solution runs out the operator walks back to the mixing tank, fills up the tank and resumes spraying. During spraying, constant agitation of the tank is essential to prevent the herbicide settling at the bottom.

The nozzle should be checked after each filling. A blocked nozzle should be cleaned with water and a brush. If the nozzle opening is damaged it will produce a bad jet resulting in disruption of spray pattern, drift problem and wastage of spray solution. In such cases, it should be replaced by a new nozzle. As the nozzle forms only a fraction of the total application cost, regular replacement is advisable. Nozzle, particularly of brass, do not last forever and hence needs replacement at least once a year for every 200 h for spraying. Spraying should be avoided when gusty wind is blowing.

Power-driven Sprayers

When a herbicide is applied to the soil by a power-driven sprayer, flooding and wide-angle or reduced-pressure cone nozzles should be used for broadcast application. Flat-fan nozzles are suitable if drift is not a major concern and the boom is stable and maintainable at a constant height. For band applications, the even-fan nozzle mounted at the proper height will provide uniform weed control across the entire band. If height is limited, nozzle tilt can also be used to obtain the proper band width.

When systemic or translocated herbicides are applied POST to the weed foliage, spray drift to the crop needs to be avoided. For a thorough coverage of the foliage, flat fan or twin-orifice flat-fan or hollow cone nozzles may be used. At times, the floodjet nozzle may also be used. The floodjet nozzle, however, may give a relatively lower penetration of the canopy than the flat-fan and hollow-cone nozzles. When crop plants are to be protected from herbicide injury, herbicide spraying may be directed to the weed foliage by using even-fan nozzles. If weeds are tall, spray shields may be used. Spray shields allow spraying close to the crop.

Cleaning the sprayer

At the end of spraying each day, the sprayer should be filled with water overnight to prevent the herbicide from drying out and forming flakes that could block the filters or nozzles. Some of the water should be sprayed out to clean the nozzle and hose. Next day, the water should be sprayed out completely and the sprayer used for applying the same herbicide. If the sprayer is to be used for applying another herbicide, it should be scrupulously cleaned out, first by rinsing with water, then washing thoroughly with a detergent, and then rinsing again with water until no trace of detergent is lelt. When sprayed with oil-based herbicides, a wash with kerosene or acetone after the wash with the detergent is useful. Similar cleaning procedure should be followed for lids, tops and other non-spraying parts of the sprayer and spray utensils.

When storing the sprayer for a long time, it should also be cleaned thoroughly as described above. It is best to keep separate sprayers for herbicides. Crop injury through traces of 2,4-D is not an uncommon occurrence. Before using the sprayer after a long storage, it is advisable to test it by spraying on susceptile weeds and observing for injury symptoms for 2 to 7 d.

CONTROLLED DROPLET APPLICATION

The technique of controlled droplet application (CDA) is used to make ultralow volume (ULV) spraying of herbicides possible. CDA is defined as the production and application of droplets that are the appropriate size both for the target and the method of delivery [20]. In the CDA technique of spraying of ultralow volume, the production of droplets is controlled within very close limits and the size is tailored to ensure that the herbicide in the spray mixture has the maximum effect on the target at which it is directed. Because of this, the amount of spray volume required is dramatically reduced to the extent that it needs to be applied only at ultralow volumes. McKinley et al. [13] reported that smaller droplets were more effective because larger droplets might become physiologically isolated.

The controlled size droplets are produced by rotary-atomizer sprayers which are now widely known as ULV sprayers or controlled droplet applications. The CDA technique is used to create a fine mist which drifts onto all exposed areas of plants, so that they become well covered with myriads of droplets of the herbicide.

In the case of herbicides, CDA is done using a ULV sprayer named 'Herbi' or 'Micron Herbi' (Fig. 14.4). Herbi is a rotary atomizer, a spinning disc, with a serrated edge which revolves at a speed of 2000 rpm to produce 250 μ droplets. The herbicide spray solution is gravity-fed to the disc from a plastic container of 2.5 L capacity. The atomizer is driven by a small electric motor which is attached to a tube carrying the batteries that provide power for the motor. The spray volume for Herbi is between 7 and 15 L ha^{-1} per hectare. If the Micron Herbi, used to apply spray in a carefully controlled swath, is held 0.5 m above the target, gravity carries the 250 m droplets to the target consistently and reliably; they need only half a second for their passage through the air to the ground [20].

Herbi covers a swath of 1.2 m wide with a walking speed of 3 to 5 km ha^{-1} or one to two paces per second. The discharge rate is approximately 1 m s^{-1}.

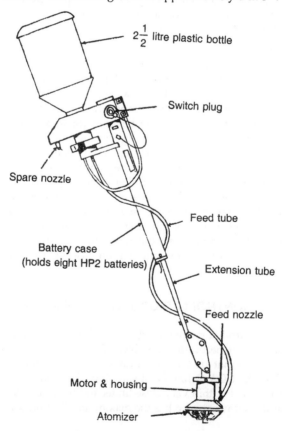

Fig. 14.4. The Micron Herbi sprayer

The CDA method is particularly useful in the case of foliage-applied systemic herbicides such as glyphostate, 2,4-D, MCPA, mecoprop, dicamba, dichlorprop, asulam, etc. However, contact herbicides such as ioxynil, bromoxynil, paraquat, etc. are less effective when applied by CDA. Taylor [21] found that the total spray deposition from CDA was higher than from conventional spraying. He observed that CDA promises more efficient spraying under a wide range of conditions than is possible with conventional systems.

DIRECT CONTACT APPLICATION (DCA)

In the DCA method, the herbicide is placed, wiped, rubbed or smeared onto the plant surface. DCA offers the following benefits [4]: greater speed and ease of operation than handweeding, freedom from concern about walking speeds as contrasted to knapsack spraying operations, ability to treat areas inaccessible to other types of cultivation or herbicide application equipment, possibility of effectively controlling weeds that are difficult or impossible to handweed due to hard-to-reach propagating organs and/or a thorny, bristly or barded nature, reduced volume of herbicide required due to treating only the target plants, and potential for local fabrication of most parts plus assembly.

DCA can be achieved by recirculating sprayer, rope wick applicator, roller application and herbicide glove.

Recirculating Sprayer

The recirculating sprayer has the advantage of the spray being directed horizontally only to weeds growing above the crop so that minimum amounts of herbicide contact the crop plants. The herbicide spray that is not deposited on weeds is collected in the recirculating sprayer and re-applied.

The recirculating sprayer consists of a conventional pumping system to provide pressure to propel the herbicide solution, a pressure regulator, cut-off valves and a hose system. The sprayer is mounted on a tractor or a high-clearance sprayer [14]. The herbicide spray not intercepted by the weeds is caught in the spray trap with an opening of 25 × 38 cm. The trap is 45 cm wide with a diagonal partition extending from one side to the other. The diagonal partition helps to prevent herbicide spray from splashing out of the trap. Herbicide spray deposited in the spray trap is returned to the spray tank with a tubing pump. The herbicide rate is expressed only on the basis of the amount of spray intercepted by the weeds. The amount of interception ranges from 20 to 40 %.

The main difference between a conventional sprayer and the recirculating sprayer is that the latter directs the spray horizontally rather than vertically. The recirculating sprayer has solid jet nozzles instead of the fan or cone nozzles normally used on conventional sprayers.

Recirculating sprayers are available in two designs. The first is designed as the box-type sprayer with two or four nozzles that spray across the rows into the box traps. These boxes trap the spray material and collect it for recirculation back into the spray tank. In the second design, the jet nozzles direct streams of fluid diagonally into a pad of rubberized fibre for collection and recirculation back into the spray tank [11]. This is known as the broadcast-type recirculating sprayer. This type of sprayer is mounted in front of the tractor or high-clearance equipment and provides full swath application to weeds that are taller than the crop.

Gebhardt [11] reported that most recirculating sprayers are designed to apply from 10 to 20 gallons per acre (100 to 200 L ha^{-1}) at 4 to 5 miles (6.4 to 8.0 km) per hour. The application rate can be changed by changing the travel speed of the sprayer or the output of the nozzle.

Rope-Wick Applicator

The rope-wick applicator, invented by J.E. Dale of the United States Department of Agriculture (USDA), Stoneville, Mississippi, USA, uses capillary action and gravitational flow to move the chemical out of a reservoir onto nylon rope wicks where the chemical is rubbed from the soaked rope onto the tall weeds [5]. This method of application not only eliminates any possibility of trash getting into the herbicide reservoir but also greatly reduces the possibility of injuring crop plants by herbicide drift, spillage and splattering [5]. The rope wick acts much like the wicking system used in kerosene lamps to deliver fuel to the flame. As the applicator moves over the field, the herbicide is applied only on weeds contacted by the soaked ropes. As a result, very little chemical is used for treating weeds on one hectare. Rope-wick application has been found particularly useful for systematic foliage herbicides such as glyphosate, 2,4-D, etc.

A rope-wick applicator offers the following advantages:

1. It is a simple and low-cost hand-operated herbicide application equipment. A farmer can build it by himself with a 15-foot (4-row) rig [5].
2. It can be used in areas less accessible for other application equipment.
3. It eliminates the need to spray at high spray volumes by conventional sprayers.
4. It can be used for spot application.
5. Herbicide residue problems, if any, are reduced.
6. It makes available some of the very expensive but effective herbicides such as glyphosate for control of perennial grasses even in sensitive crops.
7. It can be easily adapted to various weed situations, crops and crop spacings.

Rope-wick applicators appear in various designs: rope wicks (Fig. 14.5), pressurized rope wicks, carpet wicks, bunched small diameter ropes (mops), pads, rolls and numerous combinations [10].

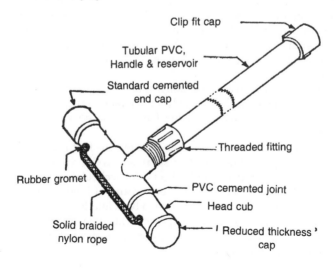

Fig. 14.5. A prototype rope wick [10].

Roller Applicator

Like the recirculating sprayer, the roller applicator applies herbicide only to weeds growing taller than the crop. The roller applicator wets weeds better than the rope-wick applicator but less effectively than the recirculating sprayer. Wind does not affect application with a roller applicator as it does with the recirculating sprayer. However, the roller applicator has a tendency to drip herbicide onto the crop when operated at an angle in hilly terrain and to pick up prickly weed seeds [22].

The roller applicator is intermediate between the recirculating sprayer and the rope-wick applicator in terms of simplicity of operation, number of trips required over a field and herbicide costs. However, it has no distinct advantage over recirculating sprayers and rope-wick applicators in terms of providing better weed control or in greater crop safety [22].

The roller applicator is particularly used for weed control in turf grass. Glyphosate is widely used through the roller applicator. When the carpet is wet, the roller applicator can dispense a considerable amount of glyphosate solution but, even so, less solution is used per unit area than with the recirculating sprayer [22]. The roller applicator is relatively heavy and also expensive.

Herbicide Glove

The herbicide glove is designed to allow safe application of specifically effective herbicides to tough-to-control weeds. It facilitates exclusive application of the herbicide to the weed. The herbicide glove consists of a plastic belt from which the container is suspended, an active glove for one hand, with flow tube, control valve, pressure bulb and sponge or foam pad, and a neutral glove for other hand (Fig. 14.6). All the parts are made of tough plastic. Below the protective palm and foam pad rests a rubber bulb which is deflated with each grip or squeeze of the glove. While walking along the crop row, the stem just below the panicle of each weed is squeezed to deposit the spray on the plant. Each squeeze of the hand releases about 1 ml of the herbicide in the palm pad. During squeezing, the grip should not be so hard as to break the plant. The optimum time of herbicide application through the glove is at panicle emergence. The container holds 2 L which is sufficient for 2,500 panicles.

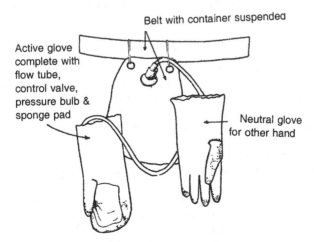

Belt with container suspended

Active glove complete with flow tube, control valve, pressure bulb & sponge pad

Neutral glove for other hand

Fig. 14.6 Herbicide glove

The herbicide glove method is slow but can help control weeds selectively and effectively. This is more useful for control of such weeds as wild oats and many tall weeds.

ELECTRODYNAMIC SPRAYING

In electrodynamic spraying, the spray liquid is atomized by an electric field. As a result, the droplet is charged and the spray liquid propelled away from the nozzle onto the target. Coffee [3] described that in this electrodynamic spraying method, the pesticide is constrained to flow through a thin-slot, high-voltage nozzle so that the liquid becomes subjected to an intense electric field upon emergence into the atmosphere. The field immediately establishes a standing wave upon the surface of the liquid, the crests of which emit jets very similar to the ligaments of a rotary atomizer. Droplets are then emitted from each jet.

A nearby earthed electrode ensures field strength at the nozzle, thus significantly improving droplet size control, and it maintains this field strength despite the constantly varying distance between the nozzle and the (earthed) target that occurs during spraying [3]. The atomization design criteria are: the electrical properties of the liquid, nozzle potential, electrode geometry, and flow rate and visco-elastic properties of the liquid. Coffee [3] observed that since the nozzle potential has the same polarity as the atomized droplets, the spray cloud is vigorously propelled away from the nozzle and onto the target. The constant-force applied to each droplet during transit to the target minimizes any effects due to extraneous forces such as gravity or air movement leading to accurate targeting and low drift.

The electrodynamic sprayer has no moving parts and this helps to reduce the need for machinery maintenance on the farm.

GRANULAR APPLICATION

Granular formulation is increasingly being used particularly in the case of PRE herbicides as it offers the following advantages against spraying: (a) water is not needed to apply the herbicide, (b) equipment cost is reduced and (c) crop injury due to spray drift is non-existent.

Granular Applicator

With the rapid increase in usage of granular formulations of PRE herbicides, there has been a corresponding increase in the popularity of granular applicators. They broadcast the granules over the entire crop row or in hands. A granular applicator consists of the hopper in which the granules are placed, a rotating disc device to get the broadcast application pattern, a long flexible discharge tube with a nozzle at the distal end and a finger-controlled mechanism for regulating the flow of granules. By rotating the handle, the granules are released into the tube through the exit hole of the hopper. The applicators are made of plastic and have a capacity of 1 to 10 kg. They are carried on the back.

The bigger type granule applicators are tractor-mounted. They consist of a band distributor or spatter plate that distributes granules laterally and uniformly over the desired band width and an easily calibrated device that meters the granules. The granule-metering devices can be calibrated by collecting and weighing the granules or by measuring the volumetric loss from carrier over a given area. The orifice setting may be adjusted on the metering device until the desire rate is achieved. In conducting a volumetric calibration, determination is made from the ratio of volume to weight, since this varies with the type of carrier and the size, kind and concentration of the herbicide impregnated on the granules.

Sand Mix Application

Granules of herbicides can be mixed in sand and applied, broadcast, on the soil. This is a widely used method for PRE and E-POST (early-POST) application of granular formulation in India and other rice-growing countries in irrigated transplanted as well as direct seeded rice.

Dry sand of fine- to medium-coarse texture is preferred for mixing the herbicide granules. Sand granules should be nearly of the size of herbicide granules to avoid the segregation problem during mixing and application. Thorough mixing of sand, 50-75 kg ha^{-1}, and the required quantity of herbicide granules is done on clean cemented ground. The herbicide sand mixture is then broadcast uniformly over the field by experienced workers. Uniformity in application enhances weed control efficacy. Herbicides such as butachlor, 2,4-D ethyl ester, anilophos, thiobencarb, propanil, and other rice herbicides can be used as sand mix. Numerous studies conducted in farmers fields showed that there was no reduction in the efficacy of weed control by using herbicide granule-sand mixture.

A variant in herbicide-sand mix is combining the EC and WP formulations with sand and applying the mixture. This is a popular farmers' practice in the wheat-growing areas of India. Isoproturon, a widely used herbicide in wheat, is applied as sand mix at PRE and E-POST. This type of sand mix is also used by rice farmers in irrigated crops.

Sand mixing reduces the cost of application to the minimum. It also helps in saving time and energy required for herbicide application.

HERBICIDE INCORPORATION

Soil-applied herbicides which have medium to high volatility require incorporation into the soil. Herbicides such as trifluralin, benefin, EPTC, vernolate, dinitramine, pebulate, bensulide, profluralin, butralin, etc. require thorough mixing with the soil for effective weed control. Incorporation not only prevents their loss by volatilization or photodecomposition by sunlight, but also reduces the soil moisture requirement and places the chemical in close contact with seed or roots for best control.

Mixing of herbicides in the soil is done by incorporating implements, e.g. spring-tooth harrow, bed conditioner, rotary hoe, rolling cultivator, pulverizing tandem disc harrow, etc. [1]. The most popular power-drawn incorporating tool in the USA is the pulverizing tandem disc harrow. In this case, the herbicide is applied from a spray boom mounted on the tractor [1]. A disc which cuts 10 to 15 cm deep provides effective incorporation on most soil types. The spring-tooth harrow is also a useful tool as it cuts as deep as 15 cm except in heavy clay soils.

The pattern of incorporation depends on soil texture depth of penetration, design of the tool, moving speed, and also rotor speed in the case of power-driven devices. Herbicides should be incorporated as uniformly and thoroughly as possible. The depth and pattern of incorporation with herbicides, crops, weeds and soil types.

HERBICIDE APPLICATION THROUGH IRRIGATION

Application of herbicides in irrigation water (**chemigation**) is a relatively new phenomenon. Herbicides may be applied through surface (furrow or flooding) irrigation, drip or trickle system and sprinkler system.

Surface Irrigation System

The surface irrigation system, also termed the gravity-flow system, is used in case of bed-planted furrow- or flood-irrigated crops. Herbicides can be applied immediately after planting, as PRE treatment, during the first irrigation. The condition of the furrow is important because it influences rate of water flow, amount of water applied and uniformity of distribution. Hence, a clod-free furrow is very helpful in controlling water flow.

In coarse-textured soils, water application of 4-5 cm is required for complete wetting of the soil, while in fine-textured soils, 8-10 cm water is needed. Very sandy soils are not well suited for this type of herbicide application. In this gravity flow system, a minimum slope of 0.2% is needed to achieve uniform water application.

Herbicides that can be used in the gravity flow irrigation system include EPTC, molinate and other PRE herbicides which require soil incorporation immediately after spraying. In this system, weeds on top of the ridge (furrow method) or bed (flooded method) will not be controlled. These require manual or mechanical removal. If the water soaks or flows completely across the ridge or bed, weed control will be very effective.

Sprinkler System

The sprinkler system is the more widely used irrigation system for herbicide application. The calibration procedure is similar to the one described for the surface irrigation system.

The sprinkler system is more useful when herbicides are applied PRE to weeds, but not necessarily to crops. Herbicides that can be used include alachlor, atrazine, benefin, bromoxynil, butylate, EPTC, metolachlor, metribuzin, pendemethalin, vernolate, certain substituted ureas and sulphonylureas, etc.

Drip or Trickle System

In this system, water is applied to the soil through orifices or emitters located at selected points along the water delivery lines. As herbicides applied through the drip or trickle irrigation systems do not move in the soil as far as the irrigation water, weeds grow in the non-treated moist area near the weed-free area around the emitter. The size of the weed-free area is affected by length of the irrigation period, rate of water application and concentration of the herbicide.

In the drip irrigation system, the amount of herbicide in solution is expressed in ppm. For deep-rooted perennial crops, herbicide concentrations of 20-100 ppm are applied over 2-4 h.

Equipment for Herbicide-Irrigation

The basic equipment for delivering the herbicide into an irrigation system includes a chemical supply tank, an injection system, and the appropriate safety and anti-siphon devices that prevent potential contamination of the water source [17]. All equipment, hoses and accessories must be resistant to all herbicides and herbicide mixtures, and formulating materials such as emulsifiers, solvents, etc.

The chemical supply tank must withstand the corrosive action of chemicals and may preferably be constructed out of stainless steel, fibreglass, polyethylene or nylon. The capacity of the tank should be large enough that an area of one acre or one hectare can be treated with a single mix. WP formulations require 8 L water for each kg of formulations, while liquid formulations require less water [16]. Most liquid formulations can also be applied directly without dilution or premixing. Agitation in the tank is required when WP, flowable and other suspended formulations are used. Mechanical and hydraulic agitation are more commonly used.

Herbicides are injected into the irrigation system by using an injection pump. An injection pump should be accurate to within 1% of maximum injection rate and easily adjustable for different injection rates [16]. Its external and internal components should be made of non-corrosive materials. Injection pumps are of two kinds: piston type and diaphragm type. Of the two, the diaphragm type is more popular because the output rates can be easily adjusted while the pump is in operation. For most of these pumps, adjustment of the injection rate is accomplished by simply turning the micrometer-type adjustment device.

The quantities of irrigation water and herbicide, injection time and delivery rate may be calibrated as shown below.

a. Quantity of herbicide needed: Area to be irrigated (ha) × herbicide formulation / ha

The herbicide may then be mixed in water

b. Injection rate (L ha^{-1}) : $\dfrac{\text{Total volume of herbicide} - \text{water mixture (L)}}{\text{Time required for irrigating the field (h)}}$

c. Calibrate delivery rate of the injection pump on the basis of injection rate

Another injection unit, based on the venturi principle, injects chemicals by generating a differential pressure across a venturi device, thereby creating a vacuum which sucks the chemical into the irrigation system [17]. It is relatively low cost compared to piston and diaphragm pumps. However, it has a disadvantage in that the chemical injection is dependent on the available differential pressure; any variation in pressure will significantly alter the rate of chemical injection. In non-pressurized or gravity irrigation (flood or furrow) system, a constant head siphon device can be used instead of an injection pump to meter herbicide into irrigation water [17].

The rate of irrigation of the herbicide, the irrigation pump, etc. are calibrated as per the manufacturer's suggestions and guidelines.

APPLICATION OF HERBICIDES IN FERTILIZERS

Fertilizers offer an alternative method of application of herbicides, particularly the soil-applied ones. Dry fertilizers can be used as a carrier for PRE herbicides such as dinitroanilines, triazines, ureas, thiocarbamates, sulfonylureas, etc. Dry fertilizers became popular in the late 1960s in the USA when thiocarbamates were impregnated on the fertilizers used. Since then, several other herbicides have been used for impregnating bulk blended fertilizers. Similarly, liquid fertilizer as a carrier for herbicides was used in the USA in the late 1950s. The combination of nitrogen solution and triazine herbicide was the first mixture to gain wide acceptance [2]. This 'weed and feed' practice led to the development of several mixtures of liquid fertilizers containing other nutrients and herbicides in the 1960s and later. As herbicide carriers, fertilizers offer several advantages. Besides saving time, energy and application cost, the fertilizer-herbicide combination enhances the efficient use of plant nutrients.

Fluid Mixtures

In a fluid mixture, liquid herbicides and fertilizers are combined. Requirements for effective use of herbicides and liquid fertilizer are not always the same [2]. The droplet size suitable for fertilizer efficacy is much larger then desired for herbicides. The two components are mixed in a closed mixing equipment, inclusive of pump, supply tank and metering device. During mixing, physical and chemical compatibility of fertilizers and herbicides needs to be determined. Features of physical incompatibility include separation of material in layers, excessive thickening or formation of precipitants. Gelling of suspensions, the physical characteristic that prevents fertilizer salts from settling, enhances the suitability of a liquid fertilizer as a herbicide carrier. Clay material, which can serve as a thickening agent, helps to maintain homogeneity of the herbicide-fertilizer mixture.

Mixtures that are physically compatible are not necessarily chemically compatible. Clay material used to thicken suspension fertilizers deactivates certain herbicides. Organic thickening agents may sometimes be used to permit the use of these herbicides in suspension fertilizers. Compatibility tests are required before a herbicide-fertilizer mixture is recommended. Generally, nitrogen fertilizers are more widely used with herbicides than the compound fertilizers.

Once a mixture is developed, it is essential to find suitable nozzles which produce normally larger drops. The higher viscosity of the mixture requires nozzles with orifices of around 4 mm or larger in diameter. Viscosity affects the pressure-nozzle flow rate relationship. The ratio of herbicide volume to fertilizer is an important criterion for developing a mixture. Normally, a fertilizer volume of 100 to 200 L ha^{-1} is desired for uniform distribution of herbicides in most fertilizer applications [2]. A suitable spray delivery system is needed for uniform distribution of the mixture. A herbicide-fertilizer mixture should be thoroughly tested from all aspects, beginning from mixing and compatibility to delivery and weed control efficacy before it can be recommended in a crop.

Dry Mixtures

In a dry mixture, a liquid herbicide formulation is impregnated or coated on a dry granular fertilizer. Impregnation of herbicides on a dry fertilizer is done in a mixing equipment, also known as a mixer or blender. The correct amount of herbicide is sprayed onto the bed of fertilizer as it is rotated in the blender and gets mixed. A rotometer or turbine flowmeter can be used to monitor herbicide flow. Horizontal cylindrical blenders require a nozzle with a 60- to 80-degree spray angle [2]. Directing the spray onto the cascading fertilizer and not onto the blender walls is important. A full-cone nozzle that delivers 12 to 20 L min^{-1} of liquid is recommended [12]. Generally, one nozzle will suffice. However, in horizontal axis cylindrical mixers, two flat-fan nozzles are required. Most herbicides can be uniformly impregnated at 500 kg ha^{-1} dry fertilizer.

After herbicide impregnation, dry mixtures are applied by using tractor-mounted centrifugal spreaders which contain one or two finned discs (spinners) that broadcast the mixture. Broadcasting of the mixture can be metered by a belt or chain through an adjustable opening. Details of application may be obtained from the equipment manu-facturers.

For application on a small area, broadcasting the dry mixture of herbicide and fertilizer may be done manually. This is a widely practiced, inexpensive method in developing countries. However, spread of the impregnated mixture may not be uni-form, although experienced workers can do a good job of uniform distribution. Uni-form application can be achieved by mixing the impregnated mixture in dry sand as explained earlier under '**Sand Mix Application**'.

Dry mixtures are also made of granular herbicides and granular fertilizers. This may not be a practical proposition because of variation in the size of granules of herbicides and fertilizers and segregation problems during mixing. However, suitable mixing technology needs to be developed as dry-granular mixtures are very useful in irrigated crops such as rice, sugarcane, vegetable crops, etc. in developing countries.

The preceding discussion indicates that soil-applied herbicides may be gainfully employed by combining them with liquid or dry (even granular) fertilizers. Of the various possibilities, impregnation of liquid herbicides in dry fertilizers and combining the granular formulation of herbicides and fertilizers offer better prospects in the de-veloping countries. If there is no decrease in the effectiveness of herbicide and fertiliz-er, combinations of the two can be used to conserve time, labour, energy and cost.

HERBICIDE COATING OF CROP SEED

Crop seeds are treated with various chemicals such as fungicides, insecticides, nematicides and herbicide safeners to protect seeds and emerging seedlings from soil-borne insects, diseases, nematodes and herbicides. Besides, seeds are also pelleted or coated with minerals, carbon protectants and inoculum of nitrogen-fixing bacteria. In view of these examples, coating of crop seeds may seem to be a convenient method of herbicide application to control weeds in the vicinity of the germinating seeds and emerging seedlings. However, this is not a safe method as the margin of herbicide tolerance and susceptibility between crop plants and weeds is so thin that an overdose may adversely affect the crop.

The success of this method is dependent on three characteristics of a herbicide: a) it should not kill the crop even at high rates, b) it should rapidly move away from the planted seeds to the zone where weeds are to be controlled and c) it should be suffi-ciently adsorbed by soil particles and retained in the soil after it has migrated from the seeds.

Dawson [6] found that alfalfa (lucerne) was highly tolerant of EPTC, a thiocarbamate herbicide, which met the last two criteria mentioned above. Alfalfa seeds coated with EPTC not only tolerated the herbicide, but also consistently controlled weeds in the alfalfa rows. Alfalfa seed was coated with commercial porous material called 'Rhizokote' and 'Gold kote' (carriers for *Rhizobia* inoculum), consisting primarily of lime and gyp-sum respectively [7, 8]. These materials were added @ 50% to the weight of the uncoated seed. The coated alfalfa seeds were mixed @ 454 g in 100 ml acetone + 27 g technical grade EPTC and stirred. As the seed was shaken gently, acetone evaporated. The saturated seed coating contained EPTC, uniformly dispersed in the porous material. The coated seed could carry 0.05 to 0.2 mg EPTC per seed without becoming sticky [7, 8].

The rate of herbicide applied to the soil through the coated seed, in the case of row planting, depends on: a) the quantity of herbicide per seed, b) spacing of the seeds in the row and c) width of the band of treated soil resulting from movement of the herbicide away from the row planted with coated seed [6]. When the herbicide-coated seed is planted broadcast, the rate of herbicide applied depends on the quantity of herbicide per seed and the seeding rate.

In a field study, Dawson [7] coated alfalfa seeds (@ 0.05 to 0.2 mg seed^{-1}) and planted them broadcast on the soil and also in rows at seed rates of 14 to 112 kg ha^{-1}, delivering 1.25 to 10.0 kg EPTC ha^{-1}. He found 95 to 100% control of perennial ryegrass (*Lolium perenne*) seedlings. He [113] further reported that a) alfalfa was inherently tolerant of EPTC, even at rates as high as 81 to 243 kg ha^{-1}, b) dry alfalfa seed was essentially immune to EPTC injury, c) germinating seeds and emerging seedlings were extremely tolerant of EPTC and d) susceptibility of alfalfa to EPTC injury disappeared once the seedlings had emerged from the soil and the hypocotyl hook straightened, by which time the massive concentration of EPTC had dispersed by diffusion.

Similar results were also obtained when seeds of field beans and snap beans were treated by applying EPTC within a porous coating [6]. The EPTC-coated bean seeds planted 3-4 cm deep and 3-6 cm apart in a row, caused complete control of annual grasses, without injuring bean seedlings appreciably.

Coating the crop seed with herbicides appears very promising as the technique allows application of herbicide and crop seed in one operation. This may lead to saving of money and time. The success of this technique at the commercial level depends on the tolerance level of a crop to a particular herbicide(s) and the rate of dispersion (by diffusion) of the herbicide in the soil after the coated seeds are planted. More work is needed to refine the technique and identify crop seeds and herbicides that can be used.

HERBICIDE ENCAPSULATION

This subject was discussed in considerable detail in Chapter 13.

HERBICIDE STORAGE

Herbicide should always be stored in the original, labelled container, with the label plainly visible. All herbicide formulations tend to deteriorate on storage. Although most formulations are subjected to rigorous storage tests during development to ensure that any deterioration is kept to a minimum, exposure to extremes of temperature and to high humidity must be avoided. Hence, herbicides must be stored in dry and well-ventilated storage rooms, godowns or warehouses specially designated for this purpose. Specific storage requirements of each herbicide product must be ascertained from the label and complied with. Liquid concentrates may crystalize or freeze if stored for prolonged periods under freezing conditions. The crystals tend to settle in a hard mass at the bottom of the container. If any crystalline deposit is seen, the formulation should not be used but returned to the supplier. High humidity may cause caking of wettable powders and granules. As far as possible, herbicide formulations should be used in the same season they are purchased. But they can be stored until the following season if the storage conditions are good.

The herbicide storage area should have a lockable entry to prevent theft and to prevent unauthorized persons from entering. The storage area should be identified with prominent water-proof signs over the entrance.

A complete inventory of all herbicides is essential. The stocks must be examined periodically for leaks, spills or any signs of deterioration. The spilled material, broken containers, etc. should be completely removed and the area decontaminated and promptly cleaned. Effective fire detection systems must be installed and fire prevention and protection measures in the storage area taken up. All workers involved in handling, transportation and storage of pesticides should have ready access to qualified medical aid and to information on the chemical, physical and toxicological properties of the herbicide products.

SAFETY IN HERBICIDE HANDLING

All herbicides are potentially toxic for human beings. They can, however, be used safely if the user has full knowledge of the hazards involved and of the procedures to be followed to avoid these hazards.

In handling herbicides, exposure to the chemical must be avoided to the extent possible. Inhalation of herbicide spray, skin contact with herbicides or their residues, injection by mouth, smoking while working with them and other avoidable exposure must be avoided. Persons engaged in the mixing and application of herbicides must wear rubber gloves, rubber boots, hat, goggles, mask or respirator and an impervious overall or coat covering the body.

In case of poisoning through swallowing, inhalation or absorption through the skin, the affected person should be given medical attention immediately.

Disposal of Herbicide Containers

Emptied pesticide containers are never completely empty and therefore they could pose serious hazard to people, animals or the environment if not properly disposed. Containers with leftover chemical may contaminate food, feed, water, soil or air and could cause damage to plants or crops. Partly full containers should be clearly labelled and stored in a lockable store. The following procedures prevent or minimize these hazards.

(a) After emptying the herbicide, the containers should be rinsed three times, each time with one litre of water for a 4-L can or 5-L water for a 20-L can. This readies the container for proper disposal and not for reuse for other purposes, such as storage and transportation of food, feed, water and other items for consumption by humans or animals.

(b) The containers which have been thoroughly drained and rinsed should be crushed and buried in a safe location on the farm premises. The burial pits should be at least 200 m away from livestock grazing areas, wells, streams and where water supplies will not be contaminated. The pit should be dug deep enough to provide for a solid cover not less than 60 cm thick. Pressurized containers must be buried without puncturing.

REFERENCES

1. Barrentine, W.L. and T.N. Jordan. 1976. Equipment for incorporating herbicides. Weeds Today **7(1)**: 20-21.
2. Broder, M.F. 1988. Application of herbicides in fertilizers. In C.G. McWhorter and M.R. Gebhardt (eds). Methods of Applying Herbicides. Weed Sci. Soc. Amer., Champaign, Illinois, USA, pp. 193-206.

3. Coffee, R.A. 1981. Electrodynamic crop spraying. Outlook on Agri. **10**: 350-356.

4. Cooper, A.S., F. Fraser, L.C. Burrill and A.E. Deutsch. 1981. Hand-held wiping devices for herbicide application. International Plant Protection Centre, Oregan State Univ., Corvallis, USA.

5. Dale, J.E. 1980. Ropewick applicator - tool with a future. Weeds Today (Spring), pp. 3-4.

6. Dawson, J.H. 1980. Selective weed control from EPTC applied with seed of alfalfa (*Medicago sativa*). Weed Sci. **28**: 607-611.

7. Dawson, J.H. 1981. Selective weed control with EPTC-treated seed of alfalfa (*Medicago sativa*). Weed Sci. **29**: 105-110.

8. Dawson, J.H. 1987. Applying EPTC and seeding alfalfa (*Medicago sativa*) simultaneously. Weed Sci. **35**: 80-88.

9. Dawson, J.H. 1988. Herbicide-treated crop seed. *In* C.G. McWhorter and M.R. Gebhardt (eds) . Methods of Applying Herbicides. Weed Sci. Soc. Amer., Champaign, Illinois, USA, pp. 253-263.

10. Deutsch, A.E., L.C. Burrill, F. Fraser and A.S. Cooper. 1981. Direct contact application of herbicides: practicability for small farm weed control. Proc. 8th Asian-Pacific Weed Sci. Soc. Conf., pp. 103-107.

11. Gebhardt, M.R. 1979. Recirculating sprayers. Weeds Today **10(2)**: 18-20.

12. Malone, A.V. 1983. Fertilizer-pesticide impregnation operations. Proc. 33rd annual Meeting of the Fertilizer Industry Roundtable. Pages 77-82. Fertilizer Industry Round table, Glen Arm, Maryland, USA.

13. McKinly, K.S., S.A. Brandt, P. Morse and R. Ashford. 1972. Droplet size and phytotoxicity of herbicides. Weed Sci. 20: 450-452.

14. McWhorter, C.G. 1970. A recirculating spray system for postemergence weed control in row crops. Weed Sci. **18**: 285-287.

15. McWhorter, C.G. 1977. Weed control in soybeans with glyphosate applied in the recirculating sprayer. Weed Sci. **25**: 135-141.

16. Ogg, A.G., Jr. and C.C. Dowler. 1988. Applying herbicides through irrigation systems. *In* C.G. McWhorter and M.R. Gebhardt (eds). Methods of Applying Herbicides. Weed Sci. Soc. Amer., Champaign, Illinois, USA, pp. 145-164.

17. Ogg, A.G., Jr., C.C. Dowler, A.R. Martin, A.H. Lange, and P.E. Heikes. 1983. Application of herbicides through irrigation systems. Cooperative Extension, U.S. Dept. of Agri. Item No. AD-FO- 2280. 8 pp.

18. Patel, S.L. 1975. Recent advances in application techniques. Pesticides Annual, pp. 22-35.

19. Patel, S.L. 1981. Spray volume. ADA (Aspee Distributors Association). News 5 (1&2): 10.

20. Rogers, E.V., G.A. Mathews and N.G. Morgan. 1976. Controlled-droplet application and ULV spraying. BP Printing, Engalnd. 28 pp.

21. Taylor, W.A. 1981. Controlled droplet application of herbicides. Outlook on Agri. **10**: 333-336.

22. Wills, G.D. and C.G. McWhorter. 1981. Developments in post-emergence herbicide applicators. Outlook on Agri. **10**: 337-341.

15

Some Prominent Weeds and Their Management

INFORMATION ON 40 PROMINENT WEED SPECIES[1]

There are over 250,000 plant species in the world. Of these, about 250 have become prominent weeds in agricultural and non-agricultural systems. They cause enormous loss and suffering to human beings by way of reduction in crop quality and quantity, wastage of human energy and resources and increased expenditure to alleviate the problems caused by them. Many of them are persistent, pernicious, obnoxious and hard-to-control. In this chapter are discussed 40 of the more prominent weeds infesting the cropped land and causing enormous loss to world economy. There may be other weed species of greater prominence, but they have to be left out due to limitation of space. Weed species predominant in an aquatic system are dealt with separately in Chapter 17.

The taxonomic and morphological characteristics of these weeds and their infestation in various crops, with particular reference to tropics and sub-tropics, are described. A brief discussion of various measures adopted to manage these weeds is also given. With respect to chemical measures, the herbicides found effective on a particular weed species are mentioned. No attempt is made here, however, to discuss whether or not a particular herbicide is safe for use in a particular crop or cropping programme. This aspect is covered to some extent in Chapter 16.

1 *Ageratum conyzoides* L. (Tropic ageratum, Goatweed)

A. conyzoides is an erect, softly hairy and 50-90 cm tall annual herb (Fig. 15.1). Leaves are opposite, soft, stalked, ovate, 2-10 cm long, 0.5-3 cm wide, with a pointed tip and the margins, regularly separated, with blunt teeth. Stem is erect, hairy and cylindrical, with enlarged nodes. Inflorescence is terminal, often axillary and made up of several branches, each bearing a number of flower heads arranged in a showy flat-topped cluster. Flowers are light blue, purple or violet. Each flower head contains 50-70 tubular flowers which are surrounded by 2 or 3 rows of narrow, pointed bracts with membranous margins. Fruit is an achene, black and ribbed or angled, with pappus of fine soft hairs on the upper end.

This weed is distributed in many tropical and subtropical countries. It grows in cultivated areas, wastelands and along roadsides. It is grouped as a winter annual. It is a common weed in tea in North-east India and South India. It is a prolific seed producer. Seeds are spread by wind and water and germinate under a wide range of conditions.

---

[1]Herbicide rate in this chapter is expressed as kilograms active ingredient per hectare, i.e., kg ha^{-1}, except when otherwise indicated. This is mentioned in parentheses after each herbicide.

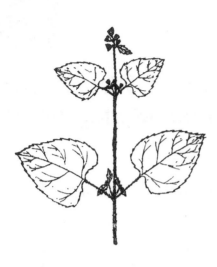

Fig. 15.1. *Ageratum conyzoides.*   **Fig. 15.2.** *Amaranthus viridus.*

Manual methods can control this weed but herbicides offer more effective control. Preemergence (PRE) application of simazine, atrazine, diuron, oxadiazon, oxyfluorfen, methazole, metribuzin, etc. provides excellent control of this weed. Postemergence (POST) application of 2,4-D controls the established infestation.

2. *Amaranthus viridus* L. (Slender amaranth).

A. viridus is a widely occurring annual broadleaf weed (Fig. 15.2) in crops such as maize, sorghum, millets, groundnut, cotton, vegetables, etc. under rain-fed as well as irrigated conditions. It is an erect summer annual growing to a height of 40-80 cm. Stem is round and ribbed. Leaves are opposite, smooth, pale green, 3-5 cm long and ovate, with blunt or shallow notch at rounded tip; margins entire. Inflorescence is long, slender, with a terminal raceme or axillary cluster. Flowers are unisexual, white, small, 1.5 cm long. Flowering time is between July and November. Fruit is one-seeded, with an urticle enclosed by perianth segments. Root system is moderately deep and spreading. *A. viridus* is a prolific producer of seeds. It is widely used as a green leafy vegetable.

Another species of *Amaranthus*, commonly found in cropping and non-cropping areas, is *A. spinosus*. It has a pair of straight spines of up to 1 cm long at the base of the petioles. Both species occur in warm areas. Maximum growth is obtained on soils of high organic matter and loamy in structure.

Amaranthus can be controlled by cultivation and hoeing. As seed germination is continuous in the season, herbicides which have longer persistence of activity are very effective. Simazine, atrazine, trifluralin, fluchloralin, dinitramine, butralin, linuron, chloramben, oxyfluorfen, oxadiazon, diuron terbacil, alachlor, etc. prevent establishment of *Amaranthus*. To control established weeds, 2,4-D, MCPA, paraquat, etc. are effective when applied on the foliage at POST.

POST application of acifluorfen (300-500 g ha^{-1}), thifensulfuron (4-10 g ha^{-1}), chlorimuron (10-15 g ha^{-1}), imazethapyr and imazaquin (both at 50-75 g ha^{-1}) also provides very good control of *Amaranthus* spp. Tank mixing a non-ionic surfactant improves weed control and reduces the optimum herbicide rate. The tank mixture of imazethapyr (25 g ha^{-1}) and bentazon (600 g ha^{-1}) is more effective than when imzethapyr (50-75 g ha^{-1}) is applied alone.

3. *Amaranthus retroflexus* L. (Redroot pigweed)

Amaranthus retroflexus L., found in 46 countries, is considered a serious weed in 16 countries including USA, Canada and Mexico. It is a roughish, somewhat pubescent annual, with a long fleshy, red or pinkish taproot and pink or white rootlets [17a]. The stems often are erect 0.1.2.0 m, high are simple or branching freely, if not crowded, greenish to reddish, with the lower part being thick and the smooth upper part often very hairy. Leaves are alternate. Long stalked, sparsely hairy, ovate to rhombic-ovate, dull green above and the lower surface with prominent white veins [32]. A vigorous plant may produce 100,000 seeds. *A. retroflexus* is commonly found in cultivated fields, orchards, waste-sites, roadsides and other open, disturbed habitats where annual weeds predominate [17a]. It seldom grows in shade. Several soil-applied and foliage-applied herbicides are effective against this weed.

4. *Anagallis arvensis* L. (Scarlet pimpernel)

A. arvensis is a procumbent, ascending or sometimes erect, glabrous annual herb (Fig. 15.3). Stem is quadrangular, weak, gland-dotted, diffusely branched from the base and 10-40 mm long. Roots are fibrous. Leaves are opposite, ovate to oval, 5–25 mm long, margins entire, obtuse to somewhat acute at tip and sessile to clasping at base. Leaf surface is glabrous, with bottom dotted with black glands. Flowers are erect, solitary, axillary on thin peduncles 1-5 cm long. Fruit is curved, membranous globose capsule, 3-5 mm across, the top falling off as a solid. Seed is 1 mm long, three-angled, brown and finely pitted.

Angallis arvensis is a moisture loving plant. It is a native of Europe but also occurs at an altitude of 1350 to 2430 m in tropical Africa as well as in the Gangetic Plain of (India) at 2130 m in the Himalayas. It grows in winter and flowers between the middle of December and March. It is seen in cultivated fields (rice, tobacco, wheat, etc.), gardens, lawns, pastures, wastelands, along roadsides and in a wide range of soils. *Anagallis* propagates by seeds, with yields as high as 900 seeds, per plant. The seeds may remain viable in the soil for as long as 10 yr. They are dormant at the time of harvesting, probably due to the presence of a water-soluble germination inhibitor. Maximum percentage of seed germination occurs at low (15° C) temperature [26]. Dormancy is broken by light or GA and with increasing

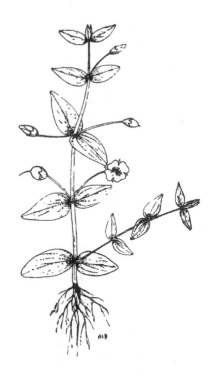

Fig. 15.3. *Anagallis arvensis.*

time of after-ripening the percentage of seeds germinating under the influence of light increases steadily [26].

Sen [26] observed that the requirement of low temperature for germination governs seasonal distribution of this species and restricts its geographic distribution to high altitudes in the tropics. Similarly, the absolute requirement of light for germination keeps deeply buried seeds dormant. Pandey [18] reported that the process of seed germination in *Anagallis* is controlled by phytochrome.

A. arvensis is a long-day plant. For flowering to occur, the plant can be sensitive to daylength at the cotyledonary leaf stage but becomes less sensitive with age. Critical daylength is about 12 h at temperatures above 10° C, but at very low temperatures the plants flower at very short daylenths. This winter annual is not a very competitive species.

Anagallis is sensitive to foliage-applied herbicides such as 2,4-D, MCPA, diclofop, mecoprop, bentazon, etc. as well as many soil-applied herbicides, e.g. triazines, ureas, carbamates, sulfonylureas etc.

5. *Argemone mexicana* L. (Mexican pricklepoppy)

A. mexicana is a prickly annual herb with alternate showy yellow flowers (Fig. 15.4). Stem is 60-90 cm tall, pithy with scattered prickles and smooth to slightly pubescent. Leaves are bluish-green, sessile somewhat clasping the stem, 10-20 cm long and pinnately lobed, with an irregularly serrate and sharply spiny margin. Veins of leaves appear greyish-white on upper surface. Flowers are sessile or on a short pedicel, 5 cm long, borne at ends of branches. They consist of 3 prickly sepals, 6 petals and numerous stamens. Fruit is spiny. Seeds are globular, reticulate and black-brown, with a prominent hilum. The plant exudes a yellow sap when cut. *A. mexicana* is adapted to a wide range of habitats in tropical and substropical regions. It germinates throughout the year, even during the dry season. It is found along roadsides, wastelands and cultivated areas. The plant is toxic to animals as well as human beings. Cattle avoid grazing this plant.

Fig. 15.4. *Argemone mexicana.*

A. mexicana can be successfully controlled in the seedling stage. 2,4-D and MCPA give moderate control when applied POST. A mixture of MCPA and 2,3,6-TBA gives better control of the weed than MCPA alone. PRE herbicides are not effective. Contact POST herbicide paraquat is effective.

6. *Avena fatua* L. (Wild oat).

A. fatua is an annual grassy weed growing to a height of 60-100 cm (Fig. 15.5). It is one of the most problematical weeds in wheat in tropical countries including India. It is considered one of the most competitive weeds in agriculture. The seeds shatter and fall to the ground before wheat or a cereal crop is harvested and persist in the soil for many years. The seeds contaminate the crop seeds and get disseminated to new places. Wild oat plants are very much like those of cultivated oat in general appearance.

Fig. 15.5. *Avena fatua.*

A. fatua is an erect annual grass with extensive fibrous root system. Leaves are flat, with broad base and acute apex, 7-20 cm long and 5-15 mm wide. Culms are smooth, erect, stout and in small tufts. Sheaths are smooth or slightly hairy on the margins on younger plants. Panicle is loose and open, slender branches ascending and rough. Spikelets are made up of 2 to 3 florets enclosed by a pair of papery bracts; each floret bears a long, abruptly bent bristle. Seed is 6-10 mm long and the most characteristic distinction from cultivated oats is the presence of long silky hairs, especially around its base. The panicles of wild oats are looser and more widely spaced than those of cultivated oats.

Wild oat plants grow on a wide range of light to heavy soil types and in both acid and alkaline soils. Freshly harvested seeds are dormant but germinate well after a prolonged storage at 20° C-25° C.

Wild oats can be controlled by cultivation after planting the crop. Cultivation should be shallow and done with care to reduce possible crop injury. The infestation can be reduced by fallowing in the summer for one season and delayed seeding in the second season.

Triallate and trifluralin, applied PRE or preplanting (PPL) give very effective control. Triallate is particularly useful in wheat and barley. These herbicides require incorporation into the soil after application.

The foliage-applied POST herbicides which are effective against wild oats are asulam, diclofop methyl, difenzoquate and benzoylprop ethyl. Diclofop methyl gives excellent control of wild oats and green foxtail (*Setaria glauca*) in wheat. Difenzoquat needs to be applied at the 3- to 5-leaf stage of wild oat plants in wheat and barley. It can be tank mixed with 2,4-D or MCPA for better results. Asulam is effective when applied at about the 3-leaf stage of the weed.

POST application of fenoxaprop. P(100-150 g ha^{-1}), flamprop (150-300 g ha^{-1}), imazamethabenz (300-500 g ha^{-1}), tralkoxydim (75-150 g ha^{-1}) and ethalfluralin (300-500 g ha^{-1}) provide good control of *A. fatua*.

7. *Axonopus compressus* (Sw.) Beauv. (Carpetgrass, Savannagrass).

A. compressus is a stoloniferous perennial grass. Leaves are linear, flat, lanceolate, 10-15 cm long, 2-15 mm wide and with broadly rounded base and blunt apex, often fringed with hairs (Fig. 15.6) Culms are erect, 20-25 cm long, laterally compressed, finely hairy along the outer margin and nodes densely pubescent. Inflorescence has 3-5 spikes, 5-7 cm long, erect and slightly digitate. Spikelets are oblong, rather acute and 2-3 mm long. Plant has short ligule fringed with short hairs. Root system is fibrous. Plant strikes roots from nodes of the stem when creeping along the ground.

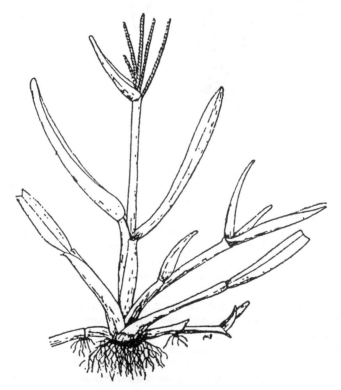

Fig 15.6. *Axonopus compressus.*

A. comprressus grows best under moist humid warm conditions. It can adapt well to both fertile soils as well as poor sandy soils if moisture is present. It grows well at the acidic side of the pH. It propagates by vegetative parts as well as by seeds. It is the most troublesome grass in plantation crops such as tea, coffee, oil palm, pineapple, etc. It also occurs in groundnut.

Mechanical methods can control this weed to some extent. But herbicides are more effective. POST applications of parquet-diuron (0.4+0.5 or 1.0) parquet-MSMA (0.4+1.0) and dalapon-MSMA (3.0+1.0) combinations give satisfactory control. When paraquat (0.4) is applied, repeat application at 10-15 d intervals gives better control. Glyphosate (0.8–1.2) gives immediate control of this weed.

8. *Bidens pilosa* L. (Hairy beggarticks, Blackjack, Cobbler's peg)

B. pilosa in an erect, branched, annual herb growing to about 60 cm in height (Fig. 15.7). Leaves are opposite, petioled, pinnate and comprise 3 ovate leaflets with acute tips and sharply serrate margins; terminal leaflet larger than lateral ones. Stem are branched, smooth, with green or brown stripes. Flower heads are about 1-1.5 cm long, borne on long stalks and arranged in branched, loose terminal inflorescence. Fruits are black, narrow, 1 cm long, ribbed and sparsely bristled.

B. pilosa is found in gardens, cultivated areas, wastelands and along roadsides. The seeds can remain dormant in the soil until conditions are favourable. It is commonly found in tea-growing areas of South India. Heavy growth is observed between July and September. PRE application of simazine, atrazine, metribuzin, diuron, etc. can keep the ground weed-free for the whole season. POST application of 2,4-D can effectively control this weed. Application of bentazon at the 2- to 4-leaf stage controls established weeds.

Fig. 15.7. *Bidens pilosa.*   **Fig. 15.8.** *Borreria hispida.*

9. *Borreria hispida* (L.) K. Schum. [B. articularis (L.f.) Will.] (Buttonweed).

B. hispida is an annual broadleaf weed. It is one of the most obnoxious weeds in tea in North-east India. In South India, *B. latifolia* is predominant. *B. hispida* is a branched herb with quadrangular stem. It is 20-60 cm tall. Stem and leaves are sparsely covered with fine hairs (Fig. 15.8). Leaves are 3-5 cm long, 2-3 cm wide, opposite, broad, elliptical and entire. Inflorescence is axillary cyme. Flowers are small and white. The plant flowers between July and October. Fruit is a capsule, globose and hairy. Root system is shallow and spreading. The weed reproduces by.seed.

B. hispida is a summer annual. It thrives well under moist, humid and heavy rainfall conditions. It grows generally in lighter soils. It grows well in acid soils with pH of 3.5–6.0. It is commonly known as an indicator plant for acid soils.

Manual methods can reduce the infestation of *B. hispida*. But continuous germination over the season makes it difficult to control by manual methods. PRE herbicides such as simazine, diuron, oxyfluorfen, oxadiazon, metribuzin and methozole are very effective in preventing its establishment. The established weed growth can be effectively controlled by POST application of 2,4-D.

10. *Celosia argentea* L. (Coxcomb)

C. argentea is a herbaceous annual weed found in many crops, e.g. groundnut, pearl millet, sorghum, maize, sesame and many dryland crops. It is widely distributed in tropical countries along the plains up to an altitude of about 1000 m.

C. argentea is an erect, glabrous, tall herb (Fig. 15.9) which may attain a height of 1–1.5 m under favourable growing conditions [26]. Leaves alternate, simple and unbranched. It propagates through seeds which germinate with the onset of rains. The inflorescence is pinkish-white and can be recognized from a distance in a crop field. It grows actively until early October and dies by the end of December. The plants have numerous lateral roots just a few centimetres below the soil surface which enable efficient absorption of nutrients from the soil. The plant has a shallow root system. Sen [25] observed two distinct forms of *C. agentea* one with lanceolate acute leaves and the other with ovate obtuse leaves.

Fig. 15.9. *Celosia argentea.*   **Fig. 15.10.** *Chenopodium album.*

Celosia is sensitive to PRE herbicides such as triazines and ureas and to POST herbicides such as phenoxy acetics.

11. *Chenopodium album* L. (Lambsquarters)

C. album is an annual, erect herbaceous weed which grows to a height of 1-1.5 m (Fig. 15.10). It is one of the most widely distributed broadleaf annual weeds in the world. Leaves are small, greenish or greyish-green, usually crowded into dense rounded clusters, simple, alternate and ovate to lanceolate and petioled. Younger leaves are covered with grey or whitish, mealy hairs. Inflorescence arises from leaf axils or at terminus of stems and branches. The weed reproduces by seed.

It is distributed over a wide range of pH and soil types. The plants occur in groups and most of the seeds are deposited near the mother plant. They are prolific producers of the seed, with each plant producing several thousands of seeds. The seeds have longer viability in the soil. *Chenopodium* can adapt to different environmental conditions. It occurs in maize, sorghum, wheat, vegetable crops, orchard crops, etc. It is used as a green leafy vegetable. [26]. *Chenopodium murale* L. grows abundantly in irrigated cultivated fields, gardens and waste places.

The seeds of *C. album* exhibit polymorphism [5, 7]. Bhati et al. [5] observed two polymorphic types in *C. album*, one producing large, heavy, black seeds are the other giving out small, light, brown seeds.

Chenopodium album is sensitive to foliage-applied herbicides such as 2,4-D, **MCPA**, mecoprop, dinoseb, paraquat, bentazon, diclofop, etc. as well as many soil-applied herbicides including triazines and ureas. PRE applications of metolachlor (2-3) + metribuzin (0.5) and alachlor (2-3 + metribuzin (0.5) give very good control of *C. album*. Effective control is also obtained by PRE applications of propachlor (4.0) chloramben (5.0). Chlorbromuron (2.0) and linuron+nitralin (1+1).

12. *Cirsium arvense* (L.) Scop. (Canada thistle)

C. arvense is an erect perennial herb, reproducing by seeds and horizontal rhizomes. Roots are very deep (up to 140 cm), with extensive system of fibrous rootlets. Stem is 40-120 cm tall, erect, grooved, branching only at the top, nearly glabrous or slightly hairy when young and increasingly hairy with maturity. Leaves are alternate, oblong or lanceolate, usually with crinkled edges and spiny margins, irregularly lobed, terminating in a spine, hairy beneath or smooth when mature; upper leaver sessile and only slightly decurrent (Fig. 15.11). Lower heads are dioecious, numerous, compact in corymbose clusters, terminal and axillary, 2-2.5 cm in diameter, with rose-purple or white disc flowers. Achene is smooth, oblong, light to dark brown, 2.5-3.5 mm long, flattened, curved or straight and apex blunt, with tubercle in centre.

C. arvense grows in many crops. It grows well in rich and heavy soils. Deep tillage can eliminate this weed. POST applications of dicamca (1.0) Picloram (1.0) glyphosate (2.0) bentazon twice (each at 2.0) and 2,4-D+dalapon (3+5) give satisfactory control of this weed.

13. *Commelina benghalensis* L. (Dayflower, Tropical spiderwort, Wandering Jew).

C. benghalensis is a broadleaf herbaceous weed widely distributed in the tropics. It grows in a wide range of situations in arable crops, particularly in moisture areas. It is a succulent or fleshy creeping annual as well as perennial herb. The leaves are alternate, simple, parallel-veined and ovate, with an entire margin contracted at the base into a narrow, stalk-like portion and a sheath enclosing the stem (Fig. 15.12). Broken pieces of stem can strike roots and grow as a separate plant.

Commelina is well adapted to moist, swampy, soil conditions with even water-logged conditions aiding rapid vigorous growth. A thick infestation can smother

Fig. 15.11. *Cirsium arvense.*   **Fig. 15.12.** *Commelina benghalensis.*

low-growing crops such as vegetables, groundnut, pulses, legumes, etc. It is also found in upland rice, sugarcane and tea.

This weed is resistant to many herbicides. The plants develop resistance with age. Soil-applied herbicides such as simazine, metribuzin, bentazon, linuron and diuron have more activity than foliage-applied herbicides. Among the POST herbicides, glyphosate controls this weed satisfactorily.

14. *Convolvulus arvensis* L. (Field bindweed)

C. arvensis is a prostrate or climbing perennial herb. Its stem, a twiner, is slender and smooth. It covers the host plant entirely due to its gregarious growth. Leaves are smooth, simple, long peotioled, alternate, 3.5-5 cm long and 2 to 3 cm wide. Flowers are funnel-shaped, solitary, pink and 3 cm long (Fig. 15.13). Flowering occurs from June to September. It propagates from deep underground roots and by seeds.

C. arvensis is a serious weed in tropical, subtropical and temperate regions. In India, it infests such as crops maize, potatoes, vegetables, rice, tea and orchards. It is extremely difficult to eradicate once it has infested an area intensively. The seeds remain viable in the soil for long periods. Repeated cultivations over a long period eradicate this weed. But this operation becomes very expensive. Among the herbicides, dicamba, MCPA, 2,4-D, glyphosate, etc. effectively control when applied POST. 2,4-D application followed by cultivation is also effective.

15. *Cuscuta chinensis* Damk. (Chinese dodder)

C. chinensis is an annual stem parasite. It reproduces by seed. It emerges along with the germinating crop seedlings and parasitizes by attaching itself to the host plant and remains parasitic until harvest. It draws nutrients from the host for sustenance and

Fig. 15.13. *Convolvulus arvensis.*

provides a dense barrier that drastically retards growth and vigour of the host plant and reduces grain yield by 35 to 50% [30]. The intensity of damage caused by *Cuscuta* depends upon its capacity to rapidly parasitize the host crop.

The stem is yellow or yellowish, slender, leafless and climbing or twining on its host.

The weed parasitizes legumes, clover, niger, forest trees and many other crops. This parasite poses a serious problem in India (in the states of Andhra Pradesh and Tamil Nadu) in pulse and legume crops, particularly in irrigated rice fallows. Seeds of pulse crops like green gram (*Phaseolus aureus* L.) and blackgram (*Phaseolus mungo* L.) are broadcast, in the standing rice crop 2-3 wk before harvesting rice. The remnant soil moisture condition allows good germination of pulse seeds. Rice crop is harvested when pulse seedlings are in 2- to 3-leaf stage. *Cuscuta* soon begins parasitizing greengram and blackgram seedlings and by the 50-60th d, the parasite gives a mat-like appearance over the crop canopy. Although several foliage-applied herbicides such as paraquat, 2,4-D and glyphosate (75-150 g ha^{-1}) and soil-applied herbicides such as chlorpropham (2-3), pronamide (1-2), ethofumesate (2-4), imazaquin (100-150), trifluralin (2-4), pendimethalin (2-4), prodiamine (2-4), etc. are known to achieve moderate to good control of *Cuscuta*, they cannot be used in the rice fallow-pulse cropping system. When pulse seeds are sown, the standing rice crop precludes PRE herbicide application, while the extreme sensitivity of pulse seedlings to foliage-applied herbicides prohibits their use at POST.

Non-chemical control methods may be useful in preventing or reducing *Cuscuta* infestation. As sowing the crop seed contaminated by *C. chinensis* seed has been the major means by which the parasitic weed is spread, planting pure crop seed could help in preventing its infestation and spread. As soon as *Cuscuta* plants appear in the field, farmers may need to uproot the weed before they produce seed, followed by drying and burning. Although insects may damage *Cuscuta*, this method is too slow to protect the host crop. The spores of *Colletotrichum gloeosporioides*, used to control *Cuscuta* selectively in soybéan ([17], may be applied uniformly over the *Cuscuta*-infested crop. The spores germinate, grow and cause a disease that suppresses *Cuscuta*.

The pulse crop can be partially protected from *Cuscuta* parasitism by growing the *Cuscuta*-resistant cluster bean (*Cyanopsis tetragonoloba*) along with green-gram (or black gram) in a mixed cropping system [21, 23]. As *Cuscuta* does not parasitize graminaceous crops, soybean, cluster bean, forage crops and tomato, these crops may be rotated with rice and pulse crops to break the life cycle of *C. chinensis*.

16. *Cynodon dactylon* (L.) Pers. (Bermudagrass, Stargrass)

C. dactylon is a perennial grass with long runners which strike roots at the nodes and extensive underground rhizomes. It occurs throughout the tropical, subtropical and semi-arid regions of the world. It is one of the world's worst weeds. It is suscepti-ble to competition and shading.

The leaves of *Cynodon* vary greatly in length from 3 to 20 cm (Fig. 15.14). There is no membranous ligule where the leaf blade joins the sheath. The flowering stems may be 15-50 cm long. The inflorescence consists of 4-5 slender purplish spikes of 10 cm long. *C. dactylon* is a variable species, with some being used as lawn grasses. It propagates vegetatively more than by seeds.

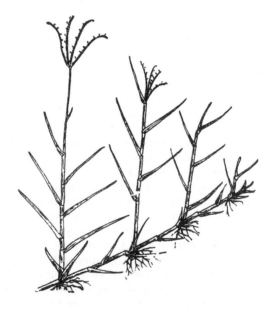

Fig. 15.14. *Cynodon dactylon.*

POST application of dalapon provide satisfactory control of *Cynodon*; it does not, however, give permanent control. Repeat applications of these herbicides may elimi-nate this weed to some extent. Glyphosate at 1.2-2.0 is extremely effective with little regrowth. *Cynodon* is sensitive to some extent to PRE application of diuron. Other herbicides which are effective against *C. dactylon* are bromacil (10.0) at PRE or POST EPTC (4.5) soil-incorporated before sowing 2,4-D+dalapon (1.0-1.5+3.0-4.5) POST and terbacil followed by dalapon at POST.

POST application of fluazifop-P (75-105 g ha^{-1}), clethodim (150-300 g ha^{-1}), fenoxaprop-ethyl (150-300 g ha^{-1}) and 40:60 ratio of (w/w) of fenoxaprop-ethyl + fluzaifop-P provides 80-95% control of *C. dactylon*. Sulfometuron (50-150 g ha^{-1}) plus 2,4-D amine (1.0-1.5) gives a very good long-term control of this persistent perennial weed.

17. *Cyperus rotundus* L. (Nugrass, Purple nutsedge)

C. rotundus is a very persistent perennial sedge. it is considered the world's worst weed as it occurs in 52 crops in 92 countries. It is a native of India. It is widely distributed throughout the tropics and sub-tropics. It grows to a height of 15-60 cm. The plant is swollen and thickened at the base. It has a triangular smooth scape, 10-60 cm in height, arising from the centre of a basal cluster of narrow grass-like leaves of 30-50 cm long and 8 mm wide (Fig. 15.15). Leaves are smooth, shiny dark green and grooved on the upper surface. Slender, underground runners grow out from the base of the stem and form series of black, irregular-shaped or nearly round tubers which may reach 2 cm in length. The tubers often sprout to produce new plants while still attached to the parent plant. The inflorescence arises from the stem apex and is suspended by a number of leaf-like bracts. It consists of a number of slender branches of unequal length, near the ends of which are clustered narrow spikelets of 1-3 cm length, brown to dark reddish-brown in colour. Each spikelet is made up of 10-30 small, closely crowded florets which ripen to form black, triangular nuts (achene).

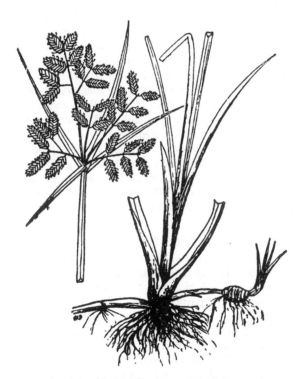

Fig. 15.15. *Cyperus rotundus.*

The roots of *C. rotundus* are fibrous and extensively branched. They spread by extensive, horizontal, slender rhizomes which are white and fleshy and covered with scale leaves when young but which turn brown and fibrous when old. The rhizomes give rise to underground tubers which proliferate profusely. The rhizomes do not give rise to new growth except through tubers. Most of these tubers grow in the top 10 cm of the soil, with none below 30 cm. Tubers store food for other parts of the plants and they are a very effective means of propagation. New tubers are produced within 3 wk of sprouting of an individual tuber [2]. Tubers have nodes, short internodes, buds and scale leaves. The scale leaves are sloughed at maturity [20]. *C. rotundus* also propagates through seeds but this is of minor importance. It is sensitive to shade and grows well in wet and dry soils and warm climates. It flowers in winter and during short photoperiods of 6 to 8 h. The emerged plant takes about 3 to 8 wk for flowering. It is a serious weed in many dryland and irrigated crops in tropical countries including India.

Mechanical methods kill only the top growth with little effect on the tubers. Herbicides which translocate rapidly into the tubers to prevent their regeneration are most effective in controlling this grass. 2,4-D and MCPA are effective at 2-5 kg ha[-1]. Arakeri [3] obtained 80% control of nutgrass with the amine salt of 2,4-D applied at 1.5 kg ae ha[-1]. He observed that application of 2,4-D or MCPA in addition to dry plough-ing and exposing tubers for desiccation was more effective than herbicides alone. Sinha and Thakur [28] found that deep cultivation in summer followed by 2,4-D sodi-um salt at 4.0 kg ha[-1] gave complete control of nutgrass. Glyphosate, a rapidly translocated herbicide, is more effective than many foliage-applied herbicides.

Amitrole-T translocates to tubers to some extent and hence can achieve control of nutgrass by repeated treatment. Paraquat kills the tops but repeated applications could deplete the tubers of the food reserves and give better control.

Of the soil-applied herbicides, atrazine (4.0) and perfluidone (3-4) give satisfactory control of this grass in crops where these herbicides are safer. EPTC, which is volatile, also gives good control of nutgrass when incorporated into the soil soon after application. Hence, EPTC needs to be used at preplanting. Alachlor is moderately effective. A mixture of alachlor and atrazine is more effective than either of them applied alone. Ray and Wilcox [22] reported that dichlobenil at PRE gave fair to complete control of *C. rotundus*, depending on the rates applied.

Soil incorporation (PPI) of imazaquin (150-250 g ha[-1]), chlorimuron (50-100 g ha[-1]) and halosulfuron-methyl (75-150 g ha[-1]) provides very good control of tuber growth, leading to satisfactory control of *C. rotundus*. Their follow up application at PRE and POST gives a much better control. POST application of chlorimuron (10-20 g ha[-1]) or imazethapyr (50-100 g ha[-1]) tank mixed with a non-ionic surfactant gives moderate to good control.

18. *Digitaria Sanaguinalis* (L.) Scop. (Crabgrass)

D. Sanguinalis is an annual grass. The clums are stout, usually decumbent at the base, smooth and 30-90 cm long when prostrate. The plant strikes roots when nodes touch the soil. Leaves are 5-15 cm long, 5-10 mm wide and somewhat hairy (Fig. 15.16). Sheath is densely hairy, particularly the lower ones. Spike is 5-15 cm long, with 3-13 digitate segments in whorls at top of stem. Spikelets are 3 mm long, paired along one side of the rachis. Although *D. sanguinalis* is an annual, it exhibits perennial growth. It flowers between July and September. It is a prolific seed producer.

Digitaria can thrive well under both ropical and temperate climates. It can grow in moist areas as well as under dry and hot weather conditions. It is found in such crops as sugarcane, sorghum, groundnut, maize, rice, tea, coffee, orchards, etc. It has an exceptional ability for spreading over an area and covering the ground very fast. In tea, it grows around the collar of the bush and protrudes through the plucking table making its control very difficult.

Once established, the weed is difficult to control with mechanical methods alone. It is tolerant to triazines which when used continuously over a number of years can create a situation where *Digitaria* becomes predominant and difficult to eliminate. It is sensitive to diuron, trifluralin, butachlor, EPTC, vernolate, metribuzin, oxadiazon, etc,

Fig. 15.16. *Digitaria sanguinalis.*

which can kill the germinating seedlings. PRE soil incorporation of butylate+atrazine, alachlor+linuron and alachlor+chloramben gives best results. Paraquat, POST, is moderately effective.

PRE application of oxadiazon (1.0-2.0), prodiamine (0.3-0.6), oryzalin (1.0-2.0), benefin + oryzalin (total rate 1.0-2.0), benefin + trifluralin (total 1.0-2.0) and dithiopyr (0.2-0.4) can provide 75-90% control of *D. sanguinalis*. POST spraying of nicosulfuron causes 90% reduction of the weed. A surfactant will enhance nicosulfuron efficacy. Sethoxydim (50-100 g ha^{-1}) in combination with atrazine (0.5-1.0), primisulfuron (20-40 g ha^{-1}), 2,4-D (0.5-1.0), bromoxynil (0.25-0.50), dicamba (0.25-0.50) or flumiclorac-pentyl (0.2-0.4) gives excellent control of this annual grass.

19. *Echinochloa crus-galli* (L.) Beauv. (Barnyardgrass)

E. crusgalli is a widely distributed annual grass throughout the warmer countries of the world. It is a major weed in rice and many field and horticultural crops. It emerges before or along with the crop and during the first several weeks it usually outgrows crop plants rapidly. It grows to a height of 30-120 cm, with thick, coarse, mostly erect, smooth and branching at the base (Fig. 15.17).

The plant has sessile leaf blades attached to a smooth sheath which encircles the stem in the absence of ligule. Leaves are rolled in flattened bud-shoot. Sheath is pale green, flattened, keeled and split, with hyaline margins. Leaf blade 10-30 cm long and 5-20 mm wide. Mid-rib is prominent. Stem is stout, glabrous, 6 mm in diameter. Culm branches at base and produce tillers. Stem ends in 10-20 cm long inflorescence, with slender spike-

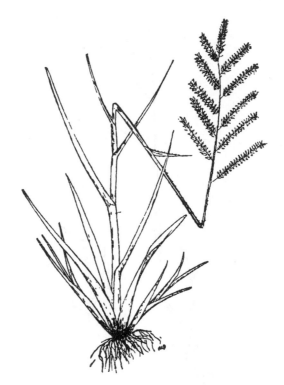

Fig. 15.17. *Echinochloa crus-galli.*

like panicle, green or purplish in colour. Lower branches of panicle wide-set are far apart from each other, while upper ones being more or less aggregate into terminal lump.

Spikelets are densely crowded in 2-4 rows on each side of stem. Spikelet is about 6 mm long and oval, with unequal pointed glumes. Seeds are strongly convex on one side, flat on the other, light orange-yellow and 2.5-3.5 mm long.

A mature barnyardgrass plant has fibrous or aventitious roots. The first adventitious roots arise from the mesocotyl (the segment between the scutellum and coleoptile) at the time of seedling emergence.

Cultivation can control barnyardgrass but it is only partially successful as the weeds continue to infest in spurts. Several rounds of manual operations or hoeing can keep the weed under check.

Herbicides which inhibit seed germination or the emerging coleoptile are more efficient in controlling this annual grass propagated by seed. The PRE herbicides alachlor, chloramben, atrazine, simazine, diuron, linuron, metolachlor, EPTC, pebulate, vernolate, etc. are particularly effective against this weed. EPTC, pebulate and vernolate need incorporation into the soil for better activity. Kaushik and Gautam [14] found that atrazine (0.25-0.50) alone was the most efficient control of barnyardgrass in pearl millet (*Pennisetum typhoides*). Chela and Gill [6] reported that benthiocarb and butachlor gave effective control of *E. crusgalli* when applied 3 d after transplanting rice.

Effective control of barnyardgrass may be achieved by fenoxaprop-ethyl (75-100 g ha^{-1}) POST, applied alone or mixed with bensulfuron (50 g ha^{-1}) or bentazon (1.0). POST application of clethodim (100 g ha^{-1}), quizalofop-P ethyl (70 g ha^{-1}), fluazifop-P

(200 g ha-1) and sethoxydim (300 g ha⁻¹), either alone or in combination with lactofen (75-150 g ha⁻¹), imazaquin (75-150 g ha⁻¹), chlorimuron (8-15 g ha⁻¹) or fomesafen (0.25-0.50) gives 80-99% control in soybean [24]. Nicosulfuron (10-25 g ha⁻¹) or primisulfuron (20-30 g ha⁻¹) or thifensulfuron (5-10 g ha⁻¹) tank mixed with an adjuvant gives good control of *E. crus-galli*.

20. *Echinochloa colonum* (L.) Link (Junglerice).

E. colonum is a slender grass growing to a height of 60-90 cm (Fig. 15.18). The stem is creeping below and erect above, with rooting at lower nodes. This weed occurs widely in tropical countries. The plant propagates mainly through seeds which have a dormancy period of about two months. It germinates readily at a soil moisture level of 20 to 30%. It can also propagate vegetatively, with nodes striking roots when in contact with soil and the new shoots, when separated, can give independent plants.

Ramakrishnan [19] observed two distinct populations of *E. colonum*: (i) the tall form, growing in very moist to waterlogged soils along the banks of ponds and drainage

Fig. 15.18. *Echinochloa colonum.*

channels and (ii) the short form, growing in comparatively drier localities. In nature, the latter form is more frequently subjected to grazing, scraping, etc. than the former. The tall form has long internodes and more vigorous growth compared to the short form.

The seeds of *E. colonum* germinate after the onset of rains, in the beginning of July. Flowering and fruiting occur between August and September, with the life cycle competing by mid-October [26].

Echinochloa colonum is sensitive to PRE herbicides such as alachlor, EPTC, trifluralin, benefin, oxadiazon, diuron, etc. At POST, it can be controlled by paraquat, glyphosate, dalapon, etc.

It may also be controlled effectively by other herbicides listed for control of *E. crusgalli*.

21. *Eleusine indica* (L.) Gaertn. (Wild fingermillet, Goosegrass)

E. indica is a coarse tufted annual grass which grows to 30-45 cm in height, with laterally flattened shoots (Fig. 15.19). Leaf blades flat or folded at base. Sheaths are flattened and keeled. Inflorescence consists of 4-8 narrow digitate spikes, 4-15 cm long, arising from the top of the stem to form a spreading umbel. Spikelets are sessile on one side of the rachis, 3-5 mm long and densely crowded in two rows along lower side of spike. Each spikelet consists of about 5 florets devoid of bristles. Roots are fibrous, quite deep and spreading. This weed flowers between June and September and reproduces by seeds and old roots.

Eleusine indica is of common occurrence in cropped and arable lands in most parts of semi-arid tropics. It is found in crops such as maize, sorghum, wheat, tea, etc.

Once established, *E. indica* is difficult to eradicate. Tillage can control this weed partially. Germinating seedlings are more susceptible to PRE applications of simazine, atrazine, monuron, diuron, oxyfluorfen, linuron, metribuzin, alachlor and other soil-applied herbicides. At POST, dalapon and paraquat are only partially effective Dalapon is useful only at the young seedling stage. Paraquat-treated plants show scorching but recover almost immediately after spraying. POST application of glyphosate alone or MSMA alternating with

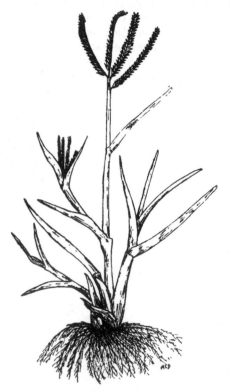

Fig. 15.19. *Eleusine indica.*

paraquat is very effective. Two POST applications of napropamide (2.2 each time) or methazole+MSMA (0.28+2.2 each time) diclofop (1.0-1.5) and diclofop (0.50-0.75)+MSMA (2.0-2.5) give satisfactory control of *E. indica*.

22. *Elytrigia repens* (L.) Nevski (Quackgrass, Couchgrass)

E. repens (earlier referred to as *Agrophyron repens*) is an aggressive erect perennial grass that propagates through rhizomes and seeds (Fig. 15.20). It is a native of Europe and a very troublesome weed in the northern USA and southern Canada. It is generally absent in the tropics and warmer regions. It is not very competitive with weeds better adapted to the tropics.

Plants have 50-120 cm tall smooth culms, with 3-5 joints. Leaves have auricles which are soft, flat and crowded, with fine ribs. Ligule is 0.5 mm long. Lower sheaths are hairy and upper ones glabrous or slightly pilose. Spike is dense or lax, 5-15 cm

Fig. 15.20. *Elytrigia repens.*

long, with 3-8 short-awned florets in compressed spikelet. The majority of quackgrass culms produce spikes.

Rhizomes are the important means of propagation this species. They regenerate when planted at a depth of 5-8 cm in the soil. Rhizomes planted at the surface show poor shoot production. The plants originating from seeds are weaker and the leaves slender. Quackgrass exhibits allelopathic effects by inhibiting the growth of several plant species and even crop plants. The inhibitor is soluble in polar solvents, partly soluble in semi-polar solvents and insoluble in nonpolar solvents.

A dense infestation of weed may have as much as 3 t acre^{-1}(7.6 t ha^{-1}) of dry quackgrass rhizomes containing about 50% carbohydrates [9]. Even after a quackgrass infestation is killed, the decay of these rhizomes can seriously deplete the supply of nitrogen available in the soil.

Due to its hardy and tenacious growth habit, it is difficult to control quackgrass by cultivation [9]. Complete quackgrass control requires not only destroying existing shoots but also killing the dormant buds on rhizomes or depleting rhizome food reserves. Deep ploughing and fork hoeing can bring the rhizomes to the soil surface and kill them. Repeat cultivations stimulate dormant buds to grow and reduce stored carbohydrates. Cultivation every 5-10 d with a heavy-duty, spring-toothed harrow is very effective.

Quackgrass can be controlled more effectively when glyphosate (0.8-1.5) is applied in short-interval, split applications rather than a single application. Spraying on actively growing quackgrass, folowed by deep ploughing and another application on regrowth effectively eliminates most of the infestation. Similar control can be obtained even at lower rates if ammonium sulphate or a lipophilic surfactant is tank mixed with glyphosate. Fluazifop-P (50-200 ha^{-1}) and sethoxydim (0.2-0.5) give moderate to satisfactory control of this perennial grass.

23. *Euphorbia hirta* L. (Garden spurge, Asthmaweed)

E. hirta is a small, prostrate, annual herb. It is 15-30 cm tall, with a taproot from which develop several-branched stems, creeping along the ground at first and becoming erect later. Branches are reddish and clothed with brownish crisp hairs. Leaves are simple, opposite, 3-4 cm long and 1 cm wide, with short stalk and finely serrate margin (Fig. 15.21). Flowers are minute and form dense, rounded, almost sessile clusters in axils of leaves. Fruits are small capsules containing 3 seeds. The main characteristics of this weed are the milky sap of the stem and sometimes purple blotched leaves with toothed margins.

E. hirta is commonly found in the tropics and subtropics, both in moist and dry environments, in cultivated lands, lawns, gardens and wastelands. It reproduces by seed. Many crops, e.g. groundnut, potato, sugarcane, maize, rice, sorghum, etc. are infested by this weed.

Cultivation, hoeing and manual methods can control this weed effectively. PRE herbicides such as triazines, ureas and thiocarbamates give better control. At POST, 2,4-D and MCPA are effective.

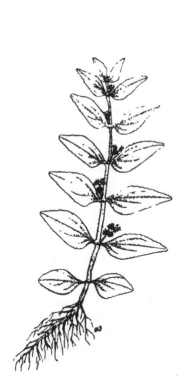

Fig. 15.21. *Euphorbia hirta.*   **Fig. 15.22.** *Imperata cylindrica.*

24. *Imperata cylindrica* (L.) Beauv. (Thatchgrass, Cogongrass, Alang-alang)

I. cylindrica a perennial grass, is one of the world's worst weeds. It has an extensive and deeply penetrating system of rhizomes. It propagates through rhizomes and seeds. The shoot grows to a height of 60-120 cm depending on growth conditions. Leaves are linear, upright, with rough sharp edges and a long tapering point (Fig. 15.22). Panicle is about 20-30 cm long, 0.5-2 cm wide, and silky with silvery white, dense, fluffy woolly hairs amidst which are concealed spikelets with purple stigma and yellow anthers. Spikelets are lanceolate to oblong. *I. cylindrica* is a prolific seed producer. The seeds travel long distances overland. The new seedlings do not develop rhizomes for 4-6 wk. The rhizomes are firmly rooted, white and succulent. Soerjani [29] estimated that a crop of one hectare of *Imperata* may produce 4.5 million shoots, more than 10 t of leaf material and 6 t of rhizomes. The rhizomes have very rapid sprouting ability. The buds of rhizomes are white, brown or dark brown. The root system is fibrous. *Imperata* flowers between May and September.

I. cylindrica occurs widely throughout the tropics. It quickly infests abandoned crop lands in low fertile soils. It can withstand dry spells in sandy soils as well as water logging situations in heavy soils. But it thrives best in wet areas of good soils. It is a most troublesome weed in the tropics in plantation crops, particularly tea, rubber, coffee and pineapple.

Imperata is used as a thatching material for homes and farm sheds. It serves as a good soil binder and thus helps to prevent soil erosion in cleared forests, in high rainfall areas, and along canal and river banks.

It is very difficult to eradicate by cultivation because of its deep rhizomes. Fork hoeing and manual removal of rhizomes may help only in reducing the infestation. Herbicides provide a more effective tool for controlling this weed. Paraquat kills the top growth and allows regeneration of rhizomes in 1 to 3 wk time. But repeated applications at 10 to 15 d intervals 4-8 times, exhaust the food reserves and hence the regenerative ability of rhizomes. Translocated herbicides such as dalapon and glyphosate are more effective. The schedule of dalapon followed one week later by paraquat and again 6 wk later by paraquat or dalapon gives more effective weed control. However, glyphosate (0.8 to 1.6), which is rapidly absorbed by the foliage and rapidly translocated to the rhizomes, gives permanent control of *Imperata*. It should be applied when the weeds are actively growing.

Elimination of *Imperata* results in an immediate infestation by other weeds, particularly the annual broadleaf weed like *Borreria hispida*. *Imperata* exhibits allelopathic effects on *Borreria hispida*, a broadleaf annual weed in tea [1].

25. Lantana camara L. (Largeleaf lantana)

L. camara, an erect perennial shrub, is a weed of cultivated land, wasteland, fence lines, roadsides, etc. It grows well in dry as well as wet regions to a height of 2-5 m and is branched. Stem is 4-angled, covered with recurved prickles. Leaves are opposite in pairs, ovate to ovate-lanceolate, 4-8 cm long and 2-5 cm wide, with serrate margins and rough upper surface (Fig. 15.23). Leaves are strongly aromatic. Petioles are 2 cm long. Flowers are small, axillary and terminal, generally yellow and pink on opening but changing to orange and red, sometimes blue or purple. Fruits are small, black berries clustered into round heads.

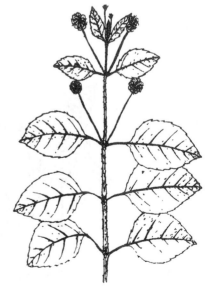

Fig. 15.23. *Lantana camara.*

L. camara is widely distributed in the tropical, subtropical and temperate regions. It is one of the most serious weeds in coffee in Indonesia, oil palms in Nigeria, cotton in Turkey and coconuts in Fiji and Trinidad. *Lantana* is a serious problem in cropped and non-cropped areas in India. The seeds are carried by birds and the plant, once established, quickly closes over open areas where it forms dense thorny thickets. When these thickets are cleared, regrowth of suckers occurs, and even if the roots are dug out, numerous seedlings will appear.

Mechanical methods such as cutting, burning and digging may result in the regrowth of an even larger number of shoots. Foliage application of herbicides kills the top growth without preventing regrowth. However, basal-bark application of 2,4-D or 2,4,5-T in diesel oil gives better results. Glyphosate (2.0-4.0) gives excellent control of *L. camara*.

Lantana can also be controlled effectively by the biological method as discussed in Chapter 11.

26. *Mikania micrantha* H.B.K. (Climbing hempvine)

M. micrantha is a fast growing perennial. It is a gregarious climber which covers up a vast area within a short time, cuts off sunlight and suppresses the host plant, eventually killing it. Stem is branched, pubescent to glabrous and ribbed. Leaves are opposite, thin, cordate, triangular or ovate; Leaf blade is 5-15 cm long and 2-10 cm wide on a petiole of 2-8 cm long (Fig. 15.24) Leaf margins are coarsely dentate and crenate. Leaf surface is glabrous. Inflorescence is a capitulum or head. Flowers are white, 4-6 mm long, surrounded by an involucre bract on a common disc (receptacle). Fruit is an achene, linear-oblong, 1.5-2 mm long, black, with tuft of hair. Flowering time is between November and January. It is propagated by seeds and vegetative parts like old rootstocks and stems.

Fig. 15.24. *Mikania micrantha.*

M. micrantha grows well in regions of high rainfall. It prefers rich damp soil and rarely grows in dry areas. It thrives in open places in forest areas, plantation crops, fallow lands, and along rivers, streams and hillsides. It makes good growth in partial shade conditions. In India, it commonly occurs in plantation crops such as tea, coffee, cardamum, etc., and in forest clearings and nurseries. In tea, it grows over the top of the bushes and suppresses shoot production; it makes plucking of tea leaf difficult.

Manual methods such as sickling and uprooting do not eradicate the weed. POST application of 2,4-D is very effective in controlling this weed. PRE application of simazine prevents new weed growth.

27. *Orobanche cernua* Loeffl (Broomrape)

O. cernua is a phanerogamic plant parasite on tobacco appearing in all the tobacco growing tracts in tropical countries including India [16]. It debilitates the plant to the maximum, resulting in stunted crop growth and very poor yields. It is endemic every year, attaining epidemic proportions in certain years.

It is a reddish-brown or yellowish-brown unbranched herb growing to a height 60 cm, with leaves scale-like, ovate to lanceolate, 6-20 mm long, acute, sessile and devoid of chlorophyll (Fig. 15.25). Stem is pubescent. Spike is loosely flowered, interrupted below and continuous above. Flowers are 1-2 cm long, with petals united into a broadly tubular corolla, 2-lobed upper lip and 3-lobed lower lip. The life cycle of *O. cernua* is completed in about 3 mon after planting [16]. In the presence of tobacco plants, *Orobanche* seeds germinate in the second week after planting and infect the roots in the third week. *Orobanche* shoots emerge from the sixth week onwards. Flowering is completed by the seventh week. Seed formation and capsule drying are completed in the eleventh and twelfth weeks.

Krishnamurty et al. [16] reported that in tobacco the incidence of *orobanche* occurs at an intensity of 4.3-7.3 shoots per diseased plant. At an infestation of 7.3 shoots per host plant, growth inhibition and yield loss in tobacco was 50%. *Orobanche* seeds germinate between November and February.

Fig. 15.25. *Orobanche cernua.*

Manual removal is the quicker method of *Orobanche* control. But this needs to be repeated many times in the season and hence it is very labour-intensive and expensive. Very few herbicides have succeeded in eradicating this weed.

The most effective method of controlling *Orobanche* is to grow trap crops in rotation with tobacco. The trap crops stimulate germination of *Orobanche* seeds, but the infection dies either during the filamentous stage (germinating seeds with whitish radicles attached to the roots of host plants) or during the subterranean button-size growth stage without subsequent emergence into flowering shoots [16]. These crops include sunflower, sesame, cotton, soybean, finger millet and others.

Orobanche may be effectively controlled by PRE and POST applications of imazethapyr (20-40 g ha^{-1}) and imazapyr (12.5-25 g ha^{-1}) in tobacco, pea and sunflower. Glyphosate (POST), due to its limited selectivity, can be used only in tobacco, lentil (*Lens culinaris* L.) and broad bean (*Vicia faba* L.), but not in pea and sunflower.

28. *Oxalis corymbosa* DC. and *O. corniculata* L. (Wood sorrel)

O. corymbosa and *O. corniculata* are creeping, stoloniferous perennial herbs. They have a short, thick, tuberous or bulbous rootstock. In *O. corymbosa*, the leaves have 20-30 cm long petioles (Fig. 15.26).

Fig. 15.26. *Oxalis corymbosa*

Peduncles of flowers are formed in centre of the base. Leaves are trifoliate and entire. Flowers are pink and 1.5-2 cm long. Flowering takes place between January and March. Root is bulbous or tuberous, formed just below leaves. Bulb consists of 70-120 bulbils which are brittle. Bulbils can grow next season into a new plant. The plant absorbs moisture from the soil and the infested soil becomes very dry.

O. corniculata is similar to *O. corymbosa* but has smaller leaves. Leaves have petioles up to 10 cm long. Leaflets are clover-shaped. Flowers are yellow, funnel-shaped and about 1-1.5 cm in diameter. Flowering occurs between February and March. These plants grow well under humid, moist and partially shaded conditions and on all types of soils.

Manual methods cannot completely eradicate the seed. However, deep cultivation or digging and overturning the soil in summer can kill the bulbs and bulbils. This however is not practicable in a permanent crop like tea. Foliage-applied herbicides such as glyphosate, which are translocated sufficiently to reach the bulbs, are very effective on these weeds. Amitrole-T is moderately effective. Repeated applications of paraquat can eradicate these weeds to a greater extent. 2,4-D is only partially effective. PRE herbicides such as diuron, metribuzin, oxadiazon, oxyfluorfen, etc. give good control of this weed.

29. *Panicum repens* L. (Torpedo grass)

P. repens is an aggressive perennial grass. It is a creeping perennial which propagates vegetatively through rhizomes. The rhizomes are knotty and swollen in appearance and

Fig. 15.27. *Panicum repens.*

send out erect culms from the rather distant nodes. Culms are clothed at base with bladeless sheaths. Leaves are 15-25 cm long and 1.5 cm wide or less, linear, flat or folded, with round base (Fig. 15.27). Panicles are 6-18 cm long, somewhat loose and open and erect or ascending. Spikelets are two flowered, pale green or pale yellow.

P. repens grows in different soils ranging from sandy to heavy clay. It is found in coastal areas and along rivers. It occurs under both irrigated and rain-fed conditions. It spreads very rapidly and becomes very persistent. In India, it is a serious weed in plantation crops such as tea, coffee, oil palm and coconut. It is also found in groundnut, cotton and sugarcane.

Tillage can control this weed only partially. Foliage-applied herbicides such as glyphosate and dalapon, which translocate readily into the rhizomes and prevent their regeneration, are more effective against this weed. Paraquat can achieve only partial control but repeated applications can exhaust the starch reserves in the roots and rhizomes and kill the weed permanently.

30. *Parthenium hysterophorus* L. (Wild carrot)

P. hysterophorus is a noxious exotic weed (Fig. 15.28) which has spread to many parts of India, covering approximately 5 million ha [11]. It flowers during most of the

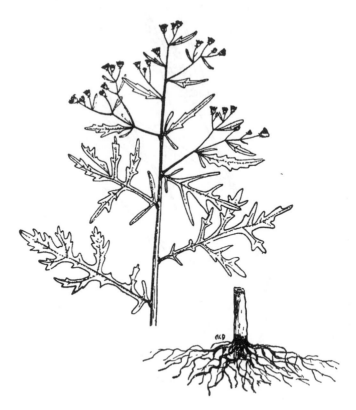

Fig. 15.28. *Parthenium hysterophorus.*

year. Kanchan and Jayachandra [11] reported that each plant produces an average of 7000 cypsella (fruits) which emerge into a dense mat of seedlings after a few showers. The young rosettes with their radial leaves closely pressed to the ground allow no other species to come up in their midst. *Parthenium* plants prefer moist, shady and organically rich habitats [8]. This weed shows remarkable adaptability to environmental extremes. It is photoperiodically and thermoperiodically neutral. It exerts allelopathic influence on neighbouring plant species. The seed leachates inhibit germination of other weed seeds. It is a rapid colonizer and competes out other vegetation in its vicinity. It causes dermatitis and other allergies in human beings.

The aboveground portions, i.e., leaf axils, stems and branches, are capable of sprouting when in touch with soil. The weed is a prolific seed producer. The seeds are extremely light weight, armed with pappus and disseminated by wind, water, birds and animals. Through its persistence and rapid spread it has become not only an agricultural weed, but a 'municipal weed' as well.

Its control by manual methods is not satisfactory, PRE herbicides such as simazine, atrazine, alachlor, metribuzin, diuron, butachlor, terbutryn, etc. can prevent seedling emergence up to 2-5 mon, depending on the rate of application. POST applications of glyphosate, 2,4-D, MSMA+2,4-D, MSMA+paraquat and 2,4-D+paraquat provide effective control of *Parthenium*.

31. *Paspalum conjugatum* Berg. (Sour paspalum, Hilograss)

P. conjugatum is a creeping, stoloniferous perennial grass. Culms are 20-40 cm long, erect, with smooth nodes. Leaves are 5-20 cm long 5-15 mm wide, lanceolate, smooth, flat and sparingly hairy, with rough or stiff hairy margins (Fig. 15.29). Spikes are two (paired), rarely three at apex of culm, widely spreading, straight or somewhat arched. Spikelets are flattened, pale green, margin fringed with long white silky hairs. Root is fibrous and shallow. Nodes stirke roots when in touch with soil. The grass reproduces by runners and seeds and spreads quickly by means of stolons. Flowering occurs between July and September.

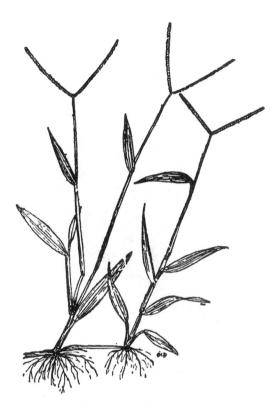

Fig. 15.29. *Paspalum conjugatum.*

It is prolific in humid tropics and a very serious weed in plantation crops such as tea, coffee, pineapple, etc. It also occurs in many annual crops in humid and high rainfall areas.

When well established, *Paspalum* is difficult to control by mechanical methods. Among the PRE herbicides, diuron, oxyfluorfen and methazole are more effective. At POST paraquat gives immediate control of the weed, with regeneration occuring after 3 to 4 wk. Paraquat-diuron and paraquat-MSMA combinations are more effective than paraquat alone. Glyphosate gives excellent control of this perennial grass.

32. *Pennisetum clandestinum* Hochst ex chiov. (Kikuyugrass)

P. clandestinum is a creeping, low, mat-forming perennial grass, with stout rhizomes below the surface and long runners above ground, which produce tough, fibrous roots and short, stout, vertical branches from the nodes. Culms are 30-100 cm long, prostrate, with rooting from the nodes. Internodes are short. Leaves are 30 cm long, narrow, spreading, blunt to pointed, initially folded and hairless or hairy, with rough margins (Fig. 15.30). The flowers are not borne at the top of a long stem, but are enclosed within the topmost leaf sheaths of short side shoots, similar in appearance to non-flowering shoots. Spikelets are 1-1.5 cm long and borne in groups of 2-4 in upper leaf sheaths.

Fig. 15.30. *Pennisetum clandestinum.*

P. clandestinum is well adapted to the humid tropics or subtropics especially in higher elevations and in soils of high fertility. It can withstand dry weather although it does not grow rapidly. The plant rarely sets seeds except in higher elevations in the tropics and subtropics. It usually propagates vegetatively.

It can be controlled by POST application of dalapon followed by repeated applications of paraquat on the regrowth or cultivation. Dalapon alone does not affect regeneration by stolons. Paraquat-MSMA combination, applied at POST, gives better control. But POST application of glyphosate gives by far the most effective control of both aerial and underground parts of *P. clandestinum.*

33. *Pennisetum pedicillatum* Trin. (Kyasumagrass)

P. pedicillatum is a tufted perennial grass with erect and woody culms and prominent nodes. Leaves are 10-25 cm long, 5-10 cm wide and accuminate. Inflorescence is a light panicle 1-3 cm long. Spikelets, 2-5 mm long, are surrounded by bristles. They are pedicelled or in groups of 2 to 5, with one sessile and the others pedicelled.

P. pedicillatum is a weed of humid tropics and subtropics. It is difficult to control by normal mechanical methods. Repeated application of paraquat or dalapon is effective. Glyphosate is most effective in controlling this weed.

34. *Phalaris minor* Retz. (Littleseed canarygrass)

P. minor (Fig. 15.31) is one of the most predominant and troublesome annual weeds of wheat fields of tropical countries including India and Pakistan. It is of Mediterranian origin. It is a major weed in Latin America also. The *Phalaris* plant is morphologically similar to the wheat plant and hence very diffi-cult to identify during the seedling stage. It flow-ers in 78 to 105 d and matures in 113 to 155 d depending on the time of planting between Oc-tober and December [4]. It produces 300 to 460 seeds per panicle and delayed sowing in De-cember causes more tiller and dry matter pro-duction than early sowing in October [4]. Verma and Yaduraju [31] reported that freshly harvest-ed seeds of *Phalaris* did not germinate for 6 mon but responded to prechilling and dehusking. Germination of one- and two-year-old *Phalaris* seeds was 96% at 15°C and 80% at 20°C. Bhan and Bhaskar Choudary [4] found that *Phalaris* seeds germinated best between 10°C and 15°C.

As *Phalaris* plants are difficult to distinguish from wheat plants at the seedling stage, herbi-cides are the only answer to control this weed. Methabenzthiazuron (0.75-1.5) metoxuron (1.0-2.0) isoproturon (0.75-1.5) applied at POST about 25 d after planting give satisfactory control of *Phalaris*. Methabenzthiazuron (0.75-1.5), pendimethalin (1.0-1.5) and terbutryn (1.0-1.5) give good PRE control of the weed.

The crop rotation involving wheat influences the population of *Phalaris*. Katyal et al. [13] re-ported that the population of *Phalaris* was high-er in a wheat plot under paddy-wheat rotation compared to cotton-wheat rotation.

Fig. 15.31 *Phalaris minor.*

35. *Portulaca oleracea* L. (Common purslane)

P. oleracea (Fig. 15.32) is an annual herb which grows erect and prostrate. It repro-duces by seed and stem fragments. The stem is succulent, fleshy, freely branched, normally forming mats when infesting intensively. Branches are numerous, often red-dish or purplish and 30 cm long. Leaves are shiny, simple, spirally arranged or oppo-site and found in clusters towards the end of branches; they are sessile, with smooth margins and broad-rounded tips, 0.5-3.0 cm long and 0.5-2 cm wide. Flowers are yellow, sessile and found solitary in the leaf axils or as clusters at the terminal end of the branch.

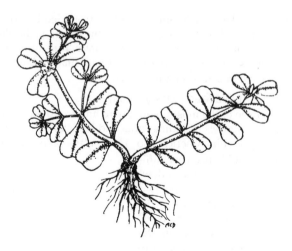

Fig. 15.32 *Portulaca oleracea*

This is a common arable weed, distributed throughout tropical and subtropical areas. It is a commonly occurring weed in different parts of India. It thrives better on moist soils and irrigated areas but requires certain temperatures before emerging. It is found in maize, sorghum, wheat and other cereal crops. It is a persistent weed as it can continue to live for weeks even after the root has been cut. The stem fragments can re-establish and survive as individual plants.

Growth regulator herbicides such as 2,4-D and MCPA are not very effective on the established plants when applied POST. The contact herbicide paraquat, is very effective on young seedlings but not on established plants.

Portulaca is sensitive to PRE herbicides such as simazine, atrazine, metolachlor, trifluralin, diuron, monuron, linuron, norflurazon, chlorpropham, metribuzin (0.5) + metolachlor (2.0-3.0) and metribuzin (0.5)+alachlor (2.0-3.0).

36. *Saccharum spontaneum* L. (Kansgrass, Wild sugarcane)

S. spontaneum (Fig. 15.33) is one of the most pernicious perennial weeds found in tropical countries. It has a creeping and penetrating root system. Roots grow 1-2 m deep. They can go even deeper in the soil and form an extensive network of root system. The weed is propagated by seeds as well as underground rhizomes. It flowers in September-October and the seeds are disseminated by wings to long distances.

S. spontaneum is found in fallow soils, along canals, ponds, marshy places and railway lines. It provides a good roof thatching material. It is also useful for making ropes, mats and brooms.

This perennial weed can be controlled by POST application of paraquat and glyphosate. Effective control with glyphosate is obtained when concentrated herbicide solution is placed on the foliage using an herbicide glove (Chapter 14).

37. *Solanum nigrum* L. (Black nightshade)

S. nigrum is an erect annual herb (Fig. 15.34) that occurs both in the tropical and temperate regions. It grows to a height of 45-60 cm. It has dark-green simple, alternate leaves, small clusters of white flowers with yellow centres and green globular berries, which normally turn black. The inflorescences arise directly from the branches.

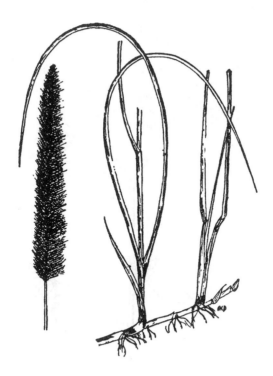

Fig. 15.33 *Saccharum spontaneum*

Solanum nigrum grows well in a moist environment but can also thrive well in low rainfall conditions. It is a prolific seed producer, with several thousands of seeds per plant. It is a major weed in sorghum, maize, cotton, vegetables, vineyards and orchard crops.

It is moderately susceptible to 2,4-D, MCPA diquat, dinoseb and other foliage-applied herbicides. *Solanum* is sensitive to herbicides applied PRE to the soil. PRE applications of chlorpropham (1.0) + propyzamide (1.0), chloropham (1.0) + metolachlor (1.5), linuron (1.0), chloramben (2.0), alachlor (2.0-4.0) monolinuron (0.5) and dinitramine (0.37) give satisfactory control of *S. nigrum*.

38. *Sonchus oleraceus* L. (Annual sowthistle)

S. oleraceus is an erect, 60-120 cm tall, soft, annual herb. Stem is smooth, pale bluish-green or powdery and exudes a milky latex when cut. Leaves are spirally arranged, up to 20 cm long and 3-6 cm wide. Basal leaves originally rosetted and have no distinct stalk. Upper leaves are sessile clasp the stem and project backwards past the stem (Fig. 15.35). Leaves are divided into broad-toothed and spiny segments, with the terminal segments largest, triangular and green on both sides. Flowering heads are numerous, stalked and borne in a loosely branched, terminal inflorescence. They consist of yellow strap-shaped florets surrounded by several rows of overlapping bracts. Fruits are flattened and crowned in a ring of long, white, simple hairs. Seeds (achenes) are 2-3 mm long, broadest towards the top, narrow at the base, ribbed lengthwise and transversely wrinkled by the presence of minute warty pumps (tubercles).

Fig. 15.34 *Solanum nigrum.*

S. oleraceus reproduces by seeds carried by wind or water. They can remain viable for several years. The weed is common in arable land, roadsides, waste land, etc. It adapts well to a number of environments as well as high and low elevations.

S. oleraceus is effectively controlled by POST applications of 2,4-D, MCPA, MCPB, mecoprop, dichlorprop, paraquat and other foliage-applied herbicides. It is also susceptible to PRE applications of diuron, monuron, simazine, atrazine, etc.

39. *Sorghum halepense* (L.) Pers. (Johnsongrass)

S. halepense is an erect perennial grass which reproduces by large rhizomes and by seeds. Root system is freely branching, fibrous, with stout rhizomes. Stem is erect, stout, 0.5-3 m tall, arising from creeping scaly rhizomes (Fig. 15.36) Leaves are alternate, simple, smooth, 20-25 cm long and 1.5-2 cm wide. Panicle is large, purplish, hairy, 15-50 cm long. Sessile spikelet is 4.5-5 cm long and ovate. Pedicillate spikelet is 5-7 cm long and lanceolate. Leaf sheath is ribbed. This grass is a heavy seed producer.

S. halepense grows well on arable land and along irrigated canals. In the tropics, it is a troublesome weed in sugarcane, maize and sorghum. Singh and Pant [27] observed 96.8 % reduction in the yield of sugarcane due to a severe infestation of *S. halepense*. They found that three applications of dalapon (3.0) or asulam (3.0) at an interval of 20 d reduced its infestation significantly and consequently increased sugarcane yield. It can be effectively eradicated by mechanical cultivation of the infested soil, followed by foliage application of glyphosate or dalapon+ammonium sulphate on regrowth and

Fig. 15.35 *Sonchus oleraceus.*

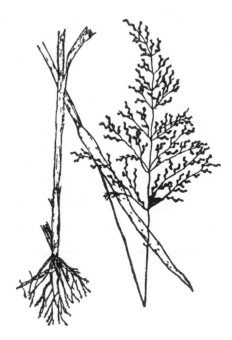

Fig. 15.36 *Sorghum halepense.*

followed, if necessary, by a soil application of atrazine+EPTC before sowing maize. Two ploughings and two discings followed by soil incorporation of trifluralin also give very good control. *S. halepense* is sensitive to PRE applications of EPTC+atrazine, alachlor, butylate, metolachlor, butralin, diuron, fluchloralin, profluralin, vernolate, trifluralin and pendimethalin. In any case, treatments in which application of these herbicides to the soil is preceded by mechanical cultivation are most effective and economical in controlling this rhizomatous perennial grass.

Good control of Johnsongrass may also be obtained by nicosulfuron (35 g ha^{-1}) and primisulfuron (40 g ha^{-1}), applied POST. Besides, PPI of butylate or EPTC, followed by POST application of nicosulfuron or primisulfuron provides greater than 90% control of seedling and rhizome Johnsongrass [12]. POST application of clethodim (100 g ha^{-1}), quizalofop-P ethyl (70 g ha^{-1}), fluazifop-P (200 g ha^{-1}) and sethoxydim (300 g ha^{-1}) either alone or in combination with lactofen, imazaquin, chlorimuron, or fomesafen gives 80-99% control in soybean [24].

40. *Striga* (Witchweed)

Striga is the most predominant root parasite in cereal crops. It establishes contact directly with the vascular system of the host plant and sucks water and nutrients from the host, resulting in crop yield loss ranging from 15 to 75% depending on the severity of infestation. When *Striga* seed germinates, the radicle emerges and grows towards the nearest host root through a chemotropic response. If the tip of the new root touches the host cells, a bell-like swelling develops and is closely attached to the host. The root cap cells are absorbed and a profuse growth of root hairs develops near the *Striga* root tip. The parasite prefers young roots. After having invaded the host roots, the parasitic roots move into the host by digesting its root tissue. The papillae spread through the host from the tip of the *Striga* root and pass between endodermis cells to enter the vascular tissues of the host and join the conducting tissues of the parasite and host. The nutrients and water pass from the conducting tissue of the host directly to that of the parasite. When *Striga* is underground, it is totally dependent on the host; once it emerges, it is able to develop chlorophyll and synthesize food materials. In that case, the parasite depends on the host only for water.

In India, five species of *Striga* are known to occur [10]: *S. lutea, S. angustifolia, S. densiflora, S. gesnerioides* and *S. sulphurea*. Of these, *S. lutea* is the cause of great economic loss in agriculture, particularly in crops like sorghum and millets.

S. lutea Lour, is an erect, stiff, 10-30 cm tall annual herb. Leaves are narrow, linear, nearly opposite or alternate, 2-3 cm long, 3-4 mm wide and sessile (Fig. 15.37). Leaf surface is rough with small prickles or hairs. Stem is slender, rigid, four-sided or grooved, simple, branched and covered with rough white hairs. Inflorescence is in terminal spikes and 10-15 cm long. Bracts in axils of which the flowers arise are longer than calyx. Flowers vary in colour—white, yellow, chrome yellow, red, pink and purple. Corolla is 2-lipped, lower lip 3-lobed and upper lip not lobed. Flowers are produced one month after emergence and seeds ripen after another month.

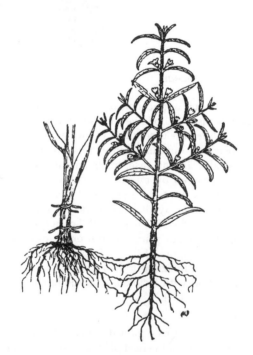

Fig. 15.37. *Striga lutea.*

S. lutea is a prolific seed producer, but the seeds remain dormant in the soil for a long time. They can only germinate when close to a suitable host and in response to a chemical stimulant secreted by the roots of host plants. This chemical is volatile and its production is influenced by temperature. It is produced before the host plant begins photosynthesis. The host roots are invaded by *Striga* within 2-3 wk after planting. The chemical stimulant is known as '**strigol**'. *S. lutea*, a native of India, is dominant in lighter soils, drier climate and low rainfall areas. It does not thrive well in wet, high rainfall areas.

Striga can be effectively controlled by stimulating germination of seeds in the soil and destroying by tillage or herbicides after they emerge. As the seeds can remain dormant in the soil for several years, repeated use of manual methods and/or herbicides is essential. Besides, proper crop rotation programme with **trap crops** which produce the chemical stimulant necessary for *Striga* seed germination but not parasitized by the witchweed and **catch crops,** which stimulate germination but, in turn, are parasitized by the weed, offers effective control of this weed. Soyabean, cotton, cowpea, groundnut, chickpea and pigeon pea are some of the trap crops which can be used in rotation with catch crops such as sorghum maize and millets.

POST application of 2,4-D, MCPA, paraquat or diesel oil can kill established plants. Since germination is continuous over a prolonged period, soil-applied herbicides which have longer persistence of activity are of greater use. Preplanting soil incorporation of fenac or 2,3,6-TBA reduces parasitism by witchweed. But these herbicides inhibit the growth of maize, sorghum and other host crops. PRE application of simazine, atrazine or propazine gives an effective control of this weed in maize, sorghum and sugarcane. Similar control is also obtained in maize with linuron and monolinuron.

REFERENCES

1. Anonymous. 1977-78. Tea Research Association, Tocklai Experimental Station, Jorhat, Assam, India. Ann. Sci. Rep. p. 24.
2. Andrews, F.W. 1960. A study of nutgrass (*Cyperus rotundus* L.). Ann. Bot. 4:177-193.
3. Arakeri, H.R. 1957. Chemicals to kill weeds. Indian Fmg. 7(3):23-27.
4. Bhan, V.M. and D.B. Bhaskar Choudary. 1976. Germination, growth and reproductive behaviour of *Phalaris minor* Retz. as affected by date of planting. Indian J. Weed Sci. 18:126-130.
5. Bhati, P.R., N. Ashraf and D.N. Sen. 1979. Ecology of Indian arid zone weeds. VII. *Chenopodium* spp. Geobios International. 6:20-23.
6. Chela, G.S. and H.S. Gill. 1980. Chemical control of *Echinochloa crus-galli* in transplanted rice (*Oryza sativa*). Indian J. Weed Sci. 12:7-14.
7. Cole, M.J. 1961. Interspecific relationships and intraspecific variations of *Chenopodium album* L. in Britain. Part II. The taxonomic delimitation of the species. Watsonia 5:47-48.
8. Gupta, S.R., J.N. Gupta and T.R. Dutta. 1977. Ecological studies on noxious weed *Parthenium hysterophorus* L. Weed Sci. Conf. Workshop in India Abstr. No. 161.
9. Harvey, R.G. 1973. Quackgrass: *Friend or Foe. Weeds Today* 4(4):8-9.
10. Hosamani, M.M. 1978. *Striga* (a Noxious Root Parasitic Weed). Royal Printers and Univ. of Agricultural Sciences, Bangalore, India. 165 pp.
11. Kanchan, S.D. and Jayachandra. 1977. *Parthenium* weed menace in India and its control. Weed Sci. Conf. Workshop in India. Abstr. No. 162.
12. Kapusta, G., K. Tomida and J.A. Bailey. 1991. Johnsongrass control in corn with preplant incorporated and postemergence herbicides. Res. Rep. North Cent. Weed Control Conf. 48:231.
13. Katyal, S.K., L.C. Godara and R.K. Bagga. 1980. Field evaluation of methabenzthiazuron for control of grassy weeds in wheat under different agro-climatic conditions. Conf. Indian Soc. Weed Sci., Abstr. no. 87.
14. Kaushik, S.K. and R.C. Gautam. 1980. Weed control studies in hybrid pearlmillet. Conf. Indian Soc. weed Sci., Abstr. No. 108.
15. Krishnamurthy, G.V.G., K. Nagarajan and G.H. Chandwani. 1976. Studies on the control of *Orobanche* on tobacco. Tobacco Research. 2(1):58-62.

446

16. Krishnamurthy, G.V.G., K. Nagarajan and R. Lal. 1977. Some studies on *Orobanche cernua* Loefl., a parasitic weed on tobacco in India. Indian J. Weed Sci. 9:95-106.

17. Li, Y.H. 1987. Parasitism and integrated control of dodder on soybean. *In* H. Chr. Weber and W. Forstreuter, (eds.). Parasitic Flowering Plants. Marburg, German, pp. 497-500.

17a. Mitich, L.W. 1997. Redroot pigweed (*Amaranthus retroflexus*). Weed Technol. 11: 199-202.

18. Pandey, S.B. 1969. Photocontrol of seed germination in *Anagallis arvensis* L. Trop. Ecol. 10:96-138.

19. Ramakrishnan, P.S. 1960. Ecology of *Echinochloa colonum* Link. Proc. Indian Acad. Sci. 52:73-90.

20. Ranade, S.K. and W. Burns. 1925. The eradication of *Cyperus rotundus*. Memoirs Dept. Agr. In India 13: 99-181.

21. Rao, P.N. and A.R.S. Reddy. 1987. Effect of China dodder on two pulses: greengram and cluster bean—the latter a possible trap crop to manage China dodder. *In* H. Chr. Weber and W. Forstreuter, (eds.). Parasitic Flowering Plants. Marburg, Germany. pp. 665-674

22. Ray, Bibhas and M. Wilcox. 1969. Chemical fallow control of nutsedge. Weed Res. 9:86-94.

23. Reddy, A.R.S. and P.N. Rao. 1987. Cluster bean—a possible herbicidal source for managing China dodder. Proc. 11th Asian-Pacific Weed Sci. Soc. Conf. Taipei, Republic of China. pp. 265-270.

24. Roy Vidrine, P., D.B. Reynolds and D.C. Blouin. 1995. Grass control in soybean (*Glycine max*) with graminicides applied alone and in mixtures. Weed Technol. 9:68-72.

25. Sen, D.N. 1977. Ecophysiological studies on weeds of cultivated fields with special reference to bajra (*Pennisetum typhoideum* Rich.) and til (*Sesamum indicum* L.) crops. 3rd US PL-480 Project Report, Jodhpur Univ., Jodhpur, India. 91 pp.

26. Sen, D.N. 1981. Ecological approaches to Indian weeds. Geobios International (Jodhpur, India) 301 pp.

27. Singh, Govindra and P.C. Pant. 1980. Control of *Sorghum halepense* (L.) Pers. In spring planted sugarcane. Conf. Indian Sci. Weed Soc. Abstr. no. 106.

28. Sinha, T.D. and C. Thakur. 1961. Control of nutgrass weed by cultivation. Indian J. Agron. 12:121-125.

29. Soerjani, M. 1970. Alang-alang, *Imperata cylindrica* (L.) Beauv., pattern of growth as related to its problem of control. BIOTROP Bulletin 1. Regional Center for Tropical Biology, Bogor, Indonesia.

30. Tosh, G.C., G.K. Patro and A. Misra. 1977. Effect of pronamide and chlorpropham on *Cuscuta* stem parasite and niger. Indian J. Weed Sci. 9:82-88.

31. Verma, K. and N.T. Yaduraju. 1980. Preliminary germination studies on wild oats and *Phalaris*. Conf. Indian Soc. Weed Sci. Abstr. No. 4.

32. Weaver, S.E. and E.L. McWilliams. 1980. The biology of Canadian weeds. 44. *Amaranthus retroflexus* L. and *A. powellii* S. and *A. hybridus* L. Can. J. Plant Sci. 60: 1215-1234.

16

Weed Management in Field and Plantation Crops

INTRODUCTION[1]

All crops grown in the field are subject to weed competition. Weed competition is one of the most important limiting factors in crop production. Successful elimination of this factor is directly linked to the production efficiency of the farm.

Weed problems vary from one crop to another, from one region to another, from one farm to another, and even from one section of the farm to another. Similarly, the weed spectrum in heavy clay soils is not always the same as in lighter soils. Weed growth is more intensive in warm, humid and high rainfall areas than in hot, dry and low rainfall areas. However, since weeds can adapt even to extreme climatic conditions, they are always competitive with crop plants in any situation.

Weeds and weed control are as old as agriculture itself. Mechanical and manual methods have always been employed for weed control. In developing countries, they still offer a viable weed control practice in field and plantation crops.

With the introduction of herbicide technology, weed management became more efficient, time-saving and less expensive. This is particularly true under intensive crop production practices involving high-yielding varieties with high fertilizer and water requirements, which weeds find more conducive for growth. The use of herbicides for weed management, however, does not eliminate the need for manual and mechanical methods and cultural practices. On the contrary, herbicide technology can enhance the efficacy of other methods of weed control as also cultural practices, and consequently the overall weed management efficiency.

Weed competition is maximum during the early stages of crop growth. However, the most critical period of weed competition varies with the crop (Fig. 16.1), growth behaviour of the crop variety, environmental conditions conducive for plant growth, weed species present and intensity of weed infestation. In some situations, late-season weed growth is as critical for crop growth as early-season weed growth. Hence, research needs to be conducted to ascertain the critical period of weed competition for each crop under different growing conditions and evolve appropriate weed management programmes.

---

[1]Herbicide rate (dose) is mentioned as kilogram active ingredient per hectare or kg ha[-1] unless otherwise specified. For example, (1-2) indicates 1.0-2.0 kg ha[-1].

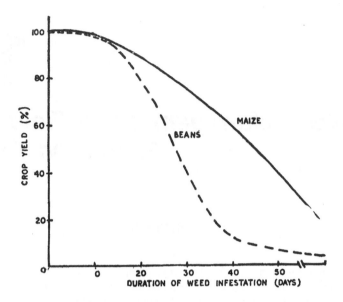

Fig. 16.1. Effect of duration of weed infestation on the yield of maize and beans, with varying weed competitive abilities.

WEED MANAGEMENT IN MAJOR FIELD CROPS

The weed management practices found effective and useful for each of the crops discussed in this chapter are based on the research conducted in various countries. For each herbicide or herbicide mixture, the dosage range at which they show generally effective weed control is given. Further work by the weed researcher is required to determine the exact rate at which they would be effective in a particular crop and weed situation.

The choice and efficacy of a weed management programme in a crop depends on the weed species present. The weed spectrum differs with soils, climatic conditions and crop. The weed spectrum of a heavier soil is different from that of a lighter soil. Similar differences in weed spectrum also exist between arid and semi-arid climates and between acid and alkaline soils. In some situations, the weed spectrum is almost identical irrespective of the crop grown.

Cereals

Rice (Oryza sativa L.)

Rice is grown by direct-seeded and transplanted methods. Weed competition is more severe in the direct-seeded crop than in the transplanted crop. The major weeds that normally infest a rice crop are *Echinchloa crus-galli, E. colonum, Panicum* spp., *Setaria glauca, Cyperus rotundus, C. difformis, Fimbristylis miliacea, Cynodon dactylon, Digitaria* spp., *Commelina benghalensis, Ageratum conyzoides, Ammania baccifera, Monochoria vaginalis, Scirpus* spp., etc. Generally, the weed spectrum in rice is dominated by grasses and sedges, and among them *Echinochloa crus-galli* and *Cyperus rotundus* are the most competitive weeds in both direct-seeded and transplanted rice crops.

Direct-seeded rice is grown under both rain-fed and flooded conditions. Transplanted rice is cultivated only under flooded conditions, but in both puddled and non-puddled situations. Weed spectrum and intensity differ according to the method under which the rice crop is grown.

Weed competition in direct-seeded rice is maximal during the first 3 wk period [4]. Yield reductions in direct-seeded upland rice ranges from 42 to 65% [24]. The most critical period, when crop losses due to weed competition were most severe, ranged from 10 to 20 d after emergence [24]. A 30-d weed-free period was required to preclude significant loss in rice yield [2]. Weed infestation depleted the soil of 24.7 kg ha^{-1} nitrogen, 5.8 kg ha^{-1} phosphorus and 63.4 kg ha^{-1} potassium in one season [10]. The loss of nitrogen by weeds ranged from 11 kg ha^{-1} in transplanted rice to 92 kg ha^{-1} in direct-seeded upland rice [14].

In direct-seeded rice under upland rain-fed conditions, weeds often emerge simultaneously with the crop and competition for soil moisture may begin early, especially in low rainfall areas. Weed seedlings, due to their greater competitive ability, grow more rapidly than rice seedlings. In transplanted rice, flooding and puddling destroy the existing weed growth before seedlings are planted. Furthermore, a new flush of weeds establishes only after 2-3 wk of transplantation, thus enabling the rice seedlings to establish well and withstand the subsequent weed competition better. The extent of yield reduction due to weeds is around 15-20% in transplanted rice, 30-35% in direct-seeded puddled rice and over 50% in direct-seeded upland rain-fed rice. Generally, the tall **indica** varieties, which grow rapidly are more competitive with weeds than the dwarf **japonica** varieties which grow slowly initially. Thus the weed problem is more acute in the high-yielding dwarf varieties.

MECHANICAL WEED CONTROL

Manual weeding by digging, sickling and pulling is the most widely used method of weed control in rice. Manual removal of weeds in rice is, however, difficult and time-consuming due to the morphological similarity between grass-type weeds and the rice crop, especially during early growth stages [24]. In the Philippines, the average labour requirement for a single manual weeding 30 d after seeding is 720 man-hours ha^{-1} [6], while in Taiwan it is 400 man-hours ha^{-1} [3]. In transplanted rice, Japan uses 500 man-hours ha^{-1} for control of weeds until 1949 and only 200 man-hours in 1962 when herbicides were used in conjunction with manual weeding [28]. Korea requires 300 man-hours ha^{-1} for a single manual weeding [11].

Inter-row cultivation by a hand rotary weeder is a useful, time-saving and more economical method of weed control than manual weeding. It uproots the weeds and buries them in the mud. In India, rotary weeding has been found to give yields within 3% of those obtained by manual weeding [12]. Manual pulling is more convenient for weed control in rice nurseries.

Preplanting tillage is an effective means of destroying the existing weed growth. This is followed by flooding and puddling. In recent years, the minimum tillage method has been favoured more than other methods. In this, the contact herbicide paraquat (0.5-0.75) is applied on the existing weed growth and followed a few days later by flooding and cultivation (or puddling). Then the crop is transplanted after a further 2-5 d or the field is drained and seeded. This method saves time on land preparation and precludes doing so when it is not economically beneficial.

CHEMICAL WEED CONTROL

As the total labour requirement for a single manual weeding ranges from 300 to 700 man-hours ha^{-1} [22], the use of herbicides could result in considerable savings to the farmer.

In rice nurseries, thiobencarb (1-2) and butachlor (1-2) may be applied 7-10 d after planting the seed. As these herbicides injure rice during seed germination, application soon after seedling emergence, but just before irrigation, is recommended. Application may be made easier if the EC or G (granular) formulation is mixed with sand (50-75 kg ha^{-1}) and broadcast over the soil surface prior to irrigation. Anilofos (0.4-0.8) and pendimethalin (1.0-1.5) may also be used in case of thiobencarb and butachlor are not available. In case sprouted seeds are sown, these PRE herbicides may be applied 4-6 d after sowing.

In irrigated transplanted crops, an effective broad-spectrum weed control can be obtained by applying the granular mixture of butachlor (5G) (0.5-1.0)+2,4-D ethyl ester (4G) (0.4-0.8) 3-7 d after transplantation of rice seedlings [17]. The granular mixture is applied as a sand mix by mixing in sand (50-75 kg ha^{-1}), This combination is very effective against many broadleaf weeds (*Marsilea quadrifolia, Eichhornia crassipes, Centella asiatica, Monochoria vaginalis, Digera arvensis, Pistia stratiotes,* etc.), grasses (*Echinochloa* spp., *Panicum maximum, Eragrostis* spp., *Brachiaria* spp., etc.), and sedges (*Cyperus rotundus, Cyperus difformis, Cyperus bulbosus, Cyperus iria, Fimbrisylis miliacea,* etc.) for 40-60 d. The granular mixture is better than the EC mixture in terms of efficacy and duration of activity. In this combination, both herbicides prevent emergence of weed seedlings, and 2,4-D kills the emerged weeds. Other granular herbicide mixtures that may be used include thiobencarb (0.5-1.0)+2,4-D EE (0.4-0.8) and anilofos (0.75-1.5)+2,4-D EE (0.4-0.8) [16]. If the weed spectrum is dominated by broadleaf weeds and sedges, 2,4-D EE granules (4G) alone (0.6-1.2) may be used as sand mix 10-15 d after rice transplantation. In case ,annual grasses predominate the field, butachlor (1.0-2.0), thiobencarb (1.0-2.0) and anilofos (0.4-0.8) alone may be applied 0-5 d after transplantation.

Other PRE herbicides, which may be used in transplanted rice when annual grasses are predominant, include oxadiazon (1.0-1.5), pendimethalin (1.0-2.0), cinmethylin (50-100 g ha^{-1}), bensulfuron (35-70 g ha^{-1}), pertilachlor (0.5-1.0), etc. As these are graminicides, combining 2,4-D EE with these herbicides will help in controlling sedges and broadleaf weed infestation. Besides, POST application of acifluorfen (0.15-0.25), bifenox (1.5-2.0), triclopyr (0.2-0.4), quinclorac (0.2-0.4) and bentazon (1.0-1.5) gives satisfactory control of several annual broadleaf weeds. Molinate (3.0-5.0) may be applied at PRE or L-PRE to control annual grasses.

In dryland rice, PRE herbicides will be effective if rain occurs within 0-3 d of their spraying. Generally, annual grasses and sedges are more predominant than broadleaf weeds in dryland rice. Herbicides (EC formulation) such as butachlor (1.0-2.0), thiobencarb (1.0-2.0), oxadiazon (1.0-1.5), anilofos (0.4-0-8), pendimethalin (1.0-2.0), etc., are applied 6-10 d after sowing paddy seeds. In case of a mixed spectrum of weeds, the combinations such as butachlor-2,4-D EE (0.75+0.5), thiobencarb-2,4-D EE (1.0+0.5) and anilofos-2,4-D EE (0.4+0.5) give satisfactory results. Quinclorac (0.25-0.50) may be used at L-PRE or E-POST to control annual grasses. The other POST herbicides that can be used in a predominantly broadleaf weed situation in dryland rice include bifenox (1.5-2.0), acifluorfen (0.15-0.25), dithiopyr (125-250 g ha^{-1}), triclopyr (0.25-0.40), bensulfuron (30-50 g ha^{-1}), fenoxaprop ethyl (50-100 g ha^{-1}) and bentazon (1.0-1.5).

As differential varietal tolerance to herbicides and herbicide mixtures exists in rice, a thorough screening of cultivars is needed to determine the suitability of a herbicide or herbicide mixture.

PPL: preplanting; PPI: preplant incorporation; PPS: preplant surface; PRE: preemergence; L-PRE: late preemergence; POST: postemergence; E-POST: early postemergence; L-POST: late postemergence

For control of algal growth in flooded rice soils, nitrofen and potassium azide are applied 2-3 wk after transplanting. The flood-water is maintained about 15 cm below the crop height for 2 wk after application. Granules can be used to facilitate easy application.

The efficacy of a herbicide depends upon the type of rice culture. Some herbicides are more suitable than others under certain cropping and weed situations. The researcher should identify the most suitable herbicides and herbicide mixtures for each of the situations. Generally, the effectiveness of herbicides in rice is dependent, to a great extent, on the water management practices and timely application of fertlizers in relation to the initial weed problems.

Wheat (Triticum aestivum L.) and Barley (Hordeum vulgare L.)

Weeds are a major problem in what and barley, the winter cereals, causing reduction in yield by 30-50%. Most of the weed competition is during the first 30-40 d after sowing the crop. Yield reduction is mostly due to lack of poor tiller development caused by weed competition.

The major weed species that generally infest wheat and barley fields are *Chenopodium album* (lambsquaters), *Avena fatua* (wild oats), *A. ludoviciana* (wild oats), *Phalaris minor* (wild canarygrass), *Fumaria* spp., *Melilotus alba, Cirsium arvense, Convolvulus arvensis,* etc.

Manual weeding once or twice in the season is effective against annual weeds but not perennial weeds. As some of the annual weeds germinate continuously during the season, repetitive treatment of manual weeding becomes essential. When manual labour is scarce and expensive, herbicides are cheaper and more efficient. Post-sowing mechanical cultivation is not feasible because of close spacing.

Broadleaf weeds in wheat and barley are successfully controlled by POST applications of 2,4-D (1-2), picloram (0.25-0.5) and the combination of dicamba (0-12-0.25)+2,4-D (0.5). The PRE application of diuron or linuron gives adequate control of annual broadleaf weeds in heavy and high organic matter soils. 2,4-D causes malformation of spikes in dwarf wheat when applied during the early stages of crop growth and hence should be applied 35 d after sowing. Bromoxynil, applied at the 3-leaf to boot stage, can control 2,4-D-tolerant weeds. Picloram shows greater selectivity when applied between the 3-leaf and tillering stage of the crop. Dicamba is combined with 2,4-D for greater activity on a wider spectrum of weeds. Diuron and linuron need to be applied to the soil soon after sowing. But they have residual activity longer than necessary and this could damage the succeeding susceptible crops in rotation.

Very effective control of several annual and perennial grasses may be obtained by POST application of fluazifop-P (50-100 g ha^{-1}), diclofop (0.5-1.0), and tralkoxydim (100-400 g ha^{-1}). Of these, diclofop has good activity against *Avena* spp. (wild oats). Herbicides such as imazamethabenz (0.3-0.45), chlorsulfuron (10-20 g ha-1), metsulfuron (40-60 g ha^{-1}), tribenuron (10-20 g ha^{-1}), prosulfuron (15-40 g ha^{-1}), triasulfuron (15-30 g ha^{-1}) and quinclorac (0.5-1.0) give good POST control of annual broadleaf weeds. Of these, triasulfuron, which has longer residual activity in soil, may injure the succeeding dicot crops for 1-3 yr after application. Quinclorac needs to be applied at the tillering stage of wheat to avoid crop phytotoxicity. Mixtures of the above-mentioned graminicides and broadleaf weed herbicides could provide very good POST broad-spectrum weed control.

Effective control of wild oats and wild canarygrass can be obtained by preplant incorporation of tirallate (1.0-1.5) and diallate (1.0-1.5). Trifluralin, PPI (0.8-1.2), gives good control of several annual grasses and broadleaf weeds. It may also be applied by injecting it in irrigation water (chemigation). The PRE herbicides effective against these and other grasses include isoproturon (1.0-1.5), methabenzthiazuron (1.0-1.5), metoxuron (1-2), terbutryn (0.5-1.0), alachlor (1.0-1.5) and pendimethalin (1.0-1.5).

Maize (Zea mays L.)

Maize (corn) is most sensitive to weed competition during its early growth period. The growth of maize plants in the first 3-4 wk is rather slow and it is during this period that weeds establish rapidly and become competitive. Maximum weed competition in maize occurs during the period of 2-6 wk after sowing, suggesting the importance of maintaining the field weed-free during this critical period of weed competition.

The wider row spacing used in maize helps in greater exposure of soil surface to sunlight, resulting in profuse weed germination and growth. A control measure, which prevents weed establishment in the early stages of crop growth will be very useful. Manual weedings and inter-row cultivations do provide weed-free environment to the crop to some extent but not very effectively. Four manual weedings are required to prevent weed growth completely but these are more expensive than simazine usage [15]. Row cultivation can destroy the weeds between rows and not within the row, thus making the operation only partially effective.

Herbicides which prevent weed establishment at least during the first 6-wk period are very useful in maize. In fact, maize is one of the most tolerant crops to many herbicides. The herbicides which are effective when applied PRE include atrazine (1-2), simazine (1-2), EPTC (2-3), butylate (2-3), alachlor (2-3), fluchloralin (1.5-2.5) metribuzin (1-2), terbutryn (1-2), etc. The combination of alachlor (1.5-2.0)+atrazine (1.0) is more effective and has activity on a wider spectrum of weeds including grasses and broadleaf weeds. Atrazine and simazine are more effective on broadleaf weeds than grasses, while alachlor has greater activity on annual grasses. Atrazine shows good POST activity when tank mixed with a phytobland oil and applied at the 2- to 4-leaf stage. Maize is exceptionally tolerant of atrazine and simazine. Under dry conditions, atrazine gives better weed control. Generally, atrazine tends to be more effective than simazine on grasses. Linuron may cause injury to maize and this needs to be applied at lower rates and only in tolerant varieties.

The other PRE herbicides which give very good control of a mixed weed spectrum include acetochlor (1.5-3.0) plus safener, dimethenamid (0.8-1.6), metolachlor (1.5-3.0), pendimethalin (1.0-1.5), oxyfluorfen (0.25-1.0), imazaquin (50-70 g ha^{-1}), thiazopyr (0.25-1.0) and halosulfuron (70-80 g ha^{-1}) plus safener. POST herbicides such as nicosulfuron (30-50 g ha^{-1}), primisulfuron (20-40 g ha^{-1}), prosulfuron (15-30 g ha^{-1}), halosulfuron (30-40 g ha^{-1}) and flumiclorac (30-60 g ha^{-1}) provide effective control of several broadleaf weeds. Imazaquin (50-70 g ha^{-1}) may also be applied POST. Herbicide combinations may need to be developed for effective broad-spectrum weed control.

Trifluralin (0.8-1.2), butylate (3.0-6.0) and EPTC (2.0-4.0), applied PPI, provide season-long control of several weeds. These herbicides control purple nutsedge (nutgrass) and many annual grasses.

Sorghum (Sorghum vulgare L.)

Grain sorghum is usually grown as a rain-fed crop. Weeds are most competitive when moisture is limiting, placing crop plants at a disadvantage. Sorghum is a poor competitor when young and hence weeds always have a competitive edge. During the first 35 d after sowing, weeds remove 46.1, 18.3 and 47.7 kg ha^{-1} N, P and K respectively, while the crop could take up only 23.8, 9.4 and 46.8 kg ha^{-1} N, P and K respectively [23]. This shows that weeds remove twice the amount of N and P, and as much as K as the crop does in the early stages of plant growth. For every 4.5 kg N, 1.5 kg P and 4.0 kg K removed by weeds, sorghum grain yields are reduced by 100 kg [23]. Thus, as in the case of maize, effective weed control measures during the first 5 or 6 wk are essential for maximizing the yields in sorghum.

If the soil is too wet in the growing season, hoeing cannot be done. Mechanical methods are difficult to use after 6-7 wk as crop plants are too tall to permit field operations. Hence, implementation of weed management practices early in the season is very essential.

The PRE herbicides found safer in sorghum are propazine, atrazine, terbutryn, alachlor and isoproturon. They are applied between 1 and 2 kg ha^{-1}. The other PRE herbicides metolachlor (1.5-3.0), pendimethalin (1.0-2.0) and fluometuron (1.0-1.5) also provide effective control of several annual grasses and some broadleaf weeds. Trifluralin (0.8-1.2) may be applied POST, via chemigation, before weed emergence. Halosulfuron (30-40 g ha^{-1}) and prosulfuron (15-30 g ha^{-1}), applied POST, are very effective against many annual broadleaf weeds.

The control of *Striga* in sorghum is discussed in Chapter 15.

Millets

Pearl Millet (Pennisetum typhoides S. & H.) and Finger Millet (Eleusine coracana Gaertn.)

Like sorghum, pearl millet and finger millet are grown mostly under rain-fed conditions in the arid and semi-arid tropics. The weeds normally found in these millet crops are *Cyperus rotundus*, *Cenchrus biflorus* (sandbur weed), *Tribulus terrestris* (puncturevine), *Tephrosia purpurea* , *Pulicaria wightiana*, *Euphorbia hirta* (garden spurge), *Phyllanthus niruri* , *Crotolaria burhia* , *Avena tomentosa* , *Heliotropium subulatum*,, etc.

Repeated cultivations in the arid and semi-arid regions aid in the loss of moisture from soil. Handweeding causes minimal soil disturbance and helps in effective control of weeds. In situations wherein chemical weed control is economically feasible, herbicides provide the best answer for weed problems in these millet crops.

PRE application of pendimethalin (0.75-1.5) gives good preventive control of a wide spectrum of weeds, without affecting pearl millet and finger millet. A follow up POST application of 2,4-D EE (0.25-0.50) controls most of the late emerging broadleaf weeds and some annual grasses. As pearl millet and finger millet are sensitive to most herbicides, mixing a safener could increase the margin of selectivity to them.

Weed management practices involving herbicides and mechanical methods will be more effective and economical than either of them applied alone.

Oilseeds

Groundnut (Arachis hypogaea L.) (Peanut)

Groundnut (peanut), normally a rain-fed crop, is sown without sufficient tillage due to frequent rains. As a result, the crop is infested with both broadleaf weeds and grasses. Yield loss in groundnut due to this kind of weed competition can be as high as 77%. Besides competing for nutrients, soil moisture and sunlight, weeds inhibit pegging in groundnut and interfere with crop harvest. Weed competition is critical up to 45 d after groundnut sowing, and weed-free environment during this period gives highest pod yield.

Manual weeding and row cultivation after the emergence of groundnut, but before peg formation, are effective in controlling established weeds. But they do not provide complete weed-free environment. Intermittent rains do not permit completing these operations on time. Once peg formation has begun, manual methods should be continued as they damage pegs and roots, and reduce crop yield. Thus, herbicides offer the most effective means of weed management in groundnut. With proper selection of suitable herbicides, chemical weed management can be made cheaper and safer than mechanical methods.

PRE herbicides provide season-long control of a wider spectrum of weeds. These include metolachlor (1.5-3.0), pronamide (1.5-3.0), napropamide (1.0-2.0), cinmethylin (0.5-1.5), thiazopyr (0.5-1.5), pendimethalin (1.0-2.0) and imazethapyr (50-70 g ha-1), which provide satisfactory PRE control of several annual grasses and broadleaf weeds. Of these, metolachlor, napropamide and cinmethylin may be applied PPI or PPS (preplant surface). The PRE, PPI and PPS treatments will be effective only if irrigation or rain followed herbicide application.

When groundnut is grown under scant rainfall conditions or when soil application of herbicides is not possible, because of heavy rains at planting time, farming problems, etc., POST application of herbicides could eliminate weed competition. Herbicides that could be applied POST include acifluorfen (0.15-0.40), imazaquin (50-70 g ha^{-1}), imazethapyr (30-50 g ha^{-1}), chlorimuron (8-15 g ha^{-1}) and pyridate (1.0-1.5). These provide satisfactory control of most of the annual broadleaf weeds and certain annual grasses. Fenoxaprop-ethyl (100-200 g ha^{-1}) and fluazifop-P (50-100 g ha^{-1}) give good POST control of many annual and some perennial grass weeds.

Sunflower (Helianthus annuus L.)

Very little information is available on the nature of weed problems and control measures in sunflower. Generally, competition from weeds is more severe and damaging during the early stage of crop growth. Maximum yields of sunflower can be obtained when the crop is left weed-free for 4-6 wk after planting [9]. Best yields can be harvested when the crop is weed-free during the first 60 d after sowing [25]. The weed species normally found in sunflower include *Anagallis arvensis* (scarlet pimpernel), *Chenopodium album* (lambsquarters), *Fumaria parviflora* , *Melilotus alba* (white sweetclover), *Vicia sativa* (common vetch), etc.

PRE application of bifenox (0.5-1.0), pronamide (1.5-3.0) and thiazopyr (0.5-1.5) give effective control of several annual grasses and broadleaf weeds in sunflower. Ethalfluralin (0.6-1.2) and trifluralin (0.5-1.0) may be applied PPI. Imazamethabenz (20-40 g ha^{-1}) gives good POST control of many annual weeds.

Safflower (Carthamus tinctorius L.)

Effective weed control in safflower is obtained by PRE application of alachlor (1-2), metoxuron (2-3) and metobromuron (2-4). EPTC (2-3), trifluralin (1-2), chlorpropham (3-5) and propham (3-4) are required to be soil-incorporated at PPL.

Sesamum (Sesamum indicum L.)

The critical period of weed competition in sesamum is between the 15th and 30th d after sowing. During the early period of crop growth, weeds remove 40, 3 and 15 kg ha^{-1} N, P and K respectively, causing 49% reduction in crop yield [27]. Alachlor (1.5-2.0) applied PRE gives very good weed control in sesamum. Other PRE herbicides found effective include dichlormate (1.0-1.5), fluchloralin (0.75-1.5) and chlorpropham (3-5).

Fibre Crops

Cotton (Gossipium spp.)

Cotton is grown under both heavy rainfall and irrigated conditions, which are conducive for heavy weed infestation. The higher level of fertilizers, applied to meet the demands of the high-yielding varieties, also create favourable conditions foe weeds to grow vigorously and establish rapidly. Weed competition causes 45-85% reduction in yield. Weeds remove 5-6 times more N, 5-12 times P and 2-5 times K than a cotton crop in the early stages of crop growth, leading to 54-85% reduction in raw cotton yields [8].

After planting, weeds germinate and begin to shade the slow-growing cotton plants. Once it rains, it will be difficult to undertake mechanical measures for weed control in the heavy cotton soils. Furthermore, the farmers are usually busy with other farm operations. As a result, the crop becomes susceptible to weed competition. Hence, chemical weed control offers the crop competitive advantage over the weeds from the very early stage.

PRE applications of metolachlor (1.5-2.0), pronamide (1.5-3.0), pendimethalin (1.0-2.0) and cinmethylin (0.5-1.5), and PPI of EPTC (2.0-4.0) achieve effective control of several annual grasses and certain broadleaf weeds. Cinmethylin may also be applied PPI. Tralkoxydim (100-400 g ha^{-1}), quizalofop-P (0.35-0.70), fluazifop-P (50-100 g ha^{-1}), sethoxydim (100-400 g ha^{-1}) and clethodim (100-300 g ha-1) provide very good POST activity against several annual and perennial grasses. Lactofen (20-30 g ha^{-1}), which is effective against several annual broadleaf weeds, may be applied as POST-directed spray to avoid possible injury to cotton.

Jute (Corchorus capsularis L.)

Weeds pose the greatest problem in jute cultivation. The conventional method of weed control in jute demands around 180 man-days ha^{-1} between the third and fifth week of crop growth, and in terms of expenditure weed management constitutes one-third the total cost of cultivation. Weed competition may reduce fibre yield to the extent of 40-60%.

As jute is a narrowly spaced crop, manual operations and row cultivation for weed control are not feasible. Hence, herbicides provide the only answer for weed problems in jute. However, jute is sensitive to most of the herbicides.

Among the herbicides, POST application of MSMA (3-4) and dalapon+MSMA (2.0+2.5) provide effective control of all grasses. Dalapon may be applied when jute plants are 3-4 wk old. PRE herbicides butachlor (1-2), thiobencarb (1.5-2.0), fluchloralin (1-2), chloramben (1.5-2.5), etc. control most of the weeds infesting jute. PPL application of propachlor (1.5-2.0) followed by one handweeding 25 d later gives excellent weed control in jute [13].

Pulse Crops

Pulse crops normally grow very short (except pigeon pea) and as a result a dense growth weeds can smother the crop plants successfully. Pulse crops such as soybean, black gram, mungbean (greengram), cowpea, etc. have all identical weed problems. The critical period of weed competition in these pulse crops is generally during the first 30 d after sowing. Thus, weed management measures during this period are essential for maximum yields.

Soybean (Glycine max Merrill)

Keeping the soybean field weed-free for one month after sowing gives virtually as much yield as when kept weed-free for the entire season. Weed competition affects both the size and number of seeds per pod.

Row cultivations can give excellent weed control but they are done after the weeds have become established. During this period, the crop is subjected to some degree of weed competition.

A wide range of soil-applied and foliage-applied herbicides provides moderate to excellent control of a wide range of weeds infesting soybean. PRE herbicides include acetochlor (1.5-3.0), dimethenamid (0.8-1.6), metolachlor (1.5-2.0), bifenox (0.5-1.0), oxyfluorfen (0.5-1.5), imazaquin (50-70 g ha^{-1}), imazethapyr (50-70 g ha^{-1}), thiazopyr (0.5-1.5), chlorimuron (30-80 g ha^{-1}) and cinmethylin (0.5-1.5). Ethalfluralin (0.6-1.2), acetochlor (1.5-3.0), dimethenamid (1.0-1.5), vernolate (2.0-3.0), imazaquin (100-150 g ha^{-1}), cinmethylin (0.5-1.5) and trifluralin (0.8-1.2) are effective when applied PPI.

Herbicides such as clethodim (100-300 g ha^{-1}), sethoxydim (100-400 g ha^{-1}), tralkoxydim (100-400 g ha^{-1}), quizalofop-P (35-70 g ha^{-1}) and fluazifop-P (35-70 g ha^{-1}) show excellent POST activity against several annual and perennial grass weeds. The POST broadleaf weed herbicides include acifluorfen 150-300 g ha^{-1}), imazaquin (50-70 g ha^{-1}), imazethapyr (50-70 g ha^{-1}), chlorimuron (8-15 g ha^{-1}) and flumiclorac (30-60 g ha^{-1}). Mixtures of graminicides and dicot herbicides provide excellent broad-spectrum weed control.

Trifluralin may be combined with imazaquin, isoxaben (0.5-1.0), clomazone (0.5-1.0), alachlor (1.5-3.0) or triallate (1.0-1.5) and applied PPI for best and season-long weed control. The PRE combination of pendimethalin (0.50-0.75)+imazethapyr (50-75 g ha^{-1}) is also very effective.

Black gram (Phaseolus mungo L.), Chickpea (Cicer arientinum L.), Green gram (Phaseolus aureus L.) and Pigeon pea (Cajanus cajan L.)

In these pulse crops, oxyfluorfen (70-150 g ha^{-1}), oxadiazon (0-5-1-0), pendimethalin (0.5-1.0), isoproturon (1.0-1.25), imazethapyr (50-100 g ha^{-1}), and pendimethalin (0.5)+imazethapyr (50-70 g ha^{-1}), all applied PRE, provide effective control of a wide range of weeds. As pulse crops are sensitive to most herbicides, the exact rate and time of application need to be determined. If required, a safener may be used to increase the margin of crop selectivity.

Other Field Crops

Tobacco (Nicotiana spp.)

Weeds adversely affect the yield and quality of tobacco. Weed problems in tobacco are acute both in the seedbed and in the transplanted field. Besides *Orobanche*, several other weeds are also predominant in tobacco crop. The management of *Orobanche* is discussed in Chapter 15.

Weed competition during the first 9 wk after transplantation reduces the yield of tobacco drastically. Hence, adequate weed control measures are essential during this period. In a transplanted crop, inter-row cultivation is feasible because of wider row spacing. Hence, one or two cultivations followed by manual weeding to remove weeds growing close to tobacco plants give adequate control. For the nursery, manual weeding is effective. Manual operations are expensive and sometimes cannot be done due to labour scarcity.

For chemical weed control in the nursery, fumigation to sterilize the soil and kill weed seeds is done by using methyl bromide (4-8 kg per 100 m^2), calcium cyanamide (40-60 kg per 100 m^2) or metham (2-4 kg per m^2). Methyl bromide, a very volatile liquid, is applied beneath a plastic gas-proof covering. Exposure below the cover should be for 1-2 d. The beds should be aerated 2-4 d before sowing seeds. The soil microorganisms as well as seeds are destroyed. Calcium cyanamide is applied 2-3 mon before sowing to avoid residual toxicity problems to the crop. The addition of urea to calcium cyanamide decreased residual toxicity to the crop. Metham is applied 3 wk before planting and watering needs to be done after application. Allyl alcohol also is a useful nursery herbicide. It should be applied as a surface drench and watered in, to penetrate 5-10 cm of soil, to control most annual weeds as they germinate. Diphenamid (2-3) and trifluralin (0.5-1.0), when PPL incorporated into the soil, control most annual weeds in a tobacco nursery.

Regarding weed control in the transplanted field, pronamide (1-2), fluchloralin (2-3) isopropalin (1.5-2.5), and isoxaben (1.5-2.5) may be applied PPI, before seedling transplantation. As tobacco is sensitive to most POST herbicides, PPI herbicides provide the best approach for effective weed control. Diphenamid (2-3) may be applied PPL or PPE to soil surface.

Sugarcane (Saccharum officinarum L.)

The duration of a sugarcane crop is 12-18 mon. Weed management in the early stages is very important since heavy infestation prevents proper bud germination. The smothering effect of weeds on the crop is observed for about 4-5 mon after planting, i.e. during tillering and elongation phases. Weed competition will ultimately have an adverse effect on cane yield and sucrose content in juice. There is a negative correlation between weed population and millable cane as well as between yield and sucrose content in juice [26]. Besides many annual broadleaf weeds and grasses, perennials such as *Cyperus rotundus, Cynodon dactylon and Sorghum halepense* are also the predominant weeds in sugarcane.

Timely mechanical methods could control weeds with moderate efficiency. Hoeing in the crop rows after planting could cause damage to the emerging cane sprouts. Two or three row cultivations besides two or three manual weedings could keep the field free of weeds until sugarcane plants reach sufficient height and have a smothering effect on weeds growing later. Mechanical methods are only partially effective against perennial grasses. The combination of mechanical methods and herbicides offers very effective weed management in sugarcane.

PRE herbicides which are effective in sugarcane include atrazine (2-3), ametryn (2-3), alachlor (1.5-2.5), metoxuron (4-6, diuron (1.5-2.5), etc. Effective PRE weed control will generally last for 8-12 wk depending on the herbicide used. Hence, in a long duration crop like sugarcane, best results are obtained by sequential applications of PRE herbicides and POST herbicides, or repeated applications of POST herbicides.

PRE application of other herbicides such as pendimethalin (1.0-2.0), metribuzin (1.0-1.5), fluometuron (1.0-1.5), imazapyr (50-150 g ha^{-1}) and thiazopyr (1.0-2.0) provides satisfactory control of several broadleaf weeds and certain annual grasses. The emerged broadleaf weeds may be eliminated by POST application of prosulfuron (15-40 g ha^{-1}) and halosulfuron (30-50 g ha^{-1}).

Glyphosate (0.8-1.6), paraquat (0.4-0.8), 2,4-D (1.0-1.5) and asulam (3-4) are the effective POST herbicides. Glyphosate controls very effectively many perennial grasses including *Cynodon dactylon, Cyperus rotundus and Sorghum halepense*. Paraquat is an excellent contact herbicide to kill the establishing weeds. 2,4-D is a very useful broadleaf weed herbicide. Among the 2,4-D formulations, isooctyl ester of 2,4-D is most effective. Ioxynil (0.2-0.5), imazapyr (50-150 g ha^{-1}) and glufosinate (0.25-0.75, directed spray) also provide good POST control of annual and perennial grasses.

Sugarbeets (Beta vulgaris L.)

In sugarbeets, PRE application of pyrazon (3.0-6.0), pebulate (3.0-6.0), and ethofumesate (0.8-1.2) is effective on most annual weeds. Of these, pebulate and ethofumesate (2.0-4.0) may also be applied PPI or PPS. The POST application of desmedipham (0.4-0.7)+ethofumesate (0.8-1.2) gives an effective broad-spectrum weed control. Sethoxydim (1.0-0.5) and diclofop (0.5-1.0), POST, control several annual and perennial grass weeds. Diclofop (0.8-1.6) may also be applied PRE or PPI.

Tuber Crops

Potato (Solanum tuberosum L.)

Potato is grown under high fertility and good soil moisture conditions and as a result, weeds pose a serious problem. Weeds get established much before the short-statured crop plants do. Irrigation or high soil moisture soon after planting of potato induces weeds to germinate and establish by the time crop plants emerge. Weed competition adversely affects sprouting of tubers, tuberization, and hence the production potential of the crop by as much as 50%. When potato crop is kept weed-free for the first 4 wk after planting, there is no significant reduction in yield [3].

During ridging, done 2-3 times in the season, some weed growth is removed, but it cannot eliminate the weeds arising in flushes between the ridging operations. Manual weeding, besides less effective, is laborious, time consuming and expensive.

PRE application of herbicides not only helps in eliminating weed competition during the early stages of crop growth but enhances the efficiency of ridging as well. Besides, perennial weeds can be controlled very effectively. Herbicides found effective at PRE include EPTC (1.5-2.0), linuron (0.75-1.25), methabenzthiazuron (0.75-1.25), simazine (0.5-1.0), pendimethalin (1.0-1.5), butachlor (1.5-2.5), oxyfluorfen (0.1-0.3), metribuzin (0.75-1.0), anilophos (0.4-0.8), diphenamid (3-4), etc. EPTC, when applied PPL and incorporated into the soil, gives very selective and most economical weed control with a cost benefit ratio of 1:10. Linuron is useful in longer duration varieties of potato and in heavy and high organic matter soils.

For POST weed control, paraquat (0.4-0.8) and propanil (1.0-1.5) are widely used. Glyphosate (0.8-1.6), dalapon (3-4) and glufosinate (0.5-1.5) give effective control of perennial grasses. Complete weed control can be obtained in a sequence of PRE herbicide followed by a POST herbicide.

Effective control in potato may also be obtained by PRE application of metolachlor (1.5-3.0), napropamide (1.0-2.0) and thiazopyr (0.5-1.5). Metolachlor and napropamide as also trifluralin (0.8-1.2) may be applied PPI. Metolachlor may also be applied POST at the hill formation stage. Its PRE application may be made just before first irrigation after planting tubers, but before tubers sprout.

Sweet Potato (Ipomoea batatus L.)

Sweet potato varieties with good top growth are highly competitive with weeds. Weed competition is, however, critical during the first 8-12 wk. Effective control of Cyperus spp., grasses and broadleaf weeds is obtained when EPTC (3-4) or vernolate is applied before planting and incorporated into the soil. A mixture of EPTC+diphenamid (2+3) applied PPL and incorporated into the soil gives most effective control of a wide spectrum of weeds without affecting the quality of potato tubers.

Vegetable Crops

Tomato (Lycopersicon esculentum L.), Onion (Allium cepa L.), Peas (Pisum sativum L.), Chillies [(Pepper) Capsicum annuum L.], Okra [Hibiscus esculentus L. or Abelomoschus esculentus (L.) Moench.], Cabbage (Brassica oleracea L.), Cauliflower, Carrot (Daucus carota L.) and Beans [French, Snap, Cluster, etc.; (Phaseolus spp.)].

Generally, vegetable crops are poor competitors of weeds. They grow very slowly and are short-statured. Dense weed growth at any stage of crop growth will have an adverse effect not only on yield but on quality of the produce as well. Weed growth harbours disease causing organisms and insects and increases their occurrence. Hence, complete elimination of weed competition, particularly during the first 30-40 d period, is a prerequisite for vegetable growing.

Manual weeding is widely practised in vegetable crops in developing countries. In spite of its high cost, it is employed because of high cash returns. However, under the high soil moisture and nutrient regimes under which vegetables grown are very conducive for intensive weed growth in a short period. Furthermore, the weeds germinate in several flushes and continue to be a problem even in the later periods of crop growth. Thus, in spite of repeated manual weedings, some weed competition always exists. A complete weed-free situation ensures maximum yields and high-quality produce. Herbicides, particularly those applied PRE are very effective in creating this weed-free situation for the crop and ensuring a better cost-benefit ratio than manual weeding.

Vegetable crops represent a diverse group of plants and almost all widely used herbicides can find a place in vegetable cultivation either alone or in combination with other herbicides. If a PPL or PRE treatment enables the crop to establish itself, a POST treatment helps in maintaining weed-free environment until crop harvest and aids in controlling troublesome perennial weeds. POST application in crop rows may create a drift hazard to young plants but this can be prevented when a proper spray-shield is used.

Napropamide (1.0-2.0), pronamide (1.5-3.0), diphenamid (2-3), isopropalin (1.5-2.5), oxyfluorfen (0.25-1.0), thiazopyr (0.5-1.5), pebulate (3.0-6.0), and cinmethylin (0.5-1.5), all applied PRE, provide very good control of a wide range of annual grasses and certain broadleaf weeds. Of these, napropamide, pebulate and cinmethylin may be effective even if applied PPI. Ethalfluralin (0.6-1.2), EPTC (2-3) and trifluralin (0.5-1.0), PPI, also give good activity against the weeds. Fluazifop-P (50-100 g ha^{-1}) provides good POST control of several annual and perennial crops in vegetable crops.

Although the above-mentioned herbicides are useful for vegetable crops, selectivity of each of the crops to them needs to be determined.

WEED MANAGEMENT IN MAJOR PLANTATION CROPS

Weed problems in plantation crops differ very much from those in field crops. Although annual weeds are predominant, perennial species, which are persistent, give greater competition to the crop. Some of the plantation crops are generally grown in arm, humid and high rainfall areas and on the hill slopes. As the weather is more conducive for plant growth, weeds make best use of it to infest an area intensively in a very short period.

Among the plantation crops, tea and coffee have serious weed problems than tree crops such as mango, coconut, citrus, rubber, etc. In tree crops, the plant spacing is very wide to allow cover crops. The shade from the canopy of tree crops and cover crops smothers weed growth. If there is any weed growth, it is controlled by manual or mechanical methods.

In situations where chemical weed control is practised in tree crops, simazine, pronamide, norflurazon, thiazopyr, fluometuron, cinmethylin and diuron are used for PRE control of many annual broadleaf weeds and some annual grasses. At POST, paraquat, 2,4-D, pyridate, fluazifop-P, glyphosate, glufosinate and dalapon are widely used. Paraquat, a contact herbicide, kills many grasses and repeated applications over a period of time can eradicate the target weed. 2,4-D is extremely effective on many established broadleaf weeds. Glyphosate, glufosinate, fluazifop-P and dalapon give good control of some perennial grasses. Trifluralin, EPTC, fluometuron and cinmethylin, all applied PPI, also provide satisfactory weed control.

Banana (Musa spp.)

Weeds are usually a problem during the early growth period of banana crop and in ratoon crops. Roots are superficial and cultivations for the purpose of weed control could cause injury. Under normal planting distances, rhizomatous and stoloniferous weeds (*Cynodon dactylon, Cyperus rotundus, Elytrigia repens*, etc.) and many broadleaf species flourish and compete severely with banana plants especially during the early stages of crop growth [2]. Weed growth could be drastically curtailed throughout the growth of banana crop by adopting high planting density (4440-6950 plants ha^{-1}) and growing initially an intercrop of cowpea [2].

PRE application of atrazine, simazine, diuron, linuron,, pronamide, cinmethylin, norflurazon, thiazopyr and fluometuron soon after planting banana keeps the ground weed-free for 3-5 mon. At POST, paraquat, dalapon, MSMA, 2,4-D, glyphosate, glufosinate and pyridate are useful in controlling many of the established perennial weeds. Application of paraquat on the weeds followed immediately by mulching with the trash of the previous crop harvest give very effective weed control. The addition of diuron or simazine to paraquat (in a mixture) will extend the period of weed control.

Pineapple (Ananas comosus L.)

In pineapple, it is very important to keep the field weed-free in the early growth period. Perennial grasses can be controlled by glyphosate (0.8-1.6), glufosinate (0.5-1.5) and dalapon (3-4) application 4-6 wk before planting and ploughing the field 2 wk before planting. PRE herbicides such as simazine (1-2), atrazine (1-2), pronamide (1.5-3.0), oxyfluorfen (0.5-1.5), norflurazon (2.5-5.0), cinmethylin (0.5-1.5) and diuron (1-2) are applied immediately after planting to prevent weed establishment. Directed spraying in the inter-rows will avoid possible injury to the young pineapple seedlings. Weed growth emerging later is treated with POST herbicides such as ametryn (2-3)+a surfactant, bromacil (2-4), terbacil (2-4), glyphosate (0.8-1.6), dalapon (3-4), paraquat (0.3-0.6), 2,4-D (0.5-1.0), glufosinate (0.35-1.25) and pyridate (1.0-1.5), depending on the weed species predominant.

Grapes (Vitis vinifera)

Young grapevines are very susceptible to weed competition. An intensive infestation of perennial grasses may result in spotty stands and stunted plants during the establishment year. Established vines can compete well with many annual grasses and broadleaf weeds.

Manual and mechanical methods are widely followed for control of weeds. However, the soil-applied PRE herbicides are very effective in preventing weed establishment. These include simazine (1-2), napropamide (1-2), oxadiazon (2-4), pronamide (1.5-3.0), fluometuron (1-2), norflurazon (2.5-5.0), thiazopy (1-2)r, oryzalin (2.5-5.0), oxyfluorfen (0.5-1.5), diuron (1-2), cinmethylin (0.5-1.5), etc. Herbicide which have longer residual activity are more economical and useful.

At POST, glyphosate (0.8-1.6) is most effective for controlling perennial weeds. Other useful POST herbicides include dalapon (3-4), paraquat (0.3-0.6), 2,4-D (0.5-1.0), glufosinate (0.35-1.5) and pyridate (1.0-1.5). Herbicide application in grapevines must be done carefully with low pressure nozzles to prevent drift.

Coffee (Coffea arabica L. and Coffea robusta L.)

Weed problems in coffee are similar to those in tea. In coffee nurseries, weeds are controlled by applying simazine (2.0-2.5) immediately after planting young coffee seedlings in polyethylene sleeves. Soil fumigation is done with methyl bromide or with a solution of allyl alcohol or metham 2-3 wk before planting.

In young and established coffee areas, simazine (1-2), atrazine (1-2), diuron (0.5-1.0), oxyfluorfen (0.5-1.5), pronamide (1.5-3.0), cinmethylin (0.5-1.5), norflurazon (2.5-5.0), etc. are used for PRE weed control. At POST, paraquat (0.3-0.6), dalapon (3-4), glyphosate (0.8-1.6), glufosinate (0.35-1.5) and pyridate (1.0-1.5) are used to control many perennial weeds. Repeated application of paraquat or one application of paraquat+diuron after the first rains followed by spot spraying of paraquat, when required, is very effective against perennial grasses.

Tea (Camelia sinensis L.)

Weeds cause 15-40% yield loss in tea. Complete elimination of an infestation of *Imperata cylindrica*, one of the world's worst weeds, in tea increases the yield by 50-80% [20].

WEED CONTROL IN TEA NURSERY

Tea is propagated by seeds and clonal cuttings on raised nursery beds or in polyethylene sleeves filled with soil. Weeds can adversely affect the emergence and establishment as well as growth of plants until they are transplanted in the main field 12-18 mon later. Weed control in the clonal nursery is done by PPL (2-3 wk before planting cuttings) application of simazine (2.0). This maintains weed-free situation for 4-6 mon after planting the young clonal plants. Later, when the clonal plants have put in two flushes (or 6 mon after planting), simazine (2.0) is applied to the soil once again after removing the existing weed growth. In the case of seed nurseries, manual removal is done for 6 mon after planting followed by simazine (2.0) application.

WEED CONTROL IN YOUNG TEA

Young tea is so termed when tea plants are 0-3 yr old from the time of their planting in the main field. Weed growth in young tea is very intensive particularly during the first two years as the ground is not completely covered by the bush canopy. Weeds remove 270 kg ha^{-1} of nitrogen in young tea [21]. Weeds also have smothering effect on tea plants in the planting year, causing stand loss [1] and inhibition of branching [2]. Infestation of perennial grasses such as *Imperata cylindrica, Cynodon dactylon, Paspalum conjugatum*, etc. can cause severe setback in tea bush growth by several years. The critical period of weed competition is 6 mon between April and September [21]. Effective weed control during this critical period results in very little growth later until the next April. Thus, weed management programme in young tea needs to be tailored with herbicide rotations to keep the ground weed-free during this period.

In young tea, simazine (1.5-2.0) is applied at PRE on clean soil after the early rains between end-March and end-April to keep the ground free of most of the annual broadleaf weeds and annual grasses. 2,4-D (0.5-1.0) is very effective when applied at POST to control broadleaf weeds such as *Borreria hispida, Ageratum conyzoides, Erechtites valerianefolia, Commelina benghalensis, Mikania micrantha*, etc.

Paraquat is the most widely used herbicide in tea. It is a broad-spectrum contact POST herbicide applied at 0.3-0.4 to control grasses such as *Paspalum conjugatum, Axonopus compressus, Digitaria sanguinalis, Imperata cylindrica*, etc. It burns the above-ground weed growth, with little effect of the underground root growth of perennial weeds, thus allowing their regeneration. Repeated applications 5-8 times, at 10-15 d intervals, could exhaust the food reserves of the underground roots and give more effective weed control. Paraquat+MSMA is used for a moderately effective control of *Paspalum conjugatum* and *P. scrobiculatum*.

Glyphosate, a broad-spectrum translocated herbicide, is most effective on perennial weeds which have an underground rhizomatous root system. It is used to effectively control perennial grasses such as *Imperata cylindrica, Cynodon dactylon, Arundinella bengalensis, Paspalum conjugatum, P. scrobiculatum, Saccharum spontaneum,* etc. and perennial broadleaf weeds such as *Polygonum chinense* and *P. perfoliatum* [19, 20]. It is also effective against ferns like *Pteridium aquilinum* [19, 20].

WEED CONTROL IN MATURE TEA

Tea above 3 yr of age is called mature tea. The plucking table of the bush is maintained at a height of 60-70 cm from the ground. A well-maintained mature tea covers the ground through leaf canopy to have a smothering effect on weed growth. However, weeds are a serious problem: a) if the leaf canopy does not adequately cover the ground, b) when there are vacancies and c) in the years following pruning (light or medium) or skiffing (deep or medium). During pruning and skiffing, done on the tea bush in the winter, the top growth of the bush is removed to promote new shoot growth and plucking points in the following year. As a result, the ground is exposed to sunlight, which breaks seed dormancy and consequently promotes intensive weed germination and growth after the early rains in March–April before the bush develops leaf canopy.

As in young tea, the critical period of weed competition in mature tea is 6 mon, from April to September. Complete prevention of weed establishment and growth during period ensures high yields of tea.

Simazine (1.5-2.0), diuron (2.0), oxyfluorfen (0.5-1.5), imazapyr (50-100 g ha-1), dithiopyr (0.25-0.50), thiazopyr (1.0-2.0) and norflurazon (2.5-5.0) are the effective PRE (with respect to weeds) herbicides used in mature tea for controlling many broadleaf annual weeds and some annual grasses. Of these, diuron is most effective. They are applied in the autumn (before rains cease) in tea areas where pruning or skiffing is done later in December-January to achieve weed-free situation in the following year [18]. As application in the spring season is less effective due to the poor contact of herbicide with the soil in the presence of pruning litter on the ground, application in autumn mitigates this problem and prevents the intensive weed growth expected in the following year.

POST herbicides paraquat, 2,4-D, paraquat+2,4-D, paraquat+diuron, paraquat+MSMA, glyphosate (0.8-1.6), imazapyr (50-100 g ha⁻¹), fluazifop-P (50-100g ha⁻¹) and glufosinate (0.35-1.25) are useful to control most of the annual and perennial weeds found in mature tea. POST application of glufosinate causes very good control of several perennial grasses and ferns, while fluazifop-P and imazapyr give excellent control of perennial grasses. Asulam (1.0-2.0), POST, provides satisfactory control of ferns.

REFERENCES

1. Anonymous. 1979-80. Tocklai Experimental Station, tea Research Association, Jorhat, Assam, India. Ann. Sci. Rep. p. 19.
2. Chacko, E.K. and Arvind Reddy. 1981. Effect of planting distance and intercropping with cowpea on weed growth in banana. Proc. 8th Conf. Asian-Pacific Weed Sci. Soc. p. 137-141.

3. Chang, W.I. 1968. Weed control effect of some herbicides in the paddy field. J. Taiwan Agri. Res. 17: 15-25.

4. Dubey, A.N., G.B. Manna and M.V. Rao. 1977. Critical period of weed competition, weed control and varietal interaction with propanil and parathion in direct-seeded rice. Indian J. Weed Sci. 9: 75-81.

5. Everaats, A.P. and Satsyati. 1977. Critical period for weed competition for potatoes in Java. Proc. 6th Asian-Pacific Weed Sci. Soc. p. 173-177.

6. International Rice Research Institute. 1966. Weed control in direct seeded rice. Ann. Rep. p. 168-173.

7. Jain, S.C. and N.K. Jain. 1980. Studies on integrated weed control approach in cotton (*Gossipium hirsutum* L.) in Madhya Pradesh. Indian J. Weed Sci. 12: 28-34.

8. Jain, S.C., B.G. Iyer, H.C. Jain and N.K. Jain. 1981. Weed management and nutrient losses in upland cotton under different ecosystems of Madhya Pradesh. Proc. 8th Asian-Pacific Weed Sci. Soc. p. 131-135.

9. Johnson, B.J. 1971. Effect of weed competition in sunflowers. Weed Sci. 19: 378-380.

10. Kaushik, S.K. and V.S. Mani. 1977. Investigations on chemical weed control in direct seeded and transplanted rice. Weed Sci. Conf. and Workshop in India, Abstr. No. 15.

11. Kim, D.M. 1969. An introduction to weed control in rice in Korea. Asian-Pacific Weed Sci. Soc. Conf. Proc. 2: 34-42.

12. Patel, J.P. 1965. Evaluating the various factors of the 'Japanese method' of rice cultivation in India. Agron. J. 57: 567-572.

13. Patro, G.K., G.C. Tosh and R.C. Das. 1978. Weed control studies on low land jute (*Corchorus trilocularis* L.) through chemical and cultural methods. Indian J. Weed Sci. 10: 49-53.

14. Pillai, K.G. V.K. Vamadevan and S.V. Subbaiah. 1976. Weed problems in rice and possibilities of chemical weed control. Indian J. Weed Sci. 8: 77-87.

15. Rajgopal, A. and S. Sankaran. 1973. Weed control in maize (var. Deccan Hybrid). Indian J. Weed Sci. 5: 50-52.

16. Rao, A.S. 1993. Studies on bio-efficacy of herbicide mixtures and sequential applications in transplanted rice. Ph.D. thesis. Benaras Hindu University, Varanasi, India.

17. Rao, V.S. 1995. Broad spectrum weed control by herbicide mixtures in rice and wheat in India. Weed Sci. Soc. Amer. Meetings, Seattle, Washington, U.S.A., Feb. 1995.

18. Rao, V.S. and B. Kotoky. 1980. Autumn application of preemergence herbicides. Two and A Bud. 27: 61-62.

19. Rao, V.S. and F. Rahman. 1978. Weed control in tea with glyphosate. Two and A Bud. 25: 71-73.

20. Rao, V.S., F. Rahman, S.N. Sarmah and H.S. Singh. 1977. Control of persistent weeds of tea. Proc. Twenty-eighth Con., Tocklai Experimental Station, Tea Research Association, Jorhat, Assam, India, p. 15-19.

21. Rao, V.S. and H.S. Singh. 1977. Effect of weed competition in young tea. Proc. Twenty-eighth Conf., Tocklai Experimental Station, Tea Research Association, Jorhat, Assam, India, p. 15-19.

22. Ray, B.R. 1973. Weed control in rice—a review. Indian J. Weed Sci. 5: 60-72.

23. Sankaran, S. and V.S. Mani. 1972. Effect of weed growth on nutrient uptake and seed yield of sorghum. (var: CSH-1). Indian J. Weed Sci. 4: 23-28.

24. Sharma, H.C. H.B. Singh and G.H. Friesen. 1977. Competition from weeds and their control in direct-seeded rice. Weed Res. 17: 103-108.

25. Singh, G. and V. Singh. 1978. Requirement of weed-free period in sunflower. Indian J. Weed Sci. 10: 79-82.

26. Srinivasan, T.R., P. Rethinam, A. Misra and A.S. Ethirajan. 1977. Relationship between the population and dry matter of weeds with the yield and quality of sugarcane under different weed control methods. Weed Sci. Conf. and Workshop in India, Abstr. No, 46.

27. Yamada, N. 1966. Recent adavnces in chemical control of weeds in Japan. International Rice Commu. Newsl. 15(2): 1-15.

17

Weed Management in Aquatic System

INTRODUCTION

Aquatic weeds are those unwanted plants which grow in water and complete at least a part of their life cycle in water. Many aquatic plants are desirable since they may play temporarily a beneficial role in reducing agricultural, domestic and industrial pollution. Letting a crop of plants grow in a lake or pond and then killing it over a period of time, and consequently releasing nutrients back into the water, may help in fish production. However, many aquatic plants are considered weeds when they deprive human beings of all facets of efficient use of water and cause harmful effects, some of which are discussed below.

1. Submersed, emersed and marginal weeds in and along irrigation canals, ditches, and drainage channels impede water flow, increase evaporation, cause damage to canals and structures, and clog grates, siphons, valves, sprinkler heads, bridge piers, pumps, etc. As a result, floods occur in the neighbouring areas. They also cause substantial losses of water through transpiration. The aquatic weeds reduce the designed flow rate by 30-40% in irrigation canals and drainage channels.

2. Floating and deep-rooted submersed weeds interfere with navigation. Some of the tougher and densely growing weeds, e.g. water hyacinth and alligator weed (*Alternanthera* spp.) become impenetrable and prevent boats and even steamers moving through.

3. Submersed and floating aquatic weeds in farm ponds, village tanks and water reservoirs reduce their utility for water storage and irrigation.

4. Aquatic weed growth also prevents or impairs the use of inland waters for fishing. The weeds assimilate large quantities of nutrients from water, thus reducing their availability for desirable planktonic algae. They cause oxygen dificiency and prevent gaseous exchange with the atmosphere, resulting in an adverse effect on fish production. Excessive growth of these weeds may provide excessive cover, resulting in an overpopulation of small fish and interference with fish harvesting.

5. Aquatic weeds provide a suitable habitat for development of mosquitoes in impounded waters, causing malaria, filariasis and encephalitis. These weeds serve as the primary vector for the disease-causing organisms.

6. Aquatic weeds reduce the recreational values of lakes, tanks, streams, etc. as the water is made turbid or dirty with an undesirable odour.

EVAPOTRANSPIRATION OF AQUATIC WEEDS

As mentioned earlier, aquatic weeds contribute to significant water losses. Otis [8] found that the highest evapotranspiration occurred in *Potamogeton nodosus* Poir. (American pondweed) and *Eichhornia crassipes* (Mart) Solms. (waterhyacinth). He calculated evapotranspiration of different aquatic weeds and found that *Typha latifolia* (common cattail), *Acorus calamus* (sweetflag), *Pontederia cordata* (pickerel weed), *Scirpus validus* (softstem bulrush) and *Nymphaea odorata* (fragrant waterlilly) caused 3.1, 2.5, 2.0, 1.2, and 1.9 times greater evaporation respectively than the evaporation from a free water surface. Timmer and Weldon [14] estimated that water evaporation of water hyacinth was 3.7 times higher than evaporation from a free water surface.

Brezny et al. [2] reported that evapotranspiration of *Trapa natans* (water-chestnut), *Pistia stratiotes* (waterlettuce), and *Ipomoea aquatica* (swamp morningglory) did not increase water losses significantly. Evapotranspiration of *Eichhornia crassipes* was 30-40 % higher, that of *Typha angustifolia* (narrowleaf cattail) was 60-70% higher and that of *Cyperus rotundus* (purple nutsedge) was 130-150% higher than evaporation from a free water surface under equivalent conditions. They also found significant positive correlations between meteorological factors (air temperature and wind velocity) and evapotranspiration from aquatic plants.

TYPES OF AQUATIC PLANTS

There are two types of aquatic plants: algae and hydrophytes.

Algae

Algae normally inhabit the surface of fresh and saline waters exposed to sunlight. While some kinds of algae are found in solid and on terrestrial surfaces exposed to air, the majority are truly aquatic and adapted to live in ponds, lakes, reservoirs, streams, swimming pools and oceans.

Freshwater algae are of two types: planktonic and flamentous. **Planktonic algae**, called **phytoplankton**, include the truly aquatic single-celled algae and the simplest filamentous or colonial forms. A heavy growth of algae may colour the water shades of green, yellow, red and black. They may also form 'water blooms' or 'scums'. They convert solar energy into food, remove CO_2 from water during photosynthesis (in daytime) and produce oxygen as a by-product. During the night or in cloudy weather, they release CO_2 in the water through respiration and consume O_2.

Certain planktonic algae are beneficial as they can maintain biotic balance in natural aquatic environment because of their ability to produce oxygen and maintain an aerobic condition. They are the original sources of food for most fish and aquatic animals. Although planktonic algae are beneficial, their overabundance may be undesirable for many domestic and commercial water uses. Excessive phytoplanktonic blooms often result in zooplanktonic (the microscopic animal forms) development that may deplete the water of oxygen and lead to overfertilization or eutrophication and destruction of fish and other aquatic wild life. Dense growth of planktonic algae will shade bottom muds sufficiently to prevent germination of seeds and growth of many species of rooted submersed weeds, thus affecting the stability of the habitat.

Generally, planktonic algae do not interfere with the use of surface waters for irrigation purposes. But some of them, the blue-green algae and green algae produce odours and scums that make water unfit for swimming. Several of the blue-green algae produce toxic substances that kill fish, birds and domestic animals.

Another group of algae, called the filamentous algae (nanoplankton) consist of single cells joined end-to-end, which may form as a single thread, branched filaments, nets, or erect stem like whorled branches or forked leaf-like forms. They do not have roots, stems or leaves as do the higher plants. They grow both in cool and warm water. They are considered a nuisance wherever they occur. The important genera of the filamentous algae are: *Chara, Nitella, Spirogyra, Hydrodictyon, Cladophora, Pithophora*, etc.

Both planktonic and filamentous algae produce undesirable odours and tastes in drinking water, secrete oily substances that interfere with domestic and industrial usage and clog sand filters in water-treatment plants. They also coat cooling towers and condensers. Filamentous algae interfere with irrigation by clogging weirs and screens and clinging to structures and concrete linings of canals. They also clog valves and spinkler heads of irrigation systems. Filamentous algae deplete ponds and lakes of nutrients required for the growth of planktonic algae used as fish food. The dense growth of filamentous algae reduce fish production and interfere with the harvesting of fish. Such dense growth may cause overcrowding of fish population and encourage production of only small fish.

Hydrophytes

The hydrophytes, which represent more than 100 families, are vascular plants. They grow wholly or partially submersed in either fresh or saline water or in plaustrine areas. They are structurally different from mesophytes and xerophytes that grow in moisture-deficient situations. The protecting and conducting tissues of hydrophytes are less developed. They also have extensive provision for aeration and buoyancy, particularly in the leaf mesophyll, ground tissue of the petiole and the cortex of the stem and root. Buoyancy is provided by aerenchyma or by air chambers. The air chambers may be either schizogenous or lysigenous, or both. Hydrophyte weeds can be grouped as submersed, emersed, marginal and floating weeds.

Submersed Weeds

Submersed weeds are mostly vascular plants that produce all or most of their vegetative growth beneath the water surface. Most submersed vascular weeds are seed plants and have true roots, stems and leaves. Abundance and density of these weeds is primarily dependent on depth and turbidity of water and physical characteristics of the bottom. A maximum depth of 3.5-4 m in clear waters is the limit for most of the submersed plants. They are capable of absorbing nutrients and herbicides through the leaves and stems as well as roots. They compete for nutrients with planktonic algae and decrease their production and a corresponding decrease in fish production.

The submersed weeds belong to the following genera: *Potamogeton, Elodea, Myriophyllum, Ceratophyllum, Utricularia, Ranunculus, Heteranthera, Alisma, Zannichellia, Lemna*, etc.

Emersed Weeds

Emersed weeds are those plants rooted in the bottom muds with serial stems and leaves at or above the water surface. They grow in situations where the water level ranges from just below ground level to about half the maximum height of the plant. They differ in leaf shape, size and point of attachment. Some of the weeds of this group have broad leaves, 5-50 cm in diameter, and others have long narrow leaves like grasses, less than 3 -15 cm or more in width; the latter are commonly called reeds. The leaves of emersed weeds do not rise and fall with water level as in the case of attached floating weeds. Some of the emersed weeds belong to the genera *Nuphar, Nelumbo, Jussiaea, Myriophyllum, Sparganium, Pontederia, Sagittaria, Rorippa, Lythrum, Epilobium*, etc.

Marginal Weeds

Most marginal weeds are emersed weeds that can grow on saturated soil above the water surface. They grow from moist shoreline areas into water up to 60-90 cm in depth. Marginal weeds vary in size, shape and habitat. Species of this group are the most widely distributed rooted aquatic plants. Plants of this group are broadleaf herbs, shrubs, trees and some grasses. The important genera to which they belong are: *Phragmitis, Typha, Polygonum, Alternanthera, Populus, Tamarix, Cephalanthus, Juncus, Scirpus*, etc.

Floating Weeds

Many water plants have leaves that float on the water surface either singly or in rosettes. They have true roots and leaves. Some are free-floating and others rooted in bottom mud have floating leaves that rise or fall with the water level. They reproduce very rapidly under favourable conditions and are among the most troublesome of aquatic weeds. Floating weeds belong to the genera *Eichhornia, Pistia, Salvinia, Lemna, Nymphaea* and *Brasenia*.

MANAGEMENT OF AQUATIC WEEDS

The degree of control of aquatic weeds depends upon the extent of damage they are causing. In situations where permanent or long-term eradication of all weed growth is necessary, the weeds are completely eradicated. These situations include serious impediments to water flow in irrigation canals and drainage channels where there is a very high risk of flooding, health problems caused by malaria, filariasis and encephalitis in impounded waters, and interference with navigation.

In situations where some weed growth is desirable for the management of fish and other wild life habitats, the excessive weed growth is removed at least for part of the year without seriously reducing the plant cover or the weeds are removed only occasionally.

Aquatic weed control is generally complicated because of lack of absolute ownership of a body of water. Frequently, approval is required from public health departments, the water purveyor, and fish and wildlife agencies. While treating a particular area with herbicides, multiple interests must be considered. For herbicide application at one place, information on the usage of water downstream is required. Interaction of various agencies is a prerequisite for the success of an aquatic weed control programme.

The various methods of control of aquatic weeds are discussed below.

Mechanical Methods

Mechanical methods employ physical forces to remove aquatic weeds or alter the environment so that the plant cannot become established or can not survive if already present. The mechanical methods include dredging, drying, mowing, handcleaning, chaining, burning, etc.

Dredging

Dredging is one of the common ways of cleaning weeds from drains and ditches. The dredge is equipped with a bucket or with a weed fork or other special tools. The bucket dredge does a good job of removing most of the weeds, but lifts out a lot of mud in the process. The weed fork drags out plant growth but leaves most of the mud behind. Generally, the more effectively the weeds are removed, the more likely the mud is removed along with them. Although dredging is efficient for complete removal of weeds, it is a slow, untidy and expensive method.

Drying

Drying is a simple, inexpensive, and satisfactory method for controlling submersed aquatic weeds. In this method, water is withdrawn from the pond or ditch, or drained from the ditch bottom and the tops of submersed weeds will dry up after several days exposure to sun and air. Drying may be repeated to control regrowth from roots or propagules in the bottom mud or sand. If clumps of weeds clog the ditch bottom, plowing may be needed to open a furrow so that the water can be drained out. Drying is ineffective against emersed weeds rooted in the bottom.

Mowing

Mowing is done to control weeds on the ditch banks. Small patches of shoreline emersed weeds are cut with scythes or swords. For clearing large ditch banks, powered equipment needs to be used where the banks are relatively smooth and not too steep. The effects of mowing are usually short-lived. It is ineffective for long-term or permanent control.

Manual Cleaning

In sparsely infested areas, manual cleaning may be the most practical method of aquatic weed control. The men cut and remove the accumulated weed growth with heavy knives and hooks. This method is effective particularly in the case of *Typha* (cattail), *Saggitaria* (arrowhead) and *Justicia* (willow) weeds.

Chaining

In this, a heavy chain is attached between two teams or tractors on opposite banks of the ditch. As these move, the chain drags over the weeds and breaks them off. The method is effective against submersed and emersed weeds. Chaining is usually not done until a ditch is severely clogged and is repeated at regular intervals. Chaining is limited primarily to ditches of uniform width, and accessible from both sides with tractors. If chaining is to be successful, a constant programme needs to be followed. Old growth must be removed at the beginning of the season. Chaining should then be done whenever new shoots rise at least 30 cm above the water level.

During chaining, the bottom mud is stirred up. It breaks up submersed aquatic weeds which then float downstream, spreading the infestation. Cleaning cannot be done near structures, particularly where reverse chaining is necessary. After chaining, more manpower is required to remove the loosened vegetation from the ditch.

Burning

Burning is used to control ditch bank weeds. Best results are obtained by first searing the green vegetation followed by complete burning after 10 to 12 d. In searing, a hot flame is passed over the vegetation at such a rate that plants wilt but are not charred. Burning can be combined with chemical or other mechanical control programmes. Mowing followed by burning the dried weeds may increase the effectiveness of the mowing.

Cutting

A mechanical weed cutter is used to cut weeds submersed at 1-1.5 m depth in the water. It consists of a sharp cutter bar operated hydraulically from a boat. The harvested weeds float to the water surface and are removed manually or by using sieve buckets. This method is inexpensive and relatively more effective.

Although some of the mechanical methods are useful, they do not provide effective and economical weed control because repetitive operations are required. The weed fragments which remain in the water can serve as a source of new infestation. However, they should be employed in situations where chemical control is not permissible or practicable.

Biological Method

The Biological method of aquatic weed control involves use of fish, snails, fungi, insects and mammals which feed on aquatic vegetation.

The chinese grass carp or white amur (*Ctenopharyngodon idella* val.) feeds primarily on submerged plants. It also feeds on small floating plants. It thrives best under cool waters although it tolerates warm waters. The small fish consume a vegetation several times their body weight. For every one gram increase in fish weight one fish needs to consume 48 g *Hydrilla*. About 75 fish can consume a vegetation of one hectare. Among the fish, *Puntius pulchellus* has shown great promise [3]. It was estimated that 10,000 fingerlings of this fish, each weighing 10 to 14 g, consume *Lemna* and *Hydrilla* weighing 25 to 50 kg day^{-1} and 9 to 18 of aquatic weeds per year.

The other fish species found useful for aquatic weed control are *Tillapia* (*T. zilli* and *T. quineensis*), silver carp (*Hypophthalmichthys molitrix*), silver dollarfish (*Metynnis roosevelti* Val.), common carp (*Cyprinus carpio* L.), goldfish (*Carassius auratus*), etc.

Among the various diseases of waterhyacinth, a thread blight caused by *Marasmiellus inoderma* (Berk.) Singh and a disease caused by *Alternaria eichhorniae* are among the potential biological agents for control of this weed.

The fungi which show promise against waterhyacinth are *Cercospora rodmanii*, *Acromonium zonatum* and *Ureds eichhornia*.

The insects found effective are flea beetle (*Agasiches hydrophilla*) on water hyacinth and *Salvinia*, and thrips (*Amgnotthrips andersoni*) and moth (*Vogtia malloi*) on other weeds. In south America the grasshopper *Paulinia acuminate* De Geer attacks *Salvinia* spp. and *Azolla* spp.

Some mammals like the sea cow or manatee (*Trichechus* spp.), a large warm-blooded marine animal known for its voracious feeding habits, consume succulent submersed vegetation. An adult manatee grows 3 m long and weighs 500 kg. It can consume 20 kg of vegetation per day. It survives in fresh, saline, brackish, muddy and clear waters. It is best suited for the tropics. A manatee lives for about 50 yr. In the absence of aquatic vegetation, it eats other vegetation, thus posing a problem for its introduction into an aquatic environment. This mammal is nearing extinction, however, because it is a source of protein.

Snails feed voraciously on submersed weeds such as pondweed, *Elodea*, coontail, etc. as well as the roots of floating weeds, e.g. waterhyacinth and *Salvinia*, and leaves of water lettuce. Over 20,000 snails ha^{-1} can consume a thick weed growth within 10-15 wk. The snails of genera *Marisa* and *Pomacea* are being increasingly used in water bodies of the USA. In India, an aquatic snail *Limnaea acuminata* was found to be a good biological agent for control of *Salvinia* [6].

Perkins [10] listed the most destructive types of insects under four groups depending on the damage they cause: (a) defoliators and external leaf feeders (*Cornops* spp.), (b) petiole borer (*Acigona infusella* Wlk.) and larvae (*Neochetina* spp.), (c) leaf tunnelers (*Orthogalumna terebrantia* Wallwork), and (d) scavenger species (*Dyscientus* spp.) which enhance the effect of attack by other insects.

Chemical Method

Control of aquatic weeds with herbicides is efficient, easier and faster, and even less expensive than mechanical control. The treated plants usually die in place and decay slowly.

The use of herbicides presents the problem of residues in the aquatic environment. However, selection of a good and a safe herbicide can mitigate this problem. further, the menace caused by aquatic weeds could outweigh this consideration and necessitates the use of herbicides. Herbicides can be effectively used to control those aquatic weeds which are inaccessible to mechanical or other control methods.

Most of the herbicides now used for control of aquatic weeds have low toxicity for human beings and other warm-blooded animals. Many herbicides are harmless to fish at concentrations necessary to control weeds. In some situations, it would sometimes become necessary to preserve some plant growth on the banks and to a lesser extent in the water to maintain a stable water course and to provide a suitable habitat for fish and wildlife. In such cases, herbicides showing selectivity are usually preferred in most situations of weed control on the banks. In the water, most submersed species will spread rapidly into space vacated by others. This makes it necessary to adopt total weed control by non-selective herbicides.

Herbicides may be applied to emersed and floating weeds by foliage application in much the same way as recommended for weeds in agricultural situations. Usually, spraying is done on only a part of the weed mat at a time. Herbicide application on the whole weed growth at one time may result in decay of all the vegetation, causing oxygen depletion and consequent fish kill. In the case of submersed weeds, herbicides are added to the main body of water to form a dilute solution. Herbicides are usually applied in two phases, the first for initial clearance and the next for subsequent maintenance.

Compared to mechanical methods, chemical method offers the following advantages:

1. The herbicide can reach the weeds otherwise inaccessible to mechanical methods.
2. One or two applications of herbicides in a season is adequate while several repetitions of mechanical methods are required.
3. Chemical weed control enables the dead weed growth to sink to the bottom avoiding the loss of plant nutrients from the water bodies.
4. The chemical method is an economical and time-saving operation.

Herbicide Application

Herbicides are applied in high volume sprays by hand, although boat-and-tractor-mounted sprayers are sometimes used. The application method varies with the properties of the herbicide. Chemicals which diffuse rapidly in water may be applied in concentrated form while others must be diluted and then sprayed evenly over the surface to ensure satisfactory distribution. The efficiency of the treatment is dependent on the length of time the plants are exposed to the herbicide. For this, continuous release of herbicide over a minimum period may be necessary. The best time for applying herbicides on emersed weeds is when the plants are well grown and causing a nuisance. In the case of submersed weeds, application on a heavy growth may cause serious deoxygenation of the water and result in fish kill. Early treatment is therefore advisable although it may not always be ideal from the point of view of efficient weed control.

Spray Calculations

EMERSED, MARGINAL AND FLOATING WEEDS

For application of herbicides to control emersed, marginal and floating weeds growing in stagnant water in ponds and lakes, the surface area needs to be considered. The quantity of herbicide can be determined as shown in the following example:

$$\text{Quantity of herbicide} = \frac{\text{Surface area (m)} \times \text{spray vol. ha}^{-1} \times \text{herbicide concen. (\%)}}{10,000 \times 100}$$

If an area of 1000 metres is to be treated with 0.5% solution of Gramoxone (paraquat dichloride) at a spray volume of 1000 L, ha^{-1}, the quantity of Gramoxone required is:

$$\frac{1000 \times 1000 \times 0.5}{10,000 \times 100} = 0.5 \text{ L}$$

SUBMERSED WEEDS IN STABLE WATER BODIES

For application on submersed weeds, the total volume of water in a pond or lake needs to be calculated. The quantity of herbicide required for treating the area can be determined as shown below:

$$\text{Quantity of herbicide} = \frac{\underset{\text{area(m)}}{\text{surface}} \times \underset{\text{water(m)}}{\text{depth of}} \times 10.11 \times \underset{\text{(ppmw)}}{\text{herbicide dose}}}{1,000,000}$$

If a pond of 5000 metres with 1.5 m deep water is treated with 0.4 ppmw of 2,4-D, the quantity of herbicide required is:

$$\frac{5000 \times 1.5 \times 10.11 \times 0.4}{10,000} = 3.03 \text{ kg}$$

Note: 0.4 ppmw is equivalent to 0.4 kg herbicide applied on a 1 km area or at 0.4 g on 10 m area.

SUBMERSED WEEDS IN FLOWING WATER BODIES

When water is flowing in a canal or channel, the herbicide calculation should be based on water discharge, which can be determined as shown under:

Discharge in cubic metres/sec (cumsec) =
width of canal (m) × depth of canal (m) × flow speed X 0.9

If a canal of 25 m wide and 1.5 m deep is flowing at a velocity of 0.8 m sec^{-1}, its discharge rate is:

$$25 \times 1.5 \times 0.8 \times 0.9 = 27 \text{ cumsec.}$$

This discharge rate should be used in determining the application (injection) rate of herbicide solution. If a canal has a discharge rate of 27 cumsec and is to be treated with a spray solution of 200 L in 90 min, the injection rate is:

$$\frac{27 \times 200}{90} = 60 \text{ L min}^{-1}$$

METHOD OF APPLICATION

The hand-operated knapsack sprayers (discussed in Chapter 14) can be used for spraying herbicides on aquatic weeds of ditchbanks. For spraying on weeds in ponds and lakes, the patches of weeds can be reached in rowing boats or floating planks and sprayed with pressurized knapsack sprayers. The boom can be fitted with more than one nozzle for faster coverage of the area.

For proper herbicide application on weeds of wider water bodies power sprayers mounted on a motor boat may be used. The spray range in this case will be 10-20 m with a discharge rate of 20-200 L min^{-1}. For injection of herbicide in stable waters, a hard pipe with several small holes is fitted. In flowing waters, the herbicide is injected from the shore itself. The herbicide is carried down the stream affecting the weeds. During spraying, the boat should maintain a speed of 2-5 km h^{-1}.

In situations where weed infestation is very dense and approach to the weeds from the ground is difficult, helicopters may be used for spraying. A single pass of a helicopter fitted with a bell-type spray boom containing many nozzles can cover a very large area. The spray delivery rate is about 1 L min^{-1} under a pressure of 2.8 kg cm^2. With an operational speed of about 60-70 kmph, a helicopter can spray up to 200-250 ha of aquatic weeds per day.

Granular formulation of herbicides is, easier to apply for the control of submersed aquatic weeds.

Herbicides Used for Aquatic Weed Management

Acrolein (acrylaldehyde)

Acrolein, known as 'Aqualin', destroys plant cells by contact action. It is effective against *Potamogeton* and *Elodea* and other submersed weed species. It is applied by pumping the spray solution into water and allowing it to move as a blanket over and through the aquatic weeds. One or two applications can provide a weed-free situation for the season. The herbicide may be introduced over a time period ranging from 30 min to 4 h in canals of up to 3 cumsec to 8 to 48 h in canals of up to 56 cumsec. Temperature of water is important. At 15° C, the dosage must be twice that at 28° C. Hence, the dosage must be adjusted to temperature and plant population. In fast-flowing canals and streams, contact of the herbicide with submerged weeds is slow and uneven, and this requires an increase in the dosage when flow is greater than 10 cumsec.

Acrolein is injected into water directly from the cylinder container through an extended acrolein resistant hose. It is applied at a concentration ranging from 4 to 7 ppmv. Acrolein-treated water is safe for irrigation. It is lost rapidly by volatilization and degradation. It is very low in toxicity for most crop plants, if the concentration in the irrigation water is below 15 ppmv. Fish are susceptible and many may be killed at concentrations of 1-5 ppmv. Hence, it should not be applied in waters where fish are grown extensively. Acrolein is a potent irritant and lachrymator (tear gas), but with proper equipment and precautions it can be safely applied. Application by licensed operators is preferable. It is inflammable.

Amitrole and Amitrole-T

Amitrole is effective on cattails (*Typha*), and amitrole-T on water hyacinth, alligator weed, and other emersed and floating species. Amitrole and amitrole-T are applied at 5-10 kg ha on the weed. They are safe on fish at rates applied for weed control

Bensulfuron

Bensulfuron may be applied PRE (50-100 g ha^{-1}) or POST (30–60 g ha^{-1}) to control many emersed and submersed broadleaf weeds and sedges. PRE application needs to be given to the soil after drawing out the water. It prevents emergence of weeds. POST application is useful to arrest the active growth of emersed and submersed weeds. Bensulfuron is effective against *Commelina* spp., *Sagittaria* spp., *Sphenoclea zeylandica, Pontederia cordata, Eleocharis* spp., etc. found in aquatic systems.

Copper Sulphate Pentahydrate

This chemical is effective against many kinds of algae including *Chara* and other species that cause pond scum. Soon after application (2-3 d), the algae colour changes to greyish-white. Copper sulphate applied continuously in sufficient quantity to maintain a concentration of 0.5-1.0 ppmw in slow moving water throughout the season gives effective control of algae and rooted submerged weeds in large canals and reservoirs carrying water for potable and industrial uses. It can be applied as a spray or by placing the crystals in a bag towed behind a boat until the chemical is dissolved, or by broadcasting crystals by hand. Copper sulphate is corrosive to commonly used sprayers. Hence, fibreglass, red brass or fibreglass equipment should be used. A concentration of up to 2.5 ppmw is considered safe for human consumption but above 1 ppmw is considered unsafe for fish. Water with 1 ppmw copper sulphate can be used as irrigation water for most crops. The activity of copper sulphate is greatly reduced in alkaline water due to precipitation.

2,4-D

2,4-D is very valuable for the control of water hyacinth, *Sagittaria* and *Alisma* spp., water lettuce (*Pistia* spp.) and other floating and emersed weeds. In *Potamogeton pectinatus*, 2,4-D moved from leaves to roots quite freely, but from roots to leaves scarcely at all [1], indicating symplastic movement with the assimilate stream. It can also control weeds such as waterhemlock (*Cicuta* spp.), milkweed (*Asclepias* spp.) and other poisonous species often found on ditch banks and in drains.

To control floating and emersed weeds, 2,4-D should be applied at 4-8 kg ha[-1] on the entire plant. The chemical should be in contact with the leaf surface for 4-10 h. The ester formulations which are non-polor (oil-soluble) are more effective because of rapid absorption by the waxy leaves of aquatic weeds, than the polar formulations. 2,4-D is effective when dissolved in water at rates of 10-30 ppmv. For this, the water-soluble formulation or granular form may be used. Addition of a wetting agent to 2,4-D solution enhances herbicide activity.

These phenoxy herbicides are toxic for some species of fish; they accumulate in the body of fish at least temporarily. Hence, fish from waters treated with 2,4-D should not be consumed before 30 d of the treatment. Ester in granular form is less hazardous to fish. 2,4-D plus TCA combination gives more effective weed control than 2,4-D alone.

Dalapon

Dalapon is very effective against grasses and cattails when applied on the foliage. Draining of water several weeks before application is advisable. It can be used in irrigation and drainage channels, lakes, ponds and ditch banks. Dalapon is applied at 15-20 kg ka[-1] (on the surface area of weed infestation) for effective control. Tank mix of dalapon (5-10 kg ha[-1]) plus amitrole-T (2 kg ha[-1]) provide excellent control of cattails. When dalapon is applied alone, a wetting agent (at 0.1%) to spray solution will enhance herbicidal activity.

Dalapon is normally harmless to fish. The treated water is, however, unsuitable for potable and irrigation purposes.

Dichlobenil

Dichlobenil is very effective when applied before weeds start growing to control *Elodea*, water milfoil, *Chara* and *Potamogeton* species. It should be applied at 5-10 kg ha[-1] on exposed bottoms where water can be drained out. This PRE treatment inhibits regeneration from roots and rhizomes. Water is let in 5 wk after treatment of the exposed bottoms.

Dichlobenil can also be applied at 10 -15 kg ha[-1] in standing waters before the weed growth begins. The treated water should not be used to irrigate crops or for human consumption. Fish grown in treated water should not be consumed before 90 d after treatment.

Diquat and Paraquat

These contact herbicides are very effective and safe for the control of many submersed weeds and algae. They should be applied (0.5-1.0 ppmv) before the weed growth reaches the water surface. They are effective against bladderwort (*Utricularia* spp.), coontail (*Ceratophyllum* spp.), elodea (*Elodea canadensis)*, water milfoil (*Myriophyllum* spp.), pondweed (*Potamogenton* spp.), *Spirogyra*, etc.

Diquat and paraquat are sprayed over or injected into the water. Application to muddy waters could result in inactivation of the chemical, leading to poor weed control. When combined with copper sulphate or triethanolamine, they kill the normally tolerant submersed weeds such as hydrilla and many algae.

Diquat and paraquat can successfully control floating weeds such as waterhyacinth and water lettuce (*Pistia)* with foliar application at 1-2 kg ha[-1].

Endothall

Endothall is very useful for controlling many submersed weeds including algae which cause scum in the stagnant waters of ponds and lakes without harming fish or wildlife. The liquid formulation of the herbicide is sprayed on the surface for control of emersed aquatic weeds or injected beneath the surface for submersed weeds. It is applied at 1-5 ppmv. The granular form may be sprinkled uniformly over the surface water. Periodic applications may be used on sprouting species.

The treated waters should not be used for potable and irrigation purposes for 15 d after application. Endothall is safer for fish. However, fish from treated waters should not be consumed for 3-5 d after herbicide application.

Fenac

Like dichlobenil, fenac is also applied to the soil after drawing out the water. It is effective in preventing regeneration of roots and rhizomes of aquatic weeds. Fenac is applied at 15-20 kg ha[-1]. It has longer persistence and hence its application is restricted to spot treatment at or around boat docks where long-term soil sterilization is desirable. Fenac pellets applied from the boat can control emersed aquatic weeds on the bottom of shallow lakes.

Fluridone

Fluridone is applied at 0.06-0.09 mg ai L[-1] (ppm ai) in ponds, 0.075-0.15 mg ai L[-1] in lakes and reservoirs, or at 2.24 kg ha[-1] of treated surface in drainage canals, irrigation canals and rivers. It can be applied to the water surface or subsurface or as bottom application just above the hydrosoil. Fluridone controls most of the submersed and emersed aquatic weed species such as *Utricularia* spp. (bladderwort), *Ceratophyllum demersum* (coontail), *Elodea canadensis* (common elodea), *Myriophyllum* spp. (water milfoil), *Najas* spp. (naiad), *Potamogeton* spp. (pondweed), *Hydrilla* spp. (hydrilla),

Brachiaria mutica (paragrass), etc. It provides season-long or longer control of some of these weeds.

Glyphosate

Glyphosate, being very effective on rhizomatous perennial weeds, controls emersed perennial grasses, *Typha* spp., (cattails), *Phragmites australis* (reed),. *Lythrum* spp., (loosestrife) and weeds that infest banks of irrigation canals and drainage ditches. It is applied on the weeds, POST, at 10-40 ml formulation (41% ai) L^{-1} water, depending on the need.

Silvex

Silvex gives effective control of surface and emersed weeds such as alligator weed (*Alternanthera philoxeroides*), waterlily (*Nymphaea* spp.), arrowhead (*Saggitaria* spp.), spikerush (*Eleocharis* spp.), etc. occurring in standing waters. It has good activity on some submersed seeds such as coontail, water milfoil, *Cabomba* spp. naiads (*Najas* spp.), etc. Silvex can be applied over the surface of standing water or injected into the water depending on the weed type. The treated water is unsafe for potable and agricultural purposes. Silvex is used in ester and potassium salt formulations.

Simazine, Diuron

Simazine and diuron, good algicides and soil sterilants, are applied to bare soil for PRE control of weeds in drainage channels and fish ponds, and on ditch-banks. As soil sterilants, they are used at 10 kg ha^{-1} and as algicides at 1-2 ppmv. They tend to be more persistent and this may result in reduced plankton growth and hence low fish population. The water of treated ponds and lakes is unfit for domestic and agricultural use for 3-5 mon.

TCA

The sodium and ammonium salts of TCA are very toxic for grasses on ditch banks and less toxic for broadleaf weeds. The herbicide is most effective when the spray is washed into the soil by rain after foliar application. It does not sterilize the soil for a long period. It is used to control *Phragmites* spp., *Sorghum halepense* (Johnsongrass), and other weeds of ditch banks and drainage systems.

PREDOMINANT AQUATIC WEEDS AND THEIR MANAGEMENT

Waterhyacinth (*Eichhornia crassipes*)

Waterhyacinth is the most predominant, persistent and troublesome aquatic weed. It exploits its freshwater habitat for efficient utilization of solar energy. It reproduces mainly vegetatively by means of slender horizontal runners called stolons. As the stolon grows, a new plant is formed at its tip and in a matter of days the parent plant is surrounded by offsprings which develop leaves and roots and send stolons by themselves. Holm et al. [5] found that two parent plants were surrounded by 300 offsprings in 23 d and by 1200 after 4 mon. A single inflorescence will have about 20 flowers and each flower produces 3000 and 4000 seeds. The seeds can remain viable for over 15 yr in the bottom soil. The seeds germinate only when the water recedes to 3-4 cm. After germination, the seedling remains attached to the mud. After some time, the plant gets detached from the mud and becomes free-floating. It then multiplies vetatatively. A single plant is capable of infesting an area of one acre in a year. The plant has blue flowers and is somewhat funnel-shaped and two-lipped. It is normally free-floating, buoyed by bladder-like inflated leaf petioles (Fig. 17.1). The leaf blade is

Fig. 17.1. *Eichhornia crassipes.*

somewhat kidney-shaped and somewhat rounded. Waterhyacinth produces greater dry matter content and converts solar energy more efficiently than many other plants (Table 17.1). Rao and Gupta [12] found that waterhyacinth produces about 800 kg dry matter ha^{-1} per day.

Table 17.1. Dry matter accumulation of many plant species and their solar energy conversion rates

Plant species	Dry matter ha^{-1} (tons ha^{-1})	Solar energy conversion rate (%)
Waterhyacinth	100-150	2.0-3.1
Sugarcane	75-112	1.6-2.4
Napier grass	106	2.2
Eucalyptus	39-54	0.8-1.1
Maize (corn)	24-37	0.8-1.8
Alfalfa (lucerne)	18-29	0.4-0.6

Chemical control of water hyacinth is more efficient and economical than mechanical methods. 2,4-D, paraquat, diquat and amitrole are applied alone or in combination on the foliage to control this weed. The amine and ester formulations of 2,4-D are more effective against waterhyacinth than sodium salt formulation. 2,4-D is applied at 2-8 kg ha^{-1} depending on mat density. Ramachandran [11] classified the mat density of water hyacinth into three groups on the basis of fresh weight with an upper limit of 13 kg sqm^{-1} as small, 23 kg sqm^{-1} as medium and 35 kg sqm^{-1} as big. However, it is simpler to apply 2,4-D at 0.3% (on the basis of formulation) concentration over the entire area. In this case, the quantity of spray solution required depends on the area and density of infestation. Addition of a wetting agent at 0.06% to spray solution (in case of sodium salt formulation) ensures better control of the weed by 2,4-D.

2,4-D should be applied on the actively, growing weed infestation. The affected mat shows epinastic symptoms characteristic of 2,4-D. The mat dies in 3-4 wk after treatment. It then decomposes and the decomposed mass sinks to the bottom in 4-8 wk. Repeat applications on the new weed growth may be given when required.

The combination of 2,4-D (4 kg ha^{-1}) paraquat (0.5 kg ha^{-1}) is more effective than 2,4-D alone as it ensures quick desiccation [12]. Similarly, 2,4-D (4 kg ha^{-1}) and diquat (1-1.5 kg ha^{-1}) combination is also more effective.

Paraquat is also applied alone. At 0.5% solution (Gramoxone formulation) with 200 L spray sol;ution per acre foot, paraquat desiccates the weed growth in 4-7 d but it does not sink. Thus, after desiccation the mat can be raked out and utilized for compost. when both desiccation and sinking are desired, urea may be added at 1% concentration to paraquat solution and the spray solution applied at 400 L per acre foot. After paraquat application, the mat shows wilting and browning appearance in 2 d. It dies and becomes a floating mass in 7-10 d and sinks to the bottom in 20-30 d. Spot application may be required on the areas missed during the first application.

Amitrole-T is applied at 0.5-1.5 % concentration. It is a fast-acting herbicide and the mat sinks more rapidly than in the case of 2,4-D alone. But it is more expensive than 2,4-D. Ametryn, a triazine herbicide, is also effective on waterhyacinth when applied at 3-4 kg ha^{-1}.

Cattails (*Typha angustata*)

Cattails are perennial, tall (2-4 m), grass-like plants with fleshy linear sheathing leaves having no midvein. The spike resembles the tail of a cat. They spread rapidly by rhizomes and by small airborne seeds. Seeds have longer viability. Each spike can produce 10,000 to 20,000 seeds. Cattails predominate in marshy areas, ditch banks, irrigation channels, water-logged fields, ponds, drainage, channels and brackish waters. Two species of *Typha*, *T. angustata* Bory and Chaub (Fig. 17.2) and *T. elephantina* Roxb. are the cattails commonly found in tropical countries.

Cattails can be controlled by dalapon and amitrole-T. Dalapon is more effective on cattails growing on ditch banks. It is applied at 2% concentration with a spray volume of 1000 to 2000 L ha^{-1}. Addition of a wetting agent at 0.06% concentration ensures better control of *Typha*. Dalapon (25-30 kg ha^{-1}) in combination with 30 L diesel and 1 kg detergent, amitrole (8 kg ha^{-1}, paraquat (1.2-1.6 kg ha^{-1}) dalapon+amitrole (15+3 kg ha^{-1}) and amitrole+TCA (5+10 kg ha^{-1}) are very effective. Paraquat gives fast top kill but heavy regrowth starts 2 mon after spray.

Mehta and Boonlia [7] showed that effective control of *Typha* was obtained when dalapon or amitrole was applied on the regrowth 3 mon after the original infestation was cut. Best control was observed when dalapon was applied in July and amitrole in September, November and January. Application on the regrowth following an initial cut could reduce the dose of herbicide required for effective control.

Mechanical control by weed bucket or weed cutter is very expensive. It was found that systematic and timely manual cutting of this weed is very economical and effective. In the areas of deep waters, three underwater cuttings done at one-month intervals during the rainy season, killed 90-95% of the *Typha*.

Ditch banks and drainage channels can be treated with a PRE herbicide such as simazine or diuron to prevent seed germination. This can be followed up later by POST application of dalapon, amitrole-T or dalapon+amitrole-T.

Pondweed (*Potamogeton* spp.)

Pondweed grows in fresh and salt water. It grows best in waters 0.5-2 m deep. The leaves of submersed plants are thin, transparent, long and slender. The emersed leaves are firm and leathery and attached to the stem by a petiole. Leaves are alternately

Fig. 17.2. *Typha angustata.*

arranged. Pondweed propagates through the buds that develop in the leaf axils, seeds and rootstocks which produce tubers. It can be controlled by 2,4-D. PRE herbicides such as simazine and diuron can also control this aquatic weed.

Hydrilla (*Hydrilla verticillata* Casp.)

Hydrilla is a submerged plant in fresh and stagnant waters. It can grow very deep. it has internodes of 3 m long. Leaves are sessile, linear, lanceolate with toothed margin arranged in whorls of 3-8. It propagates vegetatively through the aerial parts and tubers. It grows very fast even from broken fragments. The thick mat of *Hydrilla* prevents light penetration into the water and competes successfully with other submersed plants. It is sensitive to drying and hence can be controlled by draining the water from ponds and lakes, Pahuja and Sharma [9] found that *Hydrilla* required at least 5 d exposure to sun drying in April and 2 d in June. They also found that 2,4-D (2.5 ppmw), diuron (0.5 ppmw) and paraquat (1 ppmw) were effective in controlling *Hydrilla* with no regeneration.

Hydrilla is also susceptible to diquat (1 ppmw) plus copper sulphate (4 ppmw) combination. It is injected under the water surface. It is sensitive to endothall (2 ppmw) and acrolein (4 ppmw) toxicity.

Waterlettuce (*Pistia stratiotes* L.)

Water lettuce is a free-floating fern which propagates vegetatively. It has yellowish-green rosette leaves which are spongy and inflated. Its roots are filamentous, extending from the underwater rhizomes. The weed spreads very fast in stagnant or slow-moving water. Although it produces many flowers, its seed production is limited. It can be controlled effectively by diquat.

Salvinia (*Salvinia molesta* Mitchell)

Salvinia is one of the troublesome floating weeds in wetlands and rice-cultivated areas, causing problems for agriculture, water transport, irrigation, hydroelectric generation, public health, etc. It grows actively from July to November. It propagates vegetatively. It is a free-floating freshwater fern. It spreads very fast. It produces a dry matter of 650 g cm^2 yr^{-1} [4]. It forms a thick mat and prevents sunlight from reaching the water, with the result that the water becomes unfit for aquatic fauna. Although mechanical or manual removal is effective in controlling *Salvinia*, it is not economical. 2,4-D and paraquat are very effective against this weed. However, 2,4-D+paraquat combination is most effective.

Swamp Morningglory (*Ipomoea aquatica* Forsk.)

Ipomoea is a floating aquatic weed. It produces a dry matter of 24.41 t ha^{-1} [13]. This extensive weed growth hinders pisiculture, navigation, sanitation, etc. It can be successfully controlled by spraying the sodium salt formulation of 2,4-D (4 kg ha^{-1}) plus paraquat (0.5 kg ha$^{-1)}$ on activity growing weeds.

Marshy Weeds

Several marshy weeds such as alligator weed (*Alternanthera* spp.) arrowhead (*Sagittaria* spp.), spatterdock or yellow lily (*Nuphar* spp.), pickere weed (*Ponterderia cordata*), reed *(Phragmites communis)* and others hinder duck growth in duck marshes. These weeds can be controlled effectively by 2,4-D, silvex, dalapon, TCA and amitrole-T applied POST or simazine and monuron applied PRE to the bottom of ponds or lakes.

EFFECTS OF AQUATIC WEED MANAGEMENT ON FISH

The primary danger of aquatic weed control is that extensive eradication of submersed and marginal vegetation will deplete the food resources of fish. The herbicides may affect the fish population in many ways. Herbicides non-toxic to fish may kill planktonic algae and thereby decrease the fish food supply. In contrast, herbicides non-toxic to algae may increase the food supply of fish since the dead and decomposing submersed aquatic weeds release nutrients utilized by planktonic algae. However, the presence of too many weeds in a body of water may also have a long-term effect in decreasing fish population.

The herbicides mentioned earlier are relatively safer for fish but their safety can be ensured only when applied at an appropriate time. They should be used in conjunction with mechanical and manual methods. Raking, pulling and cutting may be done in small fish ponds, while mechanical weed cutters may be useful in large ponds and lakes.

UTILIZATION OF AQUATIC WEEDS FOR HUMAN BENEFIT

Aquatic weeds are a rich source of energy and are considered the most efficient converters of solar energy. For example, waterhyacinth with an annual productivity of 100 to 150 t dry weight ha^{-1} yr^{-1} has a solar energy conversion rate of 2.0-3.1% (Table 17.1) which is about 40% of the theoretical maximum. Waterhyacinth contains 26% crude protein, 26% fibre, 17% ash and 8% available carbohydrate on a dry weight basis. It can also serve as a good source of production of biogas. One kilogram of dried waterhyacinth can yield 374 L of biogas containing 70-80% methane with a fuel value of 21,000 BTU per m. Thus waterhyacinth is considered an important plant which can be used for the following purposes for the benefit of manking: (i) biogas production, (ii) manufacture of paper pulp and board because its fibre is chemically and physically similar to sugar-cane bagasse, (iii) a protein source for non-ruminant animals and human beings, (iv) a source of fertilizer and soil conditioner because of its high nitrogen and potassium content, (v) a mulching material for weed control and conserving soil moisture in crops and (vi) a material for making mats and covering house roofs.

REFERENCES

1. Aldrich, F.D. and N.E. Eto. 1959. The translocation of 2, 4-D-1-C^{14} in *Potamogeton pectinatus*, a submerged aquatic weed. Weeds 7: 295-299.
2. Brezny, O., I. Mehta and R.S. Sharma. 1973. Studies on evapotranspiration of some aquatic weeds. Weed Sci. 21: 197-204.
3. Devaraj, K.V. Manissery. 1979. Preliminary studies on the utilization of indigenous carp, *Puntius pulchellus* as one of the biological agents of weed control in ponds. Proc. Indian Soc. Weed Sci. Conf., Parbhani, Maharshtra, India, p. 82.
4. George, K. 1977. The *Salvinia* weed problem in Kerala-factors affecting growth and methods of weed control. Meeting Indian Soc. Weed Sci., Parbhani, Maharashtra, India, Abstr. No. 180.
5. Holm, L., L.W. Weldon and R.D. Blackburn. 1969. Aquatic weeds. Science 166: 699-709.
6. Ittyaverah, P.J., N.R. Nair, M.J. Thomas and P.S. John. 1979. *Limnaea acuminata*, a snail for the biological control of *Salvinia molesta*. Indian J. Weed Sci. 11: 76-77.
7. Mehta, I and D.S. Boonlia. 1980. Effect of seasonal sprays of dalapon and aminotriazole on the population fluctuations of *Typha angustata*. Meeting Indian Soc. Weed Sci., Bhubaneswar, Orissa, India, Abstr. No. 202.
8. Otis, C.H. 1914. The transpiration of emerged water plants; its measurements and relationships. Bot. Gaz. 58: 457-494.
9. Pahuja, S.S. and H.C. Sharma. 1980. Cultural and chemical control of *Hydrilla* and waterhyacinth. Meeting Indian Soc. Weed Sci., Bhubaneswar, Orissa, India, Abstr. no. 203.
10. Perkins, B.D. 1974. Arthropods that stress waterhyacinth. PANS 20: 304-314.
11. Ramachandran, V. 1969. Weed control is a must in fish culture. Indian Famg. 19: 75-77.
12. Rao, K. Narayana and K. Mahadeva Gupta. 1980. From aesthetic to pest (*Eichhornia crassipes* (Mart) Soms.). Meeting Indian Soc. Weed Sci., Bhubaneswar, India, Abstr. No. 206.
13. Seshavataram, V. and P. Venu. 1980. The problem of aquatic weeds in Kolleru lake—an appraisal. Meeting Indian Soc. Weed Sci., Bhubaneswar, Orissa, India, Abstr. no. 199.
14. Timmer, C.E. and Weldon. 1967. Evapotranspiration and pollution of waterhyacinth. Hyacinth Control J. 6: 34-37.

18

Weed Management in Forestry and Non-Agricultural Systems

WEED MANAGEMENT IN FORESTS

In forests, weed control measures are taken up in the following situations:
1. Maintaining forest tree nurseries in a weed-free environment.
2. Removal of unwanted shrub and small-tree growth in preparation for forest management.
3. Thinning of population of desired plant species.
4. Killing of cull trees and trees of unwanted species in older forests.
5. Killing or suppressing tall growing woody plants along forest roads and trails.

All forest-tree species are unwanted in one situation or another. Trees of even the most desirable species need to be removed in thinning and stand-improvement operations. All hardwoods are weeds in areas where pure stands of pine or other conifers are to be maintained. Hardwood tree species often cause weed problems because of their sprouting characteristics.

The troublesome weeds in forest nurseries are the usual broadleaf and grass weeds found in agricultural areas.

Forest Nurseries and Plantations

Weed control in forest nurseries is as important as in agricultural areas. It is achieved by manual weeding, cultivation and herbicides either alone or in combination. Handweeding in forest nurseries is very expensive. Labour is not always available. It is less effective when weed growth is very intensive in the warm, humid and high rainfall weather of the forests. Cultivation becomes impossible once the nursery beds are formed unless the seeds are sown in rows with proper spacing. If the weed problem is severe, herbicides provide the best answer.

For PRE (preemergence) weed control in nurseries, fumigation of the seed-beds with methyl bromide is done. After planting, PRE herbicides such as simazine, diuron, atrazine, linuron, oxadiazon, oxyfluorfen, monuron, alachlor, etc. may be applied to prevent establishment of weed growth.

The success of field plantings of tree species depends on the ability of planted stock to survive the first season and to dominate the associated vegetation later. In drought periods, the herbaceous weedy vegetation may deplete the soil moisture and provide strong competition, inhibiting the growth of young forest plants. Reduction of weed growth helps to conserve soil moisture for the benefit of forest plants. Preplanting application and incorporation of EPTC, trifluralin, chlorpropham, etc. give adequate control of undesired vegetation.

During establishment of forest plants in the field, weeds offer competition primarily for moisture without seriously shading or matting newly planted seedlings. PRE spraying of soil-applied herbicides (mentioned earlier) followed by one or two POST (postmergence) applications of foliage-applied herbicides provide effective weed control during the first year of planting. Elimination of weed competition in the first year is adequate for establishment of conifers under relatively severe conditions. Once established, conifers soon dominate herbaceous weeds. Later, woody plants exert adverse effect on planted conifers through shading rather than through moisture depletion. Conifers grown beneath a heavy canopy of woody and undesirable trees tend to have spindly stems that are too weak to prevent matting by deciduous foliage or clipping by animals. Conifers usually reach their most rapid growth in height somewhat later than their hardwood associates. The use of herbicides to eliminate woody trees permits the conifer seedlings to grow rapidly and become dominant. The dominance of conifers and other desirable forest trees is dependent on the sensitivity of crop species to herbicides used to suppress the brush weeds (woody plants, underwood thicket or small trees and shrubs) and the duration of beneficial effects from the treatment.

Land that needs to be planted with desired crop trees is normally covered with brush or small trees of inferior species or quality. Crops planted without eliminating these undesirable species can hardly be expected to survive. Hence, steps should be taken to (i) remove brush or herbaceous cover which interferes with the performance of planted stock, (ii) reduce physical obstacles to planting, and (iii) provide a weed-free seed-bed for direct seedling.

Mechanical methods of site preparation have some advantages over the use of herbicides. They not only remove the undesirable vegetation but prepare the soil and site for reforestation. Herbicides can kill undesirable vegetation but the dead vegetation may have to be removed by mechanical equipment or by burning to prepare soil for seeding or planting.

Usually tractors equipped with bulldozer blades or with toothed brush blades are employed in clearing burshlands for reforestation. Tractor-drawn brush cutters are very effective for site preparation.

METHODS OF MANAGING WOODY SPECIES AND OTHER WEEDS

Management of woody species and other weeds in forest areas is achieved by mechanical, biological and chemical methods.

Mechanical Methods

Cutting and Ring-barking

Cutting is done by axes, chainsaws, handsaws and winches. A tractor fitted with bulldozer blades does a quicker job. It levels the land and it is easier to work. Chaining the trees and dragging with the tractor may also be done.

Special tree-dozers which have two blades, the higher one for bending the tree and the lower one to severe roots, are most effective. Root cutters, rakes and ploughs also remove trees with roots. In slopy areas bulldozers, tree-dozers, root cutters and rakes are particularly useful.

Ring-barking or girdling is done with an axe and is the cheapest method of killing trees. It takes several months for the girdled tree to die. Time taken for a tree to die varies greatly on its health, growing conditions and species. During girdling a complete ring of cambium is removed. Girdling is done at a convenient height but as low to the ground as possible to prevent sprouting. Sprouts, if any, must be removed to prevent them from nourishing the roots. Application of herbicides in the girdled area will accelerate the death of the tree.

Burning

Burning is the cheapest and most effective method of eliminating forest brush if it is controlled under certain situations. Pines and other desirable forest trees withstand fire injury and the relative ease with which small hardwoods, shurbs and herbaceous plants can be damaged provide the basis for burning as a technique of weed control. Many brush areas are ideal for burning. If the brush is dense, burning is very effective. Many species may sprout after a burn and effective measures must be taken to control the resprouts; otherwise a much denser stand may result than that originally present. Herbicides are more useful in this situation. Burning is normally done in summer or winter. Summer fires are more destructive than winter fires. Repeated burning 2 or 3 times at intervals of 1 to 2 yr kills the larger hardwoods and inhibits regeneration so that sprouting becomes a minor problem.

Grazing

Grazing by animals, particularly goats, is employed to clear the bushes. However, it is not very effective unless supplemented with other cultural methods. Grazing is a selective process in that the animals choose the more palatable species, and overgrazing or grazing at a wrong time may cause shifts in the weed spectrum.

Chemical Method

Chemical control of brush, trees and other undesirable vegetation is more effective and economical than mechanical control. For many years, chemicals such as sodium arsenite, sodium chlorate and ammonium sulphamate were used for controlling woody plant species. However, with the discovery of phenoxyacetic herbicides and many residual soil-applied herbicides, chemical control became popular and widely accepted in forest management. Herbicides are applied in the following ways:

1. Soil application of PRE herbicides in forest tree nurseries and newly planted areas, as discussed earlier.
2. Foliar application to full-grown plants, seedlings and sprouts.
3. Dormant spray to full-grown plants.
4. Basal spray to basal cuts and girdled areas.
5. Tree injections.

Since the discovery of organic herbicides, foliar application has become the common method used to eliminate undesirable vegetation from forests. Herbicides commonly used are 2,4-D, picloram, dicamba, dalapon, glyphosate, amitrole-T, etc.

In countries such as the USA and Canada, foliar applications of herbicides are done by aerial spraying at a relatively lower cost. The total lack of selectivity is a major disadvantage in aerial foliage application. However, application at a time when hardwoods are at full leaf development and conifers have hardening new growth, will leave the conifers unharmed. Forest trees exhibit tolerance to some herbicides and display a seasonal variation in susceptibility to other herbicides. Degree of control is dependent on the quantity of herbicide translocated to various plant parts and the physiological state of the part. Actively dividing meristematic tissue is most susceptible. Satisfactory results in killing both tops and roots are obtained by slow penetration into the leaves, resulting in a slower death of the leaves and maximum translocation to root systems. Hardwood species which exhibit sprouting or suckering from the root system are particularly troublesome. When only the tops are killed, sprouting occurs on the existing hardwoods. In such cases, additional foliage treatment would become essential.

In the case of basal or stump treatments, the base of the trunk or stem must be thoroughly wetted down, to the ground level in order to reduce sprouting. The treatment is usually restricted to standing trees of 10-12 cm in diameter or to sprout clumps arising from previously treated stems. The spray solution is applied under low pressure directly to the bark surface above the ground level. There is some upward movement of the herbicide but little lateral movement or downward translocation into the root system. This method, if properly applied, effectively girdles the stem and kills dormant buds in the root-shoot transition area. For basal treatment, the mixture of 2,4-D plus picloram or 2,4-D ester alone is used at a dilution of 1-1.5 kg ai in 100 L of fuel-oil carrier. Mixing the herbicide in a straight oil gives best results.

Cut-surface treatment includes cutting trees and spraying the cut stumps. Stump spraying prevents regrowth through sprouting. Solution of 2,4-D at 1-1.5 kg ai in 100 L of fuel-oil applied to wet the cut surface and bark surface around the stump thoroughly. Picloram and glyphosate are also very effective. Hardwood stumps are effectively killed by placing ammonium sulphamate crystals. The best effect is obtained when the herbicide is applied immediately after cutting.

The herbicide is also applied around the girdled area of tree. The degree of kill obtained by mechanical girdling is variable and sometimes the trees may continue to live for three years or more. After girdling, the callus tissue formed over the girdled area provides a pathway for sufficient translocation of photosynthates to sustain life in girdled trees. Herbicides applied immediately after girdling kill trees more rapidly. Esters or amine salts of 2,4-D are very effective. The undiluted formulation of the amine salt of 2,4-D is adequate on many woody species. Glyphosate and picloram also give very effective kill of woody species. When applied, these herbicides are introduced directly into the open xylem tissues for absorption and translocation along the hydrostatic pressure gradients set up by supplying liquid to the tissues.

Herbicides are also applied by the tree-injection method. Tree-injection is accomplished by driving a chisel blade or an automatic-metering hatchet through the bark into the xylem of the tree and releasing the herbicide into the cut by a special device. Incisions are spaced 2 to 5 cm apart around the tree. For injection-treatment, the ester and amine formulations of 2, 4-D at a concentration of 2-4 kg ai, 100 L^{-1} water are used.

Besides 2,4-D, picloram, glyphosate and dicamba are extremely effective woody-plant herbicides. They are widely used in the control of woody plants along roadsides, rights-of-way, and wherever woody plants present a problem.

WEED MANAGEMENT IN NON-AGRICULTURAL SYSTEMS

Weeds in non-agricultural systems are as much of a problem as in agricultural systems. However, very little attention is paid to this matter in spite of its economic and aesthetic significance. Practically no funds are set apart in the universities, research organizations, municipal bodies and local government for solving weed problems in non-agricultural systems. The way of certain perennial weeds (e.g. *Parthenium hysterophorus*) are flourishing and becoming a major weed problem in certain tropical countries illustrates man's indifference and negligence in solving weed problems in non-agricultural systems. These include: (i) roadsides, (ii) railroads, (iii) industrial sites, (iv) power lines, (v) airfields, etc.

Roadsides

Weeds that grow along roadsides and highways are objectionable because (a) they are a good source of seed to infest adjoining cultivated lands, (b) they harbour insects and disease organisms that may spread to neighbouring crop plants, (c) they may reduce visibility at intersections and curves and (d) they may become a fire hazard to adjacent areas. Weeds normally found infesting roadsides are of many types: herbs, shrubs and woody species. They can be controlled by mechanical methods including mowing, burning, grazing, discing, hand pulling, slashing and hoeing as well as by chemical methods.

Maleic hydrazide (MH) is a good growth retardant and reduces the need for mowing herbaceous weeds. Soil applications of triazines and urea and uracil herbicides act as soil sterilants because of their long-term residual activity. In situations where the existing weed growth needs to be controlled, paraquat, picloram, amitrole-T glyphosate, asulam, 2,4,-D, dicamba, MSMA, 2,3,6-TBA, etc. may be used depending on the weed species predominant. For killing woody species, measures discussed earlier in this Chapter should be considered.

There is very little effort made by the road transport corporations and road maintenance organizations to keep the roadsides free of undesirable vegetation. It is not uncommon to see thick bushy vegetation along roadsides, providing good hideouts, for thieves, robbers and gangsters at night and thus making travel hazardous. Removal of this vegetation could make travel safe and save many lives. Replacement of this vegetation with desirable grasses maintained with regular mowing or application of growth retardants such as MH could make the roads look beautiful and travel safer.

Railroads

Like the roadsides, railroads have a weed problem, particularly from herbaccous and brush types. A railroad crosses many different major climatic zones and soil types. Railroad vegetation-control programme must be planned to meet a range of conditions and include several different types of treatment. A treatment must provide for knockdown deep-soil action, ensure adequate residual action and prove effective against a broad range of species in many environmental conditions. Spot treatments are necessary to eliminate hard-to-kill weeds.

Undesirable vegetation occurs both along the rail tracks and tracksides. Herbicides listed for roadside weed and brush control are also useful for railroads.

Industrial Sites and Airfields

Industrial sites include factories, pipelines, oil-storage tank farms, warehouses, electrical equipment sites, dams, parks, parking areas, etc. These industrial sites and airfields are infested mostly by herbaceous weeds such as perennial grasses and broadleaf weeds. To prevent this infestation, soil sterilants are most useful. This will maintain bare soil. Higher rates of triazine, urea and uracil herbicides (2-5 % solution) are particularly useful as soil sterilants. Choice of a soil-applied herbicide depends upon soil type, rainfall, weed species, etc.

Foliage-applied herbicides control the established vegetation. Herbicides such as paraquat, glyphosate, dalapon, picloram, 2,3,6-TBA, 2,4-D, amitrole-T, dicamba, MSMA, DSMA, etc. are particularly useful. They can be applied alone or in mixtures depending on the weed species present.

Vacant Areas in Villages, Towns and Cities

Weeds in vacant areas are undesirable because: (a) when dry, they are fire hazardous, (b) many weeds are poisonous and cause allergy in human beings and cattle and (c) they detract from the beauty of the living environment. These weeds are mostly neglected although they develop unsightly and dangerous growth. *Parthenium hysterophorus* is a serious menace in vacant and uncultivated areas in several tropical countries.

Parthenium flowers and seeds almost throughout the year and is a menace of dangerous proportions in many towns and cities. It poses a serious health hazard as it causes chronic skin allergy in human beings. The suffering is so unbearable and incurable that there have been instances of suicide deaths by the affected persons. Besides, its presence in agricultural fields is posing a serious problem.

Eradication of *Parthenium* can be achieved by undertaking the following measures:
(a) Prevention of seedling by destroying the existing growth before flowering with herbicides.
(b) Prevention of seed germination by using soil-applied herbicides.

As *Parthenium* plants cause allergy on contact, manual methods are not safe to man. Herbicides provide the best and effective method of controlling it. Among the PRE herbicides, atrazine (1-2), oxadiazon (1-1.5), simazine (1.5-2), monuron (1-1.5), terbutryn (1.2-1.5), neburon (1-1.5), metribuzin (1.5-2), etc. give excellent preventive control. To control the established growth, foliage applications of 2,4-D (2 -4) of both ethylamine and sodium salt formulations, and the combination of 2,4-D (1-2) +paraquat (0.8-1.2) give satisfactory results. Other herbicides which show good POST activity are picloram (1.5-2.0), dicamba (1.5-2.0), glyphosate (0.75-1.0) 2,4-D+MSMA (2+2), etc.

For control of other weeds in vacant and non-cropped areas, soil-applied herbicides mentioned earlier would give best results. The various foliage-applied herbicides which give effective control of standing weed growth are paraquat, paraquat+amitrole-T, dalapon+2,4-D, picloram, dicamba, glyhosate, etc. They are applied at 0.5-1% concentration (in the formulated product).

Combinations of petroleum oils with over 25% aromatic content and phenol herbicides such as DNBP and PCP are effective on slow growing and short-statured weeds. Herbicides are mixed in the proportion of 0.25-0.5 kg ai in 100 L petroleum oil.

Effective brush control is obtained by tebuthiuron. Promising herbicides for woody plant control are triclopyr, hexazinone and buthidazole. In the past 15 yr several old and new herbicides have been identified as useful and effective against many annual and perennial broadleaf weeds and grasses infesting forestry and non-agricultural systems. These include bromacil (PRE, 2-12), chlorimuron (POST, 15-35 g ha[-1]), dichlorprop (POST, 4-12), fosamine (POST, 10-50), imazapyr (POST, 0.5-2.5), norflurazon (PRE, 2.5-6.0), sulfometuron (PRE, POST, 70-420 g ha[-1]), tebuthiuron (POST, 0.84-4.48) and triclopyr (POST, 1-10). The activity of POST herbicides may be enhanced by tank mixing an oil adjuvant or non-ionic sufactant.

19

Weed Research Methodology—
Field Experimentation

INTRODUCTION

In order for the weed scientists to develop effective and economic recommendations for solving the multitude of weed problems in agricultural and non-agricultural systems, he or she must generate valid and reliable data from scientifically conducted field, greenhouse and laboratory research. Accurate and valid data are also essential for registration of a new herbicide compound for commercial use.

Although valuable information may be obtained from greenhouse and laboratory research, it is essential to make at least the final evaluation of a herbicide or a weed management practice on the basis of field performance of the treatments under varied soil and agroclimatic conditions. This also helps in determining the economic benefits the farmer would derive if a particular effective herbicide to weed control practice were recommended for use.

Research is becoming an increasingly expensive input in terms of money, time and effort. Therefore, the weed researcher must make every effort to conduct well-planned and well-designed experiments with minimum errors to obtain maximum information through reliable, accurate and valid data.

In this chapter, the basic principles of conducting weed research experiments beginning from planning to data reporting are discussed. In view of the greater importance of field research, special emphasis is given to the methodology related to field experiments.

PROBLEMS, OBJECTIVES, PLANNING

In planning for weed management research, the various weed problems and their magnitude need to be identified. This can be done by visiting the problem fields and discussing these problems with farmers, extention workers and scientists of other desciplines. If there are many weed problems, top priority should be given to those which are of immediate importance and whose solution will result in immense economic benefits to the farmer

For example, control of *Cyperus rotundus* and *Phalaris minor*, the worst weeds of cereal crops (rice and wheat) should get higher priority than other weed species which are not very serious. The problems of lower priority may be taken up if and when financial resources and technical personnel are available.

Before deciding to conduct experiments to solve a particular weed problem or initiating research on a particular project, adequate search into the literature must be done to ascertain the experiences of other research workers on this particular or similar problem or project. This will help in narrowing down the scope and objectives of the experiment, in executing the work more efficiently, and in avoiding unnecessary duplication. If the literature suggests the futility of using certain weed control or herbicide treatments, the researcher should then decide whether it will be justifiable to test the same treatments or herbicides under his or her conditions.

For successfully conducting the experiments, the scope of the study should be limited and objectives clearly defined to avoid confusion and ambiguity later. Some of the objectives may be as follows:

(i) Determine the herbicidal activity of a new or established compound on a particular weed and/or crop species.

(ii) Determine the optimum rate, time and method of application of a herbicide on a specific weed species or flora in a crop.

(iii) Study the harmful or beneficial effects of a herbicide or a weed control practice on the yield and quality of a particular crop or crops in a mixture or rotation.

(iv) Study the persistence of activity and residue problems, if any, of a herbicide.

(v) Determine the effect of a herbicide in different formulations.

(vi) Improve the activity of a herbicide in combination with adjuvants or other herbicides.

(vii) Study the changes in weed flora due to continuous use of a herbicide or weed control practice.

(viii) Determine the suitability of a herbicide or a weed control practice in a particular crop situation.

(ix) Determine the duration of weed competition and the time of weeding or weed control in a crop.

The success of achieving the set objectives is dependent on a number of variables such as: (i) environmental factors, (ii) agronomic factors, (iii) application of treatments, (iv) plant morphology and growth, (v) composition of tolerant strains in a susceptible weed species, (vi) design of the experiment, (vii) method of evaluation of plant response and data recording, etc.

TYPES OF FIELD EXPERIMENTS

There are different types of field experiments in weed research and each is designed to serve paticular objectives. Much of the weed research in the field includes testing and evaluation of herbicides, and it is conducted in the form of multi-stage trials.

Screening Trials

In the screening trials, newer herbicides, herbicide combinations submitted by chemical companies and herbicides found effective by other researchers are included to determine their efficacy on a specific crop and weed species. This herbicide evaluation work is conducted either in the field or greenhouse or both.

Screening is usually done in two stages. In the first stage, called **primary screening,** the herbicides are tested over a wide range of rates, preferably on a logarithmic scale, against many crops and weeds, preferably in single or double rows. The standard commercial herbicides applied at standard rates are also included to compare the performance of the products under evaluation. Primary screening trails give the estimate of general performance of a chemical against a specific crop or weed and the data obtained are highly subjective. This information helps in evaluating the selectivity of a particular herbicide in a crop or different crops and toxicity to various weeds. When there are several compounds for testing, primary screening helps to narrow down the choice of herbicides for further testing.

While evaluating a new compound about which very little is known, primary screening trials can, however, be preceded by simple observational tests with the objective of studying its efficacy on weeds and safety to crops.

In the second stage, **secondary screening**, the herbicides or herbicide combinations which performed well in the primary screening trials are further evaluated. Primary screening is, however, not a prerequisite for undertaking secondary screening trials. Chemicals which showed promise elsewhere as reported by other researchers or chemical companies are also included in secondary screening trials. The choice of chemicals to be included in these trials should be restricted to very few. They should be tested at two to four rates and at different times of application and/or growth stages of specific crop and weed species. The herbicide treatments are evaluated for selectivity towards the crop, efficacy in controlling target and subsequent weeds, and effects in changes in weed flora and crop yield. They are evaluated in comparison with standard weed control practices and/or commercial herbicides plus unweeded and complete weed-free treatments. In the case of herbicide combination treatments, each of the components at different rates should also be included. The treatments are compared with the help of qualitative as well as quantitative data.

Advanced Trials

Once secondary screening is complete, the most promising herbicides or treatments are included in advanced trials with objectives such as yield, herbicide residues, weed shifts, cost-benefits, formulation, spray volume, regional studies, cultivars, etc.

Yield Trials

Yield trials are conducted to determine: (a) the direct effect of the herbicide on the crop (through toxicity) and (b) the indirect effect of the herbicide through weed control. Both effects can be determined in one trial by including appropriate control treatments such as manually or mechanically achieved weed-free treatment and unweeded treatment to isolate the effects of the herbicide from the effects of weeds. While recording crop yield, data are also collected on weed species controlled and toxicity symptoms, if any.

Yield trials are essential for economic evaluation of a particular herbicide or weed control treatment in comparison with the standard recommended herbicide or treatment. Hence, they include only those treatments which are absolutely essential and after a thorough testing at the screening stage.

Residue Trials

Residue trials, conducted in the field or greenhouse, are useful to determine the duration of herbicide activity and effects on the following crops in rotation and weed species infesting later. They need to be conducted under different environmental and soil conditions to bring out residual toxicity problems, if any. Herbicides should be tested at 2 to 6 times higher than the normal rates and/or in multiple applications to establish the full range of residual effects. In some cases, an indicator plant should be used to prepare a standard response curve which can be used to determine residue levels in unknown soil samples.

Soil and plant samples collected from residue trails can be used for quantitative determination of herbicide residues and for understanding the transformation or degradation mechanisms of the herbicide under study. Greenhouse experiments are very useful to determine these effects under controlled environmental conditions.

Weed Shifts

Although the efforts on weed shifts can be studied to some extent in secondary screening trials, special studies are necessary for accurate evaluation of the short-term and long-term effects of the herbicide uses in a specific farming system. This type of study is not necessarily a follow-up of screening trials. It can include investigations on new or established herbicides or even a recommended weed control practice. This study can be modified to include treatments which may prove useful under the changed weed spectrum.

Herbicide Formulation

Herbicides are available in different formulations and can be compared at equivalent rates on the basis of active ingredient. Besides herbicidal effects, the convenience of their field application and performance under different agroclimatic and soil conditions and cropping practices, as well as cost benefits should be investigated.

Spray Volume

Optimum spray volume varies with the herbicide, weed species and intensity of weed infestaion. Generally, the foliage-applied contact herbicides require greater spray volume than translocated herbicides. Spray volume studies may include 2 to 4 volume rates at each of the dosage rates·to determine the interaction between the two variables.

Regional Trials

Regional trials are follow-ups of screening work to evaluate the performance of a herbicide or a treatment under diverse agroclimatic conditions, soil types and cropping practices. These trials provide a sound basis for developing weed control recommendations in different regions of a State.

Cultivars

Crop varieties differ in tolerance to herbicides. Hence, it is necessary to test all cultivars of a crop species against the herbicide under testing.

Demonstration Trials

These are basically yield trails to demonstrate or display the performance of a herbicide or a weed management treatment under field conditions for the benefit of farmers, extension workers, researchers and the public. These trials are usually conducted before a weed control treatment or a herbicide is recommended to farmers. They contain fewer treatments including those proven effective and useful in the earlier work and the standard recommended herbicide(s) and/or weed control practice(s). Demonstration trails usually have larger plots for proper comparison of treatments. Crop yields and weed control data will be useful for further evaluation of treatments. The trails should be well looked after to avoid possible crop damage by insects and diseases.

Special Trials

Weed Competition

Weed competition experiments serve a very useful purpose, namely, to study the competitiveness of different weed species or weed flora (population) as a whole in various crops and to determine the critical periods of weed competition. Such data are essential for recommending suitable herbicides or weed management programmes for different weed problems.

In these studies, certain treatments are kept weed-free for different periods of crop growth and than compared with a complete weed-free treatment (for the crop duration) and an unweeded treatment. Data are obtained on weed growth and population, crop stand, vigour and yield. Effects on disease and insect occurrence are also noted.

Application Techniques

Weed research is also conducted to standardize herbicide application techniques. These include studies for determining: (a) optimum spray delivery pressure, walking speed and delivery height, (b) proper spray nozzle, (c) optimum spray volume, (d) proper herbicide application, etc.

Other Studies

Special studies are required to determine the interaction of herbicides or weed control practices with soil fertility and nutrient levels, plant populations, row spacings, soil moisture conditions, crops and crop varieties, insect and disease occurrence, etc. These trials are usually conducted in the field.

SITE SELECTION AND WEED NURSERY

Success of a proposed experiment depends on the selection of a suitable site. The following should be taken into consideration while selecting a site for field experiments.
1. Uniformity in population and growth of target weed species. Weed growth in the previous year gives an approximate indication. If weed infestation is sparse weed seeding may be useful.
2. Suitability of the soil for experiments. The soil type should be representative of the region.
3. Fertility of the soil.
4. Objective(s) of the experiment.
5. Type of experiment.

All types of weed research trials can be conducted on the experimental farm of the institute or university. In some cases, demonstration trials and some of the advanced trials may be conducted on farmers' fields. The trials should not be located on the site used for other weed research experiments within the last 2 or 3 yr.

Pure stands of single weed species grown in special weed nurseries provide accurate date for primary and secondary screening trials designed to evaluate the performance of candidate, promising and established herbicides. These nurseries are useful for determining optimum herbicide rates, timing of application and herbicide interactions. In the case of perennial weeds, the extent and pattern of regrowth following a herbicide application or a weed control practice can be studied. The experiments conducted in weed nurseries are easier to evaluate. However, they do not provide data on crop-weed interaction in response to a particular treatment. There may also be technical problems such as obtaining weed seeds and other plant propagules.

Weed nurseries are established by making raised beds in long strips of 1.5-3 m width. Before the field is prepared for the nursery, the established weed growth, if any, should be killed by using POST herbicides. Weed seeds are planted in rows. In the case of perennial weeds, vegetative propagules such as rhizomes, bulbs, tubers, stolons, plant cuttings, etc. are used for planting. Pure stands of weed species are maintained by eliminating other weeds by hand removal or by using selective herbicides. Nurseries must be supplied with fertilizers and, if necessary, irrigation.

Maintenance of weed nurseries is expensive and difficult. However, the researcher will in turn obtain very reliable and accurate data.

Primary screening trials and herbicide residue trials may also be conducted in pots under natural conditions or under the controlled environmental conditions of the greenhouse. Weed seeds or other propagules should be planted for these experiments.

All the experimental area must be fenced in to adequately protect it from trespassing, cattle hazards, grass cutting, etc.

EXPERIMENTAL TECHNIQUES

The experimental techniques used for weed research are largely similar to those used for agronomic research.

Plot Size, Border Rows

Plot size depends on the following factors.
  (a) Type of field experiment: Primary screening trials may have smaller plots than secondary screening trials. Crop yield trials should have larger plots and demonstration trials the largest.
  (b) Uniformity in weed growth: Generally, the higher the uniformity in weed growth and infestation, the smaller the plot size.
  (c) Type of crop and weeds: Closely spaced crops may accommodate smaller plots than widely spaced crops. Trials in orchard and plantation crops usually have larger plots.

In the case of yield trials, each plot may have at least three or four rows for harvesting. It may have border rows on either side of the plot and in both directions. Sometimes, one row for two adjacent plots can serve as a common border row. The harvest rows may be used for collecting data on weed population, weed growth, crop growth, crop toxicity, crop stand, etc. Herbicide spraying should be done on the entire plot including border rows.

Arrangement of Plots, Replications

Except in weed nurseries, it is not always possible to find an experimental area with complete uniformity in weed infestation. The researcher should then arrange replications across the weed infestation gradient. For example, if weed uniformity gradient is North-South, the replications should be arranged East-West. It is not always essential to have all replications arranged in contiguous blocks. They can be located wherever the weed infestation is most uniform, so that maximum uniformity can be maintained within and between blocks. Once a replication is located, the treatments are allotted at random.

In case of primary screening trials designed to study the effects on weed growth, plots can be of different sizes arranged in different configurations. For example, pockets or patches of uniform weed growth can be marked and used to allot treatments. These patches or plots may not be always of the same size. In such cases, the spray volume differs from one patch or plot to the other, with herbicide concentration remaining constant. This arrangement is similar to growing weeds in plots and using them for experimental purposes.

It is always preferable to mark out more plots than necessary and only those which have uniform weed infestation and which meet the objective of the experiment should be used and others discarded.

In the case of crop yield trials, it is preferable to have contiguous plots within blocks. The blocks may be either arranged contiguous or not.

Control Treatments

Control treatments vary with the type and objective of the experiment conducted. They are normally: (a) unweeded or untreated with absolutely no weed control, (b) weed-free treatment with absolutely no weed growth, achieved by mechanical or manual method, (c) standard herbicide recommended for a particular weed situation and/or crop, and (d) the recommended weed control practice.

The unweeded control will be useful as a reference treatment for other treatments regarding distribution and assessment of different weed species. This is included in all the screening trials. Weed-free control treatment helps in assessing herbicide toxicity on the crop and comparing the weed control efficacy of different treatments and their influence on crop yield. The standard herbicide and weed control pratice treatments are useful to compare the effects of newer herbicides or weed control practices. They are also useful in comparing crop yield levels and determining cost-benefit ratios of new treatments which show promise.

In primary screening trials with a large number of treatments, it is desirable to have unweeded control treatments after every 5 to 6 treatments.

Plot Numbering, Labelling

Plots are numbered in different ways. The simpler method is the one which uses the three-digit numbering (e.g. 101, 102; 203, 204; 305, 306; etc.). In this, the first digit refers to the block and the second and third digits refer to plot numbers (Fig. 19.1). The treatments of the first replication may be allotted serially and those of other replications randomized (Table 19.1). Plot numbering should begin from the left side of the replications.

Plot labels should be inscribed with plot number, treatment details in abbreviated form, and time and date of application. Wooden, metal or plastic labels of smaller size (2.5-5 cm wide and 5-10 cm long) are tied to the stakes on the left side of the plot. Larger labels are fixed on the ground. In the case of advanced yield trials and demonstration trials, larger labels should be used.

101	102	103	104	105	106	107	108	109	110	111	112	113	114	115	REP I
201	202	203	204	205	206	207	208	209	210	211	212	213	214	215	REP II
301	302	303	304	305	306	307	308	309	310	311	312	313	314	315	REP III

Fig. 19.1. Numbering of plots in each replication.

Table 19.1. Treatment details and plot numbers

Treatment	Rate (kg/ha)	Time of applica.	Plot numbers		
			Rep I	Rep II	Rep III
1. Unweeded (control)	—	—	101	208	305
2. Weed-free (control)	—	—	102	204	309
3. Herbicide A	1.0	Pre	103	207	310
4. Herbicide A	2.0	"	104	201	302
5. Herbicide B	1.0	"	105	210	306
6. Herbicide B	2.0	"	106	205	308
7. Herbicide A+Herbicide B	1.0+1.0	"	107	209	303
8. Herbicide A+Herbicide B	1.0 + 2.0	"	108	206	301
9. Herbicide A+Herbicide B	2.0+1.0	"	109	203	304
10. Herbicide A+Herbicide B	2.0+2.0	"	110	202	307

HERBICIDE APPLICATION

Field application of herbicides has been dealt with in greater detail in Chapter 14 and the principles discussed there are applicable in the case of weed research experiments as well. Hence, this section briefly covers some of the special points related to herbicide application aspects in field and greenhouse experiments.

Sprayers

Herbicide formulations using water or oil as a carrier are applied by using different sprayers depending on their convenience and availability. The sprayers require an energy source to discharge the sparay solution, a spray tank to hold the liquid, a pressure regulator or gauage to control spray delivery pressure, and a spray lance or boom fitted with nozzles to distribute the spray solution over soil or plant surfaces.

The manually pumped knapsack sprayers with 5-10 L tank capacity and fitted with a pressure regulator are useful for most field trials with larger plots. Continuous pumping is necessary, however to maintain the pressure.

Smaller hand sprayers with 1-2 L capacity are useful for spraying smaller plots. At a spray volume of 500 L ha^{-1}, 50 ml spray solution will need to be delivered for 1 sq m area. For plots of 20 sq m and below, these hand sprayers are very convenient. In this type of sprayer, air is hand-pumped before spraying. It contains a 'T' shaped handle and plunger, and a discharge lance (20-25 cm long) fitted with a nozzle. In this sprayer, constant pressure cannot be maintained. This problem can be solved by pumping whenever there is a decrease in pressure.

The bicycle type of sprayers fitted with spray tanks, hoses, booms, pressure regulators, strainers and control devices to regulate spray delivery are very useful and convenient for spraying large plots in row crops. The boom height above the ground can be adjusted.

The Oxford precision sprayer is also very useful for spraying experimental plots. It requires a propellant which gives pressure. Propane gas or a refrigerant is used for this purpose. It can be carried on the back of the operator to deliver the measured quantity of spray solution.

Logarithmic sprayers (Chapter 14) are ideal for primary screening trials to determine the range of doses at which herbicide is effective.

A herbicide laboratory sprayer may also be used for precision application of herbicides to potted plants grown in the greenhouse and growth chamber. The sprayer consists of a spray both which accommodates plants up to 1 m tall. A spray cart that travels the length of the booth on an overhead track delivers the spray. The tray supporting the plant containers may be lowered or raised to spray various sizes of plants. Spray fluid is delivered to the nozzle from a small glass jar in a 100 ml pressurized container. Speed of the spray cart varies between 0.5 to 5 kmph and can be adjusted manually. This type of sprayer is extremely useful for accurate application of herbicides at different spray volumes, delivery pressures and speeds.

Spray Calibration

Spray calibration is done on the basis of area of the plot or herbicide concentration or dilution. Spraying on area basis is done in the case of plots of known areas, as shown below.

If diuron is to be applied at 2 kg ai ha^{-1} in plots of 20 sq m, the total amount of herbicide formulation required in four plots (replications) of this treatment is :

$$\frac{\text{Rate} \times 100}{\% \text{ active ingredient in herbicide formulation}} \times \frac{\text{area of 4 plots}}{10,000}$$

that is,
$$2 \times \frac{100}{80} \times \frac{20 \times 4}{10,000} = 0.02 \text{ kg or } 20 \text{ g formulation}$$

If spray volume is 400 L ha^{-1}, then the quantity of carrier (water) required for four plots can be calculated as under:

$$\frac{\text{Spray volume} \times \text{Area}}{10,000}$$

that is,
$$\frac{400 \times 20 \times 4}{10,000} = 3.2 \text{ L}$$

The herbicide formulation of 20 g should be mixed in 3.2 L water and one-fourth of this solution, i.e., 800 ml must be used for spraying each plot separately.

As some quantity of spray solution always remains in the tank at the end of spraying, calibration must be made to determine this volume. For example, if 200 ml solution is remaining at the bottom of the tank after spraying, the herbicide required for this spray volume is weighed and mixed with water. This allowance spray solution should be used only while spraying the first replication plot of each treatment. Fillings for subsequent replications should be done only with the exact volumes already calculated for each replication plot.

Spraying on volume basis is useful for POST applications on established weed growth and for applying on an unknown area. The herbicide solution is made on the basis of known spray volume and herbicide concentration. For example, if glyphosate needs to be applied at 1% concentration (on the basis of commercial formulation) at a spray volume of 400 L ha^{-1}, the quantity of commercial herbicide formulation (Roundup) to be used is determined as shown below:

$$\frac{\text{Concentration} \times \text{Spray volume}}{100}, \text{ i.e., } \frac{1 \times 400}{100} = 4 \text{ L Roundup}$$

In the case of foliage-applied herbicides, the spray volume varies with type of herbicide, weed species, stage of weed growth and intensity of weed infestation. Contact herbicides such as paraquat require higher volume than translocated herbicides to cover the same weed growth. A moderately infested area may require a lower quantity of spray solution per unit area than a heavy infestation. The exact spray volume may be determined by trial spraying the known weed-infested area (50 sq m or 100 sq m) with water and this volume maintained for all the treatments in the experiment. The herbicide rate should, however, be used on the basis of area.

Spraying Precautions

(a) Solid formulations of the herbicides must be weighed in the laboratory and brought to the field in airtight containers or envelopes. Herbicide quantities of all replications must be weighed together.

(b) Liquid formulations may be measured with a pipette (avoid mouth pipetting) or graduated cylinder directly from the herbicides container.

(c) While mixing wettable and slow-dissolving powders, the herbicide should first be added to a small quantity of water (10% or 20% of total volume of water) and mixed thoroughly. Then water is added to make up the volume. Further mixing of the entries spray solution is done by pouring it carefully form one container to the other several times before pouring into the spray tank.

(d) In the case of emulsifiable concentrates, the herbicide solution should be added to a small quantity of water and mixed thoroughly as in the case of wettable powders, before bringing the spray solution to the desired volume.

(e) Spray solution for all replications should be made up in one batch. The exact quantity of spray solution of each replication should be sprayed separately.

(f) Sprayer should be thoroughly rinsed and washed with water several times before another herbicide is applied.

(g) Generally, aqueous formulations should be applied first, followed by wettable powders and emulsifiable concentrates.

(h) Trial spraying with water for a few times in non-experimental plots will ensure uniform application in experimental plots.

(i) To avoid spray drift to adjacent plots or fields, spraying in heavy winds should be avoided. The wind movement during spraying should normally not exceed 8 kmph. Usually, spraying must be done in the direction of wind. In heavy windy areas, it is preferable to use a windshield or reduce the spray delivery pressure and delivery height from the ground.

(j) When making POST applications to weed foliage, a rain-free period of at least 3-4 h after spraying is essential.

(k) Soil incorporation by mechanical means should be done after applying soil-incorporated herbicides.

(l) In case of delayed rainfall following PPL or PRE application, irrigation may be given to ensure good herbicidal activity. This, however, depends on the objective of the experiment.

(m) The applicator should wear protective clothing (gloves, overall, etc.) to protect himself from the toxicity of the herbicides. Eye protector and respirator should be used if a diluted product of a particular herbicide is known to be toxic upon contact with the eyes or when inhaled.

(n) In case of accidental contamination of herbicide with the skin, eye or any part of the body, the exposed area must be thoroughly washed with water. It is always better to follow the guidelines given on the herbicide label by the manufacturer.

(o) All the experimental details on herbicide application, field history, weather at application, weeds, etc. must be recorded immediately after spraying.

ASSESSMENT OF PLANT RESPONSE, DATA RECORDING

Proper and unbiased assessment of plant responses to treatment is a prerequisite for the success of research. Assessment should be done at different intervals (weekly, fortnightly or monthly) or at the most appropriate time after beginning the experiment. The initial assessment should be done at shorter intervals, sometimes even daily, so that the development of herbicide toxicity on a weed and crop can be closely monitored. The method, intensity and frequency of assessment varies with the type of experiment. Treatments are usually assessed for weed population and growth as well as crop toxicity, growth and yield. Except crop yield, all other parameters can be assessed by qualitative or quantitative methods or both.

Qualitative Assessment

For simple observational tests and primary screening trials, qualitative assessment through visual observation is usually quite sufficient. In qualitative assessment, the control plots are used as reference. During visual scoring, the stand and vigour of weeds and crop are taken into consideration.

Weed Response

Visual assessment of weed response is based on such effects as weed kill, weed growth, weed population, injury to weeds, etc. by a particular treatment. This is done scoring each major weed, using the scales of 0–10, 1–9, 0–5, or 1–5. Of these, the most widely used and simpler rating scale is 0–10, where 0 represents no effect (no weed control) and 10 corresponds to complete effect (complete weed control) as given in Table 19.2. These scores may be converted later to percentage numbers. (0–10 being equal to 0–100.)

Table 19.2. Qualitative description of treatment effects on weeds and crop in the visuals croping scale of 0 to 10

Effect	Rating	Weed	Crop Description
None	0	No control	No injury, normal
Slight	1	Very poor control	Slight stunting, injury or discoloration.
	2	Poor control	Some stand loss, stunting or discoloration
	3	Poor to deficient control	Injury more pronounced but not persistent
Moderate	4	Deficient control	Moderate injury, recovery possible
	5	Deficient to moderate control	Injury more persistent, recovery doubtful
	6	Moderate control	Near severe injury, no recovery possible
Severe	7	Satisfactory control	Severe injury, stand loss
	8	Good control	Almost destroyed, a few plants surviving
	9	Good to excellent control	Very few plants alive
Complete	10	Complete control	Complete destruction

It is not always possible to find uniform infestation of all target weeds in all the plots. Some plots may not have the weed species present in the other plots and vice versa. In such a situation, one should carefully try to determine whether or not this is the effect of treatment. To eliminate this possibility, the experiment should have one unweeded or weed-free control treatment for every 5 or 6 treatments.

Visual scoring must always be done by two people separately and the data pooled. They may also do this together by standing on either side of the plot and recording means of their scores. In order to obtain unbiased assessment care should be taken not to look at the plot label until entire trial has been evaluated. After rating a few plots, it is quite common for any researcher to lose the precision and accuracy he started with. In such cases, he should check back on the weed infestation in the control plot and resume rating treatments. Sometimes, it is advisable to have a break between replications and even between treatments if the experiment is large. While rating a trail, the researcher should bear in mind that the success of the experiment and eventual recommendations largely depend upon the accuracy with which the rating is done. The data of each replication should be compared in the field immediately after scoring and if any wide variation is observed between replications in case of particular treatments, double checking should be done before leaving the site.

In a mixed weed population, weed can be grouped into broadleaf species and grasses and each group is assessed separately. In a more common method, the extent of ground cover by grasses and broadleaf weeds separately or together is rated first, followed by estimation of percentage infestation by each prominent weed species. From these data, the percentage cover by individual grass or broadleaf species in a plot is calculated as under:

$$Z = 100 - 10 \ (X) \ \frac{Y}{100}$$

where X is weed control rating in the plot, Y the % total weed infestation by area of individual weed species and Z% ground cover by individual weed species.

If a plot, for example, was rated 6 (i.e. 60% control) for grass control and *Paspalum conjugatum* accounts for 30% of the weed infestation, followed by *Axonopus compressus* 50% and *Digitaria sanguinalis* for 20%, the % group cover by each weed species is calculated thus:

Paspalum: $[100\text{-}10(6)] \ \dfrac{80}{100} = 12\%$

Axonopus: $[100\text{-}10(6)] \ \dfrac{50}{100} = 20\%$

Digitaria: $[100\text{-}10(6)] \ \dfrac{20}{100} = 8\%$

In this example, the total ground cover by grasses was 40% (as 60% were controlled), and *Paspalum, Axonopus* and *Digitaria* accounted for 12%, 20% and 8% of the infestation respectively. These data compare favourably with quantitative data from weed counts and weed weights.

Similar evaluation and calculation may be done for broadleaf weeds together and separately.

A common system or method of rating is desirable as it eliminates the wide disparities inherent when different methods are used.

Crop Response

Crop response is also rated in the scale 0–10, as shown in Table 19.2, to record herbicide toxicity on crop stand and growth.

Quantitative Assessment

Quantitative assessment of weed response gives the benefit of accurate and unbiased data. However, it does not always reflect the actual effects of the treatments. It has particularly no advantage over qualitative assessment when there is no uniformity in weed infestation. However, this method is more valid and dependable for assessing the effects on crop growth and yield. It should be used in secondary screening trials, advanced trials, yield trials and special trials either alone or in conjunction with qualitative assessment.

Weed Response

The most commonly used methods for quantitative assessment of weed response are weed counts and weed weight.

Weed counts: Weed counts may be taken by placing a quadrat in the plot at random and counting the number of plants inside the guadrat. The quadrat can be placed by throwing it into the plot at random or by placing it where weed infestation is representative of the treatment.

A quadrat is a sampling unit which has an area of definite size, the shape being rectangular, square or circular. The size of the quadrat and number of samples depends on plot size and density and uniformity of the weed population. Generally, the number of quadrats per plot should be 2 to 10 and the number of individual weeds of the species being tested and counted should be between 100 and 300 before treatment. The quadrats may differ in size, ranging from 20 cm × 20 cm to 1 m × 1 m. The area under all the quadrats should be 5 to 10% of the plot size. Increasing the number of quadrats per plot does not necessarily increase the precision of the experiment but may give a better evaluation of the variability of the weed population.

Weed population in control plots should serve as a reference to all other treatment plots. Although all the species occurring in a quadrat may be counted, meaningful comparison between treatments can be made only in the case of those species which appear consistently in all replicate control plots in sufficient numbers.

Fixed quadrats are useful for comparing the herbicidal effect before and at different times after treatment and to study regeneration of weeds and changes in weed flora. They should be used in conjunction with random quadrats to obtain more information rater than independent of them.

Although weed counts give a precise representation of the weed infestation and degree of control, they do not reflect the treatment effects on weed growth (i.e. height). This method is laborious and time-consuming.

Weed weight: In this method, weeds are cut or pulled out of the soil and weighed either fresh or after drying. This method is limited in use because it can be done only once during the season or period of crop growth, as subsequent harvesting will not give an estimation of the cumulative weed growth. It is usually employed before the experiment is terminated. In studies on perennial weeds, however, periodical harvesting is done to study the regeneration pattern.

Crop Response

Crop height: Weed control treatments affect crop growth. Hence, crop plant height is a dependable and accurate measure of treatmental effects. It is used to supplement crop yield data. The sampling size depends on size of plot and number of crop plants. Generally, 8-10% of the plants in experimental rows are used for crop height data.

Crop weight: In forage crops and other crops which are not carried to maturity or for grain yield, the crop plants are cut at different intervals and the fresh or dry weights recorded. In the case of crops grown until harvest, one experimental row is used for recording crop weight data at different intervals.

Crop yield: Yield is the most important measure of crop performance; the success of a weed control treatment and its eventual recommendation as a practice depend upon this criterion. In crops in which all plants are ready for harvesting at one time, it is easier to record yield data. However, in some vegetable crops which do not mature all at once, the period at which most of the crop can be harvested should be established. Crop yields should, when relevant, (e.g. cereal grains), be adjusted to a standard moisture percentage.

Crop stand: Herbicides frequently affect the survival of crop plants. Therefore, stand counts are useful for comparing herbicide effects with non-herbicidal treatments. Usually, all the crop rows are used for crop stand count.

BIOASSAY OF HERBICIDES

In weed research, bioassays are used to measure the biological response of a living plant to herbicide and to quantify its concentration in a substrate. Bioassays are usually conducted with sensitive plant species, also referred to as indicator or test species. Although analytical, instrumental and chemical methods may provide more accurate quantitative measurement of a herbicide concentration in soil or water, bioassay methods are relatively simple to carry out and do not require sophisticated and expensive equipment. Bioassays are often particularly valuable and relevant because they measure exactly the phytotoxic activity of the herbicide molecule and determine the phytotoxic residues in the substrate.

In a bioassay, an indicator species is grown in herbicide-treated soil or in solution of the herbicide extracted qualitatively from soil or plant tissue. This response is compared with that shown by similar plants grown in untreated soil or extract and/or in soil or extract containing known concentrations of the same herbicide selected to give responses of the sensitive indicator species ranging from nil to complete death as herbicide concentration is increased. The relationship between herbicide dose and plant response (% kill) can be compared in different ways as illustrated in Fig. 19.2 [1]

When the response is plotted against dose on an arithmetic scale a smooth curve is obtained (Fig. 19.2A). If the response is plotted against the logarithmic values of the doses, this smooth curve assumes a symmetrical sigmoid form (Fig. 19.2B). When the percentage kill is transformed to probit values and plotted against log dose values, the sigmoid curve becomes a straight line(Fig. 19.2C) which facilitates drawing the line of best fit.

The proportional increase of plant response to herbicide dose is greatest at a level where the plant produces 50% response, which is known variously as GR_{50} (does that causes 50% growth reduction), ED_{50} (equivalent dose for 50% response), LD_{50} (lethal dose for 50% kill) or ID_{50} (dose for 50% inhibition in growth). In other words, these values represent the herbicide concentration required to inhibit plant growth by 50% or achieve plant kill by 50% compared to untreated plants. This is usually the centre of the section over which the line is the steepest in the case of a sigmoid curve (Fig. 19-2B). This is the most widely accepted method of determining plant responses and is preferred to the smooth curve (Fig. 19.2A) which shows a linear graph only in a narrow dose range.

The response of a herbicide on two or more plant species can be determined by comparing the ED_{50} values [1] in a sigmoid curve (Fig. 19.3). If one species is more sensitive than the other, the ED_{50} value for the former is lower than that of the latter. Sometimes, this comparison can be made on the basis of ED_{90} values (equivalent dose for 90% response). This is particularly useful to determine the selectivity of a herbicide between a weed and a crop species.

Biological response is commonly measured by the whole plant method, and sometimes even by taking into account only the effect on shoot or root growth. The response is evaluated either qualitatively by using the visual rating system or quantitatively by using the green or dry weight of shoots. Green weights are preferable to dry weights

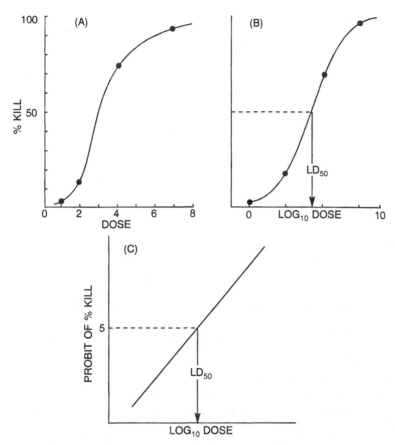

Fig 19.2. Dose-response relationships: A—dose on arithmetic scale; B—dose on logarithmic scale; C—probit of response plotted against dose on logarithmic scale[1].

as the recently dead and necrotic plants may have the same dry weight as the green healthy ones.

Once the dose-response curve is established with known concentrations of a herbicide, the plant response in soil containing unknown herbicide residue is compared with this curve and the quantity of residue is determined.

Indicator Species

The plant that can be used as an indicator species must be sensitive enough to detect even very small amounts of herbicide in the soil or another substrate. It must also show a gradual increase in susceptibility with increasing herbicide concentrations. The indicator plants should be vigorous and grow rapidly under the conditions of bioassay. The more commonly used indicator species are cucumber, oats barnyardgrass, tomato, barley, sorghum, crabgrass (*Digitaria sanguinalis*), yellow foxtail (*Setaria glauca*), etc. The ideal test species must, however, be determined from preliminary experiments with the herbicides under study.

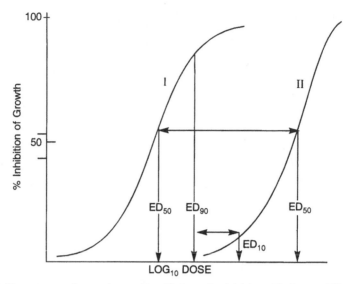

Fig 19.3. Response of weed species (I) to a herbicide with lower ED_{50} and ED_{90} values and that of crop species (II) with higher values [1].

STUDIES ON HERBICIDE PERSISTENCE AND RESIDUES IN SOIL

Bioassays are used to study the persistence of herbicides at different rates. In these studies, the indicator plants are planted in the field or in potted soil (in the case of greenhouse experiments) with known quantities of a herbicide against which the test species is very senitive. The plants are allowed to grow for a pre-determined period and harvested to record fresh weights of shoots. A dose-response curve is then established to determine the ED_{50} value. This is repeated at different intervals and at the end of the study the ED_{50} values are compared to determine the persistence pattern of the herbicide.

In studies whose objective is to determine residue levels of a known herbicide in different fields, the surface soil samples from several areas are collected and used for bioassay studies in pots under greenhouse conditions. The plant response values are compared with the ED_{50} values of the standard dose-response curve and the herbicide concentrations in the soil samples determined.

ESTIMATION OF WEED SEEDS IN SEEDBANKS

Quantifying weed seeds in the soil seedbank is of particular interest to ecologists and weed scientists is it enables them to determine the potential density, intensity and diversity of weed infestation in a particular crop or non-crop area. This information is useful in understanding and predicting the fate of weed seeds in the soil, and designing effective, viable and economic weed management programmes.

A number of methods have been described in the past for collecting soil samples and then extracting weed seeds. These include using augers, metal tubes, plastic liners and a vacuum system for collecting soil samples, and adopting the techniques of sieving, flotation in high-density solutions and centrifugation for extracting and separating the weed seeds. These methods, aside from being time consuming, are effective in estimating the densities of only some weed species in the soil.

Wiles et al. [2] described a simpler and more efficient technique of soil sampling and weed seed separation. This includes a soil sampler, elutriator and a sample flushing device, the details of which are discussed below.

The **soil sampler** includes a tube to collect cores of 5 cm in diameter and 10 cm deep, a foot rest and a sharp edge on the tube to facilitate pushing the sampler into the ground, and a rod and plunger to quickly eject an intact sample (Fig. 19.4). Cores are thrust from the sampler into plastic bags by pulling down on the spring-loaded handle. There is a short handle close to the cylinder of the sampler to provide leverage for ejecting the sample. The plastic bags containing the samples are stored at 6° C for 7-14 d until the soil can be air-dried in a greenhouse. After weighing, the air-dried samples are processed in the elutriator.

Fig 19.4. The Soil sampler with cut away view of the rod and plunger [2] (Reproduced with permission of WSSA).

The **elutriator** uses water (at pressure 275-345 kPa) to separate weed seeds from clay and silt particles in soil cores. The principle components of the elutriator include a hooded housing unit containing the spray nozzles, rotating drum, water basin and cylindrical strainers, a rheostat-controlled motor and a sediment trap (Fig. 19.5). The **rotating drum**, made of stainless steel, is mounted to the housing unit. It contains 36 capped strainers (each 6.5 cm in diameter, 17.5 cm long and lined with 318 μ stainless steel screens) which allow water to pass through while retaining weed seeds, plant debris, stones and large sand particles. The **water basin** is filled with 10 L of water before the drum starts turning. As the drum rotates, a continuous stream of water is sprayed into the strainers from six TeeJet 15006 nozzles (of Spraying Systems Co., Wheaton, Illinois, USA) mounted in the elutriator's hood. Excess water and sediment flow into the **sediment trap** through an overflow outlet in the water basin.

Using a flushing device, the sample contents in each strainer is washed (Fig. 19.6) into a propyltex fabric bag (of Teko, Inc., Elmsford, New York, USA) for drying and storage. Bags are 12 cm wide and 12 cm long on one side and 16 cm long the other side (to allow a 4 cm flap when folded down to pervent loss of the sample contents). The propyltex fabric has a mesh opening of 125 μ. The contents of the bags are dried overnight at 49° C.

Seeds are counted and identified using a parallel optical stereozoom microscope (of Nikon Corp, Melville, New York, USA), set at 7.5 X magnification and a single fibre optic illuminator (of Dolan-Jenner Industries, Woburn, Massachusets, USA). Samples with large amounts of organic matter or sand are subdivided to make counting and identifying seeds easier.

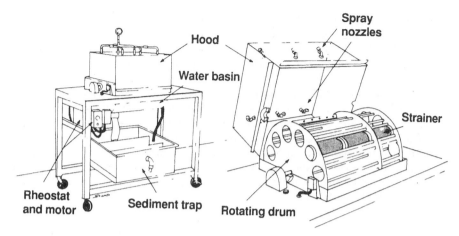

Fig. 19.5. The Elutriator and its principal components [2].

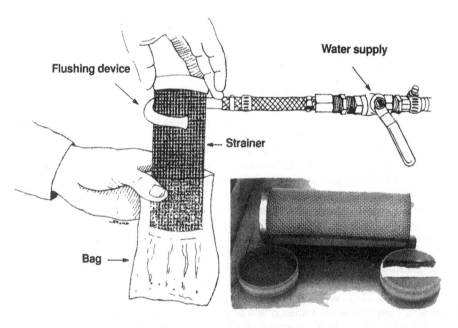

Fig. 19.6. A flushing device for transferring the sample content from a strainer to a propyltex fabric bag and a photograph of a strainer and end caps [2].

Using the above-mentioned equipment and technique, Wiles et al.[2] collected and processed 4980 samples and found the soil sampler durable, core size consistent and sampling efficient. They reported that two people could take 120 cores h⁻¹ (each approximately 200 cm³), the elutriator separated weed seeds from 36 of these cores at a time and washing required 60-75 min depending on soil texture. Seeds as small as 0.3 mm in diameter were recovered and almost 100% of the seeds were recovered from samples spiked with *Amaranthus retroflexus, Abutilon theophrasti, Echinochloa crus-galli* and *Panicum capillare*. Thus equipment like this, plus improved technology for identifying and counting seeds, possibly by image analysis, is needed to make weed seedbank studies more feasible and accurate.

Institutions involved in teaching and research of weed science will find this type of equipment extremely useful and valuable.

DATA ANALYSIS, REPORTING

Data Analysis

The weed researcher must have some knowledge of experimental statistics in order to both design and interpret experiments. Once a statistical design is chosen for an experiments, variance analysis must be done according to the design. Details of handling data for variance analysis are found in statistical textbooks and therefore these are not discussed here.

Reporting

Failure in reporting data and using them as a guidance for further work or for making out a recommendation results in complete wastage of effort and resources spent in conducting the experiment. Many a time very useful data are kept in files without ever being reported. Research is an expensive input and its continuation can be justified only if the data are made available in the form of annual reports, mimeographed reports or research publications. Once an experiment is conducted, the researcher is morally bound to report the results obtained at least for the benefit of fellow researchers.

Tabular presentation is the most common method of data reporting. The tables should have, besides data, a short and concise heading as well as information on experimental and non-experimental details for future reference purposes. Later, the data on various parameters should be separated and presented in the form of small tables, graphs or bar diagrams. Presentation of data on weed control and crop yield relationship in the form of graphs is more effective in conveying results.

REFERENCES

1. Fryer, J.D. and S.E. Evans. 1970. Weed Control Handbook, vol. 1. Principles. Blackwell Scientific Publications, Oxford, United Kingdom, 5th ed., pp. 183-185.
2. Wiles, L.J., D.H. Barlin, E.E. Schweizer, H.R. Duke and D.E. Whitt. 1996. A new soil sampler and elutriator for collecting and extracting weed seeds from soil. Weed Technol. 10: 35-41.

ABSCISSION	The shedding of fruits, leaves or stems from the parent plant.
ABSORPTION	The process of penetration into the plant tissue by roots or foliage.
ACID EQUIVALENT (ae)	The theoretical yield of parent acid from an active ingredient in acid-based herbicides.
ACEDOPHILES	The plants which can grow in acidic soils (pH range 4.5-6.5).
ACROPETAL	Towards the apex of a plant organ; generally upwards in shoots and downwards in roots. Opposite of basipetal.
ACTIVE INGREDIENT (ai)	The chemical in a product that is principally responsible for herbicidal effects.
ADJUVANT	An additive ingredient that facilitates or modifies the action of the principle ingredient.
ADP (adenosinediphosphate)	Adenosine-derived ester formed in plant cells and converted to ATP for energy storage.
ADSORPTION	The chemical or physical attraction, adhesion and accumulation of molecules at the soil-water or soil-interface, resulting in one or more ionic or molecular layers on the surface of soil particles. It can refer to gases, dissolved substances or liquids on surface of solids or liquids.
ALIPHATICS	The compounds derived from straight chain hydrocarbons.
ALLELOPATHY	The phenomenon of one plant having detrimental effect on another through the production and release of toxic chemicals.
AMBIMOBILE	The character of chemical substances (e.g. herbicides) which, when absorbed by the plant, move in both the phloem and xylem.
AMINES	A class of compounds derived from ammonia by replacing the hydrogens with organic radicals.
ANGIOSPERM	A plant having its seeds enclosed in an ovary.
ANNUAL	A plant that completes its life cycle in a year or less. It germinates from seed, and grows, flowers, produces seed and dies in the same season.
ANTAGONISM	The action of two or more chemicals in a mixture, with a total effect smaller than the most active component applied alone.

APOPLAST	The continuous, nonliving cell wall structure that surround the symplast, forming a continuous translocation system.
AQUATIC WEED	A weed that grows in water; there are three kinds: a)submersed weeds, which grow beneath the surface, b) emersed weeds, which grow roots below but extend above the surface of water and c) floating weeds, which float on the surface.
AREOLE (islet)	A small area of mesophyll delimited by intersecting veins.
AROMATICS	Compounds derived from the hydrocarbon benzene (C_6H_6).
ATP (Adenosinetriphosphate)	Adenosine-derived nucleotide that is the primary source of energy for plants through its conversion to ADP.
BAND APPLICATION	An application to a continuous restricted band (or area) such as in or along a crop row rather than over the entire field area.
BASAL APPLICATION	An application to the stems of plants at and just above the ground.
BASIPETAL	Towards the base of a plant organ; generally downwards in shoots and upwards in roots. Opposite of acropetal.
BASOPHILES	The plants which can grow in alkali soils (pH range 7.4-8.5); also known as alkali plants.
BED	A narrow-flat tipped ridge on which crops are grown with a furrow on each side for drainage of excess water or area in which seedlings or sprouts are grown before transplanting.
BIENNIAL	A plant that completes its growth in two years; produces leaves and stores food during the first year, and produces fruits and seeds during the second year.
BIOASSAY	A test method using living organisms to determine the presence of a chemical quantitatively or qualitatively.
BIOCHEMICAL ANTAGONISM	The type of antagonism that occurs when one agrochemical (the antagonist) decreases the amount of a given herbicide that would otherwise be available to its site of action in the absence of the antagonist.
BIOLOGICAL CONTROL	Controlling a pest (weed, insect nematode or pathogen) by its natural or introduced enemies.
BIOTYPE	A population within a specis that has a distinct genetuic variation.
BLANKET APPLICATION	An application (also called broadcast application) of spray or granules over an entire soil surface or weed-infested area rather than only on rows, beds or middles.
BLIND CULTIVATION	Cultivation of the field before the crop plants emerge.
BRUSH CONTROL	Control of woody plants.

CALIBRATION	Series of operations to determine the amount of solution applied per unit area of land and the amount of chemical required for a known volume of diluent.
CAMBIUM	A meristem with products of divisions arranged in an orderly fashion in parallel files. It consists of one layer of initial cells and their derivatives.
CARRIER	The liquid or solid added to a chemical compound to facilitate its application.
CHEMICAL ANTAGONISM	The type of antagonism that occurs when an antagonist reacts chemically with a herbicide to form an inactive complex.
CHLOROSIS	The loss of green colour (yellowing) in plant foliage resulting from the impairment or stoppage of the green colouring matter.
COMPATIBILITY	The nature of two compounds that permits them to be mixed without affecting the properties of either.
COMPETITIVE ANTAGONISM	The type of antagonism that occurs when an antagonist acts reversibly at the same site as the herbicide.
COMPETITION	The active acquisition of limited resources by an organism that results in a reduced supply and consequently reduced growth of other organisms in an environment.
CONCENTRATION	The amount of active ingredient in a given volume of diluent or given weight of dry material.
CONTACT HERBICIDE	The herbicide that kills primarily by contact with plant tissue rather than as a result of translocation.
CONTROL	The process of limiting an infestation.
CORTEX	The tissue region between the epidermis and vascular system of a plant.
COTYLEDON	The first leaf or pair of leaves (depending on whether the plant is monocotyledonous or dicotyledonous) of the embryo of seed plants. (Also referred as Cotyledon Leaves).
CROSS RESISTANCE	In the context of weeds, it is an expression of mechanism that confers plants the ability to resist or withstand herbicides from different classes.
CROWN	The point where stem and root join in a seed plant.
CULM	The jointed stem of a grass which is usually hollow except at the nodes or joints.
CUTICLE	A thin, continuous, non-cellular, lipoidal membrane (varnish-like layer) covering the entire plant surface (shoot), formed by oxidation of plant oils.
CUTIN	The chief structural component of cuticle, composed of polymerized long-chain fatty acids and alcohols, protecting the photosynthetic mechanisms of plant cell.
DECIDUOUS PLANTS	The plants which lose their leaves during the winter.

DEFOLIANT — A chemical that causes the foliage to abscise from a plant.

DESICCANT — A chemical that promotes or accelerats dehydration of plant tissue, causing drying; it also lowers moisture content of seeds to facilitate seed harvest.

DESORPTION — Movement of particles of chemical substances (e.g., herbicides) from the soil surface into the soil solution; the reverse of adsorption.

DICOTYLEDON (DICOT) — Any seed plant having two cotyledons.

DILUENT — Any liquid or solid material used to dilute or carry an active ingredient in the preparation of a formulation or application.

DIOECIOUS — Plants with separate male and female individuals.

DIRECTED APPLICATION — An application of a chemical directed to a restricted area, such as a row or a bed, at base of plants, minimizing contact with the crop.

DORMANCY — The condition in which seeds or other living plant organs are not dead but do not grow even under conditions (moisture, temperature and oxygen) favourable for plant growth; temporary suspension of visible plant growth.

EMERGENCE — Appearance of the first leaves of the crop plant coming through the ground.

EMULSIFIER — The material that facilitates suspension of one liquid in another.

EMULSIFYING CONCENTRATE (EC) — A single phase liquid formulation that forms an emulsion when added to water.

EMULSION — A mixture in which one liquid is suspended in minute globules in another liquid without either losing its identity (e.g., oil in water).

EPINASTY — The twisting or curling of leaves and stems caused by uneven growth of cells, especially in leaves, in which the upper surface grows faster than the lower surface, thus causing the leaf edges to bend curl downwards.

FIBROUS ROOT SYSTEM — The root system that is composed of profusely branched roots with many lateral rootlets but with no main or tap root development.

FORMULATION — A mixture of an active pesticide (herbicide) chemical with carriers, diluents or other materials; usually to facilitate handling.

FUMIGANT — A volatile liquid or gas to kill insects, nematodes, fungi, bacteria, seeds, roots, rhizomes or entire plants; usually applied to soil or in an enclosure of some kind.

GENOTYPE — The organism or an individual that possesses a separate and distinct set of genetic constitution or a set of genes.

GERMINATION — The period during which physiological processes are initiated in the seed leading to the elongation of cells and the formation of new cells, tissues and organs; morphologically observable as radicle protrusion through the seed coat.

GROWTH REGULATOR
An organic substance effective in minute amounts for controlling or modifying plant processes.

HARD WATER
Water which contains minerals, usually calcium and magnesium sulphate, chlorides or carbonates, in solution, to the extent of causing a curd or precipitate, rather than a lather, when soap is added; hard water may cause precipitation in some herbicidal sprays.

HAUSTORIUM
The major organ of parasitic weeds for attachment and penetration of the host tissue.

HERBACEOUS PLANT
A vascular plant without woody tissues.

HERBICIDE
A chemical used for killing or inhibiting the growth of plants; phytotoxic chemical (from Latin *herba*, plant and *caedere*, kill).

HERMOPHRODITE
A plant having both male and female parts in the same flower.

HORMONE
Growth-regulating substance occurring naturally in plants or animals; more correctly called growth regulator, not hormone.

HUMIDITY
Moisture or dampness in the air.

HYDROPHILIC
The character of substances having greater affinity for water and other polar solvents.

HYPOCOTYL
Portion of the stem of a plant embryo or seedling below the cotyledons.

INERT INGREDIENT
Any ingredient in a formulation which has no pesticidal (herbicidal) action.

INTEGRATED WEED MANAGEMENT
Application of many kinds of weed management technology in a mutually supportive manner.

INTERFERENCE
Total adverse effect that plants exert on each other when growing in a common ecosystem; includes competition, allelopathy, biotic interference and other modifications detrimental to plant growth.

INVERT EMULSION
One in which water is dispersed on oil (instead of oil in water); oil forms the continuous phase with the water dispersed therein.

LEAF BLADE
The expanded flat portion of a leaf.

LEACHING
Refers to downward movement of water through a soil; this may involve moving soluble chemicals and plant foods along.

LIPOPHILIC
The character of substances having greater affinity for oil and other non-polar solvents.

LOGARITHMIC SPRAYER
A sprayer devised for applying a chemical at a steadily decreasing rate.

MECHANISM OF ACTION
Precise biochemical or biophysical reaction or series of reactions that culminate into the final or ultimate effect of a herbicide; many herbicides have a primary and secondary mechanism of action.

MERISTEM
A tissue primarily concerned with protoplasmic synthesis and formation of new cells by division.

MESOCOTYL	The elongated portion of the axis between the cotyledon and the coleoptile of a grass seedling.
MISCIBLE	Two or more liquids which, when combined together, form a uniform stable mixture.
MODE OF ACTION	The sequence of events that occur from an herbicide's first contact with a plant until its final effect (often plant death) is expressed.
MONOCOTYLEDON (MONOCOT)	Any seed plant having a single cotyledon or seed leaf.
MONOECIOUS	A trait of having separate male and female flowers on the same plant.
NECROSIS	The death of plant tissue.
NEUTROPHILES	The plants which can grow in neutral soils (pH range 6.5-7.4).
NON-SELECTIVE	A characteristic that enables a herbicide to become toxic and kill all plants treated.
NOXIOUS WEED	A weed arbitrarily defined by law as being especially undesirable, troublesome and difficult to control.
OBLIGATE PARASITE	A weed never found in the wild stage, but grows only in association with another plant, as a parasite.
PERENNIAL	A plant that lives from year to year for several years; in this the shoot dies, but the roots and underground propagules persist.
PERIDERM	A part of the bark (of woody plants) that contains tightly packed suberized cells (with no intercellular spaces) which are composed of fatty acids, lignin, cellulose and terpenes; it is composed of the phellogen, the phellem (cork) and the phelloderm.
PHENOTYPE	The visual characteristic of an organism or plant as determined by the interaction of its genetic constitution and environment.
PHLOEM	The principal food-conducting living tissue of the vascular plant, basically composed of sieve tube, companion cells, fibers and sclereids; transports metabolic compounds from the site of synthesis or storage to the site of utilization.
PHOTOPHOSPHORYLATION	The process of ATP formation, in plants, in the presence of light.
PHOTOSYNTHESIS	The process that converts atmospheric carbon dioxide and water to carbohydrates, using sunlight as the source of energy, in plants.
pH VALUE	An expression of the degree of acidity or alkalinity; related to the number of hydrogen (H^+) ions in a water solution; pH values below 7.0 indicate acidity with its intensity increasing as the numbers decrease, and conversely, pH values above 7.0 indicate alkalinity with its intensity increasing as the numbers increase.

PHYSIOLOGICAL ANTAGONISM	The type of antagonism that occurs when two components of a mixture, acting at different sites, counteract each other by producing opposite effects on the same physiological process.
PLAGIOTROPIC	The term used primarily for roots, stems or branches to describe growth at an oblique or horizontal angle.
PLANT GROWTH REGULATOR	A substance used for controlling or modifying plant growth processes without severe phytotoxicity.
POLARITY	The electrical phenomenon of a molecule or an ion., on the basis of which the chemicals can be divided into polar and nonpolar compounds.
POSTEMERGENCE	Any treatment after emergence of the specified weed or crop plants.
PREEMERGENCE	Any treatment after a crop is planted but before emergence of the specified weeds or crop plants or both.
PREPLANT APPLICATION	Anytreatment applied before planting (seeding) or transplanting a crop, either as a foliar application to control the existing vegetation or as a soil application.
PREPLANT INCORPORATION	Any treatment applied and incorporated (blended) into the soil before seeding or transplanting a crop or weed; incorporation usually by tillage.
PUBESCENT	Presence of hair on stems or leaves; it may affect wetting of the foliage and retention of spray.
RADICLE	The part of the plant embryo that develops into the primary root.
RESISTANCE	The ability of a plant species to withstand the phytotoxicity of a chemical. Herbicide resistance, in the context of weeds, is defined as a characteristic of the weed species to withstand a herbicide dosage substantially higher than the wild type of the same plant species can withstand.
RHIZOME	The horizontal, slender, underground root-like stem capable of sending out roots and leafy shoots.
RIBONUCLEIC ACID (RNA)	Polymeric constituent of all living cells, consisting of a single strand of alternating phosphate and ribose units with the bases adenine, guanine, cytosine and uracil bonded to the ribose, the structure and base sequence of which are determinants of protein synthesis.
ROSETTE	A circular cluster of leaves.
SECONDARY GROWTH	Growth resulting from the formation of new cells by cambium.
SELECTIVITY	Property of differential tolerance; essential for the proper functioning of herbicides that kill some plant species when applied to a mixed population, but without serious damage to other species.
SINK	The site in a plant with a high rate of metabolic activity where food resources are used.

SOFT WATER	Water which does not contain those minerals that prevent free lathering when soap is added.
SOIL INCORPORATION	Mechanical mixing of herbicides in the soil.
SOIL INJECTION	Placement of herbicide beneath the soil, with minimum of stirring or mixing of the soil. Injection is done by an injection blade, knife or line.
SOIL PERSISTENCE	Refers to the length of time that an herbicide remains effective in the soil and exhibits some degree of phytotoxicity to some plant species.
SOIL SOLARAIZATION	The process of heating the surface soil by raising the soil temperatures.
SOIL STERILANT	Herbicide which renders the soil incapable of supporting growth of all plants; sterilization may be temporary (a few months) or relatively permanent (years).
SOLUBILITY	A measure of the amount of substance that will dissolve in a given amount of another substance.
SOLUTION CONCENTRATE	A liquide formulation, also known as soluble concentrate, that consists of a herbicide dissolved in a solvent system, which forms a solution when added to water or a carrier.
SORPTION	It refers to surface-induced removal of a chemical substance from solution by adsorption, absorption or precipitation.
SPOT TREATMENT	An application of spray to a localized or restricted area.
SPRAY DRIFT	The movement of airborne liquid spray particles outside the intended area of application.
SPREADING AGENT	A substance used to improve wetting, spreading or possibly the adhesive properties of an herbicide spray solution.
STOLON	The above-ground runner or stem that develops roots and shoots at the tip or nodes (e.g. *Cynodon dactylon*, bermudagrass).
SURFACE TENSION	A physical property of liquids, due to molecular forces, that causes them to form drops, rather than spread as a film.
SURFACTANT	A material used to pesticide (herbicide) formulations to impart spreading, wetting, dispersability or other surface-modifying properties.
SUSPENSION	Liquid or gas in which very fine solid particles are dispersed but not dissolved.
SYMPLAST	Functionally integrated unit consisting of all living cells of a multi-cellular plant; all living cells connected by plasmodesmata; including the phloem.
SYNERGISM	Action of two or more substances, in a mixture, with a total effect greater than either the sum of the two effects achieved independently or the predicted effect based on the response to each factor applied separately.

SYSTEMIC HERBICIDE	A compound which is translocated readily within the plant and has an effect throughout the entire plant system; synonymous with translocated herbicide.
TOLERANCE	In the context of herbicide tolerance, it is defined as the natural or normal variability of response to herbicides that exists within a plant species and can easily and quickly evolve.
TRANSLOCATION	Transfer of photosynthates and other materials such as herbicides from one plant part to another.
VOLATILIZATION	A process where a condensed phase, such as liquid or solid, is transformed into vapour by elevation of temperature or reduction of external pressure; also described as a measure of the tendency to evaporate or vaporize (from a liquid to a gas) at ordinary temperatures on exposure to the air.
VAPOUR	It describes a substance in the gaseous stage below its critical temperature.
VAPOUR PRESSURE	The property which causes a chemical compound to evaporate or vapourize.
VASCULAR TISSUE	A general term referring to either or both xylem and phloem.
WEED	Any plant that is objectionable or interferes with the activities and welfare of humans.
WEED CONTROL	a)The process of limiting weed infestations so that crops can be grown profitably or other operations can be conducted efficiently; b) the process reducing weed growth or weed infestation to an acceptable level.
WEED ERADICATION	The complete elimination of all live plants, plant parts, and weed seeds from an area.
WEED MANAGEMENT	Rational development of appropriate technology to minimize the impact of weeds, provide systematic management of weed problems, and optimize intended land use.
WEED PREVENTION	Stopping of weeds from invading and contaminating new area.
WETTABLE POWDER	A finely divided dry formulation (powder) that will readily form a suspension in water.
WETTING AGENT	A compound which when added to a spray solution causes it to spread over and wet plant surfaces more thoroughly due to reduction in inter-facial tensions.
WINTER ANNUAL	The plant that starts germination in the fall (after the summer), lives over the winter and completes its growth, including seed production, the following season.
WOODY PLANTS	Plants that develop woody tissues.
XYLEM	The principal water-conducting non-living tissue in vascular plants characterized by the presence of tracheids; may also contain vessels, parenchyma cells, fibres, and sclereids.

Appendices

Appendix 1. Acronyms used in the book.

ae	:	acid equivalent
ai	:	active ingredient
°C	:	degree Celsius
cm	:	centimetre
cumsec	:	cubic metre per second
cusec	:	cubic foot per second
g	:	gram
h	:	hour
ha	:	hectare
kcal	:	kilocalorie
kg ha^{-1}	:	kilogram active ingredient per hectare (or kilogram per hectore)
L	:	litre
lb acre^{-1}	:	pounds per acre
m	:	metre
M	:	mole
μg	:	microgram
μM	:	micromole
mg	:	milligram
mM	:	millimole
min	:	minute
mon	:	month
ng	:	nanogram
POST	:	postemergence (E-POST: early postemergence; L-POST: late postemergence)
ppb	:	per per billion
PPI	:	preplanting soil incorporation
PPL	:	preplanting
ppm	:	parts per million
PPS	:	preplanting surface (E-PPS: early preplanting surface)
ppt	:	parts per trillion
PRE	:	preemergence
psi	:	pounds per square inch
s	:	seconds
t	:	ton (tonne)

Appendix 2. List of weeds covered in the book, with their scientific names, common names (in English) and family names.

Scientific Name (1)	Common Name (2)	Family Name (3)
Abutilon theophrasti Medicus	velvet leaf	Malvaceae
Acacia farnesiana (L.) Willd.	huisache	Leguminosae
Acalypha ostryifolia Riddell	hornbeam copperleaf	Euphorbiaceae
Achillea millefolium L.	common yarrow	Compositae
Acorus calamus L.	sweetflag	Araceae
Acroptilon repens (L.) DC (or *Centaurea repens* L.)	Russian knapweed	Compositae
Aegilops cylindrica Host	jointed goatgrass	Gramineae
Aeschynomene virginica (L.) B.S.P.	northern jointvetch	Leguminosae
Ageratum conyzoides L.	tropic ageratum	Compositae
Agrostis gigantea Roth	redtop	Gramineae
Alisma plantago-aquatica L.	common waterplantain	Alismataceae
Allium canadense L.	wild onion	Liliaceae
Alopecurus myosuroides Huds.	blackgrass	Gramineae
Alternanthera philoxeroides (Mart.) Griseb.	alligatorweed	Amaranthaceae
Amaranthus albus L.	tumble pigweed	Amaranthaceae
Amaranthus blitoides S.Wats.	prostrate pigweed	Amaranthaceae
Amaranthus palmeri S.Wats.	Palmer amaranth	Amaranthaceae
Amaranthus retroflexus L.	redroot pigweed	Amaranthaceae
Amaranthus rudis Sauer	common waterhemp	Amaranthaceae
Amaranthus spinosus L.	spiny amaranth	Amaranthaceae
Amaranthus viridus L.	slender amaranth	Amaranthaceae
Ambrosia artemisiifolia L.	common ragweed	Compositae
Ambrosia tenuifolia Spreng.	false ragweed	Compositae
Ambrosia trifida L.	giant ragweed	Compositae
Amsinckia intermedia Fisch. & Mey	coast fiddleneck	Boraginaceae
Anagallis arvensis L.	scarlet pimpernel	Compositae
Anoda cristata (L.) Schlecht.	spurred anoda	Malvaceae
Arabidopsis thaliana (L.) Heynh.	mouse-ear cress	Cruciferae
Argemone mexicana L.	Mexican pricklepoppy	Papaveraceae
Asclepias syriaca L.	common milkweed	Asclepiadaceae
Asperugo procumbens L.	catchweed	Boraginaceae
Atriplex spp.	saltbrush	Chenopodiaceae
Avena fatua L.	wild oat	Gramineae
Avena sterilis L.	sterile oat	Gramineae
Axonopus compressus (Sw.) Beauv.	carpetgrass	Gramineae
Azolla pinnata R. Brown	pinnate mosquitofern	Azollaceae
Barbarea vulgaris R. Br.	yellow rocket	Cruciferae
Bidens pilosa L.	hairy beggarticks	Compositae
Borreris hispida (L.) K. Schum. (or *Borreria articularis* (L.f.) Willd.)	buttonweed	Rubiaceae
Brachiaria mutica (Forsk.) Stapf	paragrass	Gramineae
Brachiaria platyphylla (Griseb.) Nash	broadleaf signalgrass	Gramineae
Brasenia schreberi J.F.Gmel.	watershield	Cabombaceae
Brassica campestris L. (or *B. rapa* L.)	birdsrape mustard	Cruciferae
Brassica juncea (L.) Czern. & Coss.	Indian mustard	Cruciferae
Brassica kaber (DC.) L.C. Wheeler (or *Sinapis arvensis* L.)	wild mustard	Cruciferae
Brassica nigra (L.) W.J.D. Koch	black mustard	Cruciferae

(Contd.)

Appendix 2. *(Contd.)*

(1)	(2)	(3)
Bromus japonicus Thunb. Ex Murr.	Japanese brome	Gramineae
Bromus secalinus L.	cheat	Gramineae
Bromus tectorum L.	downy brome	Gramineae
Capsella bursa-pastoris (L.) Medicus	shepherd's purse	Cruciferae
Carthamus oxyacantha Bieb.	carthamus	Compositae
Cassia obtusifolia L.	coffee senna	Leguminosae
Cassia occidentalis L.	sicklepod	Leguminosae
Celosia argentea L.	coxcomb	Amaranthaceae
Cenchrus incertus M.A. Curtis	southern sandbur	Graminaeae
Centella asiatica (L.) Urban	pennywort	Umbelliferae
Ceratophyllum demersum L.	coontail	Ceratophyllaceae
Chenopodium album L.	common lambsquarters	Chenopodiaceae
Chorispora tenella (Pallas) DC.	blue mustard	Cruciferae
Chrysopogan aciculatus (Retz.) Trin.	love thorn, pilipiliula	Gramineae
Cirsium arvense (L.) Scop.	Canada thistle	Compositae
Commelina benghalensis L.	dayflower	Commelinaceae
Convolvulus arvensis L.	field bindweed	Convolvulaceae
Conyza bonariensis (L.) Cronq.	hairy fleabane	Compositae
Conyza canadensis (L.) Cronq. (or *Erigeron canadensis* L.)	dwarf fleabane	Compositae
Crotolaria spectabilis Roth	showy crotolaria	Leguminosae
Cucumis melo L. var. duadaim Naud.	small melon	Cucurbitaceae
Cucurbita texana (Scheele) Gray	Texas gourd	Cucurbitaceae
Cuscuta chinensis Damk.	Chinese dodder	Convolvulaceae
Cynodon dactylon (L.) Pers.	bermudagrass	Gramineae
Cyperus difformis L.	umbrella sedge	Cyperaceae
Cyperus esculentus L.	yellow nutsedge	Cyperaceae
Cyperus iria L.	flatsdge	Cyperaceae
Cyperus rotundus L.	nutgrass, purple nutsdge	Cyperaceae
Dactyloctenium aegyptium (L.) Willd.	crowfootgrass	Gramineae
Datura innoxia Mill.	sacred datura	Solanaceae
Datura stramonium L.	jimsonweed	Solanaceae
Daucus carota L.	wild carrot	Umbelliferae
Descurainia sophia (L.) Webb. Ex Prantl	flixweed	Cruciferae
Desmodium tortuosum (Sw.)	Florida beggarweed	Leguminosae
Digitaria ischaemum (Link) Radford (or *Digitaria violascens* Link)	violet crabgrass	
Digitaria sanguinalis (L.) Scop.	large crabgrass	Gramineae
Echinochloa colona (L.) Link	junglerice	Gramineae
Echinochloa crus-galli (L.) Beauv.	barnyardgrass	Gramineae
Eclipta alba (L.) Hassk.	eclipta	Compositae
Eichhornia crassipes (Mart.) Solms	waterhyacinth	Pontederiaceae
Eleocharis acicularis (L.) R. & S.	needle spikerush	Cyperaceae
Eleusine indica (L.) Gaertn.	goosegrass	Gramineae
Elytrigia repens (L.) Nevski (earlier *Agropyron repens* (L.) Beauv.)	quackgrass, couchgrass	Gramineae
Equisetum arvense L.	field horsetail	Equisetaceae
Erechtites valerianaefolia DC.	fireweed, burnweed, pilewort	Compositae
Erigeron canadensis L. (see *Conyza canadensis*)	horseweed	Compositae

Appendix 2. *(Contd.)*

(1)	(2)	(3)
Erigeron philadelphicus L.	poor-robin fleabane	Compositae
Eriochloa villosa (Thunb.) Kunth	woolly cupgrass	Gramineae
Eupatorium adenophorum Spreng.	croftonweed	Compositae
Euphorbia esula L.	leafy spurge	Euphorbiaceae
Euphorbia hirta L.	garden spurge	Euphorbiaceae
Euphorbia humistrata Engelm. Ex Gray	prostrate spurge	Euphorbiaceae
Euphorbia maculata L.	spotted spurge	Euphorbiaceae
Fimbrisylis miliacea (L.) Vahl.	globe fringerush	Cyperaceae
Galinsoga parviflora Cav.	smallflower galinsoga	Compositae
Galium aparine L.	catchweed bedstraw	Rubiaceae
Hedera helix L.	English ivy	Raliaceae
Helianthus annuus L.	common sunflower	Compositae
Helianthus tuberosus L.	Jerusalem artichoke	Compositae
Heteranthera limosa (Sw.) Willd.	ducksalad	Pontederiaceae
Hordeum glaucum Steud.	wall barley	Gramineae
Hordeum leporinum Link	hare barley	Gramineae
Hydrilla verticillata (L. f.) Royle	hydrilla	Hydrocharitaceae
Hydrocotyle javanica Thunb.	pennywort	Umberlliferae
Hypericum perforatum L.	common St. Johnswort	Hypericaceae
Imperata cylindrica (L.) Beauv.	thatchgrass, cogongrass alang-alang	Gramineae
Ipomoea aquatica Forsk.	swamp morningglory	Convolvulaceae
Ipomoea hederacea (L.) Jacq.	ivyleaf morningglory	Convolvulaceae
Ipomoea purpurea (L.) Roth	tall morningglory	Convolvulaceae
Ischaemum rugosum Salisb.	saramollagrass	Gramineae
Iva xanthifolia Nutt.	marshelder	Compositae
Jussiaea repens L. (or *Ludwigia repens* Forst.)	floating waterprimrose	Onagraceae
Kochia scoparia (L.) Schrad.	kochia	Chenopodiaceae
Lactuca serrioloa L.	prickly lettuce	Compositae
Lamium amplexicaule L.	henbit	Labiatae
Lantana camara L.	largeleaf lantana	Verbenaceae
Lepidium sativum L.	garden cress	Cruciferae
Leptochloa filiformis (Lam.) Beauv.	red sprangletop	Gramineae
Linum cartharticum L.	fairy flax	Linaceae
Lolium multiflorrum Lam.	Italian ryegrass	Gramineae
Lolium perenne L.	perennial ryegrass	Gramineae
Lolium rigidum Gaudin	rigid ryegrass	Gramineae
Lotus corniculatus L.	birdsfoot trefoil	Leguminosae
Ludwigia decurrens Walt.	winged waterprimrose	Onagraceae
Marsilea quadrifolia L.	pepperwort	Marsileaceae
Matricaria inodora L	chamomile	Compositae
Melilotus alba Medicus	white sweetclover	Leguminosae
Melilotus indicus (L.) All.	Indian sweetclover	Leguminosae
Mikania micrantha H.B.K	climbing hempweed, mile-a-minute	Compositae
Mollugo veticillata L.	carpetweed	Izoaceae
Monochoria vaginalis (Burm. f.) Kunth	monochoria	Pontederiaceae
Morrenia odorata (H. & A.) Lindl.	stranglervine	Asclepiadaceae
Muhlenbergia frondosa (Poir.) Fern.	wirestem muhly	Gramineae

(Contd.)

Appendix 2. *(Contd.)*

(1)	(2)	(3)
Myriophyllum aquaticum (Vell.) Verdc.	parrotfeather	Haloragaceae
Nelumbo lutea (Willd.) Pers.	American lotus	NelumbonaceaeN
Nymphaea odorata Ait.	fragrant waterlily	Nymphaeaceae
Opuntia imbricata (Haw.) DC.	Walkingstick cholla	Cactaceae
Orobanche cernua Loeffl.	Broomrape	Orobanchaceae
Oxalis corniculata L.	creeping woodsorrel	Oxalidaceae
Oxalis corymbosa DC.	violet woodsorrel	Oxalidaceae
Panicum capillare L.	witchgrass	Gramineae
Panicum dichotomiflorum Michx.	fall panicum	Gramineae
Panicum maximum Jacq.	guineagrass	Gramineae
Panicum miliaceum L.	wild-proso millet	Gramineae
Panicum repens L.	torpedograss	Gramineae
Parthenium hysterophorus L.	ragweed parthenium	Compositae
Paspalum conjugatum Bergius	sour paspalum, sourgrass	Gramineae
Pasaplum dilatatum Poir.	dalligrass	Gramineae
Paspalum scrobiculatum Am. auctt.	ricegrass paspalum	Gramineae
Pennisetum clandestinum Hochst. ex Chiov.	kikuyugrass	Gramineae
Pennisetum pedicillatum Trin.	kyasumagrass	Gramineae
Phalaris minor Retz.	littleseed canarygrass	Gramineae
Phragmites communis Trin.	common reed	Gramineae
Phyllanthus niruri L.	niruri	Euphorbiaceae
Pistia stratiotes L.	waterlettuce	Araceae
Poa annua L.	annual bluegrass	Gramineae
Polygonum chinens L.	smartweed	Polygonaceae
Polygonum convolvulus L.	wild buckweed	Polygonaceae
Polygonum lapathifolium L.	pale smartweed	Polygonaceae
Polygonum pensylvanicum L.	Pennsylvania smartweed	Polygonaceae
Polygonum perfoliatum L.	spiny smartweed	Polygonaceae
Polygonum persicaria L.	ladysthumb	Polygonaceae
Pontederia cordata L.	pickerelweed	Pontederiaceae
Portulaca oleracea L.	common purslane	Portulacaceae
Potamogeton nodosus Poir.	American pondweed	Potamogetonaceae
Potamogeton pectinatus L.	sago pondweed	Potamogetonaceae
Prosopis juliflora (Sw.) DC.	mesquite	Leguminosae
Pteridium aquilinum (L.) Kuhn (or *Pteris aquilina* L.)	bracken fern	Dennstaedtiaceae
Quercus stellata Wangenh.	post oak	Fagaceae
Ranunculus repens L.	birdfoot buttercup	Ranunculaceae
Raphanus sativus L.	radish	Cruciferae
Rhus radicans L. (or *Toxicodendron radicans* (L.) Ktze.	poison ivy	Anacardiaceae
Richardia scabra L.	Florida pusley	Rubiaceae
Rubus moluccanus L.	Molucca raspberry	Rosaceae
Rumex acetosella L.	red sorrel	Polygonaceae
Rumex crispus L.	curly dock	Polygonaceae
Saccharum spontaneum L.	wild sugarcane, kansgrass	Gramineae
Sagittaria montevidensis Cham. & Schlecht.	California arrowhead	Alismataceae
Salsola iberica Sennen & Pau	Russian thistle	Chenopodiaceae
Salvia reflexa Hornem.	lanceleaf sage	Labiatae

(Contd.)

Appendix 2. *(Contd.)*

(1)	(2)	(3)
Salvinia molesta Mitch.	karibaweed	Salviniaceae
Scirpus maritimus L.	bulrush	Cyperaceae
Scirpus validus Vahl	softstem bulrush	Cyperaceae
Scoparia dulcis L.	sweet broomweed	Scrophulariaceae
Senna obtusifolia (L.) Irwin and Barneby	sicklepod	Leguminosae
Senecio vulgaris L.	common groundsel	Compositae
Sesbania exaltata (Raf.) Rydb. ex A.W.Hill	Drummond rattlebush	Leguminosae
Setaria faberi Herrm.	giant foxtail	Gramineae
Setaria glauca (L.) Beauv.	yellow foxtail	Gramineae
Setaria italica (L.) Beauv.	foxtail millet	Gramineae
Setaria palmifolia (Koenig) Stapf.	palmgrass	Gramineae
Setaria viridia (L.) Beauv.	green foxtail	Gramineae
Sida spinosa L.	prickly sida	Malvaceae
Sinapis arvensis L. (see *Brassica kaber*)		
Sisymbrium altissimum L.	tumble mustard	Cruciferae
Sisymbrium officinale (L.) Scop.	hedge mustard	Cruciferae
Solanum carolinense L.	horsenettle	Solanaceae
Solanum elaeagnifolium Cav.	silverleaf nightshade	Solanaceae
Solanum nigrum L.	black nightshade	Solanaceae
Solanum ptycanthum Dun.	eastern black nightshade	Solanaceae
Solanum villosum Am. auctt.	hairy nightshade	Solanaceae
Solidago fistulosa Mill.	hollow goldenrod	Compositae
Sonchus arvensis L.	perennial sowthistle	Compositae
Sonchus oleraceus L.	annual sowthistle	Compositae
Sorghum bicolor (L.) Moench	shattercane	Gramineae
Sorghum halepense (L.) Pers.	Johnsongrass	Gramineae
Spermacoce ocymoides Burm.	buttonweed	Rubiaceae
Sphenoclea zeylandica Gaertn.	gooseweed	Campanulaceae
Stellaria media (L.) Vill.	common chickweed	Caryophyllaceae
Taraxacum officinale Weber in Wiggers	common tansy	Compositae
Thlaspi arvense L.	field pennycress	Cruciferae
Trapa natans L.	water chestnut	Trapaceae
Tribulus terrestris L.	puncturevine	Zygophyllaceae
Typha angustifolia L.	narrowleaf cattail	Typhaceae
Typha latifolia L.	common cattail	Typhaceae
Utricularia vulgaris L.	common bladderwort	Lentibulariaceae
Veronica filiformis Sm.	slender speedwell	Scrophulariaceae
Veronica persica Poir.	Persian speedwell	Scrophulariaceae
Vicia sativa L.	common vetch	Leguminosae
Xanthium pensylvanicum Wallr. (see *Xanthium strumarium*)		
Xanthium strumarium L.	common cocklebur	Compositae
Zannichellia palustris L.	horned pondweed	Zannichelliace

Sources of Information:

Composite List of Weeds. Weed Science Society of America, Champaign, Illinois, U.S.A., 112 pp.

Dutta A.C. 1983. Some Common Weeds of the Tea Estates in North-east India. Tea Research Association, Tocklai Experimental Station, Jorhat, Assam, India, 315 pp.

Appendix 3. List of herbicides with their common names, trade names, molecular weight (of acid), water solubility (acid; mg L^{-1} at 25°C and pH 7.0), LD_{50} (acute toxicity, oral, rat in mg kg^{-1} weight), toxicological use classification (GU: general use, RU: restricted use) and manufacturer (first company which discovered/developed it).

Common Name	Trade Name(s)	Mol. Wt.	Water Solubility	LD_{50}	Use Classifi.	Manufacturer
(1)	(2)	(3)	(4)	(5)	(6)	(7)
AC 263,222	Cadre	275.31	2,200	5,000	GU	American Cyanamid
Acetochlor	Harness, Guardian	269.77	223	2,148	GU	Monsanto
Acifluorfen	Blazer	361.66	250,000	1,370	GU	BASF
Acrolein	Magnacide H Herbicide	56.06	237,628	46	RU	May & Baker
Alachlor	Lasso	269.77	200	1,000	RU	Monsanto
Ametryn	Evik, Gesapax	227.33	200	1,750	GU	Ciba-Geigy
Amitrole	Amitrol-T, Weedazol	84.08	280,000	5,000	RU	Rhone-Poulenc
Asulam	Asulox	230.24	534,000	>5,000	GU	May & Baker
Atrazine	Aatrex, Gesaprim, etc.	215.69	33	5,100	RU	Ciba-Geigy
Benefin	Balan	335.28	0.01	>500	GU	DowElanco
Bensulfuron	Londax	410.40	120	>5,000	GU	Du Pont
Bensulide	Betasan, Prefar, Bensume	397.50	25	770	GU	Several
Bentazon	Basagran	240.28	500	1,100	GU	BASF
Bifenox	Mowdown	342.14	0.398	>5,000	GU	Mobil
Bromacil	Hyvar, Bromax	261.12	815	5,200	GU	Du Pont
Bromoxynil	Brominal, Buctril	276.91	130	260	RU	Rhone-Poulenc
Butachlor	Machete	311.90	23	2,000	GU	Monsanto
CGA 248757	KH-9201	403.87	850	>5,000	GU	Kumiai
Chlorimuron	Classic	414.82	1,200	3,700	GU	Du Pont
Chlorsulfuron	Glean, Telar	357.77	31,800	5,545	GU	Du Pont
Cinmethylin	Argold, Cinch	274.40	65	4,553	GU	Amer. Cyanamid, Du Pont
Clethodim	Select	359.91	NA	1,360	GU	Valent
Clomazone	Command	239.70	1,100	1,406	GU	FMC
Clopyralid	Reclaim,Stinger,Transline	293.19	1,000	4,300	GU	DowElanco
Cyanazine	Bladex,	240.70	171	334	RU	Amer. Cyan., Du Pont
Cycloate	Ro-Neet	215.35	85	2,000	GU	Zeneca (Stauffer)
Cycloxidim	Focus, Laser	325.46	85	>5,000	GU	BASF
2,4-D	Several	221.04	900	764	GU	Several
2,4-DB	Butoxone, Butyrac	249.09	46	1,960	GU	Rhone-Poulenc
Desmedipham	Betanex	300.31	7	3,720	GU	Hoechst (AgrEvo)
Diallate	Avadex	270.22	14	1,045	GU	Monsanto
Dicamba	Banvel	221.04	4,500	2,629	GU	Sandoz
Dichlobenil	Casaron	172.01	21.2	800	GU	Solvay Duphor, Uniroyal
Dichlorprop	Weedone	235.07	710	800	GU	Rhone-Poulenc
Diclofop	Hoelon, Illoxan	327.16	3,000	557	RU	Hoechst (AgrEvo)
Difenzoquat	Avenge, Superaven	249.34	817,000	270	GU	Ameri. Cyanamid
Dimethenamid	Frontier	275.79	1,174	2,400	GU	Sandoz

(Contd.)

Appendix 3. *(Contd.)*

(1)	(2)	(3)	(4)	(5)	(6)	(7)
Diquat	Diquat, Reglone, Aquacide	184.24	718,000	230	GU	Zeneca (ICI)
Dithiopyr	Dimension	401.41	1.38	3,600	GU	Monsanto, Rohm & Haas
Diuron	Karmex, etc.	233.10	42	3,400	GU	Du Pont
DSMA	DSMA	183.93	269,000	1,935	GU	ISK Biosciences
EPTC	Eptam	189.32	370	1,652	GU	Zeneca (Stauffer)
Ethalfluralin	Sonalan	333.27	0.3	10,000	GU	DowElanco
Ethametsulfuron	Muster	410.40	50	> 5,000	GU	Du Pont
Fenoxaprop	Bugle, Acclaim, Horizon	333.73	1	3,310	RU	Hoechst (AgrEvo)
Fluazifop-P	Fusilade	327.26	1.1	4,000	GU	Zeneca (ICI)
Flumetsulam	Broadstrike	325.29	5,600	> 5,000	GU	DowElanco
Flumiclorac	Resource, Sumiverde	353.74	0.189	> 5,000	GU	Valent
Fluometuron	Cotoran,	232.21	110	1,840	GU	Ciba-Geigy
Fluoroglycofen	Compete	419.70	1	1,500	GU	Rohm & Haas
Fluridone	Sonar	329.32	12	>10,000	GU	Eli Lilly
Fomesafen	Reflex, Flex	438.76	50	6,570	GU	Zeneca (ICI)
Fosamine	Krenite	153.07	1,790,000	24,400	GU	Du Pont
Glufosinate	Basta, Finale, Rely, Ignite	181.13	1,370,000	1,910	GU	Hoechst (AgrEvo)
Glyphosate	Roundup, etc.	169.07	15,700	5,400	GU	Monsanto
Halosulfuron	Permit, Sempra, Manage	434.81	1,630	8,866	GU	Monsanto
Haloxyfop	Galant	361.70	43.3	337	GU	DowElanco
Imazamethabenz	Assert	274.32	1,370	2,679	GU	Ameri. Cyanamid
Imazapyr	Arsenal,Chopper, Contain	261.28	11,272	> 5,000	GU	Ameri. Cyanamid
Imazaquin	Image, Scepter	311.34	60	> 5,000	GU	Ameri. Cyanamid
Imazethapyr	Hammer, Pursuit	289.33	1,400	> 5,000	GU	Ameri. Cyanamid
Ioxynil	Totril	370.92	130	110	GU	Rhone-Poulenc
Isoxaben	Gallery	332.40	1	NA	GU	DowElanco
Lactofen	Cobra	461.78	0.1	2,533	GU	Valent
Linuron	Lorox, Afalon, Linex	249.10	75	1,196	GU	Du Pont, Hoechst (AgrEvo)
MCPA	Chiptox, Rhomene	200.62	825	800	GU	Several
Mecoprop	Mecome, MCPP	214.65	620	650	GU	Several
Metolachlor	Dual	283.80	488	2,534	GU	Ciba-Geigy
Metribuzin	Sencor, Lexone	214.29	1,100	1,200	GU	Bayer, Du Pont
Metsulfuron	Ally, Escort	381.36	2,790	> 5,000	GU	Du Pont
Molinate	Ordram	187.3	970	584	GU	Zeneca
MSMA	Ansar, Acme, MSMA	161.95	1,040,000	2,833	GU	Several
Napropamide	Devrinol	271.36	73	> 5,000	GU	Zeneca
Naptalam	Alanap	291.31	200	> 8,200	GU	Uniroyal
Nicosulfuron	Accent	410.40	12,200	> 5,000	GU	Du Pont
Norflurazon	Predict, Solicam, Zorial	303.67	28	> 8,000	GU	Sandoz
Oryzalin	Surflan	346.36	2.6	> 500	GU	DowElanco

(Contd.)

Appendix 3. *(Contd.)*

(1)	(2)	(3)	(4)	(5)	(6)	(7)
Oxadiazon	Ronstar	345.23	0.7	> 5,000	GU	Rhone-Poulenc
Oxyfluorfen	Goal, Koltar	361.70	0.1	> 5,000	GU	Rohm and Haas
Paraquat	Gramoxone, Cyclone	257.16	620,000	138	RU	Zeneca (ICI)
Pebulate	Tillam	203.34	60	921	GU	Zeneca (Stauffer)
Pendimethalin	Pentagon, Prowl	281.31	0.275	2,679	GU	Ameri. Cyanamid
Phenmedipham	Spin-Aid, Bentanal	300.31	10	2,000	GU	Hoechst (AgrEvo)
Picloram	Tordon, Grazon	241.46	430	8,200	RU	DowElanco
Primisulfuron	Beacon	468.34	5,243	> 5,000	GU	Ciba-Geigy
Prodiamine	Barricade	350.30	0.013	15,380	GU	Sandoz
Prometon	Pramitol, Gesafram 50	225.29	720	2,276	GU	Ciba-Geigy
Prometryn	Caparol, Gesagrad	241.35	33	3,750	GU	Ciba-Geigy
Pronamide	Kerb	256.13	15	5,620	GU	Rohm and Haas
Propachlor	Ramrod, Bexton	211.69	613	1,900	GU	Monsanto
Propanil	Stam, Stampede	218.08	500	1,384	GU	Rohm and Haas
Prosulfuron	Peak	419.38	3,580	986	GU	Ciba-Geigy
Pyrazon	Pyramin	221.65	400	3,600	GU	BASF
Pyridate	Tough	378.92	1.5	4,690	GU	Agrolinz
Pyrithiobac	Staple	348.74	760	> 3,000	GU	Kumiai
Quinclorac	Facet	242.06	62	> 1,620	GU	BASF
Quizalofop-P	Assure	372.81	0.3	1,480	GU	Du Pont
Sethoxydim	Poast, Torpedo, Ultima	327.48	4,390	2,676	GU	BASF
Siduron	Tupersan	232.33	18	> 7,500	GU	Du Pont
Simazine	Princep, etc.	201.66	6.2	> 5,000	GU	Ciba-Geigy, etc.
Sulfentrazone	Authority	387.2	780	2,689	GU	FMC
Sulfometuron	Oust	364.38	300	> 5,000	GU	Du Pont
Tebuthiuron	Spike	228.31	2,500	400	GU	DowElanco
Terbacil	Sinbar	216.67	710	> 5 000	GU	Du Pont
Thiazopyr	Visor, Spindle	396.38	2.5	> 5,000	NA	Rohm and Haas
Thifensulfuron	Pinnacle	387.39	2,240	> 5,000	GU	Du Pont
Thiobencarb	Bolero, Saturn	257.78	30	920	GU	Kumiai
Triallate	Avadex, Far-Go	304.66	4	700	GU	Monsanto
Triasulfuron	Amber, Logran	401.82	815	> 5,000	GU	Ciba-Geigy
Tribenuron	Express, Granstar	395.39	2,040	> 5,000	GU	Du Pont
Triclopyr	Garlon, Grandstand	256.47	430	2,140	GU	DowElanco
Trifluralin	Treflan, etc.	335.28	0.3	500	GU	DowElanco
Triflusulfuron	Debut, Safari, Upbeat	492.43	110	> 5,000	GU	Du Pont
Vernolate	Vernam, Reward, Saverit	203.34	108	1,780	GU	Stauffer

Paraquat: Paraquat dichloride; NA : Not Available

Sources of Information:

Herbicide Handbook 1994 (Seventh Edition). Weed Science Society of America, Champaign, Illinois, U.S.A., 352 pp.

Herbicide Handbook 1998 (Supplement to Seventh Edition), Weed Science Society of America, Lawrence, Kansas, U.S.A., 104 pp.

Appendix 4. Commercialized major transgenic crops with herbicide resistance.

Transgenic Crop	Herbicide Resisted	Trademark Designation	Agrochemical or Seed Company	Commercial Status (Year)
Rice (*Oryza sativa*)	Glufosinate ammonium	LIBERTY LINK RICE	AgrEvo	USA (2000-01) Asia (2000-01)
Maize (Corn) (*Zea mays*)	Glufosinate ammonium	LIBERTY LINK CORN	AgrEvo	USA (1997)
	Glyphosate	ROUNDUP READY CORN	Monsanto DeKALB Genetics	USA (1998) Canada (1998)
	Imidazolinones	IMI CORN	American Cyanamid PIONEER, CIBA Seeds, ASGROW NORTHRUP KING	USA (1997) Australia(1998-99)
	Sethoxydim	SR CORN	BASF/DeKALB Genetics	USA (1997) Brazil (1997).
Cotton (*Gossipium hirsutum*)	Bromoxynil	BXN COTTON	Rhone-Poulenc CALGENE	USA (1997)
	Glufosinate	LIBERTY LINK COTTON	AgrEvo	USA (2000)
	Glyphosate	ROUNDUP READY COTTON	Monsanto	USA (1997)
	Sulfonylureas	19-51a COTTON	DuPont	USA (1997)
Soybean (*Glicine max*)	Glufosinate ammonium	LIBERTY LINK SOYBEANS	AgrEvo	USA (1998) Brazil (1998-99)
	Glyphosate	ROUNDUP READY SOYBEANS	Monsanto ASGROW Seeds	USA (1997) Brazil (1997) Argentina (1997)
	Sulfonylureas	STS SOYBEANS	DuPont	USA (1993)
Tobacco (*Nicotinana tabaccum*)	Bromoxynil	BXN TOBACCO	Rhone-Poulenc	Europe (1997-98)
Sugar Beet (*Beta vulgaris*)	Glufosinate ammonium	LIBERTY LINK SUGARBEET	AgrEvo	Europe (1999-2000)
	Glyphosate	ROUNDUP READY SUGAR BEET	Monsanto	Europe (1997-98)

Source of Information:
Herbicide Handbook 1998 (Supplement to Seventh Edition), Weed Science Society of America, Lawrence, Kansas, U.S.A., 104 pp.

Appendix 5. Herbicide company mergers.

1. **AgrEvo**	:	Hoechst-Roussel (Hoechst AG + Roussel Uclaf) + NOR-AM (NOR-AM Ag Products + Boots + Hercules + Fisons + Upjohn)
2. **American Cyanamid**	:	American Cyanamid + Shell International
3. **BASF Corporation**	:	BASF-Wyandotte (BASF Colors & Chemicals + Wyandotte Co.)
4. **Dow AgroSciences LLC**	:	Dow-Elanco (Dow + Elanco/Eli Lilly)
5. **FMC**	:	Food Machinery Corporation + Niagara
6. **Novartis**	:	Ciba-Geigy + Sandoz Crop Protection + Merck Crop Protection + Velsicol + International Minerals and Chemicals
7. **Rhone-Poulenc**	:	Union Carbide + Mobil + Rhodia + May & Baker + Amchem-Rhor
8. **Uniroyal**	:	Uniroyal, inc. Chemical Division + Duphar + Thomson-Hayward + Olin
9. **Valent**	:	Chevron Agric. Products + Sumitomo
10. **Zeneca Ag Products**	:	ICI + Stauffer + ISK Biosciences + Diamond Shamroc

This list includes only the companies that have undergone series of mergers or acquisitions until 1995.

Source of Information:

Herbicide Handbook 1998 (Supplement to Seventh Edition), Weed Science Society of America, Lawrence, Kansas, U.S.A., 104 pp.

Appendix 6. Conversion factors useful in field work

Area

1 acre	= 0.4047 hectare; 4047 square metres; 4840 square yards; 43,560 square feet
1 hectare (ha)	= 2.471 acres; 10,000 square metres;

Length

1 centimetre (cm)	= 0.3937 inch
1 metre (m)	= 3.28 feet; 39.37 inches; 10 decimetres
1 kilometre (km)	= 0.621 mile; 1000 metres; 3280 feet
1 inch (in)	= 2.54 centimetres;
1 foot (ft)	= 30.48 centimetres; 12 inches
1 yard (yd)	= 0.9144 metre; 91.44 centimetres; 3 feet; 36 inches
1 mile	= 1.61 kilometres; 1760 yards; 5280 feet
1 square inch	= 6.452 square centimetre;
1 square foot	= 0.093 square metre; 929 square centimetres
1 square metre	= 10.76 square feet
1 square kilometre	= 100 hectares
1 square mile	= 259 hectares; 640 acres; 2.59 square kilometres
1 mile per hour	= 1.467 feet per second; 88 feet per minute
1 kilometre per hour	= 0.278 metre per minute; 16.67 metres per minute

Weight

1 gram (g)	= 1000 milligrams; 0.0353 ounce
1 kilogram (kg)	= 1000 grams; 2.205 pounds; 35.27 ounces
1 pound (lb)	= 453.59 grams; 0.4536 kilogram
1 ounce	= 28.3495 grams
1 ton (Imperial)	= 2240 pounds; 1000 kilograms
1 ton (US)	= 2000 pounds; 907.18 kilograms

Volume (liquid)

1 litre	= 1000 millilitres; 0.22 imperial gallon; 0.2642 US gallon; 1.0567 US quarts
1 gallon (US)	= 3785.4 millilitres; 3.785 litres; 0.8333 imperial gallon; 231 cubic inches; 0.1337 cubic feet; 128 fluid ounces; 8 pints; 4 quarts; 16 cups; 256 table spoons
1 gallon (Imperial)	= 4456 millilitres; 4.456 litres; 9.825 pounds; 1.177 US gallon
1 ounce (US)	= 29.573 millilitres
1 cubic foot of water	= 62.43 pounds; 7.48 US gallon

Pressure

1 pound per square inch (psi) = 0.700 kilogram per centimetre; 700 grams per centimetre

1 pound pressure per square inch (psi) will lift water 2.31 feet

1 foot lift of water	= 0.433 pound pressure per square inch (psi)

Geometric Factors

Circumference of a circle	= $2\pi r$ or πd
Diameter of a circle	= $2r$

Temperature

Degrees Celsius (C)	= (°F -32) × 5/9
Degress Farenheit (F)	= (° C × 9/5) + 32 or (° C × 1.8) + 32

(Contd.)

Appendix 6. (*Contd.*)

°C	°F	°C	°F	°C	°F	°C	°F
–30	–22	10	50	30	86	70	158
–20	– 4	20	68	40	104	80	176
–10	14	22	72	50	122	90	194
0	32	25	77	60	140	100	212

Others

foot candle	= 10-764 lux
acre foot	= 1219.85 cu metres
	= 0.122 hectare metre
	= 1.23 million L water
cu ft	= 28.4 L water
	= 0.0284 cu metre
cu metre water	= 0.0001 hectare metre water
	= 1010.7 L water
cusec	= 0.028 cubic metre sec (sec^{-1} (cumsec)
pound $gallon^{-1}$	= 0.12 kg L^{-1}
pound $acre^{-1}$ (U.S.)	= 1.12 kg ha^{-1}

Appendix 7. Quantities of formulated herbicide required per unit area at different rates and active ingredients.

Active ingredient	Rate (kg ai/ha)	Quantity(g or ml) of formulated product/unit area (m)					
		1m	2m	5m	10m	20m	50m
20	0.2	0.10	0.20	0..50	1.00	2.00	5.00
	0.5	0.25	0.50	1.25	2.50	5.00	12.50
	1.0	0.50	1.00	2.50	5.00	10.00	25.00
25	0.2	0.08	0.16	0.40	0.80	1.60	4.00
	0.5	0.20	0.40	1.00	2.00	4.00	10.00
	1.0	0.40	0.80	2.00	4.00	8.00	20.00
35	0.2	0.057	0.114	0.286	0.571	1.143	2.857
	0.5	0.134	0.286	0.174	1.429	2.857	7.143
	1.0	0.286	0.571	1.429	2.857	5.714	14.286
40	0.2	0.050	0.100	0.250	0.500	1.00	2.500
	0.5	0.125	0.250	0.625	1.250	2.50	6.250
	1.0	0.250	0.500	1.250	2.500	5.00	12.500
50	0.2	0.04	0.08	0.20	0.40	0.80	2.00
	0.5	0.10	0.20	0.50	1.00	2.00	5.00
	1.0	0.20	0.40	1.00	2.00	4.00	10.00
72	0.2	0.028	0.056	0.139	0.278	0.556	1.390
	0.5	0.070	0.139	0.348	0.696	1.390	3.480
	0.1	0.139	0.348	0.696	1.390	2.780	6.960
80	0.2	0.025	0.050	0.125	0.250	0.500	1.250
	0.5	0.062	0.125	0.312	0.625	1.250	3.125
	1.0	0.125	0.250	0.625	1.250	2.500	6.250
85	0.2	0.024	0.047	0.118	0.235	0.471	1.176
	0.5	0.059	0.118	0.294	0.588	1.176	2.941
	1.0	0.118	0.235	0.588	1.176	2.353	5.822

Appendix 8. Quantities of formulated herbicides required per unit volume of water at different rates and active ingredients.

Active ingredient	Rate kg ai/ha	Quantity (g or ml) of formulation product/unit Value (L) of water			
		0.11	0.21	0.51	1.01
20	0.2	0.20	0.40	1.00	2.00
	0.5	0.50	1.00	2.00	5.00
	1.0	1.00	2.00	5.00	10.00
25	2.0	0.16	0.32	0.80	1.60
	0.5	0.40	0.80	2.00	4.00
	1.0	0.80	1.60	4.00	8.00
35	0.2	0.114	0.229	0.571	1.143
	0.5	0.286	0.571	1.429	2.857
	1.0	0.571	1.143	2.857	5.714
40	0.2	0.10	0.20	0.50	1.00
	0.5	0.25	0.50	1.25	2.50
	1.0	0.50	1.00	2.50	5.00
50	0.2	0.08	0.16	0.40	0.80
	0.5	0.20	0.40	1.00	2.00
	1.0	0.40	0.80	2.00	4.00
72	0.2	0.056	0.112	2.278	0.556
	0.5	0.129	0.278	0.695	1.389
	1.0	0.278	0.556	1.389	2.778
80	0.2	0.050	0.100	0.250	0.50
	0.5	0.125	0.250	0.613	1.25
	1.0	0.250	0..500	1.250	2.50
85	0.2	0.047	0.094	0.235	0.471
	0.5	0.118	0.235	0.588	1.176
	1.0	0.235	0.471	1.776	2.353

Subject Index

Abscisic Acid (ABA) 28
Absorption (penetration, uptake) of herbicides 65, 116-119
   chemical sturucture 125
   enhancement 125-127
   factors affecting 121-125
   foliage-applied herbicides 116-118
   foliar 118-119
   pH of herbicide solution 125
   soil-applied herbicides 118-119
   stem 119
   stomata 119
   surfactants 125-126
   trichomes 119
Acedophiles 24
Acetamides 68, 71-74, 127-128, 167-171, 205-206, 236, 254-255
Acetochlor 68, 71, 128
   absorption 128
   chemical degradation 265
   mechanisms of action 170
   microbial degradation 254
   photodecomposition 236
   transformations in plants 205
   translocation 128
Acetolactate synthase (ALS) 159, 160, 166
Acetyl CoA 158
Acetyl CoA carboxylase (ACCase) 158, 165, 291-293
Acetyl CoA synthetase 158
Acetyl elongase 166
Acifluorfen 68, 82
   absorption 132
   mechanisms of action 175-176
   microbial degradation 257
   photodecomposition 237
   transformations in plants 210
   translocation 132
Acrolein (acrylaldehyde) 68, 74
   aquatic weed management 472
   mechanisms of action 171
Acronyms 515

Active ingredient 60-61, 379-380
Activity, herbicide 63-66
Adaptive enzymes 252
Adaptive phase 252
Additive response (effect) 348, 359-361
Adenosine diphosphate (ADP) 146, 149, 152-153
Adenosine triphosphate (ATP) 146, 149, 151-154
Adjuvants 369-370
Adsorption, of herbicide by soil 241-248
   chemical nature of herbicides 246-248
   London van der Waals forces 242
   mechanisms 241-243
Advanced trials 489-490
AFLP (Arbitrary Fragment Length Polymorphism) 13-14
Ageratum, tropic 411-412
Age-state diversity 15
Airfields 485-486
ALA synthetase 154-155
Alachlor 68, 71
   absorption 127-128
   mechanisms of action 167-169
   microbial degration 254-255
   photodecomposition 236
   transformations in plants 205
   translocation 127-128
Alang-alang (see thatchgrass) 430-431
Algae 465-466
Aliphatics 68, 74, 128-129, 171, 206, 236, 255, 265
Allelopathy 17-23
   methodologies for studying 21-23
   potential of, in weed management 21
Allelochemicals 18-21
   chemistry 17-18
   effect 19-21
   production 19
Alleles, containment of resistant 306-307
Allozyme marker system 13

Amaranth, slender   412-413
Ametryn   70, 104
    Absorption   139
    chemical degradation   171
    leaching   248
    mechainsms of action   187-188
    microbial degradation   262-263
    translocation   139
Aminoacyl-tRNA   153
Aminoacyl-tRNA synthetase   153
δ-Aminolaevulinic acid (ALA)   154-155
Amitrole   70, 106-107
    absorption   139-140
    aquatic weed management   472
    chemical degradation   271
    mechanisms of action   188
    microbial degradation   263
    photodecomposition   240
    transformations in plants   224
    translocation   139-140
Amitrole T   70, 107
    absorption   139-140
    aquatic weed management   472
    mechanisms of action   188
    microbial degradation   263
    transformations in plants   224
    translocation   139-140
Ammonium nitrate   369
Ammonium sulphate   70, 127, 369
Ammonium sulphamate   70
Amylase, α   164
Amylase, β   164
Animals   26
Anion bonding   242
Annuals   7-8
    summer   7
    winter   7
Antagonsim (antagonistic effect/response)
    349, 361-362
    biochemical   355
    chemical   356
    competitive   356
    mechanisms of   354-356
    physiological   356
Antidotes (see safeners)
Appendices   515-529
Application, herbicide   61-63, 386-410
    band   62
    blanket   62
    controlled droplet (CDA)   397-399
    direct contact (DCA)   399-402
    granular   402-403
    rate   63
    sand mix   403

techniques   491
Applicator   402-403
    granular   402-403
    roller   401
    rope-wick   400
Aquatic plants, types of   465-467
Aquatic weed(s)   464-480
    biological method of management of
        469
    chemical method of management of
        470-475
    evapotranspiration of   465
    herbicides used for management of
        472-475
    human benefit   480
    management of   467-475
    mechanical methods of management of
        467-469
    problems of   3
    utilization of   480
Aquatic weed management   464-480
Area, conversion factors   526
Aromatic compound biosynthesis   145, 160-
    161, 166, 295-297
Aromatic compounds   332-333
Arsenicals   68, 74-75, 129, 171, 207, 236, 255,
    265
Aryloxyphenoxy propionics   69, 90-92, 135-
    136, 159, 180-181, 214-216, 259, 268
Aryl triazinones   69, 93, 136
Assessment
    plant response   496-500
    qualitative   497-498
    quantitative   498-500
Asthma weed (see spurge, garden)   430
Asulam   68, 77
    absorption   130
    chemical degradation   266
    mechanisms of action   174
    microbial degradation   256
    translocation   139
Atmospheric CO_2 48-49
Atrazine   70, 104-105
    absorption   139
    chemical degradation   270-271
    leaching   248
    mechanisms of action   187-188
    microbial degradation   262-263
    photodecomposition   240
    transformations in plants   222-224
    translocation   139
Attapulgite   383
Auxinic herbicides   293-295
Auxins   28

Azide 55

Banana 460
Barley 451
Barnyardgrass 425-427
Basophiles 24
Bean(s), cluster, dwarf, French, snap 458-459
Beggarticks, hairy 417
Benefin 68, 80
    absorption 131
    chemical degradation 266
    mechanisms of action 175
    microbial degradation 256
    photodecomposition 137
    transformations in plants 209
    translocation 131
Benoxacor 113
Bensulide 70, 110
    absorption 141
    mechanisms of action 190
    translocation 131
Bensulfuron 69, 98
    aquatic weed management 472-473
    mechanisms of action 185
    transformations in plants 218
Bentazon 68, 76
    absorption 129
    chemical degradation 266
    mechanisms of action 171
    microbial degradation 256
    photodecomposition 236
    transformations in plants 207
    translocation 129
Bentonite 383
Benzamides 68, 75, 129, 172, 207, 236
Benzoics 68, 75-76, 129, 172, 207, 236, 255-256, 266
Benzothiadiazoles 68, 76, 129, 172, 207, 236, 256, 266
Bermudagrass (stargrass) 422
Bialaphos 339
Bialaphos genes, transfer 339-340
Biennials 7-8
Bifenox 68, 82-83
    absorption 132
    mechanisms of action 175-176
    photodecomposition 237
    translocation 132
Bindweed, field 420
Bioactivation 220
Bioassay of herbicides 500-502
Biocontrol 319-324
    insect-based 319-323

nematode-based 323-324
    phytopathogenic-based 330-331
    rhizobacteria-based 326-330
Biodiversity of weeds 12-16
    age-state 15
    genetic 12-13
    habitat microsite diversity 15
    relationship of, to weed management 15-16
    somatic diversity 14-15
Bioherbicides (bioagents) 370
    enhancement of efficacy, by herbicides 340-343
    improvement of efficacy 334-340
Biological approaches in weed management 319-347
Biological method (aquatic weeds) 469-470
BIOMAL® 112, 326
Biopropagules, production of 334-335
Biorational approach 377-378
Biosynthesis
    aromatic compound 145, 160-161
    branched-chain amino acid 145, 159-160
    chlorophyll 154-155
    ethylene 146, 161-162
    fatty acid (lipid) 145, 158-159
    glutamine 146, 162-163
    nucleic acid 145, 153-154
    pigment 145, 154-158
    protein 145, 153-154
Bipyridiliums 68, 76-77, 130, 172-173, 208, 237, 256, 266
Black gram 456
Blackjack (see beggarticks, hairy) 417
Black nightshade 440-441
Border rows 492
Branched-chain amino acid biosynthesis 145, 159-160
Branching habit 122
Bromacil 70, 107-108
    absorption 140
    chemical degradation 271
    mechanisms of action 189
    microbial degradation 263
    photodecomposition 240
    transformations in plants 225
    translocation 140
Bromoxynil 69, 87
    absorption 133
    engineering for resistance to 303
    mechanisms of action 178
    microbial degradation 257
    photodecomposition 238

transformations in plants 212
translocation 133
Broomrape 433-434
Burning 44, 468, 483
Butachlor 68, 71
    absorption 128
    mechanisms of action 170
    microbial degradation 254-255
    photodecomposition 236
    transformations in plants 205
    translocation 128
Buthidazole 69, 84
    absorption 132
    mechanisms of action 177
    transformations in plants 210
    translocation 132
Buttonweed 417-418
Butylate 69, 102
    absorption 138
    mechanisms of action 186
    microbial degradation 261-262
    transformations in plants 221
    translocation 138

Cabbage 458-459
Calculus method 352
Calibration, spray 391-392, 494-495
Canada thistle 419
Canarygrass, littleseed 439
Carbamates 68, 77-78, 130, 174, 208-209, 237, 266
Carotene(s), β, ζ 156
Carotene hydroxylation 158
Carotenoid biosynthesis 156-158, 165
Carotenoid cyclization 157
Carotenoids 154
Carpetgrass (savannagrass) 416
Carrot 458-459
Carrot, wild 435-436
Cation bridging 242
Cation exchange 242
Cattails 477
Cauliflower 458-459
CDAA
    mechanisms of action 170
Cell culture
    green tissue 299
    non-photosynthetic 299
Cellulose 118
Cellulose synthase 166
Cell wall synthesis 166
Centrifugal (turbine) pumps 389-390
Cereals 448-453
CGA-185,072 115

CGA-248757 69, 85
    absorption 133
    mechanisms of action 178
    photodecomposition 238
    translocation 133
Chaining 468
Cheel how 39
Cheeling 39
Chemical factors 125
Chemical method
    aquatic weeds 470-475
    woody species, management 483-484
Chemical stability 380
Chemical structure 60, 125
Chemoinduction 27
Chick pea 456
Chillies 459-460
Chloramben
    mechanisms of action 172
    transformations in plants 207
Chlorbromuron
    microbial degradation 263-264
Chlorimuron 69, 98
    mechanisms of action 185
    transformations in plants 218
Chlorophyll 146, 154, 163
    biosynthesis 145, 154-155
    chlorophyll **a** 147, 148, 155
    chlorophyll **b** 147, 155
Chorismate 160, 161
Chorophyllide **a** 155
Chloroplast(s) 146
Chloroplast ATP synthase 147
Chlorotriazines 104
Chloroxuron
    microbial degradation 263
Chlorpropham
    absorption 130
    photodecomposition 237
    transformations in plants 208
    translocation 130
Chlorsulfuron 69, 98-99
    absorption 138
    chemical degradation 269
    mechanisms of action 185
    microbial degradation 260
    transformations in plants 218
    translocation 138
Cineoles 68, 78, 131, 209, 237
Cinmethylin 68, 78
    absorption 131
    mechanisms of action 174
    microbial degradation 257
    photodecomposition 237

transformations in plants 209
translocation 131
Citronella grass 45
Citrus 459
Clay colloid 243-244
Cleaning, manual 468
Clethodim 68, 79
absorption 131
mechanisms of action 174
photodecomposition 237
transformations in plants 209
translocation 131
Climbing hempvine 432-433
Clomazone 69, 86
absorption 137
chemical degradation 267
microbial degradation 257
transformations in plants 212
translocation 133
Clopyralid 69, 95
absorption 137
chemical degradation 269
mechanisms of action 183-184
microbial degradation 259
photodecomposition 239
transformations in plants 217
translocation 137
Cobbler's peg (see beggarticks, hairy) 417
Codistillation 234
Coffee 461
Cogongrass (see thatchgrass) 430-431
Colby's method 351-352
COLLEGO® 112, 325-326, 336, 342
Common pimpernel 413-414
Common purslane 439-440
Companion cropping 52-53
Competition, weed 29-31
Competitive crops and cultivars 51
Concentrated emulsions 382
Conjugation 199, 203-204
amino acids, with 204
glutathione 203
sugars, with 203-204
Conservation tillage 41-42
Containers, disposal of herbicide 385, 409
Control treatments 493
Controlled droplet application (CDA) 397-399
Conversion factors 526-527
Copper nitrate 70
Copper sulphate pentahydrate 473
Corn (see maize)
Cotton 454-455
Couchgrass (see quackgrass) 428-429

Coulombic forces 243
Covalent bonding 242
Coxcomb 418
CpDNA RFLP analysis 13
Crabgrass 424-425
Crop-based approaches 51-53
Cropping, companion 52-53
Crop(s)
competitive 51
cultivars 51
development of herbicide resistant 297-300
response 499-500
rotation 51-52
stand 500
weight 499
yield 499
Crop seed coating (herbicides) 407-408
Crop seeds, weed-free 37-38
Crop yield reduction 1-2
Cross enhancement 253
Cultivation
interrow 43
practices 65
Cultivator 42
Cuticle, structure 118
Cutin 118
Cutin matrix 118
Cutin wax 118
Cutting 468-469, 482
Cyanazine 70, 105
chemical degradation 271
transformations in plants 224
Cyclohexanediones 68, 79-80, 131, 159, 209, 237
Cycloxidim 68, 79
absorption 131
mechanisms of action 174
transformations in plants 209
translocation 131
Cyometrinil (CGA-43,089) 115
Cytochrome b6/f complex 147
Cytochrome P-450 204-205

2,4-D 69, 89
absorption 134
aquatic weed management 473
chemical degradation 267-268
hydroxylation 201
mechanisms of action 179-180
microbial degradation 258-259
photodecomposition 238
transformations in plants 213-214
translocation 134

Dalapon 68, 74
  absorption 128-129
  aquatic weed management 473
  mechanisms of action 171
  microbial degradation 255
  photodecomposition 236
  transformations in plants 206
  translocation 128-129
Dark phase reactions 146
Data 496, 505
  analysis 505
  recording 496
  reporting 505
Dayflower (see spiderwort, tropical) 419-420
2,4-DB 69, 89-90
  absorption 134
  chemical degradation 267-268
  conjugation 213-214
  mechanisms of action 179
  microbial degradation 258-259
  photodecomposition 238
  transformations in plants 213-214
  translocation 134
Dealkylation 199, 202
N-dealkylation 202
Deamination 199, 202
Debutoxymethylation 236
Decarboxylation 199, 202, 212
Dechlorination 236
Degradation
  chemical 265-272
  enhanced 253-254
  microbial 251-265
Dehalogenation 212
Deleterious rhizobacteria (DRB) 326-330
  weed control by 327-329
Demonstration trials 490
Desaturation 156-157
Desmedipham 68, 78
  absorption 130
  chemical degradation 266
  mechanisms of action 174
  microbial degradation 256
  transformations in plants 208-209
  translocation 130
Desorption 232
DeVine® 113, 324-326, 336
Dialkylamines 239, 262
Diallate 69, 102
  absorption 129
  mechanisms of action 186
  microbial degradation 261
  transformations in plants 221
  translocation 138

Diaphragm pumps 390
Dicamba 68, 76
  absorption 129
  chemical degradation 266
  mechanisms of action 172
  microbial degradation 255-256
  photodecomposition 236
  transformations in plants 207
  translocation 129
Dichlobenil
  aquatic weed management 474
Dichlormid (R-25788) 114, 365-366
Dichlorprop 69, 90
  absorption 135
  mechanisms of action 180
  translocation 135
Diclofop (diclofop methyl) 69, 90-91
  absorption 135
  chemical degradation 268
  mechanisms of action 180-181
  microbial degradation 259
  transformations in plants 214-215
  translocation 135
Dietholate 114
Difenzoquat 69, 93-94
  absorption 136
  mechanisms of action 182
  translocation 136
Diffusion transport 249-250
Digging 39
7,8-Dihydropteroate synthase 166
Dimethenamid 68, 72
  absorption 128
  chemical degradation 265
  microbial degradation 254-255
  transformations in plants 206
  translocation 128
Dinitramine 131
  absorption 131
  mechanisms of action 175
  microbial degradation 256
  transformations in plants 209
  translocation 131
Dinitroanilines 68, 80-82, 131-132, 175, 209-
  210, 237, 266-267
Dinoseb 69, 88
  absorption 134
  chemical degradation 267
  mechanisms of action 179
  microbial degradation 258
  photodecomposition 238
  transformations in plants 213
  translocation 134

Diphenylethers 68, 82-84, 132, 175-177, 210, 237, 257, 267
DPX-H6564 220-221
Diquat 68, 76, 130
   absorption 130
   aquatic weed management 474
   chemical degradation 266
   mechanisms of action 172-173
   microbial degradation 256
   transformations in plants 208
   translocation 130
Direct contact application (DCA) 399-402
Dispersing agents 127
Dithiopyr 69, 95-96
   absorption 137
   chemical degradation 269
   mechanisms of action 184
   microbial degradation 260
   photodecomposition 239
   translocation 137
Diuron 70, 108-109
   absorption 140
   aquatic weed management 475
   chemical degradation 263-264
   mechanisms of action 189
   microbial degradation 263-264
   photodecomposition 240
   transformations in plants 225
   translocation 140
DNA marker system(s) 13-14
Dodder, Chinese 420-422
Domain A 290
Domain B 290
Dormancy, seed 27-28
DPX-H6564 220
   transformations in plants 220
Dr.Biosedge 326
Dredging 467-468
   Drip or trickle system 404
Droplet, Controlled application (CDA) 397-399
Drying 468
DSMA 68, 75
   absorption 129
   chemical degradation 265
   mechanisms of action 171
   microbial degradation 255
   transformations in plants 207
   translocation 129

Ecological fitness 281
Ecophysiological approaches 45-49
ED_{50} (GR_{50}, ID_{50}, LD_{50}) 500-501
Electrodynamic sprayer 402

Electron flow 408
Electron transport, cyclic 151
Electron transport, non-cyclic 151
Electron transport, photosynthetic 147-149
Electron transport between PS I and PS II 150
Electron transport, PS I 150
Electron transport, PS II 148
Elicitors 341
Elutriation 380-381
Emersed weeds 466-467, 471
Emulsifiable concentrate (EC) 382
Emulsifiers 126
Encapsulated formulation 384
Encapsulation, herbicide 384, 408
Endodermis 116
Endoplasm reticulum (ER) 158
Endothall 474
Environmental factors 65, 123-124
Environmental stress 46-49
EPSP synthase 160, 166
EPTC 69, 102-103
   absorption 138
   leaching 248
   mechanisms of action 186
   microbial decomposition 261-262
   photodecomposition 239
   transformations in plants 221-222
   translocation 138
Equipment for herbicide irrigation 404-405
Ethalfluralin 68, 80
   photodecomposition 237
   transformations in plants 209
Ethofumesate 70, 110
   absorption 141
   mechanisms of action 190
   photodecomposition 240
   translocation 141
Ethylene 54
Ethylene biosynthesis 146, 161-162
Evaluation, methods of 350-354
Evapotranspiration, aquatic weeds 465
Experimental design 349
Experimental techniques 492-493
Extrachromosomal elements (ECEs) 254

Factors
   biotic 25
   climatic 23-24
   soil 24-25
Fatty acid biosynthesis 145, 158-159, 165
Fatty acid synthetase 158
Fenac 474
Fenchlorazol-ethyl (HOE-70,542) 115
Fenchlorim (CGA-123,407) 115

Fenoxaprop 69, 91
    absorption 135
    mechanisms of action 180
    microbial degradation 259
    transformations in plants 214-215
    translocation 135
Fenuron
    photodecomposition 240
Ferredoxin 151
Ferrous sulphate 70
Fertilizer additves 127
Fertilizers 369, 405-407
Fibre crops 454-55
Field bindweed 420
Field experiments (trials) 488-491
    advanced 489-490
    demonstration 490
    regional 490
    residue 489-490
    screening 488-489
    special 490-491
    yield 489
Filters 390
Finger Millet 453
Fingermillet, wild (goosegrass) 428
Fish, effects of aquatic weed management
    479
Fitness (herbicide resistance) 281-283
Floating weeds 467, 471
Flooding 44
Floodjet deflector 387
Floodjet fan nozzles 387
Flowable concentrate 383
Fluazifop-P (fluazifop butyl) 69, 91
    absorption 135
    chemical degradation 268
    mechanisms of action 180
    transformations in plants 214, 215
    translocation 135
Fluchloralin 68, 81
    absorption 131
    mechnisms of action 175
    transformations in plants 209
    translocation 131
Flumetsulam 68, 81
    absorption 131
    mechanisms of action 188-189
    microbial degradation 262
    transformations in plants 225
    translocation 140
Flumiclorac 69, 92
    absorption 136
    chemical degradation 268
    mechanisms of action 181

photodecomposition 239
    translocation 136
Fluometuron 70, 109
    absorption 140
    chemical degradation 272
    mechanisms of action 189
    microbial degradation 263-264
    photodecomposition 240
    transformations in plants 225
    translocation 140
Fluorodifen 68
    mechanisms of action 175-176
    transformations in plants 210
Fluoroglycofen 68, 83
    absorption 132
    mechanisms of action 175-176
    microbial degradation 257
    transformations in plants 210
    translocation 132
Flurazole 114
Fluridone 69, 96
    absorption 137
    aquatic weed management 474-475
    chemical degradation 269
    mechanisms of action 184
    microbial degradation 260
    translocation 137
Fluxofenim 114
Fomesafen 68, 83
    absorption 132
    mechanisms of action 175-176
    microbial degradation 257
    photodecomposition 237
    transformations in plants 210
    translocation 132
Forest nurseries, weed management in 481-
    482
Forestry, weed management in 481-486
Formulated herbicide, quantities of 528-529
Formulation(s) 335-338, 379-384
    liquid-based 336
    solid-based 336-337
    stabilization and improvement of
        337-338
    types 381-384
Fosamine 70, 110-111
    absorption 141
    chemical degradation 272
    microbial degradation 264
    photodecomposition 240
    transformations in plants 226
    translocation 141
Fumigants 384
Fungi-based herbicides 324-326

Fungicides 368-369

Garden spurge 430
Gear pumps 390
Gene flow 282-283
Genes, bialaphos 339-340
Genetic engineering 300-304
   acetolactate synthase resistance, for 300-301
   acetyl coenzyme A carboxylase resistance, for 302
   auxinic herbicide resistance, for 302-303
   photosystem II resistance, for 301-302
   resistance to other herbicides, for 303-304
Genetic manipulation 338-340
Genetic mutation 279-280
Genetic variation 279-280
Genotypic fitness 281
Germination
   phase 28
   stimulants 53-55
Gibberellic acid (GA, GA_3, Gibberellin) 28, 164
Glufosinate 70, 111, 296-297
   absorption 141
   chemical degradation 272
   engineering for resistance to 304
   mechanisms of action 190-191
   microbial degradation 265
   photodecomposition 240
   target site 296-297
   transformations in plants 226
   translocation 141
Glutamate 154-155, 162-163
Glutamate synthase 162-163
Glutamine 162-163
Glutamine biosynthesis 146, 162-163
Glutamine synthetase 162-163, 166
Glutathione 203
Glutathione conjugation 203
Glutathione-s-transferase (GSTs) 203
Glyphosate 70, 111, 295-296
   absorption 141-142
   aquatic weed management 475
   chemical degradation 272
   engineering for resistance to 303-304
   mechanisms of action 191
   microbial degradation 264-265
   photodecomposition 240
   resistant transgenic plants 296

   target site and resistance in plants 296
   transformations in plants 226
   translocation 141-142
Goatweed 411-412
Gossegrass (see fingermillet, wild) 428
GR_{50} (ED_{50}, ID_{50}, LD_{50}) 500-501
Gram, black 456
Gram, green (mung bean) 456
Grana 147
Granal lamellae 147
Granular application 402-403
Granular applicator 402-403
Granules 383-384
Grapes 460
Grazing 483
Groundnut (peanut) 453-454
Guatemala grass 45

Habitat microsite diversity 15
Hairy beggarticks 417
Halosulfuron 69, 99
   mechanisms of action 185
   microbial degradation 260
Haloxyfop-P (haloxyfop methyl) 69
   absorption 135
   chemical degradation 268
   mechanisms of action 180
   microbial degradation 259
   photodecomposition 238
   transformations in plants 214, 216
   translocation 135
Handweeding 39
Harrow 42
Haustoria (haustorium) 17
Hydrogen bonding 242
Hempvine, climbing 432-433
Herbi, Micron 398
Herbicide(s)
   absorption (uptake) 65, 116-119
   absorption enhancement 125-127
   activity 63-64
   adsorption by soil 241-248
   application 61-63, 386-410, 470, 494-496
   application in fertilizers 405-407
   application through irrigation 404-405
   aquatic weed management 472-475
   bioassay 500-501
   chemical degradation/decomposition 232, 265-272
   chemical nature• 246-247
   chemical properties 60-61

chemical structure 60
classification 67-70, 164-167
coating of crop seed 407-408
combinations 357-359
concentration 125
contact (non-systemic) 67-69
containers, disposal of 385, 409
degradation, enhanced 253
degradation, enhancement of 253
desorption 232-241
development 376, 378-384
development and formulation of 378-384
discovery 376-378
encapsulation 408
factors affecting absorption and
    translocation 120-125
factors responsible (herbicide resistance)
    305-306
foliage-applied 118-119, 120-121
formulation(s) 61, 379-384, 490
fungi-based 324-326
genetic engineering 277, 300-318
glove 401-402
granules 383-384
handling (safety in) 409
hydrophilic 125-126
incorporation 403
industry 310
information 70-112
inorganic 70
interactions 348-375
irrigation, equipment 404-405
leaching 232, 248
lipophilic 125-126
management of enhanced degradation on
    253-254
mechanisms of action 145-198
metabolism by plants 199-231
microbial 324-331
microbial degradation 232, 252-265
microbial (herbicides/agents) 324-331
mixtures 308-309
non-systemic (contact) 67-69
organic 70-112
packaging 385
persistence and behaviour of 232-276
persistence and residues in soil 502
photodecomposition 232, 235-241
phytochemical-based 332-333
placement 62
polarity 61

rate of application 63
resistance 277-318
resistance, management of 304-311
resistant crops, development of 297-300
safeners 113-115, 362-367
safety in handling 409
selection by, 280-282
selectivity 63-64
soil-applied 116-118, 120
solution, pH 125
sorption 241
spraying 386-397
storage 408
strategies (to manage resistance) 306-309
susceptibility 278
systemic (translocated) 67-70
time of application 61-62
tolerance 278
transformations of, in plants 199-231
translocation 65, 120-142
translocation, enhancement of 125-127
transport in soil 248-251
transport, diffusion 249-250
transport, mass flow 250
uptake and metabolsim by plants 233
volatilization 232
Herbicide-adjuvant interactions 369-370
Herbicide-bioherbicide interactions 356-362, 370
Herbicide-fungicide interactions 368-369
Herbicide-insecticide interactions 367-368
Herbicide-pathogen interactions 368-369
Herbicide-safener interactions 362-367
Hexazinone 70, 105
    transformations in plants 224
Hilograss (see paspalum, sour) 437
HLB 126
Hoeing 43
Hollow cone nozzles 387
Hoses 391
Host specificity 320
Human efficiency, reduced 2
Humectants 127
Hydrilla 478
Hydrogen bonding 242
Hydrolysis 199, 202
Hydrolytic enzyme activities 146, 164
Hydrophobic partitioning 242
Hydrophytes 466-467
Hydroxylation 199, 201-202

540

ID$_{50}$ (ED$_{50}$, GR$_{50}$, LD$_{50}$) 500-501
Illite 244, 383
Imazamethabenz 69, 86
    absorption 132-133
    mechanisms of action 177-178
    microbial degradation 257
    photodecomposition 238
    transformations in plants 211
    translocation 132-133
Imazapyr 69, 84-85
    absorption 132-133
    mechanisms of action 177-178
    microbial degradation 257
    photodecomposition 238
    transformations in plants 211
    translocation 132-133
Imazaquin 69, 85
    absorption 132-133
    chemical degradation 267
    mechanisms of action 177-178
    microbial degradation 257
    photodecomposition 238
    transformations in plants 211
    translocation 132-133
Imazethapyr 69, 85
    absorption 132-133
    chemical degradation 267
    mechanisms of action 177-178
    microbial degradation 257
    photodecomposition 238
    transformations in plants 211
    translocation 132-133
Imidazolidinones 69, 84, 132, 177, 210
Imidazolinones 69, 84-86, 132-133, 177-178, 211-212, 238, 257, 267
Imines 69, 86, 133, 238, 257
Imitative chemistry 377
Impellar pumps 390
Incorporation, herbicide 403
Indicator species 501-502
Inductive phase 27-28
Industrial sites 485-486
Inheritance
    Cytoplasmic 280
    Nuclear 280
Inheritance of herbicide resistance 288-289, 291, 292-293
Insect-based biocontrol 319-323
    examples of 321-323
    factors affecting 320-321
    host-specific 320
    kinds of insects 320
    mode of action 319
Insecticides 367-368

Integrated weed management 5
Interaction response, evaluation of 349-354
Intercropping 52
Interaction(s) 348-375
    bioherbicides/bioagents and herbicides 342-343, 370
    definition 349-350
    evaluation of response 349-354
    herbicide-adjuvant 369-370
    herbicide-bioherbicide 356-362
    herbicide-fertilizer 369
    herbicide-fungicide 368-369
    herbicide-insecticide 367-368
    herbicide-pathogen 368-369
    herbicide-safener 362-367
    responses 348-354
    physiological basis of herbicide 354-356
    terms 349-350
Inter-row cultivation 43
Invert emulsion 383
Ioxynil 69, 87
    absorption 133
    mechanisms of action 178
    microbial degradation 257
    photodecomposition 238
    transformations in plants 212-213
    translocation 133
Irrigation 404-405
    drip or trickle 404
    equipment 404-405
    herbicide application through 404-405
    sprinkler system 404
    surface 404
Isobole method 353
Isoleucine 159
Isopropalin
    absorption 131
    mechanisms of action 175
    microbial degradation 237
    transformations in plants 209
    translocation 131
Isoproturon 70
    absorption 140
    mechanisms of action 189
    translocation 140
Isoxaben 68, 75
    absorption 129
    mechanisms of action 172
    photodecomposition 236
    transformations in plants 207
    translocation 129
Isoxazolidinones 69, 86, 133, 212, 257, 267

Johnsongrass 442-443
Juglone 18-19
Junglerice 427-428
Jute 455

Kansgrass (sugarcane, wild) 440
Kaolinite 244, 383
Kikuyugrass 437-438
Knapsack sprayer 392
Kyasumagrass 438-439

Labelling, plot 493
Lactofen
 absorption 132
 mechanisms of action 175-176
 microbial degradation 257
 photodecomposition 237
 transformations in plants 210
 translocation 132
Lag phase 252
Lambsquarters 419
Land value, reduction in 2
Lantana, largeleaf 431-432
Laws
 quarantine 38
 weed 38
LD_{50} (ED_{50}, GR_{50}, ID_{50}) 500-501
Leaching 232
Length, conversion factors 526
Leucine 159
Ligand exchange 242
Light 45-46
Light phase 146
Light reactions 146
Linuron 70, 109
 absorption 140
 mechanisms of action 189
 transformations in plants 225
 translocation 140
Liquid-based formulations 336
Logarithmic analysis (logit analysis) 353
Logarithmic sprayer 392
London van der Waals forces 242-243
Lumen 147
Lutein 154
Lycopene 156

Maintenance phase 28
Maize (see corn) 452
Malonyl CoA 158
Management
 aquatic weeds 467-475
 biological approaches, weed 319-347
 chemical method 59-60
 enhanced herbicide degradation, of 253-254

field (major) crops, weed 448-457
herbicide resistance in weeds, of 304-311
mechanical method 40-45
plantation (major) crops, weed 459-462
programmes (implementation of herbicide resistance) 309-311
some prominent weeds, of 411-446
Marginal weeds 467, 471
Marshy weeds 479
Mass flow transport 250
Mavalonate 157
MCPA 69, 70
 absorption 134
 chemical degradation 267
 mechanisms of action 179
 microbial degradation 258-259
 transformations in plants 213
 translocation 134
MCPB 69
 absorption 134
 chemical degradation 267
 mechanisms of action 179
 microbial degradation 258-259
 transformations in plants 213
 translocation 134
Mechanical method(s) 40-45
 aquatic weed management 467-469
 woody species management 482-483
Mecoprop 69
 transformations in plants 214, 216
Melfluidide
 transformations in plants 226-227
Messenger RNA (mRNA) 153-154
Metabolism of herbicides by plants 199-231
Metflurazon 69
 mechanisms of action 182
 transformations in plants 216-217
Methabenzthiazuron 70
 absorption 140
 mechanisms of action 189
 translocation 140
Methazole 69, 88
 transformations in plants 213
Methodology, weed research 485-505
Methyltriazines 104
Methylmercapto (methylthio) triazines 104
Metolachlor 68, 72, 128
 absorption 128
 chemical degradation 265
 mechanisms of action 168, 170
 microbial degradation 254-255

photodecomposition 236
translocation 128
Metribuzin 70, 106
    absorption 136
    chemical degradation 271
    leaching 248
    transformations in plants 224
    translocation 136
Metsulfuron 69, 99
    mechanisms of action 185
    microbial degradation 260-261
    transformations in plants 218
Mexican pricklepoppy 414
MG-191 115
Microbial chemical based herbicides 333-334
Microbial degradation 232, 251-265
Microbial herbicides/agents 324-331
Micron Herbi 398
Microscopic method 380
Microtubule assemply 166
Millet 453
    finger 453
    pearl 453
Minimal tillage 41-42
Miscible oils 382
Mitochondrial activities 145, 152-153
Mitotic inhibitors 295
Mitosis 166
Mixture (herbicide)
    broad sprectrum (BSM) 308
    dry 406-407
    fluid 406
    target weed (TWM) 308-309
Modern weed science 4-6
Modified isobole method 353
Moisture, soil 246
Molinate 69, 103
    microbial degradation 261-262
    transformations in plants 222
Monogalactosyldiacylglycerol (MGDG) 183
Montmorillonite 244
Monuron 70
    chemical degradation 272
    mechanisms of action 189
    microbial degradation 263
    photodecomposition 240
    transformations in plants 225-226
Morningglory, swamp 479
Mowing 40, 468
MSMA 68, 75
    absorption 129
    chemical degradation 265
    mechanisms of action 171
    microbial degradation 255

photodecomposition 236
transformations in plants 207
translocation 129
Mulching 44-45
Multiple comparison tests 350
Mung bean (green gram) 456
Mutation 44-45
Mycoherbicide(s) 112-113, 324-326

Na (see 1-8-naphthalic anhydride) 115
1,8-Naphthalic anhydride (NA) 115
Napropamide 68, 73
    chemical degradation 265
    mechanisms of action 170
    photodecomposition 236
    transformations in plants 206
Naptalam 69, 93
    chemical degradation 268
    mechanisms of action 182
Nematode-based biocontrol 323-324
Natural products 377
Neoxanthin 154
Neutrophiles 24
Nicosulfuron 69, 99
    mechanisms of action 218-219
    transformations in plants 218-219
Nightshade, black 440-441
Nitralin 68
    chemical degradation 266
    transofrmations in plants 209
Nitrates 54
Nitriles 69, 87, 133, 212-213, 238, 257
Non-agricultural systems, weed management
    in 481-486
Norflurazon 69, 94
    absorption 136
    chemical degradation 268-269
    mechanisms of action 183
    microbial degradation 259
    photodecomposition 239
    transformations in plants 216-217
    translocation 136
No-tillage 41-42
Nozzles 387-389
Nucleic acid sysnthesis 145, 153-154
Nurseries
    forest 481-483
    weed 491-492
Nutgrass (nutsedge, purple) 423-424
Nutrient-based approaches 49-51
Nutrients
    alternative sources 50-51
    method of application 49-50
    strategies, to reduce weed competition 49-51

time of application 50
Nutsedge, purple (see nutgrass) 423-424

Oat, wild 414-415
Objectives, weed research 487-488
Oilseeds 453
Okra 458-459
Onion 458-459
Organic matter (soil) 244
Orthogonal comparison 350
Oryzalin 68
  photodecomposition 237
OTC 115
Oxabetrinil 115
Oxadiazoles 69, 87-88, 133, 179
Oxadiazolidines 69, 88, 213
Oxadiazon 69, 87-88, 199-200
  absorption 133
  mechanisms of action 179
  translocation 133
Oxidation
  α — 199
  β — 199-200
  ω — 199-200
Oxidative phosphorylation 152
Oxygen-evolving complex 147, 148
Oxyfluorfen 68, 84
  absorption 132
  mechnaisms of action 175-176
  microbial degradation 257
  photodegradation 237
  transformations in plants 210
  translocation 132

P680 147, 149
P680* 148-150
P700 147, 149-150
P700* 149-151
Paraquat 68, 77
  absorption 130
  aquatic weed management 474
  chemical degradation 266
  mechanisms of action 171-172
  microbial degradation 256
  photodecomposition 237
  transformations in plants 208
  translocation 130
Parasite 16-17
  non-obligate 16
  obligate 16
Parasitism, weed 16-17
Particle size 380-381
Paspalum, sour (hilograss) 437
Pathogen-related proteins (PR proteins) 341
Pathogens 368-369

Pathogens and fungicides 368-369
  interactions with herbicides 368-369
Pearl millet 453
Peas 458-459
Pebulate 69, 103
  absorption 138-139
  chemical degradation 270
  leaching 248
  mechanisms of action 186
  microbial degradation 261-262
  photodecomposition 239
  transformations in plants 222
  translocation 138-139
Pectin 118
Pendimethalin 68, 81
  chemical degradation 267
  mechanisms of action 175
  microbial degradation 256
  photodecomposition 237
Penetrating agents 127
Perennials 7-9
  bulbous 8
  creeping 8
  simple 8
Periderm 119
Persistence of herbicides (see herbicide persistence)
Persistence of weeds 23-25
  biotic factors 25
  climatic factors 24-25
  soil factors 23-24
pH
  herbicide solution 125
  soil 244-246
Phellem 119
Phelloderm 119
Phellogen 119
Phenmedipham 68, 78
  absorption 130
  chemical degradation 266
  mechnisms of action 174
  microbial degradation 254
  transformations in plants 208-209
  translocation 130
Phenols 69, 88, 134, 179, 213, 238, 258, 267
Phenoxyacetics 69, 88-89, 134, 179-180, 213-214, 238, 258-259, 267-268
Phenoxyalkanoic acids 69, 88-92, 134,-136, 179-181, 213-216, 238, 258-259, 267-268
Phenoxybutyrics 69, 89-90, 134, 179, 213-214, 238, 258-259, 267-268
Phenylalanine ammonia lyase (PAL) 342
N-phenylphthalimides 69, 92, 136, 181, 139, 268

Phenylpyridazines 69, 92, 136, 181, 216, 239, 259, 268
Phenyl triazinones 69, 93, 136, 182, 216
Pheophytin a 148
Phloem 120-121
Phosphinothricin (PPT; see glufosinate)
Photodecomposition 232, 235-241
Photoinduction 27
Photophosphorylation 151
Photosynthesis 145-152
Photosynthetic electron transport 147-149
Photosystem I (PS I) 147, 152, 165, 285-286
Photosystem II (PS II) 147, 152, 165, 286-289
Phthalamates 69, 93, 182, 268
Physiological differences 65-66
Phytoalexins 341
Phytochemical-based herbicides 332-333
Phytochrome 26
Phytoene 156-157
Phytofluene 157
Phytoene desaturase (PD) 156-157
Phytoene synthesis 156-157
Phytopathogenic bacteria-based biocontrol 330-331
Phytoplankton 465
Picloram 69, 95
    absorption 137
    chemical degradation 269
    mechanisms of action 183-184
    microbial degradation 259
    photodecomposition 239
    transformations in plants 217
    translocation 137
Pigment biosynthesis 145, 154-155, 165
Pigment protein complexes (PPC) 154
Pigweed, redroot 413
Pimpernel, scarlet 413-414
Pineapple 460
Piston pumps 389
Plantation crops, weed management in 459-462
Plant
    breeding 297-298
    defence mechanisms 340-342
    development, stage of 64-65
    maturity 122
    morphology 64
    population 53
    resistance to herbicides 66
    selection 297-298
    species 122
    surface 122
    uptake 232
    varieties 122

Plant response, assessment/evaluation 349
Plasmids 254
Plastohydroquinone 150
Plastoquinone 148
Plot
    arrangement 492-493
    labelling 493
    numbering 493
    size 492
Plough 42
Polymorphism 17
Polypeptides 150
Polyribosomes (polysomes) 153-154
Pondweed 477-478
Porphobilinogen (PBG) 155
Potato 458
Power-driven sprayers 393, 397
Predisposition phenomenon 360
Pressure, conversion factors 526
Pressure regulators 390-391
Pricklepoppy, Mexican 414
Primisulfuron 69, 99-100
    absorption 138
    mechanisms of action 185
    photodecomposition 239
    transformations in plants 218-219
    translocation 138
Prodiamine 68, 81
Product evaluation 378-379
Profluralin
    absorption 131
    translocation 131
Prometon
    absorption 139
    mechanisms of action 187-188
    translocation 139
Prometryn 70, 106
    absorption 139
    leaching 248
    mechanisms of action 187-188
    translocation 139
Prominent weeds 411-416
Pronamide 68, 73
    absorption 128
    chemical degradation 265
    mechanisms of action 170
    microbial degradation 254-255
    photodecomposition 236
    translocation 128
Propachlor 68, 73
    absorption 127-128
    mechanisms of action 167-169
    microbial degradation 254-255
    photodecomposition 236

transformations in plants 205-206
translocation 127-128
Propagules, vegetative 9-10
dynamics 9-10
Propanil 68, 74
absorption 128
chemical degradation 265
mechanisms of action 170-171
microbial degradation 254-255
photodecomposition 236
transformations in plants 206
translocation 128
Propazine 70
absorption 139
chemical degradation 271
mechanisms of action 187-188
transformations in plants 222
translocation 139
Propham
absorption 130
photodecomposition 237
translocation 130
Prosulfuron 69, 100
mechanisms of action 185
Protease 164
Protein biosynthesis 145, 153-154, 167
Protochlorophyllide **a** (pchlide **a**) 155
Protoplast fusion transfer 299
Protoporphyrin IX (Proto IX or PP IX) 155, 165, 176, 181
Protoporphyrinogen IX (PPG IX) 155, 176, 181
Protoporphyrinogen oxidase (Protox) 165, 176, 181
Prynachlor
mechanisms of action 167-169
Pulse crops 455-456
Pump capacity 390
Pumps 389-390
capacity 390
centrifugal or turbine 389-390
diaphragm 390
gear 390
impellar 390
piston 389
Purslane, common 439-440
Pyrazoliums 69, 93, 136, 181
Pyrazon 69, 94
absorption
mechanisms of action 182-183
microbial decomposition 259
photodecomposition 239
transformations in plants 216-217
translocation 136

Pyridate 69, 92
absorption 136
chemical degradation 268
mechanisms of action 181-182
microbial degradation 259
photodecomposition 239
transformations in plants 216
translocation 136
Pyridazines 136
Pyridazinones 69, 94, 136, 157, 159, 182-183, 217, 239, 259, 268
Pyridines 69, 95-96, 137, 184, 218, 239, 260
Pyridinecarboxylic acids 69, 95, 137, 183-184, 217-218, 239, 259, 269
Pyridinones 69, 96, 137, 184, 260
Pyrimidinylthio-benzoates 69, 97, 137, 184
Pyrithiobac 69, 97
absorption 137
mechanisms of action 184
translocation 137
Pyruvate 159

Quackgrass (couchgrass) 429-430
Qualitative assessment 497-498
Quantitative assessment 497-498
Quaratine Laws 38
Qinclorac 69.97
absorption 138
mechanisms of action 184-185
transformations in plants 218
translocation 138
Quinolinecarboxylic acids 69, 97, 138, 184, 218
Quizalofop-P 69, 91-92
absorption 135
transformations in plants 214, 216
translocation 135

R-25788 (see dichlormid)
R-29148 114, 363-365
Railroads 485
Rainfall, effect of
herbicide transport 250-251
Random screening 377
RAPD analysis 13-14
Recirculating sprayer 399-400
Regional trials 489-490
Regression estimate analysis 352
Regression techniques 352-355
Reproduction, vegetative 28-29
Residue trials 489-490
Resistance (herbicide) 277-311
acetolactate synthase (ALS) inhibitors, to 289-291

acetyl coenzyme A carboxylase inhibitors, to 291-293
aromatic compound biosynthesis inhibitors, to 295-297
auxinic herbicides, to 293-295
cross 283, 290-291
development 278-285, 294-295
factors responsible (for development) 305-306
fitness 281-282
inheritance 280
management of 304-311
mechanisms (resistance development) 285-297
mitotic inhibitors, to 295
multiple 283
non-target site, cross 284-285
photosystem I (PS I) inhibitors, to 285-286
photosystem II (PS II) inhibitors, to 286-289
target site, herbicide 285, 287-290, 292
strategies to manage herbicide 306-309
spread of 282-283
target site 283-284
Resistant alleles 278, 306-307
Response 497-500
crop 498-500
weed 497-499
Rhizobacteria 326-330
deleterious (DRB) 326-330
weed control by 327-329
Rhizobacteria-based biocontrol agents 326-330
Rhizobacteria, integration into weed management 329-330
Ribosomes 153-154
Rice, weed management in 448-451
Ridge-till 41-42
Ring barking 482-483
Ring cleavage 199-203
RNA (ribonucleic acid) 167
mRNA 153-154
rRNA 154
tRNA 153-154
Roadsides 485
Roller applicator 401
Ropewick applicator 400

Safener(s) 362-367
herbicide-safener interaction 362-367
interaction responses 366-367
mode of action of 364-366

Safety, herbicide handling 409
Safflower 454
Salvinia 479
SAN 9785
mechanisms of action 182-183
Sand mix application 403
Savannagrass (see carpetgrass) 416
Screening trials 488-489
primary 488-489
secondary 489
Sedimentation method 381
Seedbank, weed 10-11, 305-307, 502, 505
Seedbed sale 43-44
Seed(s)
bank 10-11, 305-307, 502, 505
certification 38
dissemination 25-26
dormancy 27-28
estimation, in seedbank 502-505
germination 26-27
Selectivity, herbicide 63-66
Selectivity, sulfonylurea herbicides 221
Sesamum 454
Sethoxydim 68, 79
absorption 131
mechanisms of action 174-175
photodecomposition 237
transformations in plants 209
translocation 131
Shikimate 160
Sickling 40
Siduron
mechanisms of action 189
photodecomposition 240
Sigmoid curve 500-501
Silvex 475
Simazine 70, 106
absorption 138
aquatic weed management 475
chemical degradation 271
leaching 248
mechanisms of action 187-188
microbial degradation 262-263
photodecomposition 240
transformations in plants 222, 224
translocation 138
Site selection 491-492
Sodium
arsenite 70
borate 70
chlorate 70
chloride 70
metaborate 70
nitrate 70
tetraborate 70

Soil
  clay colloid type   243-244
  factors   124
  moisture   246
  organic matter   244
  pH   244-246
  solarization   47-48
  temperature stress   47
  water content   249
  water stress   46-47
Solarization, soil   47-48
Solid-based formulations   336-337
Soluble powder   381
Solution concentrate   382-383
Somatic diversity   14-15
Sorghum   452-453
Sorption   241
Sorrel, wood   434
Sour paspalum (see Hilograss)   437
Sowthistle, annual   441-442
Soybean   455-456
Special trials   490-491
Species, indicator   501-502
Spiderwort, tropical (dayflower, wandering
  Jew)   419-420
Spray
  boom   391
  calibration   391-392, 494-495
  calculations   471
  density   393-395
  drift   395
  droplet   393-394
  droplet size   393-395
  lance   391
  mixing   395-396
  number median diameter (NMD)   393-394
  pattern   388
  precaution   395-397
  tanks   386-387
  volume   490
  volume median diameter (VMD)
    393-394
Sprayer(s)   392-393, 494
  cleaning   397
  electrodynamic   402
  hand-operated   396-397
  knapsack   392
  logarithmic   392
  nozzles   387-389
  power-driven   393, 397
  recirculating   399-400
  types of   392-393
Spraying
  equipment   387-391

  precautions   395-397, 495-496
  procedure   396-397
  ultralow volume   397
Spreading agents   126
Sprinkler system   404
Spurge, garden (asthma weed)   430
Stability, chemical   380
Stabilization of formulation   337-338
Stale seedbed   43-44
Stargrass (see bermudagrass)   422
Steroids   333
Sticking agents   127
Stimulants   54-55
  dormancy, weed seed   53-55
  effective utilization of   55
  germination, weed seed   26-27, 53-55
Stomata   119
Storage, herbicide   408
Strainers   390
Stress
  environmental   46-48
  temperature   47
  water   46-47
Strigol   55
Stroma   146
Stromal lamellae   147
Sulfentrazone   69, 93
  absorption   136
  mechanisms of action   182
  transformations in plants   216
  translocation   136
Sulfonylures   69, 97-101, 138, 185, 218-221,
    239, 260-261
Submersed weeds   466, 471
Sugarbeets   457-458
Sugarcane   457
Sugarcane, wild (see kansgrass)   440
Sulfometuron   69, 100
  absorption   138
  mechanisms of action   185
  translocation   138
Sulphoxidation (S-oxidation)   201
Sunflower   454
Suppressors   341-342
Surface irrigation system   404
Surfactants   125-126
  hydrophilic   125-126
  lipophilic   125-126
Swamp morningglory   479
Sweet potato   458
synergism (synergistic effect/response)
    349, 359-361
  mechanisms of   354

Target site resistance 283-284
TCA 475
Tea 461-462
Tebuthiuron 70, 109-110
Temperature, conversion factors 526-527
Temperature stress 47
Terbacil 70, 108
　absorption 140
　chemical degradation 272
　mechanisms of action 189
　microbial degradation 263
　transformations in plants 225
　translocation 140
Terbutryn
　absorption 139
　mechanisms of action 187-188
　microbial degradation 262-263
　translocation 139
Terpenoids 333
Tetrahydropyrimidinones 69, 101, 157
　mechanisms of action 186
Thatchgrass (alang-alang, cogongrass) 430-431
Thermoinduction 27
Thiazopyr 69, 96
　absorption 137
　chemical degradation 269
　mechanisms of action 184
　microbial degradation 260
　photodecomposition 239
　tranformations in plants 218
　translocation 137
Thickening agents 127
Thifensulfuron 69, 100
　mechanisms of action 185
Thiobencarb 69, 103
　absorption 138-139
　chemical degradation 270
　mechanisms of action 187
　microbial degradation 261-262
　photodecomposition 239
　translocation 138-139
Thiocarbamates 69, 102-104, 138-139, 186-187, 221-222, 239. 261-262, 270
Thistle, Canada 419
Threonine 159
Thylakoid membrane 146-147
Tillage 40, 42
　minimal till (minimal tillage) 41-42
　no-till (no-tillage) 41-42
　practices 53-54
　ridge-till (ridge tillage) 41-42
　zone-till 41
Tissue culture 299-300

Tobacco 456-457
Tomato 458-459
Torpedograss 434-435
Tralkoxydim 68, 80
　absorption 131
　mechanisms of action 174
　translocation 131
Transfer RNA (tRNA) 153-154
Transformations in plants, herbicide 199-231
　conjugation 199, 203-204
　cytochrome P-450-mediated 199, 204-205
　decarboxylation 199, 202
　N-dealkylation 199, 202
　hydrolysis 199, 202
　hydroxylation 199, 201-202
　oxidation 199-200
　pathways 199-205
　ring cleavage 199-203
　sulphoxidation 199, 201
Translocation, herbicides 120-142
　factors affecting 121-125
　foliage-applied 120-121
　soil-applied 120
Transport of herbicides in soil 248-251
　diffusion 249-250
　mass flow 250
　rainfall, effect of 250-251
Transposans 254
Triallate 69, 103-104
　absorption 138
　mechanisms of action 187-188
　microbial degradation 261
　transformations in plants 221
　translocation 138
Triasulfuron 69, 101
　chemical degradation 269
　mechanisms of action 185
　transformations in plants 221
Triazines 70, 104-106, 139, 186-187, 222-224, 240, 263, 271
　chloro 104
　methoxy 104
　methylmercapto (methylthio) 104
Triazinones 70, 106, 224, 271
Triazoles 70, 106-107, 139-140, 157, 188, 224-225, 240, 263, 271
Triazolopyrimidine sulfonanilides 70, 107, 140, 188-189, 225, 263
Tribenuron
　mechanisms of action 185
Trichomes 119
Triclopyr 69, 95
　absorption 137

chemical degradation 269
mechanisms of action 183-184
microbial degradation 259
photodecomposition 239
transformations in plants 217
translocation 137
Tridiphane 70, 112
Trifluralin 68, 81
absorption 131-132
chemical degradation 267
mechanisms of action 175
microbial degradation 256
photodecomposition 237
transformations in plants 209
translocation 131-132
Triflusulfuron
mechanisms of action 185
photodecomposition 239
transformations in plants 218-219
Trigger phase 28
Triggering agent 28
Tuber crops 458

UCC-C4243 70, 108
mechanisms of action 189
Ultralow volume spraying 397
Unclassified herbicides 70, 110-112, 141-142,
190-191, 226-227, 240, 264-265
Uncoupler(s) 167
Uracils 70, 107-108, 140, 189, 240, 271-
272
Ureas 70, 108-110, 140, 189-190, 240, 263-
264, 272
UV irridation 232

Vacant areas 486
Volatilization 232
Valine 159
Vapour pressure 233
Vegetable crops 458-459
Vegetative reproduction 28-29
Vernolate 69
chemical degradation 270
microbial degradation 261-262
Viooxanthin 154
Visual scoring 497
Volatilization of herbicides 233-235
Volume, conversion factors 526
Volume median diameter (VMD) 393-394

Wandering Jew (see spiderwort, tropical)
419-420
Waterhyacinth 475-477
Water lettuce 479
Water stress 46-47

Water table 232
Weed(s)
annuals 7-8
aquatic 465, 467-472, 475-480
biennials 7-8
biodiversity 12-16
biology 7-23
bulbous perennials 8
classification 7-8
competition 29-31, 490-491
counts 499
creeping perennials 8
ecology 23-31
emersed 466-467
eradication 38
floating 467
genetic diversity 12-13
germination 26-27, 53-55
harmful effects caused by 1-3
Law(s) 38
management, biological approaches in
319-347
management in major field crops
447-459
management in major plantation crops
459-462
marginal 467
nursery 491-492
parasitism 16-17
perennials 7-9
persistence 23-25
prevention 37-38
propagation 8-11
resistance 285-286
response 498-499
seedbank 10-11, 502-505
seed dormancy 27-28, 53-55
shifts 490
simple perennials 7-8
some prominent weeds 411-446
submersed 466
summer annuals 7
survival mechanisms 25-29
vegetative propagules, dynamics 9-
10
weight 499
Weed biology and ecology 7-35
Weed competition 29-30
characteristics 30-31
Weed control by deleterious rhizobacteria
326-330
Weed ecology 23-31
Weed management
aquatic systems 464-480

biological approaches to 319-343
cereals, in 448-453
chemical 59-66
CO_2, atmospheric 48-49
crop-based approaches in 51-53
ecophysiological approaches, in 45-49
environmental stress 46-48
evolution 3-4
fibre crops, in 454-455
field and plantation crops, in 447-463
forest nurseries and plantations, in 481-482
forestry, in 481-486
integrated 5
integration of rhizobacteria into 329-330
light 45-46
manual methods 39-40
mechanical methods 40-45
non-agricultural systems, in 481-486
nutrient-based approaches, in 49-51
oilseeds, in 453-454
plantation crops, in 459-462
potential of allelopathy in 21
post-infested 38-45
preventive 37
pulse crops, in 455-456
relationship of biodiversity, to 15-16
traditional approaches in 36-45
tuber crops, in 458
vegetable crops, in 459-462
weed-free crop seed 37-38
Weed research 487-505
methodology 497-505
planning 487-488
problems 487-488
Weed science, modern 4-6
Weed seed(s) 502-505
estimation in seedbanks 502-505
Weight, conversion factors 526
Wettable powder 381
Wetting agents 126
Wheat 451-452
Wild carrot 435-436
Wild oat(s) 414-416
Witchweed 443-445
Wood sorrel 434
Woody species, managing 482-484

Xanthophyll(s) 154, 156
Xylem 120-121

Z-scheme 149

Weed Index

Abutilon theophrasti 11, 15, 20, 31, 46, 48, 54, 81, 84-87, 92, 97, 99-100, 105-107, 212, 288, 328, 330-331, 333, 504

Acacia farnesiana 361

Acalypha ostryifolia 93

Achillea spp. 9

Acorus calamus 465

Acroptilon repens 328

Aegilops cylindrica 100, 328-330

Aeschynomene virginica 103, 112, 325, 338, 342

Ageratum conyzoides 84, 411-412, 448, 461

Agrostis gigantea 9

Alisma spp. 466-467

Allium canadense 2

Alopecurus myosuroides 288-289, 295, 305

Alternanthera philoxeroides 475

Alternanthera spp. 479

Amaranthus albus 45

Amaranthus blitoides 48

Amaranthus palmeri 295

Amaranthus retroflexus 3, 8, 20, 47, 50, 54, 72-75, 78, 80, 110, 129, 132, 225, 285, 289, 328, 331, 359, 413, 504

Amaranthus rudis 290

Amaranthus spinosus 2, 20, 51, 412

Amaranthus spp. 26, 31, 80, 82-85, 88, 92-94, 97-98, 100-101, 103, 105

Amaranthus viridus 412-413

Ambrosia artemisiifolia 47, 75, 78, 82, 83, 86, 92, 93, 95, 100-101, 106, 133, 330-331

Ambrosia spp. 2, 83, 85, 98, 105

Ambrosia tenuifolia 100

Ambrosia trifida 133

Ammania spp. 98

Amsinckia intermedia 212

Anagallis arvensis 413-414, 454

Anoda cristata 85

Arabidopsis thaliana 294, 298, 301

Arceuthobium spp. 16

Arctotheca calendula 286

Argemone mexicana 2, 3, 414

Arundinella bengalensis 462

Asclepias syriaca 25

Asperugo procumbens 86

Atriplex spp. 31

Avena fatua 2, 3, 8, 29, 45, 86, 91, 102-103, 291, 293, 328, 415-416, 451

Avena sterilis 52, 292

Avena tomentosa 453

Axonopus compressus 84, 111, 416-417, 498

Azolla anabena 53

Azolla pinnata 53

Barbarea vulgaris 11

Bidens pilosa 417

Borreria hispida (B. articularis) 20, 27, 84, 417-418, 461

Borreria latifolia 417

Brachiaria mutica 96, 475

Brachiaria platyphylla 103, 109

Brachiaria spp. 72, 80-81, 450

Brasenia spp. 467

Brassica campestris 302

Brassica juncea 284

Brassica kaber (see *Sinapis arvensis*) 82-83, 86-87, 101, 105, 107, 207, 284, 293-295

Brassica napus 12

Brassica nigra 8

Brassica spp. 98-99, 101

Bromus japonicus 328

Bromus secalinus 328

Bromus tectorum 91, 110, 327-329

Cabomba spp. 475

Calamagrostis canadensis 328-329

Camelina alyssum 18-19

Capsella bursa-pastoris 87

Cassia obtusifolia 109, 333, 337, 342

Cassia occidentalis 93, 136, 216, 337

Celosia argentea 418

Cenchrus biflorus 453

Cenchrus incertus 88

Cenchrus spp. 80

Centella asiatica 450

Cephalanthus spp. 467
Ceratophyllum demersum 76, 96, 474
Ceratopteris richardii 286
Chara spp. 466
Chenopodium album 11, 14, 20, 54, 71, 73, 75, 78, 80, 82, 84-87, 92-93, 98-101, 103, 105-110, 137, 171, 188, 219, 279, 288, 331, 419, 451, 454
Chrysopogan aciculatus 26
Chorispora tenella 87
Cirsium arvense 19, 87, 94-95, 98-99, 110, 217, 330, 331, 361, 419, 451
Commelina alyssum 19
Commelina benghalensis 24, 84, 419-420, 448, 461
Commelina spp. 98, 103
Convolvulus arvensis 4, 9, 19, 44, 48, 97, 226, 296
Conyza bonariensis 11, 285-286
Conyza canadensis (see *Erigeron canadensis*)
Crassocephalum crepidioides 285
Crotolaria burhia 453
Crotolaria specatbilis 338
Cucumis melo 93
Cucurbita texana 336, 342
Cuscuta chinensis 420-422
Cuscuta spp. 16
Cynodon dactylon 2, 24, 42, 44, 48, 87, 91-92, 107, 111, 135, 422, 448, 458, 460-461
Cyperus bulbosus 450
Cyperus difformis 101, 186, 289, 450
Cyperus esculentus 9, 10, 28, 71-72, 102, 128, 205, 326, 359-360
Cyperus iria 450
Cyperus rotundus 2, 9, 30, 48, 50-51, 75, 87, 98, 102, 139, 141, 172, 207, 322, 423-424, 448, 450, 453, 457, 460, 465, 487
Cyperus serotinus 101, 186

Datura innoxia 290
Datura stramonium 20, 54, 82-83, 85, 95, 98, 105, 106, 171, 338
Daucus carota 3, 87, 305
Descurainia sophia 86
Desmodium tortuosum 337, 342
Digera arvensis 450
Digitaria ischaemum 216
Digitaria sanguinalis 20, 24, 46, 49, 72-73, 77, 84, 94, 103, 110-111, 223, 361, 424-425, 498, 501

Echinochloa colonum 51, 54, 98, 103, 427-428
Echinochloa crus-galli 20, 29, 31, 46, 51, 71-75, 80, 83-86, 91, 96-98, 100-103, 105, 107-111, 176, 186, 187, 206, 219, 221, 334, 342, 360, 425-427, 448, 504

Eclipta alba 98
Eichhornia crassipes 3, 74, 77, 326, 342, 450, 465, 475-476
Eleocharis acicularis 101, 186
Eleocharis spp. 333, 475
Eleusine indica 49, 54, 71, 74, 77, 84, 88, 94, 96, 106, 109-111, 289, 292, 295, 428
Elodea canadensis 474
Elodea spp. 77, 96
Elytrigia repens 9, 13, 24, 28, 42, 74, 84, 87, 91-92, 99-100, 102, 111, 127, 130, 132, 138, 141-142, 170, 210, 216, 331, 360, 428-429, 460
Epilobium ciliatum 285
Epilobium spp. 467
Equisetum arvense 140-141
Eragrostis spp. 450
Erechtites valerianefolia 461
Erichloa villosa 99
Erigeron canadensis (see *Conyza canadensis*) 285, 286
Erigeron philadelphicus 285
Eupatorium adenophorum 322-323
Euphorbia esula 81, 87, 109, 217-218, 328, 330
Euphorbia hirta 430, 453
Euphorbia humistrata 81
Euphorbia maculata 92
Euphorbia spp. 85, 96

Fimbristylis miliacea 448, 450
Fumaria parviflora 454
Fumaria spp. 451

Galinsoga spp. 71
Galium aparine 175
Galium spp. 86

Hedera helix 51
Helianthus annuus 20
Helianthus tuberosus 330
Heliotropium subulatum 453
Heteranthera limosa 98, 322
Heteranthera spp. 466
Hordeum glaucum 285, 286
Hordeum leporium 285
Hydrilla verticillata 3, 96, 478
Hydrocotyle spp. 77
Hypericum perforatum 322

Imperata cylindrica 2, 20, 24, 25, 28, 74, 91, 111, 127, 223, 322, 360, 361, 369, 430-431, 461
Ipomoea aquatica 3, 465, 479
Ipomoea hederacea 93
Ipomoea purpurea 103, 226
Ipomoea spp. 82, 83, 85, 94, 98, 105, 106, 109
Iva xanthifolia 95

Juncus spp. 467
Jussiaea spp. 74
Justicia spp. 468

Kochia iberica 283
Kochia scoparia 46, 80, 82, 85, 87, 98-101, 105, 107, 110, 283, 289-291, 305, 307

Lactuca sativa 291
Lactuca serriola 10, 108, 290-291, 307, 331
Lamium amplexicaule 101, 108
Lamium spp. 110
Lemna spp. 466-467
Lantana camara 321, 322, 431-432
Lepidium sativum 186
Lepidium spp. 26
Leptochloa filiformis 54
Leptochloa spp. 103
Linum catharticum 218
Lolium multiflorum 292, 307
Lolium perenne 87, 219, 283-284, 288-289, 296, 298, 408
Lolium rigidum 50, 283, 288-289, 291-292, 295, 305-307
Lolium spp. 91, 108
Loranthus spp. 16
Lotus corniculatus 296
Ludwigia decurrens 338
Lupinus albus 20
Lythrum spp. 475

Malva pusilla 112, 326
Marsilea quadrifolia 450
Melilotus alba 451, 454
Matricaria inodora 219
Mikania micrantha 2, 111, 432-433, 461
Mollugo verticillata 72-73, 75, 88
Monochoria vaginalis 29, 101, 186, 448, 450
Morrenia odorata 113, 324
Muhlenbergia frondosa 92
Myriophyllum spp 96, 474.

Najas spp. 96, 474, 465
Nelumbo spp. 467
Nuphar spp. 479
Nymphaea odorata 465
Nymphaea spp. 3, 475

Opuntia aurantiaca 321
Opuntia dellenii 322
Opuntia imbricata 321
Opuntia inermis 321
Opuntia monocantha 321
Opuntia streptocantha 321
Opuntia stricta 321
Opuntia tomentosa 321

Opuntia vulgaris 322
Orobanche cernua 17, 433
Oxalis corniculata 434
Oxalis corymbosa 434
Oxalis spp. 28, 88

Panicum capillare 31, 72, 105, 505
Panicum dichotomiflorum 72, 73, 75, 80, 85, 94, 100, 102, 105, 109, 223
Panicum maximum 111, 450
Panicum miliaceum 91, 361
Panicum repens 111, 434-435
Parthenium hysterophorus 2, 285, 323, 435-436, 486
Paspalum conjugatum 84, 111, 360, 437, 462, 498
Paspalum dilatatum 75, 111
Paspalum scrobiculatum 462
Pennisetum clandestinum 437-438
Pennisetum pedicillatum 438-439
Phalaris minor 26, 305, 439, 451, 487
Phoradendron spp. 16
Phragmites australis 475
Phragmites communis 479
Phragmites spp. 475
Phyllanthus niruri 453
Pistia stratiotes 74, 450, 465, 479
Plantago spp. 98
Poa annua 87, 88, 285, 328, 331
Poa spp. 86
Polygonum chinense 111
Polygonum convolvulus 86, 87, 105, 109
Polygonum lapathifolium 289
Polygonum pennsylvanicum 20
Polygonum perfoliatum 2
Polygonum persicaria 331
Polygonum spp. 82, 95, 98, 100, 105, 107, 109, 110
Pontederia cordata 98, 465, 479
Populus spp. 467
Portulaca oleracea 8, 20, 71, 73, 84, 87, 103, 105, 109, 360, 439-440
Portulaca spp. 31, 94, 109
Potamogeton nodosus 465
Potamogeton pectinatus 473
Potamogeton spp. 3, 96, 474, 477
Prosopis juliflora 361
Pteridium aquilinum 77, 111, 130, 174
Pteridium spp. 24
Puccinallia spp. 24
Pulicaria wightiana 453

Quercus stellata 369

Ranunculus spp. 466

Richardia scabra 88
Rhus radicans 2
Rorippa spp. 467
Rubus spp. 328
Rumex acetosella 24
Rumex crispus 100
Rumex spp. 26

Saccharum spontaneum 25, 111, 440
Saggitaria montevidensis 289
Saggitaria pygmaea 101, 186
Saggitaria spp. 475, 479
Salsola iberica 31, 46, 289, 291
Salvia reflexa 97
Salvinia spp. 467, 469
Salvinia molesta 3, 479
Scirpus juncoides 101, 186
Scirpus validus 465
Scoparia dulcis 27
Senecio vulgaris 188, 286
Senna obtusifolia 136, 216
Sesbania exaltata 82, 97, 330, 335, 337-338, 342
Setaria anceps 47
Setaria faberi 11, 14, 54, 93, 140, 140, 171, 221, 223, 361
Setaria glauca 416, 448, 501
Setaria italica 298, 302, 361
Setaria lutescens (see *Setaria glauca*) 171
Setaria palmifolia 111, 361
Setaria viridis 29, 84, 88, 292, 295, 298, 302, 307, 328, 331
Sida spinosa 93, 97, 106, 338
Sinapis arvensis (see *Brassica kaber*)
Sisymbrium altissinum 8, 328
Sisymbrium officinale 54
Sisymbrium spp. 86
Solanum carolinense 137
Solanum nigrum 8, 46, 72, 78, 80, 82-83, 85, 93, 100, 103, 107, 110, 218, 221, 285, 300, 437-438
Solanum ptycanthum 45, 46, 338
Solanum spp. 83, 87
Solanum villosum 103
Solidago fistulosa 225
Sonchus arvensis 78, 100
Sonchus oleraceus 289, 441-442
Sorghum halepense 9, 31, 44, 48, 75, 77, 81-82, 85, 91-92, 94, 99-100, 102, 107, 111, 135, 139, 220-221, 334, 337, 442-443, 457, 475
Sparganium spp. 467
Spermacoce ocymoides 27
Sphenoclea zeylandica 98
Spirogyra spp. 466, 474

Stellaria media 108, 110, 216, 218, 289-290, 294
Stiga angustifolia 18, 444
Stiga asiatica 17, 18
Stiga aspera 18
Striga bilabiata 18
Striga baumanii 18
Striga brachycalyx 18
Striga crysantha 18
Striga curviflora 18
Striga densiflora 18, 444
Striga baumanii 18
Striga brachycalyx 18
Striga crysantha 18
Striga curviflora 18
Striga densiflora 18, 444
Striga elegans 18
Striga forbesii 18
Striga fulgen 18
Striga gesnerioides 18, 444
Striga hallaie 18
Striga hermonthica 18
Striga junodii 18
Striga klingii 18
Striga latericea 18
Striga ledormannii 18
Striga linearifolia 18
Striga lutea 18, 444-445
Striga macrantha 18
Striga masuria 18
Striga multiflora 18
Striga parviflora 18
Striga passargei 18
Striga primuloides 18
Striga publiflora 18
Striga sulphurea 18, 444

Tamarix spp. 467
Taraxacum officinale 25, 84, 87
Tephrosia purpurea 453
Thlapsi arvense 86
Trapa natans 465
Tribulus terrestris 107, 453
Typha angustata 477-478
Typha angustifolia 3, 465
Typha elephantina 477
Typha latifolia 465

Urtica spp. 88
Utricularia spp. 76, 86, 474

Veronica filiformis 9
Veronica persica 220
Vicia sativa 454

Viscum spp.   16

Xanthium pensylvanicum   129
Xanthium spinosum   338
Xanthium spp.   14, 83

Xanthium strumarium   11, 45-46, 48, 75, 82, 85, 87, 93, 95, 97-100, 105, 109, 121, 283, 286, 289-290, 328, 331, 338

Zannichellia spp.   466

The Author

Dr. Vallurupalli Sivaji Rao, an eminent weed scientist credited with significant long-lasting contributions to the field of weed science for over 35 years, was born on December 6, 1940 in Gudlavalleru, Andhra Pradesh, India. He received Bachelors degree in Agricultural Science from Karnataka University, Dharwar, India in 1961 and Masters degree in Agronomy from Osmania University, Hyderabad, India in 1963. After working as Research Fellow of Council of Scientific and Industrial Research, Government of India and Agronomist for the Rockefeller Foundation, India, he went to Cornell University, Ithaca, NY, USA in 1969 and graduated with Ph.D. degree in 1973, majoring in Weed Science. He followed it up with a three-year stint as Research Associate in the Department of Seed and Vegetable Sciences at Cornell.

After returning to India in October 1975, Dr. Rao worked as Agronomist (Weed Control) at Tocklai Experimental Station, Tea Research Association, Jorhat, Assam, India for over six years. During this period, he built a strong weed research programme and developed numerous and far-reaching weed management practices for the benefit of tea industry in India. In March 1982, he became Director of Fredrick Institute of Plant Protection and Toxicology, near Madras, India. After leading this pesticide/herbicide contract research laboratory for over three years, Dr. Rao became a professional weed scientist advising national and state governments, herbicide industry and research institutions on matters concerning herbicide technology and its transfer to farmers. During this period (1985-94), Dr. Rao served as Adviser/Consultant to herbicide companies including Mascot Agrochemicals Pvt. Ltd., Bangalore and Hoechst India Ltd., Bombay. He also served as a member of Agronomy Panel of Indian Council of Agricultural Research and Herbicide Panel of Central Insecticides Board, Ministry of Agriculture, New Delhi. Besides, he was also a member of numerous national-level and state-level government committees related to weed management.

After relocating to USA with his family in March 1995, Dr. Rao continued to pursue his interest in his field. Currently, he is a Weed Scientist in a private pesticide laboratory, besides being a Consultant (Weed Science) to certain herbicide companies.

Dr. Rao is an author of over 80 research papers and reports in weed science, plant physiology and agronomy. The first edition of his book 'PRINCIPLES OF WEED SCIENCE', published in 1983, has been a very popular reference-cum-textbook in weed science in India. Dr. Rao has also made a significant contribution to the growth of weed science profession in India. He served as Executive Vice President of Indian Society of Weed Science and Editor-in-Chief of Indian Journal of Weed Science, each for a four-year term, during 1982-90. He is a member of Indian Society of Weed Science for life and Weed Science Society of America for over 25 years. He is presently located at 3575 Lehigh Dr., House # 2, Santa Clara, California 95051, USA.

T - #0447 - 101024 - C0 - 234/156/30 - PB - 9781578080694 - Gloss Lamination